Pearls and Pitfalls in

EMERGENCY RADIOLOGY

Variants and Other Difficult Diagnoses

Pearls and Pitfalls in EMERGENCY RADIOLOGY

Variants and Other Difficult Diagnoses

Edited by

Martin L. Gunn

Associate Professor
Department of Radiology
University of Washington
Seattle, WA, USA

CAMBRIDGE UNIVERSITY PRESS
Cambridge, New York, Melbourne, Madrid, Cape Town,
Singapore, São Paulo, Delhi, Mexico City

Cambridge University Press
The Edinburgh Building, Cambridge CB2 8RU, UK

Published in the United States of America by Cambridge University Press, New York

www.cambridge.org
Information on this title: www.cambridge.org/9781107021914

First published 2013

Printed and bound in the United Kingdom by the MPG Books Group

A catalog record for this publication is available from the British Library

Library of Congress Cataloging in Publication data

Gunn, Martin L.
Pearls and pitfalls in emergency radiology : variants and other difficult diagnoses /
Martin L. Gunn.
p. ; cm.
Includes bibliographical references and index.
ISBN 978-1-107-02191-4 (Hardback)
I. Title.
[DNLM: 1. Radiography–methods–Case Reports. 2. Diagnosis, Differential–Case Reports. 3. Emergencies–Case Reports. WN 445]
616.07′572–dc23

2012025997

ISBN 978-1-107-02191-4 Hardback

Contents

Contributors

Robert B. Carr, MD
Clinical Instructor in Neuroradiology
Department of Radiology
Massachusetts General Hospital
Harvard Medical School
Boston, MA, USA

Joel A. Gross, MD, MS
Associate Professor
Director of Emergency Radiology
Department of Radiology
University of Washington
Seattle, WA, USA

Martin L. Gunn, MB ChB, FRANZCR
Associate Professor
Department of Radiology
University of Washington
Seattle, WA, USA

Ramesh S. Iyer, MD
Assistant Professor
Department of Radiology
Seattle Children's Hospital
University of Washington
Seattle, WA, USA

Randy K. Lau, MD
Acting Instructor
Department of Radiology
University of Washington
Seattle, WA, USA.

Ken F. Linnau, MD, MS
Assistant Professor
Department of Radiology
University of Washington
Seattle, WA, USA

Michael J. Modica, MD
Fellow in Neuroradiology
Department of Radiology
Massachusetts General Hospital
Harvard Medical School
Boston MA, USA

Matthew H. Nett, MD
Clinical Instructor in Musculoskeletal Radiology
Department of Radiology
University of Wisconsin
Madison, WI, USA

Claire K. Sandstrom, MD
Acting Assistant Professor of Radiology
Department of Radiology
University of Washington
Seattle, WA, USA

Preface

"To see what is in front of one's nose needs a constant struggle."
George Orwell, from the essay *In Front of Your Nose*, 1946.

The rise of Emergency Radiology as a discrete subspecialty over the last 25 years, and in particular in the last decade, has unquestionably led to the more timely recognition of life-threatening illnesses and injuries, expedited appropriate management, and improved clinical outcomes.

The need for rapid decision making and management in patients who present with acute illnesses makes it especially important that radiologists recognize not only the typical findings of disease, but also mimickers and unusual diagnoses. As with others in this series, the aims of this book are to help the radiologist to discern the abnormal from the normal, to identify imaging artifacts that simulate disease, and to make subtle but important diagnoses that might otherwise go unrecognized.

The growth of imaging utilization in the emergency department over the past 10 years has been enormous. This is particularly so for MDCT. CT is now used in the emergency department not only as a way to resolve clinical enigmas, but also as a screening tool for a number of conditions, such as cervical spine trauma, pulmonary embolism, and suspected appendicitis in adults. Quite rightly, the expectations our clinical colleagues have of radiologists continue to grow with their increased use of complex imaging for acute patient care. Radiologists are now expected to provide real-time, accurate diagnoses, 24 hours per day, 365 days per year, and in many cases, discuss cases before and after images are acquired.

Unfortunately, false-positive interpretations in radiology have been identified as a significant cause of error, leading to unnecessary investigation and treatment, increased healthcare costs, delays in appropriate management, and litigation. And that is not to mention the embarrassment of getting a case wrong in front of one's colleagues.

This book follows the outstanding structure set by Dr. Fergus Coakley in the first book of this series: *Pearls and Pitfalls in Abdominal Imaging*. Each "case" includes a short synopsis of the imaging appearances, importance, clinical scenario, and differential diagnosis, plus a key teaching point. This is followed by a number of images that illustrate the pearl, pitfall, or differential diagnosis. Where there are several variants on a theme (for example, accessory ossicles), images were chosen to form a mini-atlas of variants.

The book is intended more as a bench-side reference than a book that should be read from cover-to-cover. As such, some of the descriptions are brief and to the point. We have tried to include cases that represent the gamut of diagnoses that might be encountered in a busy emergency radiology practice that includes a high volume of trauma and pediatrics. However, by necessity, cases that are either too rare, of limited clinical importance, or unlikely to cause diagnostic confusion have been excluded.

In summary, unlike most reference books on the topic of Emergency Radiology, this book should provide guidance when you see a puzzling case, and help you determine if it is simply a borderland of normal, or an important but less recognized diagnosis.

Acknowledgments

Since I arrived at the University of Washington, I have been extremely lucky to work with a group of energetic and talented academic radiologists and learn from some of the outstanding fellows and residents who pass through our program. As an "outsider" coming from New Zealand, I have received an immense amount of support and nurturing over the years by the Department in my adopted home of Seattle.

For this book, I want to highlight the outstanding work done by the contributors, many of whom are early in their academic careers. The remarkable quality of their chapters in this book reflects their effort, insights, and talent better than I can express with words of gratitude.

In preparing the book I had significant help from several of the staff at Cambridge University Press. Deborah Russell conceived and managed the project with aplomb. Caroline Mowatt oversaw the book through many of its stages. Anne Kenton did an extremely astute job reading the proofs and finding unexpected errors and contradictions.

Our current emergency radiology fellows at the University of Washington, R. Travis Clark and Claudia Zacharias, and our always good-humored Emergency Radiology Program Coordinator, Daniel Willems helped with proof reading. Bindi, my wife, seemed to instinctively understand that this book project would accompany us on our family vacations and occupy countless evenings and weekends, increasing her work as the parent of our two children.

Thanks to you all.

Martin L. Gunn

Isodense subdural hemorrhage

Michael J. Modica

Imaging description

Subdural hematoma is the most common extra-axial collection and is present in up to 5% of trauma patients. Subdural hemorrhages are located between the arachnoid and inner layer of the dura and typically appear as a crescent-shaped fluid collection. As a rule, subdural hemorrhages can cross sutures, but cannot breach the dural attachments. Subdural hemorrhage will displace the cortical vascular structures medially and compress and mildly displace the underlying brain. The typical imaging appearance of subdural hemorrhage can vary depending on acuity. Hemorrhage can be classified as acute (6 hours to 3 days), subacute (3 days to 3 weeks), and chronic (>3 weeks). The CT density decreases by approximately 1.5 Hounsfield Units (HU) per day as the hemorrhage resolves.

Subdural hemorrhage that is isodense to brain parenchyma is typically subacute. Rarely, isodense subdural hemorrhage can be acute in patients who are anemic (serum hemoglobin <8–10 g/dL) (Figure 1.1A). Either case can easily be missed because small collections may be difficult to see when they are the same density as the underlying cortex [1].

Contrast-enhanced CT can detect isodense subdural hemorrhage. Enhancement of the dura, displacement of the cortical veins away from the skull, and cortical enhancement all help make the hemorrhage more conspicuous (Figure 1.1B) [2].

Based on the paramagnetic properties of blood at various stages of breakdown, MRI has become an excellent tool for the evaluation of isodense subdural hemorrhage (Table 1.1) [3]. Subacute blood products are composed of methemoglobin and are hyperintense on T1- and T2-weighted MR (Figure 1.2).

Importance

Missing subdural hematomas is potentially fatal. Small acute and mixed age or acute-on-chronic subdural hematomas can grow over time, with increasing mass effect if untreated. Use wide window settings (125–200 HU) to detect small subdural fluid collections.

Typical clinical scenario

Subdural hemorrhage is commonly acquired following deceleration injury, often from a motor vehicle collision. It is more common in the elderly. Isodense subdural hemorrhage is commonly subacute; however, it may present acutely in the anemic patient. The clinical presentation can vary from asymptomatic to loss of consciousness. Patients may experience a "lucid interval" following acute extra-axial hemorrhage. This refers to a trauma patient initially awake who shows neurologic decline a few hours after the event.

Differential diagnosis

The differential diagnosis for isodense subdural hemorrhage is narrow and includes:

- *Subdural hygroma*: crescent-shaped subdural collection composed of simple to slightly proteinacous fluid that typically displaces the cerebral cortex medially. Hygromas are isodense to cerebrospinal fluid (CSF) on non-contrast CT and follow CSF signal on all MR pulse sequences (Figure 1.3).
- *Dural thickening*: This can be patchy or diffuse. CT is nonspecific and often normal. MR demonstrates a characteristic dural signal that is a dark band on T2-weighted images between the calvarium and subarachnoid CSF, bright on FLAIR, and strongly enhances on post-contrast images (Figure 1.4) [4].

Teaching point

Isodense subdural hemorrhage can be difficult to detect on non-contrast CT but can grow rapidly if undetected. Consider contrast-enhanced CT to detect subtle subdural hemorrhages or isodense hemorrhage in anemic patients. MRI will help characterize the acuity of the subdural hemorrhage and differentiate it from subdural hygroma or dural thickening.

Table 1.1. MRI signal of blood degradation products

	Blood products	T1 signal	T2 signal	Time
Hyperacute	Oxyhemoglobin	Intermediate	Bright	< 24 hours
Acute	Deoxyhemoglobin	Intermediate	Dark	1–3 days
Subacute	Extracellular methemoglobin	Bright	Bright	> 3 days
Chronic	Hemosiderin	Dark	Dark	> 7 days

REFERENCES

1. D'Costa DF, Abbott RJ. Bilateral subdural haematomas and normal CT brain scans. *Br J Clin Pract*. 1990;**44**(12):666–7.
2. Tsai FY, Huprich JE, Segall HD, Teal JS. The contrast-enhanced CT scan in the diagnosis of isodense subdural hematoma. *J Neurosurg*. 1979;**50**(1):64–9.
3. Wilms G, Marchal G, Geusens E, *et al*. Isodense subdural haematomas on CT:MRI findings. *Neuroradiology*. 1992;**34**(6):497–9.
4. Tosaka M, Sato N, Fujimaki H, *et al*. Diffuse pachymeningeal hyperintensity and subdural effusion/hematoma detected by fluid-attenuated inversion recovery MRI in patients with spontaneous intracranial hypotension. *AJNR Am J Neuroradiol*. 2008;**29**(6):1164–70.

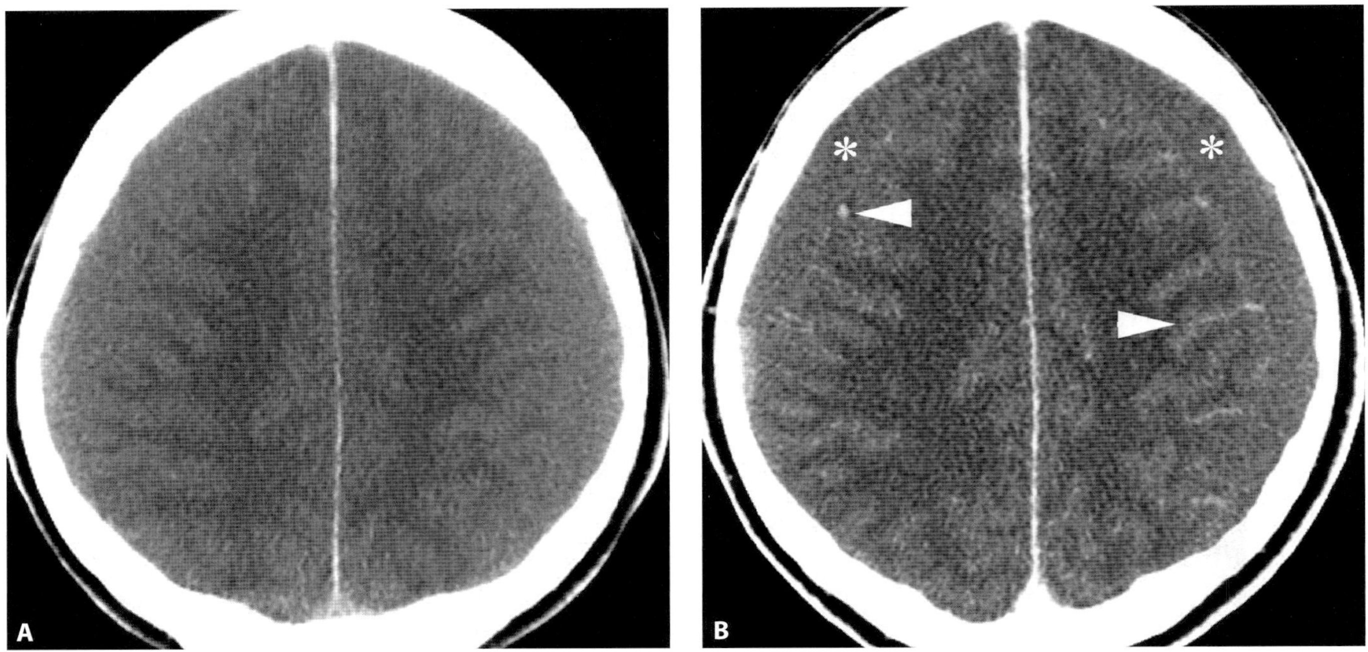

Figure 1.1 A. Axial non-contrast head CT from a 69-year-old woman involved in a high-speed motor vehicle collision shows subtle diffuse sulcal indistinctness. The patient's hemoglobin concentration at the time of the study was 8.4 g/dL. **B.** Axial contrast-enhanced CT shows normal enhancement of the underlying cortex and medial displacement of the cortical veins (arrowheads). The contrast study makes the large bilateral acute isodense subdural hematomas more conspicuous (asterisks).

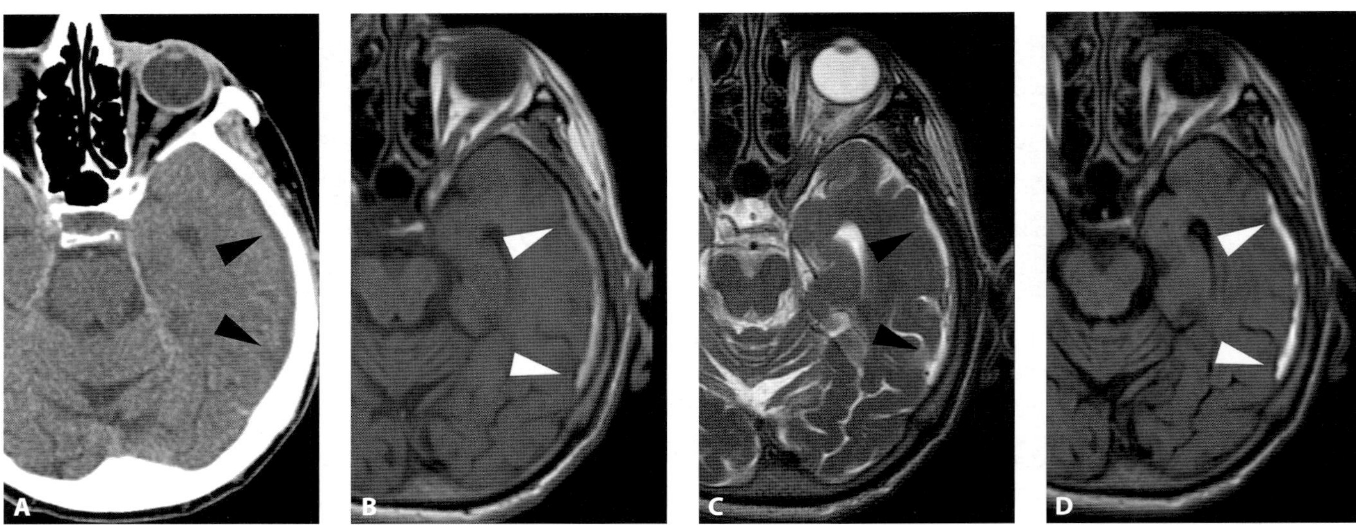

Figure 1.2 A. Axial non-contrast head CT from an 80-year-old man with altered mental status shows an iso- to slightly hypodense subdural collection beside the left inferior temporal lobe (arrowheads). The patient did not report a history of trauma and this collection was thought to be a small hygroma or widened subarachnoid space secondary to cerebral atrophy. Axial T1-weighted (**B**), T2-weighted (**C**), and FLAIR (**D**) MR images from the same patient show signal characteristics typical of a subacute subdural hemorrhage: bright signal on T1-weighted and T2/FLAIR images (arrowheads).

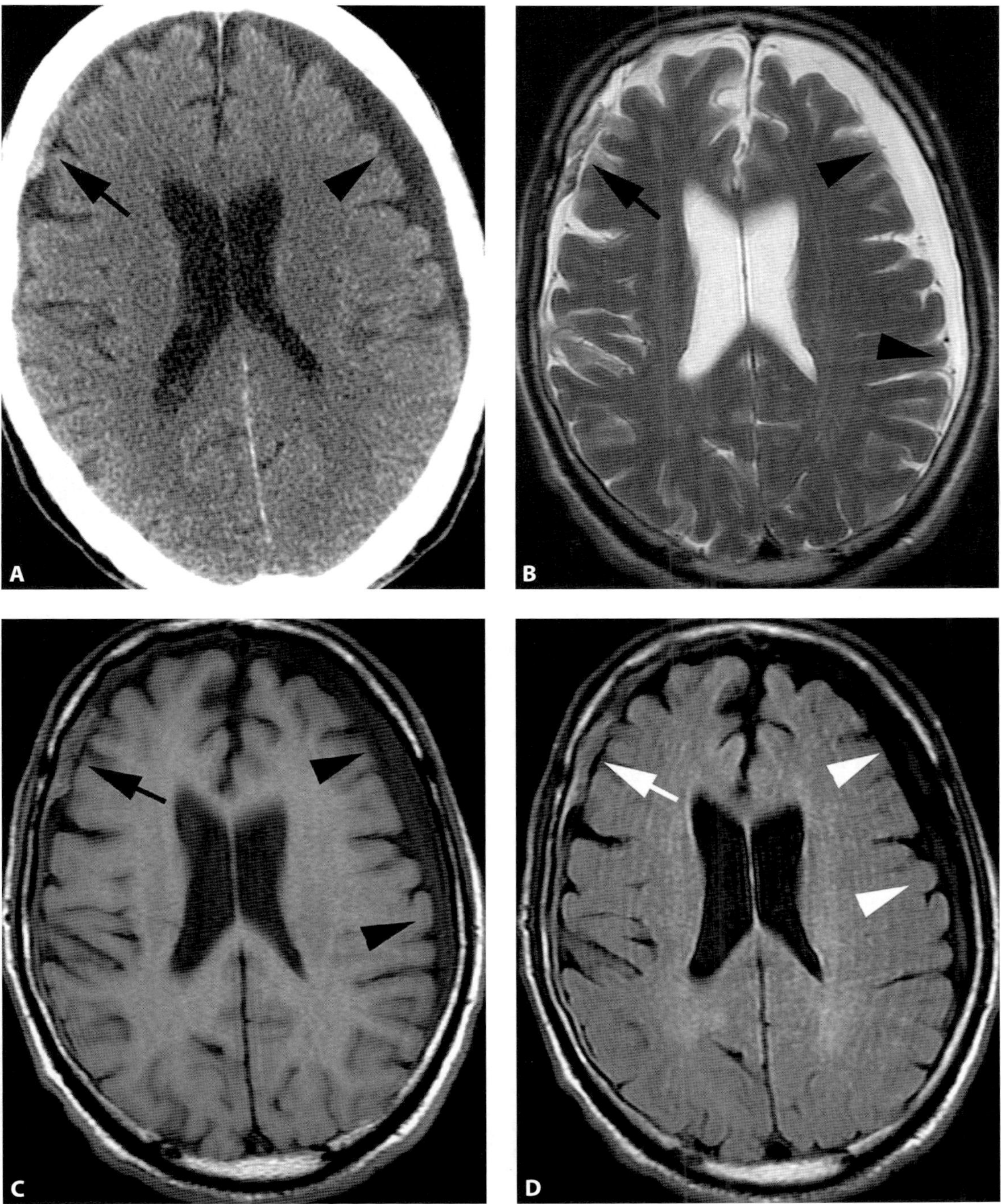

Figure 1.3 A. Axial non-contrast head CT from a 64-year-old man after a ground-level fall shows a hypodense extra-axial collection along the left frontal convexity (arrowhead) suggestive of a chronic subdural hemorrhage. Additionally, there is a hyperdense collection along the right frontal convexity consistent with acute subdural hemorrhage (arrow). Axial T2-weighted (**B**), T1-weighted (**C**), and FLAIR (**D**) MR images from the same patient show characteristic signal pattern seen in subdural hygroma along the left frontal convexity. The hygroma displaces the neighboring parenchyma medially and follows CSF signal on all pulse sequences (arrowheads). Notice the MR also characterizes the acute subdural hemorrhage along the right frontal convexity (arrow) with intermediate signal on T1- and corresponding dark signal on T2-weighted images.

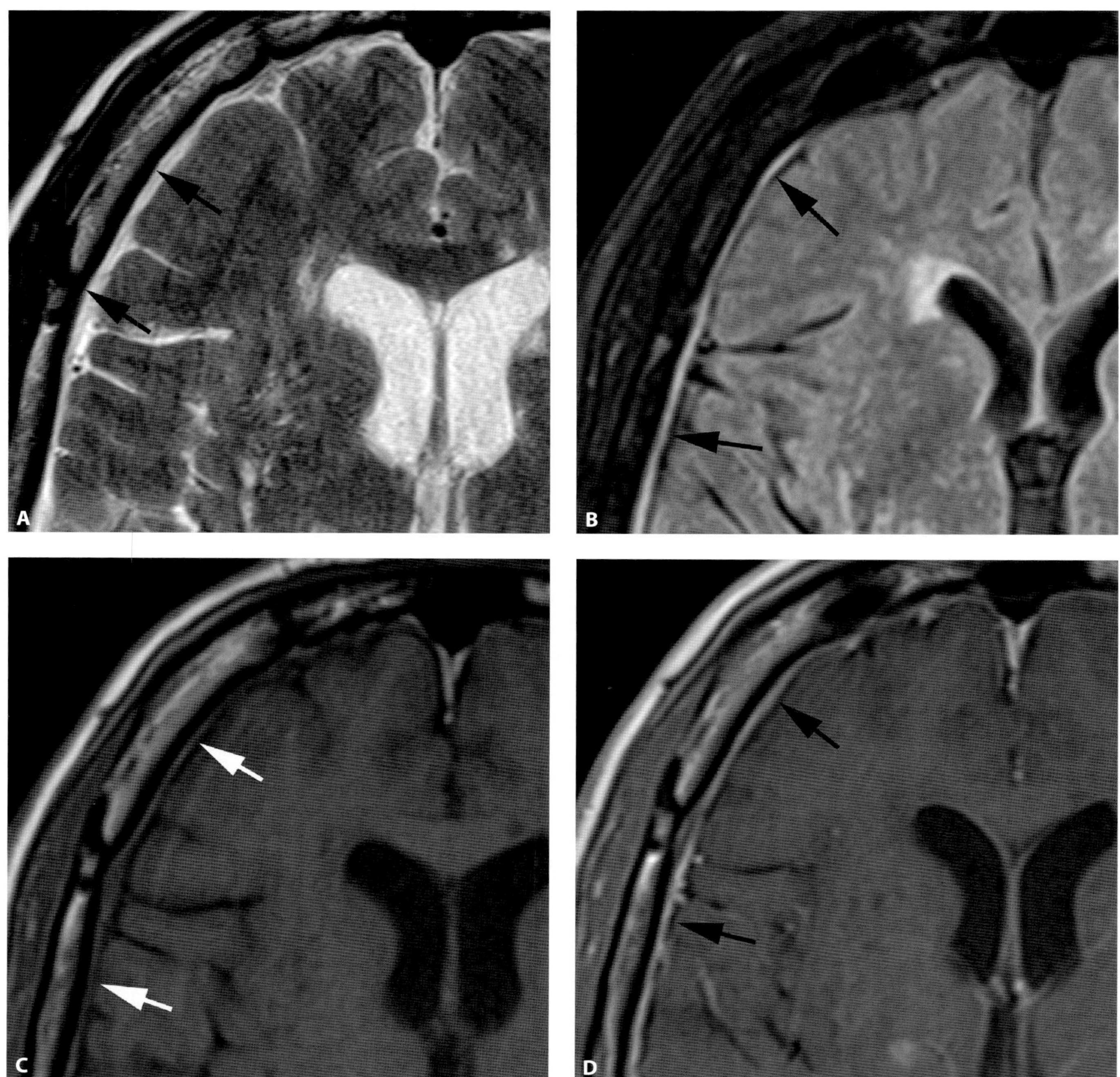

Figure 1.4 Axial T2-weighted (**A**), FLAIR (**B**), T1-weighted (**C**), and T1-weighted post-contrast (**D**) MR images from a 66-year-old woman with altered mental status show thickening of the dark dural signal on T1 and thick bright signal on FLAIR. There is intense post-contrast enhancement, characteristic for diffuse dural thickening.

Non-aneurysmal perimesencephalic subarachnoid hemorrhage

Robert B. Carr

Imaging description

Most cases of non-traumatic subarachnoid hemorrhage (SAH) are caused by aneurysm rupture. However, approximately 15% of patients will have no identifiable cause on CT angiography (CTA). In these patients, it is important to assess the pattern of SAH.

A subset of patients with CTA-negative SAH will have a pattern known as non-aneurysmal perimesencephalic subarachnoid hemorrhage (NAPH) (Figure 2.1). Criteria have been established for this, and include the following [1, 2]:

- Subarachnoid hemorrhage within the perimesencephalic cisterns, centered anterior to the midbrain.
- Possible extension into the posterior aspect of the anterior interhemispheric fissure, but not completely filling the anterior interhemispheric fissure.
- Possible extension into the medial aspects of the Sylvian fissures, but no extension laterally within the fissures (Figure 2.2).
- Possible small amounts of layering intraventricular hemorrhage sedimentation, but no frank intraventricular hemorrhage.
- No intraparenchymal hemorrhage.

There are other patterns of hemorrhage which have been described as compatible with NAPH. Hemorrhage centered anterior to the pons has been referred to as "pretruncal" and is considered a variant of NAPH [3]. There is also a variant described in which the hemorrhage is centered within the quadrigeminal plate cistern [4].

Importance

Patients with NAPH have a much more favorable outcome and can be managed less aggressively. They usually demonstrate complete recovery without complications from vasospasm or hydrocephalus [5]. Their risk of repeat hemorrhage is equivalent to the general population. Thus, identification of this pattern is important for risk stratification.

Identification of NAPH is also important in directing further imaging. A study of 93 patients with a CT pattern of NAPH and a negative CTA demonstrated negative digital subtraction angiography (DSA) in all patients. The authors concluded that cases which fulfill the criteria of NAPH and demonstrate no cause of hemorrhage on CTA do not require further investigation with DSA. Other patterns of non-traumatic CTA-negative SAH, including diffuse SAH and SAH limited to the peripheral sulci, require further evaluation with DSA [6].

A study evaluating the inter- and intra-observer agreement in CT characterization of NAPH found a small level of disagreement [7]. Thus, there should be a high level of confidence that the pattern fulfills the established criteria before suggesting NAPH as the diagnosis.

Typical clinical scenario

NAPH represents approximately 5% of all causes of non-traumatic SAH. The patients are more likely to be younger and less likely to be hypertensive than those presenting with aneurysmal hemorrhage.

Differential diagnosis

The differential diagnosis for non-traumatic CTA-negative SAH includes occult vascular abnormalities such as an aneurysm or vascular malformation, particularly if the SAH is in a pattern that does not fulfill the criteria of NAPH. These patients require DSA for further evaluation.

Teaching point

Identification of NAPH is important in determining the prognosis and need for follow-up imaging. A confident diagnosis of NAPH may preclude the need for DSA.

REFERENCES

1. Rinkel GJ, Wijdicks EF, Vermeulen M, *et al.* Nonaneurysmal perimesencephalic subarachnoid hemorrhage: CT and MR patterns that differ from aneurysmal rupture. *AJNR Am J Neuroradiol.* 1991;**12**(5):829–34.
2. Velthuis BK, Rinkel GJ, Ramos LM, Witkamp TD, van Leeuwen MS. Perimesencephalic hemorrhage. Exclusion of vertebrobasilar aneurysms with CT angiography. *Stroke.* 1999;**30**(5):1103–9.
3. Wijdicks EF, Schievink WI, Miller GM. Pretruncal nonaneurysmal subarachnoid hemorrhage. *Mayo Clin Proc.* 1998;**73**(8):745–52.
4. Schwartz TH, Mayer SA. Quadrigeminal variant of perimesencephalic nonaneurysmal subarachnoid hemorrhage. *Neurosurgery.* 2000;**46**(3):584–8.
5. Hui FK, Tumialan LM, Tanaka T, Cawley CM, Zhang YJ. Clinical differences between angiographically negative, diffuse subarachnoid hemorrhage and perimesencephalic subarachnoid hemorrhage. *Neurocrit Care.* 2009;**11**(1):64–70.
6. Agid R, Andersson T, Almqvist H, *et al.* Negative CT angiography findings in patients with spontaneous subarachnoid hemorrhage: When is digital subtraction angiography still needed? *AJNR Am J Neuroradiol.* 2010;**31**(4):696–705.
7. Brinjikji W, Kallmes DF, White JB, *et al.* Inter- and intraobserver agreement in CT characterization of nonaneurysmal perimesencephalic subarachnoid hemorrhage. *AJNR Am J Neuroradiol.* 2010;**31**(6):1103–5.

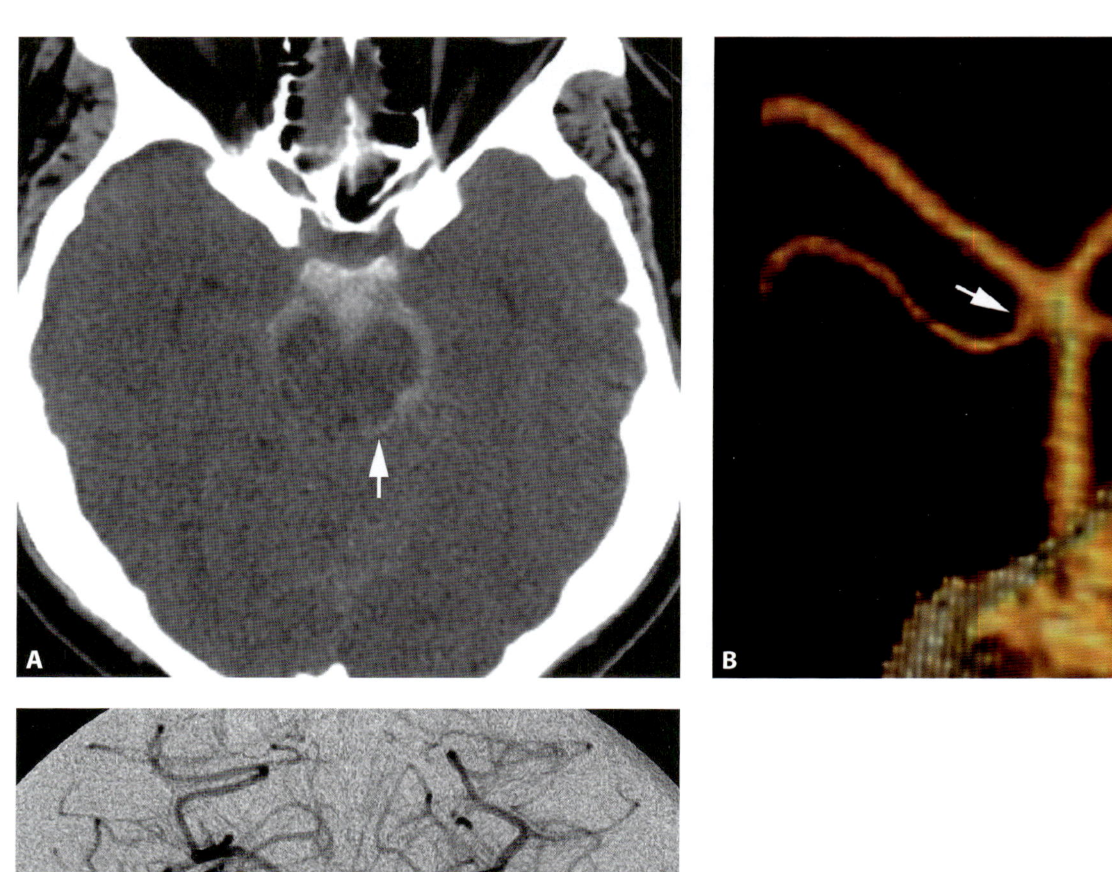

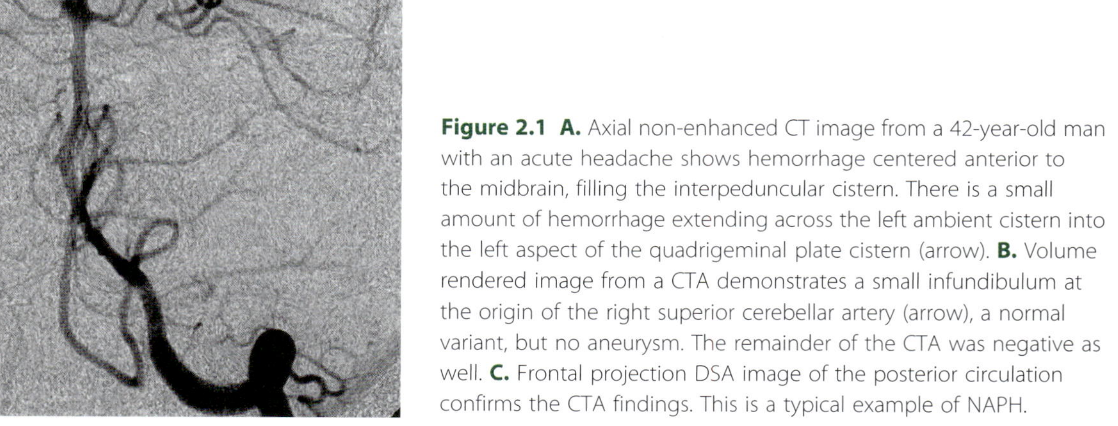

Figure 2.1 A. Axial non-enhanced CT image from a 42-year-old man with an acute headache shows hemorrhage centered anterior to the midbrain, filling the interpeduncular cistern. There is a small amount of hemorrhage extending across the left ambient cistern into the left aspect of the quadrigeminal plate cistern (arrow). **B.** Volume rendered image from a CTA demonstrates a small infundibulum at the origin of the right superior cerebellar artery (arrow), a normal variant, but no aneurysm. The remainder of the CTA was negative as well. **C.** Frontal projection DSA image of the posterior circulation confirms the CTA findings. This is a typical example of NAPH.

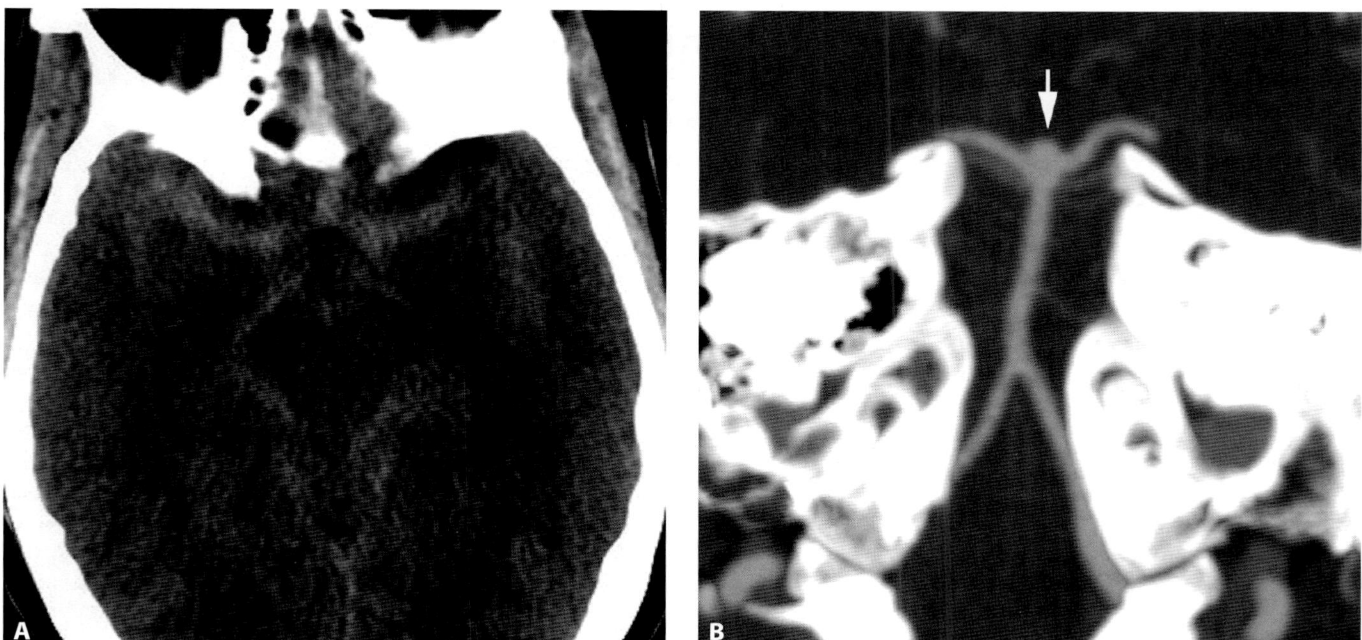

Figure 2.2 A. Axial non-enhanced CT image from a 48-year-old woman with an acute headache shows SAH in the perimesencephalic cisterns, but there is more hemorrhage within the Sylvian fissures. Also notice that the hemorrhage extends laterally within the Sylvian fissures. CTA is recommended for further evaluation. Even if the CTA were normal, DSA would be required because this does not fulfill the criteria for NAPH. **B.** Coronal maximum intensity projection (MIP) image of the basilar artery demonstrates a small basilar tip aneurysm (arrow), which is responsible for the SAH.

Missed intracranial hemorrhage

Robert B. Carr

Imaging description

Traumatic head injuries may result in intraparenchymal, intraventricular, subarachnoid, subdural, or epidural hemorrhage. Acute hemorrhage is characterized by hyperattenuation on CT, and the classic appearances of the various types of hemorrhage are well known. However, certain types of hemorrhage may be overlooked, especially subdural and subarachnoid hemorrhages.

Images from a head CT are routinely reviewed in the axial plane. However, important findings may be missed on axial images alone. In particular, hemorrhages oriented in a horizontal plane are prone to volume-averaging effects which may result in false-negative results. This is especially true of hemorrhages which occur adjacent to bone, such as the floor of the anterior and middle cranial fossae, where volume-averaging with adjacent bone leads to decreased detection (Figure 3.1). This issue is compounded by the fact that hemorrhages have a tendency to occur adjacent to bony structures in certain mechanisms of injury [1].

The addition of coronal and sagittal reformations may improve the diagnostic accuracy by reducing both false-negative and false-positive results (Figure 3.2). A study of 109 patients with intracranial hemorrhage found that the addition of coronal reformations resulted in a change in interpretation in approximately 25% of cases, compared with axial images alone [2].

Another cause of missed hemorrhage involves the use of inappropriate window and level values (Figure 3.3). If the window is too narrow, a small subdural hemorrhage may be difficult to distinguish from the adjacent bone. Optimal values for the window and level will vary among scanners, but a reasonable starting point may be a window of 200 and a level of 50.

Small quantities of subarachnoid hemorrhage may be overlooked. Subarachnoid hemorrhage tends to accumulate in dependent areas. Thus, the occipital horns of the lateral ventricles and the interpeduncular cistern should also be specifically evaluated, as these are sites where a tiny amount of hemorrhage may be seen (Figure 3.4).

Importance

The detection of intracranial hemorrhage is important for accurate treatment and appropriate follow-up of the patient.

Typical clinical scenario

Traumatic brain injury is very common, and accounts for up to 10% of healthcare costs [3]. Many of these patients will have associated intracranial hemorrhage.

A study of radiology residents found that the types of hemorrhages most often missed were subdural hemorrhages (especially frontal and parafalcine hemorrhages) and subarachnoid hemorrhages (especially interpeduncular cistern hemorrhage) [4]. Thus, when reviewing images, it is prudent to search specifically for signs of subdural and subarachnoid hemorrhage.

Differential diagnosis

Imaging artifacts may occasionally be misinterpreted as intracranial hemorrhage. Experience with interpreting head CTs allows one to predict the usual location and appearance of such artifacts. Reformatted images are also useful.

Teaching point

The additional of coronal and sagittal reformations, and the use of appropriate window and level values, can lead to a more accurate diagnosis of intracranial hemorrhage.

REFERENCES

1. Hardman JM, Manoukian A. Pathology of head trauma. *Neuroimaging Clin N Am.* 2002;**12**(2):175–87, vii.
2. Wei SC, Ulmer S, Lev MH, *et al.* Value of coronal reformations in the CT evaluation of acute head trauma. *AJNR Am J Neuroradiol.* 2010;**31**(2):334–9.
3. Lee B, Newberg A. Neuroimaging in traumatic brain imaging. *NeuroRx.* 2005;**2**(2):372–83.
4. Strub WM, Leach JL, Tomsick T, Vagal A. Overnight preliminary head CT interpretations provided by residents: locations of misidentified intracranial hemorrhage. *AJNR Am J Neuroradiol.* 2007;**28**(9):1679–82.

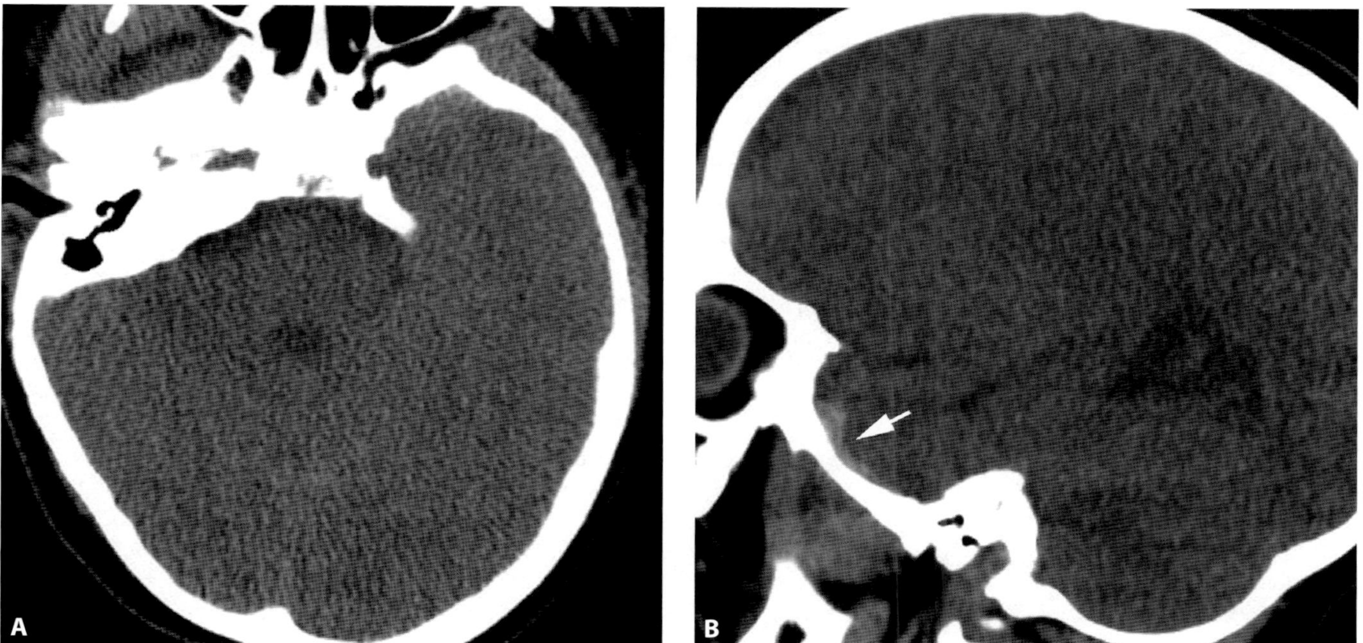

Figure 3.1 A. Axial non-enhanced CT image from a 9-year-old boy with head trauma from a motor vehicle accident shows no obvious hemorrhage. This was a low-dose exam, reducing the signal-to-noise ratio. **B.** Sagittal reformatted CT image shows a subdural hematoma extending along the floor of the left middle cranial fossa (arrow). This hemorrhage was initially overlooked on the axial images.

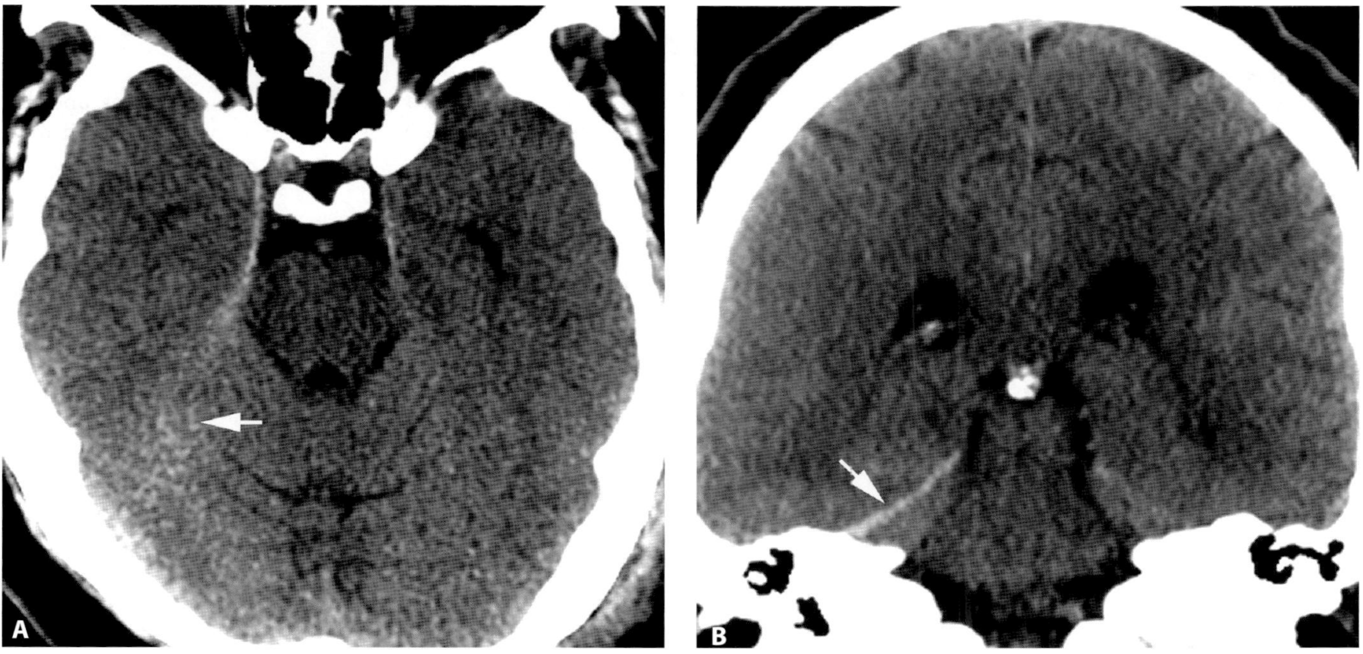

Figure 3.2 A. Axial non-enhanced CT image from a 43-year-old man shows subtle increased density along the right aspect of the tentorium (arrow). However, it is difficult to be certain that this represents hemorrhage. **B.** Coronal reformatted CT image shows obvious thickening and increased density of the right aspect of the tentorium, consistent with a tentorial subdural hematoma. This case illustrates the importance of reformatted images.

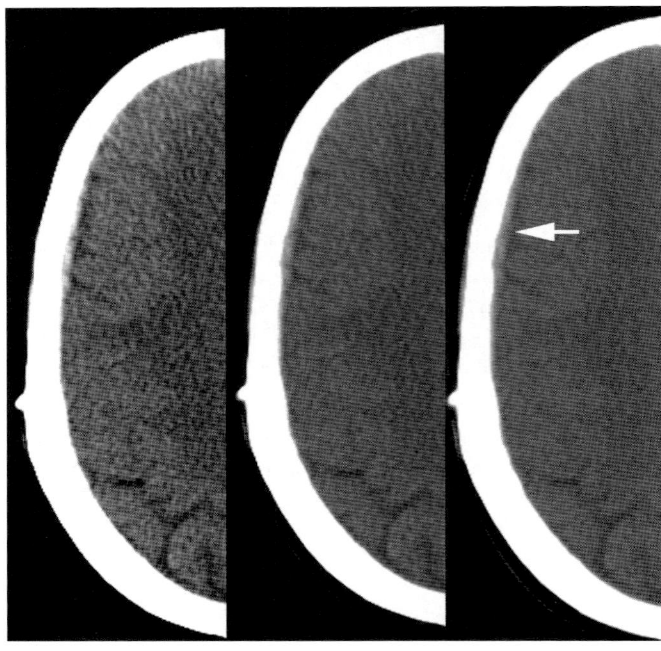

Figure 3.3 Montage of three axial non-enhanced CT images from a 22-year-old man shows a small right frontal subdural hemorrhage (arrow). The image on the left has a window of 80 and level of 40; the middle image has a window of 150 and level of 50; the right image has a window of 200 and level of 50. Notice that as the window widens, the subdural hemorrhage becomes more distinct from the adjacent bone.

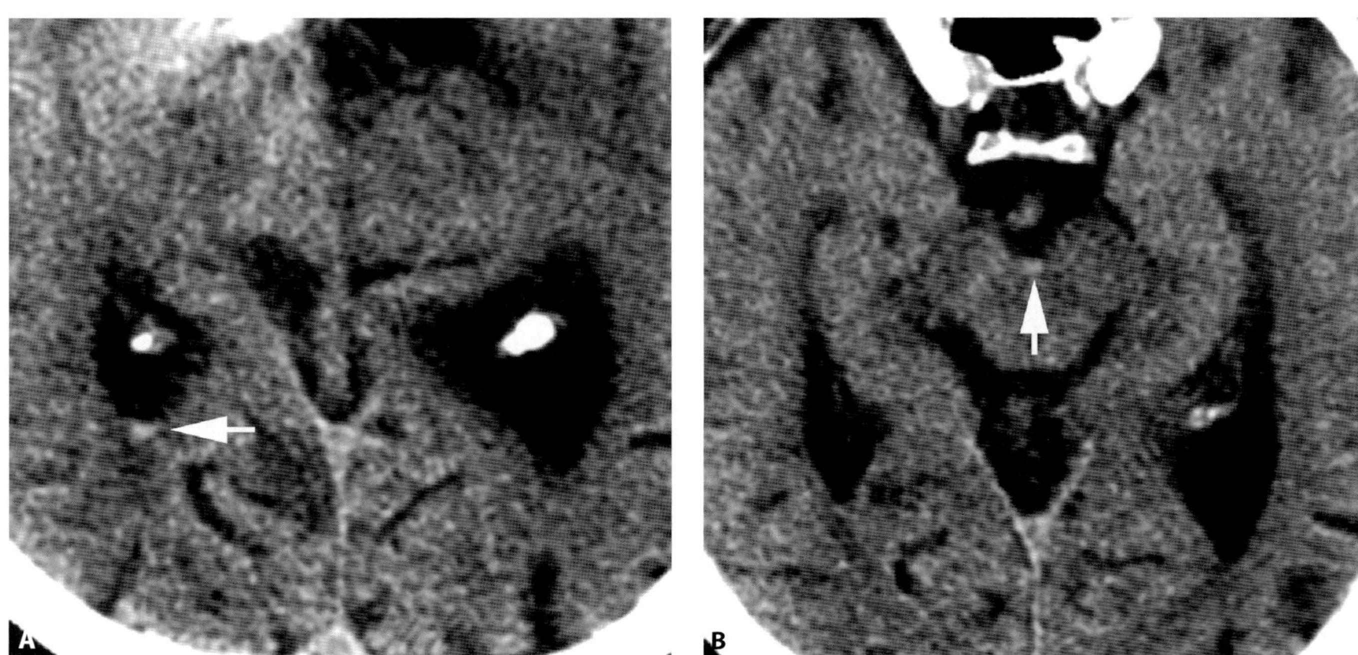

Figure 3.4 **A.** Axial non-enhanced CT image from a 67-year-old woman with head trauma shows a tiny amount of hemorrhage within the right occipital horn (arrow). **B.** Axial non-enhanced CT image in the same patient shows a tiny amount of hemorrhage within the interpeduncular cistern (arrow). These were the only foci of hemorrhage, and could have been missed if these sites were not specifically evaluated.

Pseudo-subarachnoid hemorrhage

Robert B. Carr

Imaging description

Pseudo-subarachnoid hemorrhage (pseudo-SAH) refers to increased attenuation within the basal cisterns and subarachnoid spaces that mimics subarachnoid hemorrhage (SAH), but has a different etiology. The causes of pseudo-SAH include diffuse cerebral edema, meningitis, and intrathecal contrast [1].

Diffuse cerebral edema is the most common cause of pseudo-SAH. Cerebral edema leads to decreased attenuation of the brain parenchyma. There is also compression of the dural venous sinuses, which may lead to venous congestion and engorgement of the superficial veins. The combination of decreased brain attenuation and venous engorgement is postulated to be the etiology of pseudo-SAH in the setting of cerebral edema (Figure 4.1) [2].

The measured attenuation of the subarachnoid spaces will be lower than that seen with true SAH. Venous engorgement will demonstrate attenuation coefficients of 30–42 HU. SAH will demonstrate higher attenuation. Therefore, if accurate measurements can be made, the distinction of pseudo-SAH from true SAH can be made in the setting of cerebral edema [3]. When cerebral edema is caused by a hypoxic event, there may be loss of the gray–white matter differentiation, especially involving the basal ganglia (Figure 4.2).

Exudative meningitis leads to increased protein content within the subarachnoid space. This may rarely produce a pattern of pseudo-SAH [4]. Similar findings may be seen along the pachymeninges (Figure 4.3).

Importance

Pseudo-SAH is often misdiagnosed as true SAH. However, post-mortem studies have proven that the appearance is not caused by the presence of SAH [5]. Therefore, it is important to be aware of this pitfall in order to avoid misdiagnosis. Additionally, it has been shown that patients with post-resuscitation encephalopathy and findings of pseudo-SAH may have a poorer prognosis than those without pseudo-SAH [3].

Typical clinical scenario

The clinical scenario is helpful in arriving at the correct diagnosis. Patients with pseudo-SAH often have a history of an anoxic event, such as cardiac arrest. The rare cases of meningitis that may cause pseudo-SAH will usually have supporting clinical signs and symptoms.

Differential diagnosis

The differential diagnosis is primarily the distinction between pseudo-SAH and true SAH.

Teaching point

Pseudo-SAH is a rare but important pitfall that is most commonly encountered in the setting of diffuse cerebral edema.

REFERENCES

1. Given CA, 2nd, Burdette JH, Elster AD, Williams DW, 3rd. Pseudo-subarachnoid hemorrhage: a potential imaging pitfall associated with diffuse cerebral edema. *AJNR Am J Neuroradiol.* 2003;**24**(2):254–6.
2. Avrahami E, Katz R, Rabin A, Friedman V. CT diagnosis of non-traumatic subarachnoid haemorrhage in patients with brain edema. *Eur J Radiol.* 1998;**28**(3):222–5.
3. Yuzawa H, Higano S, Mugikura S, *et al.* Pseudo-subarachnoid hemorrhage found in patients with postresuscitation encephalopathy: characteristics of CT findings and clinical importance. *AJNR Am J Neuroradiol.* 2008;**29**(8):1544–9.
4. Cucchiara B, Sinson G, Kasner SE, Chalela JA. Pseudo-subarachnoid hemorrhage: report of three cases and review of the literature. *Neurocrit Care.* 2004;**1**(3):371–4.
5. Chute DJ, Smialek JE. Pseudo-subarachnoid hemorrhage of the head diagnosed by computerized axial tomography: a postmortem study of ten medical examiner cases. *J Forensic Sci.* 2002;**47**(2):360–5.

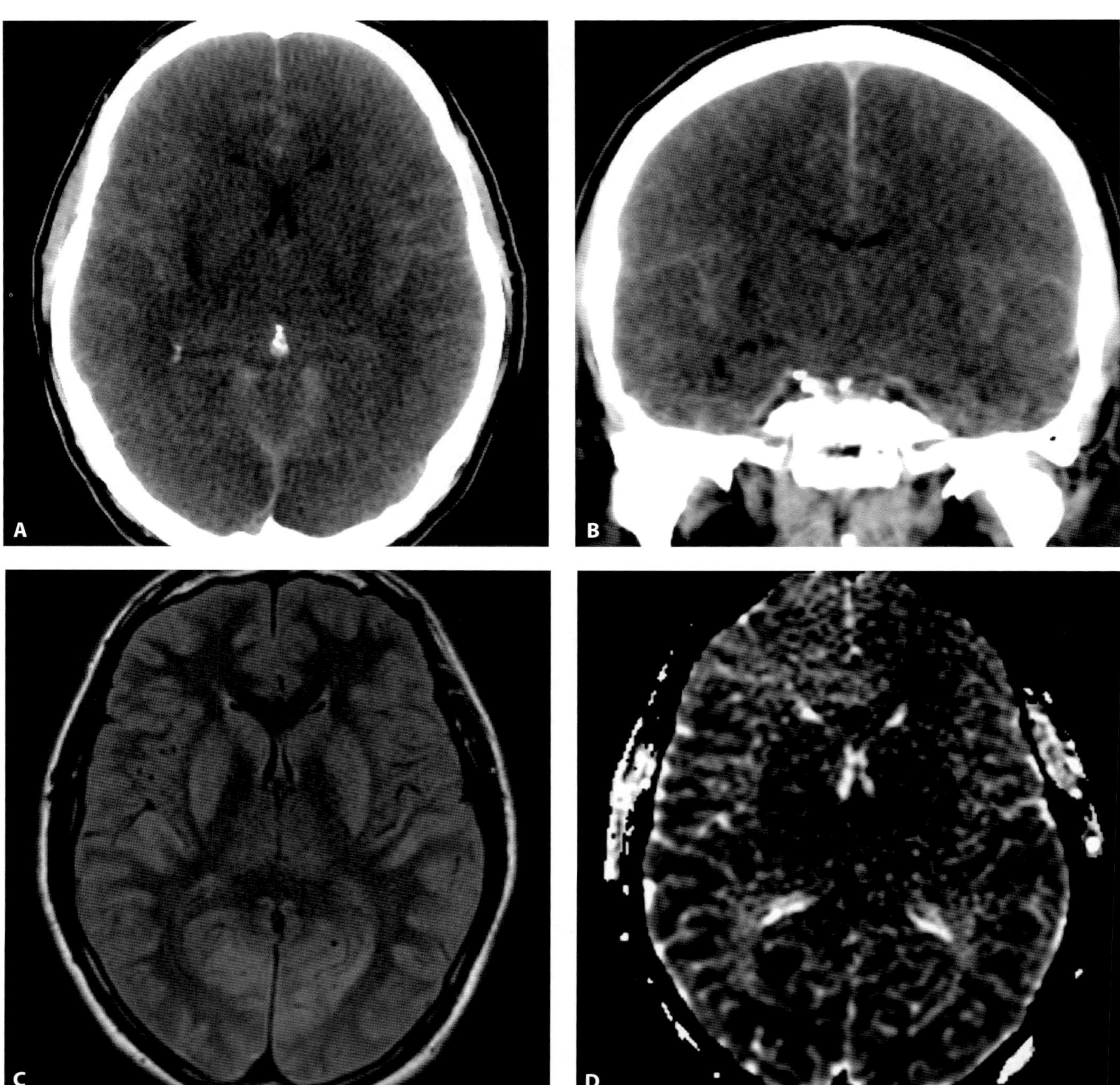

Figure 4.1 **A.** Axial non-enhanced CT image from a 35-year-old man with cardiac arrest shows diffusely increased attenuation within the subarachnoid spaces. Notice the decreased attenuation of the basal ganglia. **B.** Coronal reformatted CT image again demonstrates diffusely increased attenuation within the subarachnoid spaces, especially prominent in the Sylvian fissures. **C.** Axial FLAIR image shows no evidence of sulcal hyperintensity to suggest the presence of SAH. The subarachnoid spaces appeared normal in signal on all MR sequences. **D.** Axial apparent diffusion coefficient map shows decreased diffusivity diffusely, most pronounced centrally within the basal ganglia. These findings illustrate the presence of pseudo-SAH in the setting of anoxic injury.

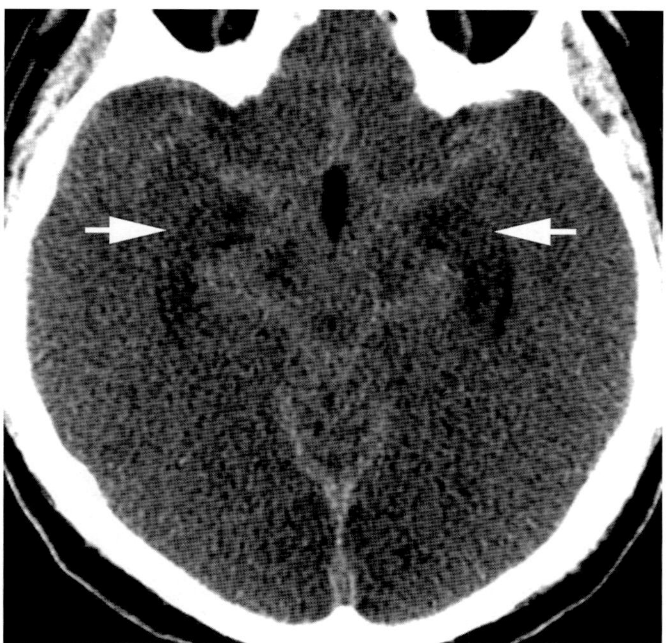

Figure 4.2 Axial non-enhanced CT image from a 48-year-old man two days after an acute myocardial infarction shows increased attenuation within the basal cisterns and Sylvian fissures. Notice the decreased attenuation within the medial temporal lobes (arrows), as well as the diffusely decreased gray–white matter differentiation. The attenuation coefficient of the subarachnoid space is 34HU, lower than expected for true SAH. These findings are typical of pseudo-SAH.

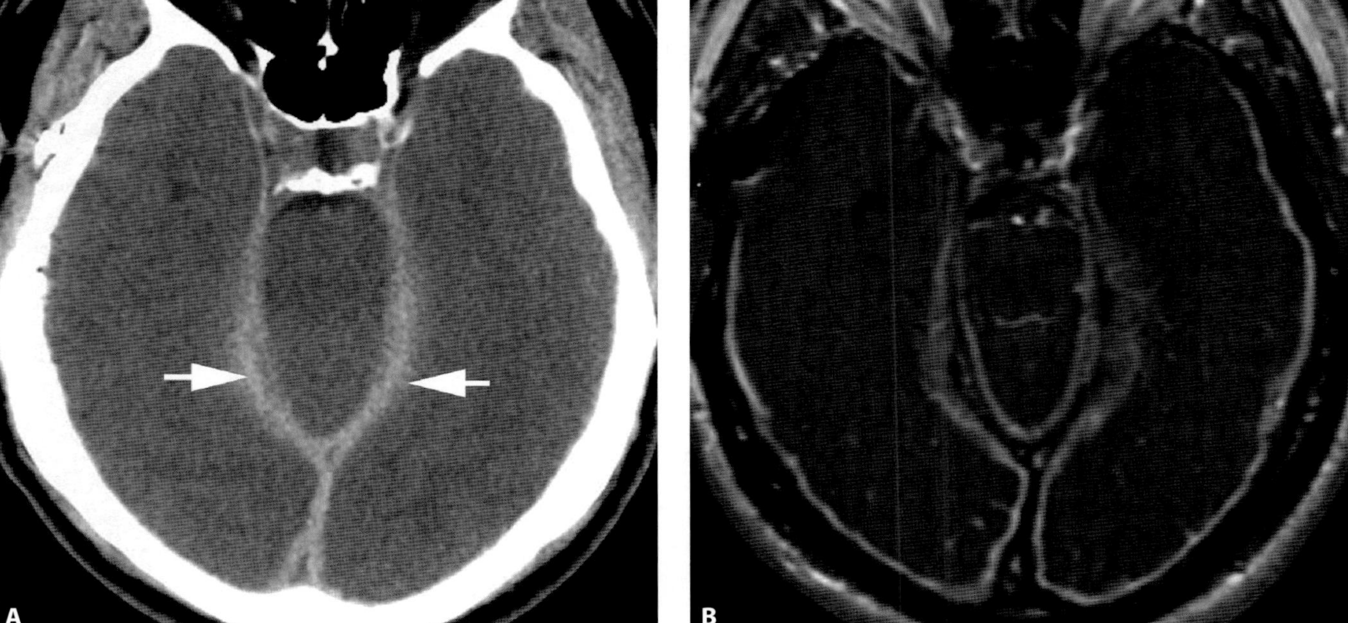

Figure 4.3 **A.** Axial non-enhanced CT image from a 58-year-old man with tuberculous meningitis shows increased attenuation along the tentorium which simulates the appearance of subdural hemorrhage (arrows). **B.** Axial contrast-enhanced T1-weighted image shows diffuse pachymeningeal enhancement and thickening. The findings on CT are analogous to pseudo-SAH.

Arachnoid granulations

Robert B. Carr

Imaging description

Arachnoid villi represent the normal sites of cerebrospinal fluid (CSF) resorption from the subarachnoid space into the venous sinuses. The villi are not visible radiologically, but they may enlarge over time due to distension with CSF. This causes progressive penetration of the arachnoid membrane into the dura, beneath the vascular endothelium of the venous sinus. The result is formation of an arachnoid granulation. These granulations tend to increase in size and number with age.

Arachnoid granulations are most commonly located within the transverse sinuses, superior sagittal sinus, and parasagittal venous lacunae [1]. They usually range in size from 2 to 8 mm [2], though may be larger than 10 mm at which time they are referred to as "giant" arachnoid granulations.

Contrast-enhanced CT demonstrates a round or oval filling defect within a dural venous sinus. An arachnoid granulation typically occurs where a superficial vein drains into the venous sinus, which is thought to induce a focal weakness in the sinus wall. There may be a smooth corticated erosion of the adjacent calvarium. A granulation will never demonstrate hyperattenuation on non-enhanced CT, unlike dural venous sinus thrombosis, which commonly demonstrates increased attenuation (Figures 5.1 and 5.2).

Usually, an arachnoid granulation will follow CSF signal on all MR sequences (Figure 5.3) [3]. However, several studies have demonstrated that this rule does not always hold true, especially with giant granulations. For instance, the signal may be higher than CSF on both T1- and T2-weighted images, and suppression on FLAIR images may be incomplete. In these cases, the characteristic shape, location, and lack of solid enhancement are helpful clues to the correct diagnosis [4].

Importance

Arachnoid granulations are common. Autopsy series have reported their occurrence in up to two-thirds of people. Imaging studies report a widely varying incidence. However, with the increased use of high-resolution MRI, they are encountered more commonly. A study of 90 patients found 433 arachnoid granulations using contrast-enhanced magnetization prepared rapid acquisition gradient-echo (MPRAGE) sequences [5]. It is important to recognize an arachnoid granulation as a normal structure, and not to misinterpret it as an extra-axial mass or venous thrombosis.

Typical clinical scenario

Arachnoid granulations are almost always incidental findings. Rarely, a giant arachnoid granulation may functionally occlude a dural venous sinus, resulting in venous hypertension. In such cases, if there is no other finding to explain symptoms of increased intracranial pressure, venous sinus pressure measurements proximal and distal to the lesion may be obtained to assess for the presence of obstruction [6].

Differential diagnosis

The primary differential considerations include dural venous sinus thrombosis (Figure 5.4) and an extra-axial mass, such as a metastasis or meningioma. Most dilemmas will arise in the setting of a giant arachnoid granulation, which may not follow CSF on all MR sequences. However, they will always be similar to CSF attenuation on CT and should not demonstrate enhancement. The characteristic location at the site of superficial vein penetration is also helpful.

Teaching point

Arachnoid granulations demonstrate hypoattenuation on CT and often follow CSF signal on MRI. However, larger arachnoid granulations may differ in signal from CSF, and other findings may be required for the correct diagnosis.

REFERENCES

1. VandeVyver V, Lemmerling M, De Foer B, Casselman J, Verstraete K. Arachnoid granulations of the posterior temporal bone wall: imaging appearance and differential diagnosis. *AJNR Am J Neuroradiol.* 2007;**28**(4):610–12.
2. Ikushima I, Korogi Y, Makita O, *et al.* MRI of arachnoid granulations within the dural sinuses using a FLAIR pulse sequence. *Br J Radiol.* 1999;**72**(863):1046–51.
3. Roche J, Warner D. Arachnoid granulations in the transverse and sigmoid sinuses: CT, MR, and MR angiographic appearance of a normal anatomic variation. *AJNR Am J Neuroradiol.* 1996;**17**(4):677–83.
4. Trimble CR, Harnsberger HR, Castillo M, Brant-Zawadzki M, Osborn AG. "Giant" arachnoid granulations just like CSF?: NOT!! *AJNR Am J Neuroradiol.* 2010;**31**(9):1724–8.
5. Liang L, Korogi Y, Sugahara T, *et al.* Normal structures in the intracranial dural sinuses: delineation with 3D contrast-enhanced magnetization prepared rapid acquisition gradient-echo imaging sequence. *AJNR Am J Neuroradiol.* 2002;**23**(10):1739–46.
6. Kan P, Stevens EA, Couldwell WT. Incidental giant arachnoid granulation. *AJNR Am J Neuroradiol.* 2006;**27**(7):1491–2.

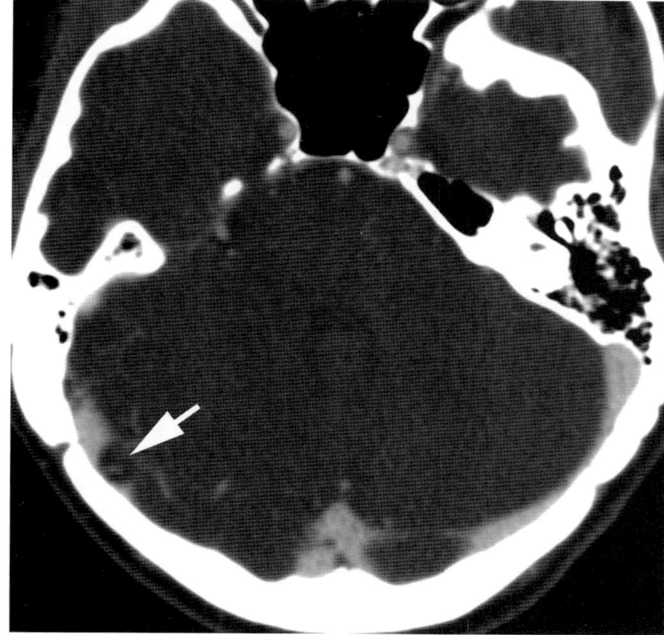

Figure 5.1 Axial contrast-enhanced CT image from a 42-year-old man with an incidental arachnoid granulation shows a round low-attenuation filling defect within the right transverse sinus (arrow). The central focus of enhancement represents a penetrating superficial vein. This is the characteristic appearance of an arachnoid granulation.

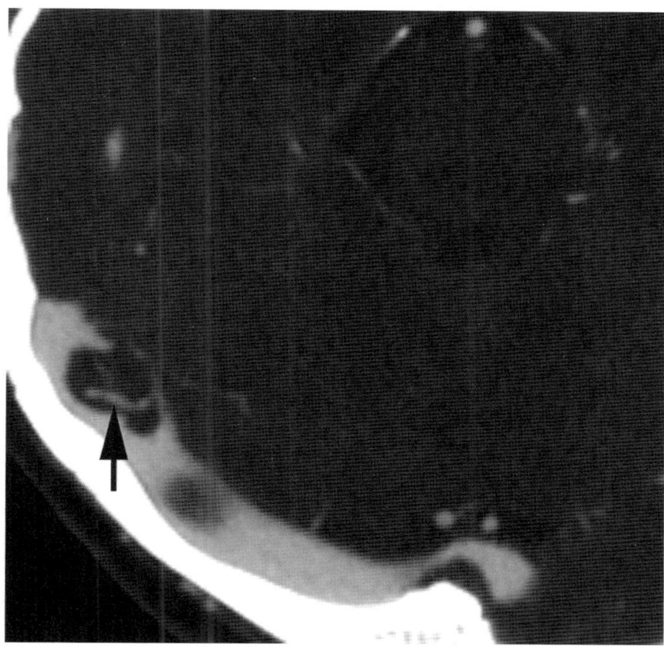

Figure 5.2 Axial contrast-enhanced CT image from a 48-year-old woman with an incidental arachnoid granulation shows a lobular, oval, low-attenuation filling defect within the right transverse sinus. Notice the prominent penetrating superficial vein coursing through the granulation (arrow).

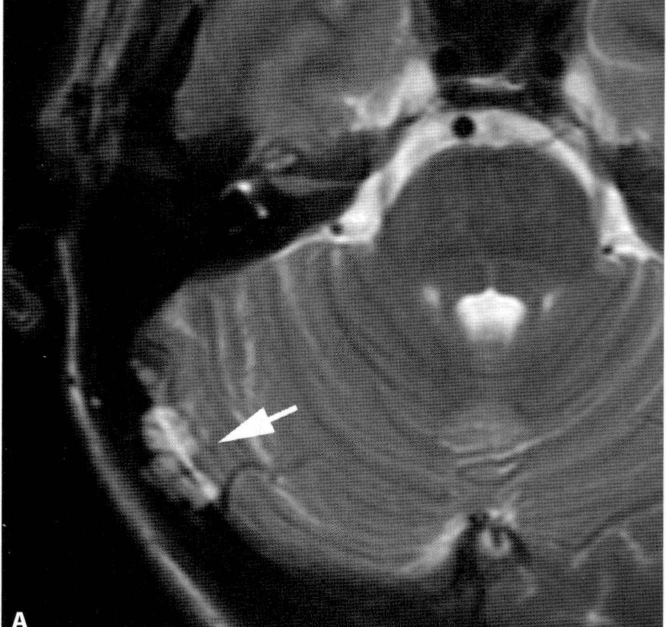

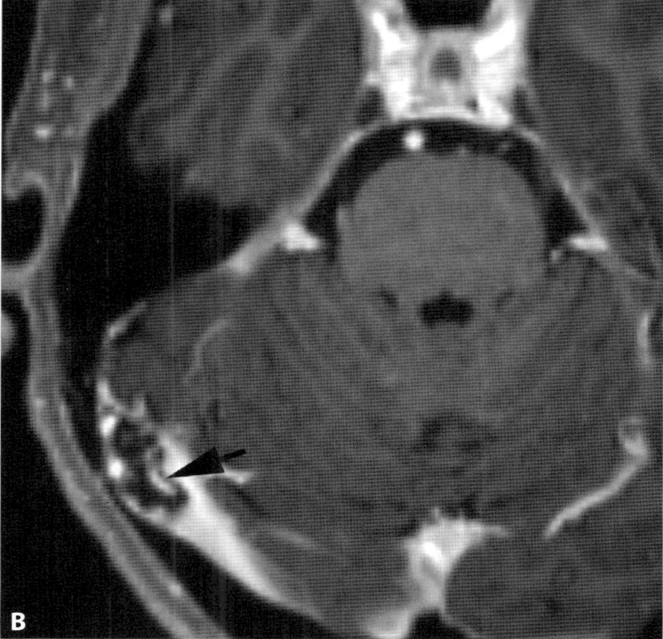

Figure 5.3 **A.** Axial T2-weighted MR image from a 51-year-old man with an incidental arachnoid granulation of the right transverse sinus (arrow). Notice that the granulation is isointense to CSF. **B.** Axial contrast-enhanced T1-MPRAGE image also demonstrates isointensity with CSF. Notice the enhancing penetrating superficial vein (arrow). The granulation itself does not enhance.

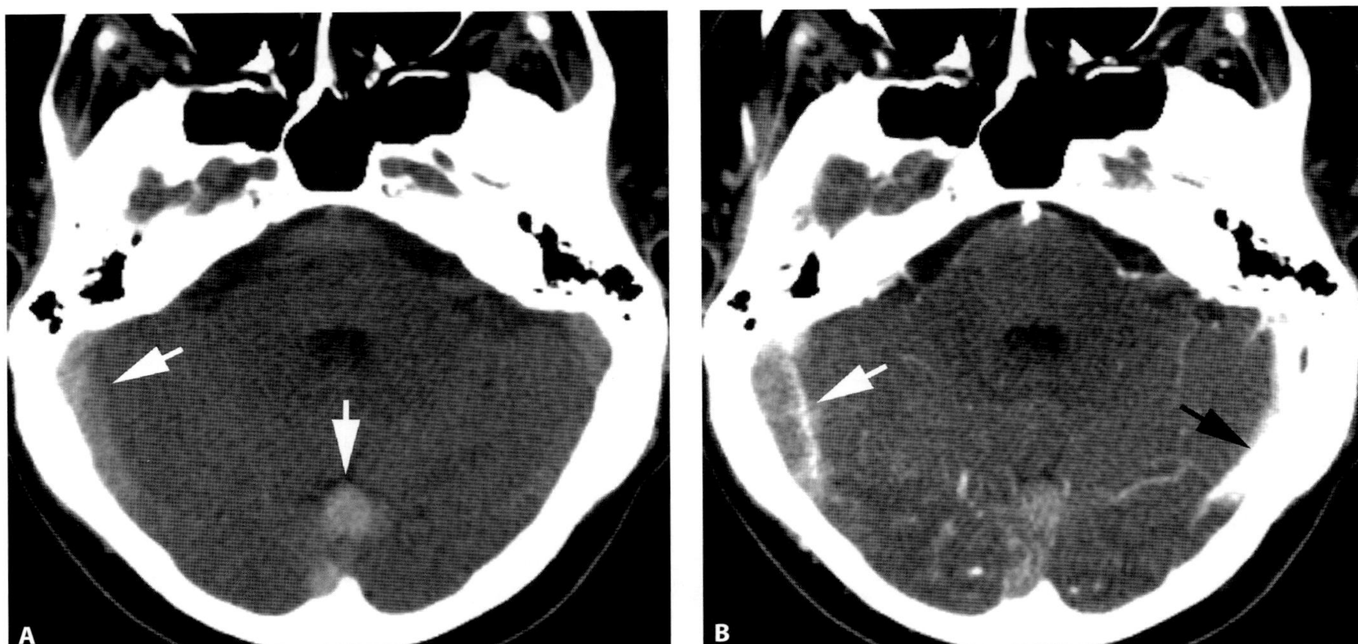

Figure 5.4 A. Axial non-enhanced CT image from a 24-year-old female on oral contraceptives shows hyperattenuation within the right transverse and straight sinuses (arrows). An arachnoid granulation is never hyperdense. **B.** Axial contrast-enhanced CT image shows only thin peripheral enhancement of the right transverse sinus (white arrow). The left transverse sinus enhances normally (black arrow). This is a typical example of dural venous sinus thrombosis.

Ventricular enlargement

Robert B. Carr

Imaging description

Enlarged ventricles can be caused by hydrocephalus or parenchymal loss. Hydrocephalus is classified as non-communicating (obstructive) or communicating. It is important to try and distinguish among these different patterns, as this will direct further workup and management.

Non-communicating hydrocephalus results from obstruction of the ventricular outflow of cerebrospinal fluid (CSF). Frequent causes include neoplasms, aqueductal stenosis, and intraventricular hemorrhage. The site of obstruction can be implied by which ventricles are enlarged.

Communicating hydrocephalus usually results from obstruction of CSF resorption at the arachnoid granulations. Less common causes include overproduction of CSF and compromised venous outflow. Normal-pressure hydrocephalus (NPH) is a specific form of communicating hydrocephalus which is associated with the clinical triad of dementia, ataxia, and urinary incontinence.

Hydrocephalus may lead to transpendymal edema, caused by transpendymal resorption of CSF. This will produce a rim of decreased attenuation (CT) or high T2/FLAIR signal (MRI) around the lateral ventricles (Figure 6.1). This usually indicates an acute enlargement of the ventricles [1].

Some apparent cases of communicating hydrocephalus are caused by fourth ventricular outflow obstruction. An apparent obstructive lesion may not be evident; CT and conventional MRI may miss small webs which obstruct outflow at the foramina of Luschka and Magendie. The addition of 3D constructive interference in the steady state (CISS) images may allow for improved detection of small membranes at the suspected site of obstruction (Figure 6.2) [2].

Parenchymal volume loss can result in focal or diffuse ventricular enlargement. Diffuse ventricular enlargement caused by volume loss may sometimes be difficult to distinguish from hydrocephalus. Several findings may be helpful in making the distinction [3], including the following, which are features of hydrocephalus:

- Widening of the third ventricular recesses
- Thinning and displacement of the corpus callosum
- Decrease in the mammillopontine distance
- Depression of the fornix.

Normal-pressure hydrocephalus is a unique form of communicating hydrocephalus which can be difficult to distinguish from parenchymal atrophy. The findings described above may be useful in making the distinction. An additional finding which is sometimes present in NPH is focal dilation of the subarachnoid spaces over the convexity or medial surface of the cerebrum in the setting of compressed subarachnoid spaces elsewhere. There may also be isolated enlargement of the Sylvian fissures or basal cisterns (Figure 6.3) [4].

Importance

It is important to distinguish true hydrocephalus from atrophy, as this will direct clinical management. Patients with hydrocephalus may require surgical intervention. In particular, patients with NPH will often improve after ventricular shunting [5].

Typical clinical scenario

The clinical presentation varies widely based upon the underlying abnormality. Hydrocephalus may be associated with symptoms of increased intracranial pressure, such as headache, nausea, vomiting, seizures, and visual changes. NPH may be associated with the clinical triad of dementia, ataxia, and urinary incontinence. Parenchymal loss may be associated with other forms of dementia, such as Alzheimer's dementia or vascular dementia (Figure 6.4).

Differential diagnosis

The differential diagnosis of enlarged ventricles includes communicating hydrocephalus (including NPH), non-communicating hydrocephalus, and ex vacuo dilation due to parenchymal loss. Features that help distinguish between these entities are described above.

> **Teaching point**
>
> It may not always be possible to distinguish hydrocephalus from enlarged ventricles due to atrophy. It is important to use clinical signs and symptoms to help direct the diagnosis in difficult cases.

REFERENCES

1. Ulug AM, Truong TN, Filippi CG, *et al.* Diffusion imaging in obstructive hydrocephalus. *AJNR Am J Neuroradiol.* 2003;**24**(6):1171–6.
2. Dincer A, Kohan S, Ozek MM. Is all "communicating" hydrocephalus really communicating? Prospective study on the value of 3D-constructive interference in steady state sequence at 3T. *AJNR Am J Neuroradiol.* 2009;**30**(10):1898–906.
3. Segev Y, Metser U, Beni-Adani L, *et al.* Morphometric study of the midsagittal MR imaging plane in cases of hydrocephalus and atrophy and in normal brains. *AJNR Am J Neuroradiol.* 2001;**22**(9):1674–9.
4. Kitagaki H, Mori E, Ishii K, *et al.* CSF spaces in idiopathic normal pressure hydrocephalus: morphology and volumetry. *AJNR Am J Neuroradiol.* 1998;**19**(7):1277–84.
5. Katzen H, Ravdin LD, Assuras S, *et al.* Postshunt cognitive and functional improvement in idiopathic normal pressure hydrocephalus. *Neurosurgery.* 2011;**68**(2):416–19.

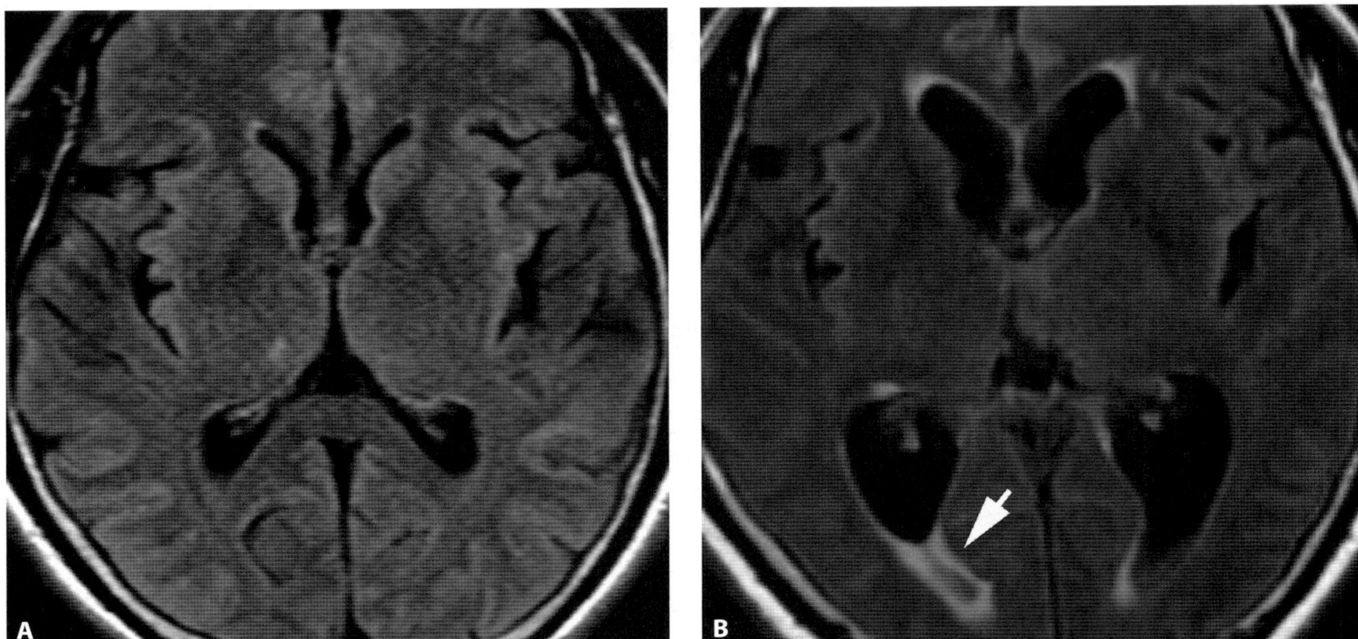

Figure 6.1 A. Axial FLAIR image from a 58-year-old man with a posterior fossa tumor (not shown) shows a normal size of the lateral ventricles. **B.** Axial FLAIR image obtained six months later shows interval enlargement of the ventricles, with transependymal edema (arrow). The posterior fossa tumor had enlarged and obstructed CSF outflow from the fourth ventricle.

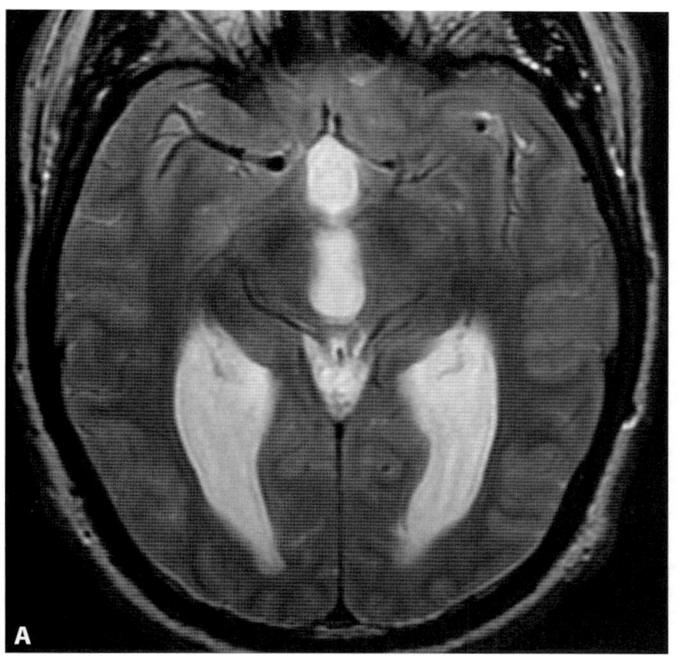

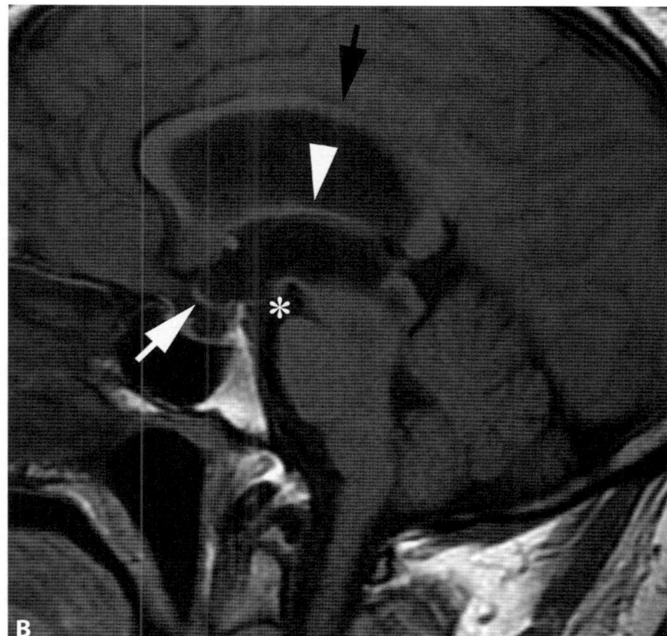

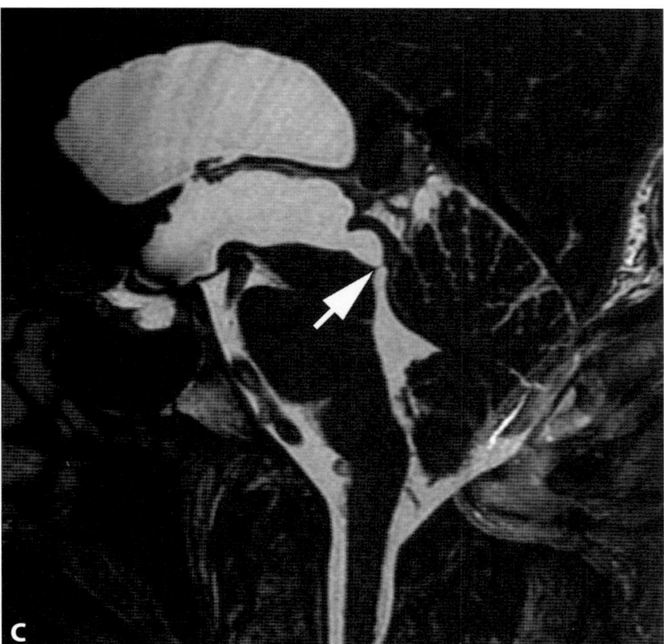

Figure 6.2 **A.** Axial T2-weighted image from a 40-year-old man with headaches and visual changes shows enlargement of the lateral and third ventricles which is out of proportion to the size of the sulci. **B.** Sagittal T1-weighted image demonstrates the typical findings of hydrocephalus, including upward displacement and thinning of the corpus callosum (black arrow), downward displacement of the fornices (arrowhead), enlargement of the anterior third ventricular recesses (white arrow), and decrease in the mammillopontine distance (asterisk). **C.** Sagittal 3D-CISS image demonstrates a thin web within the cerebral aqueduct (arrow) which is responsible for the obstructive hydrocephalus.

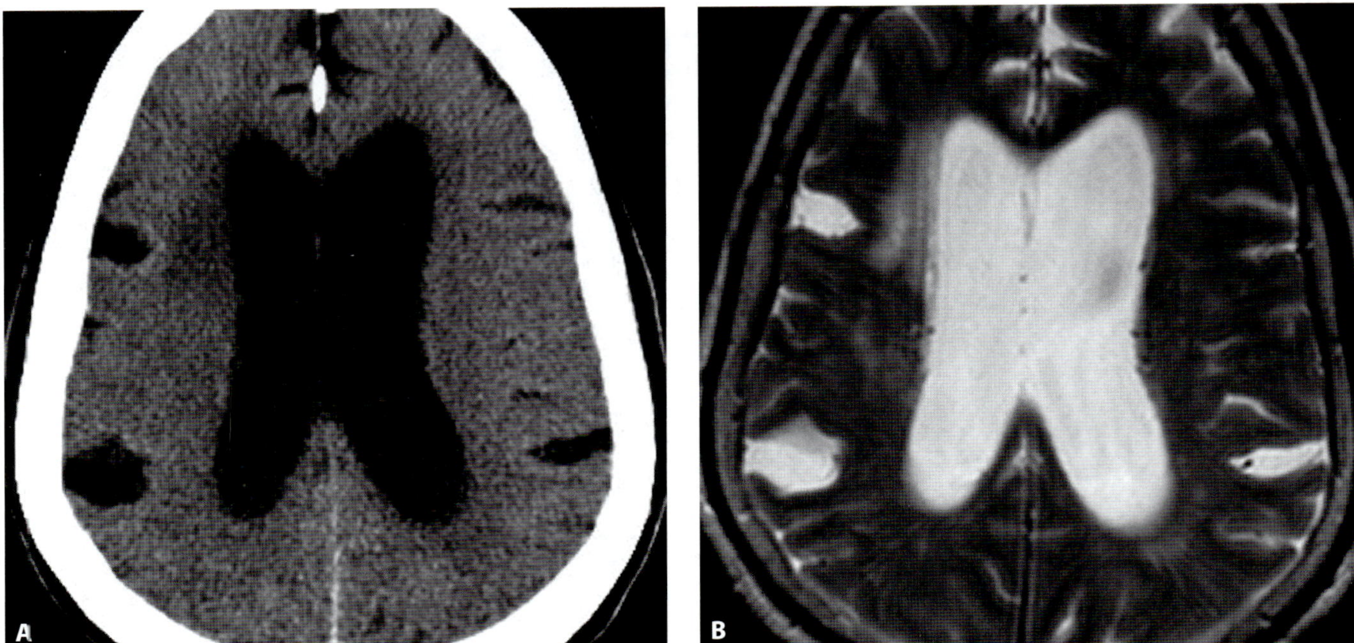

Figure 6.3 A. Axial non-enhanced CT image from a 69-year-old woman with clinical symptoms of NPH demonstrates enlargement of the ventricles and focal enlargement of a few isolated sulci. **B.** Axial T2-weighted image confirms the findings seen on CT. This pattern of isolated sulcal enlargement is often associated with NPH.

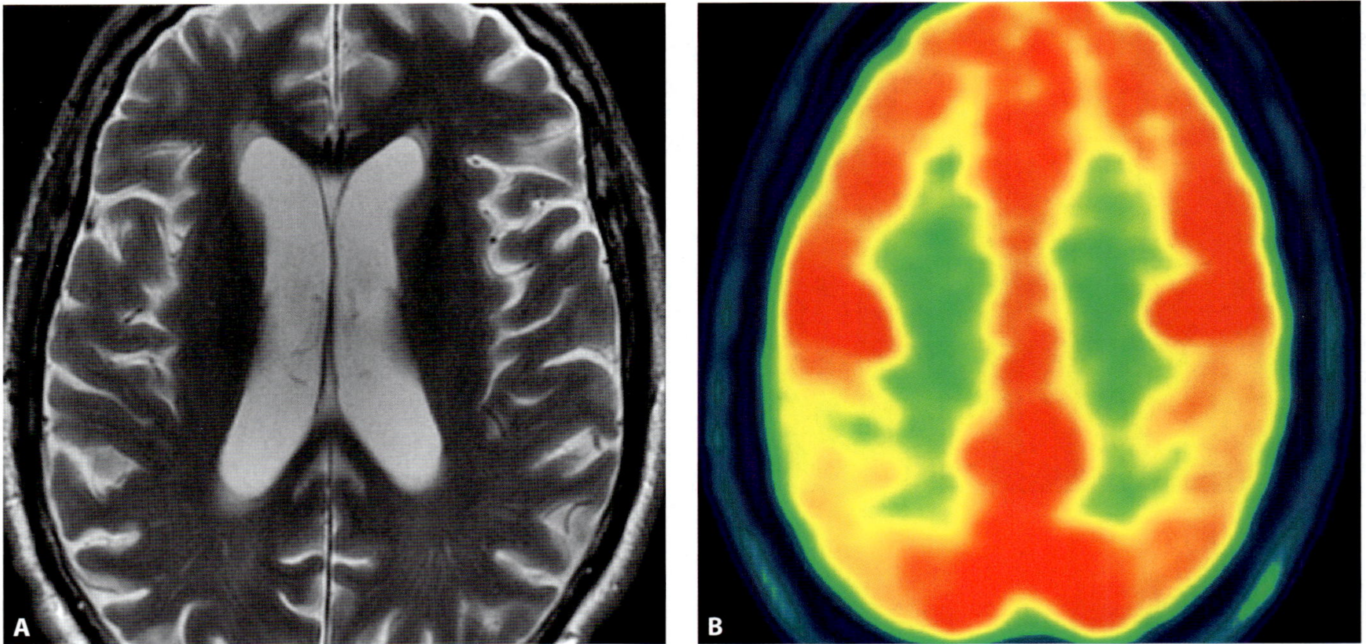

Figure 6.4 A. Axial T2-weighted image from a 71-year-old man with symptoms of Alzheimer's dementia demonstrates mild enlargement of the lateral ventricles and mild diffuse sulcal enlargement. The ventricular enlargement is not out of proportion to the sulcal enlargement. **B.** Axial PET image demonstrates hypometabolism in the parietal lobes bilaterally, a pattern supportive of the clinical diagnosis of Alzheimer's dementia.

CASE 7

Blunt cerebrovascular injury

Robert B. Carr

Imaging description

Blunt cerebrovascular injury (BCVI) may present as a range of clinical and radiologic manifestations. While there is controversy regarding the accuracy of CT angiography (CTA) compared with digital subtraction angiography (DSA), most cases will be first identified on CTA.

The various imaging manifestations of BCVI include minimal intimal injury, raised intimal flap, dissection with intramural hematoma, occlusion, pseudoaneurysm, transection, and arteriovenous fistula. The carotid artery is most often injured in the cervical segment, just inferior to the skull base. The vertebral artery is most often injured in the cervical segment. Multi-vessel injury is common (Figure 7.1) [1].

A grading system was developed by Biffl *et al.*, based upon DSA findings. Five grades are defined, and with regards to the carotid artery, a higher grade is associated with a higher likelihood of cerebral infarction (Table 7.1) [2]. In patients who do not exhibit clinical symptoms of BCVI, there are several imaging findings that warrant vascular screening. In general, any severe facial or cervical injury can be associated with BCVI. Several specific findings which warrant vascular screening have been grouped together as the Denver criteria. These include Le Fort II or III facial fractures; carotid canal fractures; diffuse axonal injury with a Glasgow Coma Scale (GCS) less than 6; near hanging with anoxic brain injury; cervical fractures of C1 through C3 or involving a transverse foramen; and cervical spine subluxation (Figure 7.2) [3].

Importance

There is a high rate of morbidity and mortality from BCVI related to brain infarction [4]. Treatment with antithrombotic medications and carotid artery stenting in eligible patients has been shown to be effective in reducing cerebral infarction [5]. Therefore, it is important to recognize these injuries and understand when vascular imaging is warranted.

There is controversy regarding the accuracy of CTA in detecting BCVI. Some studies have found CTA inferior to DSA [6]. However, CTA will remain an important screening and diagnostic tool, and the majority of cases will be initially detected by CTA.

Table 7.1. Grades of blunt cerebrovascular injury

Grade
Grade I: Minimal intimal irregularities or intramural hematoma with luminal narrowing less than 25%
Grade II: Intramural hematoma with greater than 25% luminal narrowing, intraluminal thrombus, and hemodynamically insignificant arteriovenous fistula
Grade III: Pseudoaneurysm
Grade IV: Vessel occlusion
Grade V: Hemodynamically significant arteriovenous fistula

Typical clinical scenario

The incidence of BCVI was reported in a large series to be 1.1% of all blunt trauma victims [7]. The diagnosis is often suspected clinically based upon the presence of focal neurologic deficits. However, some patients will be asymptomatic or may develop delayed symptoms related to subsequent cerebral infarction [8].

Differential diagnosis

Atherosclerotic plaque may mimic findings of BCVI. The presence of calcification and a characteristic location help in the distinction. Beam-hardening artifact may also be confused with injury.

Teaching point

The morbidity and mortality of BCVI are high, but early treatment is effective. Therefore, there should be a relatively low threshold for screening at-risk patients.

REFERENCES

1. Sliker CW. Blunt cerebrovascular injuries: imaging with multidetector CT angiography. *Radiographics*. 2008;**28**(6):1689–708; discussion 709–10.
2. Biffl WL, Moore EE, Offner PJ, *et al.* Blunt carotid arterial injuries: implications of a new grading scale. *J Trauma*. 1999;**47**(5):845–53.
3. Cothren CC, Moore EE. Blunt cerebrovascular injuries. *Clinics (Sao Paulo)*. 2005;**60**(6):489–96.
4. Cothren CC, Moore EE, Ray CE, Jr., *et al.* Screening for blunt cerebrovascular injuries is cost-effective. *Am J Surg*. 2005;**190**(6): 845–9.
5. Edwards NM, Fabian TC, Claridge JA, *et al.* Antithrombotic therapy and endovascular stents are effective treatment for blunt carotid injuries: results from longterm followup. *J Am Coll Surg*. 2007;**204**(5):1007–13; discussion 14–15.
6. DiCocco JM, Emmett KP, Fabian TC, *et al.* Blunt cerebrovascular injury screening with 32-channel multidetector computed tomography: more slices still don't cut it. *Ann Surg*. 2011;**253**(3):444–50.
7. Schneidereit NP, Simons R, Nicolaou S, *et al.* Utility of screening for blunt vascular neck injuries with computed tomographic angiography. *J Trauma*. 2006;**60**(1):209–15; discussion 15–16.
8. Biffl WL, Moore EE, Offner PJ, Burch JM. Blunt carotid and vertebral arterial injuries. *World J Surg*. 2001;**25**(8):1036–43.

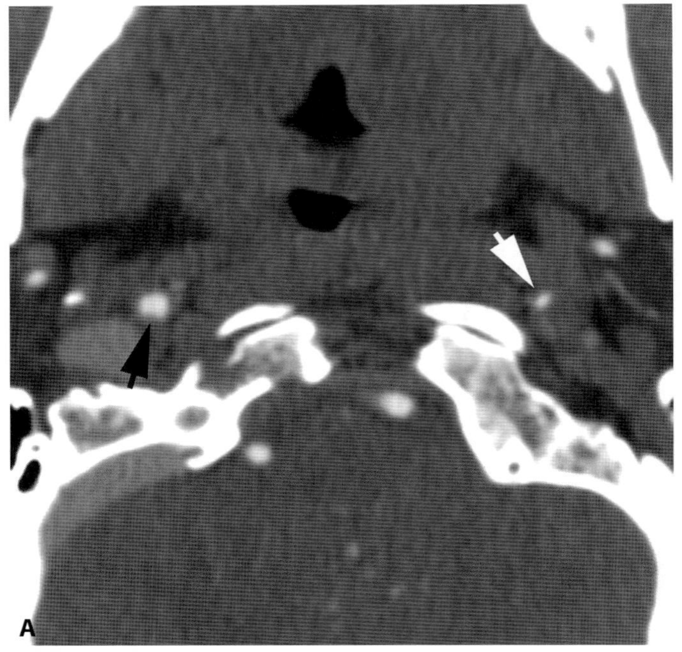

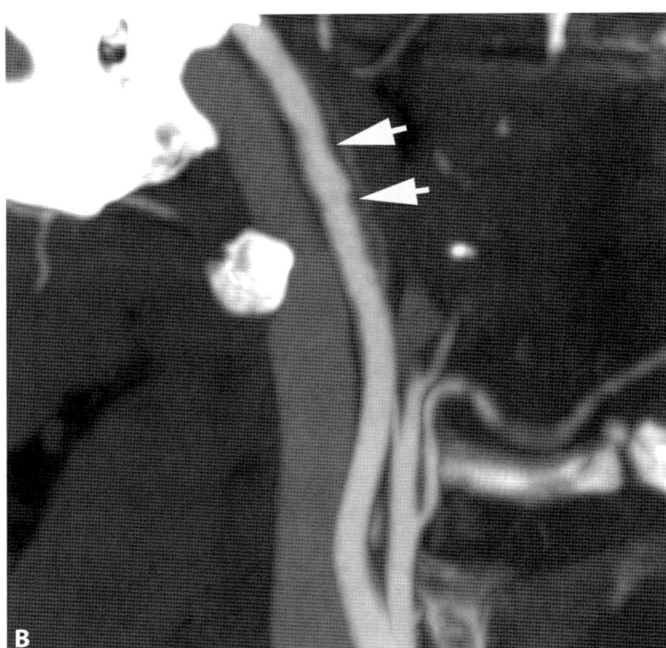

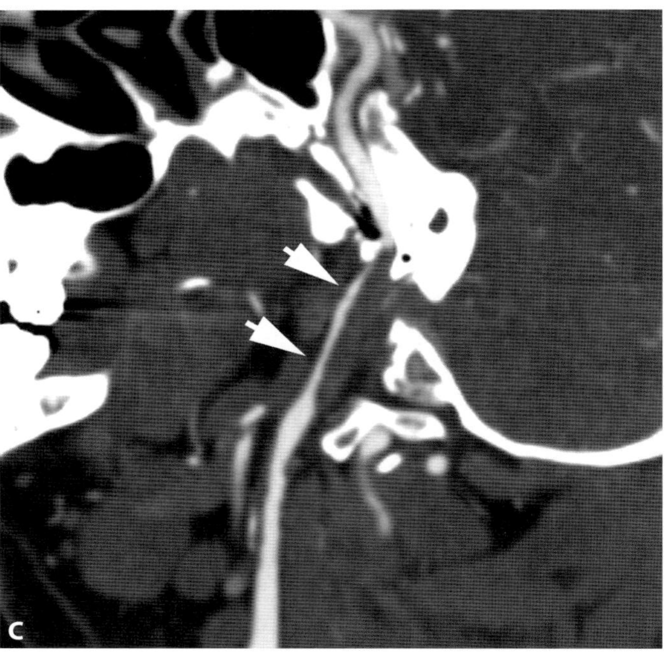

Figure 7.1 **A.** Axial CTA image from a 36-year-old man who was involved in a motor vehicle collision shows mild irregularity of the right internal carotid artery (black arrow) and greater than 50% narrowing of the left internal carotid artery (white arrow). **B.** Curved reformatted image of the right internal carotid artery demonstrates mild luminal irregularities (arrows). This is a young patient, and there are no findings to suggest the presence of atherosclerosis. This would be classified as a grade I injury in the Biffl classification. **C.** Curved reformatted image of the left internal carotid artery shows irregular narrowing over a long segment of the artery (arrows). This would be classified as a grade II injury. Multi-vessel injury is common, so injury in one vessel should prompt a careful evaluation of the other vessels.

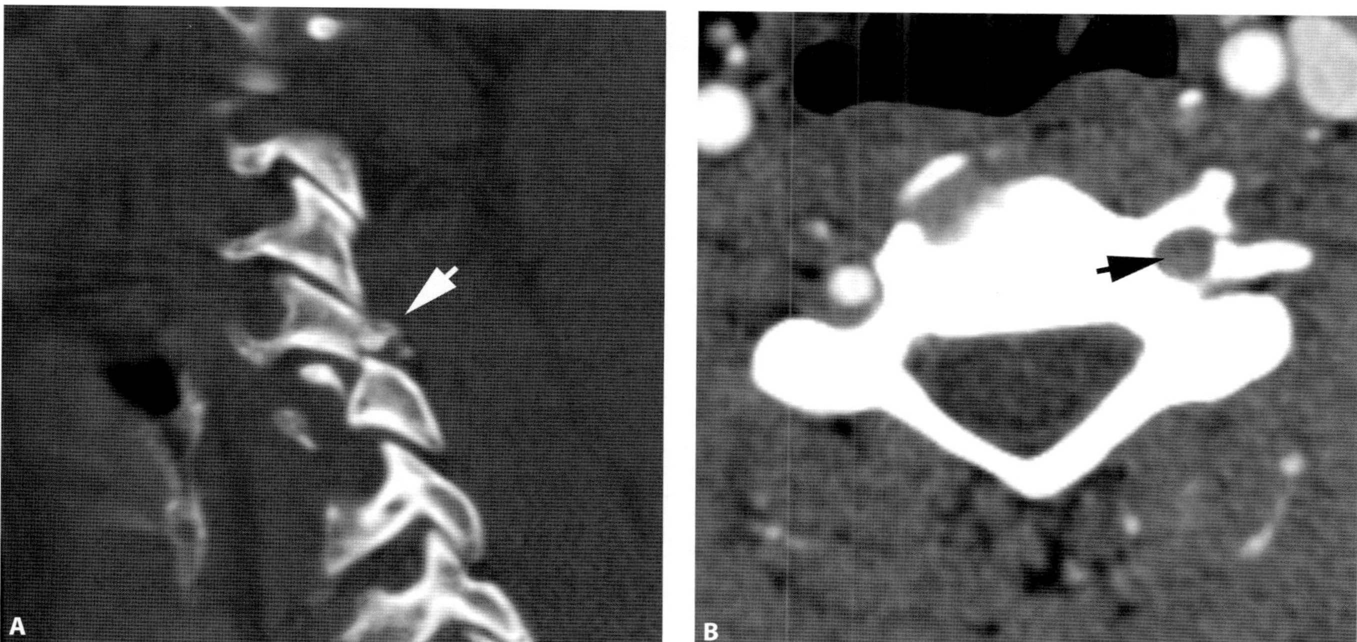

Figure 7.2 A. Sagittal non-enhanced CT image of the cervical spine from a 29-year-old man demonstrates a comminuted left-sided facet fracture with subluxation (arrow). This injury should prompt further evaluation with CTA. **B.** Axial CTA image at the level of the fracture demonstrates occlusion of the left vertebral artery (arrow). This is a grade IV injury.

Internal carotid artery dissection presenting as subacute ischemic stroke

Michael J. Modica

Imaging description

Internal carotid artery (ICA) dissection may be clinically unsuspected as a cause of subacute ischemic stroke, but there are imaging manifestations that suggest this diagnosis. A watershed distribution of hypodensity (Figure 8.1A) in a patient presenting with acute neurologic deficit is highly suspicious for ischemic infarct from an ipsilateral ICA dissection. MRI typically demonstrates restriction (bright signal) on diffusion-weighted images and decreased signal on the apparent diffusion coefficient (ADC) map which corresponds to acute cytotoxic edema in the areas of hypodensity (Figure 8.1B). Additionally, finding intrinsic high T1 and FLAIR signal outlining the affected cortex (Figure 8.1C and 8.1D) is consistent with laminar necrosis, as these signal abnormalities take 10–14 days to appear after the ischemic event. This indicates subacute ischemia. Corresponding gyriform or superficial enhancement in the affected cortex and subcortical white matter (Figure 8.1E and 8.1F) can help differentiate the process from other diseases. Gyriform enhancement is usually caused by vascular or inflammatory processes and is rarely neoplastic.

Finding bright signal within the ipsilateral carotid artery on T1-weighted images, corresponding to loss of the normally visualized signal void, suggests slow flow or thrombus as the etiology for the acute ischemic event (Figure 8.1G). If the vascular flow voids are normal on the brain MR study, MR angiography (MRA) of the neck vasculature will often reveal the source of the ischemic event.

In the majority of cases, carotid artery dissection involves the extracranial ICA, sparing the bifurcation. Time-of-flight MRA will show signal loss in the ipsilateral ICA (Figure 8.1H), consistent with a flow-related abnormality secondary to dissection and possible thrombosis. Three-dimensional maximum intensity projection images will usually demonstrate flow-related signal loss or smooth tapering of the ICA lumen distal to the bifurcation. On CTA, ICA dissection is characterized by eccentric luminal narrowing with an increase in the external diameter of the artery from mural thrombus, producing a target appearance [1].

Typical clinical scenario

ICA dissection has two typical presentations: spontaneous or the sequelae of blunt trauma. Spontaneous ICA dissection can present with ipsilateral frontotemporal headache that can mimic the headache associated with subarachnoid hemorrhage [2, 3]. This is accompanied by Horner's syndrome [3] in less than half of patients. These symptoms can be followed by cerebral or retinal ischemia, reported in 50–95% of patients, in hours to days after symptom onset [3, 4].

Importance

Stroke due to traumatic or spontaneous ICA dissection is a common cause of ischemic events in young adults, occurring in approximately 60% of reported cases [5]. Ischemia is caused by either thromboembolic events or hypoperfusion. Early recognition of the clinical and imaging manifestations of ICA dissection is critical for early intervention and differentiation from other disease processes. MRI has demonstrated sensitivity and specificity of 84% and 99% respectively compared with conventional angiography for the diagnosis of ICA dissection [6]. Once ICA dissection is recognized, immediate anticoagulation is the first-line treatment to prevent thromboembolism and allow restoration of vessel anatomy.

Endovascular stent placement is an option in the following situations:

- Failure of anticoagulation – recurrent transient ischemia attacks (TIAs), fluctuating neurologic status or neurologic deterioration,
- Impending stroke,
- Contraindication to anticoagulation,
- Symptomatic thromboembolic occlusion of cerebral vessels.

Differential diagnosis

The imaging manifestations of acute stroke can have considerable variation, and can resemble other vascular, neurologic, or neoplastic processes.

The CNS abnormalities caused by ICA dissection and subsequent watershed stroke can resemble:

- *Arteriosclerosis*: commonly referred to as microvascular ischemic disease. Scattered, bilateral multifocal lesions, predominantly in the subcortical and periventricular white matter without a predilection for the watershed zone. Confluent lesions are usually found around the atria of the lateral ventricles and are common in hypertensive patients. No enhancement on post-contrast images and typically spares the cortex.
- *Infiltrating CNS neoplasm*: presents as a dominant mass, typically with mass effect on surrounding tissues with or without necrosis. Diffusion-weighted imaging will show a vasogenic pattern of edema – diffusion restriction (bright signal) and increased signal on ADC map.
- *Posterior reversible encephalopathy (PRES)*: often subcortical occipital, sparing the basal ganglia. Typically lacks diffusion restriction.

- *Encephalitis/cerebritis*: imaging is often non-specific. Non-contrast CT is often negative. MRI typically shows abnormal T2 hyperintensity involving both cortical gray and subcortical white matter. Can present as a large, poorly defined area of involvement. Diffusion-weighted imaging shows a vasogenic rather than a cytotoxic pattern of edema. Late cerebritis often demonstrates ring enhancement. Clinical history is often helpful.

Arterial dissection must be differentiated from other causes of arterial wall thickening or irregularity including:

- *Atherosclerosis*: occurs in the elderly and most commonly involves the carotid bifurcation and carotid bulb.
- *Beam-hardening artifact*: often originates from the skull base or dental amalgam. Can simulate an intimal flap near the distal cervical ICA.
- *Fibromuscular dysplasia (FMD)*: non-atherosclerotic, non-inflammatory disorder primarily occurring in young to middle-age women and frequently occurring in the renal arteries and craniocervical arteries. Bilateral involvement is found in 60% of cases. Symptoms are related to tight stenosis, subarachnoid hemorrhage, or craniocervical artery dissection. FMD is angiographically present in approximately 15% of patients with spontaneous craniocervical ICA dissection. CTA and MRA usually demonstrate focal or long, tubular, multifocal stenosis with adjunct dilation, the so-called "string of beads" sign.
- *ICA dysgenesis*: can mimic occlusion or long segment stenosis of the ICA. Diagnosis is strongly suggested when the carotid canal is absent or hypoplastic.

Teaching point

A young patient presenting with headache, Horner's syndrome, or focal neurologic deficits with initial imaging showing ischemia in a watershed distribution should raise the suspicion for ICA dissection. The presence of intrinsic ribbon-like cortical high signal on T1weighted images with corresponding gyrifom cortical contrast enhancement suggests a subacute ischemic or inflammatory process. Neck MRA will often show ICA occlusion or restricted flow from intimal disruption.

REFERENCES

1. Petro GR, Witwer GA, Cacayorin ED, *et al.* Spontaneous dissection of the cervical internal carotid artery: correlation of arteriography, CT, and pathology. *AJR Am J Roentgenol*, 1987;**148**(2):393–8.
2. Silbert PL, Mokri B, Schievink WI. Headache and neck pain in spontaneous internal carotid and vertebral artery dissections. *Neurology*. 1995;**45**(8):1517–22.
3. Schievink WI. Spontaneous dissection of the carotid and vertebral arteries. *N Engl J Med*. 2001;**344**(12):898–906.
4. Patel RR, Adam R, Maldjian C, *et al.* Cervical carotid artery dissection: current review of diagnosis and treatment. *Cardiol Rev*. 2012; **20**(3):145–52. doi: 10.1097/CRD.0b013e318247cd15.
5. Lee VH, Brown RD, Jr., Mandrekar JN, Mokri B. Incidence and outcome of cervical artery dissection: a population-based study. *Neurology*. 2006 Nov 28;**67**(10):1809–12.
6. Levy C, Laissy JP, Raveau V, *et al.* Carotid and vertebral artery dissections: three-dimensional time-of-flight MR angiography and MRI versus conventional angiography. *Radiology*. 1994;**190**(1):97–103.

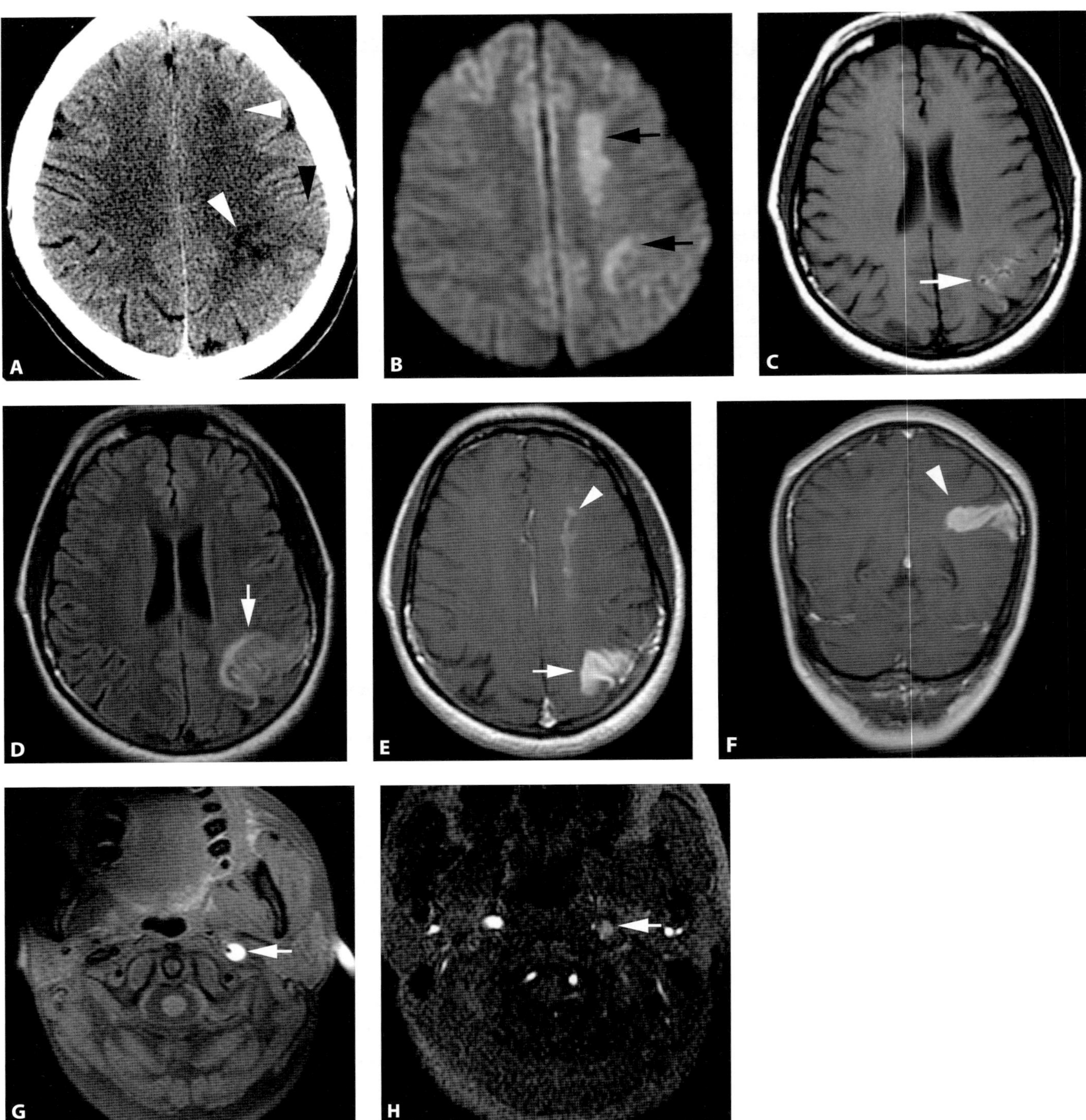

Figure 8.1 A. Axial non-contrast CT of the head from a 42-year-old woman with headache and upper extremity weakness shows poor gray–white matter differentiation in the left posterior parietal lobe (black arrowhead) as well as ill-defined foci of hypodensity in the left frontal and parietal subcortical white matter (white arrowheads) corresponding to a watershed infarct. **B.** Axial diffusion-weighted MR image shows increased signal in the left frontal and parietal subcortical white matter (arrows), corresponding to the hypodensities seen on non-contrast CT. Axial T1-weighted **(C)**, axial FLAIR **(D)**, axial **(E)** and coronal **(F)** post-contrast contrast T1-weighted MR images show ribbon-like intrinsic T1 hyperintensity outlining the affected cortex (arrow) with intense contrast enhancement in the left posterior parietal lobe (arrows) and left frontal subcortical white matter (arrowheads). These findings represent laminar necrosis and early gyriform enhancement of subacute ischemic infarct in a watershed distribution. **G.** Axial fat-saturated T1-weighted images from MRA demonstrates bright signal in the left internal carotid artery consistent with slow or no flow (arrow). **H.** Corresponding axial 2D time-of-flight MR image shows focal signal loss corresponding to absent cephalad blood flow in the ipsilateral ICA. These findings are consistent with left internal carotid artery dissection.

CASE **9**

Mimics of dural venous sinus thrombosis

Michael J. Modica

Imaging description

Evaluation for cerebral dural venous sinus thrombosis is generally undertaken with one of many imaging modalities: MRI, non-contrast CT, time-of-flight MR venography (TOF MRV), contrast-enhanced MRV and CT venography.

Absence of blood flow secondary to sinus aplasia or hypoplasia can be confused with venous sinus thrombosis on non-contrast CT of the brain. Sinus hypoplasia or atresia is one of the most common anatomic variations of the dural venous sinuses. In most patients, the right transverse sinus is larger than the left [1]. Studies using conventional TOF MRV have shown the left transverse sinus to be atretic or severely hypoplastic in 20–39% of people, with the medial aspect of the left transverse sinus being the most significantly affected (Figure 9.1). When transverse sinus hypoplasia or aplasia is found, the ipsilateral sigmoid and jugular sinuses are usually also hypoplastic or aplastic [2].

Clues that suggest an etiology other than sinus hypoplasia or aplasia include:

- Secondary signs of thrombosis or injury e.g. cerebral infarct, edema or hemorrhage (Figure 9.2)
- Collateral vessel filling or recanalization
- Intrinsic high T1 signal within a dural sinus, which suggests thrombosis (Figure 9.3).

When these findings are present, examination of the sigmoid sinus and jugular sinus for thrombus is mandatory.

In difficult cases, TOF or contrast-enhanced MRV should be obtained to evaluate for thrombus (Figure 9.4). A pitfall of MRV is T1 hyperintense subacute thrombus resembling normal flow. Evaluation of pre-contrast MR sequences and source images will help avoid this. Phase contrast MRV sequences are not limited by T1 hyperintense thrombus and should be obtained in difficult cases. Additionally, contrast-enhanced MRV will evaluate small-vein detail and collaterals much better than 2D TOF MRV.

Arachnoid granulations are normal structures that invade the dural sinus lumen, mimicking a focal sinus thrombus on contrast-enhanced CT venogram or contrast-enhanced MR. They typically, but do not always, have signal intensity and attenuation similar to cerebrospinal fluid (CSF). They are most typically seen in the far lateral transverse sinuses at the junction of the vein of Labbe and lateral tentoral sinus; however, they occur with equal frequency in the sagittal and sigmoid sinuses, and have a typical imaging appearance [3]. Arachnoid granulations are covered thoroughly in Case 5.

High or asymmetric bifurcation of the venous sinuses at the confluence can resemble an intraluminal thrombus, particularly on contrast-enhanced CT examinations. This variant has been termed "pseudo-empty delta sign" [4, 5]. The empty delta sign refers to a filling defect surrounded by enhancing dura at the sinus confluence that is seen' on contrast-enhanced CT or contrast-enhanced MR [5]. This sign should not be described on non-contrast CT, as hypodensity at the confluence surrounded by normal hyperdense dura is commonly a normal finding. Difficult cases should be evaluated with CT or MR venography for full evaluation of the venous sinuses.

Importance

The wide range of variants of intracranial veins and venous sinuses makes identification of venous thrombosis problematic. Variations are generally subdivided into three major categories: variants that mimic venous occlusion, asymmetric or variant sinus drainage, or normal sinus filling defects. Incorrect diagnosis of venous thrombosis can lead to unnecessary patient anxiety, treatment, and complications of anticoagulation.

Typical clinical scenario

Venous sinus thrombosis mimics are generally encountered incidentally. The clinical presentation of venous sinus thrombosis is extremely variable and can be non-specific. Headache, nausea, and vomiting are the most common symptoms; however, altered mental status, coma, and death are possible in patients with large cerebral infarcts.

Differential diagnosis

Sinus hypoplasia and aplasia and arachnoid granulations generally have characteristic imaging findings. Dural venous sinus thrombosis is the major differential consideration in cases where there is absence of flow or a filling defect in one of the venous sinuses.

> **Teaching point**
>
> Transverse venous sinus aplasia and arachnoid granulations are commonly mistaken for venous sinus thrombosis. Knowledge of their typical appearance and lack of additional findings associated with sinus thrombus can help avoid this pitfall.

REFERENCES

1. Zouaoui A, Hidden G. Cerebral venous sinuses: anatomical variants or thrombosis? *Acta Anat (Basel).* 1988;**133**(4):318–24.
2. Alper F, Kantarci M, Dane S, *et al.* Importance of anatomical asymmetries of transverse sinuses: an MR venographic study. *Cerebrovasc Dis.* 2004;**18**(3):236–9.

3. Leach JL, Jones BV, Tomsick TA, Stewart CA, Balko MG. Normal appearance of arachnoid granulations on contrast-enhanced CT and MR of the brain: differentiation from dural sinus disease. *AJNR Am J Neuroradiol.* 1996;**17**(8):1523–32.

4. Yeakley JW, Mayer JS, Patchell LL, Lee KF, Miner ME. The pseudodelta sign in acute head trauma. *J Neurosurg.* 1988; **69**(6):867–8.

5. Lee EJ. The empty delta sign. *Radiology.* 2002;**224**(3):788–9.

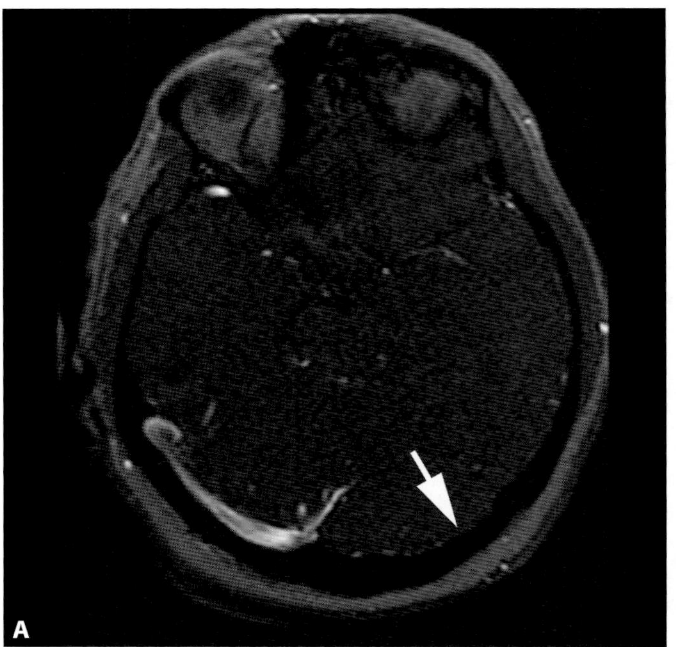

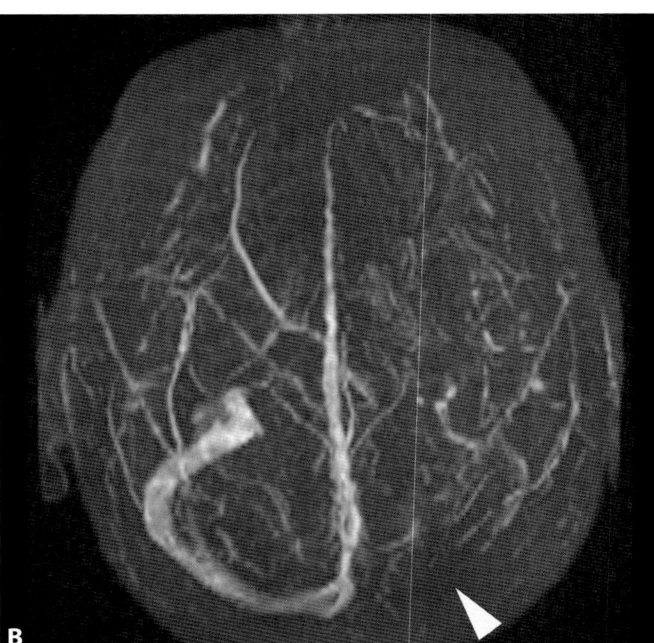

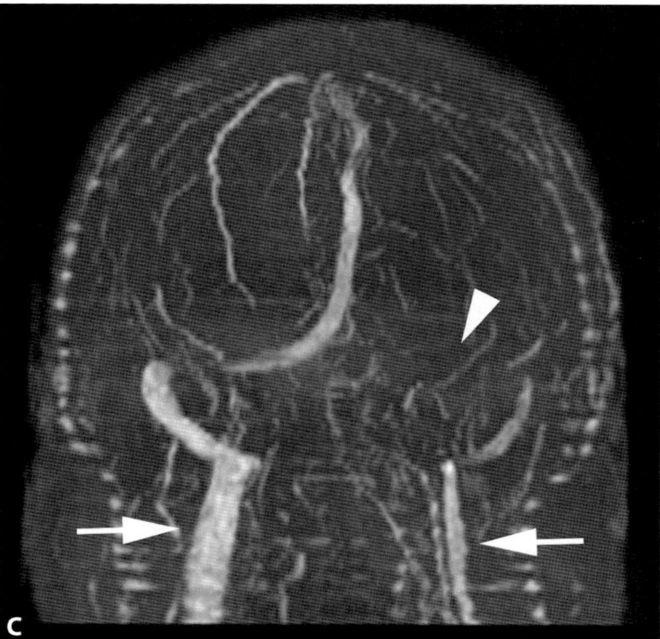

Figure 9.1 **A.** Axial image from 2D TOF MRV in a 71-year-old woman with headache shows absence of flow within the left transverse sinus (arrow). **B.** Axial maximum intensity projection (MIP) confirms complete absence of flow within the left transverse sinus (arrowhead). **C.** Coronal MIP image compares the caliber of the left and right sigmoid and jugular sinuses. Transverse sinus aplasia/hypoplasia (arrowhead) is commonly accompanied by varying degrees of ipsilateral sigmoid and jugular sinus hypoplasia (arrows). T1-weighted images were normal, further supporting the diagnosis of left transverse sinus aplasia.

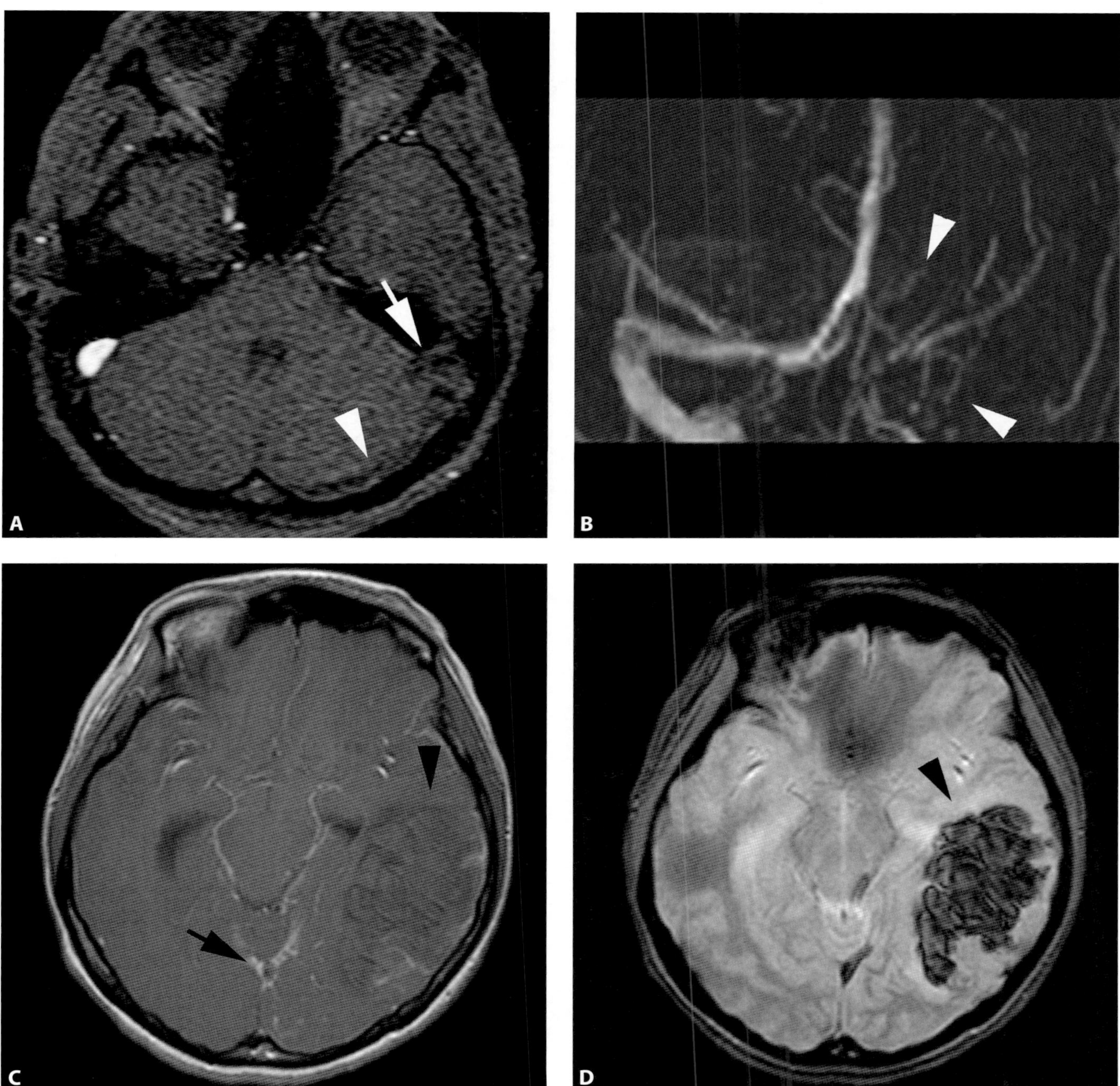

Figure 9.2 A. Axial 2D (TOF) image from an MR venogram in a 52-year-old man with headache and weakness shows absence of contrast and complex heterogeneous hypointense signal in the enlarged left transverse (arrowhead) and sigmoid (arrow) venous sinuses. Normal flow enhancement is seen in the right transverse sinus. **B.** Coronal MIP image from a TOF MR venogram shows absence of flow within the left transverse sinus and recanalization of subjacent venous collaterals (arrowheads). **C.** Axial contrast-enhanced T1-weighted image shows clot extending into the deep venous system (arrow) with a large parenchymal hypointensity consistent with extensive temporal lobe venous infarct (arrowhead). **D.** Axial gradient-echo MR image shows blooming of thrombosed blood in both the temporal lobe (arrowhead) and deep venous system.

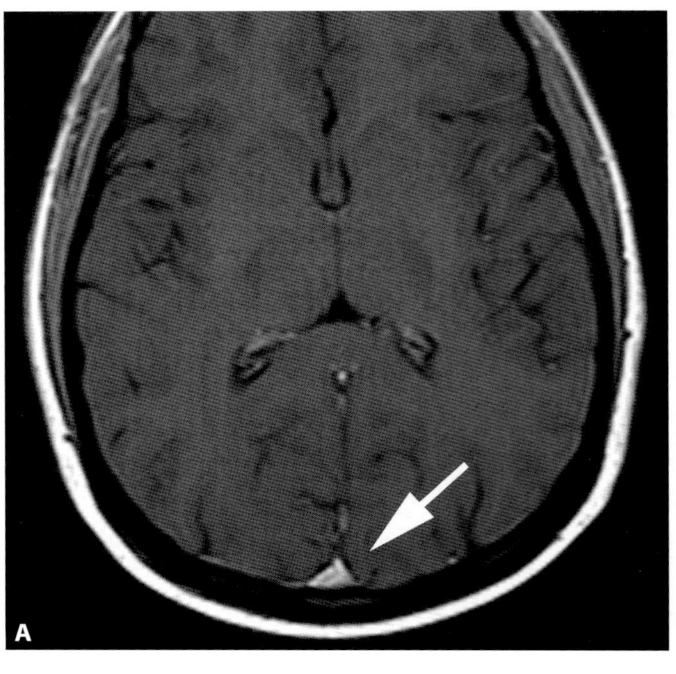

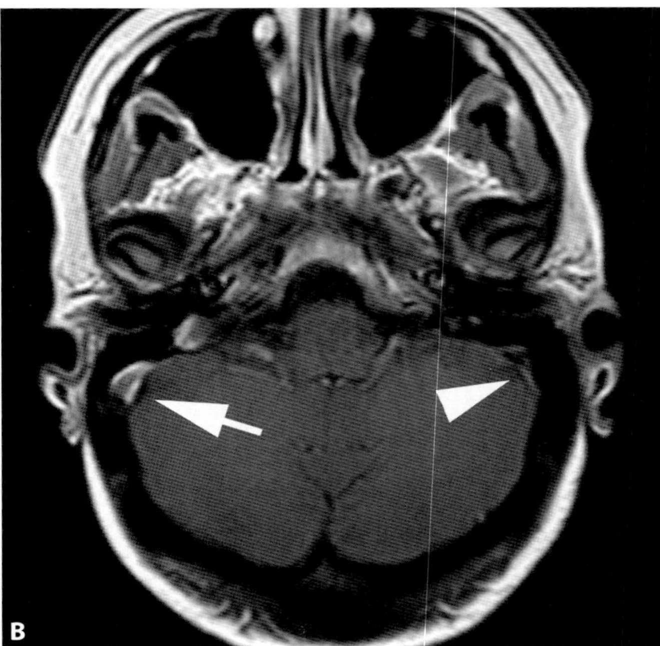

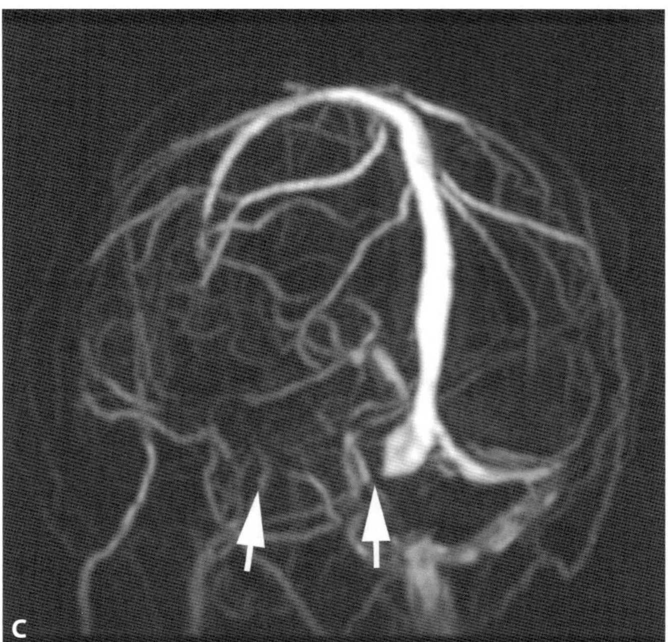

Figure 9.3 A. Axial T1-weighted MR image from a 35-year-old postpartum woman with headache shows intrinsic high signal at the sinus confluence (arrow). **B.** More inferiorly, at the junction of the right transverse and sigmoid sinuses, there is high signal filling the sinus lumen (arrow), suggesting thrombus. Note the normal flow void on the contralateral side (arrowhead) **C.** Oblique coronal MIP image from a TOF MRV shows absent flow within the right transverse sinus (arrows), confirming right transverse sinus thrombosis.

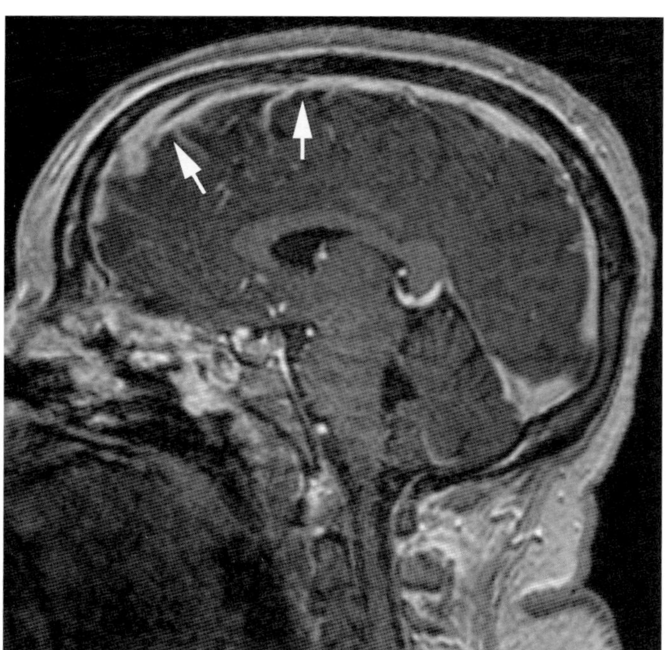

Figure 9.4 Sagittal contrast-enhanced MR venogram from a 56-year-old with headache demonstrates a long, linear filling defect within the anterior superior sagittal sinus (arrows) consistent with superior sagittal sinus thrombus.

CASE

10 Pineal cyst

Robert B. Carr

Imaging description

Pineal cysts are common incidental findings detected on CT and MRI. They are round or oval in shape, and may be unilocular or multilocular. On CT, the cyst will demonstrate hypoattenuation compared with brain parenchyma, and there may be pineal calcifications adjacent to the cyst periphery. Most cysts are 2–15 mm in diameter. Thin-section images and sagittal reformatted images are often helpful in their evaluation (Figures 10.1 and 10.2).

Sometimes it is difficult to confirm the benign nature of a cystic pineal lesion on a routine head CT, so a MRI is obtained. Features of a benign pineal cyst include thin walls and lack of a solid internal component. The cyst walls will normally enhance, but the enhancement should be smooth and linear. The cyst contents will often differ in signal from cerebrospinal fluid (CSF), and FLAIR images often show non-suppression of signal.

High-resolution MR sequences, such as balanced steady state free precession (SSFP) and constructive interference steady state (CISS) sequences, may demonstrate internal architecture such as thin internal septations and smaller internal cysts. These findings should not be viewed as suspicious for malignancy [1].

A cyst may enlarge over time due to hemorrhage or accumulation of fluid. This may result in local mass effect. Compression of the superior colliculus may result in upward gaze palsy (Parinaud syndrome), and effacement of the cerebral aqueduct may lead to obstructive hydrocephalus (Figure 10.3). Thus, it is important to assess the relationship of the cyst to adjacent structures, specifically the tectal plate and cerebral aqueduct.

Importance

There have been several studies designed to determine the natural history of pineal cysts. These studies suggest that follow-up of cysts with benign features is not necessary, in both adults [2, 3] and children [4].

Another study evaluated pineal cystic lesions which were not clearly benign, and thus characterized as indeterminate. Follow-up was obtained in 26 lesions over a course ranging from seven months to eight years. There was no change in size or appearance of these indeterminate lesions on follow-up imaging [5].

However, there are no standard guidelines on follow-up, and follow-up remains controversial. Some authors recommend follow-up of cysts that are larger than 10 mm [6]. If there are atypical features, such as nodular enhancing internal components, follow-up may be prudent (Figure 10.4).

Typical clinical scenario

Autopsy series have shown the incidence of pineal cysts to be 25–40%. A study of healthy adults revealed an incidence of 23% as seen on high-resolution MRI [7]. Therefore, it is clear that pineal cysts are common entities.

Small cysts are incidental findings and unlikely to be related to symptoms. Cysts have been reportedly associated with headaches [8]. However, given the common occurrence of headache and the common occurrence of pineal cysts, a causal relationship is difficult to verify.

Rarely, a patient will present with ocular gaze disturbance or symptoms related to hydrocephalus. In such cases, the cyst will be large with local mass effect.

Differential diagnosis

The primary differential consideration is a cystic pineal neoplasm. However, if high-quality MRI is performed, a pineocytoma will not demonstrate the features of a benign cyst [9]. Other cystic lesions may occur in the pineal region, most commonly an arachnoid cyst (Figures 10.5 and 10.6). Rarely, an epidermoid will be encountered [10].

Teaching point

Small benign-appearing pineal cysts require no follow-up. Larger cysts (>10 mm) probably require no follow-up, but given the lack of consensus, follow-up may be recommended at the discretion of the radiologist.

REFERENCES

1. Pastel DA, Mamourian AC, Duhaime AC. Internal structure in pineal cysts on high-resolution magnetic resonance imaging: not a sign of malignancy. *J Neurosurg Pediatr*. 2009;**4**(1):81–4.

2. Al-Holou WN, Terman SW, Kilburg C, *et al*. Prevalence and natural history of pineal cysts in adults. *J Neurosurg*. 2011;**115**(6):1106–14.

3. Barboriak DP, Lee L, Provenzale JM. Serial MR imaging of pineal cysts: implications for natural history and follow-up. *AJR Am J Roentgenol*. 2001;**176**(3):737–43.

4. Al-Holou WN, Maher CO, Muraszko KM, Garton HJ. The natural history of pineal cysts in children and young adults. *J Neurosurg Pediatr*. 2010;**5**(2):162–6.

5. Cauley KA, Linnell GJ, Braff SP, Filippi CG. Serial follow-up MRI of indeterminate cystic lesions of the pineal region: experience at a rural tertiary care referral center. *AJR Am J Roentgenol*. 2009;**193**(2):533–7.

6. Smith AB, Rushing EJ, Smirniotopoulos JG. From the archives of the AFIP: lesions of the pineal region: radiologic-pathologic correlation. *Radiographics*. 2010;**30**(7):2001–20.

7. Pu Y, Mahankali S, Hou J, *et al.* High prevalence of pineal cysts in healthy adults demonstrated by high-resolution, non-contrast brain MR. *AJNR Am J Neuroradiol.* 2007;**28**(9):1706–9.

8. Seifert CL, Woeller A, Valet M, *et al.* Headaches and pineal cyst: a case-control study. *Headache.* 2008;**48**(3):448–52.

9. Fakhran S, Escott EJ. Pineocytoma mimicking a pineal cyst on imaging: true diagnostic dilemma or a case of incomplete imaging? *AJNR Am J Neuroradiol.* 2008;**29**(1):159–63.

10. Osborn AG, Preece MT. Intracranial cysts: radiologic-pathologic correlation and imaging approach. *Radiology.* 2006;**239**(3):650–64.

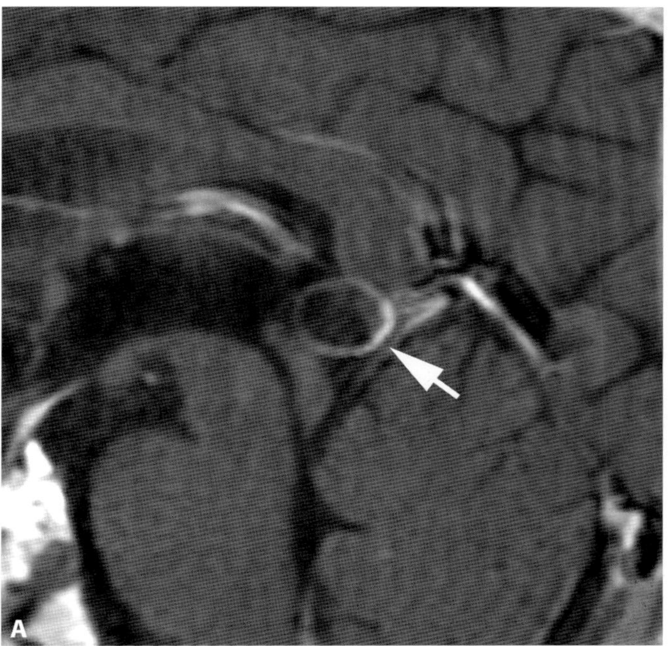

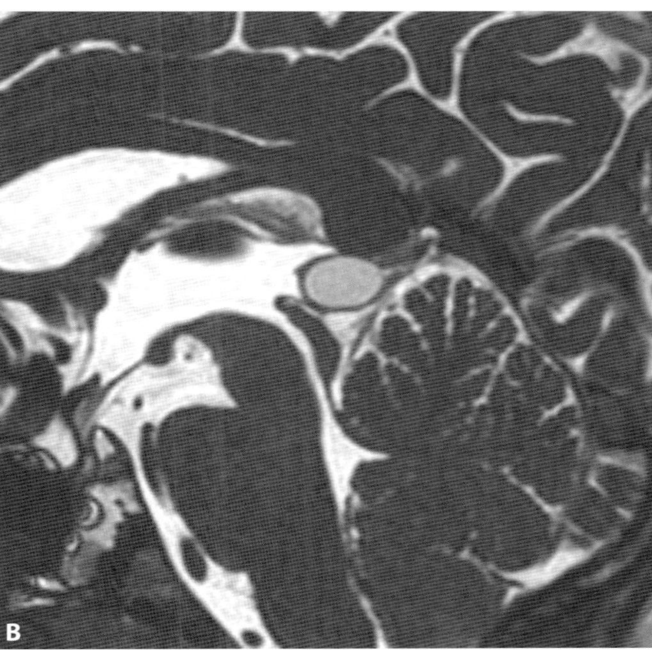

Figure 10.1 A. Sagittal contrast-enhanced T1-weighted image from a 36-year-old woman shows an incidental pineal cyst. Notice the thin peripheral enhancement (arrow), an expected finding. **B.** Sagittal CISS image demonstrates the cystic nature of the lesion. Notice that the signal differs from CSF. This lesion requires no follow-up.

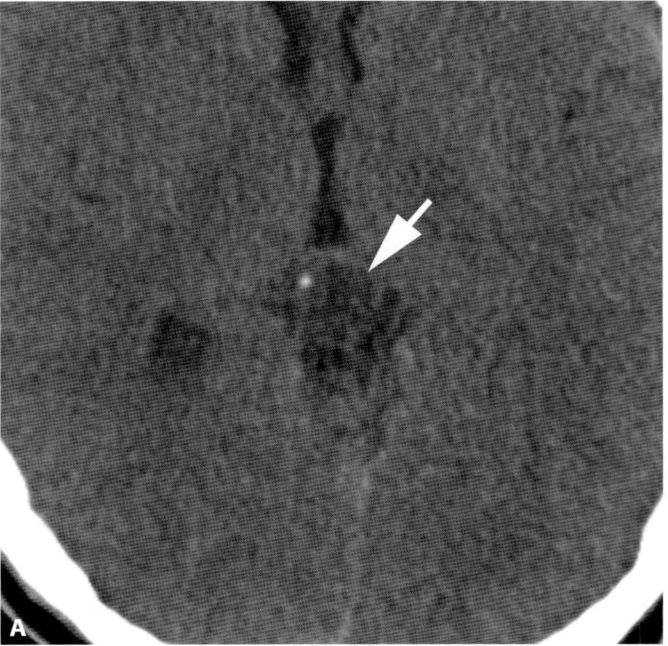

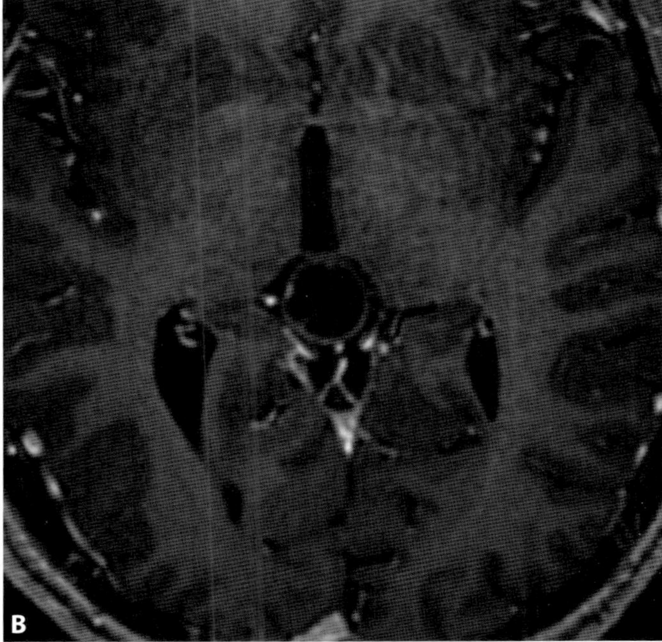

Figure 10.2 A. Axial non-enhanced CT image from a 31-year-old man shows an incidental low-attenuation pineal lesion (arrow). The lesion could not be fully characterized on CT, so an MRI was obtained. **B.** Axial contrast-enhanced T1-weighted image shows the typical appearance of a benign pineal cyst. Follow-up was not suggested.

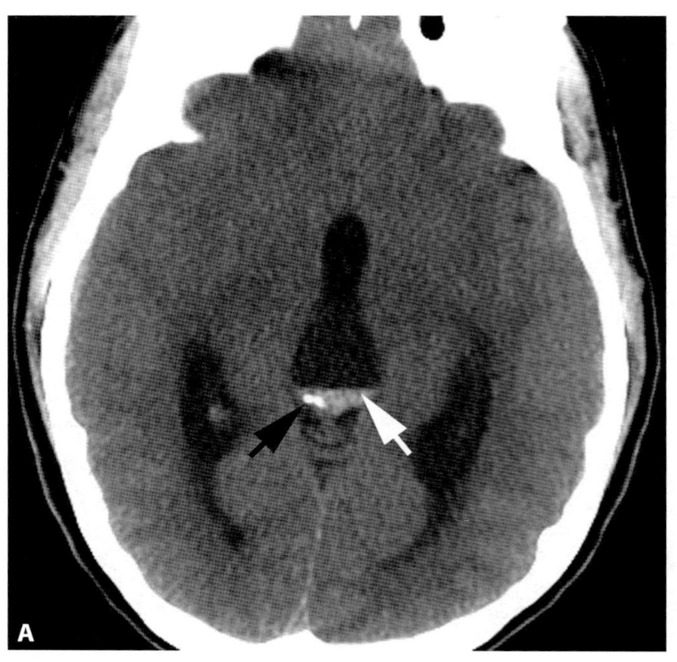

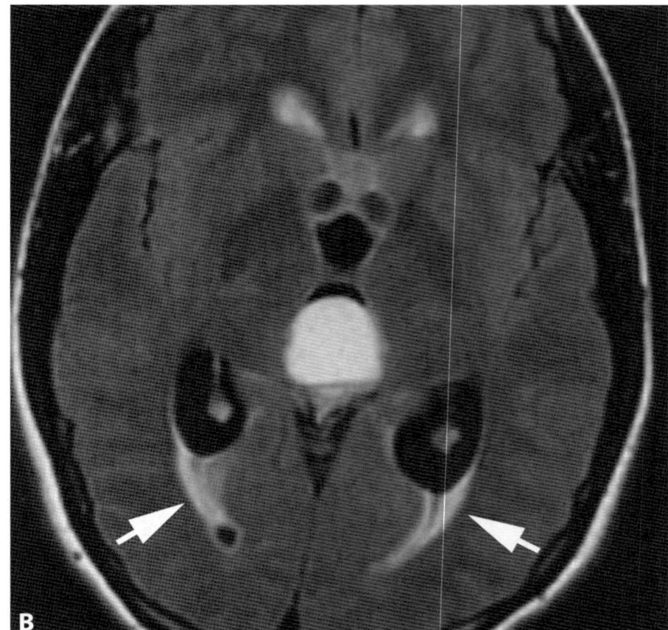

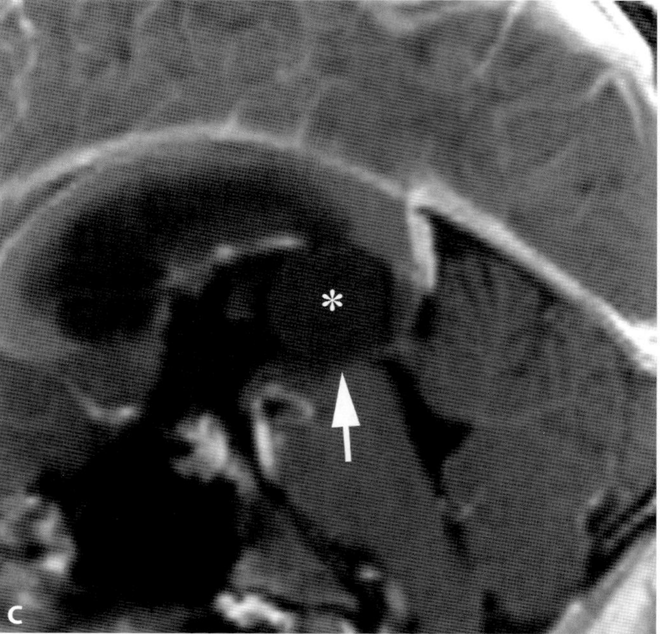

Figure 10.3 **A.** Axial non-enhanced CT image from a 32-year-old woman with headaches shows a low-attenuation lesion within the pineal gland with a dense fluid level layering dependently (white arrow). Also notice the adjacent pineal calcifications (black arrow). The lateral and third ventricles are enlarged. **B.** Axial FLAIR image shows a hyperintense pineal lesion with acute obstructive hydrocephalus resulting in transpendymal flow of CSF (arrows). **C.** Sagittal contrast-enhanced T1-weighted image shows no evidence of nodular enhancement within the cyst (asterisk). There is complete effacement of the cerebral aqueduct (arrow). This represents a benign pineal cyst which underwent hemorrhage, increased in size, and caused acute obstructive hydrocephalus.

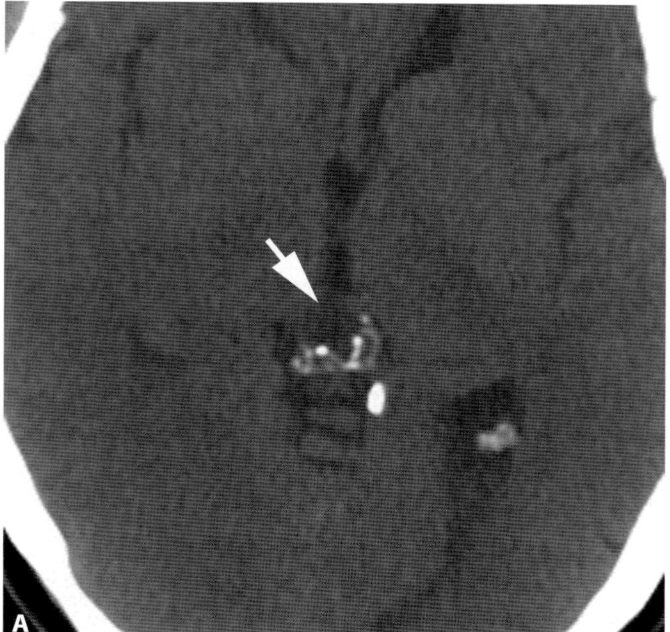

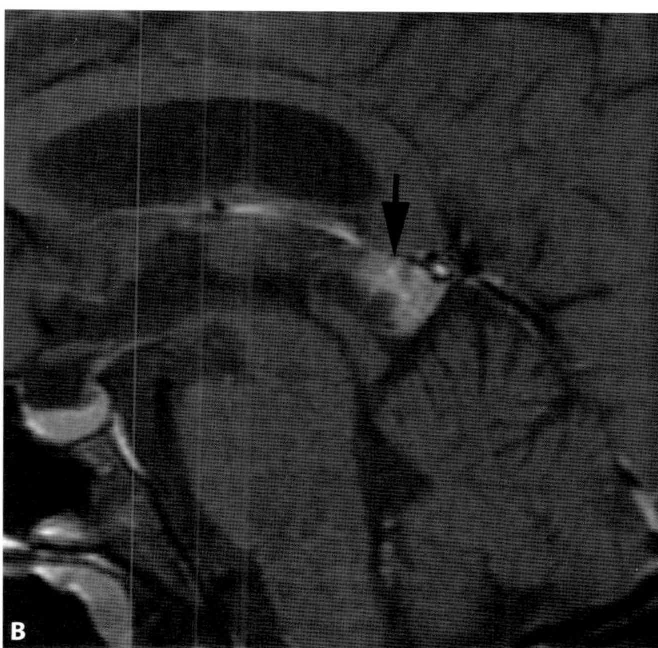

Figure 10.4 A. Axial non-enhanced CT image from a 26-year-old man with headaches shows a low-attenuation lesion within the pineal gland (arrow), with prominent pineal calcifications posteriorly. **B.** Sagittal contrast-enhanced T1-weighted image shows a large area of nodular enhancement (arrow) adjacent to an irregular cyst. These findings are not compatible with a benign cyst, and follow-up was suggested.

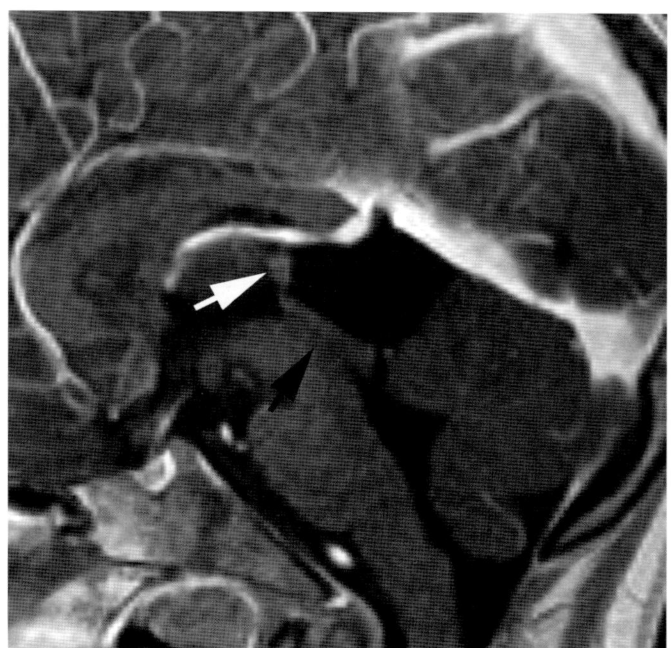

Figure 10.5 Sagittal contrast-enhanced T1-weighted image from a 24-year-old woman shows an arachnoid cyst in the quadrigeminal plate cistern. Notice that the cyst is posterior to the pineal gland (white arrow). Also notice the effacement of the cerebral aqueduct (black arrow).

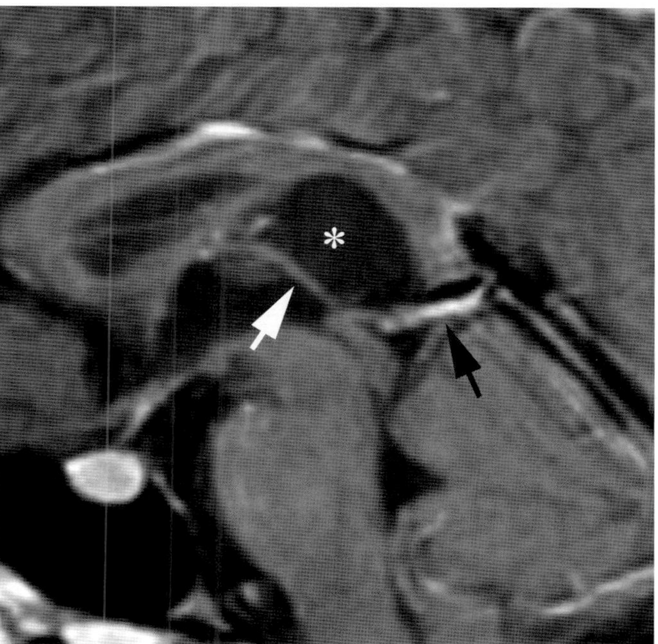

Figure 10.6 Sagittal contrast-enhanced T1-weighted image from a 33-year-old man shows an arachnoid cyst of the cistern for the velum interpositum (asterisk). This space extends anteriorly from the quadrigeminal plate cistern, superior to the pineal gland. Notice the inferior displacement of the fornix (white arrow) and internal cerebral vein (black arrow).

CASE 11

Enlarged perivascular space

Michael J. Modica

Imaging description

Virchow-Robin (VR) spaces, also commonly referred to as perivascular spaces, are pia-lined extensions of the subarachnoid space surrounding the walls of cerebral vessels. Different theories have been proposed for their mechanism of expansion; however, it remains unknown. The typical appearance varies depending on the patient's age. Small (<2 mm) VR spaces appear in all age groups. Larger (>2 mm) VR spaces are found with increasing frequency with advancing age (Figure 11.1) [1].

Generally, VR spaces are well-defined, round or oval fluid-filled structures with smooth margins and they usually measure 5 mm or less [2]. They are typically found in the inferior third of the anterior perforated substance and basal ganglia [2].

On non-contrast CT, VR spaces are isodense with cerebrospinal fluid (CSF). On MR, the signal intensities of the VR spaces are similar to those of CSF on all pulse sequences (Figure 11.2). There is no restriction on diffusion-weighted imaging (DWI). The dilated spaces do not enhance on post-contrast images [3].

Dilated VR spaces typically occur in three characteristic locations, which determine their type (Table 11.1).

Type I and occasionally, type II VR spaces are commonly seen on CT whereas all three types are commonly seen on MRI.

Atypical appearances of VR spaces have been reported. They can present as clusters of type II enlarged spaces in one hemisphere or unilateral spaces high in one hemisphere (Figure 11.4). Rarely, VR spaces can be markedly enlarged, cause mass effect, and assume bizarre cystic configurations. This manifestation is most common with type III VR spaces [4].

Importance

Recognition of their typical location and imaging appearance will allow them to be dismissed as an anatomic variant, avoiding misdiagnosis.

Typical clinical scenario

Virchow-Robin spaces are usually encountered as an incidental finding or during a work-up for a non-specific symptom. Occasionally, a patient presents with non-specific, non-localizing symptoms and an atypical enlarged VR lesion is reported as a bizarre multi cystic neoplasm.

Differential diagnosis

Perivascular spaces mimic several other diseases that present with focal parenchymal signal abnormality.

Table 11.1. Virchow-Robin spaces

Type I: Along the lenticulostriate arteries entering the basal ganglia through the anterior perforated substance
Type II: Along the path of the perforating medullary arteries as they enter the cortical gray matter over the high cerebral convexities and extend into the white matter
Type III: In the midbrain (Figure 11.3) [1]

- *Lacunar infarcts*: found in end-artery territory: the upper two-thirds of the anterior perforated substance, basal ganglia, and thalamus. These tend to be larger than VR spaces and wedge-shaped on CT [5]. On MRI, acute lacunar infarcts demonstrate increased T2 and FLAIR signal and decreased T1 signal. High signal is present on DWI with corresponding low signal on ADC map. Chronic lacunar infarcts lack increased signal on DWI and appear as low signal with a hyperintense surrounding rim on FLAIR (Figure 11.5) which represents gliosis. Enhancement can be seen up to 8 weeks after the acute event.
- *Microvascular ischemic disease*: typically found in the periventricular and subcortical white matter. Appear as bilaterally symmetric confluent foci of increased T2-weighted and FLAIR signal abnormality. The findings are usually subtle on T1-weighted images (Figure 11.6).
- *Multiple sclerosis*: lesions are typically oval and commonly occur in the periventricular and subcortical white matter, similar to type II VR spaces. The lesions are arranged perpendicular to the ventricular walls (finger-like projections) and demonstrate T1 isointensity/mild hypointensity to brain parenchyma. The lesions can have rim or solid enhancement, depending on the degree of lesion inflammation (Figure 11.7) [6].
- *Cystic neoplasms*: typically located in the pons, cerebellum, and thalamus. The signal does not exactly follow CSF and there are usually associated parenchymal signal abnormalities (edema). Cystic neoplasms may have an enhancing component or nodule. Generally neoplasms demonstrate low signal on DWI with corresponding high ADC values.
- *Neurocysticercosis*: cysts are less than 1 cm and have a scolex. Cyst walls can enhance and there is often surrounding parenchymal edema present.
- *Arachnoid cysts*: intra-arachnoid CSF-containing cysts that do not communicate with the ventricular system. Common locations include the middle cranial fossa, the perisellar cisterns and the subarachnoid space over the convexities. They follow CSF signal on CT and MR and do not enhance. They can be differentiated from VR spaces by their typical location.

Teaching point

Perivascular spaces are typically found in the inferior third of the anterior perforated substance and can range in size from 2 mm up to 10 mm in size. They follow CSF on all pulse sequences, allowing differentiation from focal acute lacunar infarcts and chronic microvascular ischemic disease.

REFERENCES

1. Heier LA, Bauer CJ, Schwartz L, *et al.* Large Virchow-Robin spaces: MR-clinical correlation. *AJNR Am J Neuroradiol.* 1989;**10**(5):929–36.

2. Jungreis CA Kanal E, Hirsch WL, Martinez AJ, Moossy J. Normal perivascular spaces mimicking lacunar infarction: MR imaging. *Radiology.* 1988;**169**(1):101–4.

3. Braffman BH, Zimmerman RA, Trojanowski JQ, *et al.* Brain MR: pathologic correlation with gross and histopathology. 1. Lacunar infarction and Virchow-Robin spaces. *AJR Am J Roentgenol.* 1988;**151**(3):551–8.

4. Ogawa T, Okudera T, Fukasawa H, *et al.* Unusual widening of Virchow-Robin spaces: MR appearance. *AJNR Am J Neuroradiol.* 1995;**16**(6):1238–42.

5. Brown JJ, Hesselink JR, Rothrock JF. MR and CT of lacunar infarcts. *AJR Am J Roentgenol.* 1988;**151**(2):367–72.

6. McDonald WI, Compston A, Edan G, *et al.* Recommended diagnostic criteria for multiple sclerosis: guidelines from the International Panel on the diagnosis of multiple sclerosis. *Ann Neurol.* 2001;**50**(1):121–7.

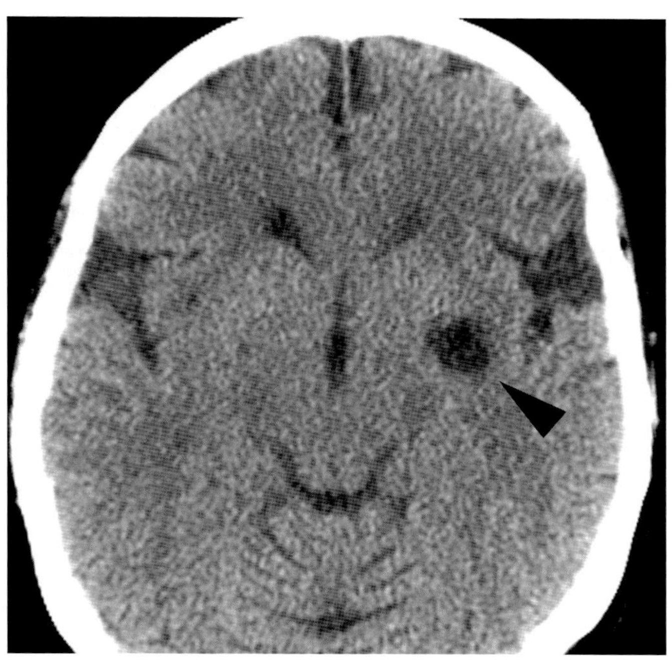

Figure 11.1 Axial non-contrast CT examination from a 67-year-old man with right upper extremity weakness shows a focal 7 mm oval enlarged perivascular space in the inferior third of the left anterior perforated substance (arrowhead) which extends down to the basal cisterns. Note the density of this lesion is similar to CSF seen in the neighboring Sylvian fissure.

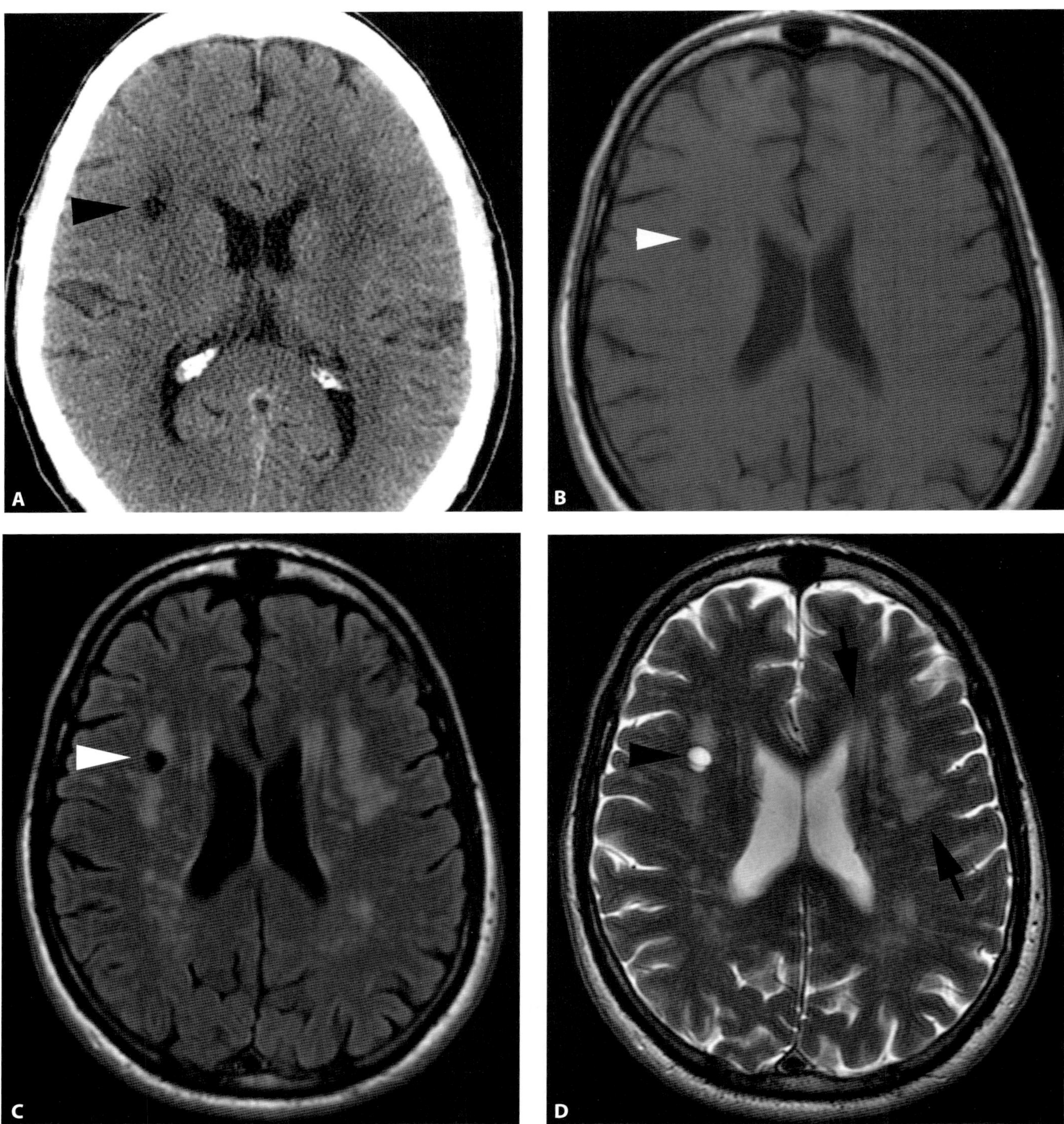

Figure 11.2 Axial non-contrast CT from a 60-year-old woman with transient upper extremity weakness shows a 5 mm round enlarged perivascular space in the right frontal subcortical white matter (arrowhead). There is ill-defined hypodensity in the bilateral frontal periventricular white matter. Axial T1-weighted (**B**), FLAIR (**C**), and T2-weighted (**D**) MR images from the same patient show the perivascular space (arrowhead) follows CSF signal on all sequences. Note on (D) how the confluent periventricular/subcortical abnormal T2 signal (black arrows) is slightly less intense than CSF, consistent with chronic microvascular ischemic change.

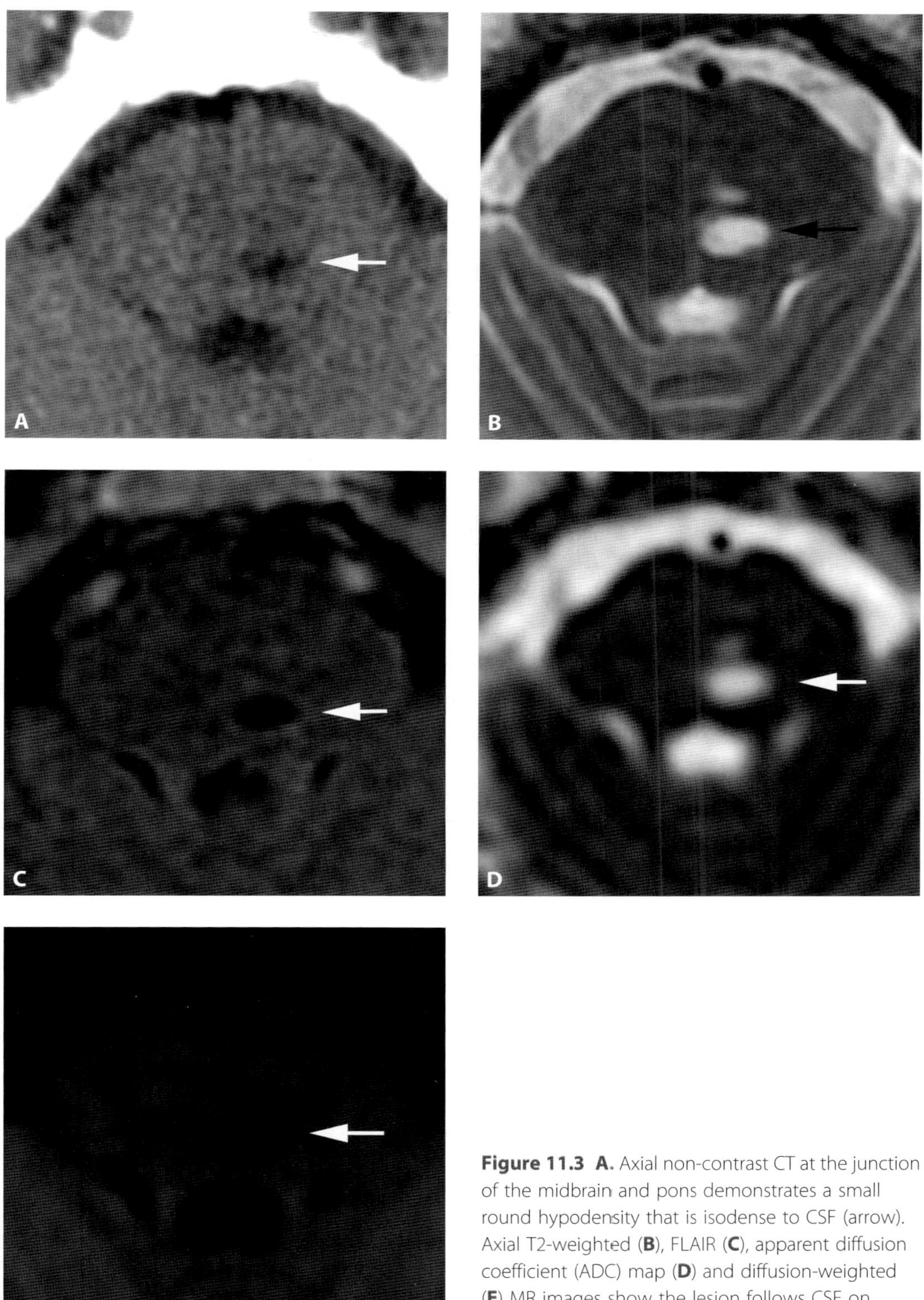

Figure 11.3 A. Axial non-contrast CT at the junction of the midbrain and pons demonstrates a small round hypodensity that is isodense to CSF (arrow). Axial T2-weighted (**B**), FLAIR (**C**), apparent diffusion coefficient (ADC) map (**D**) and diffusion-weighted (**E**) MR images show the lesion follows CSF on all pulse sequences (arrow), characteristic of a type III perivascular space.

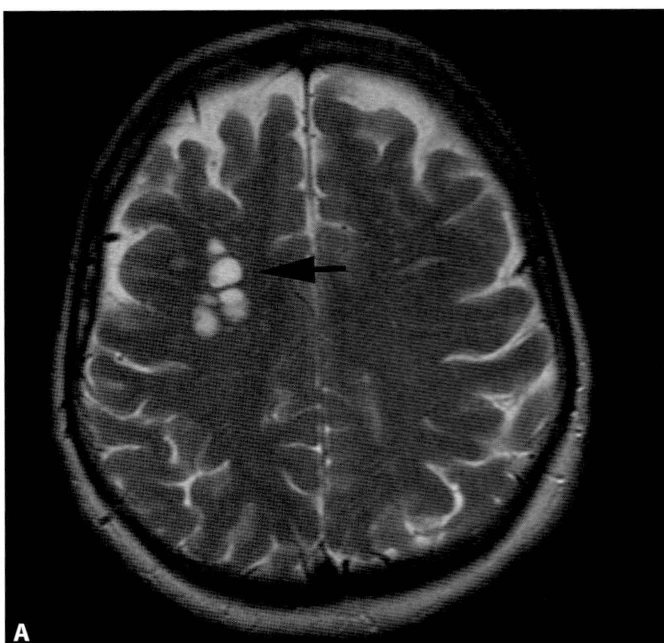

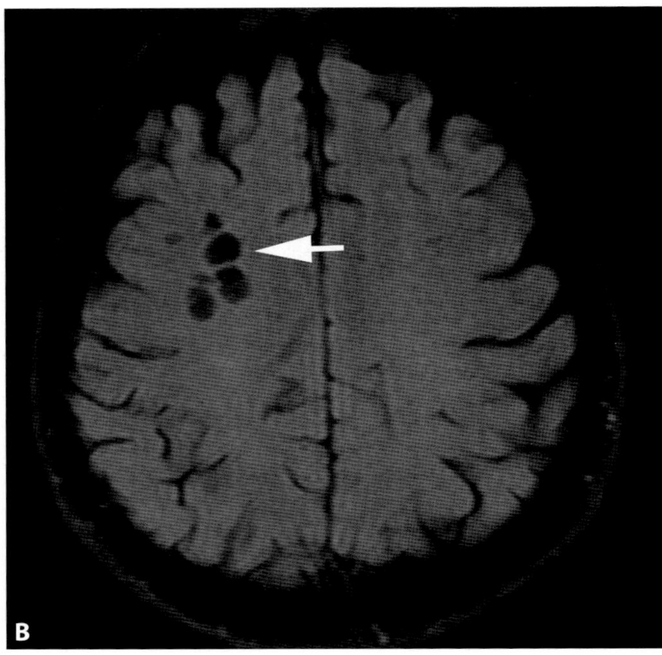

Figure 11.4 Axial T2-weighted (**A**) and FLAIR (**B**) MR images from a 68-year-old man who was found down show cystic lesions in the right frontal subcortical white matter that follow CSF signal on all pulse sequences (arrows). The surrounding brain parenchyma is normal. These represent atypical type II perivascular spaces.

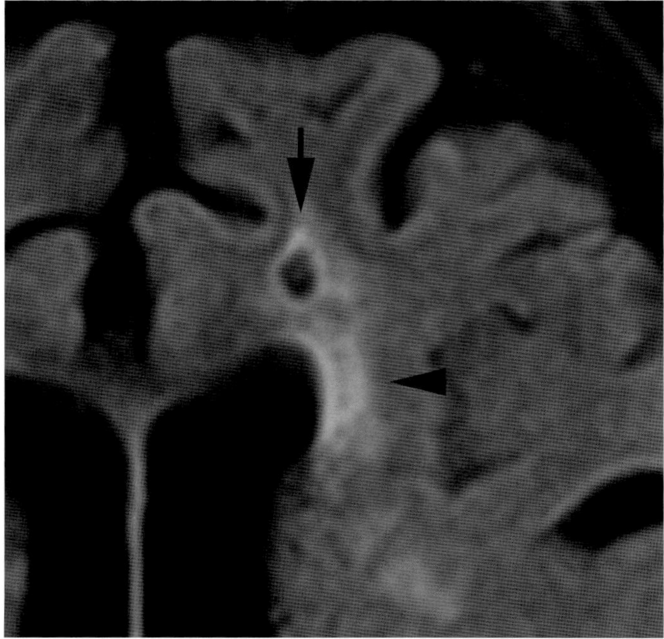

Figure 11.5 Axial FLAIR MR image from a 72-year-old man with confusion shows the typical appearance of a chronic lacunar infarct (arrow) in the left frontal periventricular white matter. Notice the oval hyperintense lesion is surrounded by increased FLAIR signal representing gliosis. The arrowhead depicts microvascular ischemic white matter change.

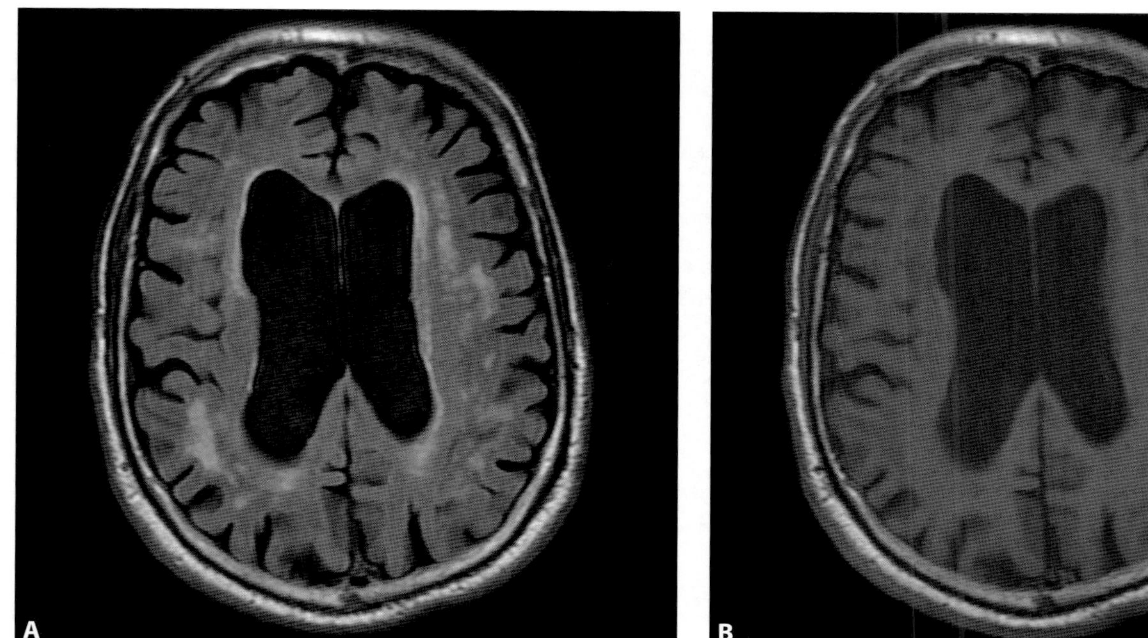

Figure 11.6 Axial FLAIR (**A**) and T1-weighted (**B**) MR images from an 88-year-old man with upper extremity weakness show typical findings of diffuse microvascular ischemic white matter change. There is confluent abnormally increased FLAIR signal in the subcortical and periventricular white matter bilaterally. The white matter changes are subtle iso- to slightly hypointense on T1-weighted images.

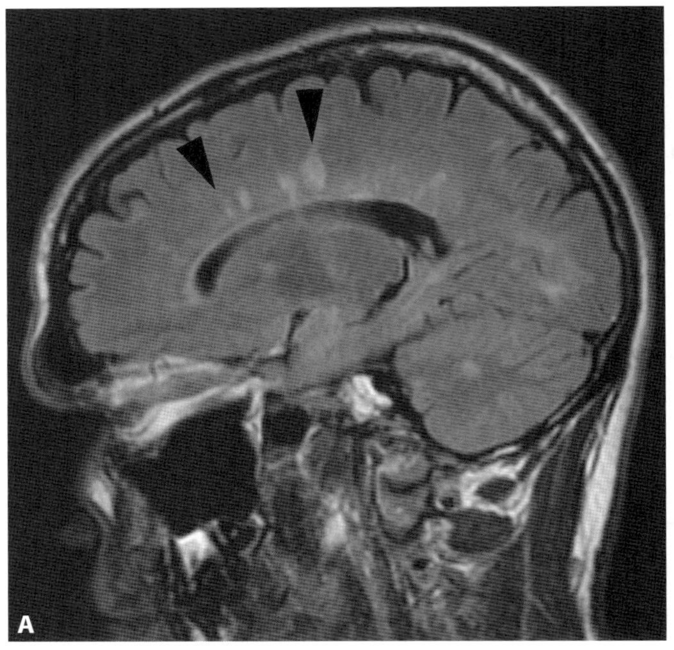

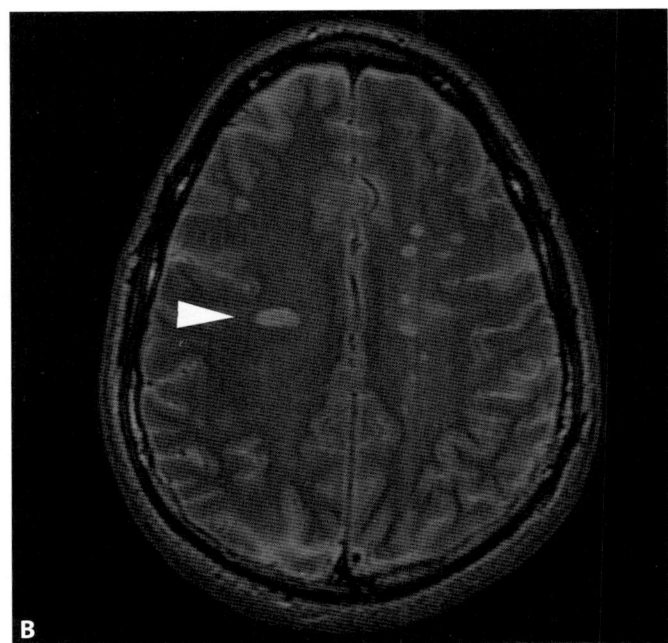

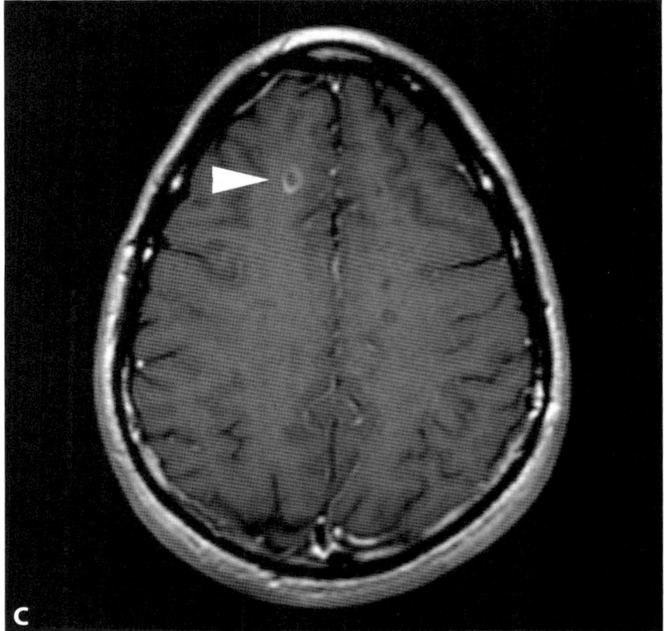

Figure 11.7 Sagittal FLAIR (A) and axial T2-weighted (B) images from a 42-year-old woman show the typical appearance of multiple sclerosis plaques. The oval plaques typically project from the corpus callosum and show increased signal on T2-weighted and FLAIR images with corresponding hypointensity on T1-weighted images (arrowheads). C. Axial T1-weighted post-contrast image demonstrates peripheral enhancement after the administration of gadolinium consistent with an actively demyelinating lesion (arrowhead).

CASE **12** # Tumefactive multiple sclerosis

Michael J. Modica

Imaging description

Non-contrast head CT is often the first study for evaluation of acute cognitive decline in the emergent setting. Tumefactive multiple sclerosis (TMS) typically shows a large area of confluent hypodensity in the periventricular white matter, often extending from the body of the corpus callosum. This area of hypodensity can extend to the subcortical white matter and, in some cases, exhibit mass effect on the neighboring ventricle, distorting the overlying cortex. While tumefactive lesions can be large enough to exhibit some mass effect, the mass effect is often less than would be expected for the size of the lesion. Post-contrast images can show subtle, discontinuous edge enhancement (Figure 12.1A).

Further characterization with MRI (Figure 12.1B–E) usually reveals the characteristic imaging findings associated with TMS that differentiate it from other diseases. The confluent hypodensity seen on CT will correlate to a similar territory of intrinsic hypointensity on T1-weighted images. There will be corresponding central hyperintensity with a thin edge of hypointensity on T2-weighted images. Post-contrast MRI yields a characteristic horseshoe-shaped leading edge of enhancement that is open towards the cortex. Corresponding restriction on diffusion-weighted imaging (DWI) occurs in the region of enhancement (Figure 12.1F) [1, 2].

Physiologic (permeability) and hemodynamic (blood volume) parameters on perfusion CT have been used to discriminate TMS from high-grade neoplasms. Tumefactive multiple sclerosis typically lacks neo-angiogenesis and vascular endothelial proliferation, and thus shows a low permeability surface-area-product and low blood volume on CT perfusion. These perfusion studies can also demonstrate vascular structures, usually veins, within the lesions, only present in TMS [3]. MR spectroscopy can also help distinguish TMS by demonstrating elevation of glutamate/glutamine peak on short echo time MR spectroscopy [4].

Importance

Tumefactive demyelination is a spectrum of disease that includes multiple sclerosis and acute disseminated encephalomyelitis (ADEM) [4]. These lesions can present a diagnostic dilemma to both the clinician and radiologist. Prompt recognition or suspicion can prevent unnecessary surgical biopsy or intervention. These patients often respond well to high-dose corticosteroids, with regression of the demyelinating lesion.

Tumefactive demyelination can share imaging findings with high-grade intracranial neoplasms and infections: contrast enhancement, perilesional edema, and mass effect. Correlation with clinical history, patient demographics, and diagnostic imaging is valuable to differentiate these diseases. When present, recognition of the typical features of TMS is critical: periventricular lesion centered predominantly within white matter with less mass effect than expected for size and a leading edge of enhancement. Despite presenting with typical imaging findings, TMS can mimic a necrotic neoplasm or infection.

Typical clinical scenario

The typical patient is between the ages of 16 and 45 years and presents with acute cognitive decline, seizure, or acute onset of hemiparesis or weakness.

Differential diagnosis

- *Glioblastoma multiforme*: irregular mass with thick enhancing rim surrounding central necrosis (Figure 12.2). Extensive peritumoral edema and mass effect. Typically involves and crosses the corpus callosum. Diffusion-weighted imaging typically does not show restriction. Elevated flow is demonstrated during dynamic contrast-enhanced MR consistent with neovascularity or hypervascularity. MR spectroscopy characteristically shows decreased N-acetylaspartate and myo-inositol.
- *Cerebral abscess*: poorly marginated region of hypodensity on CT or hypointensity on T1-weighted MR representing edema surrounding a central area of even further decreased density/signal. Smooth thin rim of enhancement that is typically thinnest on the ventricle side. Typically in an anterior and middle cerebral artery distribution. Restriction on DWI is present (Figure 12.3). MR spectroscopy of the central necrosis will show presence of acetate, lactate, alanine, pyruvate, and amino acids.

Teaching point

Tumefactive multiple sclerosis should be considered when leading-edge enhancement is seen in an aggressive appearing lesion. Diffusion-restriction can help differentiate this from a high-grade neoplasm. Difficult cases should be studied with perfusion imaging and spectroscopy for better characterization before surgical biopsy.

REFERENCES

1. Lucchinetti CF, Gavrilova RH, Metz I, *et al.* Clinical and radiographic spectrum of pathologically confirmed tumefactive multiple sclerosis. *Brain.* 2008;**131**(Pt 7):1759–75.
2. Dagher AP, Smirniotopoulos J. Tumefactive demyelinating lesions. *Neuroradiology.* 1996;**38**(6):560–5.
3. Eklund A. The contents of phytic acid in protein concentrates prepared from nigerseed, sunflower seed, rapeseed and poppy seed. *Ups J Med Sci.* 1975;**80**(1):5–6.
4. Malhotra HS, Jain KK, Agarwal A, *et al.* Characterization of tumefactive demyelinating lesions using MR imaging and in-vivo proton MR spectroscopy. *Mult Scler.* 2009;**15**(2):193–203.

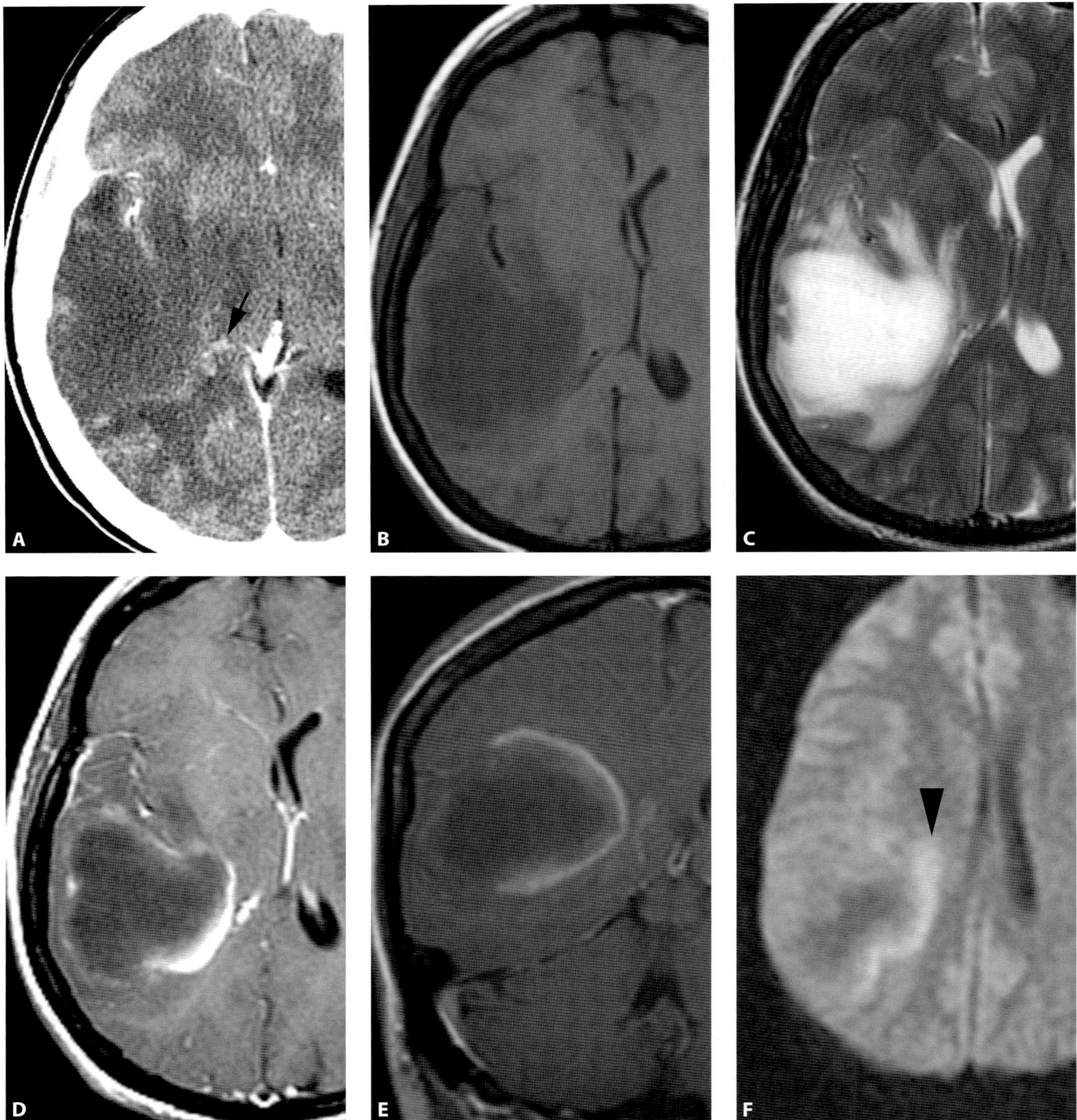

Figure 12.1 A. Axial post-contrast head CT from a 38-year-old woman with weakness and confusion, transferred to a referral center for evaluation of presumed astrocytoma. There is a large confluent area of subcortical hypodensity in the right temporal lobe. Hypodensity extends into the cortex and there is mild right to left midline shift. There is subtle peripheral enhancement along the medial edge of the lesion (arrow) when compared to the pre-contrast scan (not shown). **B.** Axial T1-weighted MR image shows confluent hypointensity in the right temporal lobe corresponding to the abnormality seen on non-contrast head CT. Axial T2-weighted (**C**) images show corresponding large confluent focus of abnormal signal in the right temporal lobe. Axial (**D**) and coronal (**E**) T1-weighted post-contrast images demonstrate medial "leading edge" peripheral enhancement. Notice the horseshoe-shaped enhancement pattern open to the cortex on the coronal image. Axial diffusion-weighted MR image (**F**) shows decreased signal within the body of the lesion with increased signal (arrowhead) corresponding to the edge enhancement seen on post-contrast images. The enhancement pattern and diffusion characteristics are typical for tumefactive demyelination.

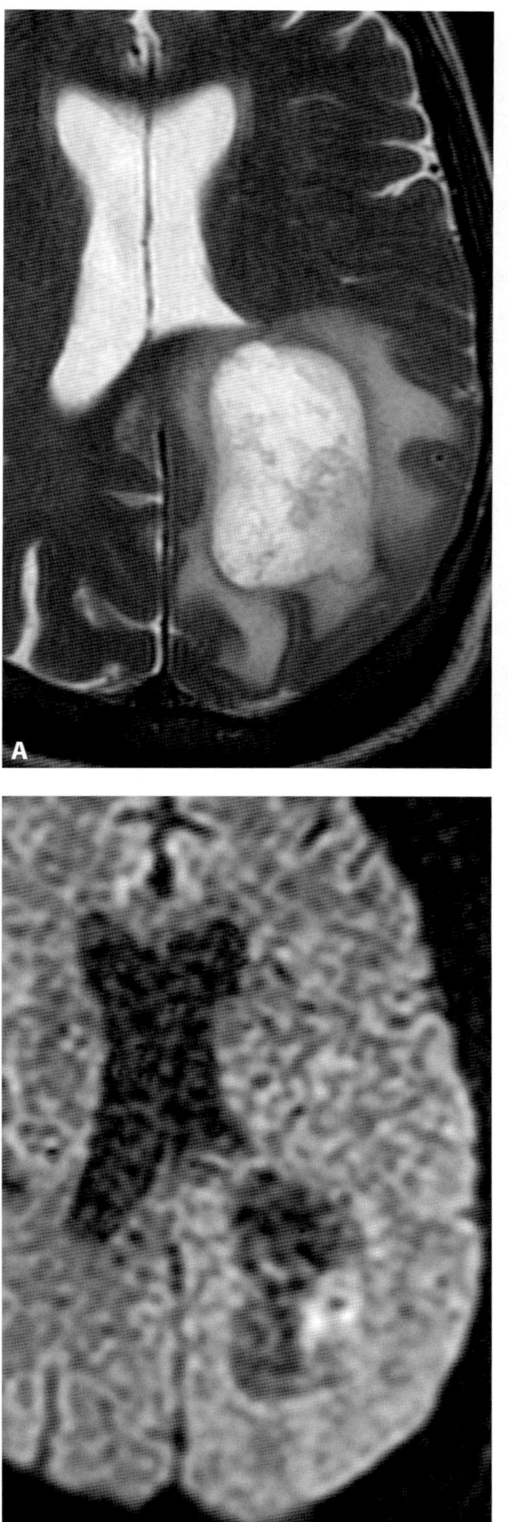

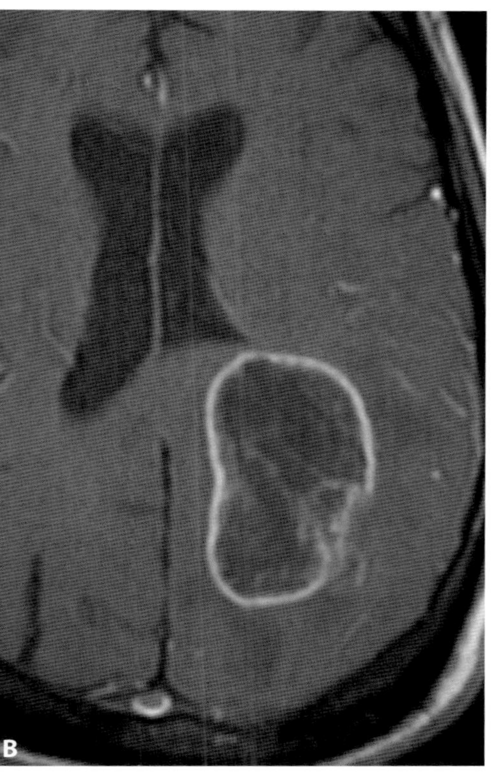

Figure 12.2 MR images of the brain from a 61-year-old man with seizures. **A.** Axial T2-weighted image shows a confluent area of hyperintensity in the left parietal subcortical white matter, in continuity with the spenium of the corpus callosum that is heterogeneous with high signal centrally and is associated with subcortical edema. **B.** Axial T1 post-contrast image shows rim enhancement with non-enhancing central necrosis. **C.** Axial diffusion-weighted image shows the lesion does not demonstrate diffusion restriction. These findings are typical of a high-grade CNS neoplasm, in this case a glioblastoma multiforme.

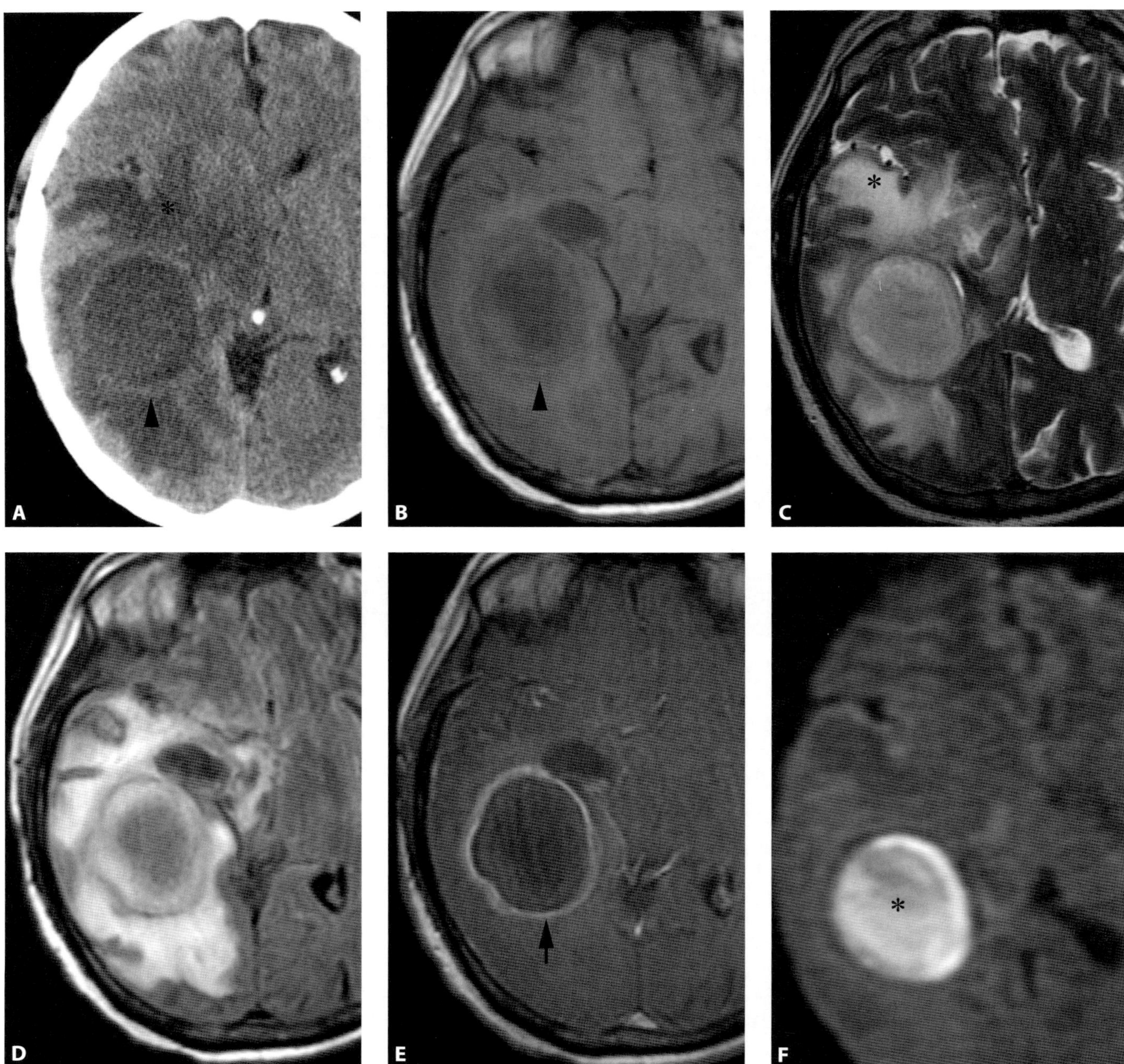

Figure 12.3 Images from a 40-year-old male with altered mental status demonstrating the typical imaging findings for cerebral abscess. **A.** Axial non-contrast CT reveals a round hypodense lesion in the right temporal lobe (arrowhead) with surrounding subcortical white matter hypodensity (asterisk). **B.** T1-weighted MR image shows a large round area of hypointense signal in the right temporal lobe (arrowhead) corresponding to the lesion seen on CT. **C.** Axial T2-weighted MRI image shows abnormally increased signal within the lesion with surrounding hyperintensity (asterisk) indicating edema. **D.** Axial FLAIR image shows central hypointensity suggesting necrosis. **E.** Axial T1-weighted post-contrast image shows typical ring enhancement (arrow) with central hypointensity. **F.** Axial diffusion weighted image shows bright focus of restriction (asterisk) which corresponded to low signal on the apparent diffusion coefficient map.

CASE 13 Cavernous malformation simulating contusion

Michael J. Modica

Imaging description

As cavernous malformations have a propensity to undergo repeated intralesional hemorrhage, they can have an imaging appearance that overlaps with cerebral contusions [1].

Cavernous malformations typically present as round or oval hyperdense lesions on non-contrast CT and have a variable appearance on MR, depending on the stage of hemorrhagic degradation (Figure 13.1). On MR they commonly appear as discrete, lobulated masses – "popcorn ball" – with variable-sized foci of blood products at different stages of evolution. On T2 they typically show a mixed signal intensity core with a complete hypointense hemosiderin rim. There is corresponding hypointense rim signal on gradient-recalled echo (GRE) sequences owing to hemosiderin susceptibility effect, otherwise known as "blooming." Lesions with subacute hemorrhage may show surrounding FLAIR hyperintensity from edema, although this is uncommon. Diffusion-weighted images are typically normal [2, 3].

The second most common presentation of cavernous malformation is numerous punctate foci of hypointense "black dots" scattered throughout both cerebral hemispheres on GRE images (Figure 13.2). These lesions are usually occult on other imaging sequences primarily due to their small size [2–4].

Cavernous malformations are associated with developmental venous anomalies. These should be sought whenever a cavernous malformation is detected.

An MRI-based classification of cavernous malformations has been proposed based on clinical symptomatology of the lesions. Type I and type II lesions are more prevalent in symptomatic patients.

Zabramski *et al.* proposed a classification of cavernous malformations (Table 13.1) [5].

Type I cavernous malformations have the potential to mimic acute contusion whereas type III and IV lesions can resemble chronic contusions and diffuse axonal injury.

Importance

Cavernous malformations are benign vascular hamartomas of the central nervous system composed of immature blood vessels without intervening neuronal tissue [6]. They typically undergo spontaneous intralesional hemorrhage and can produce focal neurologic deficits or seizures or can mimic hemorrhagic cerebral contusions in the acute trauma patient.

If deemed necessary surgical resection is the treatment mainstay for cavernous malformation due to their propensity for bleeding and re-bleeding. When a contusion is diagnosed on CT or MR, follow-up is helpful to evaluate for changes in size and/or resolution. A non-resolving contusion should raise the suspicion for an underlying cavernous malformation.

Table 13.1. Classification of cavernous malformations

Type I: Subacute hemorrhage appearing hyperintense on T1 and hyper- or isointense on T2
Type II: Classic popcorn ball lesions appearing as mixed signal intensity on T1 and T2 representing degrading hemorrhage of various stages (Figure 13.3)
Type III: Chronic hemorrhage appearing hypo- to isointense on T1- and T2-weighted imaging
Type IV: Numerous punctate microhemorrhages ("black dots"), not seen on any sequence other than GRE

Typical clinical scenario

Solitary cavernous malformations can be discovered incidentally, such as in patients with a history of recent trauma. Patients are typically between the ages of 40 and 60 years; however, they occur in any age group. Cavernous malformations may present with a range of clinical symptoms including a ground-level fall secondary to seizure, progressive neurologic deficit from intralesional hemorrhage, or they may be completely asymptomatic. Multiple (familial) cavernous malformation syndrome is an autosomal dominant disease primarily affecting the Hispanic population and tends to present earlier than sporadic cavernous malformations [5].

Differential diagnosis

Cerebral contusions often involve the anterior-inferior temporal lobes and frontal lobes. Acute contusions can be difficult to differentiate from cavernous malformations based on CT alone. On non-contrast CT contusions often appear as non-specific cortical and subcortical hyperdensities, and may be multiple. They are typically accompanied by surrounding perihemorrhagic edema. Contusions and surrounding subcortical edema can increase in size on follow-up scans, which may lead to mass effect on neighboring parenchyma [7].

Contusions can be differentiated from cavernous malformations based on the typical MRI appearance. Acute contusions will generally lack a surrounding ring of hemosiderin with central "popcorn ball" appearance on MR that is found with cavernous malformation, instead having surrounding hyperintense signal on T2-weighted and FLAIR images from cortical edema. They may show predominantly intra-lesional rather than peripheral ring-like hypointense hemorrhagic "blooming" on GRE images (Figure 13.4) [8, 9].

Chronic contusions can be differentiated from type III cavernous malformations by their typical imaging pattern. Chronic contusions appear as focal frontal or temporal lobe or diffuse cerebral encephalomalacia, findings not present with type III or IV cavernous malformations [8]. MR will show some non-focal residual parenchymal hemosiderin staining that appears hypointense on FLAIR and GRE images, but signal abnormality will not be focal or rounded in configuration like type III cavernous malformations [8].

Diffuse axonal "shear" injury (DAI) is another type of traumatic brain injury that can be confused with type IV cavernous malformations. DAI lesions appear as multifocal punctate foci of hypointensity on GRE images secondary to susceptibility from hemorrhage. However, in distinction to type IV cavernous malformations, DAI is accompanied by hyperintensity on T2/FLAIR images and they characteristically occur within the corticomedullary junction, corpus callosum, and deep gray matter nuclei [8, 10]. Diffuse axonal injury is covered in detail in Case 14.

Teaching point

Cavernous malformations can mimic cerebral contusions. Both can be hyperdense on non-contrast CT. MR features that suggest cavernous malformations include a mixed signal intensity lesion with a hemosiderin rim and lack of perilesional edema.

REFERENCES

1. Maraire JN, Awad IA. Intracranial cavernous malformations: lesion behavior and management strategies. *Neurosurgery.* 1995;**37**(4):591–605.

2. Rigamonti D, Hadley MN, Drayer BP, *et al.* Cerebral cavernous malformations. Incidence and familial occurrence. *N Engl J Med.* 1988;**319**(6):343–7.

3. Ide C, De Coene B, Baudrez V. MR features of cavernous angioma. *JBR-BTR.* 2000;**83**(6):320.

4. Barker CS. Magnetic resonance imaging of intracranial cavernous angiomas: a report of 13 cases with pathological confirmation. *Clin Radiol.* 1993;**48**(2):117–21.

5. Zabramski JM, Wascher TM, Spetzler RF, *et al.* The natural history of familial cavernous malformations: results of an ongoing study. *J Neurosurg.* 1994;**80**(3):422–32.

6. Voigt K, Yasargil MG. Cerebral cavernous haemangiomas or cavernomas. Incidence, pathology, localization, diagnosis, clinical features and treatment. Review of the literature and report of an unusual case. *Neurochirurgia (Stuttg).* 1976;**19**(2):59–68.

7. Alahmadi H, Vachhrajani S, Cusimano MD. The natural history of brain contusion: an analysis of radiological and clinical progression. *J Neurosurg.* 2010;**112**(5):1139–45.

8. Gentry LR, Godersky JC, Thompson B. MR imaging of head trauma: review of the distribution and radiopathologic features of traumatic lesions. *AJR Am J Roentgenol.* 1988;**150**(3):663–72.

9. Akiyama Y, Miyata K, Harada K, *et al.* Susceptibility-weighted magnetic resonance imaging for the detection of cerebral microhemorrhage in patients with traumatic brain injury. *Neurol Med Chir (Tokyo).* 2009;**49**(3):97–9; discussion 99.

10. Parizel PM, Ozsarlak, Van Goethem, JW, *et al.* Imaging findings in diffuse axonal injury after closed head trauma. *Eur Radiol.* 1998;**8**(6): 960–5.

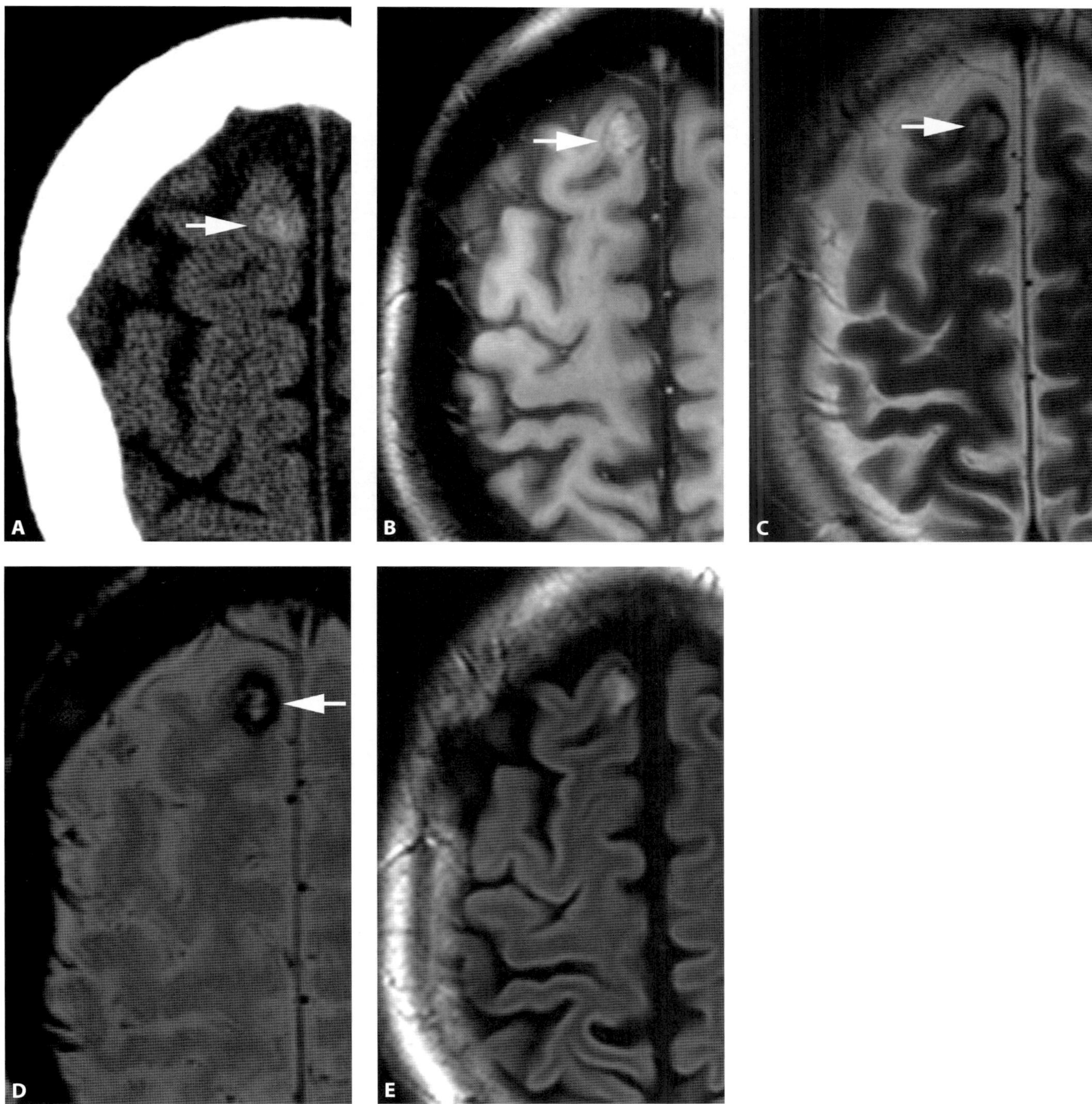

Figure 13.1 A. Axial non-contrast head CT from a 41-year-old woman after a high-speed motor vehicle collision shows a hyperdense lesion in the superior aspect of the right frontal lobe (arrow). Note the lack of peri-lesional edema. Axial T1- (**B**) and T2- (**C**) weighted MR images show a well-defined hyperintense lesion in the right frontal lobe, near the junction of the cortical and subcortical white matter (arrow). **D.** Axial GRE MR image shows prominent susceptibility effect represented by hypointense blooming around the hyperintense lesion (arrow). This is a classic Type I cavernous malformation. **E.** Axial FLAIR MR image shows the lack of perilesional edema, further differentiating this lesion from a hemorrhagic contusion.

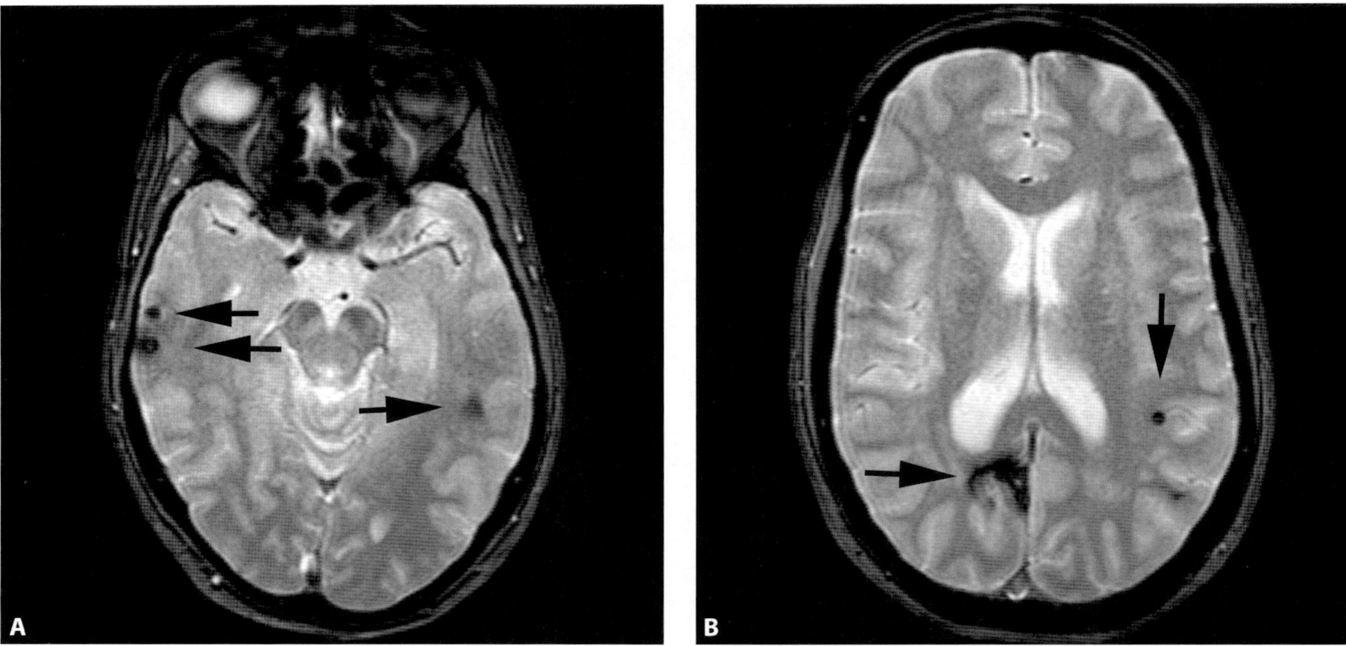

Figure 13.2 A and **B.** Axial GRE images from a brain MR from a 38-year-old woman with headache demonstrate multiple round foci of susceptibility throughout the brain (arrows). No abnormalities were seen on any other imaging sequence. Non-contrast CT was negative. This is a typical imaging pattern for type IV cavernous malformations.

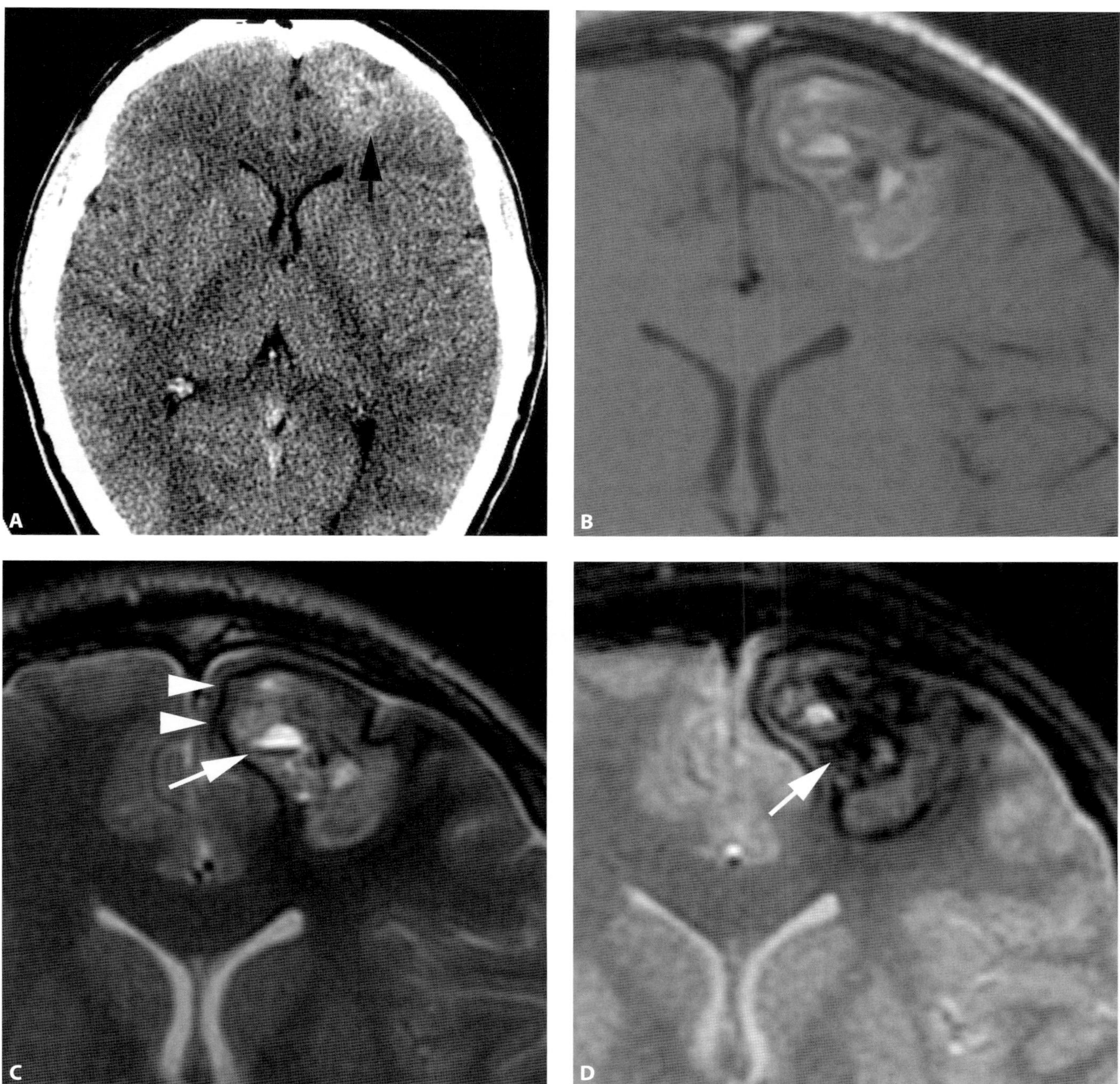

Figure 13.3 A. Axial non-contrast head CT from a 45-year-old man who was assaulted shows an ill-defined hyperdense lesion in the left frontal gyrus. This lesion was initially thought to represent a hemorrhagic contusion. Axial T1- (**B**) and T2- (**C**) weighted MR images show a mixed signal intensity lesion. Notice the fluid-fluid level (arrow) within the lesion representing acute-on-chronic hemorrhage, classic for type II cavernous malformations. There is a complete hypointense rim on T2 (arrowheads) representing the characteristic hemosiderin rim. There is a lack of surrounding edema, confirmed on FLAIR. **D.** Axial GRE MR image shows prominent susceptibility effect represented by hypointense blooming centrally (arrow) due to abundant methemoglobin.

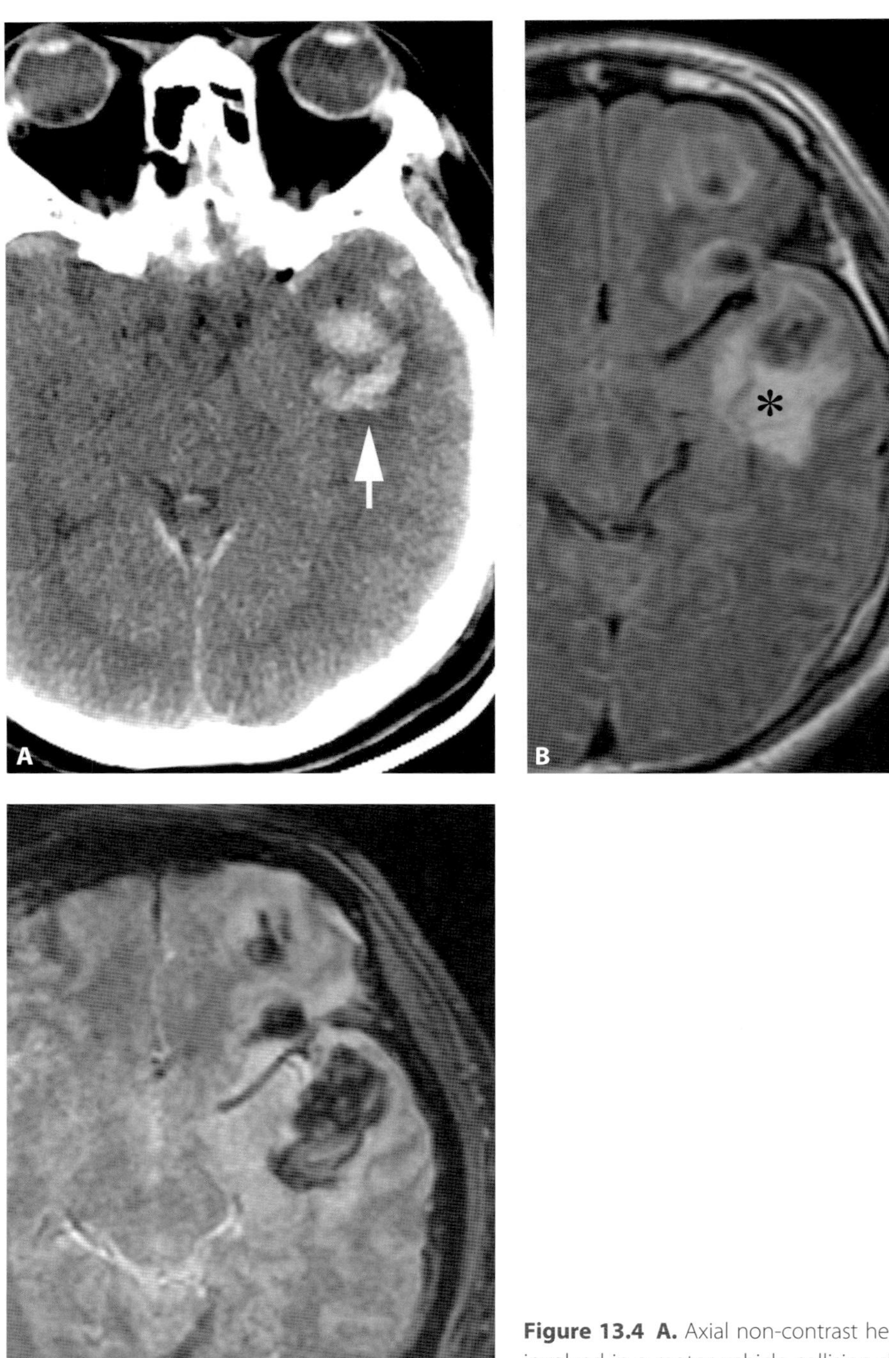

Figure 13.4 **A.** Axial non-contrast head CT from an 18-year-old male involved in a motor vehicle collision shows a hyperdense lesion in the left anteroinferior temporal lobe (arrow). Note the surrounding hypodensity indicating surrounding edema. **B.** Axial FLAIR image from an MR examination from the same patient demonstrates multiple foci of hypointensity in the left inferior frontal and temporal lobes with surrounding edema (asterisk). **C.** Axial GRE T2* image shows intralesional susceptibility artifact corresponding to hemorrhage.

CASE **14**

Diffuse axonal injury

Robert B. Carr

Imaging description

Diffuse axonal injury (DAI) is caused by shearing forces that occur during rapid acceleration or deceleration of the brain. This results in tearing of the axons. Most lesions are small and multiple. Characteristic locations include the gray–white matter junctions (Figure 14.1), splenium of the corpus callosum, basal ganglia, internal capsules, and dorsolateral brainstem [1, 2].

It has been previously reported that most DAI lesions are non-hemorrhagic. However, both pathologic literature and imaging studies with improved techniques suggest that more lesions are hemorrhagic than previously thought [3]. Lesions may be seen on CT if there is sufficient hemorrhage or edema to produce discernible hyperattenuation or hypoattenuation, respectively. However, CT is very insensitive to the detection of DAI, and this limitation should be realized when imaging a patient with traumatic brain injury.

MRI is more sensitive for the detection of DAI, and will detect many lesions which are not visible on CT [4]. FLAIR and T2-weighted images will depict DAI lesions as foci of increased signal. However, these sequences are also relatively insensitive. Since most blood products are paramagnetic (including deoxyhemoglobin, intracellular methemoglobin, and hemosiderin), they produce susceptibility effect on gradient-recalled echo (GRE) images. GRE images are therefore sensitive to the identification of microhemorrhages and detect more foci of DAI than conventional MRI sequences (Figure 14.2). Susceptibility-weighted imaging (SWI) has been shown to detect even more lesions than GRE images [5].

Importance

DAI may have profound consequences, and its detection is important for the evaluation, treatment, and prognosis of patients with traumatic brain injury. It is important to realize that CT is very insensitive to the detection of DAI. MRI is much more sensitive, specifically GRE and SWI sequences.

Typical clinical scenario

DAI occurs from acceleration or deceleration forces related to traumatic brain injury. The patient will often be comatose. The clinical findings may be out of proportion to the imaging findings. In such cases, MRI with a GRE or SWI sequence may be beneficial in identifying lesions.

Differential diagnosis

In the appropriate clinical setting, the differential diagnosis is limited and includes cerebral contusions. It may be difficult to ascribe a lesion as a focus of shear injury or contusion. However, DAI is usually multifocal and located in characteristic locations as described.

Teaching point

CT has a low sensitivity for the detection of DAI. MRI, specifically GRE and SWI sequences, is much more sensitive and should be obtained in appropriate cases.

REFERENCES

1. Arfanakis K, Haughton VM, Carew JD, *et al.* Diffusion tensor MR imaging in diffuse axonal injury. *AJNR Am J Neuroradiol.* 2002;**23**(5):794–802.
2. Giugni E, Sabatini U, Hagberg GE, Formisano R, Castriota-Scanderbeg A. Fast detection of diffuse axonal damage in severe traumatic brain injury: comparison of gradient-recalled echo and turbo proton echo-planar spectroscopic imaging MRI sequences. *AJNR Am J Neuroradiol.* 2005; **26**(5):1140–8.
3. Scheid R, Preul C, Gruber O, Wiggins C, von Cramon DY. Diffuse axonal injury associated with chronic traumatic brain injury: evidence from T2*-weighted gradient-echo imaging at 3 T. *AJNR Am J Neuroradiol.* 2003;**24**(6):1049–56.
4. Topal NB, Hakyemez B, Erdogan C, *et al.* MR imaging in the detection of diffuse axonal injury with mild traumatic brain injury. *Neurol Res.* 2008;**30**(9):974–8.
5. Tong KA, Ashwal S, Obenaus A, *et al.* Susceptibility-weighted MR imaging: a review of clinical applications in children. *AJNR Am J Neuroradiol.* 2008;**29**(1):9–17.

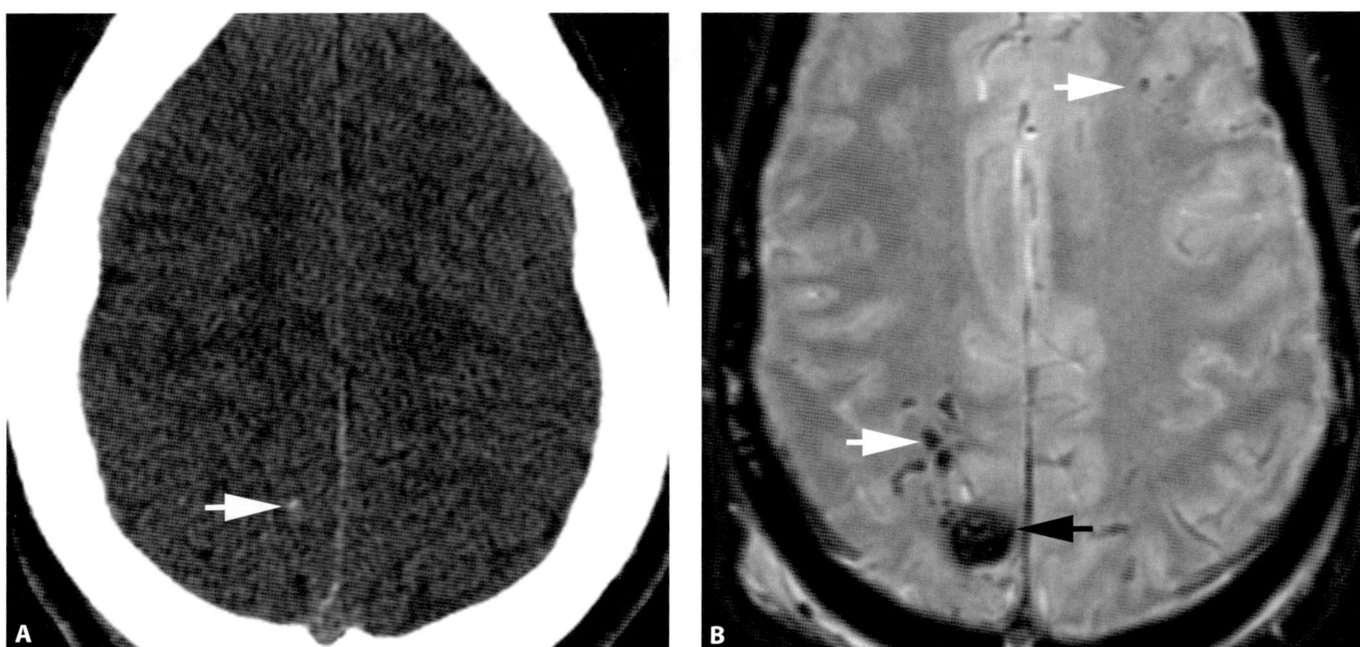

Figure 14.1 **A.** Axial non-enhanced CT image from a 34-year-old man who sustained a motor vehicle accident shows a single small focus of increased attenuation within the right parietal lobe (arrow). **B.** Axial GRE image obtained soon afterwards demonstrates a much larger area of hemorrhage (black arrow), with numerous additional lesions (white arrows). This case demonstrates the insensitivity of CT to the detection of DAI.

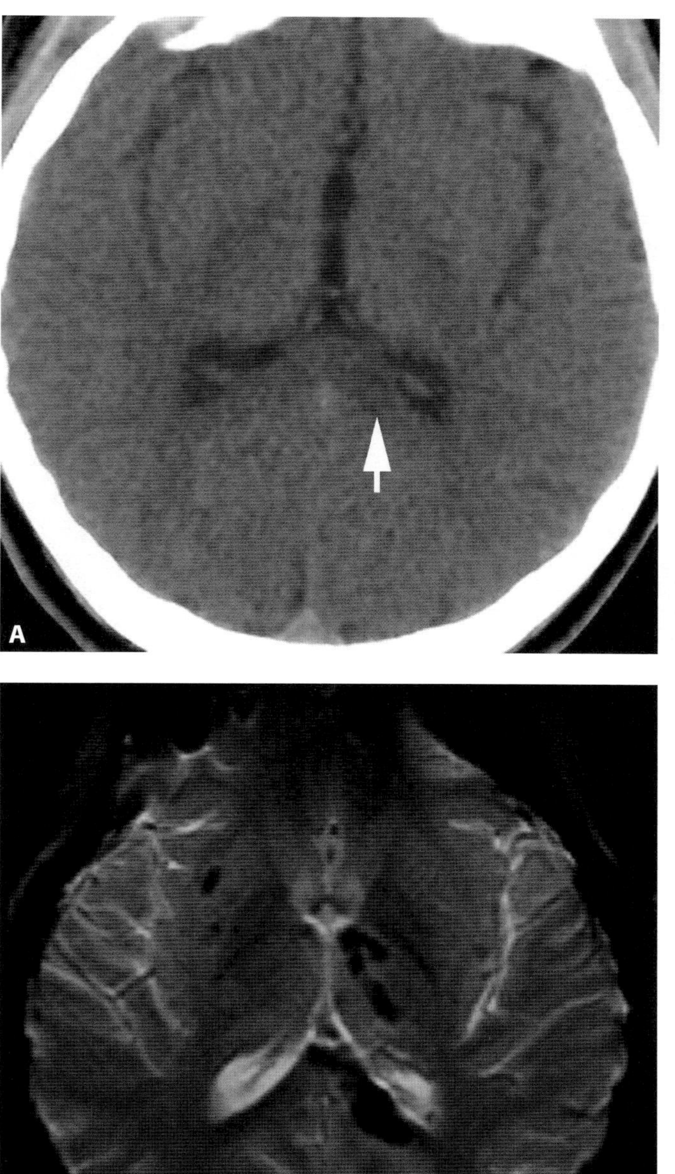

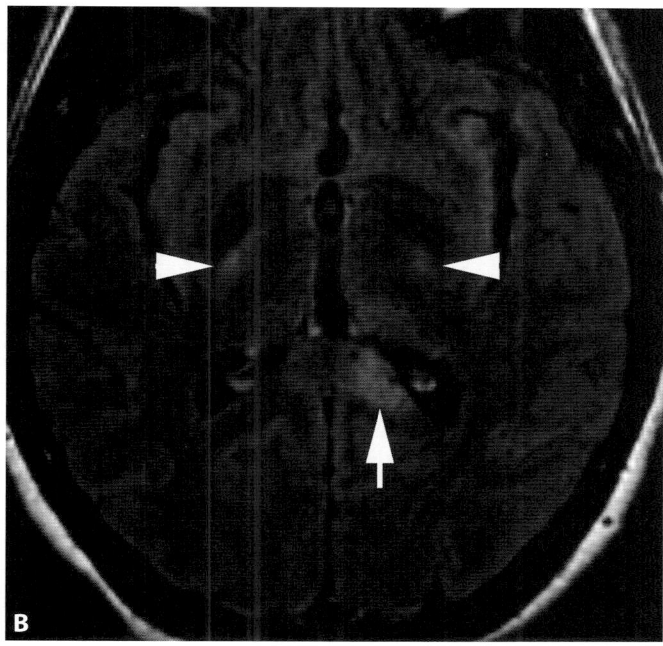

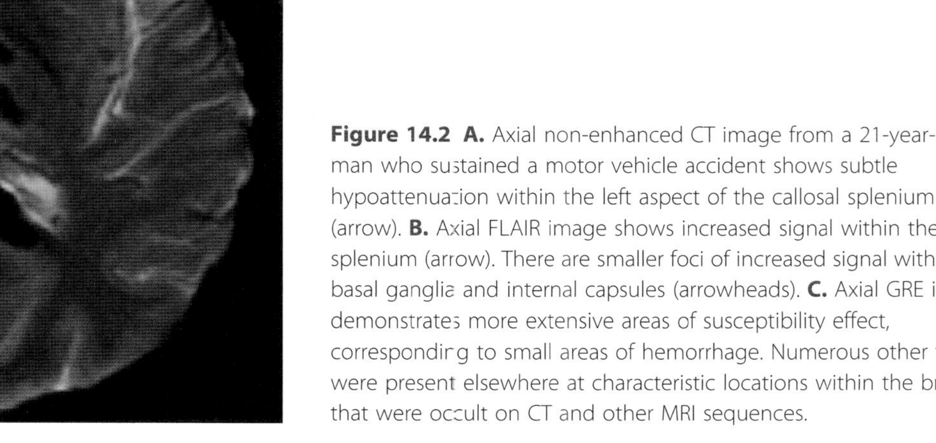

Figure 14.2 A. Axial non-enhanced CT image from a 21-year-old man who sustained a motor vehicle accident shows subtle hypoattenuation within the left aspect of the callosal splenium (arrow). **B.** Axial FLAIR image shows increased signal within the splenium (arrow). There are smaller foci of increased signal within the basal ganglia and internal capsules (arrowheads). **C.** Axial GRE image demonstrates more extensive areas of susceptibility effect, corresponding to small areas of hemorrhage. Numerous other foci were present elsewhere at characteristic locations within the brain that were occult on CT and other MRI sequences.

CASE

15 Orbital infection

Robert B. Carr

Imaging description

The evaluation of an orbital infection should seek to define the extent of infection, the source of infection, and the presence of complications.

An imperative distinction is preseptal (periorbital) versus postseptal (orbital) cellulitis. The orbital septum is a thin fibrous layer of the eyelids that blends with the periosteum of the bony orbit. The septum cannot be specifically delineated on conventional imaging, but its position can be inferred. Inflammatory changes posterior to the septum, such as fat stranding, muscle enlargement, and abscess formation, imply the presence of orbital cellulitis (Figures 15.1 and 15.2). Inflammatory changes entirely anterior to the septum are classified as periorbital cellulitis [1].

The most common cause of an orbital infection is sinusitis, especially of the ethmoid sinus. Therefore, the sinuses should be scrutinized for signs of inflammation (Figure 15.1). Infection spreads through the bony walls via perivascular routes [2]. Other sources of infection include orbital foreign objects, adjacent dermal infection, and septicemia. An uncommon source is odontogenic infection (Figures 15.3 and 15.4), which usually arises from the maxillary dentition and spreads through the paranasal sinuses, premaxillary tissues, or infratemporal fossa. The most common signs of dental infection include periapical lucency, indistinctness of the lamina dura, and widening of the periodontal ligament space [3].

Complications include abscess formation, meningitis, and cavernous sinus thrombosis [4]. A phlegmon or abscess will often form in the subperiosteal space due to adjacent sinusitis. Any soft tissue or fluid collection immediately adjacent to the bony margin in the setting of infection is suspicious for a subperiosteal phlegmon or abscess (Figure 15.5). Infection may spread intracranially, resulting in meningitis, epidural abscess, or subdural abscess. Evaluate the dura and meninges for areas of abnormal enhancement or small foci of restricted diffusion on MRI, suggesting the presence of an abscess. Venous extension into the cavernous sinus, often via the superior ophthalmic vein, may result in cavernous sinus thrombosis (Figure 15.6). Typical imaging findings include lack of normal enhancement within the sinus and convex bowing of the lateral margin [5].

Importance

Orbital infections are usually bacterial in origin. Recognizing postseptal extension of infection is of paramount importance, as this may lead to intracranial involvement or visual loss. Patients with postseptal infections will be treated more aggressively. While periorbital cellulitis can typically be managed with oral antibiotics, orbital cellulitis often requires intravenous antibiotics [6]. Identifying the cause of infection, as with odontogenic disease, will also facilitate appropriate clinical management.

Typical clinical scenario

Orbital infections affect all age groups, but are more common in young children. The patient may present with pain, edema, erythema, and chemosis. The diagnosis is often apparent clinically, but the extent of infection may not be known. Imaging is obtained to define the extent of infection and presence of complications.

Differential diagnosis

Orbital inflammatory pseudotumor and orbital lymphoma may demonstrate inflammatory changes similar to orbital infection [7]. In the absence of ancillary findings such as sinusitis and abscess formation the distinction may be difficult. The clinical scenario and response to treatment should guide the diagnosis. However, in difficult cases, the symptoms may overlap. Therefore, if the diagnosis is not clear, inflammatory pseudotumor and lymphoma should be considered.

Teaching point

The key roles of imaging in orbital infection are to define the extent of infection (preseptal versus postseptal), identify a possible source of infection, and identify the presence of complications.

REFERENCES

1. LeBedis CA, Sakai O. Nontraumatic orbital conditions: diagnosis with CT and MR imaging in the emergent setting. *Radiographics*. 2008;**28**(6):1741–53.
2. Ludwig BJ, Foster BR, Saito N, *et al.* Diagnostic imaging in nontraumatic pediatric head and neck emergencies. *Radiographics*. 2010;**30**(3):781–99.
3. Caruso PA, Watkins LM, Suwansaard P, *et al.* Odontogenic orbital inflammation: clinical and CT findings – initial observations. *Radiology*. 2006;**239**(1):187–94.
4. Eustis HS, Mafee MF, Walton C, Mondonca J. MR imaging and CT of orbital infections and complications in acute rhinosinusitis. *Radiol Clin North Am*. 1998;**36**(6):1165–83, xi.
5. Schuknecht B, Simmen D, Yuksel C, Valavanis A. Tributary venosinus occlusion and septic cavernous sinus thrombosis: CT and MR findings. *AJNR Am J Neuroradiol*. 1998;**19**(4):617–26.
6. Seltz LB, Smith J, Durairaj VD, Enzenauer R, Todd J. Microbiology and antibiotic management of orbital cellulitis. *Pediatrics*. 2011;**127**(3):e566–72.
7. Kapur R, Sepahdari AR, Mafee MF, *et al.* MR imaging of orbital inflammatory syndrome, orbital cellulitis, and orbital lymphoid lesions: the role of diffusion-weighted imaging. *AJNR Am J Neuroradiol*. 2009;**30**(1):64–70.

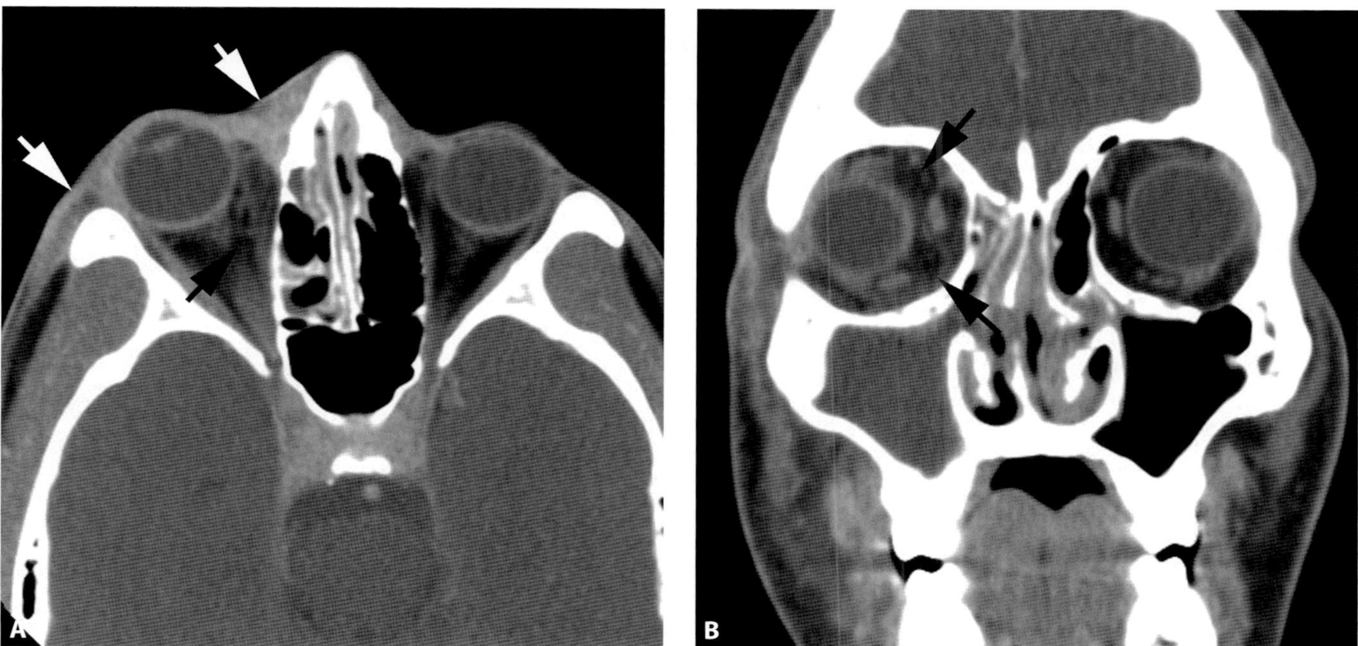

Figure 15.1 A. Axial contrast-enhanced CT image from a 20-year-old man with orbital cellulitis shows inflammation of the medial and lateral preseptal soft tissues (white arrows), as well as postseptal extension (black arrow). **B.** Coronal contrast-enhanced CT image shows inflammation within the right postseptal soft tissues (arrows). Compare with the normal left orbit. Also notice the extensive sinus inflammation, indicating a probable sinogenic origin of the orbital cellulitis.

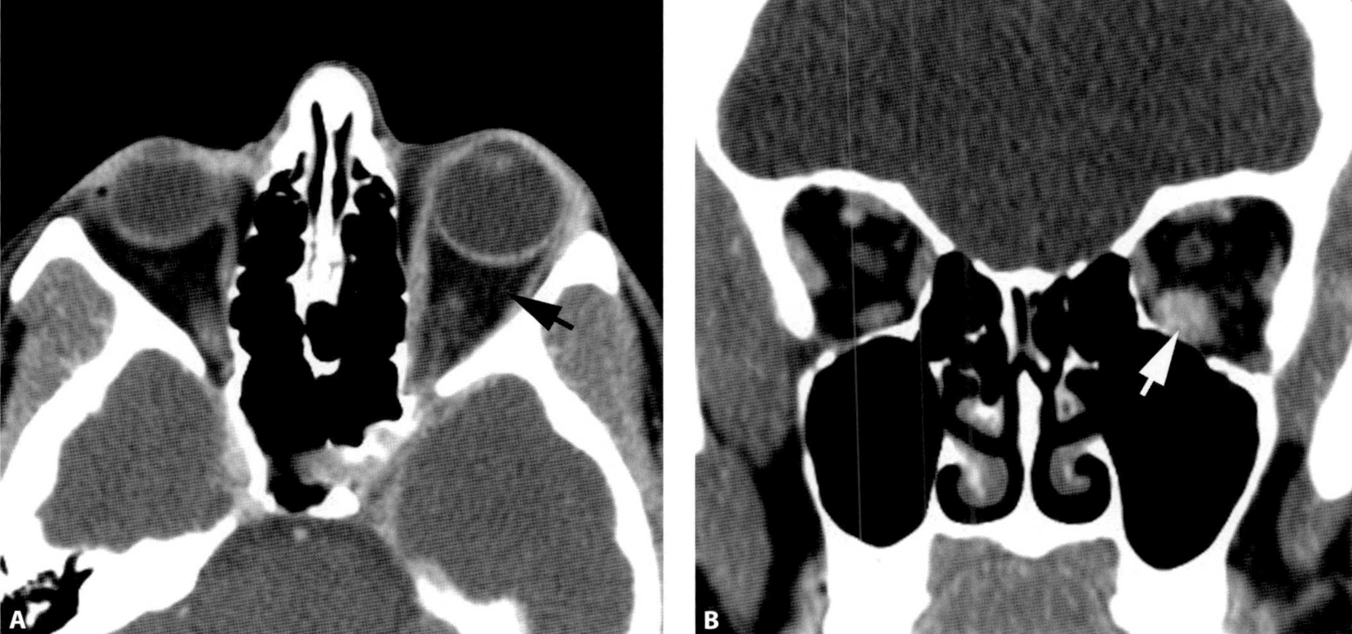

Figure 15.2 A. Axial contrast-enhanced CT image from a 23-year-old woman with orbital cellulitis shows very mild increased attenuation within the left intraconal fat (arrow) compared with the right. **B.** Coronal contrast-enhanced CT image shows enlargement and increased enhancement of the left inferior rectus muscle (arrow). Mild fat stranding and increased attenuation is again seen. The coronal images are often very helpful when evaluating the orbits.

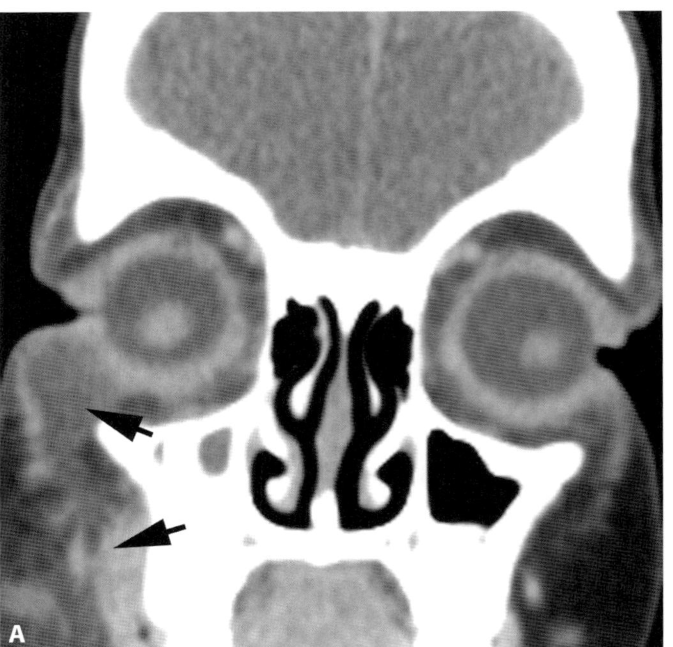

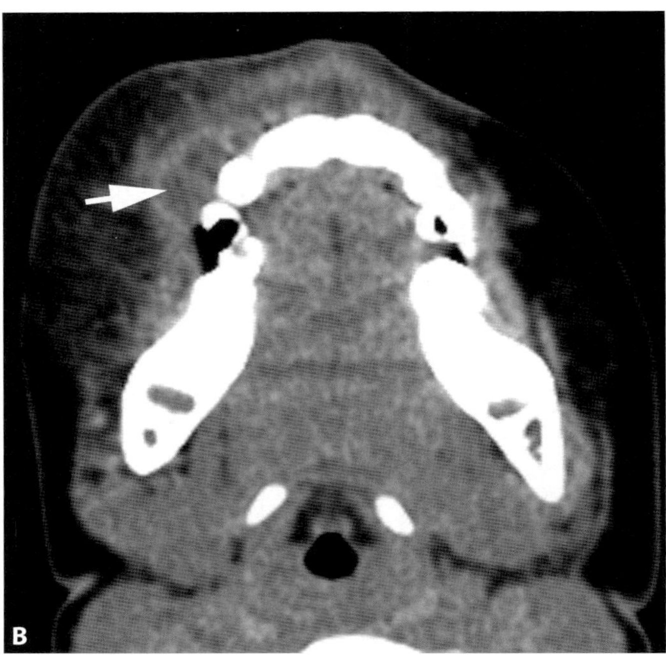

Figure 15.3 A. Coronal contrast-enhanced CT image in a 17-year-old man with periorbital cellulitis shows inflammation of the right preseptal soft tissues (arrows). **B.** Axial contrast-enhanced CT image shows a dental abscess, with inflammatory changes ascending through the right premaxillary soft tissues (arrow). This confirms an odontogenic origin of the preseptal cellulitis.

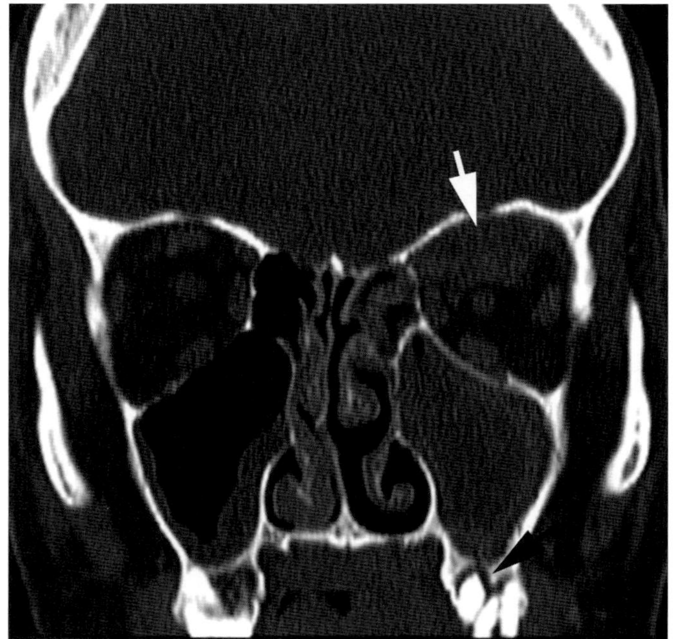

Figure 15.4 Coronal contrast-enhanced CT image with bone algorithm from a 24-year-old man with odontogenic orbital cellulitis shows periapical lucency and widening of the periodontal ligament space around a left maxillary tooth (black arrow). The infection has ascended into the orbit, producing a large abscess (white arrow).

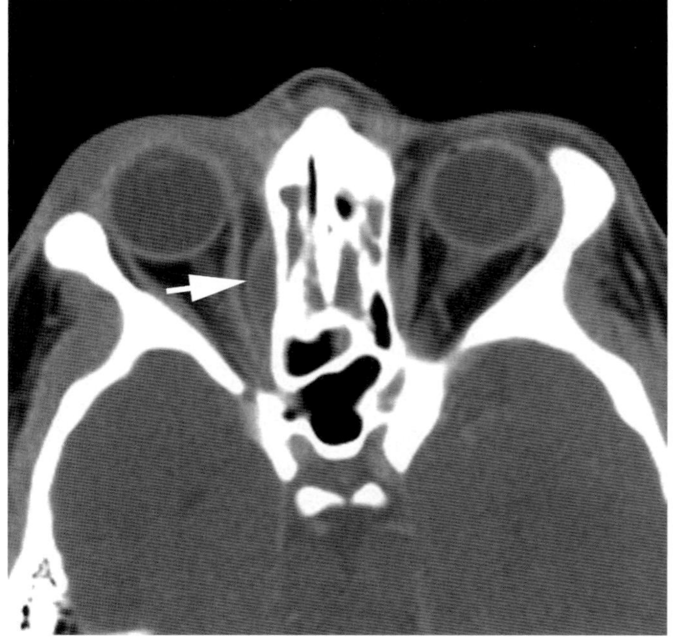

Figure 15.5 Axial contrast-enhanced CT image from a 20-year-old man with sinusitis shows a linear fluid collection along the inner margin of the right lamina papyracea (arrow). This is a typical appearance for a subperiosteal abscess.

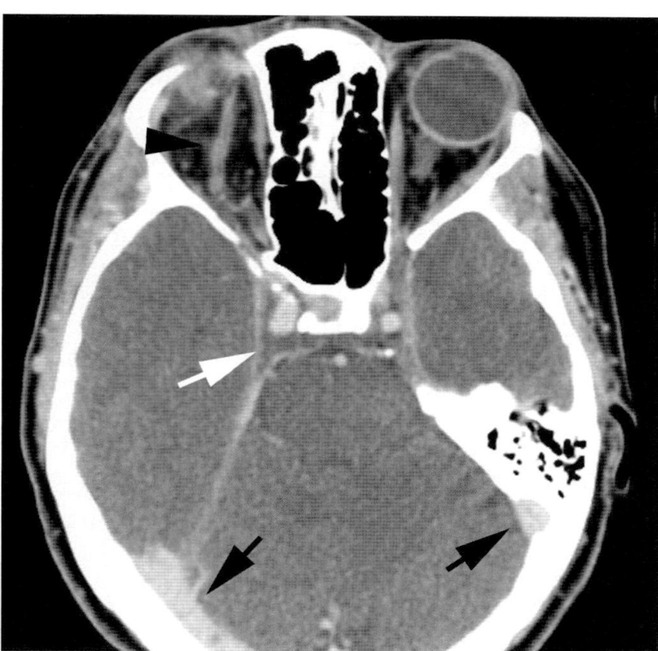

Figure 15.6 Axial venous phase contrast-enhanced CT image from a 34-year-old woman with orbital cellulitis and cavernous sinus thrombosis shows non-opacification of the cavernous sinuses (white arrow). Notice the normal opacification of the right transverse and left sigmoid sinuses (black arrows). There is also thrombosis and mild expansion of the right superior ophthalmic vein (black arrowhead).

CASE

16 Globe injuries

Robert B. Carr

Imaging description

Globe injuries often present with subtle or confusing appearances on CT. A systematic approach is useful, paying particular attention to the anterior chamber, the lens, the vitreous body, the shape of the globe, and the presence of foreign objects.

The anterior chamber should be scrutinized with respect to size and attenuation. Decreased depth of the anterior chamber may be caused by a full-thickness corneal laceration or by anterior dislocation of the lens (Figure 16.1). Increased depth of the anterior chamber may be seen with a posterior globe rupture [1]. The change in depth may be subtle, and it is most helpful to compare with the contralateral globe. Increased attenuation within the anterior chamber indicates the presence of hemorrhage, known as a hyphema (Figures 16.2 and 16.3).

Injury to the zonular attachments of the lens may result in posterior (more common) or anterior lens dislocation, and dislocations may be partial. Trauma to the lens capsule may result in the influx of fluid, leading to hypoattenuation of the lens; this is known as a traumatic cataract (Figure 16.4).

The posterior chamber may rupture, producing deformity along the posterior margin of the globe. There may also be detachment of the vitreous, choroid, or retina (Figure 16.5). Each type of detachment demonstrates a different morphology. Vitreous detachment usually begins posteriorly and crosses the optic disk. Choroid detachment extends anteriorly to the margin of the lens, and diverges posteriorly as it approaches the optic disk (Figure 16.6). Retinal detachment extends anteriorly to the ora serrata, and converges posteriorly on the optic disk (Figure 16.7).

Irregularity or flattening of the globe contour may indicate globe rupture. If there is retrobulbar hemorrhage or inflammation, the globe may be proptotic. It is important to evaluate the insertion of the optic nerve for evidence of posterior globe tenting (Figure 16.8). It has been reported that a posterior globe angle of less than 120 degrees is associated with poor visual outcome [2].

Importance

There are numerous methods by which the globe may be injured. Many of these injuries are apparent clinically. However, physical examination may be limited by overlying soft tissue edema. Ultrasound is commonly employed, but is contraindicated in the setting of suspected globe rupture [3]. Therefore, CT may be the primary mode of initial evaluation, and it is important not to overlook or misinterpret important findings.

There are data that suggest that the change in anterior chamber size is the most important single predictor of globe injury. A change in anterior chamber size associated with other findings, such as intraocular hemorrhage or a change in globe contour, further increases the sensitivity [4]. Therefore, it is important to closely evaluate the anterior chamber and compare it with the contralateral globe.

It is important to remember the limitations of imaging. A normal appearance of the globe does not exclude globe rupture [5].

Typical clinical scenario

Orbital trauma accounts for approximately 3% of all emergency room visits [6], and may be caused by blunt or penetrating injuries. There will often be other injuries such as orbital wall and facial fractures.

Differential diagnosis

The differential diagnosis for globe injury would include a globe mass, such as melanoma, that may simulate hemorrhage. Silicone injection within the vitreous may be confused with hemorrhage. Abnormal globe shape may be seen with a staphyloma or coloboma. However, characteristic findings in the appropriate clinical setting are unlikely to be confused with other entities.

Teaching point

Globe injuries may be subtle or confusing. Therefore, it is important to have a systematic approach when evaluating the globe for trauma.

REFERENCES

1. Weissman JL, Beatty RL, Hirsch WL, Curtin HD. Enlarged anterior chamber: CT finding of a ruptured globe. *AJNR Am J Neuroradiol.* 1995;**16**(4 Suppl):936–8.
2. Dalley RW, Robertson WD, Rootman J. Globe tenting: a sign of increased orbital tension. *AJNR Am J Neuroradiol.* 1989;**10**(1):181–6.
3. Kubal WS. Imaging of orbital trauma. *Radiographics.* 2008;**28**(6):1729–39.
4. Kim SY, Lee JH, Lee YJ, *et al.* Diagnostic value of the anterior chamber depth of a globe on CT for detecting open-globe injury. *Eur Radiol.* 2010;**20**(5):1079–84.
5. Hoffstetter P, Schreyer AG, Schreyer CI, *et al.* Multidetector CT (MD-CT) in the diagnosis of uncertain open globe injuries. *Rofo.* 2010;**182**(2):151–4.
6. Bord SP, Linden J. Trauma to the globe and orbit. *Emerg Med Clin North Am.* 2008;**26**(1):97–123, vi–vii.

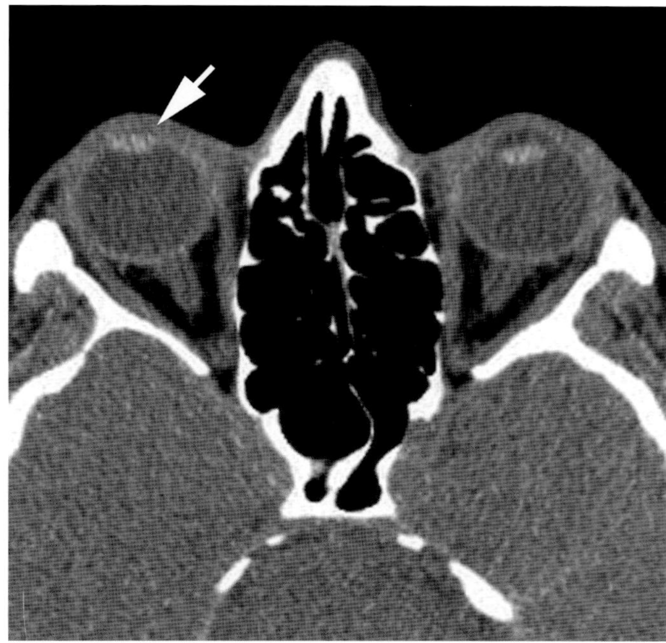

Figure 16.1 Axial non-enhanced CT image from a 33-year-old man who sustained trauma to the right globe shows near complete loss of the right anterior chamber (arrow). Compare with the normal contralateral side. This was caused by a full-thickness corneal laceration.

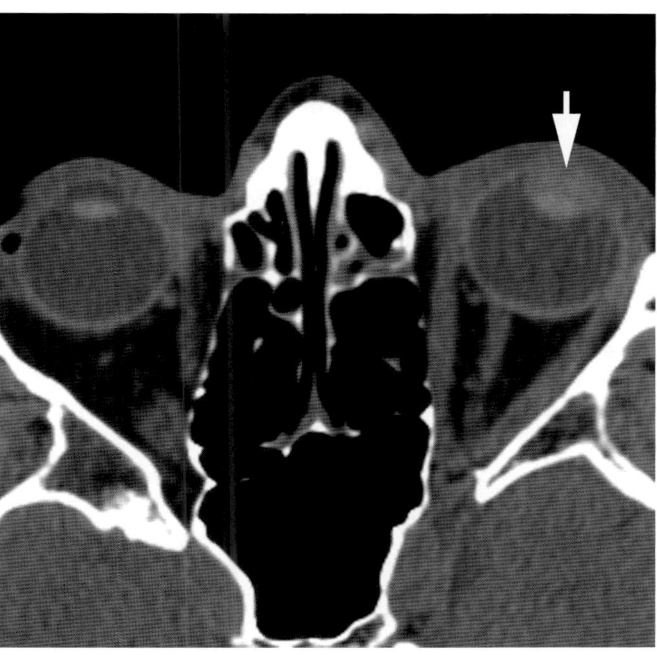

Figure 16.2 Axial non-enhanced CT image from a 42-year-old man who sustained blunt trauma to the left orbit shows increased attenuation within the left anterior chamber (arrow). This represents a hyphema with hemorrhage completely filling the anterior chamber.

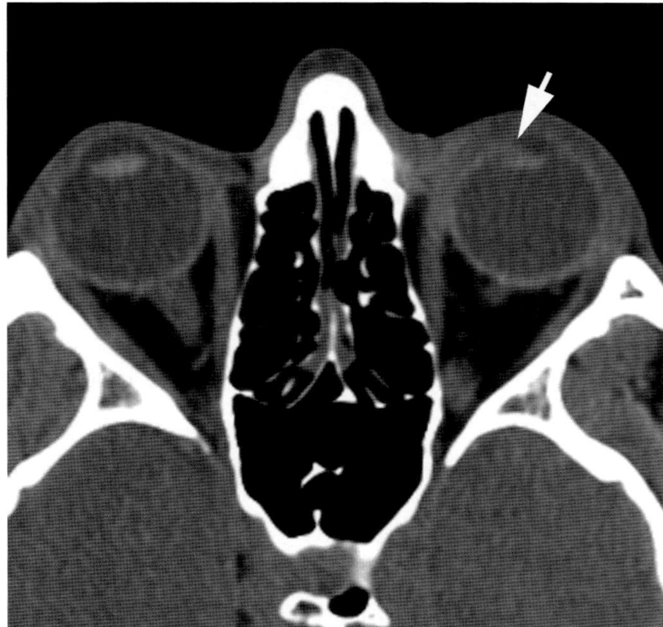

Figure 16.3 Axial non-enhanced CT image from a 23-year-old woman who sustained trauma to the left orbit shows a small amount of increased attenuation layering dependently within the left anterior chamber (arrow), consistent with a small hyphema.

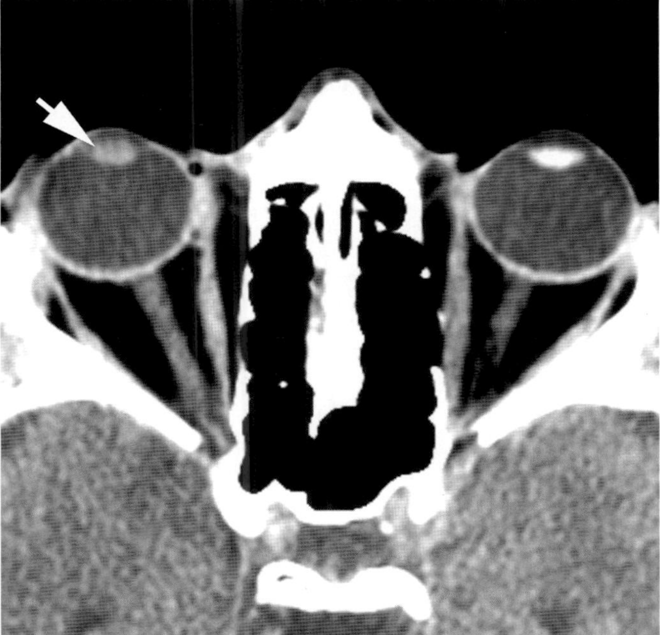

Figure 16.4 Axial non-enhanced CT image from a 58-year-old man with penetrating right globe injury showing decreased attenuation of the right lens (arrow) compared with the left. This is consistent with a traumatic cataract.

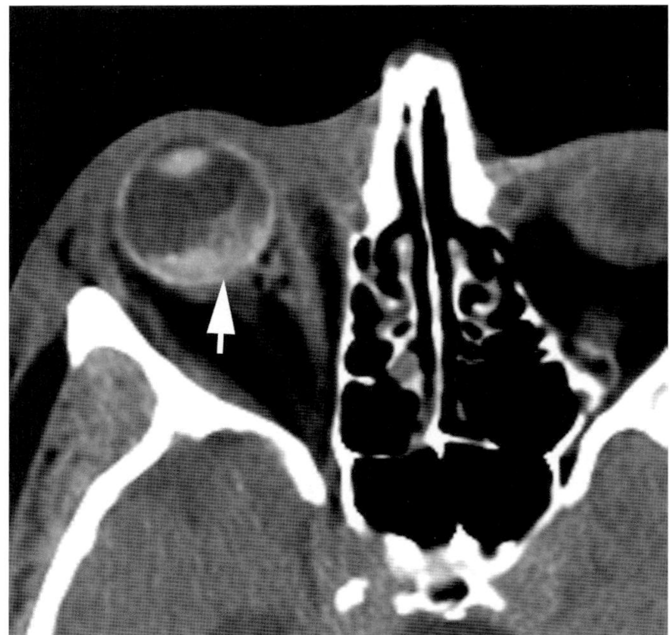

Figure 16.5 Axial non-enhanced CT image from a 47-year-old man with blunt right orbital trauma shows increased attenuation within the vitreous body which crosses the optic disk (arrow). This is a typical configuration of vitreous hemorrhage. Also notice the extensive periorbital edema.

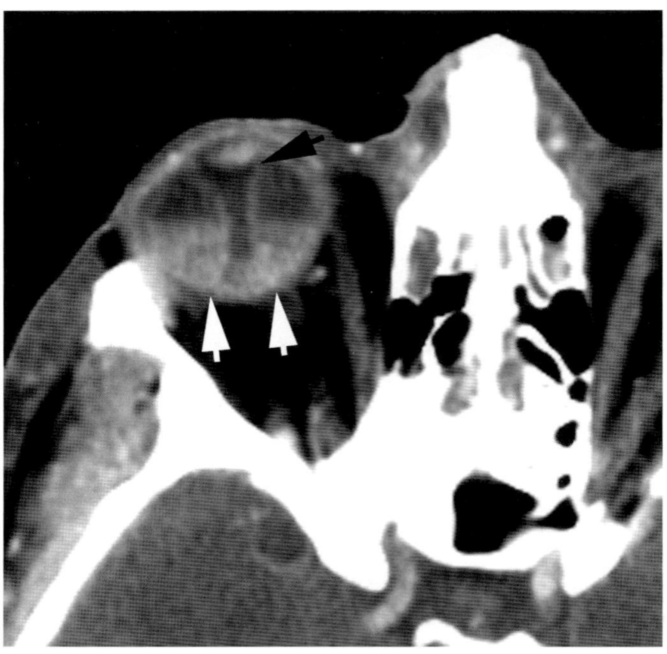

Figure 16.6 Axial non-enhanced CT image of a 61-year-old woman with acute right orbital pain shows a detachment of the choroid that extends anteriorly to the level of the lens (black arrow) and diverges posteriorly (white arrows). Notice the fluid-fluid level caused by layering of blood products.

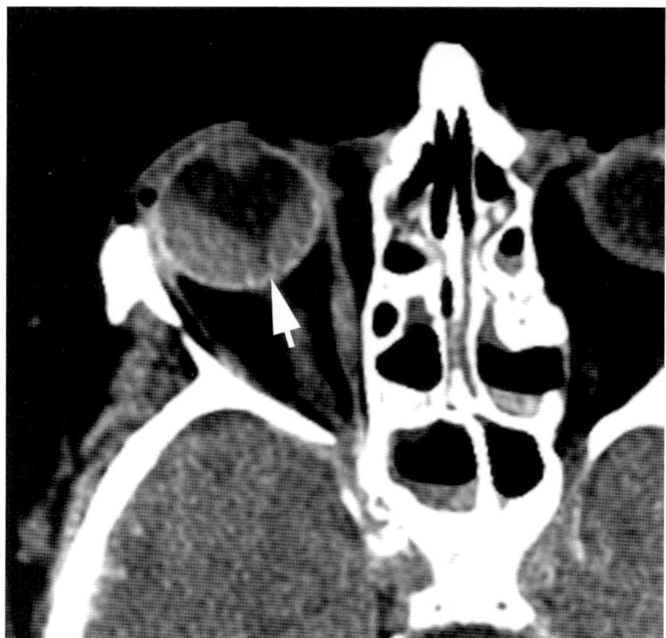

Figure 16.7 Axial non-enhanced CT image of a 68-year-old man with acute vision loss in the right eye shows a hemorrhagic retinal detachment. Notice that the retina converges posteriorly at the optic disk (arrow).

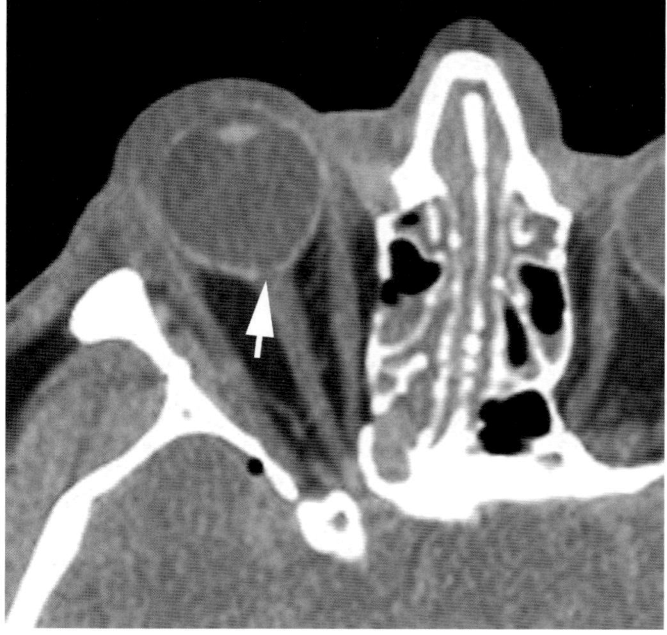

Figure 16.8 Axial non-enhanced CT image from a 37-year-old man in a motor-vehicle accident shows proptosis of the right globe and tenting at the optic disk (arrow). This is an important finding to convey, as severe tenting is associated with vision loss.

CASE 17

Dilated superior ophthalmic vein

Robert B. Carr

Imaging description

The superior ophthalmic vein (SOV) is formed by the confluence of the angular, supraorbital, and supratrochlear veins. Anteriorly, it is located medially near the trochlea. As it passes posteriorly, it courses beneath the superior rectus muscle and curves laterally. It passes outside the muscular annulus and enters the cavernous sinus [1, 2]. The vein is valveless throughout its course.

The diameter of the normal SOV ranges from 1 to 2.9 mm as measured on coronal MR images, with a mean diameter of approximately 2 mm [3]. When evaluating an SOV that appears enlarged it is important to assess for symmetry and enhancement characteristics. However, note that mild asymmetry can be normal.

Enlargement of the SOV has been described in several conditions. In the acute setting, enlargement may be caused by a cavernous-carotid fistula (CCF) or increased intracranial pressure (ICP).

A CCF may develop from trauma, surgery, or spontaneously. Fistulas can be described according to the Barrow classification, forming via a direct connection from the internal carotid artery or indirectly via the internal and/or external carotid arteries [4]. Radiologic findings include enlargement of the ipsilateral SOV, often with arterialized enhancement and signal characteristics on CT angiography (CTA) and MR angiography (MRA). The globe is often proptotic (Figure 17.1). Evaluation with conventional angiography is usually required to delineate the sites of fistula formation and for treatment [5].

Enlargement of the SOV can occur with increased ICP, and typically affects both sides (Figure 17.2). A study of 69 patients showed a positive correlation between the ICP and the SOV diameter. Only 3% of the patients with an SOV <1 mm had increased ICP, whereas 58% of patients with an SOV >2.5 mm had increased ICP [6].

Various other conditions associated with enlargement of the SOV have been described. These include Grave's disease, orbital pseudotumor, and varices. An orbital varix may result in enlargement of the SOV that fluctuates with venous pressure. At rest, the SOV may be normal or only mildly enlarged. However, with breath-holding or Valsalva, the vein may dilate significantly (Figure 17.3) [7].

Importance

When an enlarged SOV is encountered, it is important to evaluate for other imaging findings that may suggest the diagnosis. In the acute setting, increased ICP and CCF are the most important diagnoses that should not be overlooked. A CCF may lead to blindness or intracranial hemorrhage [8], and often requires endovascular treatment.

It is important to remember that mild enlargement of the SOV can be a normal incidental finding. Thus, it is necessary to correlate the finding with clinical and other radiologic findings.

Typical clinical scenario

Most patients with a CCF will present with clinical signs and symptoms such as chemosis, proptosis, and vision changes. Patients with increased ICP may present with headache, nausea, papilledema, and cranial nerve palsies. Patients with orbital varices often present with proptosis that occurs with stress maneuvers, such as coughing or bending over.

Differential diagnosis

The various causes of an enlarged SOV have been discussed above. Other lesions which may simulate a markedly enlarged SOV include an orbital hemangioma and lymphatic malformation. Careful evaluation should allow distinction in most cases.

Teaching point

Enlargement of the SOV may be an important finding suggesting the diagnosis of a CCF or increased ICP.

REFERENCES

1. Rene C. Update on orbital anatomy. *Eye (Lond)*. 2006;**20**(10):1119–29.
2. Cheung N, McNab AA. Venous anatomy of the orbit. *Invest Ophthalmol Vis Sci*. 2003;**44**(3):988–95.
3. Ozgen A, Aydingoz U. Normative measurements of orbital structures using MRI. *J Comput Assist Tomogr*. 2000;**24**(3):493–6.
4. Barrow DL, Spector RH, Braun IF, *et al.* Classification and treatment of spontaneous carotid-cavernous sinus fistulas. *J Neurosurg*. 1985;**62**(2):248–56.
5. LeBedis CA, Sakai O. Nontraumatic orbital conditions: diagnosis with CT and MR imaging in the emergent setting. *Radiographics*. 2008;**28**(6):1741–53.
6. Lirng JF, Fuh JL, Wu ZA, Lu SR, Wang SJ. Diameter of the superior ophthalmic vein in relation to intracranial pressure. *AJNR Am J Neuroradiol*. 2003;**24**(4):700–3.
7. Smoker WR, Gentry LR, Yee NK, Reede DL, Nerad JA. Vascular lesions of the orbit: more than meets the eye. *Radiographics*. 2008;**28**(1):185–204; quiz 325.
8. Kirsch M, Henkes H, Liebig T, *et al.* Endovascular management of dural carotid-cavernous sinus fistulas in 141 patients. *Neuroradiology*. 2006;**48**(7):486–90.

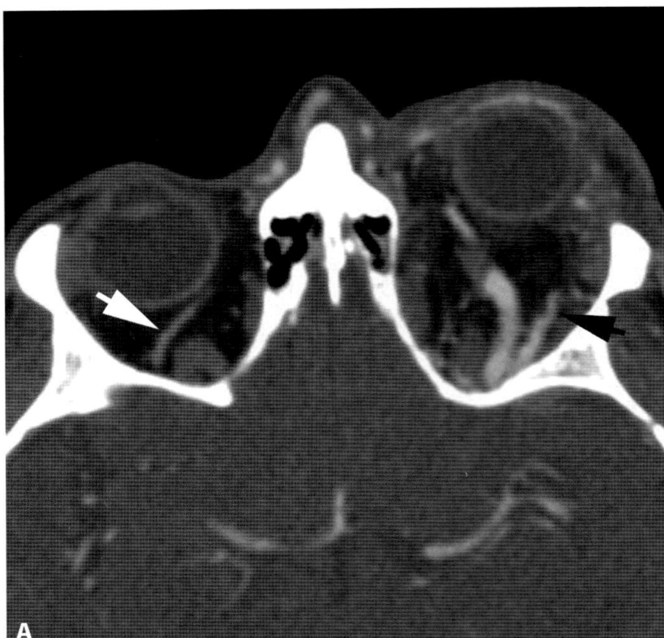

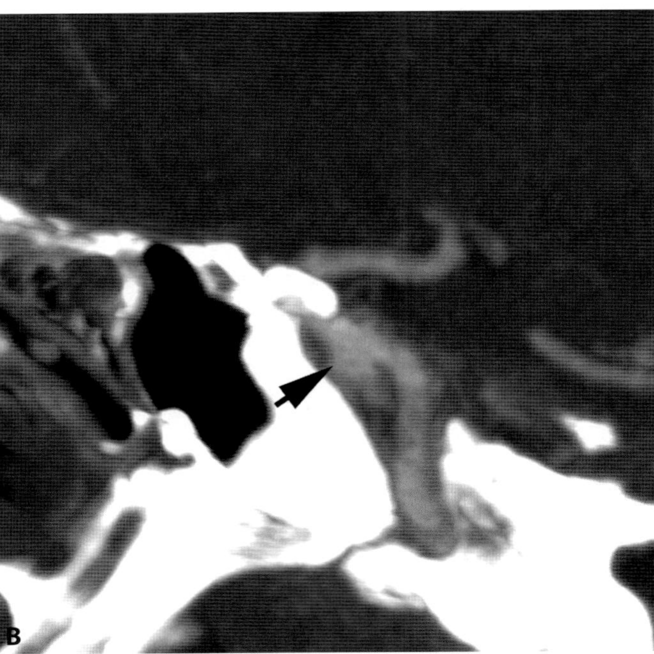

Figure 17.1 A. Axial contrast-enhanced CT image from a 37-year-old man with a recent history of trauma shows significant enlargement of the left SOV; compare with the normal right SOV (white arrow). There is also enlargement of the left lacrimal vein (black arrow), which usually drains into the SOV. The globe is proptotic and inflammatory changes are present. **B.** Reformatted image from the CT angiogram shows an eccentric outpouching of the left cavernous carotid artery (arrow). This represents a post-traumatic pseudoaneurysm, which has resulted in a cavernous-carotid fistula.

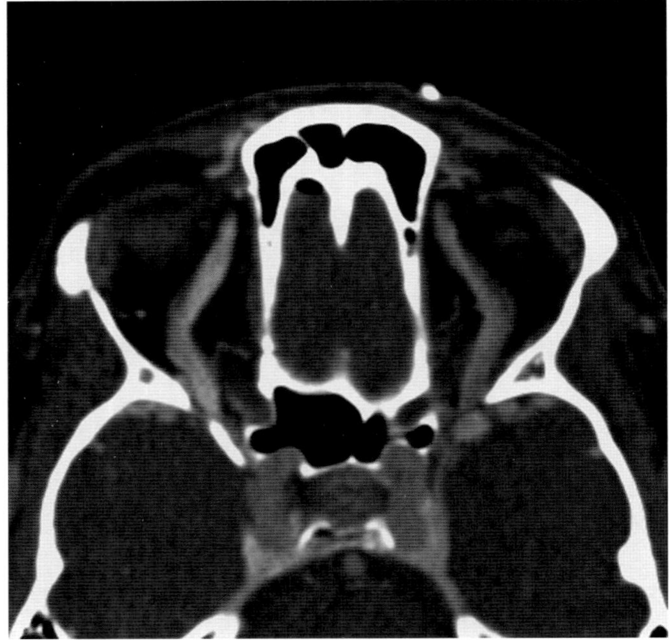

Figure 17.2 Axial contrast-enhanced CT image from a 48-year-old woman with meningitis shows significant symmetric enlargement of both SOVs, which is a result of increased intracranial pressure.

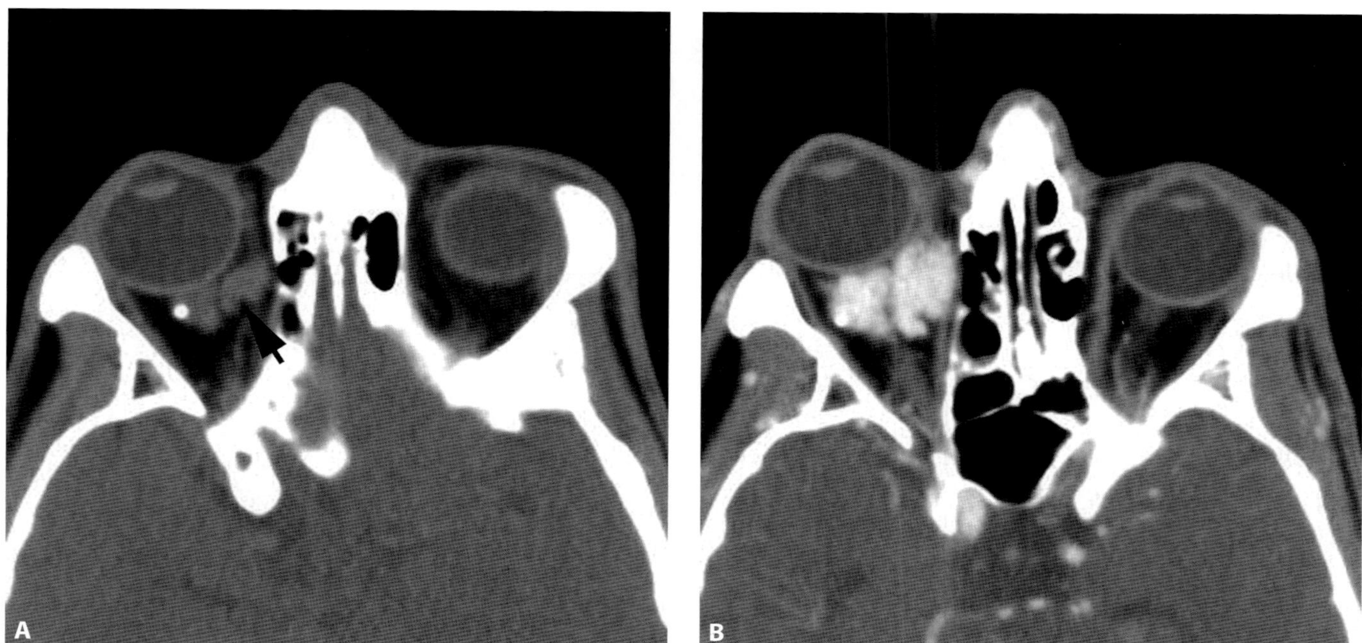

Figure 17.3 A. Axial non-enhanced CT image from a 52-year-old woman shows a lobular mass within the right orbit (arrow) which represents an ectatic SOV. Notice the adjacent punctate calcification. **B.** Axial contrast-enhanced CT image obtained during a Valsalva maneuver demonstrates marked enlargement and enhancement of the mass with resultant proptosis. These findings are typical for a varix of the SOV.

CASE 18

Orbital fractures

Robert B. Carr

Imaging description

Blunt trauma to the orbit often results in an orbital wall fracture. The predominant fracture patterns are different between adults and pediatric patients. In adults, the medial orbital wall and the orbital floor are the most common sites of fracture (Figure 18.1). In children, especially those less than seven years of age, the most common orbital fracture involves the orbital roof. This is explained by the prominence of the frontal bone relative to the size of the face in children. Also, the frontal sinus does not develop until the age of seven years; thus, there is lack of the normal cushioning effect from the sinus, and frontal bone fractures tend to extend into the orbital roof [1].

A trapdoor fracture may occur in children and young adults (Figure 18.2). In this type of injury a linear orbital floor fracture results in inferior bony displacement. However, due to the elasticity of the floor, the bone fragment swings back to the normal position in a hinge-like manner. These types of fractures may be subtle, but they are associated with a high rate of tissue entrapment [2]. If there is any evidence of orbital fat inferior to the orbital floor, a fracture must be presumed present. Another useful finding is hemorrhage within the maxillary sinus, which is often but not always associated with an orbital floor fracture (Figure 18.3).

In the setting of orbital trauma, always evaluate the shape and position of the extraocular muscles, especially the inferior rectus muscle. An abnormal rounded configuration and downward displacement of the inferior rectus muscle indicates disruption of the fascial sling (Figure 18.4). A normal flattened configuration of the muscle and a normal position suggest that the fascial sling is intact, minimizing the possibility of entrapped orbital tissue [3].

Fractures may extend through the orbital apex, involving the optic canal. It is important to evaluate for the presence of these fractures, as they may be associated with traumatic optic neuropathy (Figure 18.5). MRI may demonstrate increased T2 signal within the optic nerves [4]. A recent study suggests that diffusion tensor imaging may also be of value [5].

Importance

Orbital fractures may be associated with soft tissue injury to the globe, muscles, or optic nerve. Therefore, it is important to closely evaluate these structures.

Complications of fractures include diplopia, visual loss, and cosmetic deformities such as enophthalmos. Fractures involving more than half of the orbital floor usually result in a cosmetic and/or functional deformity [6]. Thus, it is important to describe the size and extent of the fracture.

Typical clinical scenario

Most orbital fractures are caused by blunt trauma, such as assault or motor vehicle accident. Common signs and symptoms include diplopia, restricted range of eye movement, vision loss, and enophthalmos.

Differential diagnosis

Many fractures do not meet criteria for surgical repair. Therefore, untreated chronic fractures are not uncommon. They can be distinguished from acute fractures by the lack of inflammatory changes in the soft tissues and the lack of sharp fracture lines.

Teaching point

Rounding and displacement of the inferior rectus muscle suggests disruption of the fascial sling. Orbital fat inferior to a normal-appearing orbital floor indicates a trapdoor injury.

REFERENCES

1. Alcala-Galiano A, Arribas-Garcia IJ, Martin-Perez MA, *et al.* Pediatric facial fractures: children are not just small adults. *Radiographics.* 2008;**28**(2):441–61; quiz 618.
2. Grant JH, 3rd, Patrinely JR, Weiss AH, Kierney PC, Gruss JS. Trapdoor fracture of the orbit in a pediatric population. *Plast Reconstr Surg.* 2002;**109**(2):482–9; discussion 90–5.
3. Hopper RA, Salemy S, Sze RW. Diagnosis of midface fractures with CT: what the surgeon needs to know. *Radiographics.* 2006;**26**(3):783–93.
4. Kubal WS. Imaging of orbital trauma. *Radiographics.* 2008;**28**(6):1729–39.
5. Yang QT, Fan YP, Zou Y, *et al.* Evaluation of traumatic optic neuropathy in patients with optic canal fracture using diffusion tensor magnetic resonance imaging: a preliminary report. *ORL J Otorhinolaryngol Relat Spec.* 2011;**73**(6):301–7.
6. Joseph JM, Glavas IP. Orbital fractures: a review. *Clin Ophthalmol.* 2011;**5**:95–100.

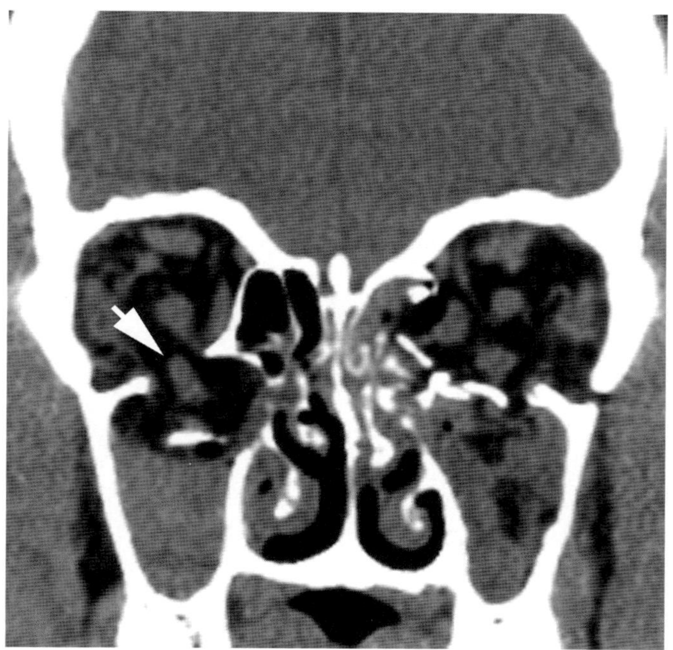

Figure 18.1 Coronal non-enhanced CT image from a 38-year-old man with facial trauma shows fractures of the orbital floors bilaterally and the left medial orbital wall. Hemorrhage and orbital fat are present within both maxillary sinuses. Also notice the downward herniation of the right inferior rectus muscle (arrow).

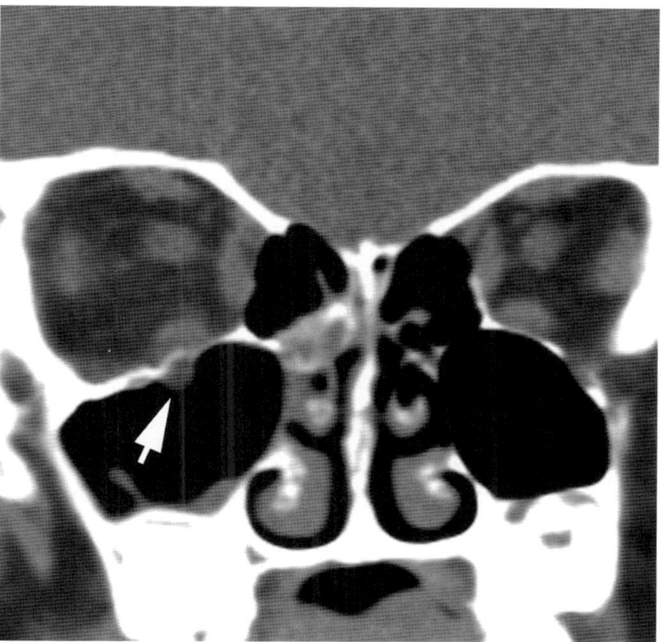

Figure 18.2 Coronal non-enhanced CT image from an 11-year-old child with right orbital trauma shows a small amount of orbital fat inferior to the orbital floor (arrow). This finding confirms the presence of a trapdoor fracture.

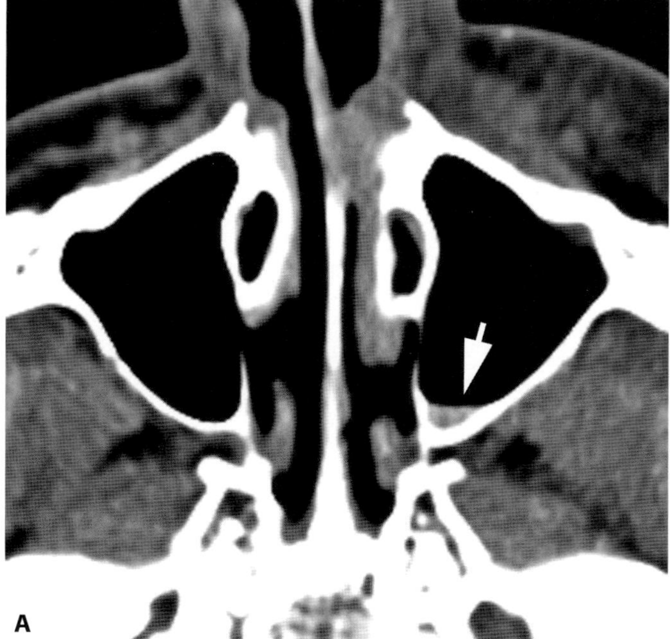

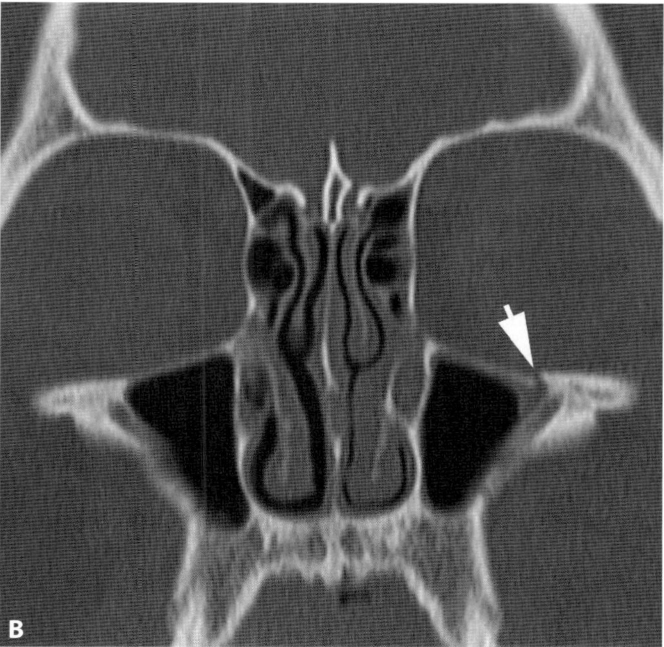

Figure 18.3 A. Axial non-enhanced CT image from a 27-year-old man with blunt trauma to the left orbit shows a small amount of hyperdense fluid layering within the left maxillary sinus (arrow), consistent with hemorrhage. This finding should prompt a careful inspection of the orbital floor. **B.** Coronal non-enhanced CT image demonstrates a minimally displaced linear fracture of the left orbital floor (arrow).

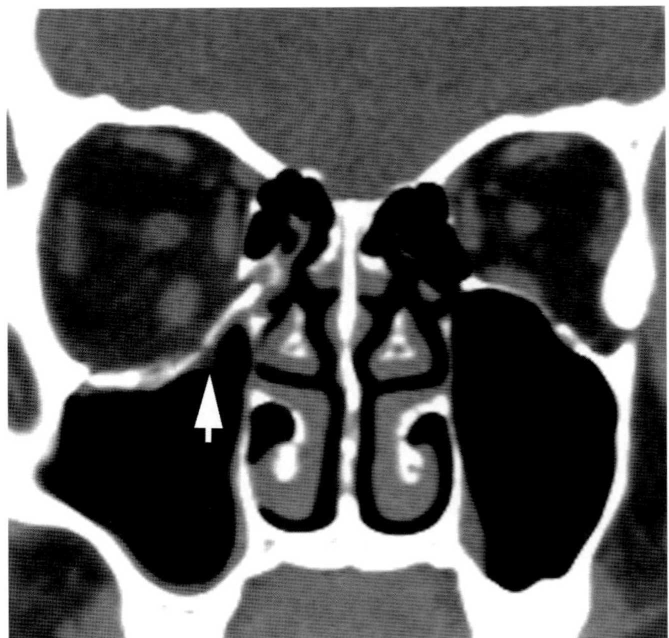

Figure 18.4 Coronal non-enhanced CT image from a 21-year-old man with right orbital trauma shows a small amount of fat inferior to the orbital floor (arrow), confirming the presence of a non-displaced fracture. Also notice the rounded configuration of the right inferior rectus muscle, indicating disruption of the fascial sling; compare with the normal contralateral side.

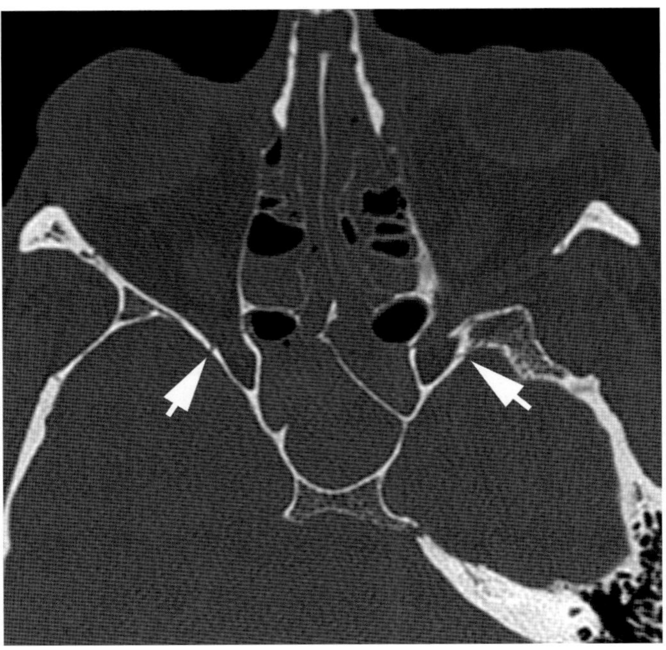

Figure 18.5 Axial non-enhanced CT image from a 51-year-old man who after a motor vehicle collision shows a skull base fracture which extends through both orbital apices (arrows). This is an important finding which may be associated with traumatic optic neuropathy.

Variants of the upper cervical spine

Robert B. Carr

Imaging description

There are several skeletal variants of the C1 and C2 vertebrae which may be confused with injury. Knowledge of the normal development of these vertebrae is essential for distinguishing anatomic variants from pathology.

The atlas typically develops from three primary ossification centers: one anterior arch and two neural arches. Two neurocentral synchondroses separate the anterior arch from the neural arches (Figure 19.1). A single posterior synchondrosis separates the neural arches. The atlas is normally fused by eight years of age [1].

The axis typically develops from five primary ossification centers: two odontoid centers, two neural arches, and one centrum. The odontoid centers usually fuse prior to birth (Figure 19.2). The remaining primary centers are usually fused by six years of age. A secondary center of ossification, known as the os terminale, forms at the odontoid tip and usually fuses by 12 years of age [2].

Incomplete fusion of the atlas may result in a cleft, usually at the site of a synchondrosis. The cleft will usually demonstrate a smooth margin at a characteristic location, and should not be mistaken for a fracture (Figure 19.3). If there is nondevelopment of the anterior arch, the neural arches may overgrow and attempt to fuse anteriorly, resulting in an anterior midline cleft (Figure 19.4).

Variants related to the axis include an unfused os terminale and an os odontoideum. An unfused os terminale will appear as a round corticated bony structure at the tip of the odontoid process (Figure 19.5). It lies superior to the atlantodental articulation and transverse atlantal ligament. An os odontoideum is much larger than the unfused os terminale. The etiology is uncertain, though many cases are likely due to prior trauma [3]. There may be hypertrophy of the anterior arch of C1 (Figure 19.6). While this is not an acute injury, the patient may have chronic symptoms and the spinal canal and alignment should be closely examined.

Importance

It is important to be aware of the normal ossification patterns of the C1 and C2 vertebrae. This will allow accurate interpretation of pediatric cervical spine imaging and prevent misdiagnosis of a fracture in an older patient with a persistent cleft.

An os odontoideum does not represent an acute injury. However, there may be associated instability and the patient may be symptomatic. Therefore, it is important to describe the relationship of the os odontoideum to the skull base, C1, and C2, and to evaluate for evidence of spinal canal narrowing. If there is clinical concern that the lesion is producing symptoms, surgery may be warranted [4].

Typical clinical scenario

These findings are usually incidental. However, more advanced congenital anomalies and os odontoideum may result in instability with pain and neurologic symptoms.

Differential diagnosis

The differential diagnosis primarily involves acute fractures (Figure 19.7). Coexistant posterior and anterior clefts may especially simulate a Jefferson burst fracture [5].

Teaching point

There are many variants of the upper cervical spine. Knowledge of the normal ossification patterns will allow identification of many of these variants.

REFERENCES

1. Junewick JJ, Chin MS, Meesa IR, *et al.* Ossification patterns of the atlas vertebra. *AJR Am J Roentgenol.* 2011;**197**(5):1229–34.
2. Lustrin ES, Karakas SP, Ortiz AO, *et al.* Pediatric cervical spine: normal anatomy, variants, and trauma. *Radiographics.* 2003;**23**(3): 539–60.
3. Sankar WN, Wills BP, Dormans JP, Drummond DS. Os odontoideum revisited: the case for a multifactorial etiology. *Spine (Phila Pa 1976).* 2006;**31**(9):979–84.
4. Arvin B, Fournier-Gosselin MP, Fehlings MG. Os odontoideum: etiology and surgical management. *Neurosurgery.* 2010;**66**(3 Suppl): 22–31.
5. Mellado JM, Larrosa R, Martin J, *et al.* MDCT of variations and anomalies of the neural arch and its processes: part 2 – articular processes, transverse processes, and high cervical spine. *AJR Am J Roentgenol.* 2011;**197**(1):W114–21.

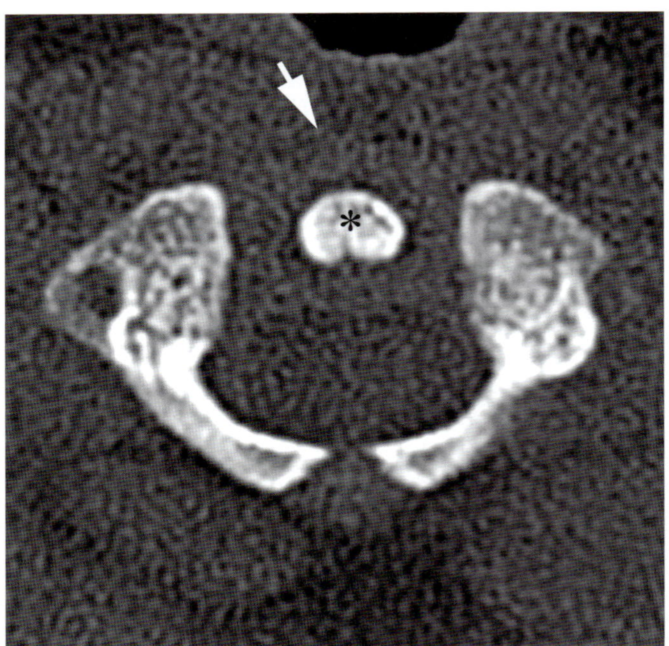

Figure 19.1 Axial non-enhanced CT image from an 11-month-old child shows the normal ossification pattern of the atlas. Notice that there are two neural arches. The single anterior arch has not yet ossified (arrow). Also notice the normal position of the dens (asterisk).

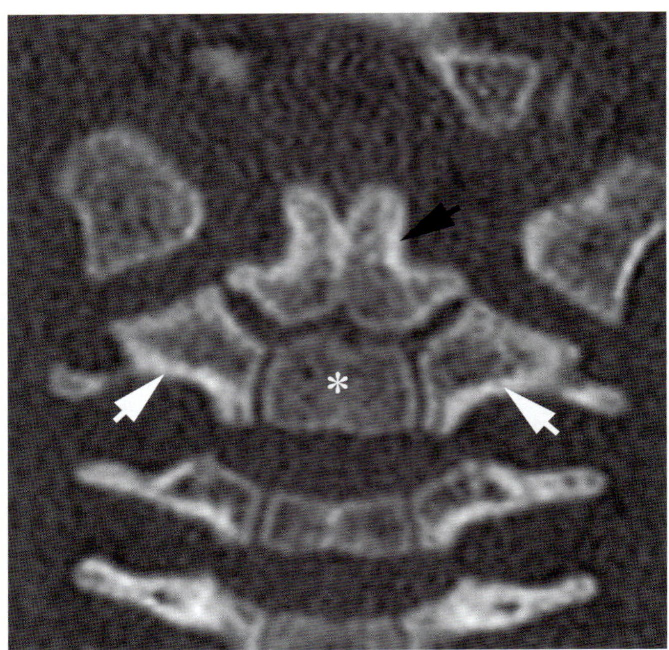

Figure 19.2 Coronal non-enhanced CT image from a 1-year-old child shows the normal primary ossification centers of the axis (C2). The two odontoid centers usually fuse before birth (black arrow). The two neural arches are seen laterally (white arrows) and the centrum is seen centrally (asterisk).

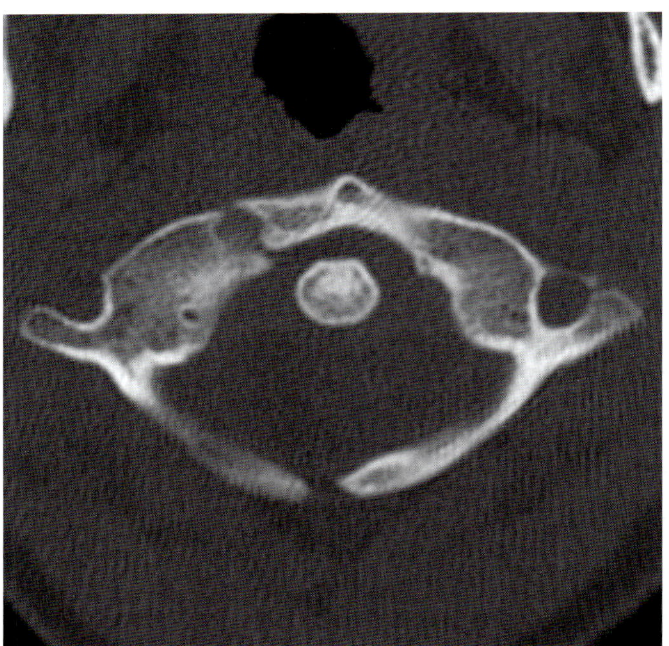

Figure 19.3 Axial non-enhanced CT image from a 29-year-old man with neck pain shows clefts from incomplete bony fusion at the posterior and right neurocentral synchondroses. Notice the smooth margins and characteristic locations, features that argue against the presence of acute fractures.

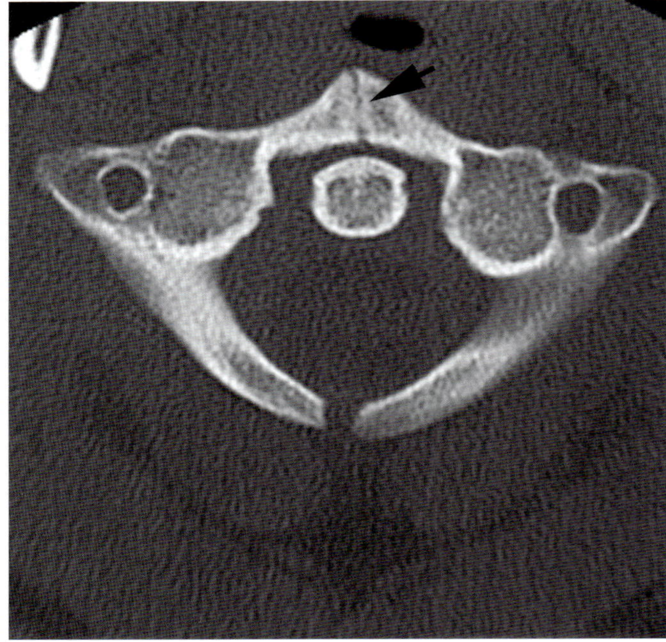

Figure 19.4 Axial non-enhanced CT image from a 34-year-old man with acute trauma shows an anterior midline cleft (arrow). This is caused by non-formation of the anterior arch, with overgrowth and attempted fusion of the neural arches. Also notice the posterior cleft.

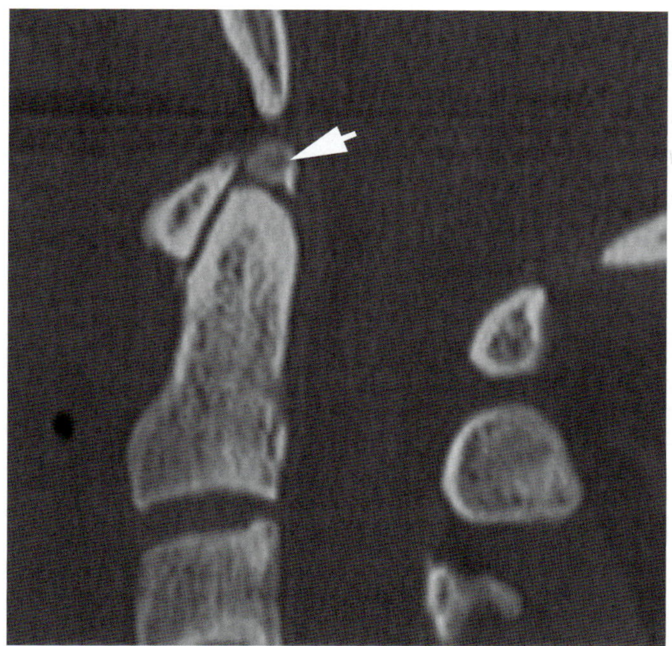

Figure 19.5 Sagittal non-enhanced CT image in a 30-year-old woman shows a small corticated ossicle at the tip of the odontoid process (arrow). This is the characteristic appearance of an unfused os terminale. Notice that the os terminale is located superior to the atlantodental articulation.

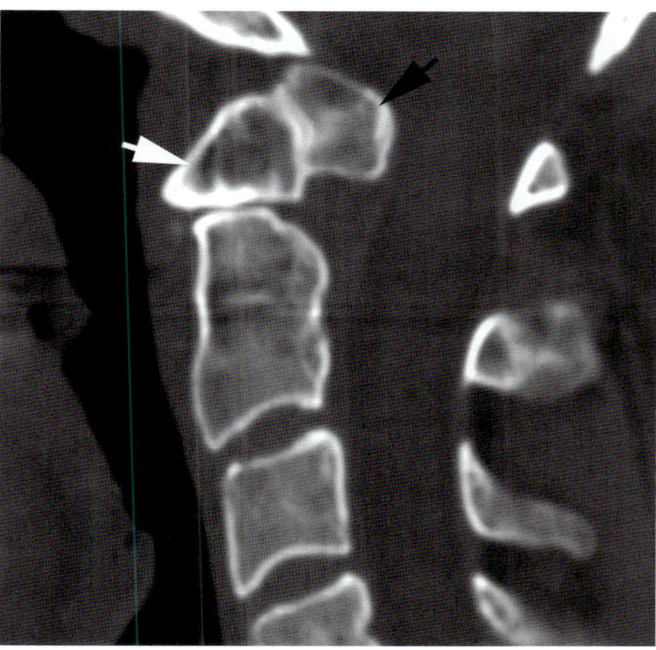

Figure 19.6 Sagittal non-enhanced CT image in a 22-year-old man with neck trauma shows an enlarged ossicle (black arrow) closely associated with the anterior arch of C1 (white arrow). The anterior arch of C1 is markedly hypertrophied. The odontoid process is foreshortened, and there is malalignment of C1 and C2 anteriorly. These are typical findings of an os odontoideum.

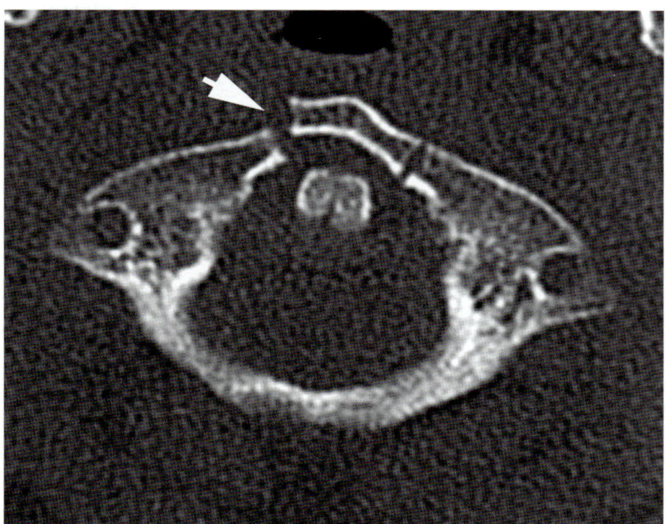

Figure 19.7 Axial non-enhanced CT image from a child with trauma shows asymmetric displacement of the anterior ossification center along the right neurocentral synchondrosis (arrow), suggesting a fracture through the synchondrosis. Findings were confirmed with MRI.

Atlantoaxial rotatory fixation versus head rotation

Claire K. Sandstrom

Imaging description

Open mouth odontoid radiographs or cervical spine CT demonstrate asymmetric position of the dens between the lateral masses of C1 [1]. The C1 lateral mass contralateral to the direction of head rotation is subluxed anteriorly relative to the C2 lateral mass, while the ipsilateral C1 lateral mass might be posteriorly displaced. This asymmetry can be normal during head rotation, and is very frequently encountered on imaging. If, however, the asymmetric relationship persists when the head is then turned to the contralateral side, atlantoaxial rotatory fixation (AARF) is diagnosed.

Importance

At our institution, far more trauma patients are transferred from outside institutions for further evaluation for AARF than are ultimately diagnosed with it. Differentiation between AARF and acute torticollis, however, is important because torticollis responds to analgesics and conservative management, while AARF requires reduction. Furthermore, reduction has been more successful when performed early in the course of the condition.

C1 rotates on C2 when the head is turned, with a normal range of motion of 29–44 degrees, and this relationship allows much of the axial rotation achieved in the cervical spine (Figure 20.1) [2]. This relationship occurs in both voluntary rotation of the head and involuntary fixation, such as seen with torticollis and AARF [3]. If the C1–C2 relationship is fixed, or at least partially restrained, when the patient attempts to turn his head to the contralateral side AARF is diagnosed (Figure 20.2) [4].

Typical clinical scenario

A patient, typically a child, presents with painful, fixed rotation of the head with a lateral tilt, or "cock-robin position," with inability to rotate the head to the contralateral side [2]. Onset may be preceded by minor trauma, surgery, or upper respiratory infection. The cause of rotary subluxation is debated, possibly relating to entrapment of inflamed synovium or of torn capsular ligaments at the articular facets of C1–C2.

Differential diagnosis

When a static CT or radiograph reveals the rotation of C1 on C2 such that AARF is considered, the other two conditions in the differential should include acute torticollis due to muscular spasm and normal head rotation. The latter can usually be resolved clinically, when the patient is found to have no restriction to head motion.

If, however, the patient has a fixed rotation of the head and complains of pain with movement, acute torticollis and AARF are the primary considerations. The typical imaging protocol for diagnosis or exclusion of AARF involves dynamic CT. Initially, CT is obtained with the head in the presenting position. CT is also performed with the head turned as far to the contralateral side as the patient can tolerate, and some protocols suggest a third series with the head in a neutral position [5]. If adequate head rotation cannot be achieved in the conscious patient because of pain, dynamic CT under anesthesia has been recommended [3].

Teaching point

Detection of subluxation of the C1 lateral masses and asymmetric positioning of the dens should raise consideration for AARF, though in practice head rotation in an otherwise normal patient or in a patient with acute torticollis is much more commonly encountered.

REFERENCES

1. Fielding JW, Hawkins RJ. Atlanto-axial rotatory fixation. (Fixed rotatory subluxation of the atlanto-axial joint). *J Bone Joint Surg.* 1977;**59A**(1):37–44.
2. Kowalski HM, Cohen WA, Cooper P, Wisoff JH. Pitfalls in the CT diagnosis of atlantoaxial rotary subluxation. *AJR Am J Roentgenol.* 1987;**149**:595–600.
3. Been HD, Kerkhoffs GMMJ, Maas M. Suspected atlantoaxial rotatory fixation-subluxation: the value of multidetected computed tomography scanning under general anesthesia. *Spine*. 2007;**32**(5):E163–7.
4. Pang D. Atlantoaxial rotatory fixation. *Neurosurgery*. 2010;**66** (3 Suppl):161–83.
5. Pang D, Li V. Atlantoaxial rotatory fixation: Part 2 – New diagnostic paradigm and a new classification based on motion analysis using computed tomographic imaging. *Neurosurgery*. 2005;**57** (5):941–53.

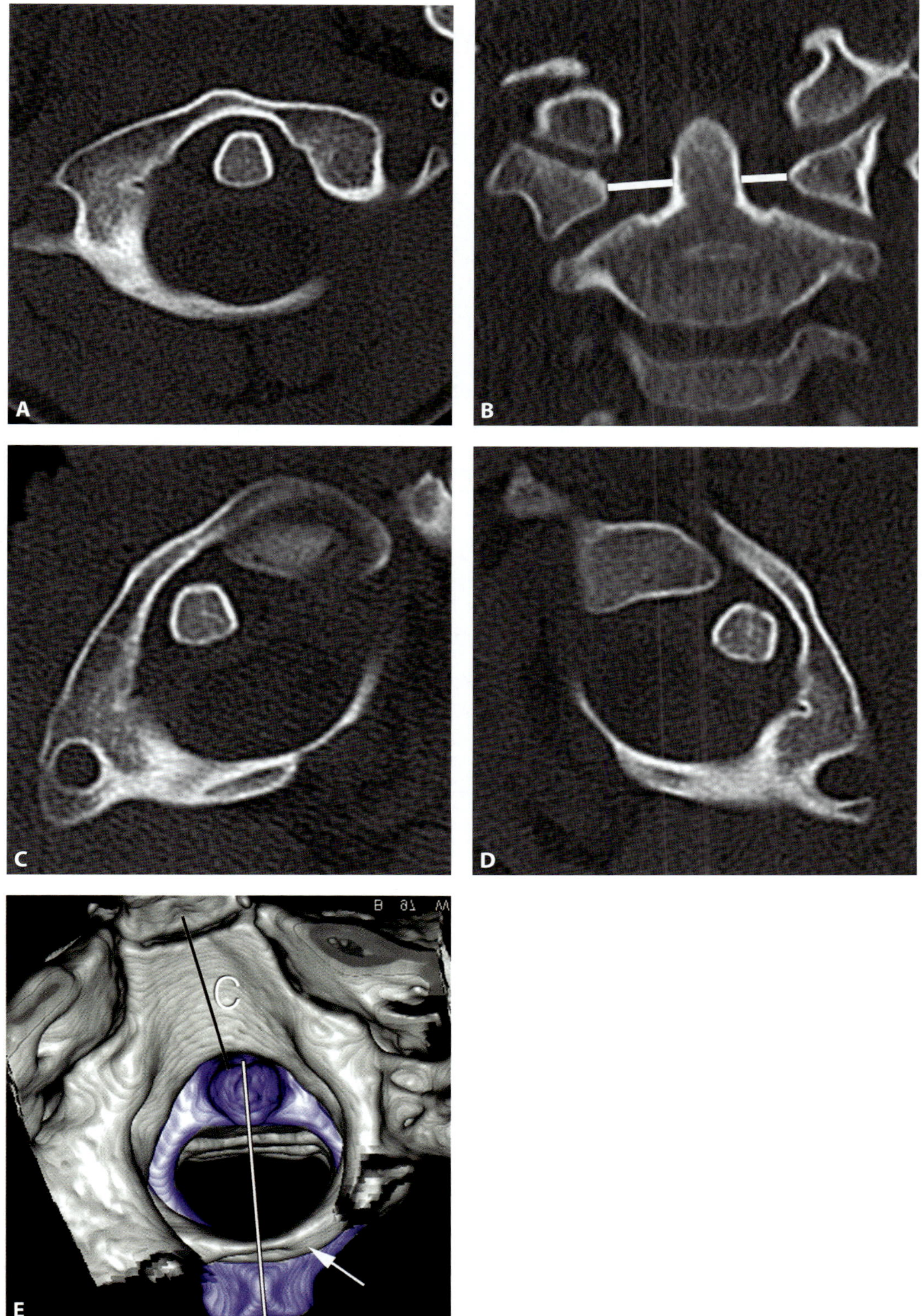

Figure 20.1 **A.** Axial image from CT of the cervical spine in bone window from a 5-year-old boy with acute neck pain after reaching for a book shows mild asymmetry of the distance between the dens and right compared to left ateral masses of C1. There is mild rightward rotation of the head on this sequence, but it was unclear if this was the cause of the asymmetry. **B.** Coronal reformation from the same axial sequence confirms the asymmetry (white line). Axial CT images were then obtained with the head fully rotated toward the right (**C**) and toward the left (**D**). These dynamic CT images showed normal motion of C1 relative to the dens without evidence of atlantoaxial rotatory fixation. **E.** 3D surface rendering of the initial CT sequence with the head in "neutral" position shows the angle between the clivus (C, black line defining axis of midline) and C2 (blue shading, white line denoting midline odontoid to spinous process) due entirely to head rotation. The white arrow identifies the posterior arch of C1, but the anterior arch is obscured by the skull base.

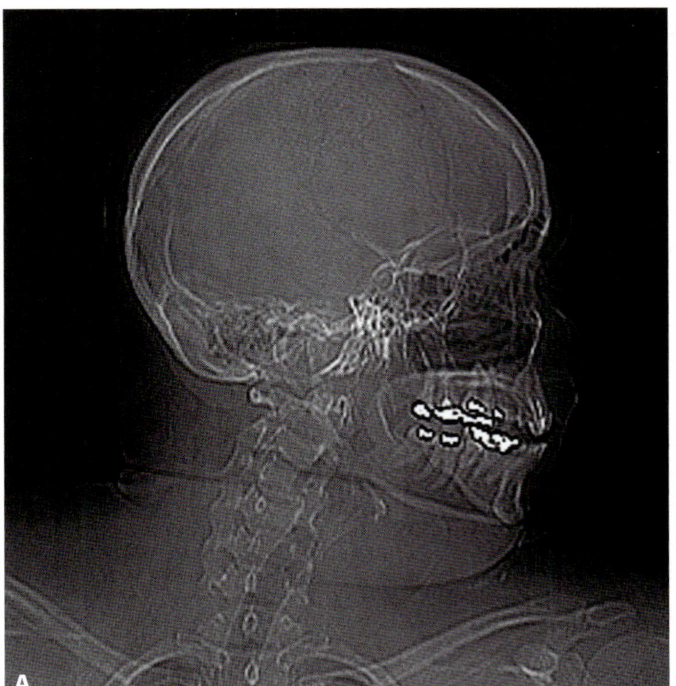

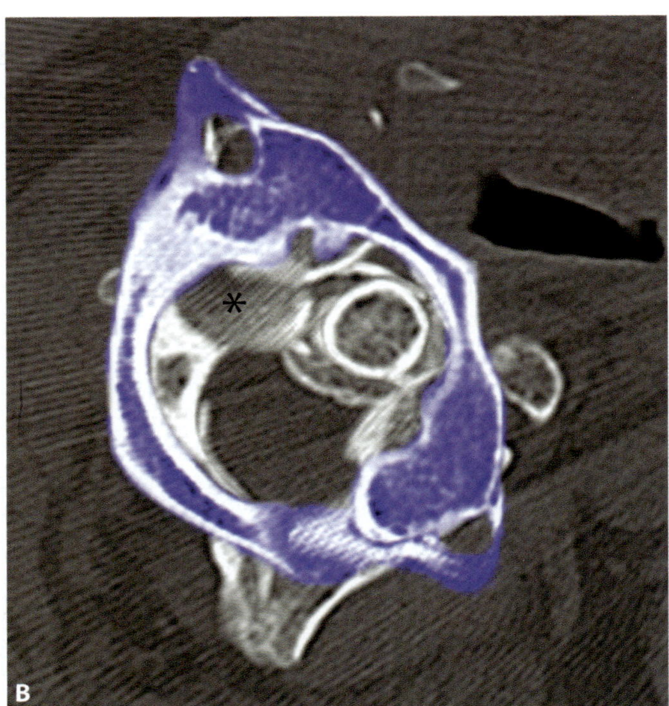

Figure 20.2 A. Posterior–anterior topographic image of the cervical spine CT in a 36-year-old woman with achondroplasia shows the head rotated to the left, the position in which she was fixed since waking that morning. **B.** Overlay of selected static axial CT images in bone windows shows subluxation of C1 (blue shading) relative to C2. The left lateral masses of C1 and C2 are mildly offset, while there is significant subluxation of the right C1 lateral mass relative to the right C2 facet (asterisk). A diagnostic dynamic CT was not performed, but rather the patient was empirically treated for AARF with traction and recovered fully within two days.

CASE

21 Cervical flexion and extension radiographs after blunt trauma

Ken F. Linnau

Imaging description

Flexion and extension radiographs of the cervical spine can be used in the setting of blunt trauma to assess for stability if the initial imaging with cervical spine radiography or CT is normal. Patients with a normal Glasgow Coma Score (GCS) who are not intoxicated, but who complain of persistent posterior midline tenderness may undergo this dynamic test as an alternative to MRI [1].

Dynamic evaluation of the cervical spine has been performed for many decades and numerous radiographic findings have been described. Despite this long history of use, specific validated measurable criteria to determine what constitutes a normal or abnormal radiographic flexion-extension (FE) study do not exist. The usual practice is to assess intervertebral body motion (subluxation and angulation) to indirectly determine presence or absence of an unstable cervical spine injury (Figure 21.1). The criteria used are mostly derived from cadaver models with very limited validation data for the trauma setting [2]. A wide variation exists in the degree of physiologic range of motion between cervical levels, among individuals, at different ages, and between genders [3, 4].

Subluxation (translation)

Subluxation is usually defined as the distance from the posterior margin of one vertebral body to the posterior margin of the next caudal vertebral body. A measurable anterior (more common) or posterior displacement of more than 2–3 mm between flexion and extension in any of the cervical segments is usually considered abnormal [5, 6]. Radiographic signs of anterior subluxation are usually accentuated in flexion and diminished in extension.

Abnormal localized kyphosis (angulation)

Angulation (also called "rotational displacement") describes a vertebral body in abnormal localized kyphosis. Localized kyphosis can be assessed either on the flexion or the extension view by drawing a line through the endplate of the vertebral body in question and the endplates above and below it (Figure 21.1). If the angulation of a motion segment in relation to the vertebral body below and above it exceeds 11 degrees, angulation is considered to be abnormal [7].

Importance

Flexion and extension radiographs of the cervical spine have been used since the 1930s [2] and have been advocated in the acute setting after blunt trauma since the 1960s. Use of this technique to evaluate the injured cervical spine assumes: the existence of unstable ligamentous cervical spine injury or very subtle but unstable fractures, and radiographic signs on FE studies that allow identification of instability.

None of the proposed radiographic findings for flexion and extension radiography have been validated prospectively with clearly defined variables and clearly defined outcome measures [2]. Prior series persistently report very low numbers of positive FE studies [1]. The low number of positive examinations in these clinical case series limits the calculation of sensitivity within an acceptably narrow confidence interval.

Flexion and extension radiographs are even less reliable in the presence of spondyloarthropathy and some authors advise against their use in the elderly [1].

Current practice recommendations for clearance of the cervical spine, including those of the Eastern Association for the Surgery of Trauma [8], suggest performing FE studies as an alternative to MRI in awake and alert patients who complain of neck pain if cervical radiography and CT scanning are normal (Figure 21.2). In this setting, voluntary and painless flexion and extension must exceed 30 degrees in each direction from the neutral position to be considered adequate. In patients with a range of motion of less than 60 degrees total, a repeat FE study is performed two weeks after the initial exam when muscle spasm has subsided, with cervical spine collar protection in the interval [8].

In intubated patients, passive FE fluoroscopy (i.e., where the provider moves the head of the patient) is not recommended due to cases of iatrogenic complications following passive FE fluoroscopy [3].

Typical clinical scenario

A patient presents with persistent neck pain or midline cervical tenderness after blunt trauma, such as a motor vehicle collision, and the initial radiographic evaluation with static radiographs (cervical spine series [3–6 views] or CT shows no abnormality to suggest fracture or malalignment and the patient has a reliable clinical examination.

Differential diagnosis

The primary differential diagnosis is between a significant ligamentous injury without abnormal translation or angulation during flexion and extension radiographs, or underlying degenerative cervical spondylolisthesis due to facet joint arthrosis, or disk degeneration. Degenerative cervical spondylolisthesis is most common at the C3–4 and C4–C5 levels [9]. Cervical spine ligamentous laxity may also occur in connective tissue disorders, genetic syndromes, and inflammatory arthritis, in particular rheumatoid arthritis.

In normal children, pseudosubluxation at the C2–C3 level may be observed. This is discussed in Case 22.

Teaching point

Although recommended in some guidelines, and supported by limited evidence, the use of flexion and extension radiographs for the assessment of stability of the cervical spine following normal radiographic or CT evaluation has not been systematically validated.

REFERENCES

1. Khan SN, Erickson G, Sena MJ, Gupta MC. Use of flexion and extension radiographs of the cervical spine to rule out acute instability in patients with negative computed tomography scans. *J Orthop Trauma*. 2011;**25**(1):51–6.

2. Knopp R, Parker J, Tashjian J, Ganz W. Defining radiographic criteria for flexion-extension studies of the cervical spine. *Ann Emerg Med*. 2001;**38**(1):31–5.

3. Dvorak J, Panjabi MM, Grob D, Novotny JE, Antinnes JA. Clinical validation of functional flexion/extension radiographs of the cervical spine. *Spine (Phila Pa 1976)*. 1993;**18**(1):120–7.

4. Lin RM, Tsai KH, Chu LP, Chang PQ. Characteristics of sagittal vertebral alignment in flexion determined by dynamic radiographs of the cervical spine. *Spine (Phila Pa 1976)*. 2001;**26**(3):256–61.

5. Hammouri QM, Haims AH, Simpson AK, Alqaqa A, Grauer JN. The utility of dynamic flexion-extension radiographs in the initial evaluation of the degenerative lumbar spine. *Spine (Phila Pa 1976)*. 2007;**32**(21):2361–4.

6. White AP, Biswas D, Smart LR, Haims A, Grauer JN. Utility of flexion-extension radiographs in evaluating the degenerative cervical spine. *Spine (Phila Pa 1976)*. 2007;**32**(9):975–9.

7. White AA, 3rd, Johnson RM, Panjabi MM, Southwick WO. Biomechanical analysis of clinical stability in the cervical spine. *Clin Orthop Relat Res*. 1975;(**109**):85–96.

8. Como JJ, Diaz JJ, Dunham CM, *et al.* Practice management guidelines for identification of cervical spine injuries following trauma: update from the eastern association for the surgery of trauma practice management guidelines committee. *Trauma*. 2009;**67**(3):651–9.

9. Jiang SD, Jiang LS, Dai LY. Degenerative cervical spondylolisthesis: a systematic review. *Int Orthop*. 2011;**35**(6):869–75.

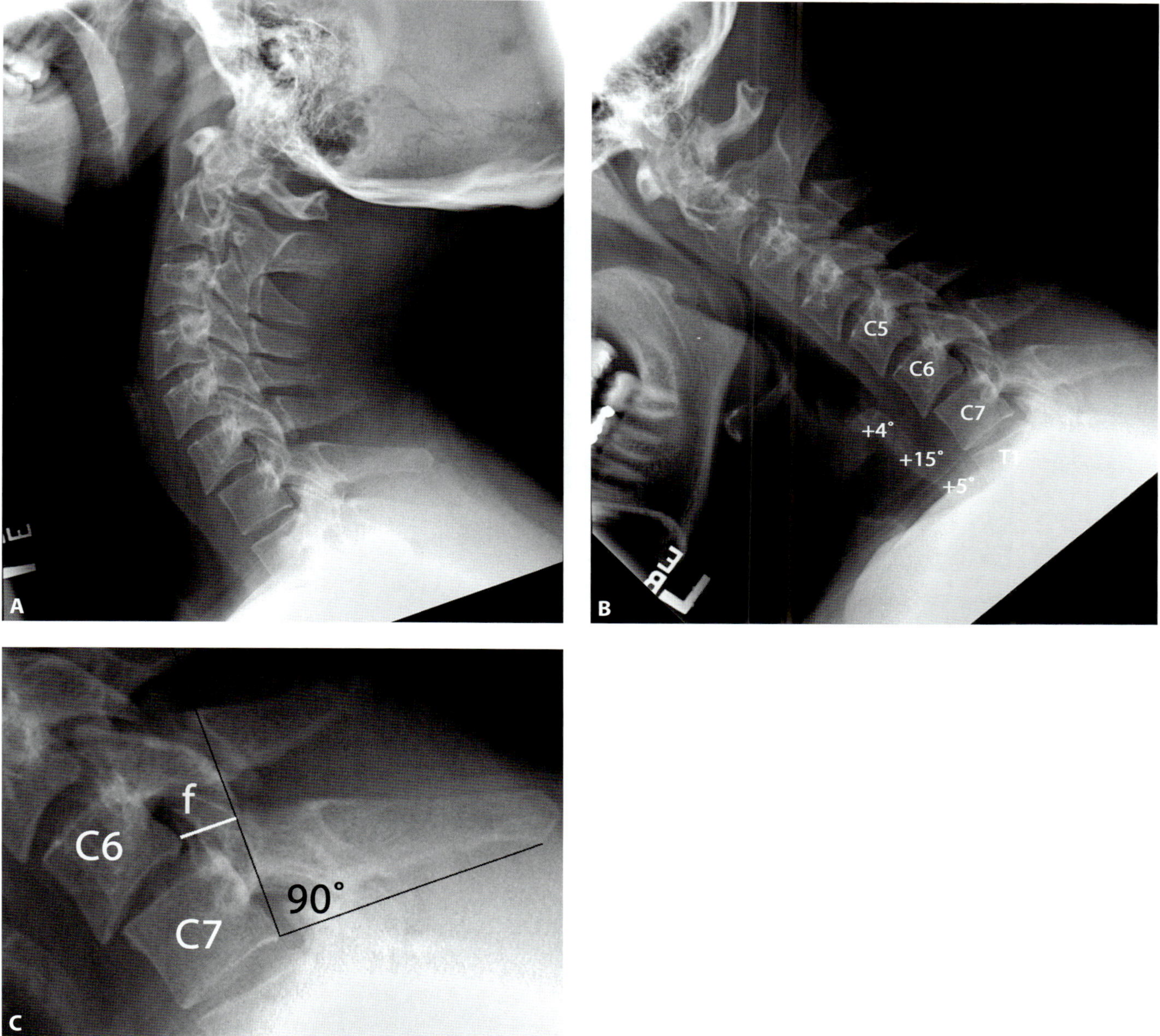

Figure 21.1 A. The extension radiograph of this 27-year-old man who was involved in a motor vehicle collision shows minimal anterolisthesis of C6 on C7. **B.** On the flexion view, the anterolisthesis increases to about 4 mm, and focal kyphosis of about 15 degrees between C6 and C7 becomes apparent. **C.** Collimated view of the radiograph in flexion illustrates how subluxation can be objectively assessed on both flexion and extension views: a perpendicular line is raised from a line connecting the posterior inferior corner of the lower vertebra (C7) and the subluxation (f) is measured at the inferior posterior corner of the upper vertebra (C6). Due to spinal instability, the patient underwent cervical fusion.

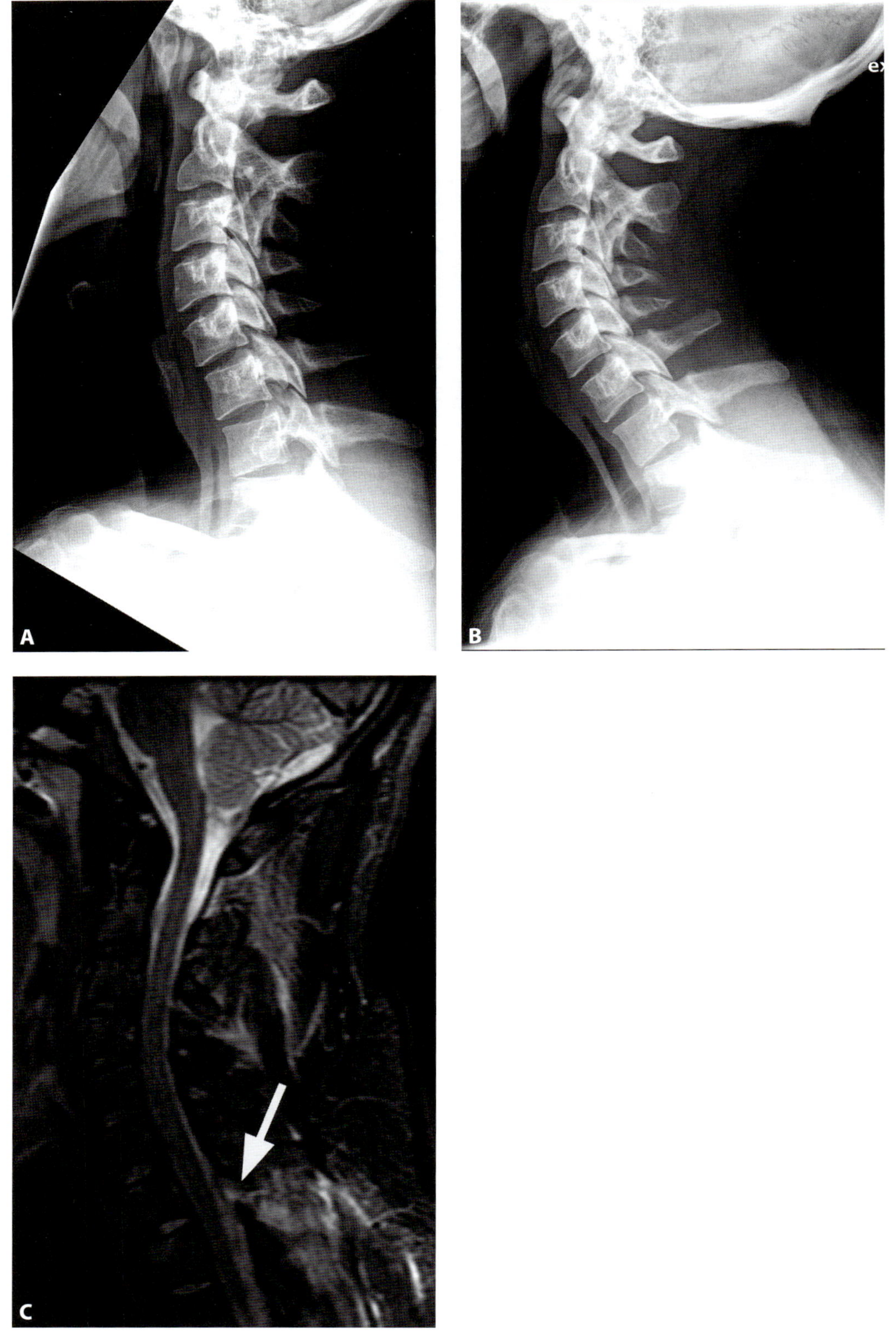

Figure 21.2 Upright flexion (**A**) and extension (**B**) lateral radiographs of the cervical spine in a 25-year-woman pedestrian who was hit by a bus. A cervical spine CT, reported as normal, had been performed earlier the same day. There is less than 60 degrees of motion in the series, limiting assessment for translation. However, on both views there is widening of the C7–T1 facet joints and interspinous distance, and narrowing of the anterior C7–T1 disc space. **C.** Sagittal short tau inversion recovery (STIR) MRI clearly shows the flexion distraction injury, with the additional findings of disruption of the ligamentum flavum (arrow), and significant edema in the posterior soft tissues. A further FE radiographic series performed nearly one month after the injury was similar in appearance to the images performed on the day of the injury, confirming the limited value of the dynamic evaluation. The patient underwent operative C7–T1 spinal fusion.

Pseudosubluxation of C2–C3

Robert B. Carr

Imaging description

Pseudosubluxation refers to physiologic anterior spondylolisthesis of C2 on C3, caused by ligamentous laxity and a more horizontal position of the facet joints compared with adults. It is seen in children less than 16 years of age, with most patients less than eight years of age. Rarely, it may be seen in an adult patient [1].

Lateral radiographs will reveal anterior displacement of the C2 vertebral body relative to C3. Displacement is most conspicuous during flexion, and may resolve during extension. A posterior cervical line may be drawn between the anterior cortex of the C1 and C3 posterior arches. This line, referred to as Swischuk's line, should pass within 2 mm of the anterior cortex of the C2 posterior arch (Figure 22.1) [2]. If it does not, injury should be suspected [3].

CT will reveal similar findings to those seen on radiography. However, one may more confidently exclude a fracture of the axis in the setting of malalignment. If there is concern for ligamentous injury, MRI should be obtained. The absence of ligamentous edema is reassuring, and further suggestive of the normal variant of C2–C3 pseudosubluxation (Figure 22.2).

An important discriminator is the age of the patient. Pseudosubluxation of C2–C3 is much more common in children less than eight years of age. As the age increases beyond eight, this variant becomes much less common. Therefore, if malalignment at C2–C3 is identified in an older child or adult, it should be viewed with a much higher suspicion of injury.

Importance

Cervical spine injuries occur in 1–3% of pediatric trauma patients. Children less than eight years of age usually sustain injury to the C1 through C3 levels, as the fulcrum of movement is located at C2–C3 in this age group [4]. This is the same age group that most commonly presents with pseudosubluxation of C2–C3. Therefore, it is imperative to distinguish these entities in order to prevent unnecessary treatment and avoid complications of an untreated injury.

Typical clinical scenario

Many pediatric trauma patients will present with multiple injuries, which may raise the suspicion for cervical injury. However, a study of 138 pediatric patients admitted with polytrauma demonstrated C2–C3 pseudosubluxation on admission radiographs in 30 (21.7%) patients, with no significant association with injury status or outcome [5]. Therefore, the presence of this variant should not be alarming, even in the setting of polytrauma.

Differential diagnosis

Ligamentous injury and traumatic C2 spondylolisthesis (Hangman's fracture) may result in abnormal anterolisthesis of C2 on C3. While chronic degenerative spondylolisthesis may also occur, this would not be present in the pediatric population.

Teaching point

Pseudosubluxation of C2–C3 is a normal variant that occurs in children, usually less than eight years of age, that should not be confused with acute injury.

REFERENCES

1. Curtin F, McElwain J. Assessment of the "nearly normal" cervical spine radiograph: C2–C3 pseudosubluxation in an adult with whiplash injury. *Emerg Med J*. 2005;**22**(12):907–8.
2. Swischuk LE. Anterior displacement of C2 in children: physiologic or pathologic. *Radiology*. 1977;**122**(3):759–63.
3. Lustrin ES, Karakas SP, Ortiz AO, *et al*. Pediatric cervical spine: normal anatomy, variants, and trauma. *Radiographics*. 2003;**23**(3):539–60.
4. Egloff AM, Kadom N, Vezina G, Bulas D. Pediatric cervical spine trauma imaging: a practical approach. *Pediatr Radiol*. 2009;**39**(5):447–56.
5. Shaw M, Burnett H, Wilson A, Chan O. Pseudosubluxation of C2 on C3 in polytraumatized children – prevalence and significance. *Clin Radiol*. 1999;**54**(6):377–80.

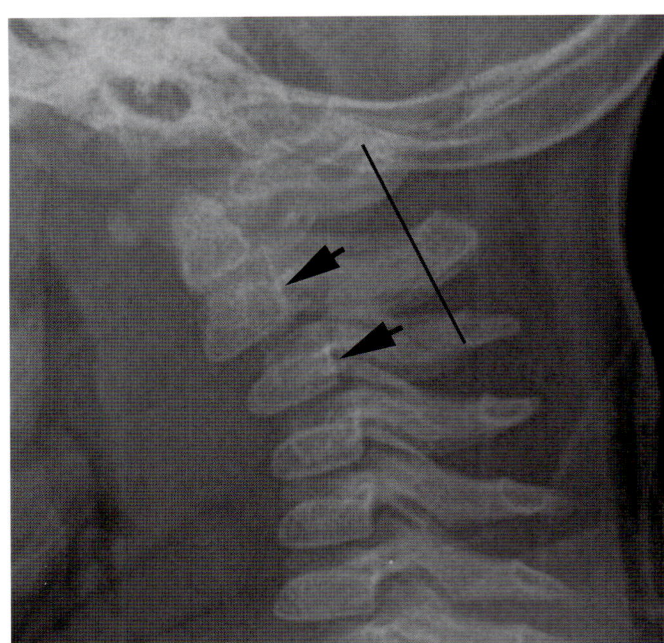

Figure 22.1 Lateral radiograph of the cervical spine in a 1-year-old boy who fell while playing shows anterior displacement of C2 relative to C3 (black arrows). The posterior cervical line of Swischuk is normal (black line). This is a typical appearance of pseudosubluxation.

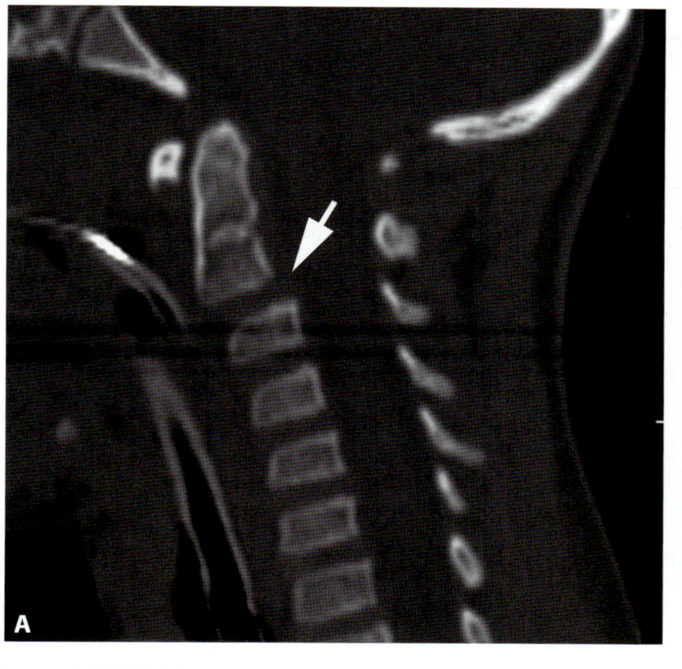

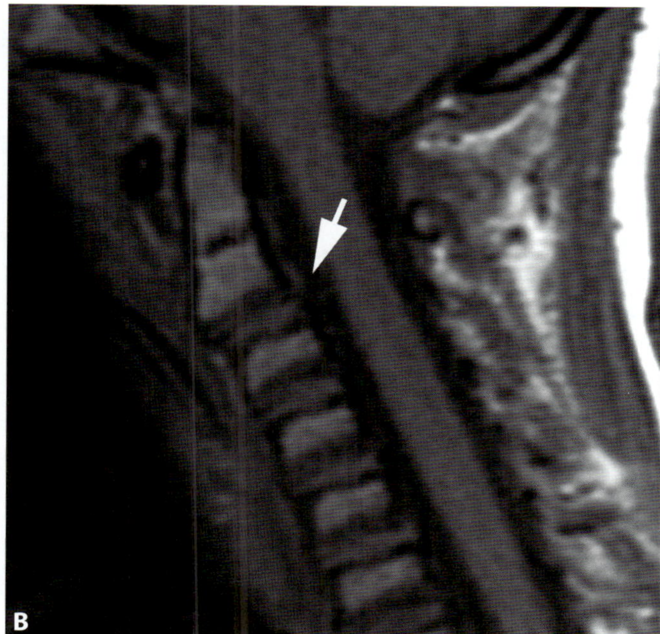

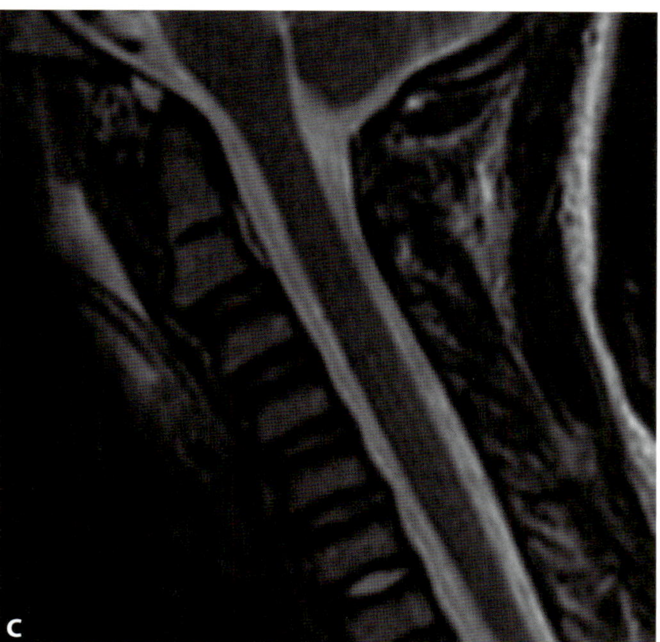

Figure 22.2 **A.** Sagittal non-enhanced CT image from a 7-year-old boy who sustained a motor-vehicle accident shows 3 mm of anterolisthesis of C2 relative to C3 (arrow). Skull base fractures were present, but no spine fractures were identified. **B.** Sagittal T1-weighted MR image redemonstrates the offset at C2–C3 (arrow). **C.** Sagittal T2-weighted MR image reveals no evidence of injury, including no findings to suggest ligamentous injury.

Calcific tendinitis of the longus colli

Matthew H. Nett

Imaging description

The longus colli muscle lies anterior to the cervical spine in the prevertebral space, covered by the prevertebral layer of the deep cervical fascia. It extends from the level of the anterior tubercle of the atlas (C1 vertebra) to the level of the T3 vertebral body in the superior mediastinum. Although the superior tendon fibers at C1–C3 are classically affected in acute calcific tendinitis, theoretically calcific tendinitis could occur in any of the tendon fibers, and there are reports of this process occurring in the inferior tendon fibers as well [1].

Calcific tendinitis of the longus colli was first reported on radiography in 1964 [2]. Although originally reported on radiographs, CT offers improved visualization and localization of the calcium deposits due to its superior spatial resolution. CT and radiographic findings of calcific tendinitis of the longus colli include calcifications anterior to the C1–C3 vertebral bodies with associated soft tissue edema (Figure 23.1A–E). If the calcifications are subtle, however, these may be missed with radiography. Typically MR shows edema, indicating inflammation in and around the longus colli tendon fibers (Figure 23.1F) [3]. A retropharyngeal fluid collection may be seen but should cause smooth enlargement of the retropharyngeal space as opposed to a septic fluid collection (Figures 23.1C, E, F) [4]. As with radiography, the MR depiction of calcifications in fibers of the longus colli muscle is often inferior to CT. Contrast-enhanced cross-sectional imaging may be helpful if fluid collections are present as these will lack wall enhancement, thereby excluding an abscess (Figure 23.1C) [5]. Lack of cervical lymphadenopathy is also a helpful sign to differentiate calcific tendinitis from an infection [1].

Importance

When calcific tendinitis occurs in the longus colli, in the perivertebral or prevertebral space, it may be misinterpreted as a neoplasm, fracture dislocation, or infection. The acute symptoms may overlap with these diagnoses, so it cannot be easily distinguished on clinical parameters alone. If not recognized by the radiologist, this could lead to unnecessary or inadequate workup and treatment, possibly with invasive procedures such as biopsy, aspiration, or surgery.

Typical clinical scenario

Calcific tendinitis is an inflammatory condition resulting in deposition of hydroxyapatite crystals, which may or may not be symptomatic [6]. It most commonly occurs in the third to sixth decades of life and typically affects the shoulder, but can take place in any tendon. The symptoms are non-specific, and may include neck pain, dysphagia, odynophagia, limited range of motion, and fever [1]. Onset is typically acute and progresses over several days. There may be mild elevation of the erythrocyte sedimentation rate and mild leukocytosis [3].

Treatment of the calcific tendinitis of the longus colli is conservative and typically involves rest and non-steroidal anti-inflammatory medications [3]. Signs and symptoms will typically abate within a few weeks.

Differential diagnosis

The differential diagnosis for calcific tendinitis of the longus colli includes trauma, neoplasm, or infection. An acute fracture should have an associated osseous defect, and the calcification will not be within tendon fibers. Neoplasm will typically have an associated soft tissue mass, enhancement, or lymphadenopathy. If a fluid collection is present, infection may be considered, but this should have post-contrast rim enhancement typical of abscess, and often lymphadenopathy as well.

Teaching point

Although the clinical symptoms may overlap with more serious conditions, calcific tendinitis of the longus colli has a characteristic imaging appearance of soft tissue swelling and amorphous calcification in the prevertebral space, anterior to the C1–C3 vertebrae..

REFERENCES

1. Park SY, Jin W, Lee SH, *et al.* Acute retropharyngeal calcific tendinitis: a case report with unusual location of calcification. *Skeletal Radiol.* 2010;**39**(8):817–20.
2. Hartley J. Acute cervical pain associated with retropharyngeal calcium deposit. A case report. *J Bone Joint Surg Am.* 1964;**46**:1753–4.
3. Offiah CE, Hall E. Acute calcific tendinitis of the longus colli muscle: spectrum of CT appearances and anatomical correlation. *Br J Radiol.* 2009;**82**(978):e117–21.
4. Eastwood JD, Hudgins PA, Malone D. Retropharyngeal effusion in acute calcific prevertebral tendinitis: diagnosis with CT and MR imaging. *AJNR Am J Neuroradiol.* 1998;**19**(9):1789–92.
5. Omezzine SJ, Hafsa C, Lahmar I, Driss N, Hamza H. Calcific tendinitis of the longus colli: diagnosis by CT. *Joint Bone Spine.* 2008; **75**(1):90–1.
6. Ring D, Vaccaro AR, Scuderi G, Pathria MN, Garfin SR. Acute calcific retropharyngeal tendinitis. Clinical presentation and pathological characterization. *J Bone Joint Surg Am.* 1994;**76**(11):1636–42.

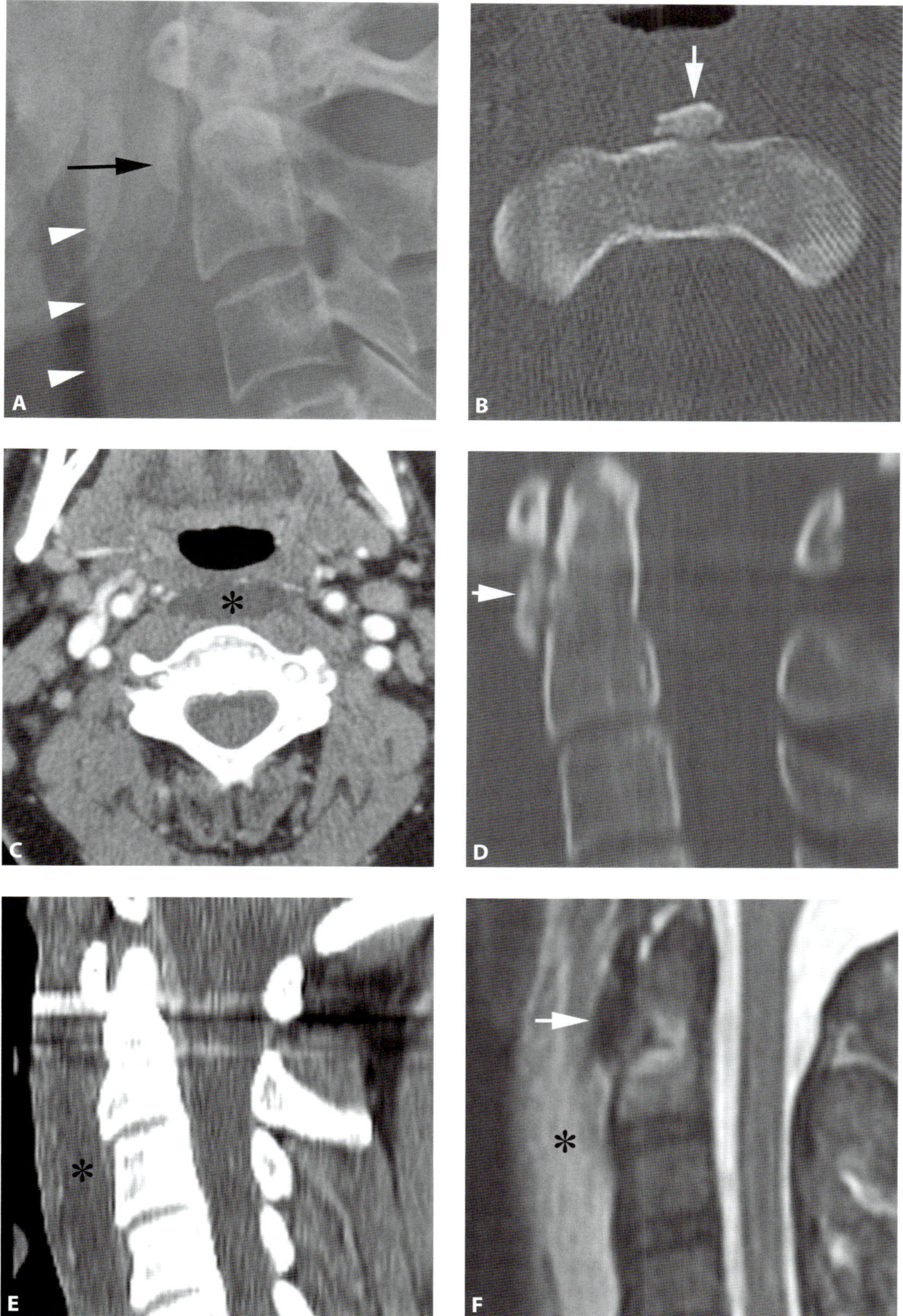

Figure 23.1 **A.** Lateral radiograph from a 32-year-old female with severe neck pain demonstrates a calcification anterior to the dens (arrow) with marked prevertebral soft tissue swelling (arrowheads). **B.** Axial CT in the bone window more clearly defines the calcification (arrow) and shows an intact adjacent C2 cortex excluding fracture or osseous lesion. **C.** Axial CT with contrast in the soft tissue window shows an associated soft tissue fluid collection (asterisk) without rim enhancement or lymphadenopathy. **D.** and **E.** Sagittal CT images in the bone and soft tissue windows demonstrate the extent of the calcification in the longus colli (arrow) and the soft tissue edema and fluid (asterisk). **F.** Sagittal short tau inversion recovery image from a corresponding MR shows the smoothly marginated fluid collection (asterisk) and longus colli calcification (arrow).

Motion artifact simulating spinal fracture

Ken F. Linnau

Imaging description

Many patients undergoing spine evaluation in the emergency department are intubated or otherwise unable to hold still while undergoing CT. Misregistration is a CT artifact caused by patient or physiologic motion during helical or sequential image acquisition [1–3]. Slight movement may cause blurring on axial CT images (Figure 24.1). If greater physical displacement occurs, motion artifact can result in double edges on images or ghosting (Figure 24.2A). The cervical spine is the most mobile portion of the spinal column, and images of the cervical spine are prone to motion artifact.

CT images of the spine degraded by patient motion can be mistaken for fractures. When an equivocal fracture is suspected on axial views confirmation of the finding on multiplanar reformations (usually obtained in the sagittal and coronal planes) is advised. Multiplanar image reformations improve clinical accuracy for the detection of spine fractures [4, 5], intracranial hemorrhage [6], and pulmonary embolism [7]. In the spine, orthogonal image reformation in the sagittal and coronal plane increases sensitivity, when compared to axial image review alone [4]. Thin axial image reconstruction (1 mm) also improves sensitivity to 95% for the detection of spine fractures compared to thicker slice reconstruction in the osteoporotic patient [5].

On sagittal CT reformations, motion artifacts may extend beyond the osseous portion of the spine and into soft tissues. Patient motion may cause "stair-step" artifact in the patient's skin, tubes, lines and air–soft tissue interfaces running orthogonal to the plane of motion (Figure 24.2B).

CT scanner manufacturers aim to minimize the effect of motion artifacts using over- or under-scanning modes and built-in software corrections [1]. Motion artifacts also decrease with reduced scan time and shorter rotation times (higher temporal resolution). Currently, the highest temporal resolution can be achieved with dual-source CT scanners [8].

At times, CT reformations may not allow confident confirmation of artifact and findings remain equivocal. To distinguish motion artifact from fracture or malalignment, a single collimated lateral radiograph, or repeat CT, of the spinal portion in question can be useful for clarification (Figure 24.2C).

Importance

Overdiagnosis of spine fractures may lead to unnecessary immobilization or follow-up imaging.

Typical clinical scenario

A patient, typically after blunt trauma, cannot undergo spine clearance based on clinical signs and symptoms alone, and requires imaging of the spine for clearance. Due to pain, a reduced level of consciousness, intoxication, or unwillingness of the patient to lie still, misregistration artifact occurs during the CT acquisition. Examining the CT images while the patient remains on the CT table will assist with early detection of this artifact, and judicious use of rescanning of the area in question before the patient leaves the scanner may assist. Occasionally, sedation, intubation, and paralysis may be necessary for adequate imaging of combative patients [9].

Differential diagnosis

This diagnostic pitfall can usually be avoided by careful inspection for other clues of patient motion and the use of multiplanar reformations. However, in some cases, osteophytes arising from vertebral body endplates can yield a similar imaging appearance to endplate fractures. In this setting, thin-section reconstructions in orthogonal planes to the endplate assists in discriminating between fracture and osteophyte.

Teaching point

Image interpretation of the spine should be performed with sagittal and coronal reformations to avoid misinterpretation of motion artifacts as pseudofractures. Thin slice reconstruction thickness (1 mm) improves sensitivity for the detection of spine fractures..

REFERENCES

1. Barrett JF, Keat N. Artifacts in CT: recognition and avoidance. *Radiographics*. 2004;**24**(6):1679–91.
2. Bushberg J, Seibert J, Leidholdt EJ, Boone J. *The Essential Physics of Medical Imaging*, 2nd edn. Philadelphia: Lippincott Williams & Wilkins; 2002.
3. McCollough CH, Bruesewitz MR, Daly TR, Zink FE. Motion artifacts in subsecond conventional CT and electron-beam CT: pictorial demonstration of temporal resolution. *Radiographics*. 2000;**20**(6):1675–81.
4. Gross EA. Computed tomographic screening for thoracic and lumbar fractures: is spine reformatting necessary? *Am J Emerg Med*. 2010;**28**(1):73–5.
5. Bauer JS, Muller D, Ambekar A, *et al.* Detection of osteoporotic vertebral fractures using multidetector CT. *Osteoporos Int*. 2006;**17**(4):608–15.
6. Wei SC, Ulmer S, Lev MH, *et al.* Value of coronal reformations in the CT evaluation of acute head trauma. *AJNR Am J Neuroradiol*. 2010;**31**(2):334–9.
7. Wu C, Sodickson A, Cai T, *et al.* Comparison of respiratory motion artifact from craniocaudal versus caudocranial scanning with 64-MDCT pulmonary angiography. *AJR Am J Roentgenol*. 2010;**195**(1):155–9.
8. Donnino R, Jacobs JE, Doshi JV, *et al.* Dual-source versus single-source cardiac CT angiography: comparison of diagnostic image quality. *AJR Am J Roentgenol*. 2009;**192**(4):1051–6.
9. Redan JA, Livingston DH, Tortella BJ, Rush BF, Jr. The value of intubating and paralyzing patients with suspected head injury in the emergency department. *J Trauma*. 1991;**31**(3):371–5.

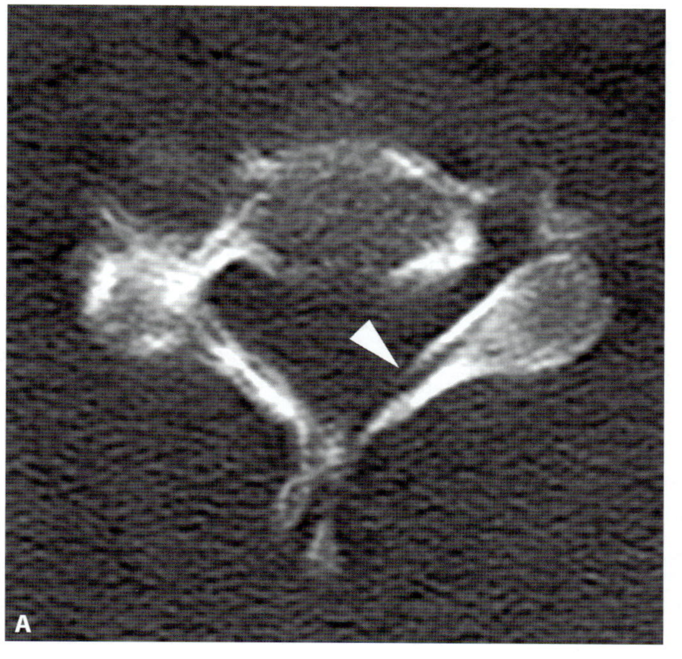

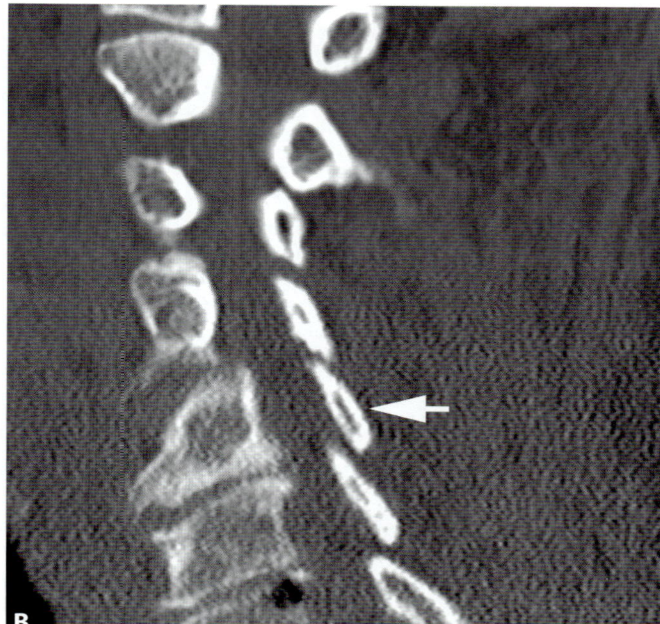

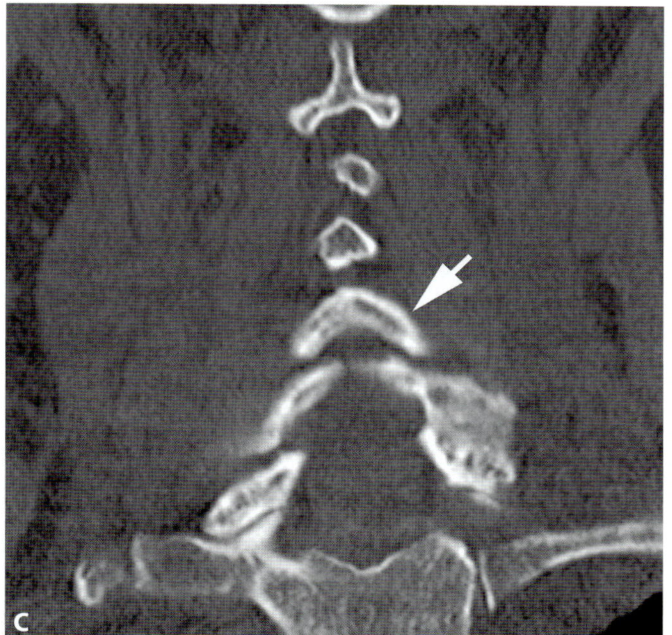

Figure 24.1 A. Axial non-contrast CT of the cervical spine in a 52-year-old man who fell shows blurring of the cortices of C5, suggestive of motion artifact. Additionally, there is apparent discontinuity of the anterior cortex of the left lamina (arrowhead). suspicious for fracture. **B.** Left paramedian sagittal reformation of the same CT shows no fracture line through the left lamina (arrow), which is confirmed on coronal reformations (**C**, white arrow) discriminating the axial image finding (**A**) as motion artifact.

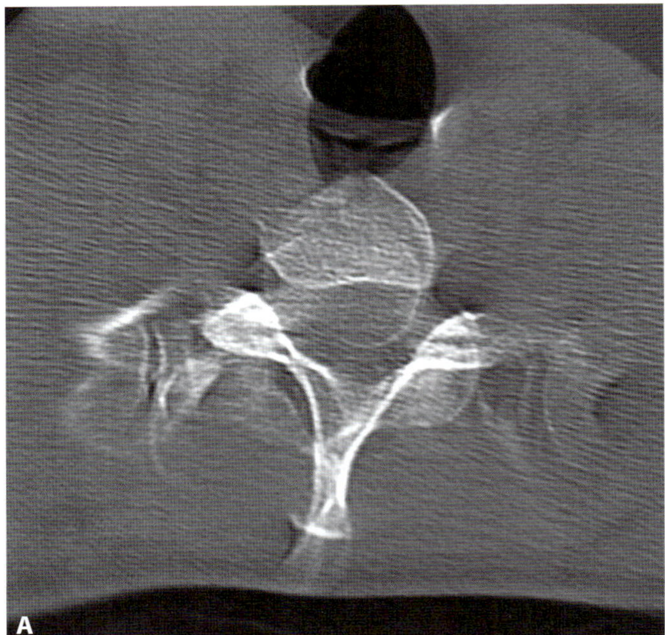

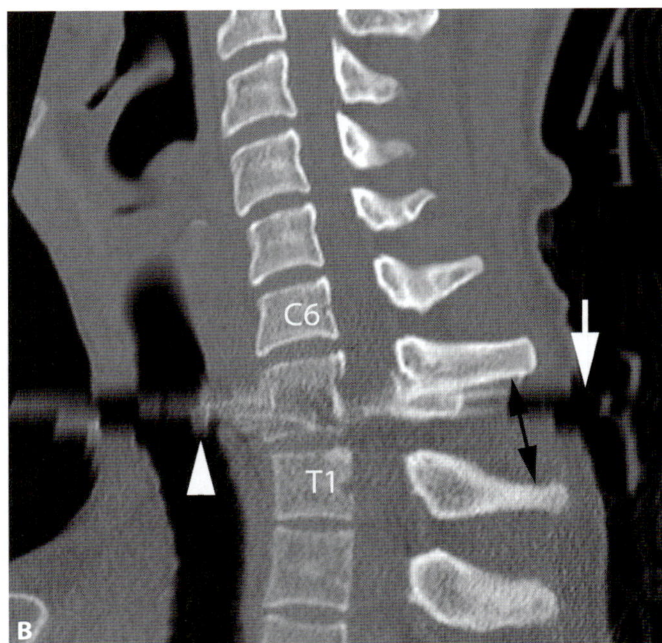

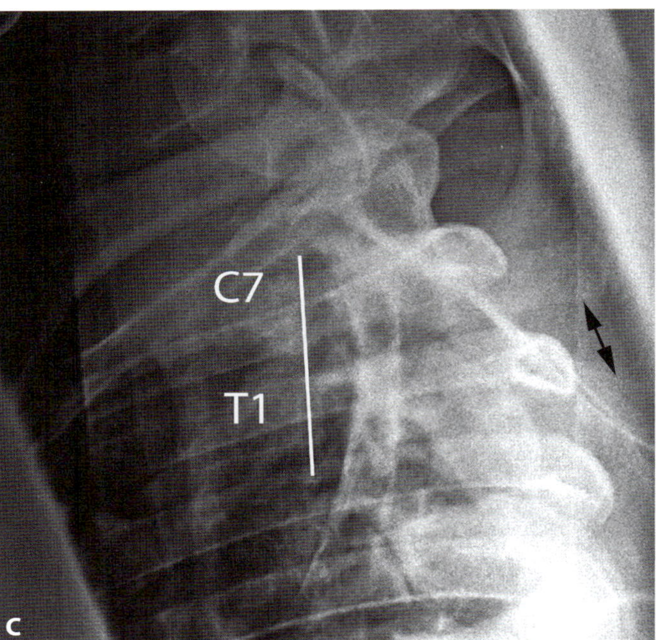

Figure 24.2 **A.** Axial cervical spine CT image from a 24-year-old man who fell during a seizure shows double contours (ghosting) of C7, suggestive of misregistration artifact due to substantial patient motion. **B.** Mid-sagittal reformation image confirms motion artifact, which extends through the anterior and posterior soft tissues. There is stair-step artifact in the patient's skin and cervical collar (white arrow) as well as the trachea (white arrowhead). Despite confirmation of motion artifact, spinal alignment at C7–T1 remains somewhat suspicious for a flexion-distraction injury (black double-arrow). **C.** Subsequently obtained swimmer's view of the cervical spine shows normal alignment of the C7–T1 vertebral bodies (white line) without widening of the interspinous distance (black double-arrow), which allowed radiographic clearance of the cervico-thoracic junction.

Pars interarticularis defects

Michael J. Modica

Imaging description

The pars interarticularis (or pars) is a short segment of the vertebra located between the superior and inferior facets of the articular process. Spondylolysis, also commonly referred to as a pars defect, is a unilateral or bilateral osseous defect in the pars interarticularis and is most common at the L5 vertebral body level (Figure 25.1) [1]. Pars defects usually result from dysplastic pars at birth exposed to chronic repetitive stress.

The radiographic appearance of pars defects varies with the age of the lesion.

Acute traumatic pars defects occurring in a non-dysplastic vertebral level are rare and result from high-energy trauma [2]. They are usually hyperextension injuries, and can be missed on plain radiographs [3]. Findings that suggest an acute injury include irregular bony edges, lack of soft tissue calcification, and associated fractures (Figure 25.1).

In contrast, chronic injuries typically have smooth, rounded, and corticated edges (Figures 25.2 and 25.3). Fibrocartilaginous material will develop in the gap, subsequently replaced by hypertrophic bone (Gill's nodules) [4]. Other imaging signs of a chronic unilateral spondylolysis include deviation of the spinous process and sclerosis of the contralateral pedicle [5]. Collimated lateral radiographs of the region of concern can help in questionable cases, but unless the beam is tangential to the defect, it may be missed [1, 4]. A five-view radiographic series (which includes 45 degree obliques) has a 96.5% sensitivity for the detection of pars defects [4].

Multidetector row CT with multiplanar reformations is the best way to detect the bony defect [4]. Wide sclerotic margins indicate chronic fracture with no potential for healing while narrow, non-sclerotic margins favor an acute fracture that could benefit from immobilization [6]. MRI will detect the edematous stress response in adolescents before the fracture line develops [4].

Spondylolisthesis is present in 80% of patients when spondylolysis is initially diagnosed. Progression of spondylolysis to spondylolisthesis is most likely to occur in patients younger than 16 years old (Figure 25.4). This often results in an increased sagittal diameter of the spinal canal.

Importance

Accurate diagnosis is important in the emergency setting for two reasons. Firstly, chronic pars defects can be mistaken for acute fractures, and secondly, they are a common cause for back pain in adolescents involved in certain sporting activities.

Pars defects are common, affecting 6% of the asymptomatic general population, but the prevalence rises to 13–47% in certain adolescent athletes [7]. They are rare in children less than five years old [4].

As a common etiology of back pain in adolescents, pars defects should be sought and acuity should be determined. Acute fractures are amenable to immobilization. Chronic non-union can be symptomatic when motion is detected on flexion-extension radiographs. Normal facet joints can mimic pars defects in the axial plane on CT and when in question sagittal images should be reviewed for further clarification. Orthopedic management decisions are based on the patient's age, symptoms, and degree of spondylolisthesis, if present.

Progression in older patients from stable spondylolysis to spondylolisthesis is thought to be secondary to disk degeneration and results in a narrowed sagittal diameter.

Typical clinical scenario

Acute spondylolysis is rare, and results from high-energy trauma, usually, but not always, from hyperextension.

Pars defects are asymptomatic or present as non-specific low back pain and represent the most common identified cause of low back pain in children and adolescents more than 10 years of age [8].

Differential diagnosis

The differential diagnosis for suspected symptomatic pars defects in children and adolescents includes acute fracture, osteoid osteoma, and osteomyelitis.

Teaching point

Pars defects commonly cause adolescent back pain, particularly in the young athlete. Acute pars fractures in the setting of a non-dysplasic pedicle result from high-energy trauma. CT and MRI are valuable in determining the age of the defect, and guiding management.

REFERENCES

1. Amato M, Totty WG, Gilula LA. Spondylolysis of the lumbar spine: demonstration of defects and laminal fragmentation. *Radiology*. 1984;**153**(3):627–9.
2. El Assuity WI, El Masry MA, Chan D. Acute traumatic spondylolisthesis at the lumbosacral junction. *J Trauma*. 2007;**62**(6):1514–16; discussion 1516–17.
3. Reinhold M, Knop C, Blauth M. Acute traumatic L5-S1 spondylolisthesis: a case report. *Arch Orthop Trauma Surg*. 2006;**126**(9):624–30.
4. Leone A, Cianfoni A, Cerase A, Magarelli N, Bonomo L. Lumbar spondylolysis: a review. *Skeletal Radiol*. 2011;**40**(6):683–700.

5. Ravichandran G. A radiologic sign in spondylolisthesis. *AJR Am J Roentgenol.* 1980;**134**(1):113–17.
6. Smith JA, Hu SS. Management of spondylolysis and spondylolisthesis in the pediatric and adolescent population. *Orthop Clin North Am.* 1999;**30**(3):487–99, ix.
7. Nayeemuddin M, Richards PJ, Ahmed EB. The imaging and management of nonconsecutive pars interarticularis defects: a case report and review of literature. *Spine J.* 2011;**11**(12):1157–63.
8. Afshani E, Kuhn JP. Common causes of low back pain in children. *Radiographics.* 1991;**11**(2):269–91.

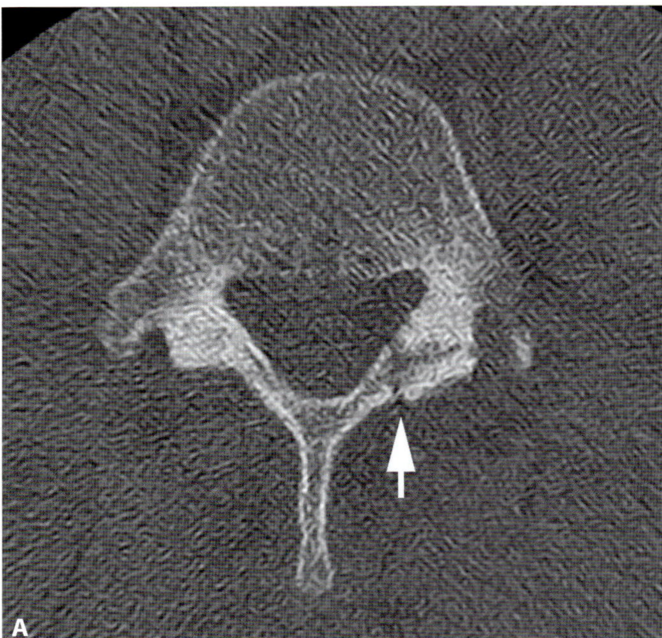

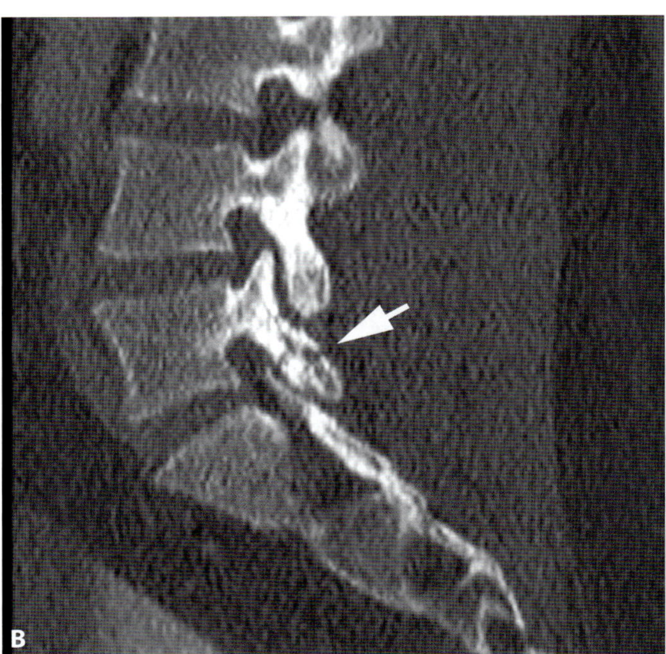

Figure 25.1 A. Axial CT from a 32-year-old man involved in a high-speed motor vehicle collision who sustained multiple organ injuries including left transverse process fractures of all the lumbar vertebrae. A comminuted fracture extending through the lamina, interior facet, and pars interarticularis is present (arrow). **B.** Sagittal reformation shows the plane of the fracture (arrow). Note the absence of cortication along the fracture line.

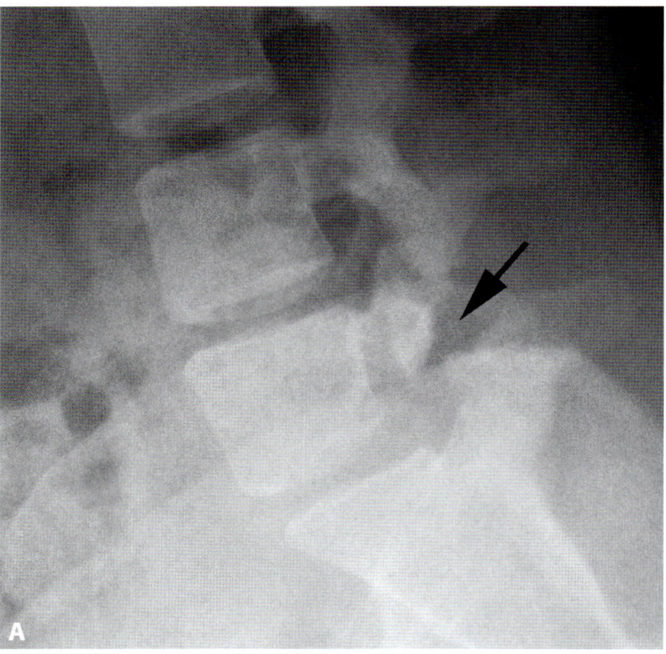

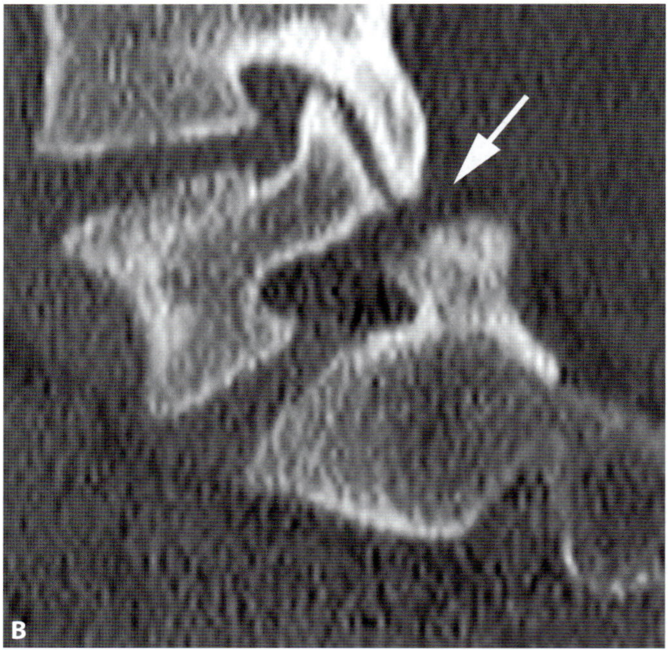

Figure 25.2 A. Collimated lateral radiograph of the lumbar spine in a 17-year-old man shows a wide defect in the pars interarticularis at the L5 level (arrow). **B.** Sagittal CT of the lumbar spine in the same patient shows the pars defect (arrow).

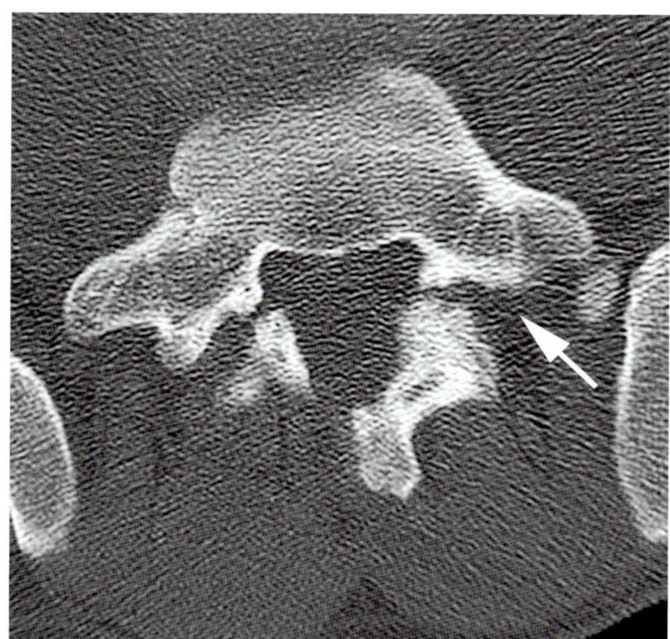

Figure 25.3 Axial CT image of the lumbar spine at the L5 level shows a defect in the left pars interarticularis (arrow) with irregular sclerotic margins.

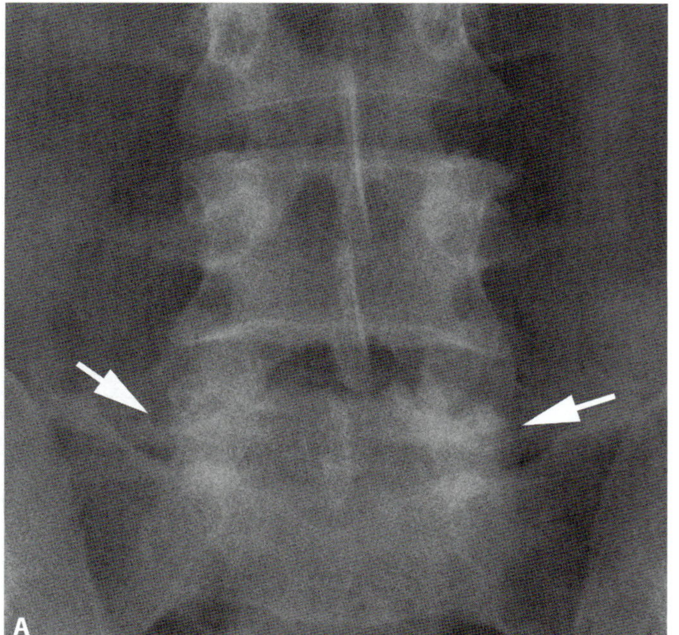

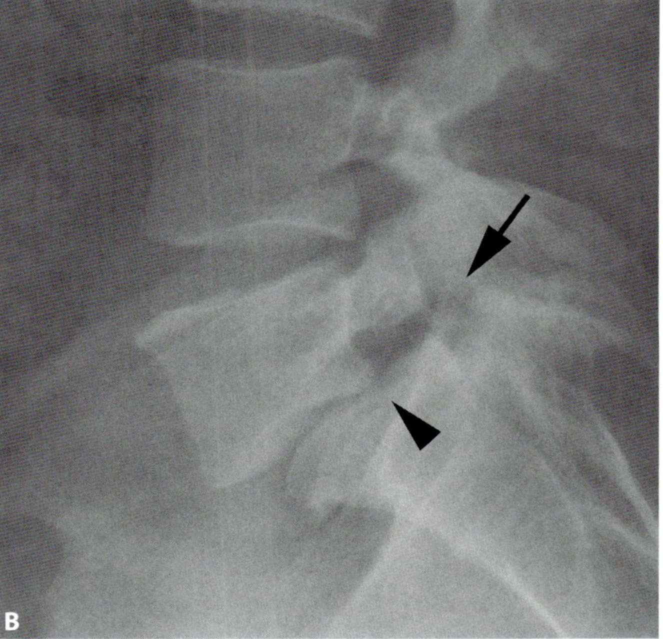

Figure 25.4 **A.** Anterior-posterior radiograph of the lumbar spine in a 20-year-old gymnast with low back pain shows bilateral L5 pedicle sclerosis (arrows). **B.** Lateral radiograph from the same patient shows spondylolysis (arrow) that has progressed to grade II spondylolisthesis (arrowhead) at the L5–S1 level.

Limbus vertebra

Joel A. Gross

Imaging description

A limbus vertebra (LV) demonstrates separation of a segment of the rim of the vertebral body. This was first described by Schmorl in 1927 [1], and is caused by intraosseous penetration of disk material at the junction of the cartilaginous endplate and the bony rim during childhood or adolescence [2]. An oblique radiolucent defect extends from the vertebral endplate to the outer surface of the vertebral body, separating off a small segment of bone (Figure 26.1). In adults, this is typically triangular in shape and has sclerotic margins. This helps distinguish the lesion from an acute fracture. In children, the separate fragment may not be ossified, and not visible on radiographs, and only a lucent defect in the vertebral body may be evident.

Limbus vertebrae most commonly occur at the anterosuperior margin of a single lumbar vertebra [3], followed by the anteroinferior margin of a lumbar vertebra, and far less commonly at the posteroinferior corner of a lumbar vertebra or in the thoracic spine [1].

Radiographs are usually adequate to make the diagnosis, but if findings are not characteristic, the improved visualization of findings on CT can help make a confident diagnosis [4]. The etiology of the lesion, intraosseous herniation of the nucleus pulposus, has been demonstrated in the past using injection of contrast into the disk [5], but similar findings can now be less invasively visualized incidentally on MRI performed for other reasons.

Importance

A limbus vertebra may be mistaken for degenerative changes, acute fracture, or infection, but the well-defined cortical sclerotic margins of a limbus vertebra fragment along with the associated osseous defect will differentiate it from these other possibilities in adults. In children, the ossified fragment may not be visible.

A limbus fracture is defined as a "traumatic separation of a segment of bone from the edge of the vertebral ring apophysis at the site of anular attachment" in the Combined Task Force on the Classification of Lumbar Disk Pathology [6]. In our experience, this terminology is not commonly used, and should be avoided, or used with caution, to prevent confusion when referring to the more commonly described limbus vertebrae, as described in this case.

Typical clinical scenario

Limbus vertebrae may be identified incidentally, as part of torso imaging, or may be identified during dedicated spine imaging. It is important to avoid mistaking this lesion for significant spine pathology.

Differential diagnosis

Acute vertebral body fractures do not have well-defined sclerotic margins, and may demonstrate adjacent hematoma.

Osteomyelitis and diskitis demonstrate loss of cortex (typically of both adjacent vertebral body endplates), and may have an adjacent soft tissue mass.

Anterior osteophyte fragments lack a matching defect in the vertebral body, as do calcified disk herniations.

Teaching point

Limbus vertebrae in adults have a typical corticated triangular fragment with a matching vertebral body defect.

They typically involve the anterosuperior margin of lumbar vertebrae and should not be mistaken for fractures, infection, or other significant spine pathology.

REFERENCES

1. Henales V, Hervas JA, Lopez P, *et al.* Intervertebral disc herniations (limbus vertebrae) in pediatric patients: report of 15 cases. *Pediatr Radiol.* 1993;**23**(8):608–10.
2. Resnick D. *Diagnosis of Bone and Joint Disorders.* Philadelphia: Saunders; 2002.
3. Kumar R, Guinto FC, Jr., Madewell JE, Swischuk LE, David R. The vertebral body: radiographic configurations in various congenital and acquired disorders. *Radiographics.* 1988 May;**8**(3):455–85.
4. Yagan R. CT diagnosis of limbus vertebra. *J Comput Assist Tomogr.* 1984 Feb;**8**(1):149–51.
5. Ghelman B, Freiberger RH. The limbus vertebra: an anterior disc herniation demonstrated by discography. *AJR Am J Roentgenol.* 1976 Nov;**127**(5):854–5.
6. Fardon DF, Milette PC. Nomenclature and classification of lumbar disc pathology. Recommendations of the Combined Task Forces of the North American Spine Society, American Society of Spine Radiology, and American Society of Neuroradiology. *Spine (Phila Pa 1976)* 2001:**26**(5): E93–113. Available from: http://www.asnr.org/spine_nomenclature/ (accessed January 17, 2012).

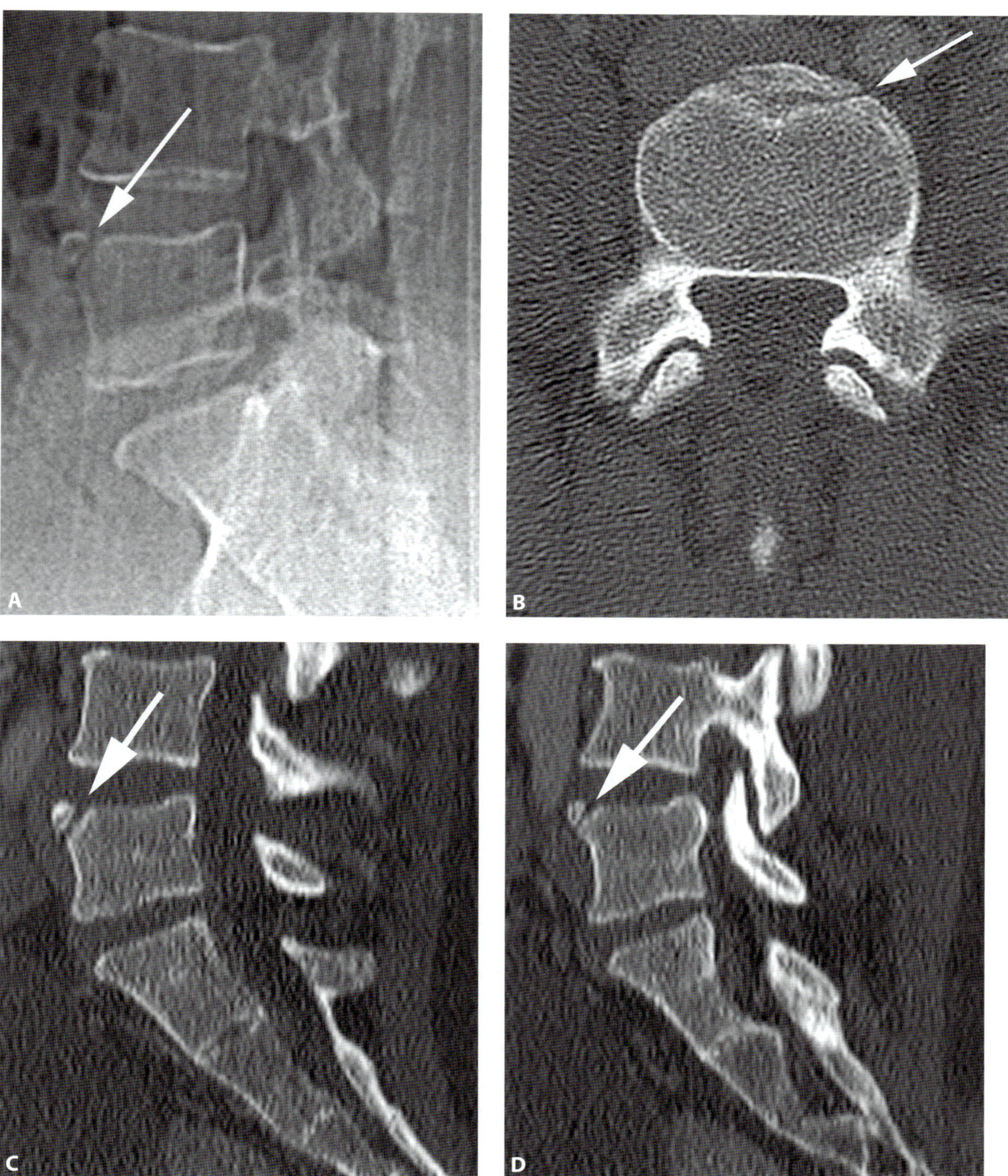

Figure 26.1 Lumbar spine CT from a 42-year-old woman following a motor vehicle crash. **A.** Lateral topogram demonstrates typical findings similar to those expected on radiographs, with triangular corticated fragment arising from the superior endplate of L5 (arrow). **B.** Axial CT image demonstrates fragment arising from the anterior superior endplate of L5 (arrow). Corticated edges may be appreciated, depending on image quality, slice thickness, and plane of the axial image with respect to the vertebra. This feature helps distinguish the fragment from an acute fracture. **C.** and **D.** Sagittal reformations more clearly demonstrate the triangular fragment with corticated edges (arrow) of both the fragment and the "donor site."

CASE 27

Transitional vertebrae

Joel A. Gross

Imaging description

Transitional vertebrae are fairly frequently identified in the lumbosacral region, where they are referred to as lumbosacral transitional vertebrae (LSTV). Lumbosacral transitional vertebrae have a reported prevalence of 4–30%, and can reflect sacralization or lumbarization [1]. Sacralization is more common [2].

In sacralization, there is assimilation (fusion) of the lowest lumbar segment into the sacrum. In lumbarization, the S1 segment is not fused with the remainder of the sacrum, and appears as a separate "lumbar type" vertebral body. Partial sacralization or lumbarization often occurs in which there is partial fusion or separation of the transitional vertebra (Figure 27.1 and 27.2).

While transitional vertebrae may be identified on radiographs, CT with multiplanar reformations (MPRs) provides more optimal evaluation of the vertebrae, pseudarthroses, and fusion.

Castellvi [3] described a classification system for LSTV, ranging from type I to type IV anomalies, with A and B subtypes for unilateral or bilateral involvement respectively.

Importance

The lucency between the transitional vertebra and the sacrum may be mistaken for infection or fracture.

Lumbosacral transitional vertebrae may result in numbering errors, when the radiologist counts up from the last lumbar appearing vertebra to identify and enumerate the vertebra on a lumbar spine study. A sacralized L5 may cause them to mistakenly number T12–L4 as L1–L5, or a lumbarized S1 may cause them to mistakenly number L2–S1 as L1–L5. Errors in enumeration of the vertebrae can result in surgery or procedures being performed at the wrong levels. Spine imaging of all other levels should be reviewed to ensure a correct enumeration. If this is not possible (for example if the thoracic spine was not imaged), a clear description of your method for numbering the spine should be provided, to avoid ambiguities leading to errors.

Transitional vertebrae are incidental asymptomatic findings in many patients, but a relationship between low back pain and LSTV (known as the Bertolotti syndrome) has been debated in the literature. The type and site of pathology of the transitional segment should be accurately identified to help guide treatment. In young adults, degeneration of the disk immediately superior to the LSTV level is more common than in patients without a LSTV. These differences become less evident with age as typical degenerative changes occur [4].

Typical clinical scenario

Transitional vertebrae are frequently incidentally noted on torso CTs, but are of most significance in spine imaging obtained following trauma or for pain or neurologic symptoms thought to be related to the spine.

Differential diagnosis

Acute fractures of the sacrum demonstrate absence of cortex across the acute fracture site. In contrast, the lucencies between transverse processes and the sacrum, or across diarthrodial joints, should have well-defined cortices bordering them. Axial images may not clearly demonstrate these corticated edges and the anatomy of the surrounding structures. Multiplanar reformatted images, especially coronal reformations, are helpful for accurate evaluation and characterization.

Diskitis and osteomyelitis may also present with lucencies around the lumbosacral junction, but cortical endplate destruction and the absence of appropriate anatomic variants (enlarged transverse processes, fusion or pseudarthrosis at the lumbosacral junction) prevent mistaken diagnosis of a transitional vertebra.

Teaching point

Lucencies around transitional vertebrae can simulate fractures or spondylodiskitis. Coronal CT reformations often help to correctly identify them as transitional vertebrae.

Transitional vertebrae may lead to incorrect enumeration of vertebrae, and result in procedures or surgery being performed at the wrong levels.

REFERENCES

1. Konin GP, Walz DM. Lumbosacral transitional vertebrae: classification, imaging findings, and clinical relevance. *AJNR Am J Neuroradiol.* 2010;**31**(10):1778–86.
2. Köhler A, Zimmer E-A, Freyschmidt J, *et al. Freyschmidts' "Koehler/Zimmer". Borderlands of Normal and Early Pathological Findings in Skeletal Radiography*. Stuttgart: Thieme; 2003.
3. Castellvi AE, Goldstein LA, Chan DP. Lumbosacral transitional vertebrae and their relationship with lumbar extradural defects. *Spine (Phila Pa 1976).* 1984;**9**(5):493–5.
4. Bron JL, van Royen BJ, Wuisman PI. The clinical significance of lumbosacral transitional anomalies. *Acta Orthop Belg.* 2007;**73**(6):687–95.

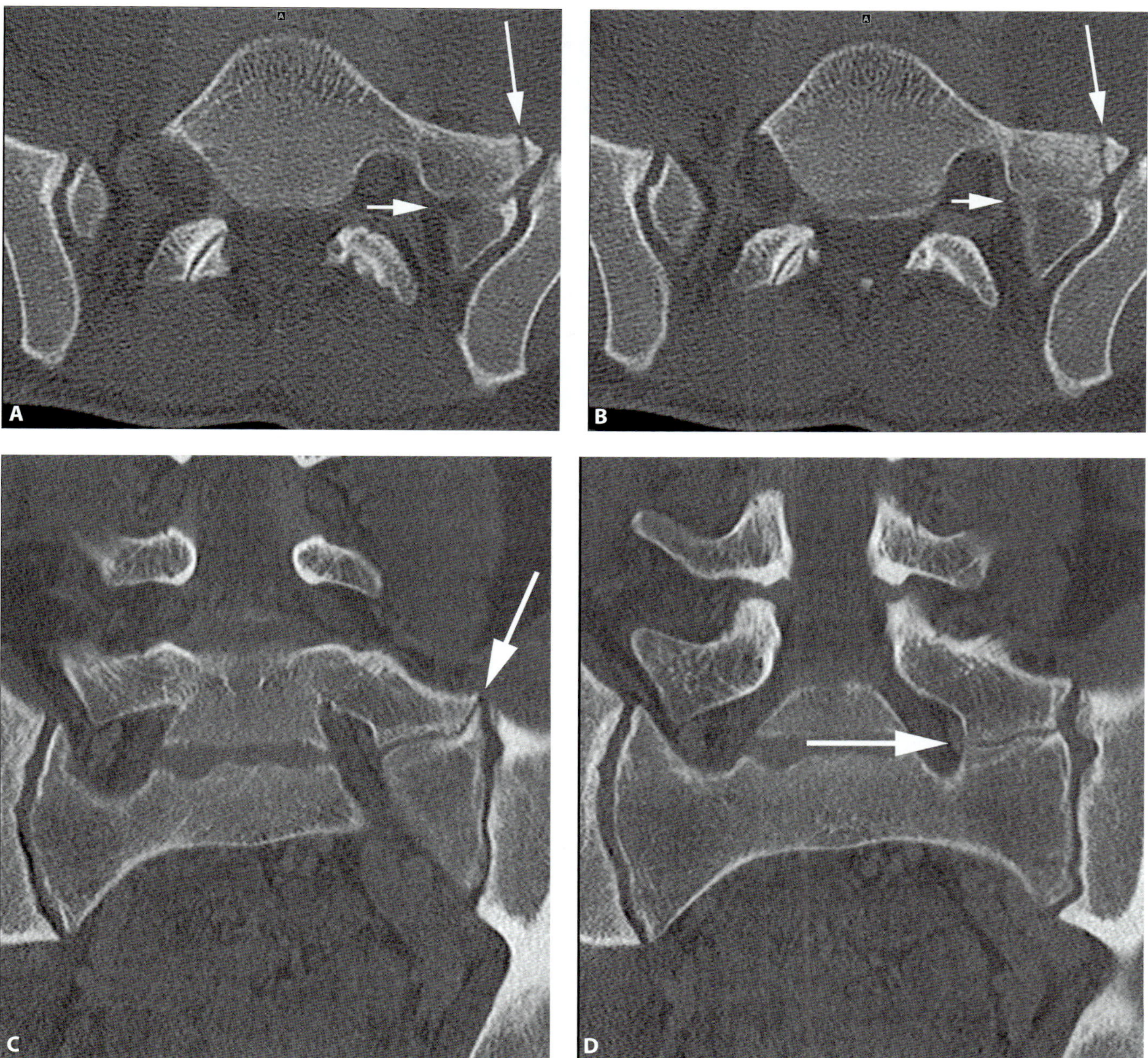

Figure 27.1 Lumbar spine CT from a 21-year-old man following a motor vehicle collision. **A** and **B.** Axial CT demonstrates a sagittally oriented lucency (long arrow) and transversely oriented lucency (short arrow) without well-defined cortical margins, thought to represent left sacral fractures. **C.** and **D.** Coronal reformations clearly demonstrate the left pseudarthrosis between L5 and S1 (arrows), with well corticated margins. While the right L5 transverse process is enlarged, representing partial right sided sacralization, there is no associated pseudarthrosis or fusion.

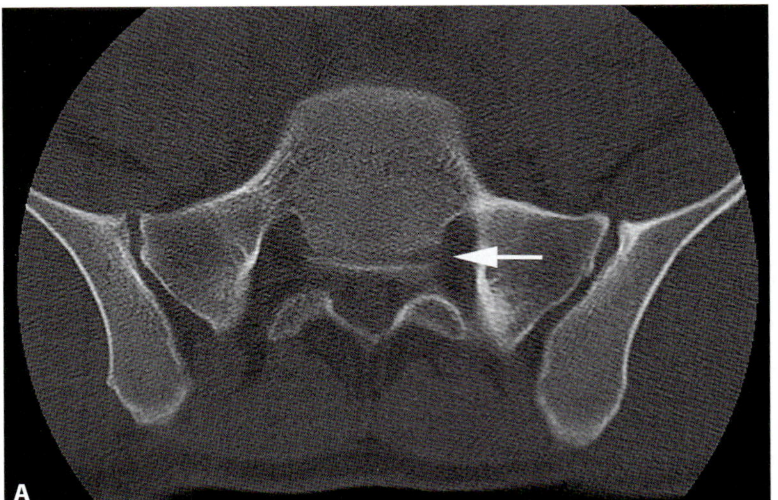

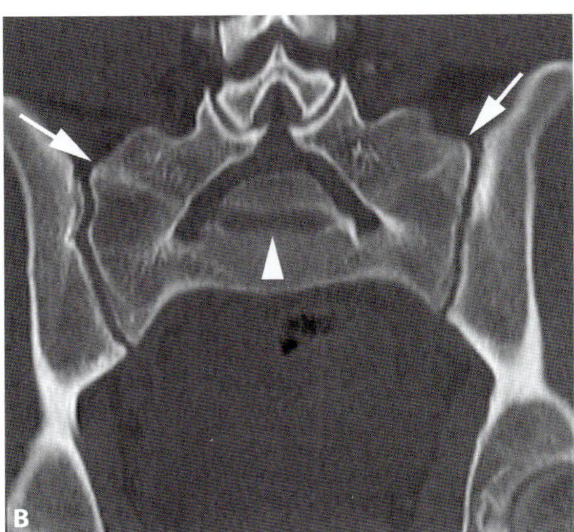

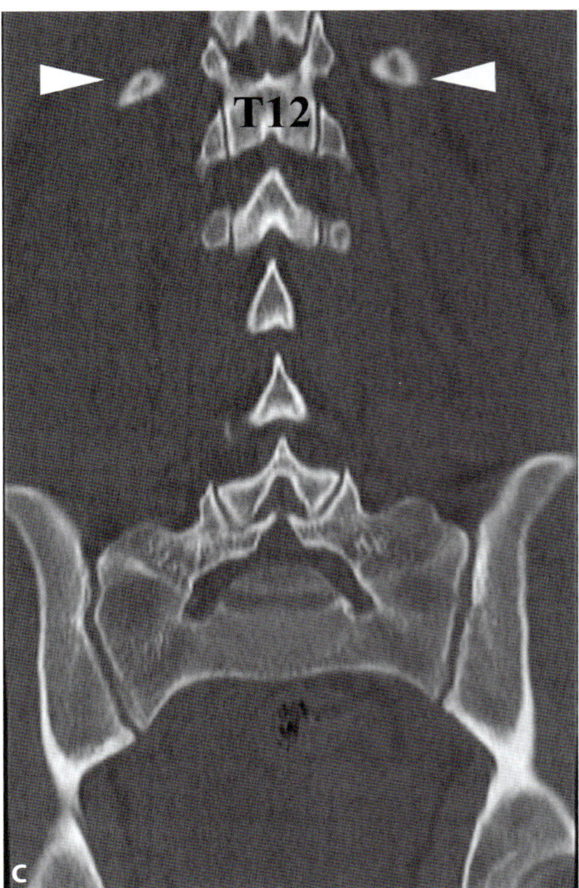

Figure 27.2 Lumbar spine CT from a 25-year-old man following a gunshot wound to the abdomen. **A.** Axial CT demonstrates sacralization of L5, with bilateral lateral osseous fusion with S1, and a persistent L5–S1 disk (arrow). **B.** Coronal reformation again demonstrates the disk at L5–S1 (arrowhead), but also demonstrates the "scar" at the site of osseous fusion between L5 and S1 (arrows), which was not evident on axial images due to its oblique course. **C.** Coronal reformation demonstrates the inferior most ribs (arrowheads) arising from the T12 vertebral body (T12). T12 level was confirmed by review of thoracic spine radiographs (not shown), confirming that the transitional vertebra is a sacralized L5 rather than a lumbarized S1.

Subtle injuries in ankylotic spine disorders

Matthew H. Nett

Imaging description

Ankylotic spine disorders, including seronegative spondyloarthropathy (especially ankylosing spondylitis) and diffuse idiopathic skeletal hyperostosis (DISH), result in osseous fusion of the vertebral elements which restrict physiologic motion of the vertebral column and predispose the spine to atypical fractures even with minimal trauma.

Radiographs may not delineate the subtle fractures often seen in ankylotic disorders (Figure 28.1). It can also be difficult on radiographs to determine the full extent of the fracture (Figure 28.2), differentiate incomplete ankylosis from fracture, or evaluate the soft tissues. For these reasons, when ankylosis is encountered, the radiologist must have a high level of suspicion for an occult fracture, and CT or MR should be utilized liberally.

Injuries in the setting of ankylosis often involve all three spinal columns [1]. MR is superior for the evaluation of the disk space, soft tissues, and spinal cord. Soft tissue trauma including ligament tears and transdiskal injuries may be diagnosed on CT by subtle widening of the intervertebral space compared to adjacent levels [2]. CT angiography or MR angiography may also be necessary to evaluate for vascular injury if the fracture extends to the foramen transversarium.

Importance

Ankylosing spinal disorders, including the seronegative spondyloarthropathies and DISH, are common conditions in the adult population and often go undiagnosed, especially DISH. These patients are prone to significant, yet often subtle, fractures of the spinal column, even with relatively minor trauma. The altered mechanics of the spinal column also result in unusual fracture patterns compared to the normal population, which can lead to significant neurologic impairment.

Typical clinical scenario

A recent review of the literature on spinal fractures in ankylotic disorders found patients were typically in the fifth to sixth decade of life, and approximately two-thirds of these patients had relatively minor trauma, most commonly a fall from sitting or standing [3]. Hyperextension was the most frequent mechanism, and fractures mostly occurred in the cervical spine. Neurologic deficits are common and may be acute or delayed.

Differential diagnosis

Differentiation between fracture and incomplete fusion can be difficult. Helpful features to exclude fracture include smooth margins. Even displaced fracture fragments should fit together when extrapolated to their donor sites. In difficult cases an MR can demonstrate marrow or soft tissue edema if a fracture has occurred. Edema related to a disk injury could be mistaken for an infection on MR; however, infection should not widen the disk space, and the patient should have clinical signs and symptoms to support this diagnosis. Lack of post-contrast enhancement also helps to exclude infection.

Teaching point

In ankylotic spine disorders, subtle fractures may occur with minor trauma. MR and CT imaging are important as radiographs may be falsely negative.

REFERENCES

1. Wang YF, Teng MM, Chang CY, Wu HT, Wang ST. Imaging manifestations of spinal fractures in ankylosing spondylitis. *AJNR Am J Neuroradiol.* 2005;**26**(8):2067–76.
2. Campagna R, Pessis E, Feydy A, *et al.* Fractures of the ankylosed spine: MDCT and MRI with emphasis on individual anatomic spinal structures. *AJR Am J Roentgenol.* 2009;**192**(4):987–95.
3. Westerveld LA, Verlaan JJ, Oner FC. Spinal fractures in patients with ankylosing spinal disorders: a systematic review of the literature on treatment, neurologic status and complications. *Eur Spine J.* 2009;**18**(2):145–56.

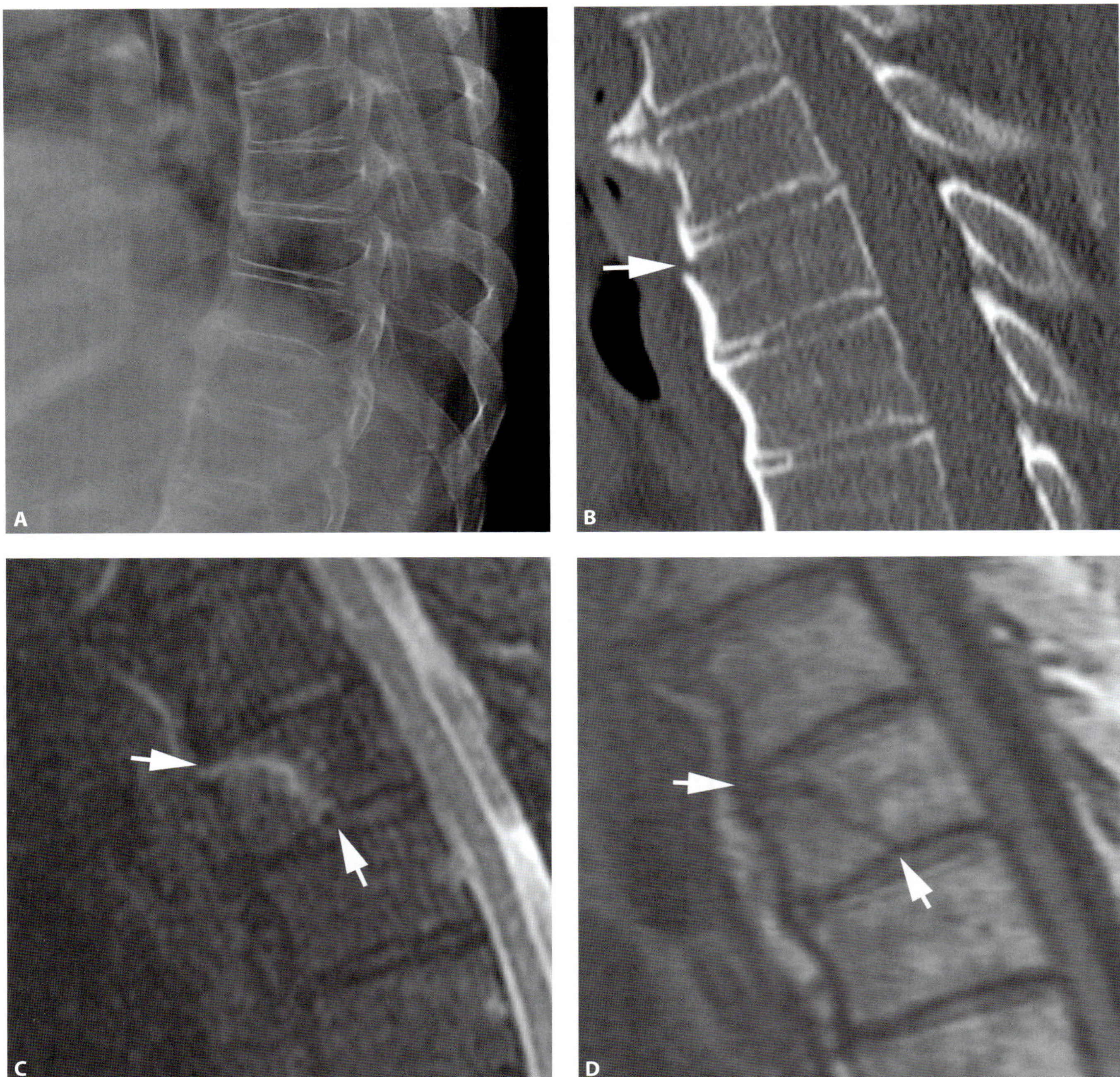

Figure 28.1 A. Lateral radiograph from a 74-year-old man who fell from a ladder demonstrates ankylosis of the thoracic spine consistent with DISH, but does not show evidence of fracture. **B.** Sagittal non-contrast CT image again shows DISH with a subtle break in the ankylosis anteriorly (arrow) indicating an extremely subtle fracture. Sagittal short tau inversion recovery **(C)** and T1 MRI **(D)** images demonstrate a non-displaced hyperextension fracture of the anterior vertebral body (arrows), which is much more conspicuous than on the radiographs and CT.

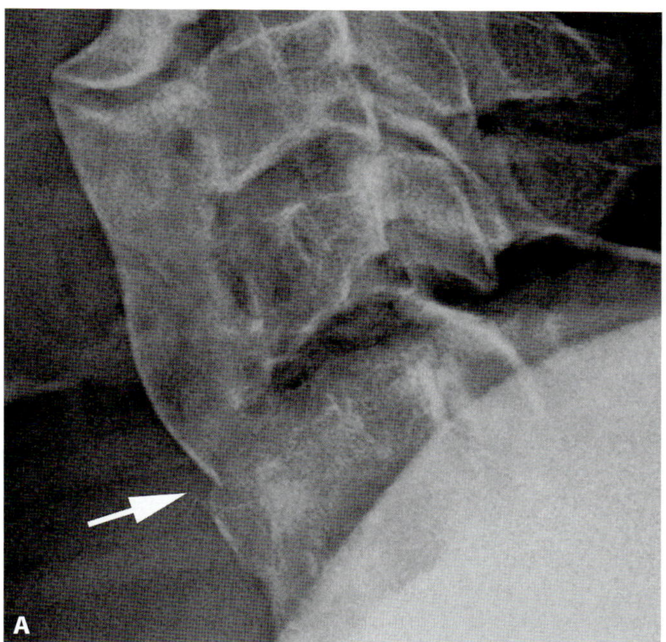

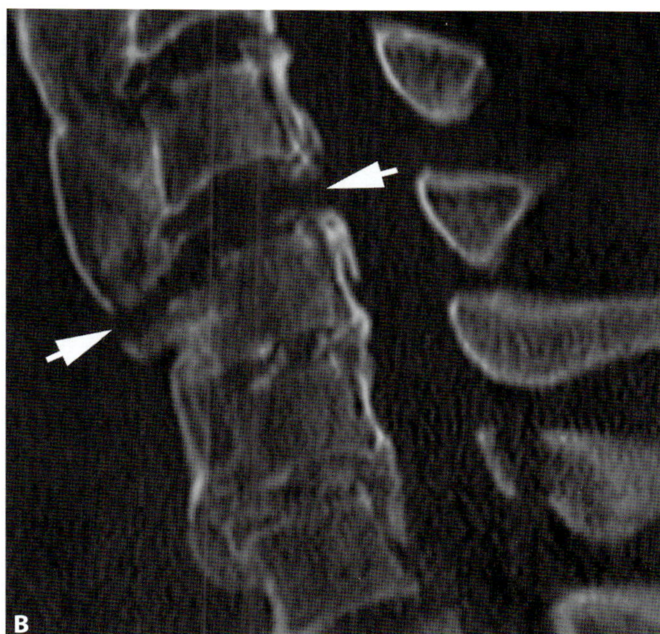

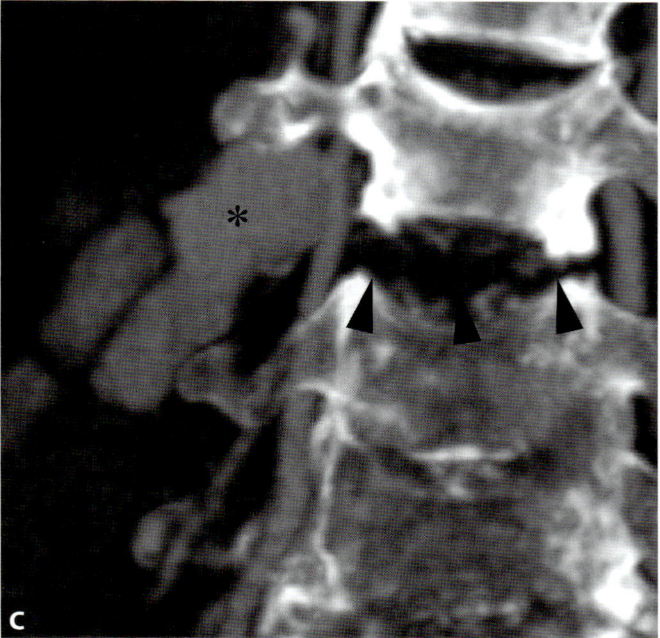

Figure 28.2 A. Lateral cervical spine radiograph from an 83-year-old man who fell from standing demonstrates flowing anterior osteophytosis consistent with DISH with a subtle break in the anterior cortex (arrow), which appears minor and could easily be overlooked. **B.** Sagittal CT image shows the fracture extents through the vertebral body and disk (arrows) and involves at least the anterior and middle columns. **C.** Coronal CT angiogram demonstrates a large pseudoaneurysm of the right vertebral artery (asterisk) adjacent to the fracture (arrowheads).

Spinal dural arteriovenous fistula

Michael J. Modica

Imaging description

Spinal dural arteriovenous fistula (DAVF) is the most common spinal vascular malformation. It is an intradural-extramedullary lesion commonly found in the distal cord or at the conus medullaris and is composed largely of distended intradural draining veins. MRI typically shows a distal spinal cord which is enlarged and edematous due to venous congestion, with corresponding hypointensity on T1- and hyperintensity on T2-weighted images. In the setting of venous hypertensive myelopathy, the cord edema can spare the periphery and the central edema can be "flame shaped" at its superior and inferior margins; these findings typically are best appreciated on T2 series (Figure 29.1). Close inspection will often show multiple abnormal vessel flow voids on the pial surface of the cord.

When a spinal DAVF is suspected on MRI, thoracic spine MRI or MR angiography (MRA) can be considered as the next diagnostic modality for evaluation of the extent of spinal involvement [1]. However, catheter angiography should be considered in all cases as this will confirm the diagnosis and help identify the exact level of the vascular shunt to plan for endovascular embolic therapy [2].

Importance

The clinical findings for DAVF are typically non-specific and the diagnosis requires a high level of suspicion. Duration from onset of symptoms to diagnosis is usually delayed. However, if undiagnosed, DAVF can lead to permanent neurologic dysfunction, including paraplegia and bowel or bladder dysfunction. Endovascular treatment of the fistula with embolic agents provides permanent occlusion and up to 60% of patients experience symptom improvement [2].

Typical clinical scenario

Patients are typically more than 40 years of age and typically present with acute back pain or have a stuttering clinical course that may be accompanied by lower extremity weakness and bowel or bladder dysfunction. Symptoms may be exacerbated by exercise. Patients may have a clinical diagnosis of spinal stenosis or claudication. Often, these patients are sent to MR to evaluate for "cauda equina syndrome" or spinal stenosis. If unexplained edema is encountered in the cord in this setting, further imaging is warranted. Early diagnosis and management is important, as this may prevent further clinical deterioration.

Differential diagnosis

The differential diagnosis of spinal DAVF includes normal cerebrospinal fluid (CSF) pulsations and tortuous redundant cauda equina nerve roots (Figures 29.2 and 29.3). Normal CSF pulsations typically appear on T2-weighted imaging as globular, ill-defined signal loss dorsal to the cord while the cord signal remains normal. This artifact, which is caused by time-of-flight loss effects, is accentuated by faster CSF flow velocity, thinner slices, longer echo time, and imaging perpendicular to the direction of CSF flow [3]. In lumbar spinal stenosis, the cauda equina nerve roots can appear clumped above and below the level of stenosis, resulting in serpentine areas of low signal on sagittal T2-weighted MRI. The signal of the cord usually is normal [4, 5].

Teaching point

Spinal DAVF should be suspected in patients presenting with back pain and lower extremity weakness, bowel or bladder dysfunction. Distal spinal cord edema present on lumbar spine MR may be the only clue to the presence of DAVF. Consider catheter angiography in these cases, as early treatment is associated with improved outcomes.

REFERENCES

1. Farb RI, Kim JK, Willinsky RA, *et al.* Spinal dural arteriovenous fistula localization with a technique of first-pass gadolinium-enhanced MR angiography: initial experience. *Radiology*. 2002; **222**(3):843–50.
2. Patsalides A, Santillan A, Knopman J, *et al.* Endovascular management of spinal dural arteriovenous fistulas. *J Neurointerv Surg*. 2011;**3**(1):80–4.
3. Lisanti C, Carlin C, Banks KP, Wang D. Normal MRI appearance and motion-related phenomena of CSF. *AJR Am J Roentgenol.* 2007;**188** (3):716–25.
4. Duncan AW, Kido DK. Serpentine cauda equina nerve roots. *Radiology.* 1981;**139**(1):109–11.
5. Pau A, Viale ES, Turtas S, Viale GL. Redundant nerve roots of the cauda equina. *Surg Neurol.* 1981;**16**(4):245–50.

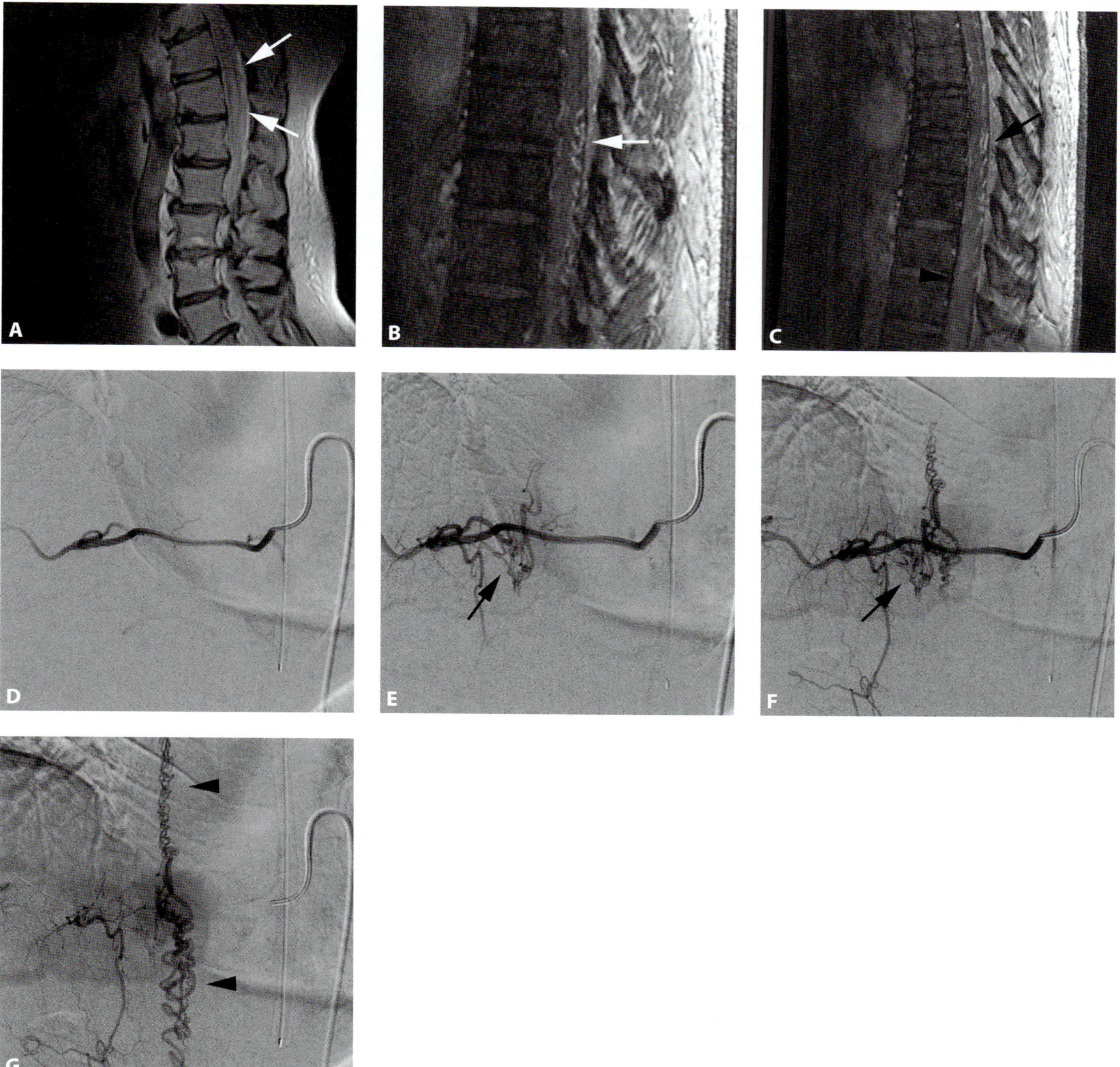

Figure 29.1 **A.** Sagittal T2-weighted MRI of the lumbar spine in a 66-year-old man with rapidly progressive lower extremity weakness and back pain shows an abnormally hyperintense conus medullaris (arrows). Notice the flame-shaped signal abnormality spares the extreme periphery of the cord, consistent with extensive edema. This finding was missed on initial reading and the patient's back pain and weakness was attributed to advanced disk degeneration **B.** Subsequent sagittal T1-weighted fat-saturated contrast-enhanced images of the thoracic spine show enhancing corkscrew-like vessels posterior to the spinal cord (arrow). These represent dilated collateral vessels from venous congestion. **C.** Sagittal T1-weighted fat-saturated contrast-enhanced image of the thoracic spine centered at midline shows scattered enhancing dorsal collateral vessels (arrow). Notice the diffuse enhancement throughout the lower thoracic cord (arrowhead). **D–G.** Catheter angiography images following injection of the right T8 segmental artery show direct communication between the radicular-pial arterial network (arrow) and the dilated dorsal dural vein (arrowheads) consistent with a dural arteriovenous fistula.

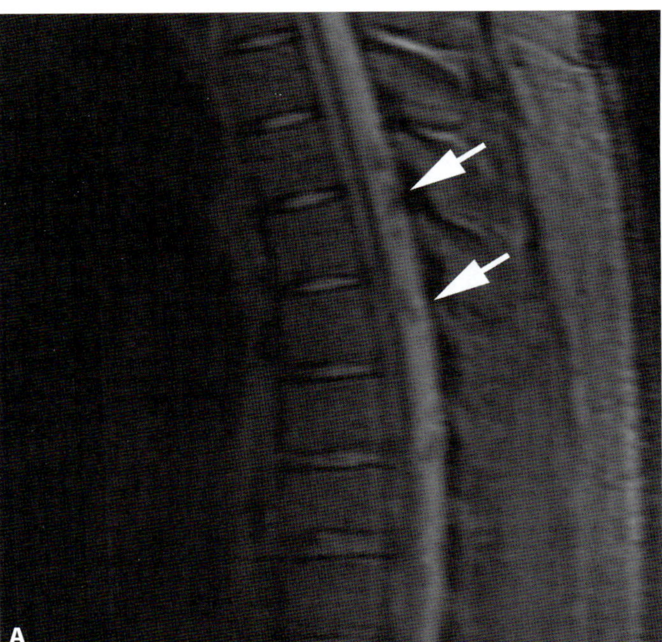

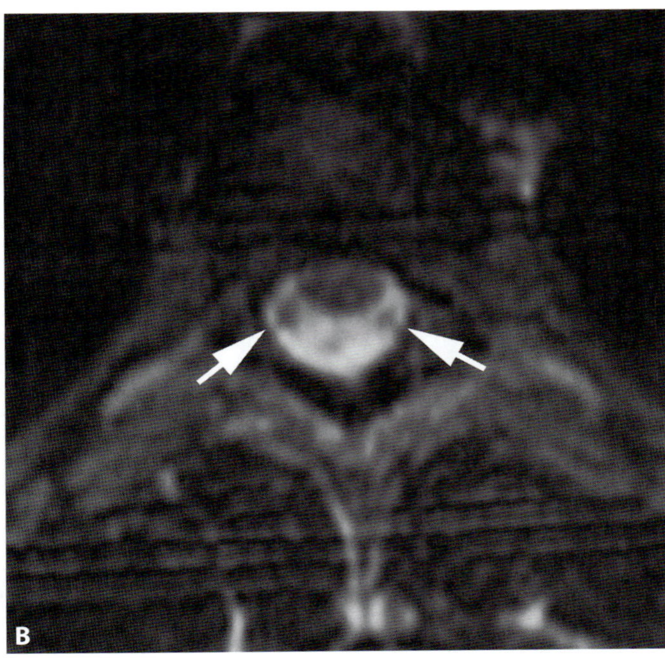

Figure 29.2 **A.** Sagittal short tau inversion recovery (STIR) MRI of the thoracic spine shows irregular hypointense lesions in the dorsal CSF (arrows) in a patient with a congenitally small caliber thoracic cord. **B.** Axial STIR MRI demonstrates similar foci of hypointensity within the CSF, dorsal to the cord (arrows). This is the typical appearance of CSF pulsation artifact.

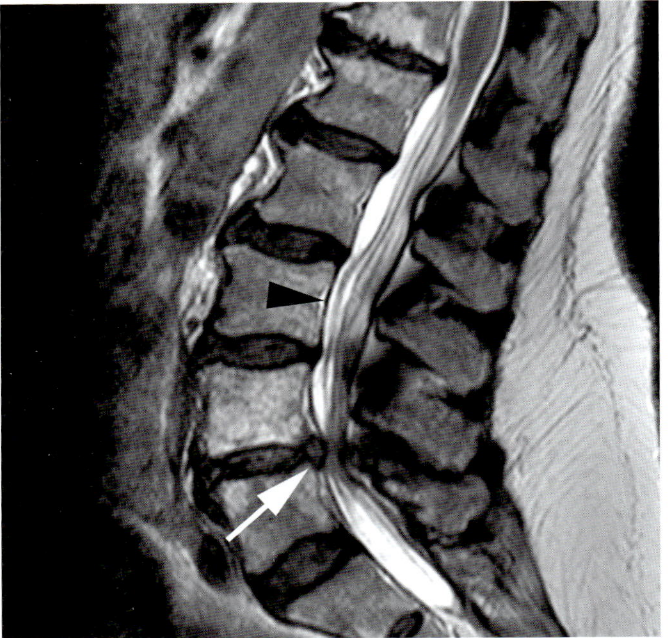

Figure 29.3 Sagittal T2-weighted image of the lumbar spine shows anterior and posterior dural compression from a posterior disk bulge and ligamentum flavum hypertrophy at the lumbosacral intervertebral disk level (arrow). There is a wavy, serpentine appearance of the cauda equina nerve roots above the level of stenosis (arrowhead). This bunched appearance of the nerve roots should not be mistaken for anomalous vasculature.

Pseudopneumomediastinum

Randy K. Lau

Imaging description

Pseudopneumomediastinum can result from Mach bands, normal anatomic structures, or a paratracheal air cyst (tracheal diverticulum).

Mach bands, first described by Ernst Mach in 1865, are optical edge-enhancement phenomena secondary to lateral inhibition in the retina [1]. Mach bands appear as a region of lucency adjacent to structures with convex borders and can be distinguished from true pneumomediastinum as Mach bands do not have an opaque line (Figures 30.1–30.3). For a more detailed description of the Mach effect, see Case 79.

Normal anatomic structures can occasionally mimic pneumomediastinum. As described by Zylak and colleagues, the superior aspect of the major fissure or the anterior junction line may appear as a white line, mimicking pneumomediastinum. This tends to occur with lordotic positioning [2].

Tracheal diverticula are present in 3–4% of the US population [3]. These benign entities are typically located on the right at the level of the thoracic inlet. Communication between the trachea and diverticulum is seen in only 8–35% of cases [3, 4]. The rounded shape and characteristic location will prevent confusion of tracheal diverticulum with pneumomediastinum (Figures 30.4 and 30.5).

Importance

Pneumomediastinum has many causes and can be benign, requiring no further treatment or diagnostic workup. However, pneumomediastinum can also be significant, for example pneumomediastinum associated with tracheobronchial injury. Recognizing causes of pseudopneumomediastinum can help avoid misdiagnosis, unnecessary workup and treatment.

Typical clinical scenario

Pneumomediastinum is often asymptomatic, but can present with chest pain or dyspnea [5]. However, because pneumomediastinum is often asymptomatic, it can be an incidental finding on chest radiographs or CT of the chest ordered for any one of a myriad of reasons.

Differential diagnosis

The primary differential consideration is pneumomediastinum. Potential sources of pneumomediastinum include gas tracking from the tracheobronchial tree, esophagus, lung, pleura, head and neck, retroperitoneum, and peritoneal space [2]. Perhaps the most common cause of mediastinal air in the acute setting is introduction during central venous catheter placement.

Teaching point

Causes of pseudopneumomediastinum include Mach bands, superior aspect of the major fissure, anterior junction line, and paratracheal air cyst/tracheal diverticulum.

REFERENCES

1. Chasen MH. Practical applications of Mach band theory in thoracic analysis. *Radiology*. 2001;**219**(3):596–610.
2. Zylak CM, Standen JR, Barnes GR, Zylak CJ. Pneumomediastinum revisited. *Radiographics*. 2000;**20**(4):1043–57.
3. Buterbaugh JE, Erly WK. Paratracheal air cysts: a common finding on routine CT examinations of the cervical spine and neck that may mimic pneumomediastinum in patients with traumatic injuries. *AJNR Am J Neuroradiol*. 2008;**29**(6):1218–21.
4. Goo JM, Im JG, Ahn JM, *et al.* Right paratracheal air cysts in the thoracic inlet: clinical and radiologic significance. *AJR Am J Roentgenol*. 1999;**173**(1):65–70.
5. Bejvan SM, Godwin JD. Pneumomediastinum: old signs and new signs. *AJR Am J Roentgenol*. 1996;**166**(5):1041–8.

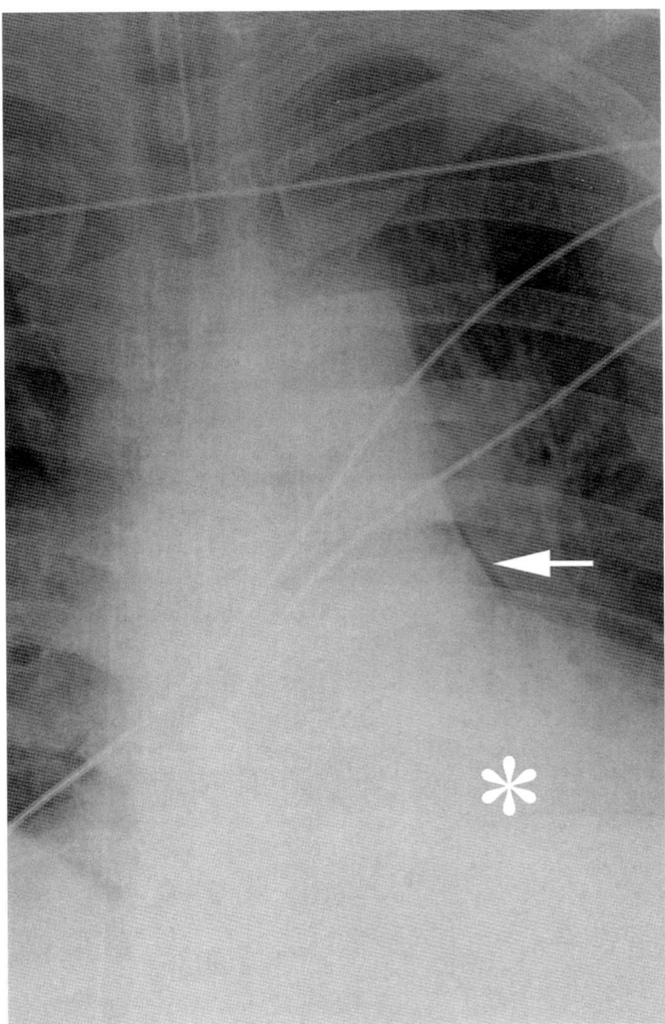

Figure 30.1 Chest radiograph from a 49-year-old man with new onset seizures and hypoxia requiring intubation demonstrating linear lucency (arrow) from a Mach band adjacent to convex margins of the left atrium. Notice absence of a thin white line representing the mediastinal parietal pleura that would be seen in true pneumomediastinum. This patient also had bilateral basilar pulmonary consolidation (asterisk) from aspiration.

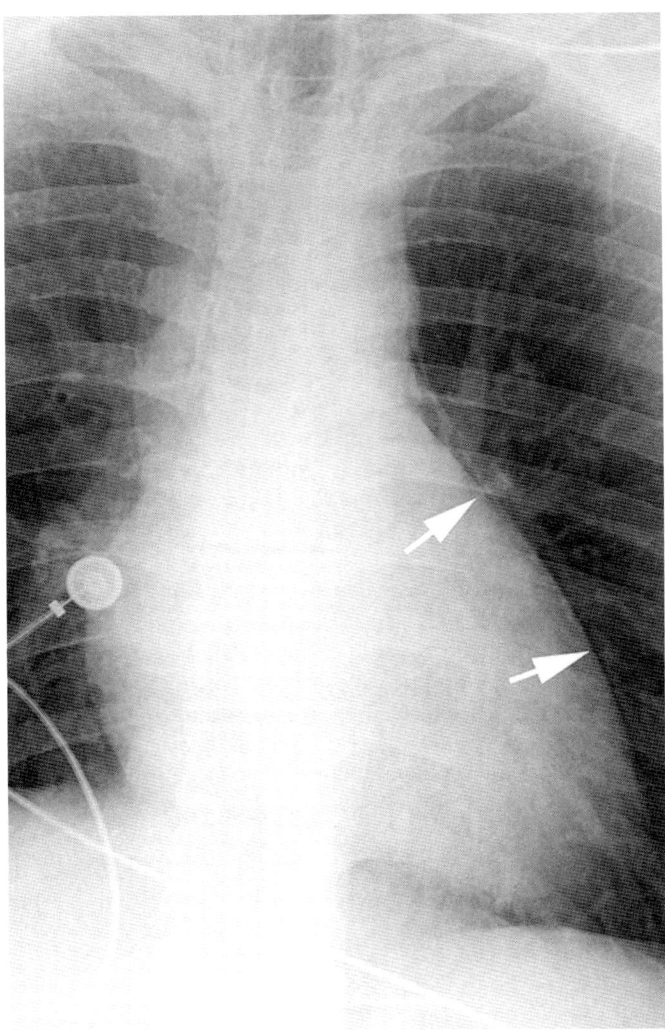

Figure 30.2 Chest radiograph from a 32-year-old man who was assaulted demonstrates linear lucency (arrows) adjacent to the left heart border related to Mach band with absence of thin white line of the mediastinal parietal pleura seen in true pneumomediastinum.

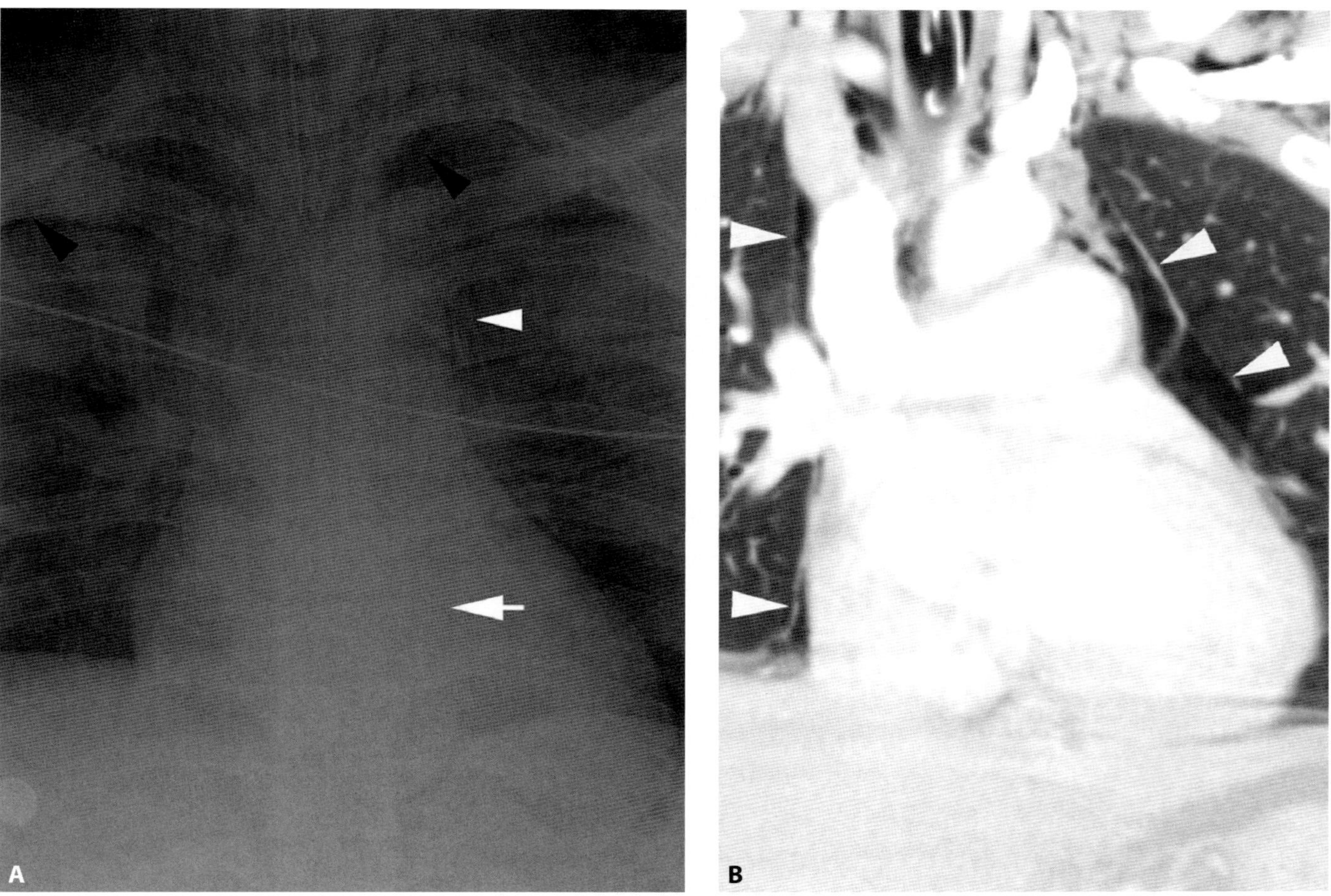

Figure 30.3 A. Chest radiograph from an intubated 18-year-old man in a high-speed motor vehicle collision demonstrates bilateral pulmonary contusions, bilateral apical pneumothoraces (black arrowheads), pneumomediastinum which extends into the neck, and significant subcutaneous emphysema. The left mediastinal parietal pleura is seen as a thin white line (white arrowhead) outlined by mediastinal air and adjacent lung parenchyma. Also note the linear vertical lucency (white arrow) in the left paraspinal region and along the descending thoracic aorta. **B.** Corresponding coronal contrast CT demonstrating subcutaneous emphysema and pneumomediastinum with the mediastinal parietal pleura (arrowheads) clearly demarcated by the pneumomediastinum. Pneumothoraces had resolved following insertion of chest tubes (not shown).

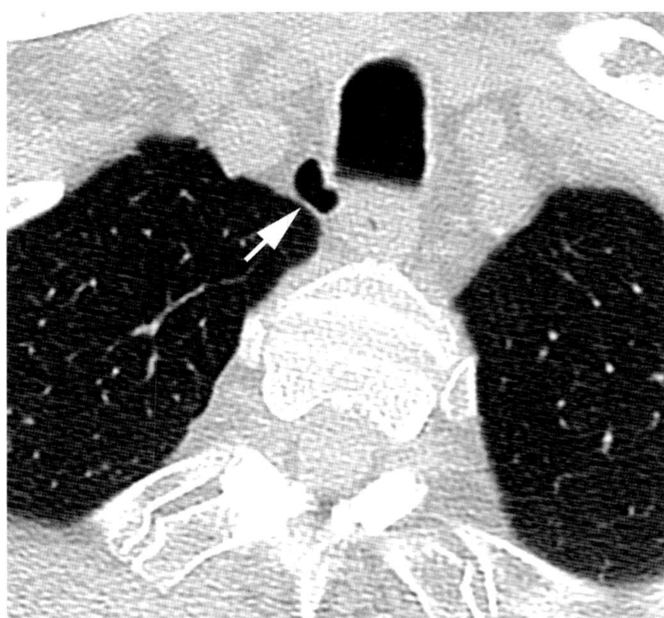

Figure 30.4 Lung windows of a non-contrast CT of the cervical spine in a 79-year-old man who fell, demonstrating rounded lucency (arrow) along the right posterior aspect of the trachea in a location typical for a tracheal diverticulum.

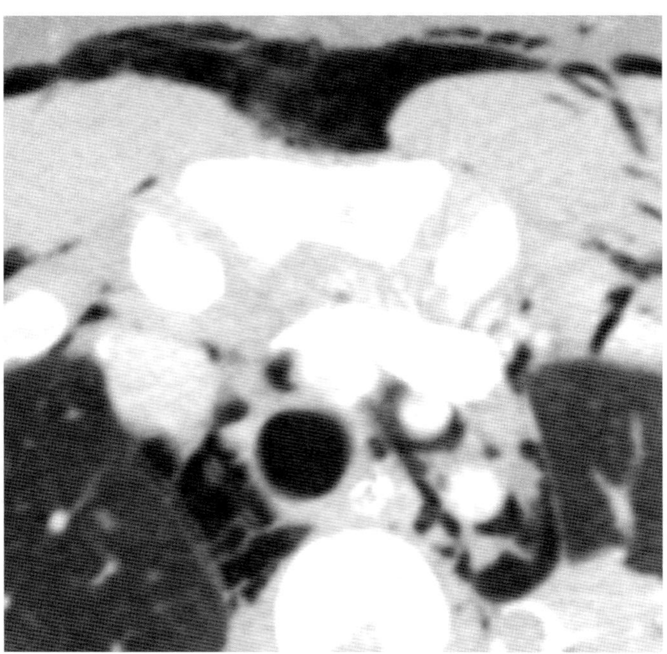

Figure 30.5 Axial contrast CT from an 18-year-old man involved in a high-speed motor vehicle collision with subcutaneous emphysema and pneumomediastinum. Notice the sharp, angular margins of pneumomediastinum, which also typically extends into the neck and chest wall. This is in contrast to the rounded margins of a tracheal diverticulum.

Traumatic pneumomediastinum without aerodigestive injury

Randy K. Lau

Imaging description

Pneumomediastinum occurs in up to 10% of patients following blunt trauma. However, only the minority of patients have an aerodigestive injury. In a recent study of 136 patients with pneumomediastinum identified by CT following blunt thoracic trauma, 27 patients had CT findings suspicious for aerodigestive tract injury, and of these only 10 patients required operative intervention [1]. No patient had a significant injury that was not suggested by the CT. In this same series, only 20 of the 136 cases of pneumomediastinum were identified on radiographs.

Suspicious CT findings of aerodigestive injury when pneumomediastinum is identified in blunt trauma patients include airway irregularity, disruption of the cartilage or tracheal wall (Figure 31.1), focal thickening or indistinctness of the trachea or main bronchi, laryngeal disruption, and concurrent pneumoperitoneum on CT of the abdomen. Massive pneumomediastinum despite adequate tube drainage of pneumothoraces is also considered a suspicious finding for an aerodigestive tract injury. If all these findings are absent, further investigation is likely unwarranted [1].

The most common cause of pneumomediastinum in blunt trauma is thought to be the Macklin effect. The Macklin effect occurs when elevated alveolar pressure leads to alveolar rupture, dissection of air along peribronchovascular interstitial sheaths, interlobular septa, and visceral pleura. From here, gas may spread to the mediastinum. On CT, the Macklin effect can be identified as pulmonary interstitial emphysema (PIE) in these locations (Figure 31.2). This finding should be interpreted with some caution, as the finding of PIE does not exclude aerodigestive injury [2].

CT should be considered for the evaluation of stable patients with transmediastinal penetrating trauma [3–5]. Pneumomediastinum in the setting of penetrating trauma should be treated with suspicion, especially in the setting of transmediastinal injury, where pharyngoesophagography with water-soluble contrast media and/or bronchoscopy should be considered [4]. Gas from penetrating trauma to the pharynx and hypopharynx may also extend to the mediastinum.

Pneumomediastinum detected on chest radiographs has been termed "overt pneumomediastinum", while pneumomediastinum detected only on CT scans is termed "occult pneumomediastinum" [6].

Importance

Patients with pneumomediastinum without aerodigestive injury tend to be more severely injured than those patients with no pneumomediastinum on CT [2, 6]. However, in the absence of specific signs of aerodigestive injury, excessive investigation is probably unwarranted in both adults and children following blunt trauma [1, 2, 6, 7]. Conversely, patients with pneumomediastinum from penetrating injuries require a thorough investigation for injury to the airways or esophagus. This usually includes bronchoscopy and esophagography.

Typical clinical scenario

The classic presentation is in patients with blunt thoracic trauma.

Differential diagnosis

Pseudopneumomediastinum secondary to Mach band, normal anatomic structures, and tracheal diverticula. These are fully described in the preceding chapter.

Pneumomediastinum may also be due to gas tracking from other contiguous compartments, such as the retroperitoneum, chest wall, face and the soft tissues of the neck. Gas may also be introduced directly into the mediastinum during penetrating injury or central venous line placement. Pulmonary interstitial emphysema may also be the consequence of alveolar rupture in the setting of positive pressure ventilation.

Teaching point

Most cases of pneumomediastinum following blunt trauma are not due to aerodigestive injury, and do not require an extensive workup, or intervention unless localizing signs of airway or esophageal injury are present.

REFERENCES

1. Dissanaike S, Shalhub S, Jurkovich GJ. The evaluation of pneumomediastinum in blunt trauma patients. *J Trauma*. 2008; **65**(6):1340–5.
2. Wintermark M, Schnyder P. The Macklin effect: a frequent etiology for pneumomediastinum in severe blunt chest trauma. *Chest*. 2001; **120**(2):543–7.
3. Stassen NA, Lukan JK, Spain DA, *et al.* Reevaluation of diagnostic procedures for transmediastinal gunshot wounds. *J Trauma*. 2002; **53**(4):635–8; discussion 638.
4. Hanpeter DE, Demetriades D, Asensio JA, *et al.* Helical computed tomographic scan in the evaluation of mediastinal gunshot wounds. *J Trauma*. 2000;**49**(4):689–94; discussion 694–5.
5. Ibirogba S, Nicol AJ, Navsaria PH. Screening helical computed tomographic scanning in haemodynamic stable patients with transmediastinal gunshot wounds. *Injury*. 2007;**38**(1):48–52.
6. Rezende-Neto JB, Hoffmann J, Al Mahroos M, *et al.* Occult pneumomediastinum in blunt chest trauma: clinical significance. *Injury*. 2010;**41**(1):40–3.
7. Neal MD, Sippey M, Gaines BA, Hackam DJ. Presence of pneumomediastinum after blunt trauma in children: what does it really mean? *J Pediatr Surg*. 2009;**44**(7):1322–7.

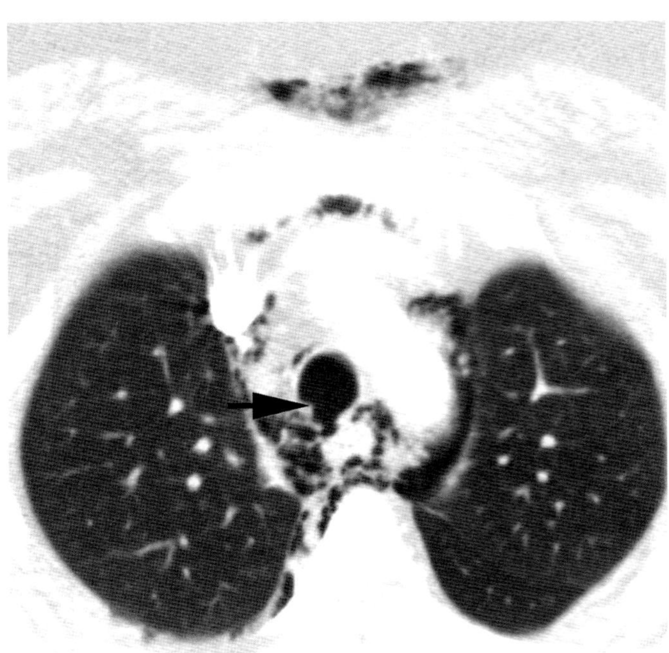

Figure 31.1 CT from a 45-year-old woman following trauma. CT demonstrates a large posterior tracheal defect (arrow), consistent with tracheal injury. Note the extensive pneumomediastinum.

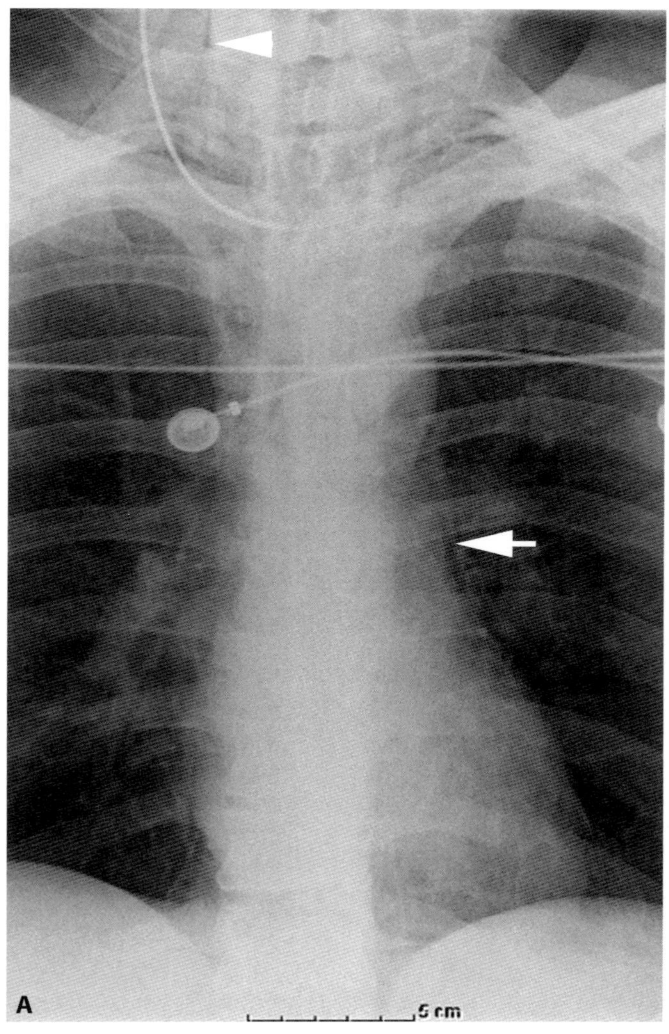

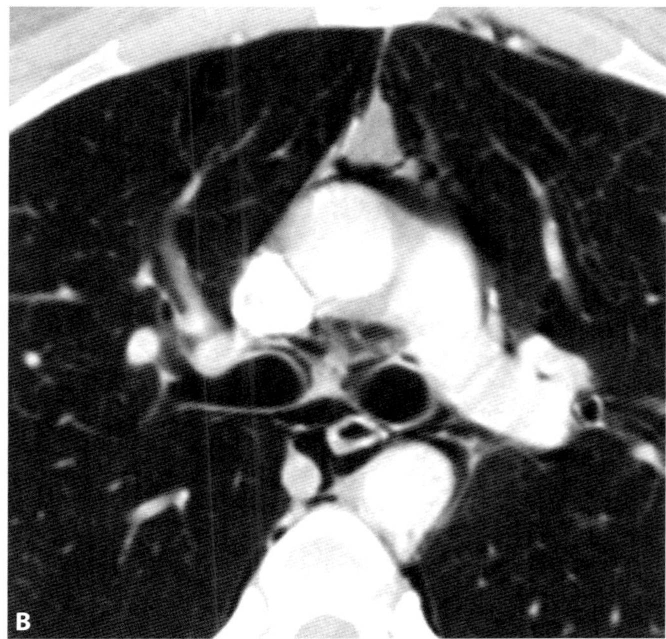

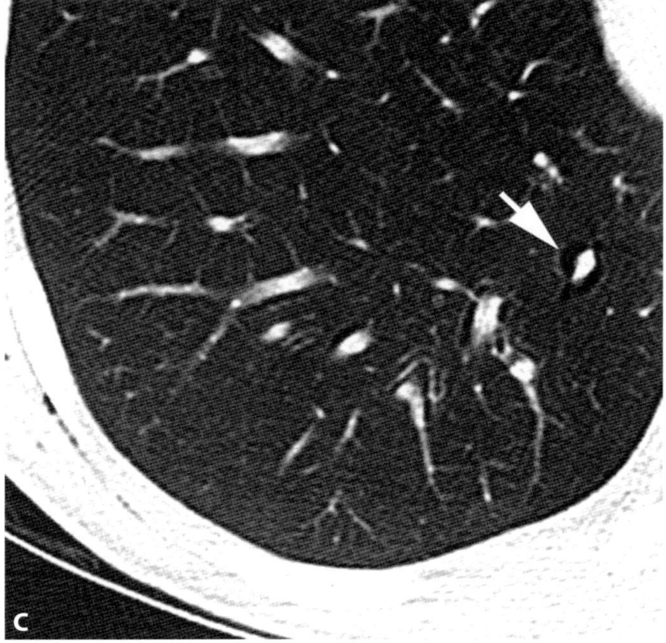

Figure 31.2 **A.** Chest radiograph from a 20-year-old man following a bicycle collision demonstrates a pneumomediastinum (arrow) which extends into the neck tissues (arrowhead). **B.** Contrast-enhanced chest CT shows extensive pneumomediastinum with gas surrounding the proximal mainstem bronchi and esophagus. **C.** Magnified CT image of the right lower lobe shows air tracking along the peri-bronchovascular sheaths (arrow). This illustrates the Macklin effect. The patient had normal esophagography and bronchoscopy, and was discharged without complication.

CASE

32 Pseudopneumothorax

Randy K. Lau

Imaging description

Various artifacts can be mistaken for pneumothoraces. These include skin folds, giant bullous emphysema, calcified pleural plaque, folds of blankets or clothing, lateral edges of breast tissue, and the medial border of the scapula [1–4].

Skin folds mimicking pneumothoraces can be distinguished from true pneumothoraces by looking for lung markings beyond the skin fold and discriminating between the abrupt interface or edge caused by skin folds and the thin white visceral pleural line of a pneumothorax (Figure 32.1). In addition, interfaces or edges related to skin folds may extend beyond the thoracic cavity.

Giant bullous emphysema mimicking tension pneumothorax can be distinguished from true tension pneumothorax by the lack of hemodynamic instability in giant bullous emphysema, lack of re-expansion after thoracostomy tube placement, and septations and vessels within bullous emphysema on CT (Figure 32.2).

Calcified pleural plaques seen tangentially can mimic the visceral pleural line of a pneumothorax. This pitfall can be recognized by identifying a white line that is thicker than that typical for the visceral pleural line of a pneumothorax. In addition, the white line created by a calcified pleural plaque will not follow the expected contour of the lung. The presence of other calcified pleural plaques is another clue that this pitfall may be present.

Air located between the folds of clothing or blankets is another potential pitfall mimicking a pneumothorax. This artifact can be distinguished from a true pneumothorax when the white line from the folds extends beyond the thoracic cavity (Figure 32.3). In addition, lung markings may be identified lateral to the white lines of folded clothing or blankets.

Air along the edges of backboards may decrease the attenuation of the costophrenic angle and create a false impression of a "deep sulcus sign", a sign of a pneumothorax on a supine radiograph (Figure 32.4).

Lateral edges of breast tissue extending from the axilla to the breast can also simulate the deep sulcus sign on a supine radiograph.

Mediastinal or chest wall gas may extend between the parietal pleura and chest wall, simulating pneumothorax, especially adjacent to the apex of the thoracic cavities. Sometimes this can be distinguished by the presence of fibers crossing the potential space (Figure 32.5).

On CT, beam-hardening and streak artifact related to concentrated intravenous contrast within the brachiocephalic vein can mimic a small pneumothorax, especially when a low kilovoltage technique is used. This same pitfall may also be seen when patients are scanned with their arms placed at their sides. Pseudopneumothorax from streak artifact and beam hardening can be distinguished from pneumothorax by the extension of linear hypodensity beyond anatomic boundaries and the presence of adjacent highly attenuating substances (Figure 32.6).

In contrast to pseudopneumothorax, the appearance of pneumothorax on chest radiography is influenced by gravity, lung recoil, and anatomy of the pleural recesses. On a posterior-anterior chest radiograph, visualization of the visceral pleural line is a sine qua non of a pneumothorax. The visceral pleural line will appear as a thin white line with air on either side of it. If there is parenchymal disease or a pleural effusion, a pneumothorax will appear as an interface or edge rather than a thin white line [5].

Importance

Placement of a chest tube is a low-risk procedure, but is not risk free. Complications of tube thoracostomy include: nerve injuries, cardiac and vascular injuries, esophageal injuries, broncho-pleural fistula, lung herniation through the site of thoracostomy, chylothorax, empyema, wound infections, and cardiac dysrhythmias [6].

Typical clinical scenario

Pneumothoraces may be spontaneous or traumatic (including iatrogenic) [7, 8]. Pneumothorax is seen in 15–40% of all patients with blunt chest trauma [9, 10].

Spontaneous pneumothoraces can be further classified as primary or secondary. Primary spontaneous pneumothorax is one that occurs in patients without clinically apparent lung disease, while secondary spontaneous pneumothorax occurs in patients with underlying lung disease [8].

With the rising use of CT in acute chest trauma, the entity of occult pneumothorax has been increasingly identified. Occult pneumothorax is a pneumothorax that was not suspected clinically nor was evident on plain radiographs but was identified using more sensitive techniques, usually CT [11]. Due to the higher sensitivity of sonography compared to supine radiography for the detection of traumatic pneumothorax, occult pneumothorax may also be diagnosed with sonography [12]. Occult pneumothoraces are present in about 5% of injured patients [13]. Limited evidence suggests that small to moderate-sized occult pneumothoraces in patients not undergoing mechanical ventilation do not require treatment. However, at present, the data do not allow us to determine which patients with occult pneumothoraces who will undergo positive pressure ventilation require tube thoracostomy [13].

Differential diagnosis

The differential diagnosis lies between a true pneumothorax, artifact, and underlying lung or pleural pathology simulating a pneumothorax.

Teaching point

Many causes of pseudopneumothorax exist, and should be considered prior to tube thoracostomy.

Pneumothoraces that can be identified by CT or sonography but not by radiography are termed occult pneumothoraces. Current evidence suggests that small and moderate-sized pneumothoraces in trauma patients not undergoing mechanical ventilation do not require tube thoracostomy in most cases.

REFERENCES

1. El-Gendy KA, Atkin GK. Calcified pleural plaque mimicking a traumatic pneumothorax. *Emerg Med J*. 2009;**26**(12):914.
2. Waseem M, Jones J, Brutus S, *et al*. Giant bulla mimicking pneumothorax. *J Emerg Med*. 2005;**29**(2):155–8.
3. Silva FR. Shirt fold mimicking pneumothorax on chest radiograph: accurate diagnosis by ultrasound. *Intern Emerg Med*. 2007;**2**(3):236–8.
4. Lee CH, Chan WP. Skin fold mimicking pneumothorax. *Intern Med*. 2011;**50**(16):1775.
5. Tocino I. Pneumothorax in the supine patient: radiographic anatomy. *Radiographics*. 1985;**5**:557–86.
6. Kesieme EB, Dongo A, Ezemba N, *et al*. Tube thoracostomy: complications and its management. *Pulm Med*. 2012;**2012**:256878.
7. Haynes D, Baumann MH. Pleural controversy: aetiology of pneumothorax. *Respirology*. 2011;**16**(4):604–10.
8. Baumann MH. Pneumothorax. *Semin Respir Crit Care Med*. 2001;**22**(6):647–56.
9. Mayberry JC. Imaging in thoracic trauma: the trauma surgeon's perspective. *J Thorac Imaging*. 2000;**15**(2):76–86.
10. Miller LA. Chest wall, lung, and pleural space trauma. *Radiol Clin North Am*. 2006;**44**(2):213–24, viii.
11. Omar HR, Abdelmalak H, Mangar D, *et al*. *J Trauma Manag Outcomes*. 2010;**4**:12.
12. Rowan KR, Kirkpatrick AW, Liu D, *et al*. Traumatic pneumothorax detection with thoracic US: correlation with chest radiography and CT – initial experience. *Radiology*. 2002;**225**(1):210–14.
13. Ball CG, Kirkpatrick AW, Feliciano DV. The occult pneumothorax: what have we learned? *Can J Surg*. 2009;**52**(5):E173–9.

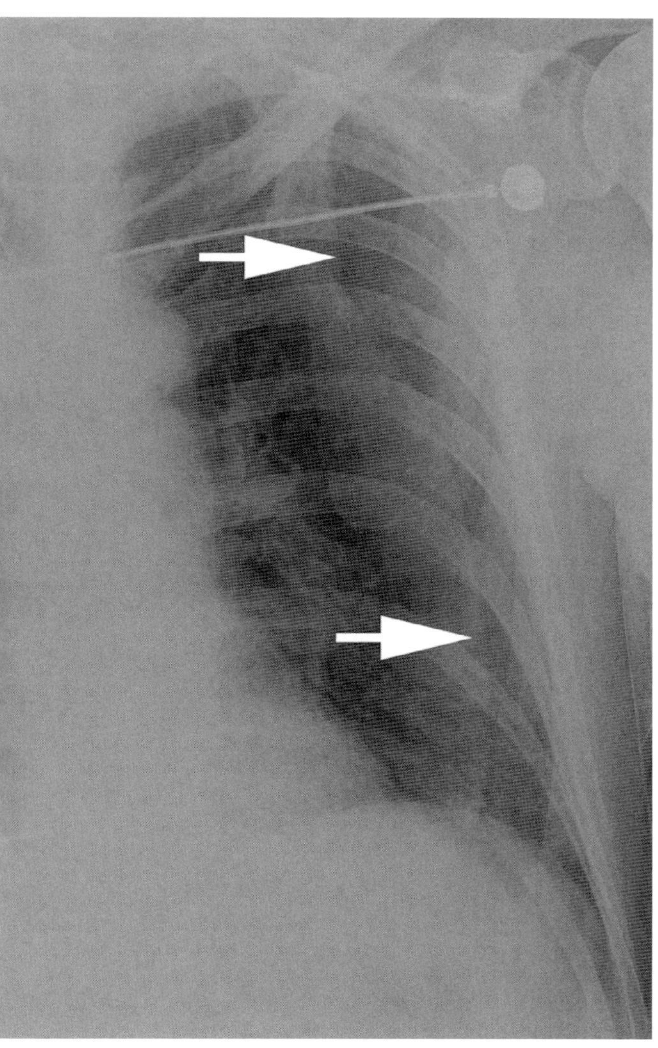

Figure 32.1 Chest radiograph from an 89-year-old man who fell into a ditch demonstrates a skin fold on the left mimicking a pneumothorax. The key to diagnosis is the presence of an interface or edge (white arrows) rather than the thin white pleural line of a pneumothorax. In addition, lung markings are present lateral to the skin fold.

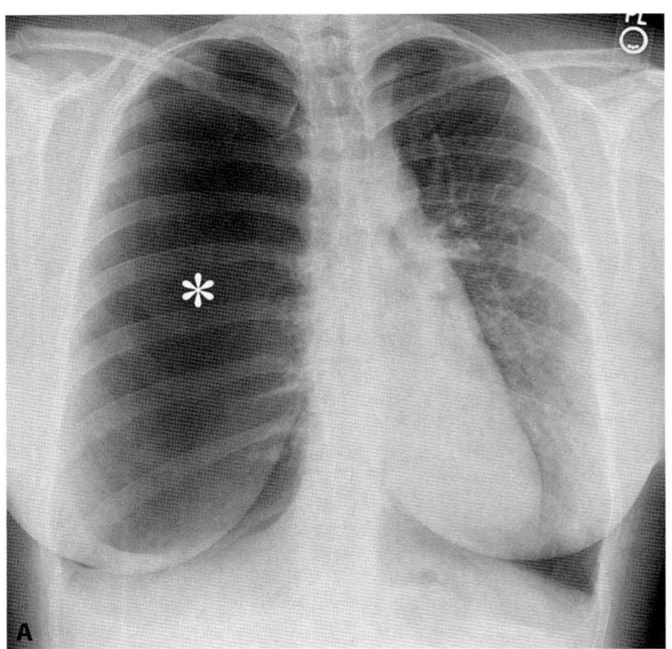

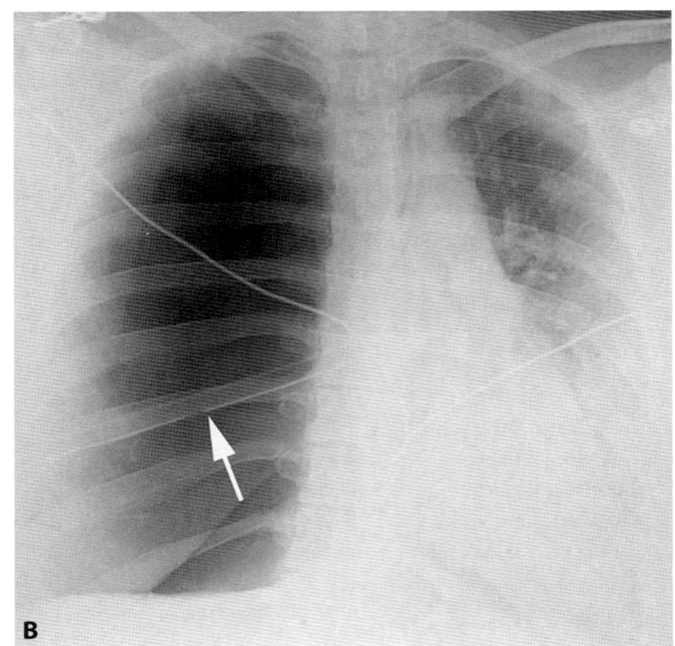

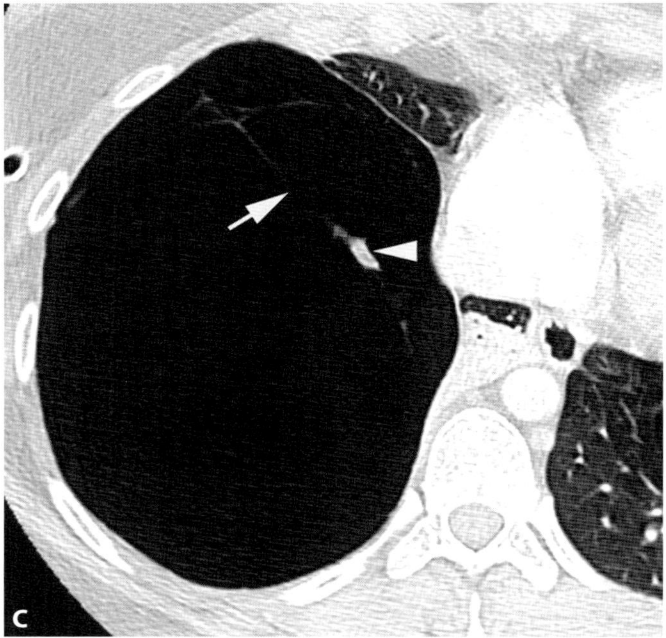

Figure 32.2 A. Chest radiograph in a 23-year-old woman who presented with shortness of breath and chest tightness after an airplane flight. There is an apparent right peumothorax with increased lucency (white asterisk) and slight mediastinal shift to the left. This was interpreted as a tension pneumothorax. **B.** Follow-up chest radiograph after placement of chest tube (white arrow) demonstrates no change in lucency of the right hemithorax or mediastinal deviation and no re-expansion of the right lung. **C.** Axial contrast CT demonstrates thin septations (white arrow) and a pulmonary vessel (white arrowhead). This is an example of giant bullous emphysema mimicking a right tension pneumothorax.

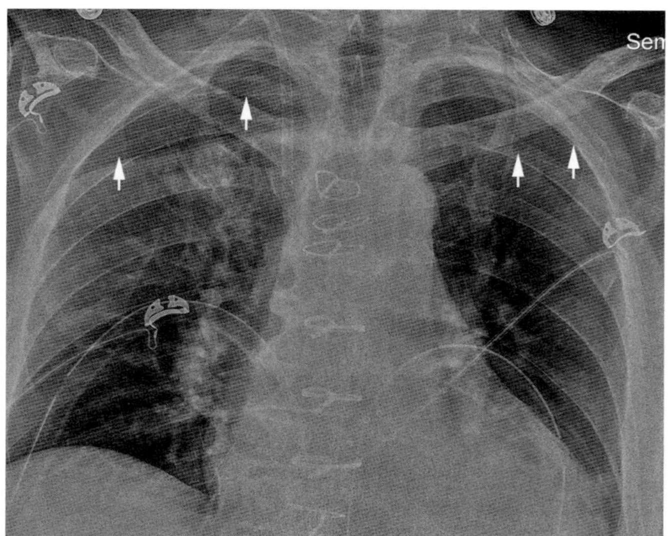

Figure 32.3 Semi-upright chest radiograph from a 73-year-old man following recent aortic valve replacement. There are many thin white lines (arrows) that mimic the thin white pleural lines of a pneumothorax. However, these thin white lines extend beyond the thoracic cavity and are artifacts related to air located between folds of a blanket.

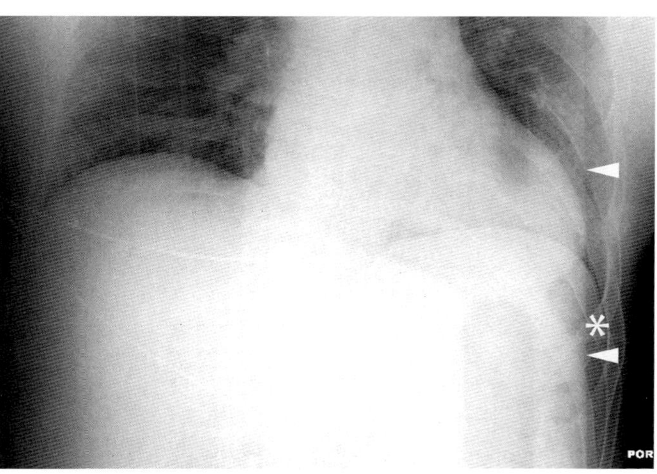

Figure 32.4 Chest radiograph from a 48-year-old man demonstrates an apparent left deep sulcus sign (white asterisk). Notice the edge of the backboard extending from the left upper quadrant and projecting over the left thoracic cavity (white arrowheads).

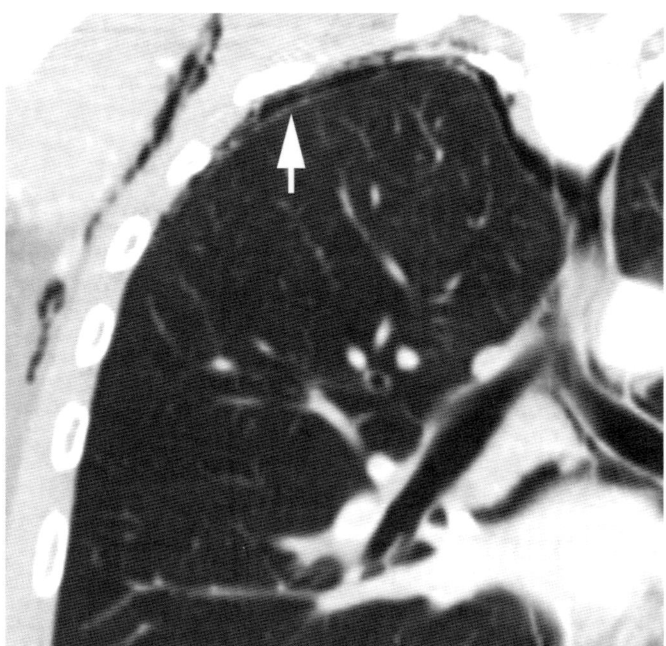

Figure 32.5 Coronal reformation of a CT of the chest in a man with a tracheal injury and pneumomediastinum from a bicycle collision. Note how the extrapleural air tracks over the apex of the right lung (arrow), within the extrapleural space, and the fibers between the parietal pleura and the upper chest wall.

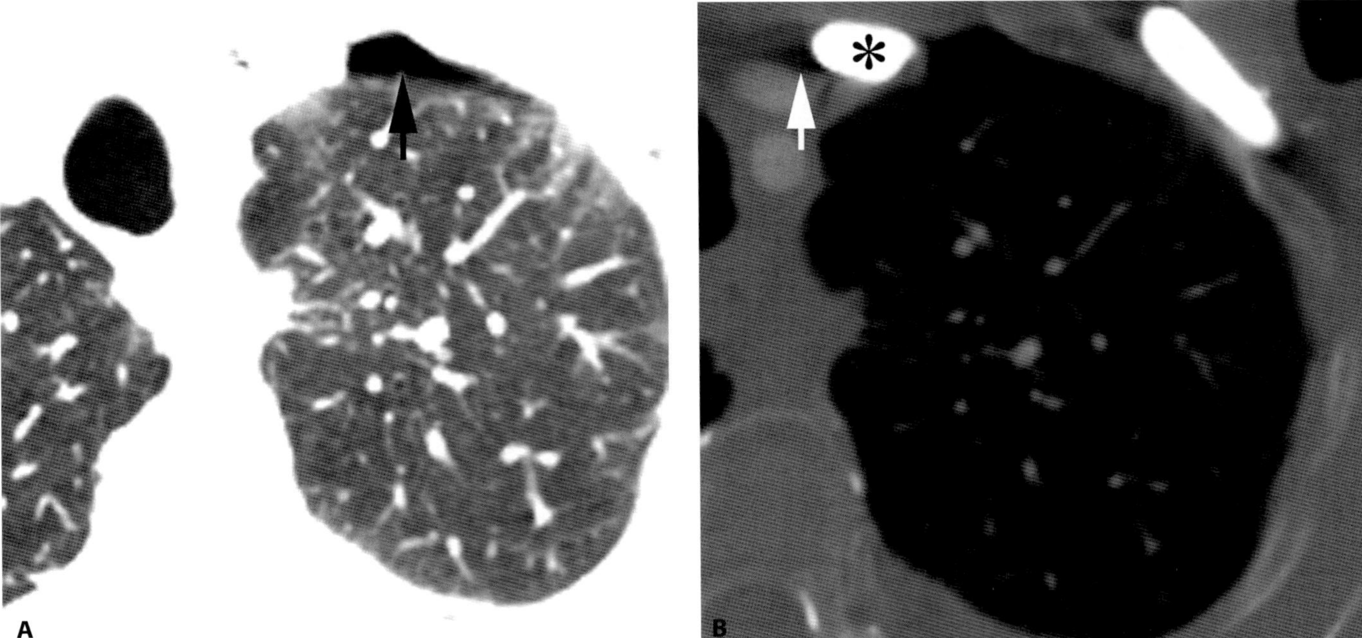

Figure 32.6 **A.** Lung windows of an axial contrast CT from an 80-year-old woman with chest pain from an intramural aortic hematoma demonstrating a left apical pseudopneumothorax (arrow). **B.** Axial contrast CT at the same level in bone windows demonstrating concentrated, dense contrast within the left brachiocephalic vein (black asterisk), which is the source of the beam hardening and streak artifact simulating the appearance of a pneumothorax. Note that the artifact causing the pseudopneumothorax extends beyond the lung (white arrow).

Subcutaneous emphysema and mimickers

Randy K. Lau

Imaging description

Subcutaneous emphysema on chest radiographs appears as radiolucent striations that outline muscle fibers or as irregular, ill-defined lucencies.

Importance

Subcutaneous emphysema itself does not require treatment. However, subcutaneous emphysema may be the main or initial presenting sign of serious underlying pathology that does require treatment (Figure 33.1).

Subcutaneous emphysema can also spread along fascial planes and extend to other parts of the body including the head, neck, extremities, and abdomen. Hence, subcutaneous emphysema in one body region may occur from injury to another body region that might not be initially imaged.

Typical clinical scenario

Subcutaneous emphysema is a common radiographic finding in emergency patients. Causes of subcutaneous emphysema of the chest wall include chest wall infection, blunt trauma with injury to the respiratory or gastrointestinal tract, and penetrating injuries [1]. In one series by Marti de Gracia and colleagues, the most common etiology was trauma [2]. Chest wall subcutaneous emphysema is commonly associated with pneumomediastinum and pneumothorax. Thus, signs of pneumomediastinum and pneumothorax should be sought if there is subcutaneous emphysema. There are also iatrogenic causes of subcutaneous emphysema, such as surgery, and placement of tubes and intravenous catheters. A focused clinical history is necessary to determine the iatrogenic cause of subcutaneous emphysema.

Differential diagnosis

Gas trapped in skin folds can be mistaken for subcutaneous emphysema. Skin folds are more likely to occur in patients who have loose skin and soft tissue. These patients may be elderly, have lost weight, or have deep lacerations. Skin folds can occur anywhere, but a typical location is in the axilla and along the lateral chest walls (Figures 33.2–33.4).

Other mimics of subcutaneous emphysema include gas interspersed between folds of clothes, blankets, sheaths, and dressings [2].

Superimposition of external structures such as long hair may also mimic subcutaneous emphysema.

Teaching point

Subcutaneous emphysema may be the main or initial presenting sign of serious underlying pathology requiring treatment.

Subcutaneous emphysema in one region of the body may be the result of injury to a distant body region.

Mimickers of subcutaneous emphysema include skin folds, clothes, or blankets; sheaths, dressings, or even long hair.

REFERENCES

1. Ho ML, Gutierrez FR. Chest radiography in thoracic polytrauma. *AJR Am J Roentgenol.* 2009;**192**(3):599–612.

2. Marti de Gracia M, Gutierrez FG, Martinez M, Duenas VP. Subcutaneous emphysema: diagnostic clue in the emergency room. *Emerg Radiol.* 2009;**16**(5):343–8.

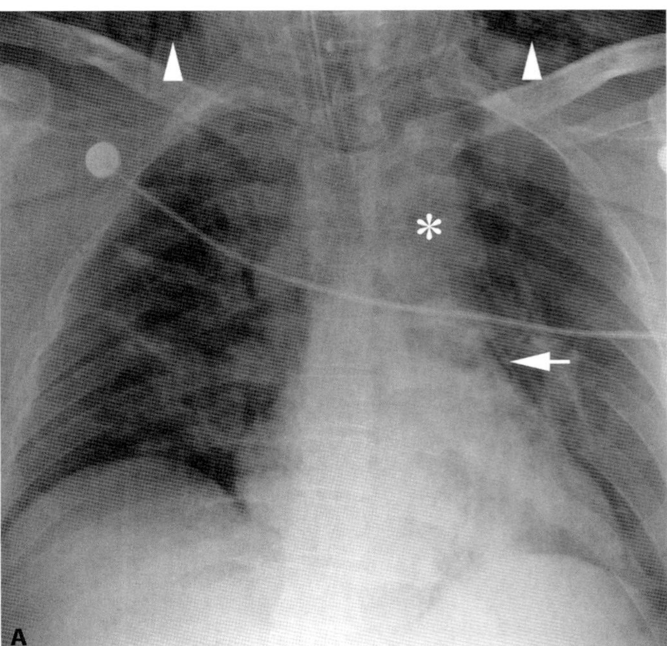

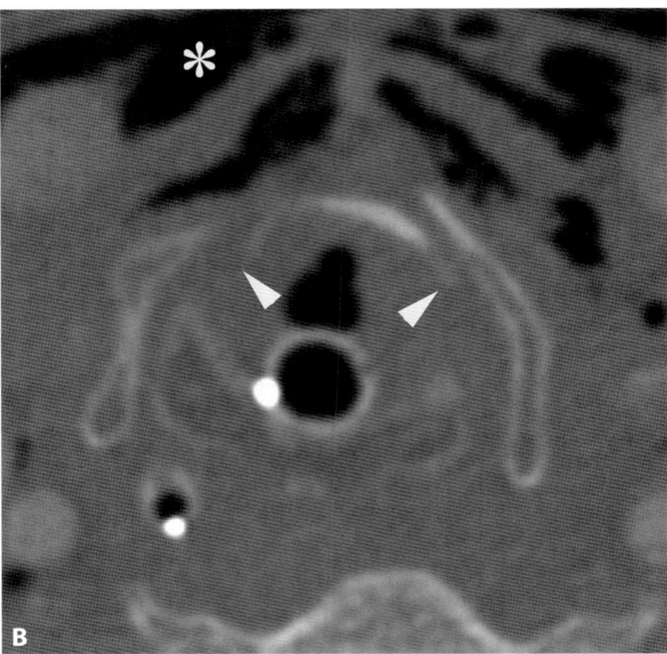

Figure 33.1 A. Chest radiograph from an intubated 43-year-old man involved in a high-speed motor vehicle collision who has extensive subcutaneous emphysema (arrowheads), pneumomediastinum (arrow), and abnormal mediastinum with indistinct aortic arch (asterisk) suggestive of blunt traumatic aortic injury. **B.** Bone window axial image from CT angiography demonstrates a fractured thyroid cartilage (arrowheads). The subcutaneous emphysema (asterisk) was caused by a laryngeal injury in this patient. No aortic injury was identified.

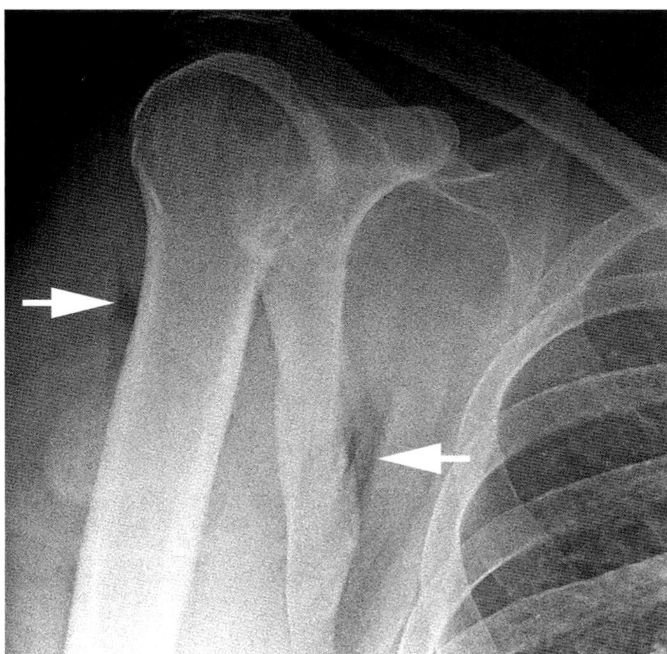

Figure 33.2 Right shoulder anterior-posterior radiograph from a 37-year-old man who suffered a fall while he was being seen in the emergency room for a heroin overdose. There are lucencies (arrows) within the right axilla and lateral to the right humeral diaphysis from skin folds that mimic subcutaneous emphysema. The patient had no soft tissue injuries.

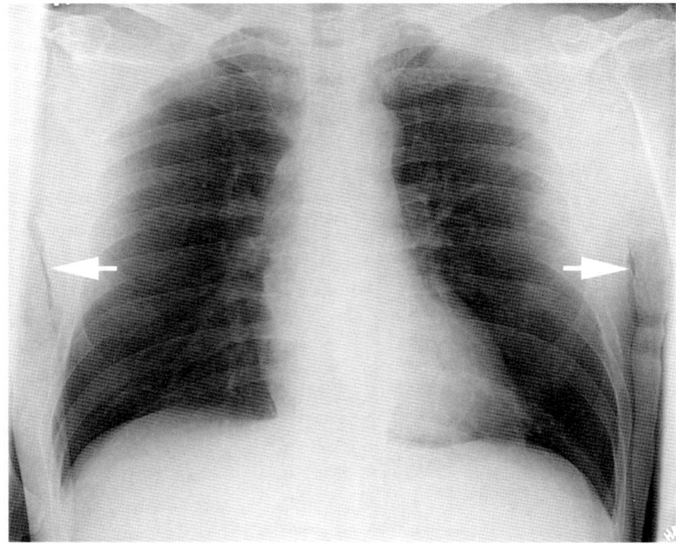

Figure 33.3 Chest radiograph from a 52-year-old man who was assaulted demonstrates linear lucencies (arrows) along the left hemithorax and in the right axilla secondary to skin folds that mimic subcutaneous emphysema. There are no associated osseous injuries.

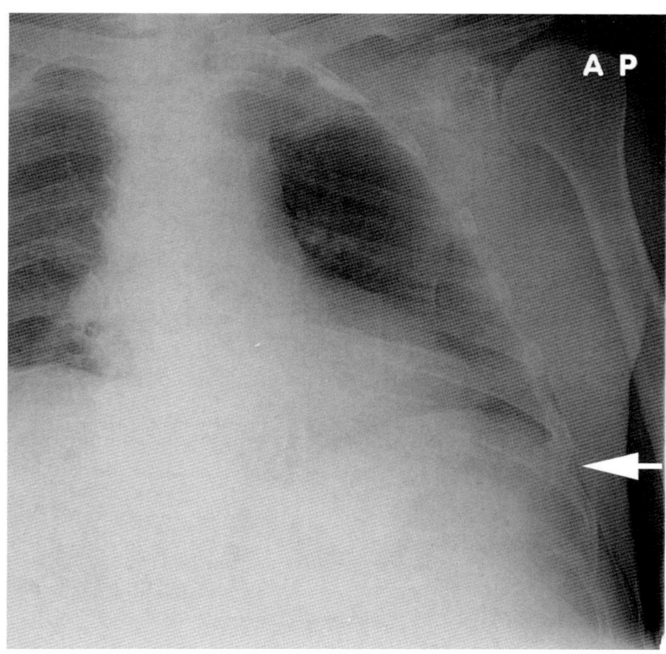

Figure 33.4 Chest radiograph from a 70-year-old man with unrelenting nausea and vomiting demonstrates a linear lucency along the left hemithorax due to a skin fold (arrow).

CASE **34** Tracheal injury

Randy K. Lau

Imaging description

Blunt thoracic tracheobronchial injuries usually occur within 2.5 cm of the carina [1]. In contrast, most penetrating injuries occur in the cervical trachea [2, 3]. In blunt trauma, bronchial lacerations are more common than tracheal lacerations. Chest radiographs may demonstrate pneumomediastinum (60%) and pneumothorax (70%) [4]. CT may identify the site of tracheobronchial injury in 70–100% of cases [5, 6].

Suspicious CT findings of aerodigestive injury include airway irregularity, disruption of the cartilage or tracheal wall, focal thickening or indistinctness of the trachea or main bronchi, and laryngeal disruption. Massive pneumomediastinum despite adequate tube drainage of pneumothoraces is also considered suspicious finding for aerodigestive tract injury (Figure 34.1) [7].

Cervical and thoracic subcutaneous emphysema is the most consistent radiographic finding in tracheobronchial injuries [8]. Other recognized findings include over-distention of the endotracheal tube balloon (>28 mm), an oval-shaped balloon, and herniation of the balloon into the tracheal defect (Figures 34.1 and 34.2) [5]. Although it is difficult to rupture the trachea with an endotracheal tube balloon, rupture from intubation does occur (Figure 34.3) [5, 9].

The "fallen lung sign" of bronchial transection is based on a radiographic case report in 1970 by Kumpe *et al.*, who describes how the lung falls away from the mediastinum when there is complete bronchial transection without vascular injury at the hilum [10, 11]. This is in contrast to a pneumothorax without bronchial injury where the lung usually collapses against the mediastinum. In our clinical practice, we have not identified a fallen lung sign.

Importance

Late complications of untreated tracheobronchial injury include: bronchial stenosis, recurrent pneumonia, and bronchiectasis [12].

Typical clinical scenario

A large proportion of patients with tracheobronchial injury (30–80%) die before reaching the hospital [11]. Acute tracheobronchial injury is rare following blunt trauma, with an incidence of approximately 3% [4]. At one academic level I trauma center over an 8-year period, 16 out of 12 789 trauma patients (0.13%) had tracheobronchial injuries of which 10 (63%) were secondary to penetrating trauma [13].

The most common clinical signs and symptoms are subcutaneous emphysema (35–85%) and hemoptysis (14–25%) [2].

Differential diagnosis

Paratracheal air cyst, otherwise known as tracheal diverticulum. Tracheal diverticula are present in 3–4% of the US population [14]. These benign entities characteristically occur on the right at the level of the thoracic inlet. Communication between the trachea and diverticulum is seen in 8–35% of cases [14, 15].

Other causes of pneumomediastinum are discussed in Case 30.

Teaching points

Blunt tracheobronchial injuries occur within 2.5 cm of the carina, whereas penetrating injury is most common in the cervical trachea.

CT identifies the site in 70–100% of blunt tracheobronchial injuries, and can be used to select which patients with pneumomediastinum need bronchoscopy.

REFERENCES

1. Kaewlai R, Avery LL, Asrani AV, Novelline RA. Multidetector CT of blunt thoracic trauma. *Radiographics*. 2008;**28**(6):1555–70.
2. Rossbach MM, Johnson SB, Gomez MA, *et al.* Management of major tracheobronchial injuries: a 28-year experience. *Ann Thorac Surg*. 1998; **65**(1):182–6.
3. Kiser AC, O'Brien SM, Detterbeck FC. Blunt tracheobronchial injuries: treatment and outcomes. *Ann Thorac Surg*. 2001;**71**(6):2059–65.
4. Karmy-Jones R, Avansino J, Stern EJ. CT of blunt tracheal rupture. *AJR Am J Roentgenol*. 2003;**180**(6):1670.
5. Chen JD, Shanmuganathan K, Mirvis SE, Killeen KL, Dutton RP. Using CT to diagnose tracheal rupture. *AJR Am J Roentgenol*. 2001; **176**(5):1273–80.
6. Scaglione M, Romano S, Pinto A, *et al.* Acute tracheobronchial injuries: impact of imaging on diagnosis and management implications. *Eur J Radiol*. 2006;**59**(3):336–43.
7. Dissanaike S, Shalhub S, Jurkovich GJ. The evaluation of pneumomediastinum in blunt trauma patients. *J Trauma*. 2008; **65**(6):1340–5.
8. Meislin HW, Iserson KV, Kaback KR, *et al.* Airway trauma. *Emerg Med Clin North Am*. 1983;**1**(2):295–312.
9. Sippel M, Putensen C, Hirner A, Wolff M. Tracheal rupture after endotracheal intubation: experience with management in 13 cases. *Thorac Cardiovasc Surg*. 2006;**54**(1):51–6.
10. Burke JF. Early diagnosis of traumatic rupture of the bronchus. *JAMA*. 1962 Aug **25**;181:682–6.
11. Kumpe DA, Oh KS, Wyman SM. A characteristic pulmonary finding in unilateral complete bronchial transection. *Am J Roentgenol Radium Ther Nucl Med*. 1970;**110**(4):704–6.

12. Chu CP, Chen PP. Tracheobronchial injury secondary to blunt chest trauma: diagnosis and management. *Anaesth Intensive Care*. 2002; **30**(2):145–52.
13. Huh J, Milliken JC, Chen JC. Management of tracheobronchial injuries following blunt and penetrating trauma. *Am Surg*. 1997; **63**(10):896–9.
14. Buterbaugh JE, Erly WK. Paratracheal air cysts: a common finding on routine CT examinations of the cervical spine and neck that may mimic pneumomediastinum in patients with traumatic injuries. *AJNR Am J Neuroradiol*. 2008;**29**(6):1218–21.
15. Goo JM, Im JG, Ahn JM, *et al*. Right paratracheal air cysts in the thoracic inlet: clinical and radiologic significance. *AJR Am J Roentgenol*. 1999;**173**(1):65–70.

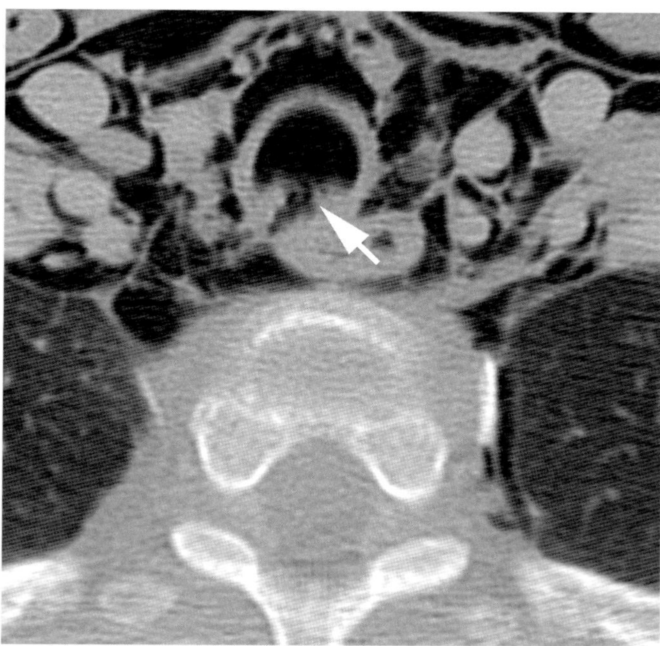

Figure 34.1 Non-contrast axial CT from a teenage male whose bicycle collided with his sister's bicycle. The handlebar struck the patient's lower neck. CT demonstrates extensive subcutaneous emphysema and pneumomediastinum. There is a tracheal defect along the posterior membrane (arrow).

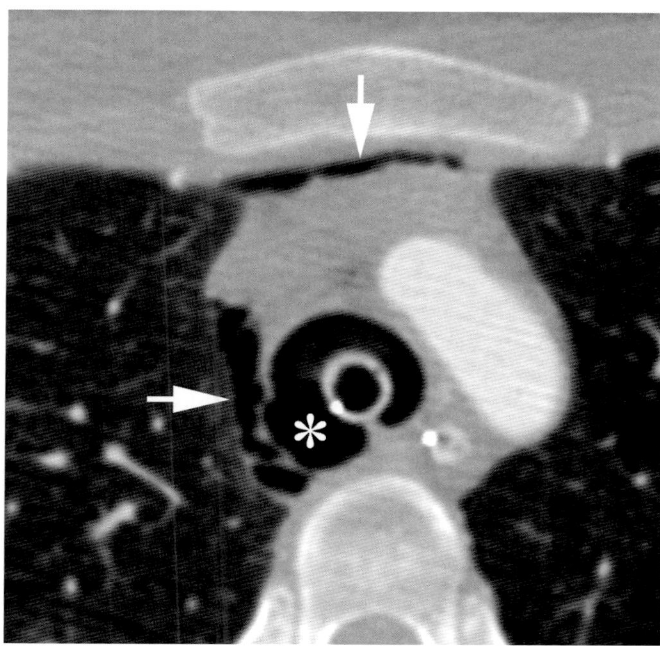

Figure 34.2 Axial contrast CT from a 15-year-old female who was an unrestrained passenger in a high-speed motor vehicle collision demonstrating pneumomediastinum (arrows). Note the contour abnormality (asterisk) due to herniation of the endotracheal tube balloon into the site of the tracheal laceration.

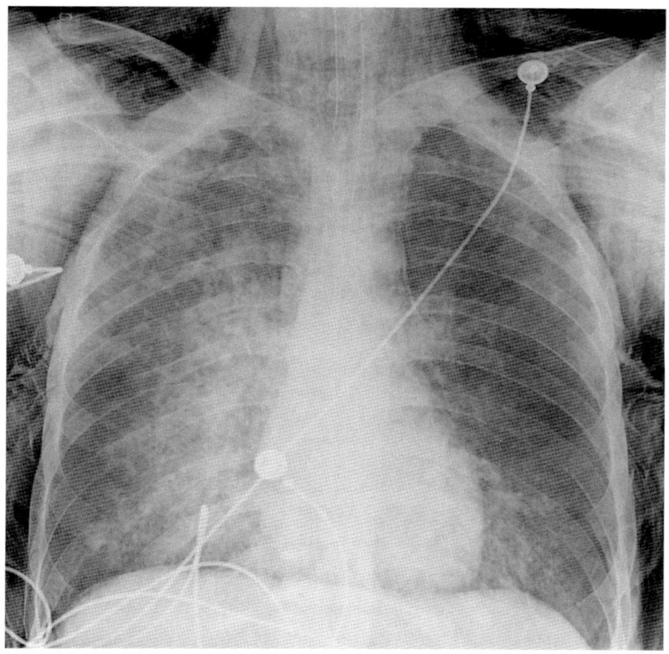

Figure 34.3 Chest radiograph from a 25-year-old male with loss of consciousness and a difficult intubation who developed massive subcutaneous emphysema. The patient had a tracheal rupture diagnosed by bronchoscopy. The cause was thought to be trauma during endotracheal intubation.

Pulmonary contusion and laceration

Randy K. Lau

Imaging description

Pulmonary contusions, aspiration, and pneumonia have overlapping imaging appearances, and can be difficult to distinguish in the acutely injured patient. However, both the imaging and appearance and time course can assist in discriminating between these entities. On radiographs, contusions appear as geographic areas of air-space opacification, and are usually located adjacent to bony structures and hence are peripherally located [1].

Pulmonary contusions may not be evident on initial radiographs but may appear on chest radiographs acquired up to 6 hours after injury. Development of pulmonary opacities 24 hours or more after injury suggests nosocomial pneumonia, atelectasis, or aspiration. However, caution is still warranted. Series have shown that the appearance of a pulmonary opacity has a specificity of only 27–35% for pneumonia in a patient on mechanical ventilation [2]. Specific signs that suggest ventilator-associated pneumonia (VAP) include rapid or progressive cavitation of the pulmonary opacity, an airspace process abutting a fissure, and an air bronchogram [2].

Pulmonary contusions typically resolve within 3–10 days [3]. However, both the presence of pulmonary contusions and endotracheal intubation for $\geq$5 days are strong risk factors for the development of VAP pneumonia [4–6]. In patients with a pulmonary contusion, persistence of consolidation beyond a few days after injury should be considered suspicious for VAP.

Aspiration is typically segmental in distribution and commonly involves the posterior segments of the upper lobes and the superior segments of the lower lobes. Alternatively, aspiration may present with confluent, irregular perihilar opacities with relative peripheral sparing. This is in contrast to contusions, which tend to be more peripheral in location [7]. Like pneumonia, and unlike contusions, aspiration follows the segmental anatomy of the lung. Aspiration or pneumonia can be distinguished from atelectatic lung by the relative decrease in attenuation of the lung compared to atelectatic lung on contrast-enhanced CT.

On CT, contusions appear as patchy airspace opacities or consolidation with ill-defined margins in a non-segmental distribution. Subpleural sparing may be seen.

Pulmonary lacerations are typically accompanied by contusion. Lacerations occur when there is a tear within the lung parenchyma. Due to the elastic recoil of normal lung tissues, the lung will usually recoil upon itself, and the laceration will form a sphere [3]. Hence pulmonary lacerations characteristically appear ovoid on CT (Figures 35.1 and 35.2). They can be filled with blood (hematocele, pulmonary hematoma), air (pneumatocele) or a combination of the two. Lacerations can be obscured on radiographs, but on CT performed within hours of injury appear as an air-filled or air–fluid-filled cavity within an area of pulmonary contusion. In days or weeks, they may fill with blood or fluid, and then take up to months to resolve. Their appearance soon after injury helps differentiate pneumatoceles from cavitation within nosocomial pneumonia.

Importance

Although the pulmonary contusion itself is not usually the cause of significant morbidity, it is a marker of high-energy trauma and is associated with other serious injuries. However, the finding of pulmonary contusion on CT but not chest radiographs is not a risk factor for significant morbidity in either children or adults [8, 9]. In contrast, VAP, the most common type of nosocomial pneumonia, is associated with a mortality rate of up to 50% [2].

Typical clinical scenario

Pulmonary contusions and lacerations can be seen with either penetrating or blunt trauma. Pulmonary contusions are the most common lung injury following blunt chest trauma [10, 11]. Pulmonary lacerations are especially common in children and young adults, given their more pliable chest wall.

Differential diagnosis

The differential diagnosis of airspace opacities in the injured patient includes contusion, pneumonia, aspiration, atelectasis and fat embolism (Figure 35.3). Occasionally a pneumothorax contained within a fissure mimics a contusion (Figure 35.4).

Teaching point

Contusions may not be present on initial radiographs and may present up to 6 hours after trauma. However, if pulmonary opacities develop 24 hours after trauma, or persist for 5–7 days, then nosocomial pneumonia, aspiration, or atelectasis must be considered.

REFERENCES

1. Ho ML, Gutierrez FR. Chest radiography in thoracic polytrauma. *AJR Am J Roentgenol.* 2009;**192**(3):599–612.
2. Koenig SM, Truwit JD. Ventilator-associated pneumonia: diagnosis, treatment, and prevention. Clin Microbiol *Rev.* 2006;**19**(4):637–57.
3. Kaewlai R, Avery LL, Asrani AV, Novelline RA. Multidetector CT of blunt thoracic trauma. *Radiographics.* 2008;**28**(6):1555–70.
4. Croce MA, Fabian TC, Waddle-Smith L, Maxwell RA. Identification of early predictors for post-traumatic pneumonia. *Am Surg.* 2001;**67**(2):105–10.
5. Hanes SD, Demirkan K, Tolley E, *et al.* Risk factors for late-onset nosocomial pneumonia caused by *Stenotrophomonas maltophilia* in critically ill trauma patients. *Clin Infect Dis.* 2002;**35**(3):228–35.

6. Joseph NM, Sistla S, Dutta TK, Badhe AS, Parija SC. Ventilator-associated pneumonia: a review. *Eur J Intern Med.* 2010;**21**(5):360–8.
7. Franquet T, Gimenez A, Roson N, *et al.* Aspiration diseases: findings, pitfalls, and differential diagnosis. *Radiographics.* 2000;**20**(3):673–85.
8. Kwon A, Sorrells DL, Jr., Kurkchubasche AG, *et al.* Isolated computed tomography diagnosis of pulmonary contusion does not correlate with increased morbidity. *J Pediatr Surg.* 2006;**41**(1):78–82; discussion 78–82.
9. Deunk J, Poels TC, Brink M, *et al.* The clinical outcome of occult pulmonary contusion on multidetector-row computed tomography in blunt trauma patients. *J Trauma.* 2010;**68**(2):387–94.
10. Wagner RB, Crawford WO, Jr., Schimpf PP. Classification of parenchymal injuries of the lung. *Radiology.* 1988;**167**(1):77–82.
11. Cohn SM. Pulmonary contusion: review of the clinical entity. *J Trauma.* 1997;**42**(5):973–9.

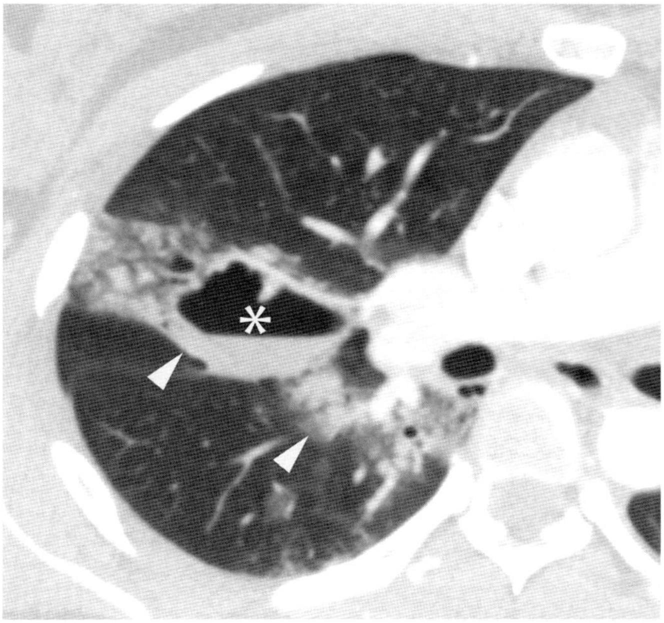

Figure 35.1 Axial contrast CT from a 23-year-old woman in a high-speed motor vehicle collision demonstrating typical ovoid appearance of a large right upper lobe pulmonary laceration (asterisk) due to recoil of lung tissue. The laceration in this case is filled with both blood and air. The laceration is accompanied by adjacent lung contusion (arrowheads), which is non-segmental, extending across the major fissure to involve the superior segment of the right lower lobe.

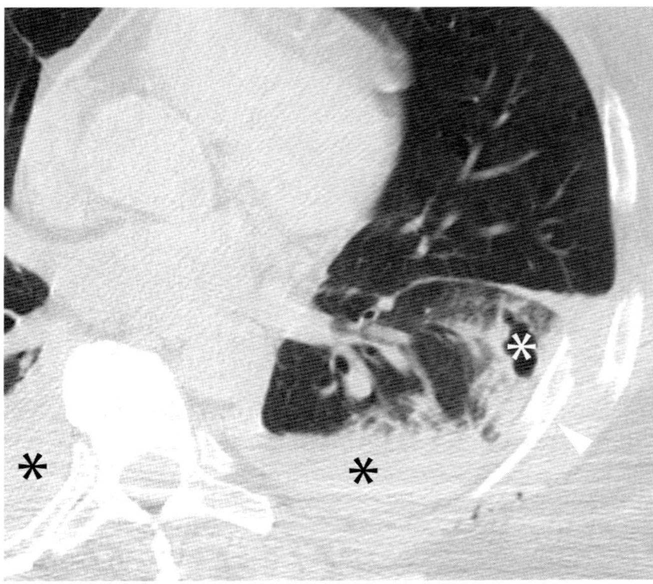

Figure 35.2 Axial contrast CT from a 57-year-old male in a bicycle accident demonstrates another left lower lobe pulmonary laceration (white asterisk) with adjacent left rib fracture (arrowhead). There are also bilateral hemothoraces (black asterisks).

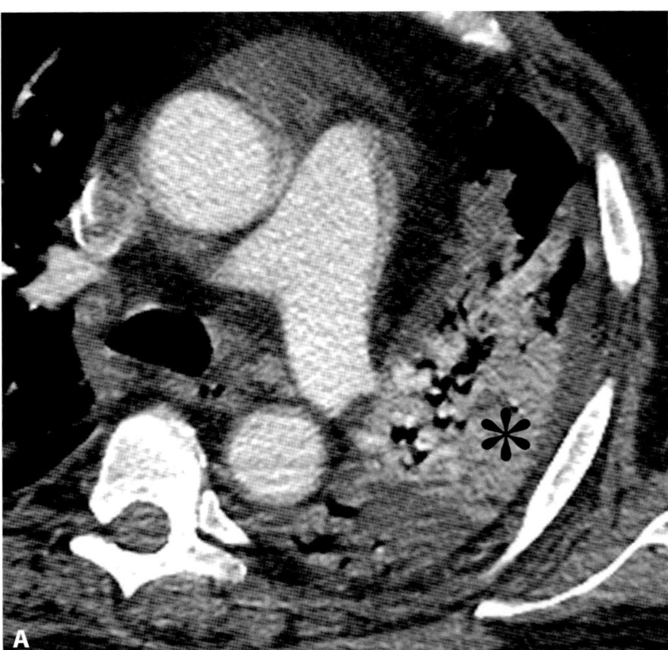

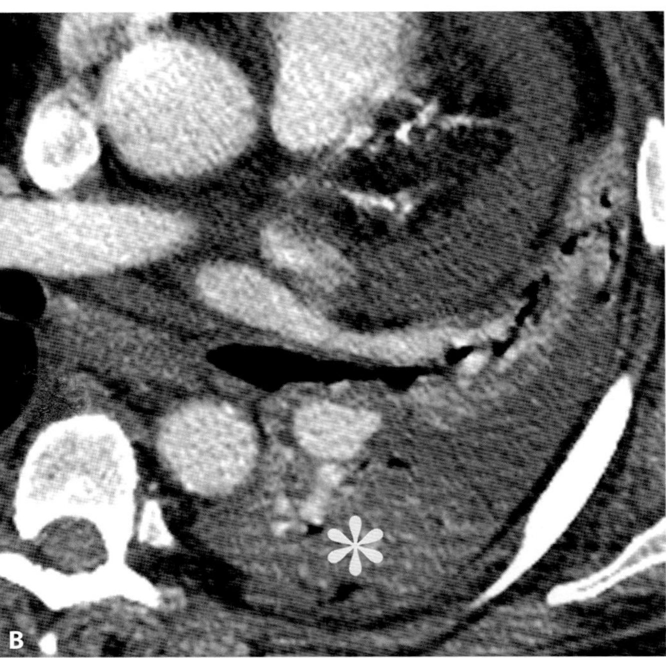

Figure 35.3 Atelectasis and pneumonia. **A.** Axial contrast CT from a 56-year-old man with acute myeloid leukemia demonstrates higher attenuation in atelectatic left upper lobe (asterisk) with Hounsfield units of 91. There is also a left pleural effusion and a pericardial effusion. **B.** In comparison, the same patient demonstrates lower attenuation in the left lower lobe consolidation (asterisk) related to aspiration and/or pneumonia with Hounsfield units of 46.

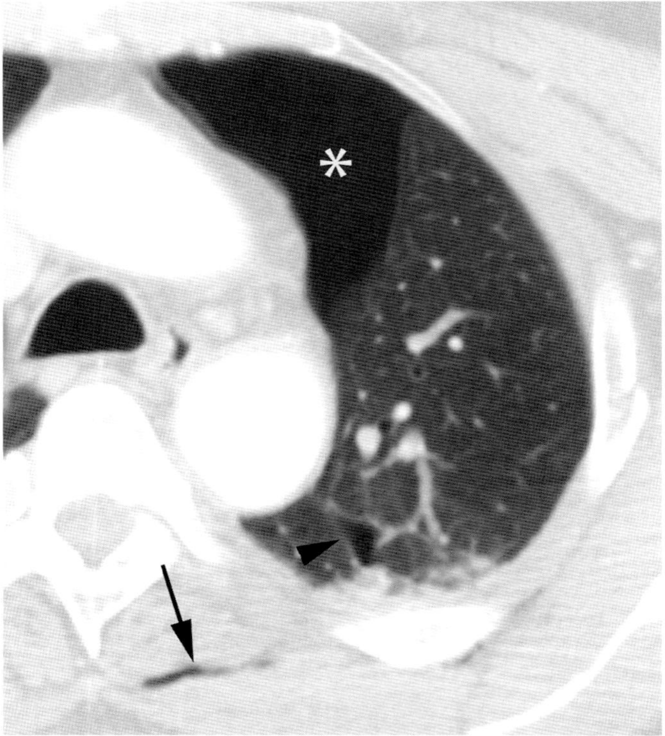

Figure 35.4 Axial contrast CT from a 67-year-old male who fell off a horse demonstrates a left pneumothorax (asterisk) with air within the left major fissure (arrowhead), which could mimic a pulmonary laceration. However, this collection of air is not ovoid in shape as is typical for pulmonary lacerations. There is adjacent atelectasis. Note the subcutaneous emphysema along left paraspinal musculature (arrow).

Sternoclavicular dislocation

Randy K. Lau

Imaging description

On chest radiographs, a difference in the relative craniocaudal positions of the medial clavicles of greater than 50% of the width of the clavicular heads is suggestive of sternoclavicular joint (SCJ) dislocation (Figure 36.1) [1]. While a number of additional specific radiographic views including the serendipity (Rockwood) (Figure 36.2), Hobb, Kattan, and Heinig views have been proposed, CT is the best study to perform when SCJ dislocation is suspected. CT helps characterize dislocation as either anterior or posterior and is useful for assessing associated injuries [2]. If a posterior dislocation is suspected, we perform CT angiography of the mediastinum as part of the evaluation (Figure 36.3).

Sternoclavicular joint dislocations are rare injuries [3]. Less than 2–3% of all dislocations involving the pectoral girdle involve the sternoclavicular joint [4]. Anterior or presternal displacement is much more common than posterior displacement [5].

Importance

Although anterior dislocations have limited skeletal morbidity, they indicate high-energy trauma and more than two-thirds of anterior dislocations are associated with other serious injuries including pneumothorax, hemothorax, and pulmonary contusions [5].

Posterior dislocations are more frequently associated with life-threatening complications because of their proximity to the great vessels and aerodigestive tracts. Approximately 25% of patients with posterior dislocations will have significant soft tissue complications in the upper chest [4]. These include subclavian vessel compression and injury, pneumothorax, esophageal rupture, myocardial conduction abnormalities, brachial plexopathy, tracheal tear, and late thoracic outlet syndrome [6–10].

Treatment of anterior dislocations is controversial as most anterior dislocations have little long-term functional impact [11]. Conversely, attempts are made at closed reduction of posterior dislocations under general anesthesia provided a cardiothoracic surgeon is available in the event of complications, and the patient presented within 7–10 days of injury [10].

Typical clinical scenario

Most SCJ dislocations are due to high-energy motor vehicle collisions. They can also be caused by athletic injuries or falls [12]. The dislocation can result from direct force to the clavicle or indirect force to the shoulder. The vector of force against the shoulder determines whether the dislocation is anterior or posterior.

Differential diagnosis

In many patients under the age of 25 years, the SCJ dislocation may in fact be a physeal plate injury (Salter–Harris I or II) as the medial clavicular physis may not close until after this age [12].

Teaching point

Sternoclavicular joint dislocations are rare, but should be suspected when there is a difference in relative craniocaudal positions of medial clavicles on chest radiographs.

If a SCJ dislocation is suspected clinically, perform a CT with multiplanar reformats, rather than evaluating with plain radiographs. Perform CT angiography (CTA) if a posterior sternoclavicular dislocation is suspected.

Sternoclavicular joint dislocation is a marker of high-energy trauma. Complications of posterior SCJ dislocation include pneumothorax, esophageal and tracheal injury, and compression of vascular and neural structures.

REFERENCES

1. McCulloch P, Henley BM, Linnau KF. Radiographic clues for high-energy trauma: three cases of sternoclavicular dislocation. *AJR Am J Roentgenol.* 2001;**176**(6):1534.
2. Restrepo CS, Martinez S, Lemos DF, *et al.* Imaging appearances of the sternum and sternoclavicular joints. *Radiographics.* 2009; **29**(3):839–59.
3. Cope R, Riddervold HO, Shore JL, Sistrom CL. Dislocations of the sternoclavicular joint: anatomic basis, etiologies, and radiologic diagnosis. *J Orthop Trauma.* 1991;5(3):379–84.
4. Rogers LF. *Radiology of Skeletal Trauma.* Philadelphia: Churchill Livingstone; 2002.
5. Rockwood CA, Wirth M. Injuries to sternoclavicular joints. In: Rockwood CA Green DP, Bucholtz RW, Heckman JD, eds. *Fractures in Adults*, 4th edn. Philadelphia: Lippincott-Raven; 1996: 1415–71.
6. Asplund C, Pollard ME. Posterior sternoclavicular joint dislocation in a wrestler. *Mil Med.* 2004;**169**(2):134–6.
7. Jougon JB, Lepront DJ, Dromer CE. Posterior dislocation of the sternoclavicular joint leading to mediastinal compression. *Ann Thorac Surg.* 1996;**61**(2):711–13.
8. Mirza AH, Alam K, Ali A. Posterior sternoclavicular dislocation in a rugby player as a cause of silent vascular compromise: a case report. *Br J Sports Med.* 2005;**39**(5):e28.
9. Noda M, Shiraishi H, Mizuno K. Chronic posterior sternoclavicular dislocation causing compression of a subclavian artery. *J Shoulder Elbow Surg.* 1997;**6**(6):564–9.

10. Wirth MA, Rockwood CA, Jr. Acute and chronic traumatic injuries of the sternoclavicular joint. *J Am Acad Orthop Surg.* 1996;**4**(5):268–78.
11. Bicos J, Nicholson GP. Treatment and results of sternoclavicular joint injuries. *Clin Sports Med.* 2003;**22**(2):359–70.
12. Macdonald PB, Lapointe P. Acromioclavicular and sternoclavicular joint injuries. *Orthop Clin North Am.* 2008;**39**(4):535–45, viii.

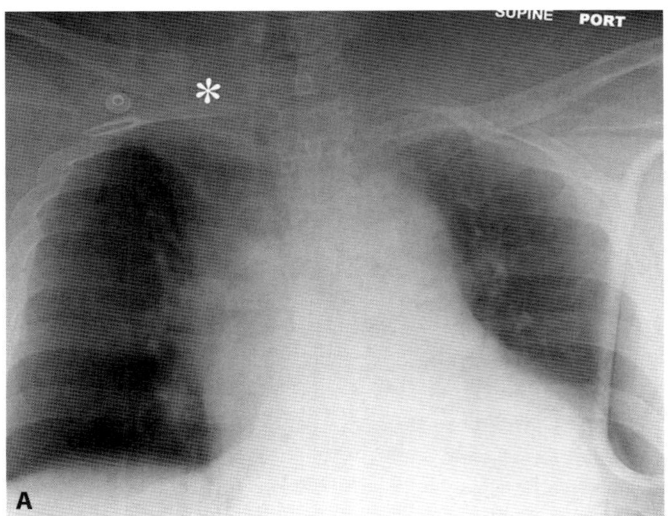

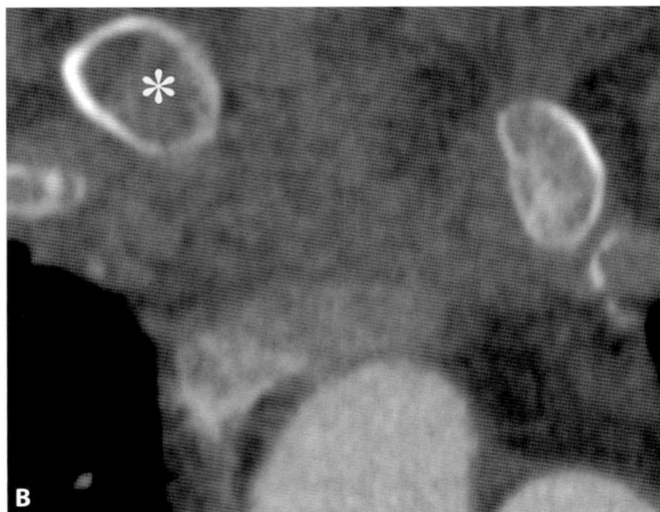

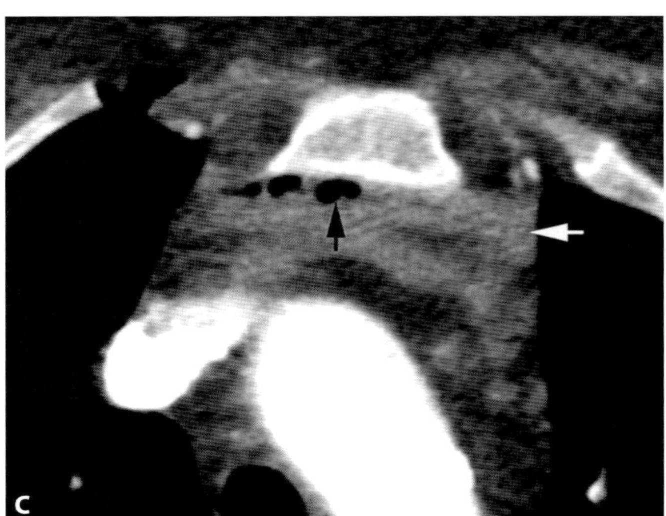

Figure 36.1 A. Chest radiograph from a 64-year-old woman after a motor vehicle collision demonstrating sternoclavicular joint asymmetry with superior displacement of the medial right clavicle (asterisk) by more than 50%, suggesting sternoclavicular joint dislocation. There is also widening of the mediastinum worrisome for mediastinal hematoma. **B.** CT angiogram with a coronal reformat confirms superior dislocation of the medial right clavicle at the sternoclavicular joint (asterisk). **C.** Axial image demonstrates associated mediastinal hematoma (white arrow) and pneumomediastinum (black arrow).

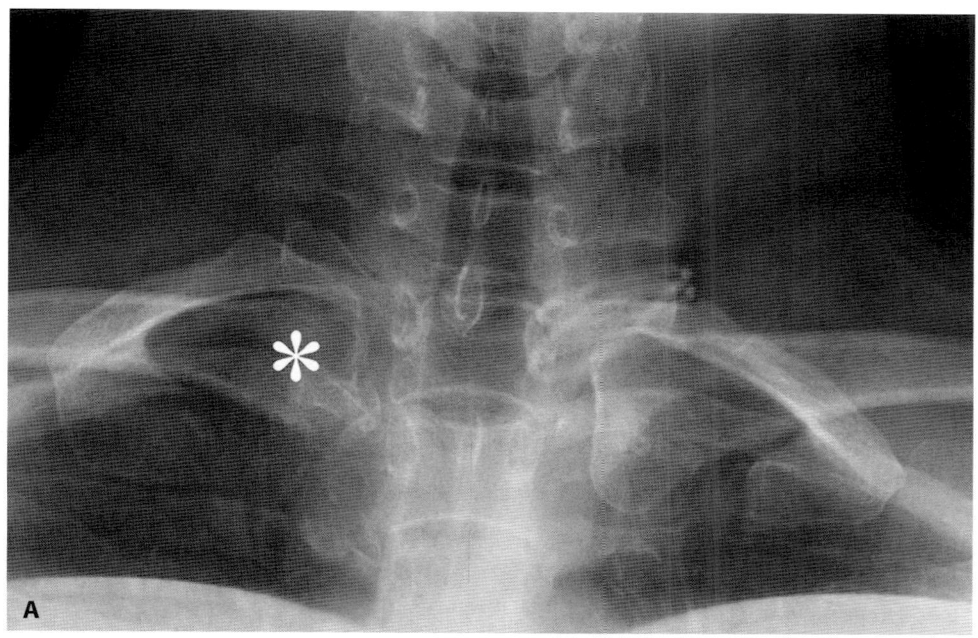

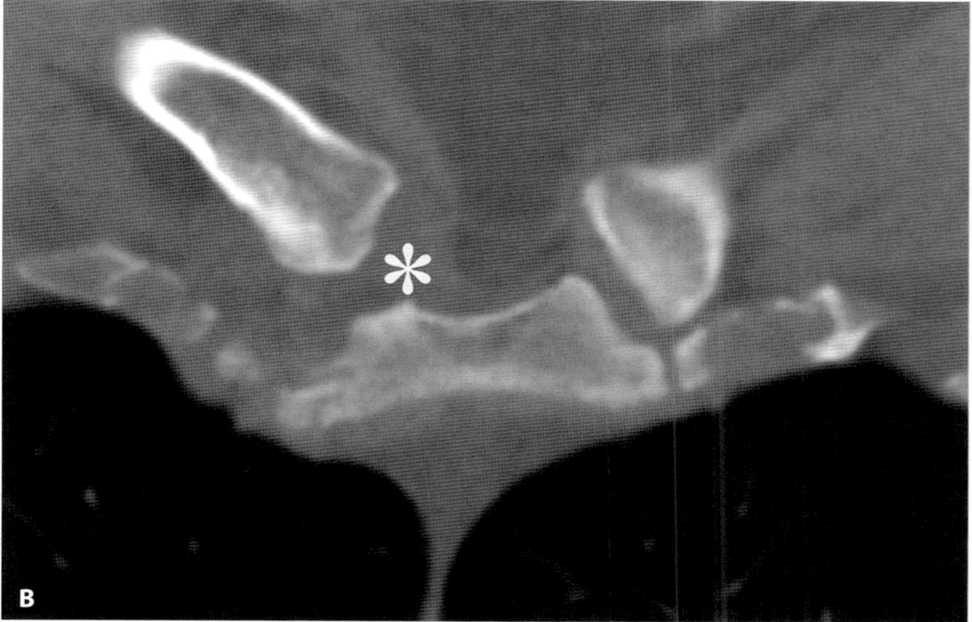

Figure 36.2 **A.** Serendipity view of the sternoclavicular joints in a 44-year-old woman who was assaulted demonstrates asymmetry of the sternoclavicular joints (asterisk) with superior displacement of the right medial clavicle suggestive of dislocation. **B.** Coronal non-contrast CT confirming dislocation of the right sternoclavicular joint (in this case, anterior) with superior displacement (asterisk).

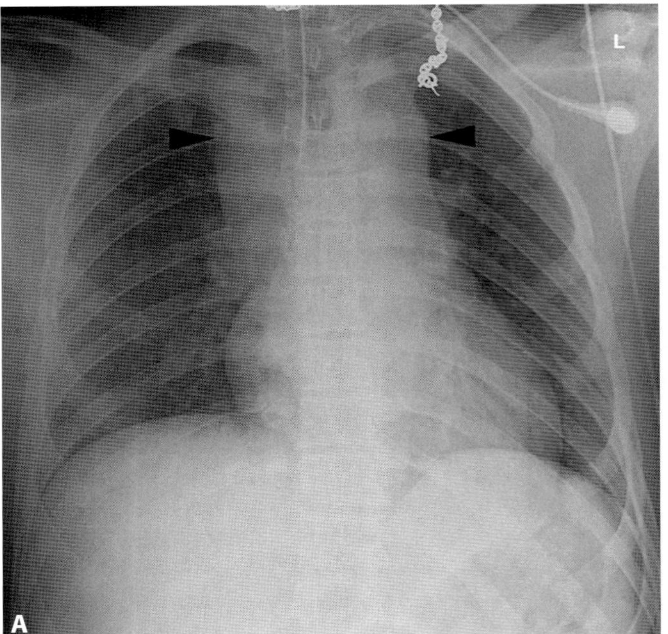

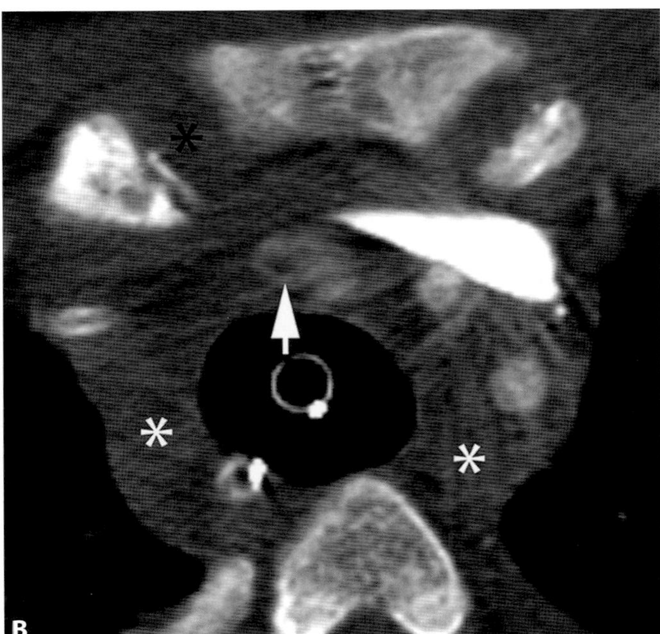

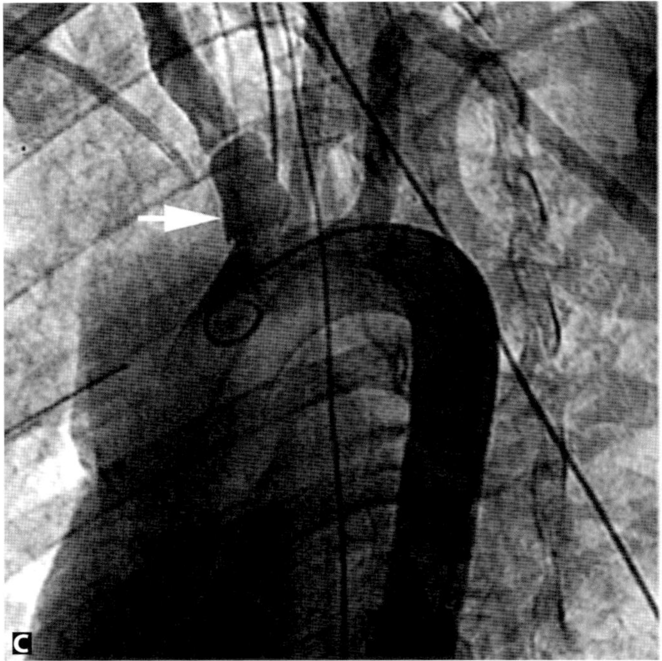

Figure 36.3 **A.** Chest radiograph from an intubated 20-year-old man involved in a high-speed motor vehicle collision demonstrating an abnormally widened mediastinum (arrowheads) with aortic arch indistinctness. Minimal asymmetry of the sternoclavicular joints is present, not identified prospectively. **B.** CT angiography demonstrates posterior dislocation and widening of the right sternoclavicular joint (black asterisk). There is also a minimally displaced fracture of the medial right clavicular physis. Mediastinal widening seen on the chest radiograph is secondary to mediastinal hematoma (white asterisks). There is dilation of the proximal right brachiocephalic artery with intraluminal thrombus (arrow) secondary to a vascular injury. Note the beam-hardening artifact from intravenous contrast in the left brachiocephalic vein. This is not ideal. In cases of suspected great vessel injury, inject through the right upper extremity where possible. **C.** Angiography confirming pseudoaneurysm (arrow) of the proximal right brachiocephalic artery.

CASE **37**

Boerhaave syndrome

Randy K. Lau

Imaging description

Boerhaave syndrome represents the clinical syndrome associated with spontaneous esophageal rupture from retching and vomiting. The factor most often associated with a high mortality is a delay in diagnosis as this can result in significant mediastinal infection and tissue destruction [1–3].

Surgical repair remains the mainstay of therapy. However, there is increasing evidence that spontaneous esophageal rupture can be managed successfully conservatively, or with endoscopic stenting, in carefully selected patients who have an early diagnosis [2].

Although radiographic findings may be non-specific or subtle, early identification of these findings and a low threshold for further investigation in the absence of definitive radiographic findings is important.

Rupture of the esophagus from Boerhaave syndrome tends to occur in the distal left posterior wall. This will result in the classic radiographic findings of pneumomediastinum, left pneumothorax, and left pleural effusion (Figure 37.1) [4]. However, the effusion and pneumothorax are not always left-sided, and the chest radiograph may be normal [5]. Evidence of pneumomediastinum may include a white line adjacent to the mediastinum, linear gas in the mediastinal or cervical soft tissues, a continuous diaphragm sign, and a "V"-shaped air lucency in the left lower mediastinum (of Naclerio) [6].

Findings of Boerhaave syndrome on CT include esophageal wall thickening, pneumomediastinum, and extravasation of oral contrast into the mediastinum or pleural space [7]. CT is also particularly useful to assess the degree of mediastinitis, fluid collections, or abscess formation associated with delayed diagnosis.

Although esophagography remains the mainstay of diagnosis, this test may be falsely negative in as many as 10% of patients [7]. We perform the examination with water-soluble contrast, and follow this with dilute barium if no leak is seen (Figure 37.2) [8]. The tear can usually be seen in the left posterior wall of the esophagus, near the left diaphragmatic hiatus [9]. Tears are typically 2 cm long and are located 3–6 cm above the diaphragm [10]. Other findings on esophagogram may include extravasation of contrast, submucosal collections of contrast, and an esophagopleural fistula from extensive tissue destruction [5, 11].

Importance

Early diagnosis and treatment is critical as mortality rate can be as high as 40% if diagnosis is delayed, compared to 6.2% if diagnosed and treated within 24 hours [3].

Typical clinical scenario

Spontaneous esophageal rupture is more common in middle-aged men. Fifty percent will have a history of alcoholism or heavy drinking [7]. Typically the patient will have a history of severe retching or vomiting, followed by epigastric pain, chest pain, back pain, dyspnea, or shock. Nowadays, spontaneous rupture only represents about 15% of cases of esophageal rupture, the remainder occurring during diagnostic or therapeutic procedures (Figure 37.3) [12]. Unfortunately the signs and symptoms of esophageal rupture may mimic other causes of acute chest pain, such as aortic dissection, myocardial infarction, pneumonia, lung abscess, or acute pancreatitis.

Differential diagnosis

This includes Mallory–Weiss syndrome, epiphrenic diverticulum, esophageal injury, and malignant fistula.

Mallory–Weiss syndrome is a partial thickness tear of the distal esophagus that may extend into the gastric fundus. This syndrome typically occurs following violent retching. Because the tear is only partial thickness, there are usually no discernible findings on chest radiograph or CT. Diagnosis using esophagography is difficult. Occasionally a longitudinal, intramural collection of contrast may be seen [9].

Epiphrenic diverticula are mucosa-lined outpouchings of the distal esophagus that usually occur in the setting of esophageal dysmotility or distal esophageal obstructions. Pneumomediastinum and surrounding inflammatory changes are absent in epiphrenic diverticulum.

Malignant esophageal fistulas can mimic Boerhaave syndrome radiographically. However, the clinical context will help differentiate the two.

> **Teaching point**
>
> Early diagnosis of esophageal rupture is critical as a delay in diagnosis and repair of even just 24 hours results in a high mortality rate. Radiographic findings may be non-specific or negative.

REFERENCES

1. Port JL, Kent MS, Korst RJ, Bacchetta M, Altorki NK. Thoracic esophageal perforations: a decade of experience. *Ann Thorac Surg*. 2003;**75**(4):1071–4.
2. Abbas G, Schuchert MJ, Pettiford BL, *et al*. Contemporaneous management of esophageal perforation. *Surgery*. 2009;**146**(4):749–56.
3. Shaker H, Elsayed H, Whittle I, Hussein S, Shackcloth M. The influence of the 'golden 24-h rule' on the prognosis of oesophageal perforation in the modern era. *Eur J Cardiothorac Surg*. 2010;**38**(2):216–22.
4. Young CA, Menias CO, Bhalla S, Prasad SR. CT features of esophageal emergencies. *Radiographics*. 2008;**28**(6):1541–53.
5. Gimenez A, Franquet T, Erasmus JJ, Martinez S, Estrada P. Thoracic complications of esophageal disorders. *Radiographics*. 2002;**22** Spec No: S247–58.

6. Sinha R. Naclerio's V sign. *Radiology*. 2007;**245**(1):296–7.
7. Katabathina VS, Restrepo CS, Martinez-Jimenez S, Riascos RF. Nonvascular, nontraumatic mediastinal emergencies in adults: a comprehensive review of imaging findings. *Radiographics*. 2011;**31** (4):1141–60.
8. Gollub MJ, Bains MS. Barium sulfate: a new (old) contrast agent for diagnosis of postoperative esophageal leaks. *Radiology*. 1997; **202**(2):360–2.
9. Kanne JP, Rohrmann CA, Jr., Lichtenstein JE. Eponyms in radiology of the digestive tract: historical perspectives and imaging appearances. Part I. Pharynx, esophagus, stomach, and intestine. *Radiographics*. 2006;**26** (1):129–42.
10. de Schipper JP, Pull ter Gunne AF, Oostvogel HJ, van Laarhoven CJ. Spontaneous rupture of the oesophagus: Boerhaave's syndrome in 2008. Literature review and treatment algorithm. *Dig Surg*. 2009;**26**(1):1–6.
11. Ghanem N, Altehoefer C, Springer O, *et al*. Radiological findings in Boerhaave's syndrome. *Emerg Radiol*. 2003;**10**(1):8–13.
12. Sepesi B, Raymond DP, Peters JH. Esophageal perforation: surgical, endoscopic and medical management strategies. *Curr Opin Gastroenterol*. 2010;**26**(4):379–83.

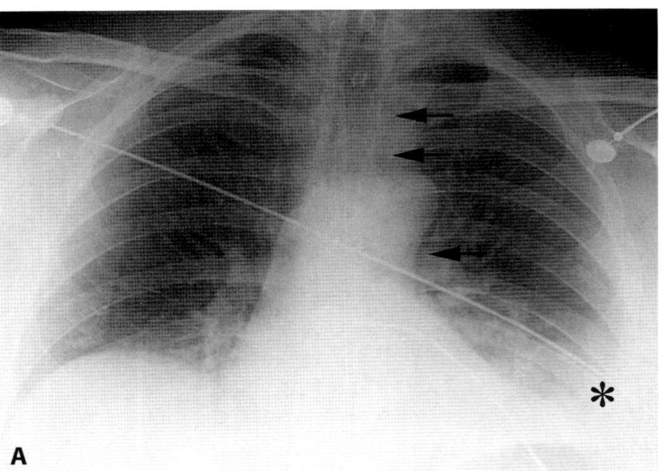

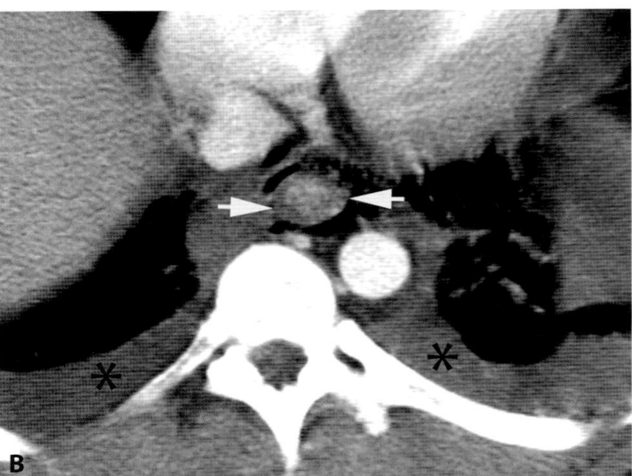

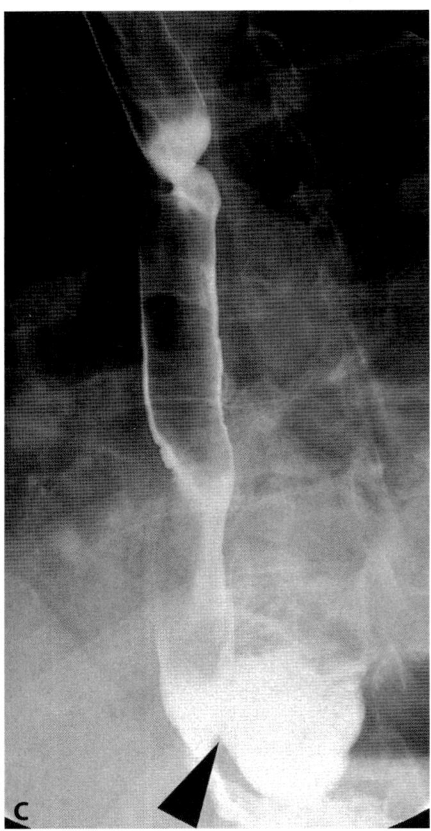

Figure 37.1 Boerhaave syndrome. **A.** A 37-year-old man who experienced retching during a meal followed by acute, sharp epigastric pain. Chest radiograph demonstrates extensive pneumomediastinum (arrows) and left pleural effusion (asterisk). **B.** Axial contrast-enhanced CT shows circumferential esophageal wall thickening (arrows) surrounded by pneumomediastinum and bilateral pleural effusions (asterisks). **C.** Esophagogram with water-soluble contrasts demonstrates extravasation from left posterior aspect of distal esophagus (arrowhead). At surgery a 4cm defect was repaired.

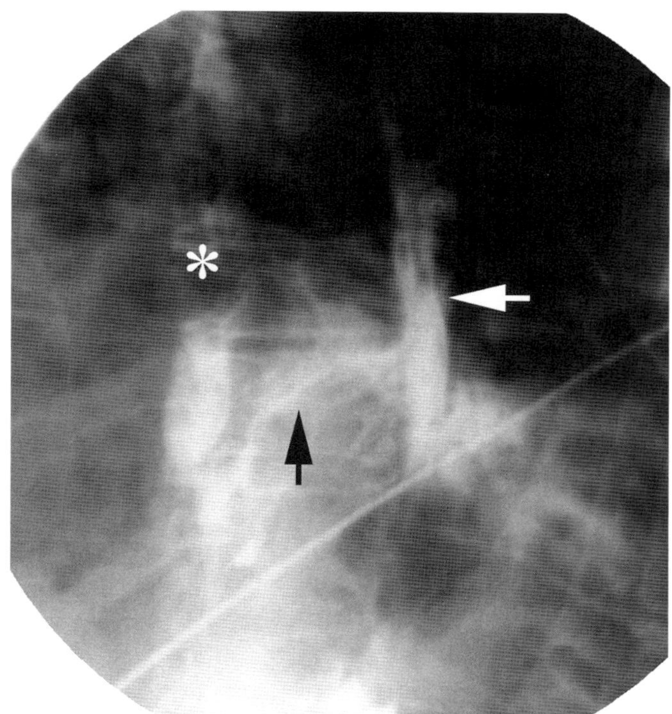

Figure 37.2 Boerhaave syndrome with leakage of contrast into the left pleural space. Selected image from esophagogram performed with water-soluble contrast in a 70-year-old man with epigastric and lower thcracic pain after vomiting demonstrates extravasation of oral contrast from the left lower esophagus (asterisk) into the mediastirum (black arrow) and left pleural space (white arrow).

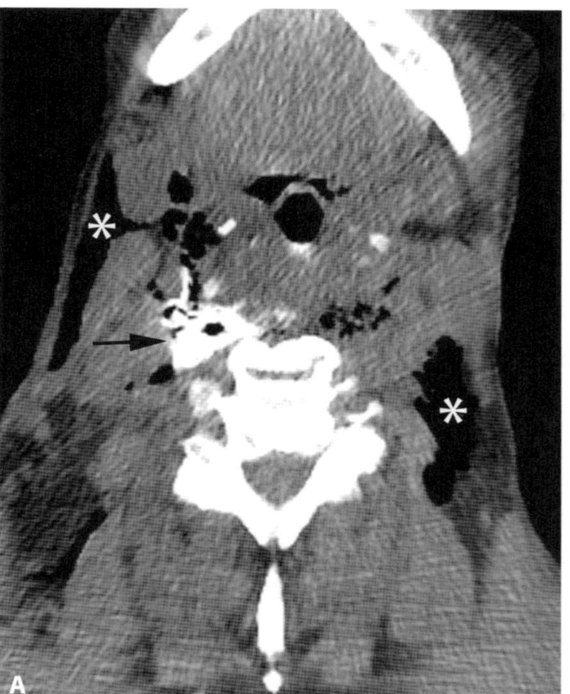

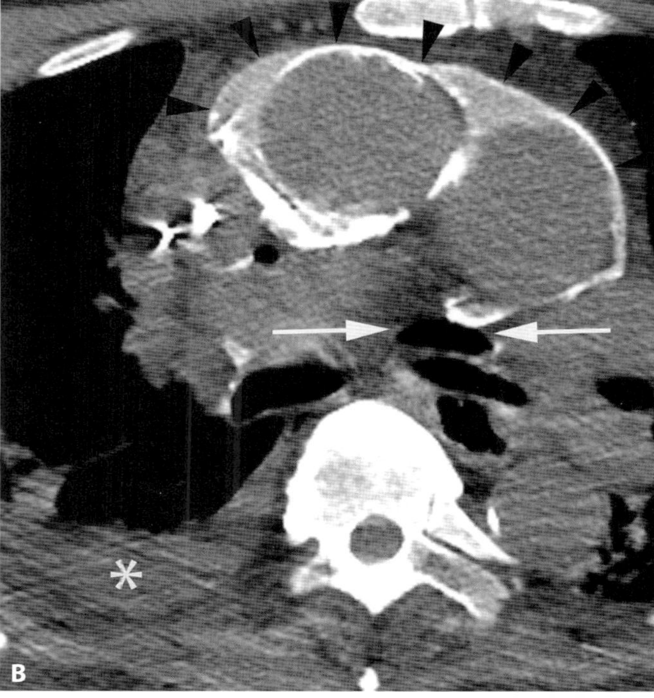

Figure 37.3 Axial CT from a patient following ingestion of oral contrast with iatrogenic esophageal perforation after transesophageal echocardiogram. **A.** There is extensive subcutaneous emphysema (asterisks). There is also extravasation of oral contrast from the hypopharynx (arrow). This is in contrast to Boerhaave syndrome, in which the tear typically occurs in the distal esophagus, near the gastroesophageal junction. **B.** Another slice from the same CT demonstrates pneumomediastinum (arrows) and oral contrast within the pericardial space representing esophagopericardial fistula (arrowheads). Note right pleural effusion (asterisk).

Variants and hernias of the diaphragm simulating injury

Randy K. Lau

Imaging description

Most blunt diaphragmatic ruptures are longer than 10 cm and occur along the posterolateral aspect of the left hemidiaphragm [1]. Imaging findings of diaphragmatic rupture on chest radiography include an intrathoracic location of abdominal viscera (with or without the "collar sign") a nasogastric tube above the left hemidiaphragm, distortion or obliteration of hemidiaphragm outline, contralateral mediastinal shift, and marked elevation of the left hemidiaphragm (>4 cm) compared to the right [1–3]. CT findings of diaphragmatic injuries include segmental diaphragm non-visualization, intrathoracic herniation of viscera, collar sign, dependent viscera sign, and a thickened diaphragm (Figure 38.1) [3]. Although sensitive for injury, focal thickening of the diaphragm in the absence of other signs of diaphragmatic injury is not specific [4].

Diagnostic pitfalls for diaphragmatic injury include hernias (Bochdalek, Morgagni, and hiatal) and discontinuity of the diaphragm between crura and lateral arcuate ligaments [5].

Foramen of Bochdalek hernias

The foramen of Bochdalek is a 2 cm opening in the posterior fetal diaphragm that normally closes by the eighth week of gestation. The left foramen closes later than the right. Hence, 85% of Bochdalek hernias occur on the left [6]. Most symptomatic Bochdalek hernias present in the neonatal period whereas asymptomatic foramina and hernias are detected incidentally later in life, during imaging for other reasons. Posterolateral diaphragmatic defects are found in 6% of asymptomatic adults, are more common on the left side, and are thought to represent a type of Bochdalek hernia [1]. These defects do not allow organ herniation but contain intra-abdominal fat. This is also a common site of diaphragmatic rupture. The absence of organ herniation, presence of only fat, smoothly tapering diaphragmatic margins, and lack of associated thoracic or abdominal visceral injury favor persistent foramina of Bochdalek over diaphragmatic injury (Figure 38.2).

Foramen of Morgagni hernias

Morgagni hernias are rarer than Bochdalek hernias and occur through defects in the right anteromedial diaphagm in the retrosternal region. The defect is bounded medially by muscle fibers from the xiphoid process and laterally by muscle fibers from the costal cartilages [7]. The typical location of a Morgagni hernia is key to its distinction from diaphragmatic injury, which is typically posterolateral (Figure 38.3).

Diaphragmatic eventration

Diaphragmatic eventrations can mimic blunt traumatic diaphragmatic injury. However, thin multiplanar reformations will confirm diaphragmatic continuity (Figure 38.4).

Lateral arcuate ligament

The lateral arcuate ligament typically blends imperceptibly with the diaphragmatic crus. However, in 11% of patients, there may be discontinuity between the lateral arcuate ligament and diaphragmatic crus, and this should not be mistaken for diaphragmatic injury [8]. Smooth margins, the typical imaging appearance, and absence of organ herniation or associated thoracic or abdominal injury is key to correct diagnosis (Figure 38.5).

Importance

Diaphragmatic injuries will not heal spontaneously, and even tiny diaphragmatic ruptures may require surgical treatment [9]. Complications from missed diaphragmatic rupture include bowel obstruction or infarction from herniation of the abdominal contents. These complications may increase morbidity and mortality.

Typical clinical scenario

Diaphragmatic injury may be the result of blunt or penetrating trauma. Blunt diaphragmatic rupture represents an uncommon injury with an incidence of 0.16–5% [10–13]. Blunt diaphragmatic rupture is even more uncommon in the pediatric population with a reported incidence in only 0.07% of pediatric patients admitted over a 21-year period in one study [14]. Blunt diaphragmatic rupture is highly associated with other high-energy injuries such as injuries to the thoracic aorta, spleen, liver, and pelvis fractures [10, 12, 15–17].

Diaphragmatic laceration should always be assumed when there is a sign of contiguous penetrating injury on either side of the diaphragm (Figure 38.6) [18].

Differential diagnosis

Concurrent injuries such as pulmonary contusion, atelectasis, and pleural effusion may obscure the diaphragm on chest radiography. Consequently, diaphragmatic injuries are missed in 12–66% of blunt trauma patients on radiography [19]. The differential diagnosis of CT findings includes congenital mimickers of the diaphragmatic injury that are described here.

Teaching point

A smooth defect in the posterolateral diaphragm between the lateral arcuate ligament and the diaphragmatic crus is most commonly a developmental variant and not a diaphragmatic injury. Another common posterolateral diaphragmatic defect is due to persistence of the fetal foramen of Bochdalek.

REFERENCES

1. Iochum S, Ludig T, Walter F, *et al.* Imaging of diaphragmatic injury: a diagnostic challenge? *Radiographics.* 2002;**22** Spec No:S103–16; discussion S116–18.
2. Gelman R, Mirvis SE, Gens D. Diaphragmatic rupture due to blunt trauma: sensitivity of plain chest radiographs. *AJR Am J Roentgenol.* 1991;**156**(1):51–7.
3. Sliker CW. Imaging of diaphragm injuries. *Radiol Clin North Am.* 2006;**44**(2):199–211, vii.
4. Nchimi A, Szapiro D, Ghaye B, *et al.* Helical CT of blunt diaphragmatic rupture. *AJR Am J Roentgenol.* 2005;**184**(1):24–30.
5. Restrepo CS, Eraso A, Ocazionez D, *et al.* The diaphragmatic crura and retrocrural space: normal imaging appearance, variants, and pathologic conditions. *Radiographics.* 2008;**28**(5):1289–305.
6. Nyhus LM Condon RE. *Hernia.* Philadelphia: J.B. Lippincott; 1995.
7. Lee GH, Cohen AJ. CT imaging of abdominal hernias. *AJR Am J Roentgenol.* 1993;**161**(6):1209–13.
8. Naidich DP, Megibow AJ, Ross CR, Beranbaum ER, Siegelman SS. Computed tomography of the diaphragm: normal anatomy and variants. *J Comput Assist Tomogr.* 1983;**7**(4):633–40.
9. Sandstrom CK, Stern EJ. Diaphragmatic hernias: a spectrum of radiographic appearances. *Curr Probl Diagn Radiol.* 2011;**40**(3):95–115.
10. Mihos P, Potaris K, Gakidis J, *et al.* Traumatic rupture of the diaphragm: experience with 65 patients. *Injury.* 2003;**34**(3):169–72.
11. Barsness KA, Bensard DD, Ciesla D, *et al.* Blunt diaphragmatic rupture in children. *J Trauma.* 2004;**56**(1):80–2.
12. Rodriguez-Morales G, Rodriguez A, Shatney CH. Acute rupture of the diaphragm in blunt trauma: analysis of 60 patients. *J Trauma.* 1986;**26**(5):438–44.
13. Tarver RD, Conces DJ, Jr., Cory DA, Vix VA. Imaging the diaphragm and its disorders. *J Thorac Imaging.* 1989;**4**(1):1–18.
14. Koplewitz BZ, Ramos C, Manson DE, Babyn PS, Ein SH. Traumatic diaphragmatic injuries in infants and children: imaging findings. *Pediatr Radiol.* 2000;**30**(7):471–9.
15. Athanassiadi K, Kalavrouziotis G, Athanassiou M, *et al.* Blunt diaphragmatic rupture. *Eur J Cardiothorac Surg.* 1999;**15**(4):469–74.
16. Kearney PA, Rouhana SW, Burney RE. Blunt rupture of the diaphragm: mechanism, diagnosis, and treatment. *Ann Emerg Med.* 1989;**18**(12):1326–30.
17. Gunn ML. Imaging of aortic and branch vessel trauma. *Radiol Clin North Am.* 2012;**50**(1):85–103.
18. Bodanapally UK, Shanmuganathan K, Mirvis SE, *et al.* MDCT diagnosis of penetrating diaphragm injury. *Eur Radiol.* 2009;**19**(8):1875–81.
19. Sangster G, Ventura VP, Carbo A, *et al.* Diaphragmatic rupture: a frequently missed injury in blunt thoracoabdominal trauma patients. *Emerg Radiol.* 2007;**13**(5):225–30.

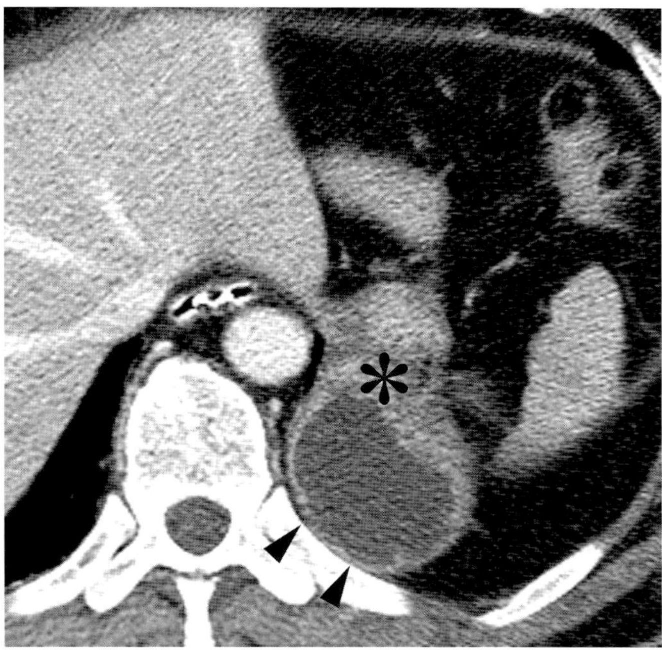

Figure 38.1 Axial contrast-enhanced CT from a 55-year-old man involved in a high-speed motor vehicle collision demonstrating the collar sign (asterisk) and dependent viscera sign (arrowheads) typical of diaphragmatic injury.

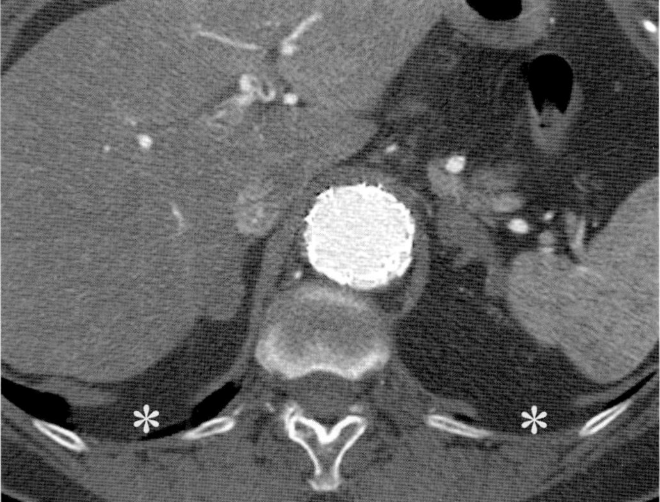

Figure 38.2 Axial contrast-enhanced CT from a 77-year-old man with a history of repaired thoraco-abdominal aneurysm demonstrating bilateral, small foramen of Bochdalek defects in the diaphragm, containing only fat (asterisks). Notice the smooth tapered margins.

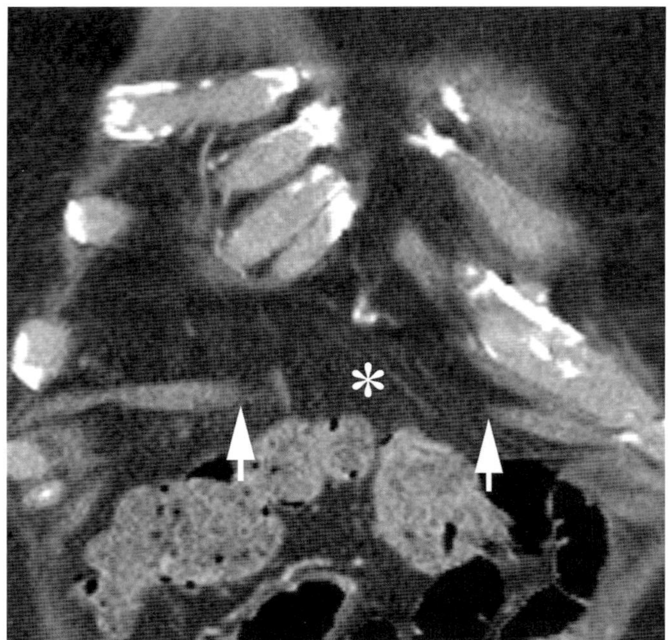

Figure 38.3 Anterior coronal image from contrast-enhanced CT in an 87-year-old man with a large Morgagni hernia shows herniation of omental fat (asterisk) between diaphragmatic defects (arrows). The anterior location is typical for Morgagni hernias in contrast to the posterolateral location typical for diaphragmatic rupture.

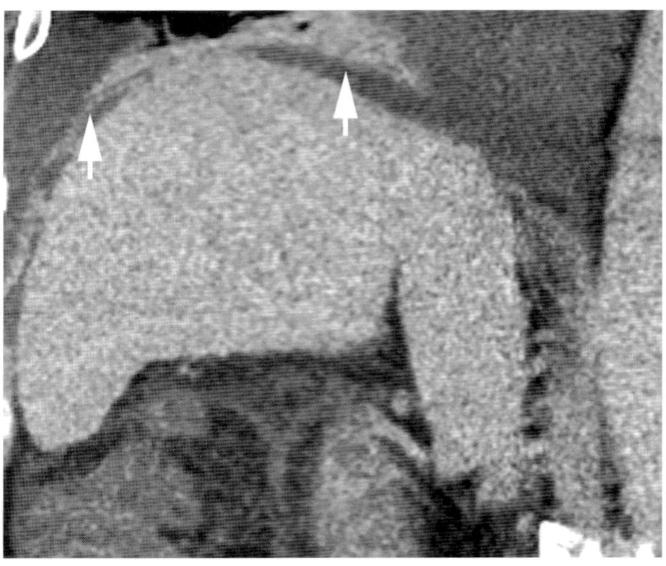

Figure 38.4 Coronal image from a contrast-enhanced CT in a 58-year-old man with hepatitis C-related cirrhosis demonstrating thinning of the diaphragm (arrows) outlined by ascites and a right pleural effusion. Diaphragmatic eventration can be distinguished from diaphragmatic injury through use of thin multiplanar reformats.

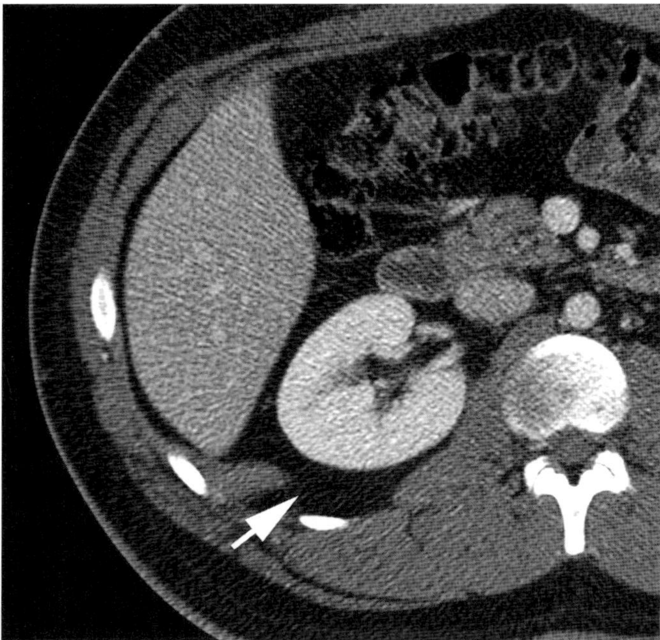

Figure 38.5 Thick lateral arcuate ligament in a 28-year-old man. Axial contrast-enhanced CT demonstrates variable appearance of the lateral arcuate ligament. This appearance can mimic a diaphragmatic defect. Notice abdominal fat invaginating behind the ligament, but remaining contained within the abdominal cavity.

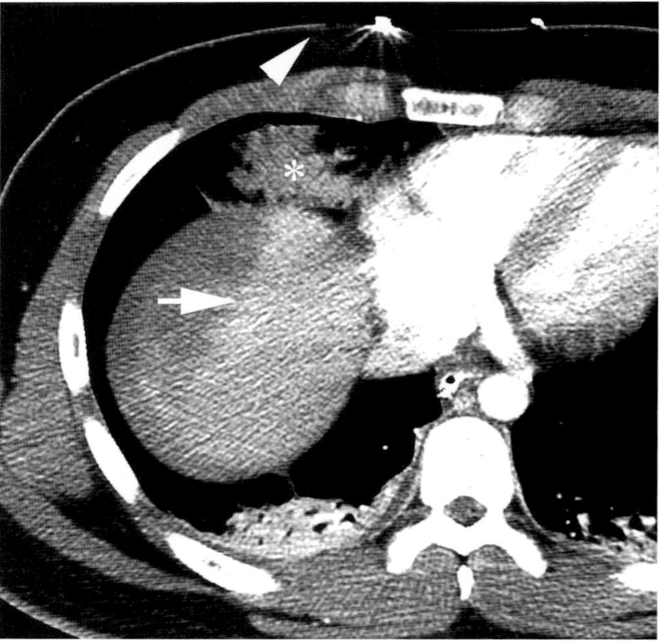

Figure 38.6 Axial contrast-enhanced CT from a 17-year-old teenager shot in the anterior chest with contiguous injuries above and below the right hemidiaphragm, indicating a diaphragm laceration. Part of the entry site through the skin can be identified (arrowhead). The trajectory included the right pleural space, right middle lobe (asterisk) and right lobe of the liver (arrow). Note the sharp anterior contour to the pulmonary hemorrhage in the right middle lobe, consistent with a pneumothorax. The diaphragm laceration was confirmed and repaired by surgery.

Aortic pulsation artifact

Randy K. Lau

Imaging description

A common artifact that can mimic blunt traumatic aortic injury (BTAI) or an aortic dissection is pulsation artifact. This artifact occurs most commonly in the ascending aorta, but it may occur elsewhere in the thoracic aorta, including the aortic isthmus ([1–4]. Pulsation artifacts have been identified in up to 92% of non-ECG-gated CT scans [3]. Blunt traumatic aortic injuries involving the ascending aorta and root are extremely rarely encountered in the emergency department as these patients almost always die before reaching the hospital [4–6]. In the vast majority of cases, pulsation artifact affects the left anterior and right posterior aspects of the aortic circumference [3]. This artifact, which can also simulate aortic dissection, is described in more detail in Case 43. Important differentiating features are extension of the linear hypodensity of pulsation artifact into the mediastinal fat, and similar "pseudoflaps" in the main pulmonary artery and superior vena cava at the same slice level (Figure 39.1). Multiplanar reformations can be very useful. Other helpful differentiating features include the absence of periaortic hematoma (in suspected BTAI) and the presence of similar artifacts on adjacent structures such as tubes and lines (Figure 39.2).

Recent work has demonstrated considerable physiologic variation in the diameter, shape, and length of the normal thoracic aorta during the cardiac cycle, especially in younger patients who have more elastic vessel walls [7, 8]. During non-gated aortic CT angiography (CTA), these changes in aortic contour and size can mimic aortic injury. Smooth variation in the shape of the thoracic aorta in young patients should not be misinterpreted as a sign of aortic injury [1].

Importance

There is high morbidity and mortality associated with untreated thoracic aortic transections. Only 20% of patients with acute traumatic aortic injuries survive for more than 1 hour and most of these deaths occur before the victim reaches the hospital [9]. Consequently, BTAIs are considered vascular emergencies. Misinterpretation of pulsation artifacts has been identified as a cause of unnecessary hospital transfer and management.

Typical clinical scenario

Blunt traumatic aortic injury is caused by high-energy trauma, most commonly by high-speed motor vehicle collisions but also falls from great height, pedestrian–motor vehicle collisions, and crush injuries [10]. They are very rare in children <10 years of age, and almost never occur in children under the age of 5. Clinical signs and symptoms are non-specific and poor for exclusion of BTAIs. In fact, clinical signs may be absent in up to 33% of patients [11]. Thoracic aortic dissection related to blunt trauma is almost never seen. To discriminate between the artifact and an aortic dissection, always consider half-scan reconstruction, a follow-up cardiac-gated CTA of the thoracic aorta or transesophageal echocardiography.

Differential diagnosis

The differential diagnosis for pulsation artifact simulating BTAI includes: intramural hematoma, aortic dissection, and pseudoaneurysms.

Teaching point

Pulsation artifact within the aorta is a common artifact and may mimic BTAI or aortic dissection. To distinguish from a true injury examine the adjacent pulmonary artery and superior vena cava for "pseudoflaps." The artifact may extend into the mediastinal fat or to adjacent lines and tubes. In equivocal cases, consider ECG-gated CTA or transesophageal echocardiography.

REFERENCES

1. Gunn ML. Imaging of aortic and branch vessel trauma. *Radiol Clin North Am.* 2012;**50**(1):85–103.
2. Kuhlman JE, Pozniak MA, Collins J, Knisely BL. Radiographic and CT findings of blunt chest trauma: aortic injuries and looking beyond them. *Radiographics.* 1998;**18**(5):1085–106; discussion 1107–8; quiz 1.
3. Ko SF, Hsieh MJ, Chen MC, *et al.* Effects of heart rate on motion artifacts of the aorta on non-ECG-assisted 0.5-sec thoracic MDCT. AJR *Am J Roentgenol.* 2005;**184**(4):1225–30.
4. Pretre R, LaHarpe R, Cheretakis A, *et al.* Blunt injury to the ascending aorta: three patterns of presentation. *Surgery.* 1996;**119**(6):603–10.
5. Steenburg SD, Ravenel JG, Ikonomidis JS. Blunt traumatic injury of the ascending aorta: multidetector CT findings in two cases. *Emerg Radiol.* 2007;**13**(4):217–21.
6. Symbas PJ, Horsley WS, Symbas PN. Rupture of the ascending aorta caused by blunt trauma. *Ann Thorac Surg.* 1998;**66**(1):113–17.
7. Morrison TM, Choi G, Zarins CK, Taylor CA. Circumferential and longitudinal cyclic strain of the human thoracic aorta: age-related changes. *J Vasc Surg.* 2009;**49**(4):1029–36.
8. Ganten M, Krautter U, Hosch W, *et al.* Age related changes of human aortic distensibility: evaluation with ECG-gated CT. *Eur Radiol.* 2007; **17**(3):701–8.
9. Parmley LF, Mattingly TW, Manion WC, Jahnke EJ, Jr. Nonpenetrating traumatic injury of the aorta. *Circulation.* 1958;**17**(6):1086–101.
10. Steenburg SD, Ravenel JG, Ikonomidis JS, Schonholz C, Reeves S. Acute traumatic aortic injury: imaging evaluation and management. *Radiology.* 2008;**248**(3):748–62.
11. Kram HB, Wohlmuth DA, Appel PL, Shoemaker WC. Clinical and radiographic indications for aortography in blunt chest trauma. *J Vasc Surg.* 1987;**6**(2):168–76.

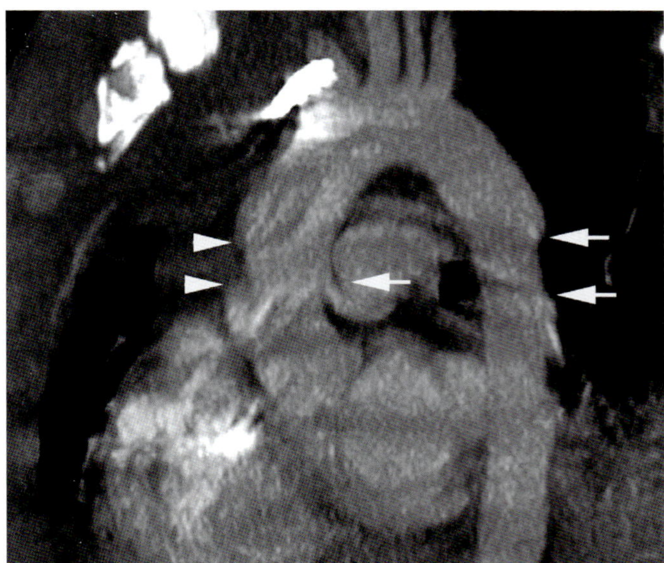

Figure 39.1 Oblique maximum intensity projection image from contrast CT of the thoracic aorta in an 18-year-old man in a high-speed motor vehicle collision demonstrates pulsation and motion artifact mimicking blunt traumatic aortic injury of the ascending aorta (arrowheads). Note that the same artifact is seen in adjacent structures including the pulmonary artery and proximal descending thoracic aorta (arrows).

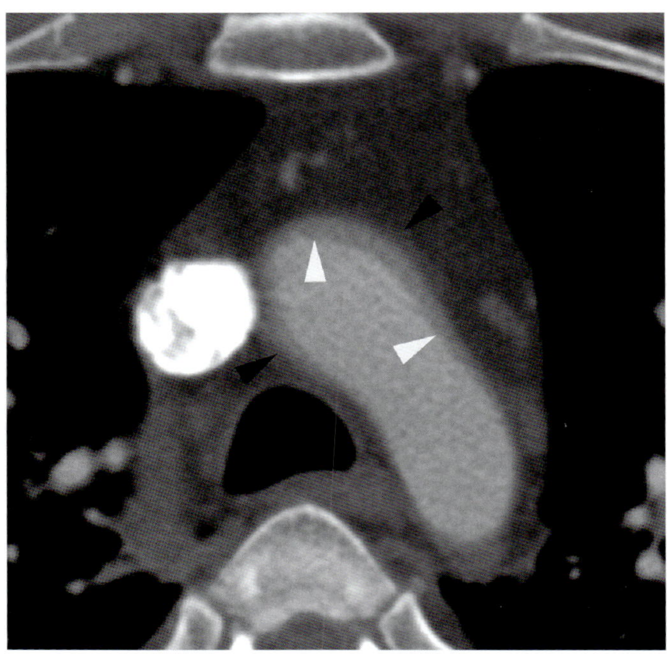

Figure 39.2 Axial contrast CT from a 39-year-old man in a high-speed motor vehicle collision shows pulsation artifact (white arrowheads) within the ascending aorta mimicking blunt traumatic aortic injury. There is extension of the linear hypodensity of pulsation artifact into the mediastinal fat (black arrowheads). Note the absence of indirect signs of aortic injury such as mediastinal hematoma.

Mediastinal widening due to non-hemorrhagic causes

Randy K. Lau

Imaging description

The approach to non-traumatic, non-hemorrhagic causes of mediastinal widening on emergency radiographs can be divided into broad diagnostic categories. Imaging findings will vary depending on the cause, but CT will invariably be diagnostic.

Mediastinal widening is perhaps the best known radiographic sign of blunt thoracic aortic injury (BTAI). However, the definition of a widened mediastinum varies. Quantitatively, it refers to a mediastinal width of 8 cm at the level of the aortic arch on a supine (or erect) chest anterior-posterior radiograph [1]. Due to variation in patient size, mediastinum to chest-width ratios of (>0.25 [and >0.38]) have been suggested as more accurate measures, but these have not been found to be consistently sensitive [2]. Supine radiography has a poor specificity for aortic injury [2]. Although it offers greater specificity, an erect chest radiograph often cannot be obtained in unstable trauma patients and in the setting of a potential spine injury. No single radiographic sign has adequate specificity or sensitivity for the diagnosis or exclusion of aortic injury in patients with a BTAI. The specificity and sensitivity of radiographic evaluation is increased by considering a combination of other suggested signs. These include an abnormal aortic knob contour, shift of the tracheal wall to the right of the T4 transverse process, rightward deviation of the nasogastric tube, an apical pleural cap, widened paraspinal lines, or depression of the left main bronchus [2]. For a detailed discussion of the signs of aortic injury see Case 41.

Mediastinal lipomatosis is a well-recognized benign cause of mediastinal widening. It is associated with obesity, steroid use, and Cushing's syndrome [3]. Mediastinal widening due to mediastinal lipomatosis is usually bilateral and demonstrates smooth, sharply defined linear contours (Figure 40.1).

Vascular anomalies such as a persistent left superior vena cava (SVC), azygous continuation of the inferior vena cava, aberrant right subclavian artery with a diverticulum of Kommerell, and cervical aortic arches can also cause mediastinal widening (Figures 40.2 and 40.3). Tortuous brachiocephalic arteries mimicking mediastinal widening are usually associated with atherosclerotic or hypertensive disease of the aorta and are usually seen in older patients. Contrast-enhanced CT will be diagnostic.

Atelectasis can occasionally cause apparent mediastinal widening. Radiographic signs of atelectasis include increased density of the affected lung, displacement of fissures, displacement of mediastinum structures, bronchovascular crowding, and elevation of affected hemidiaphragm with a possible juxtaphrenic peak sign.

Technical factors can also result in apparent mediastinal widening. These include: inadequate inspiration, patient rotation, short target-film distance, and supine or lordotic positioning (Figure 40.4) [4].

In young children, the thymus obscures the thoracic aortic contour. However, BTAI is quite uncommon in children <10 years of age, and vanishingly rare in children <5 years [5].

Importance

Chest radiography is the first imaging examination performed in patients with symptoms referable to the chest and an important part of the diagnostic workup in trauma patients. While there are many pathologic causes of an abnormal mediastinum, there are also normal variants and technical factors that can mimic an abnormal mediastinum, causing confusion in trauma patients.

Typical clinical scenario

The typical clinical scenario will vary widely depending on the cause of mediastinal widening.

Differential diagnosis

Within the anterior mediastinum, prevascular masses that may widen the mediastinum include lymphadenopathy, retrosternal thyroid goiter, thymic masses, and germ cell tumors [6]. Other more rare lesions include hemangiomas and lymphangiomas. Lymphadenopathy will have lobulated borders. Retrosternal thyroid goiters will demonstrate deviation of the trachea.

Within the middle mediastinum, non-traumatic, non-hemorrhagic causes of mediastinal widening include lymphadenopathy, enlarged pulmonary arteries, and aortic abnormalities. Aortic abnormalities include aortic aneurysms and acute aortic syndromes.

Posterior mediastinal causes of non-traumatic, non-hemorrhagic mediastinal widening include esophageal lesions, neurogenic tumors, paraspinal abscess, foregut duplication cysts, and aortic abnormalities.

Teaching point

Mediastinal widening as a radiographic sign alone is non-specific for traumatic aortic injury. Benign causes of mediastinal widening include atelectasis, inadequate inspiration, mediastinal lipomatosis, congenital vascular anomalies, tortuous vessels, and radiographic technique.

REFERENCES

1. Gunn ML. Imaging of aortic and branch vessel trauma. *Radiol Clin North Am.* 2012;**50**(1):85–103.

2. Mirvis SE, Bidwell JK, Buddemeyer EU, *et al.* Imaging diagnosis of traumatic aortic rupture. A review and experience at a major trauma center. *Invest Radiol.* 1987;**22**(3):187–96.

3. Homer MJ, Wechsler RJ, Carter BL. Mediastinal lipomatosis. CT confirmation of a normal variant. *Radiology.* 1978;**128**(3):657–61.

4. Reeder MM, Felson B. *Reeder and Felson's Gamuts in Radiology: Comprehensive Lists of Roentgen Differential Diagnosis* 4th edn. New York, Springer Verlag 2003.

5. Barmparas G, Inaba K, Talving P, David JS, Lam L, Plurad D, *et al.* Pediatric vs adult vascular trauma: a National Trauma Databank review. *J Pediatr Surg.* 2010;**45**(7):1404–12.

6. Whitten CR, Khan S, Munneke GJ, Grubnic S. A diagnostic approach to mediastinal abnormalities. *Radiographics.* 2007;**27**(3):657–71.

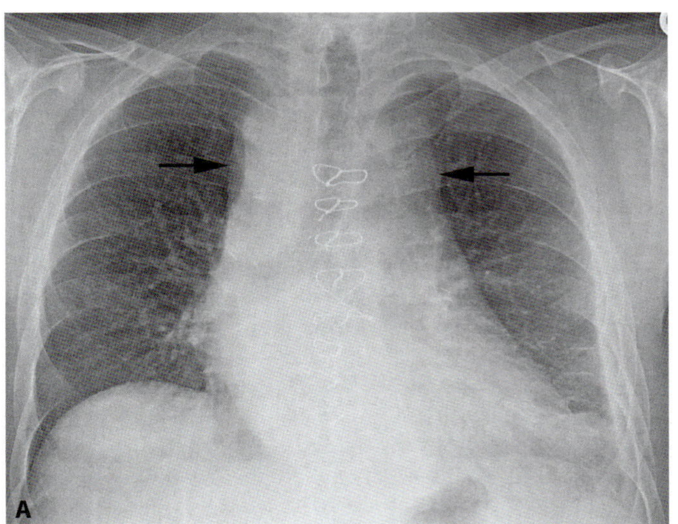

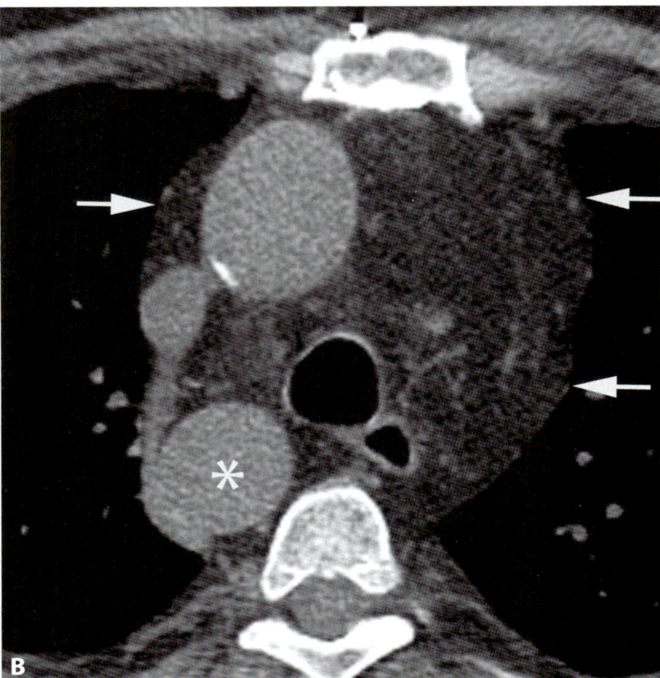

Figure 40.1 A. A 61-year-old man after combined heart–lung transplantation with apparent mediastinal widening (arrows). Note the relative lucency of the mediastinal widening that is typical for mediastinal lipomatosis. **B.** Axial contrast CT from the same patient showing mediastinal lipomatosis (white arrows). Incidental note is made of a right aortic arch (asterisk).

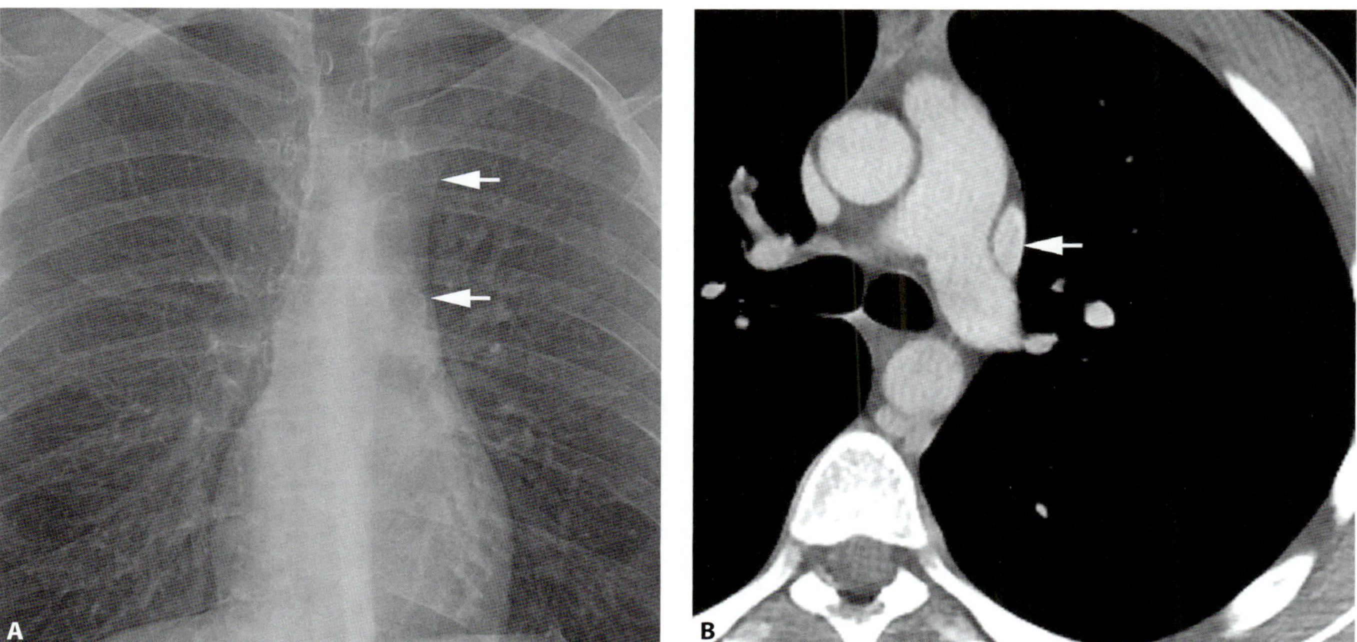

Figure 40.2 **A.** An 18-year-old man with chest radiograph demonstrating widening of left superior mediastinum (arrows). Note, the smooth linear nature of the mediastinal widening related to its vascular nature. **B.** Axial contrast CT demonstrating an abnormal vessel lateral to the main and left pulmonary artery (arrow) representing persistent left SVC.

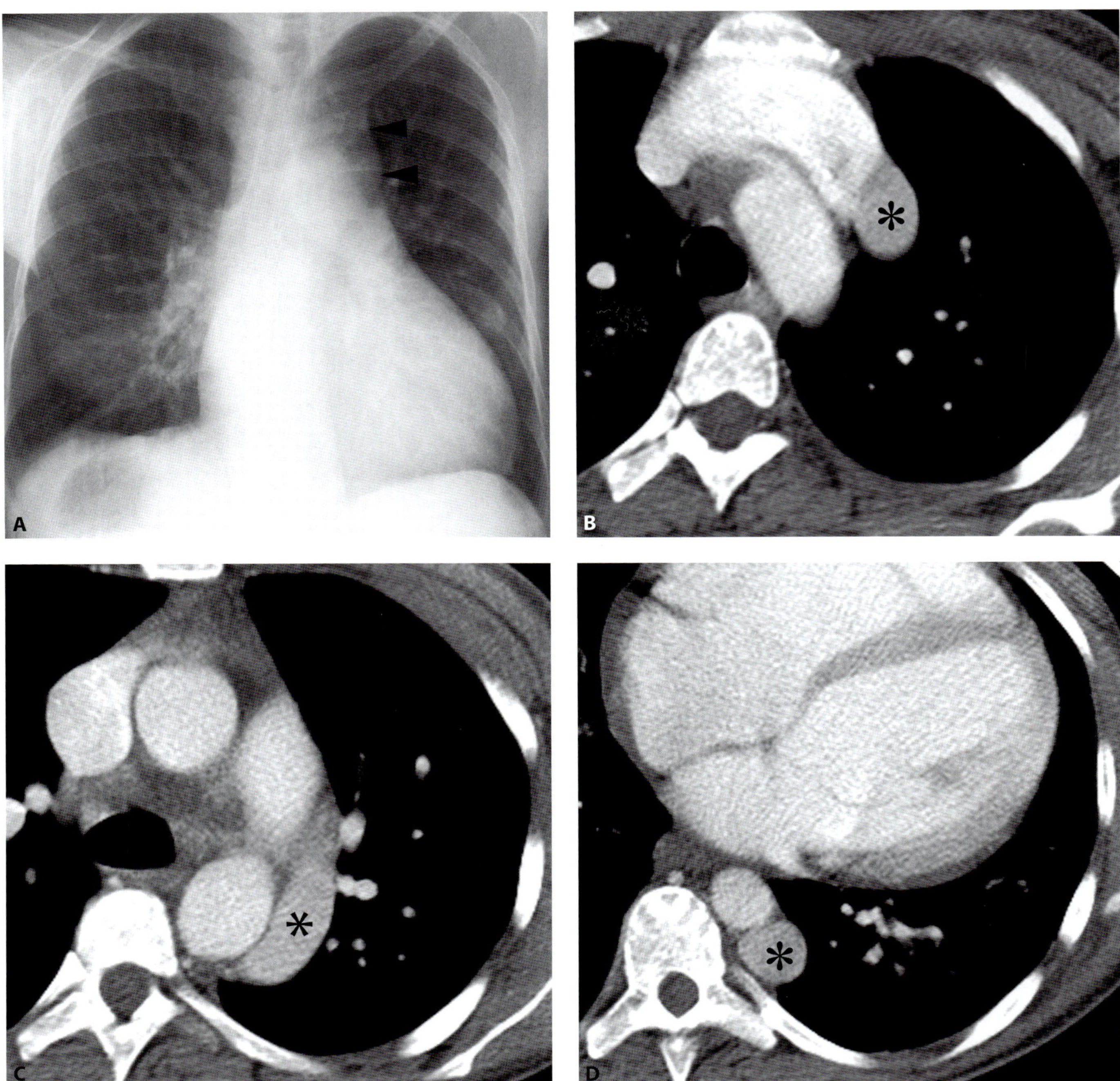

Figure 40.3 **A.** Chest radiograph from a 32-year-old woman with sickle cell anemia shows abnormal widening of the left superior mediastinum (arrowheads). **B–D.** There is a congenital venous anomaly with drainage of venous blood (asterisk) from the abdomen via the hemi-azygous vein, left superior intercostal vein, and into the left brachiocephalic vein.

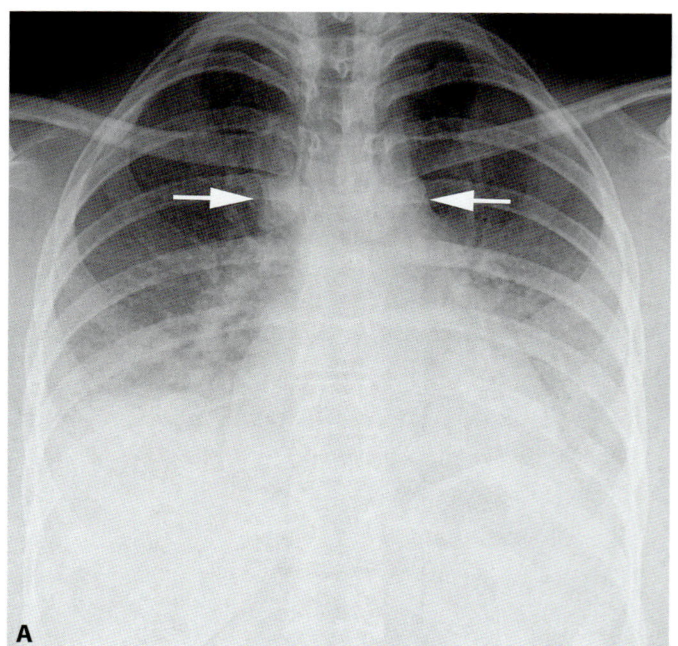

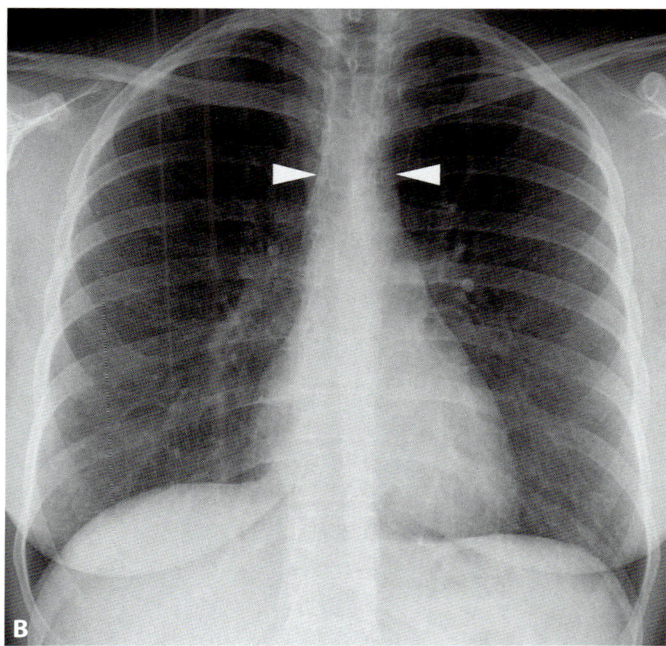

Figure 40.4 A. A 22-year-old woman with chest pain and right shoulder pain after an upper respiratory infection. A chest radiograph performed on end expiration demonstrates low lung volumes and apparent mediastinal widening (arrows). **B.** Chest radiograph repeated at end inspiration demonstrates normal mediastinum (arrowheads) with larger lung volumes.

CASE

41 Aortic injury with normal mediastinal width

Martin L. Gunn

Imaging description

In many trauma centers, chest radiography is used as the initial diagnostic test to evaluate for blunt thoracic aortic injury (BTAI). Radiography is usually performed using a supine portable anterior-posterior (AP) projection. Acute BTAI is Frequently associated with mediastinal hematoma, which may widen the superior mediastinal contour. Mediastinal widening on chest radiography refers to a mediastinal width of 8 cm at the level of the aortic arch on a supine (or erect) chest AP radiograph. Due to variation in patient size, a mediastinum to chest-width ratio of (>0.25 [or >0.38]) has been suggested as a more accurate measure. However, a subjective assessment of mediastinal width is usually used in practice [1].

Reports of a high negative predictive value of a normal chest radiograph (>98%) simply reflect the low incidence of patients with BTAI in the populations being assessed. The chest radiograph may be normal in 7–11% of patients with BTAI [2–4]. Moreover, radiographs have a low specificity for thoracic aortic injury. This is due to several other potential causes of mediastinal widening, which are discussed in Case 40. In order to increase specificity, consider using a combination of radiographic signs of BTAI (Figures 41.1 and 41.2, Table 41.1).

Table 41.1. Radiographic signs of BTAI

Widened mediastinum
Loss of the aortopulmonary window
Loss of definition of the descending thoracic aorta
Widened right paratracheal stripe
Tracheal shift to the right of the T4 spinous process
Left mainstem bronchus depression
Nasogastric tube displaced to the right
Widened paraspinal stripe
Left apical cap sign*

*Almost never seen without other signs of BTAI.

Importance

Despite changes in the pattern of vehicular intrusion and advances in automobile safety technology, the incidence of BTAI does not appear to be declining. Blunt thoracic aortic injury is a lethal injury. Up to 80% of patients with BTAI from motor vehicle collisions will die at the scene. However, prompt identification and management with beta blockade and repair (either endovascular or open) is associated with an excellent outcome. In the $AAST_2$ trial, mortality in patients with BTAI who presented to hospital was 13% [5]. There are multiple reasons for recent improvement in the outcome for patients with aortic injury. One is likely to be the widespread availability and use of CT angiography (CTA) as a screening tool. Both the sensitivity and specificity of CTA are in the 95–100% range [1, 6]. Accordingly, it has almost completely replaced transesophageal echo and catheter angiography as the primary diagnostic test.

Typical clinical scenario

Blunt thoracic aortic injury typically occurs as a consequence of high energy trauma such as motor vehicle collisions. The mechanism is usually due to rapid deceleration, or near-side impact. Other causes include motorcycle and aircraft crashes, pedestrian injuries, falls from height, and crush injuries. Blunt aortic injuries do not result from falls from standing.

Differential diagnosis

The several potential causes other than aortic injury that can cause a widened or abnormal mediastinal contour are discussed in Case 40. As there is an association between BTAI and first rib fractures, these fractures were previously considered predictive of aortic and great vessel injury. However, first and second rib fractures are quite frequently identified on imaging and on their own they do not indicate the need for angiography. Associated injuries should be considered in all cases of suspected aortic injury. In one large autopsy study, victims of BTAI had extrathoracic injuries in 96% of cases [7]. Patients with BTAI are significantly more likely to have cardiac injury, diaphragmatic lacerations, hemothorax, upper rib fractures, pelvic fractures, and intra-abdominal injuries.

Patients with posterior sternoclavicular dislocation (Case 36) and great vessel injuries may have a normal mediastinal contour.

Minimal aortic injury, which is an injury to the thoracic aorta without external contour irregularity and usually measures less than 10 mm in size, can occur in the absence of periaortic hematoma detectable by CT (Figure 41.3) [8, 9].

Teaching point

Aortic injury can present without mediastinal widening on plain radiographs. If an aortic injury is suspected based on the mechanism of injury, CTA of the thoracic aorta should be obtained even in the absence of mediastinal widening.

REFERENCES

1. Gunn ML. Imaging of aortic and branch vessel trauma, *Radiol Clin North Am.* 2012;**50**(1):85–103.

2. Ekeh AP, Peterson W, Woods RJ, *et al.* Is chest x-ray an adequate screening tool for the diagnosis of blunt thoracic aortic injury? *J Trauma* 2008;**65**(5):1088–92.

3. Woodring JH. The normal mediastinum in blunt traumatic rupture of the thoracic aorta and brachiocephalic arteries. *J Emerg Med* 1990;**8** (4):467–76.

4. Benjamin ER, Tillou A, Hiatt JR, *et al.* Blunt thoracic aortic injury. *Am Surg* 2008;**74**(10):1033–7.

5. Fabian TC, Richardson JD, Croce MA, *et al.* Prospective study of blunt aortic injury: multicenter trial of the American Association for the Surgery of Trauma. *J Trauma* 1997;**42**(3):374–80; discussion 380–3.

6. Steenburg SD, Ravenel JG. Acute traumatic thoracic aortic injuries: experience with 64-MDCT. *AJR Am J Roentgenol* 2008; **191**(5):1564–9.

7. Teixeira PG, Inaba K, Barmparas G, *et al.* Blunt thoracic aortic injuries: an autopsy study. *J Trauma* 2011;**70**(1):197–202.

8. Starnes BW, Lundgren RS, Gunn M, *et al.* A new classification scheme for treating blunt aortic injury. *J Vasc Surg.* 2012;**55**(1):47–54.

9. Malhotra AK, Fabian TC, Croce MA, *et al.* Minimal aortic injury: a lesion associated with advancing diagnostic techniques, *J Trauma.* 2001;**51**(6):1042–8.

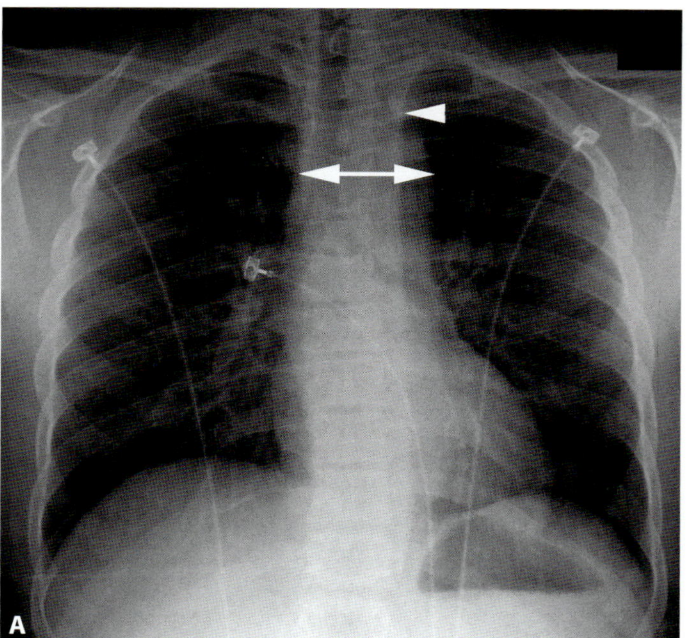

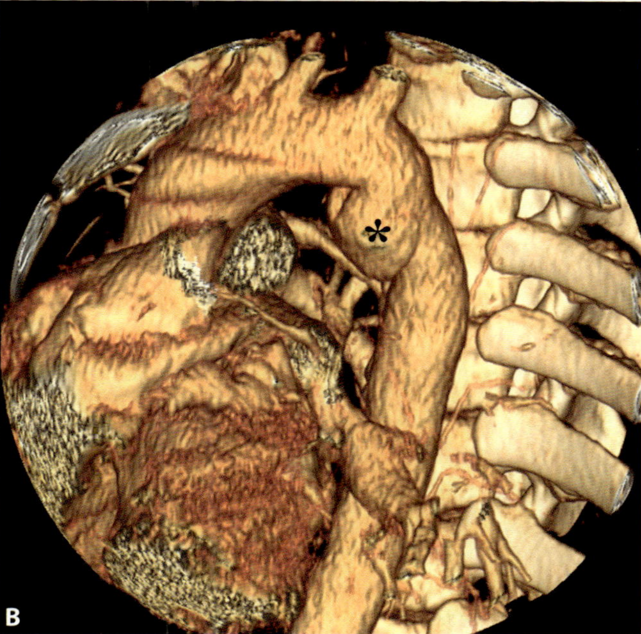

Figure 41.1 **A.** Portable chest radiograph from a 17-year-old man who fell from a high bridge. The mediastinal width measures 6.5 cm, and has a normal contour (white double-arrow). Subtle widening of the left paravertebral stripe is visible (arrowhead), but was not identified prospectively. **B.** Volume rendered reconstruction from a CTA of the thoracic aorta reveals a large pseudoaneurysm arising from the aortic isthmus (asterisk). This was treated successfully with a thoracic endovascular stent graft.

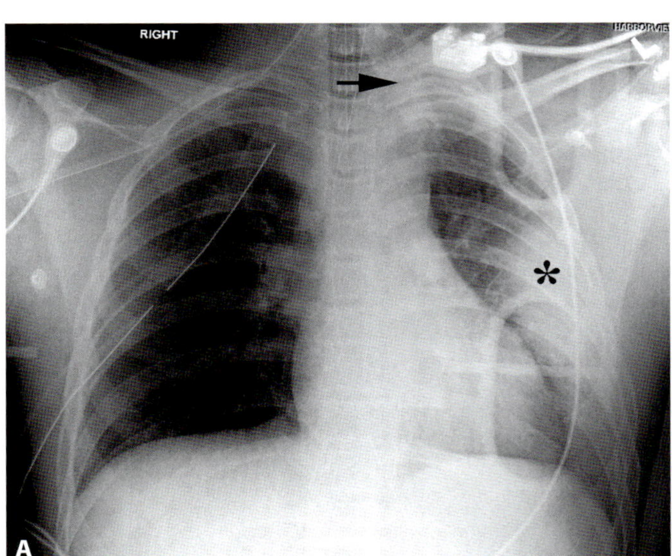

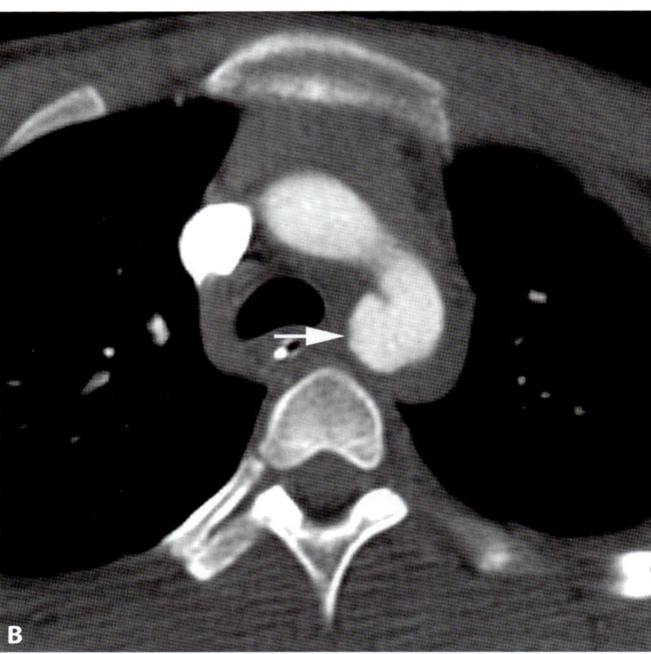

Figure 41.2 A. Portable chest radiograph from a 16-year-old man involved in a head-on motor vehicle collision. The mediastinum is not widened, although there is a left "apical pleural cap" (arrow), and large left pulmonary contusion (asterisk). Note the absence of rib fractures (although the scapula and left humeral head are fractured [not shown]. **B.** Chest CTA reveals a pseudoaneurysm arising from the aortic isthmus (arrow).

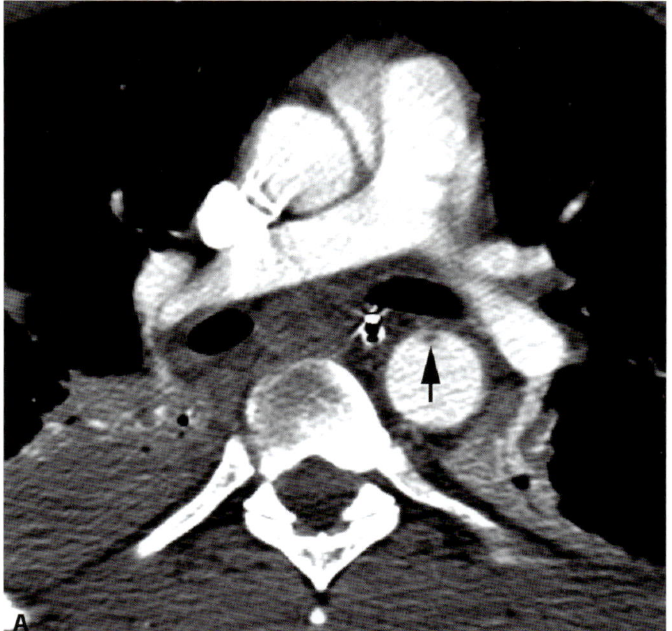

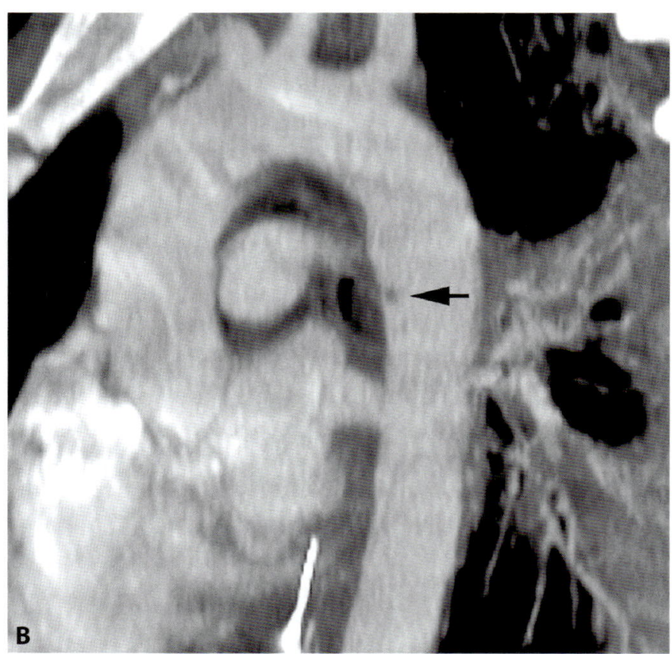

Figure 41.3 Axial (**A**) and sagittal (**B**) images from a CTA of the chest in a 24-year-old man following a high-speed motor vehicle collision. A minimal aortic injury (MAI) is present (arrow) in the proximal descending thoracic aorta. Note the absence of periaortic hematoma. The mediastinal contour on the chest radiograph (not shown) was abnormal due to bilateral pulmonary atelectasis and the right hemothorax. The MAI had resolved on follow-up CTA without intervention.

Retrocrural periaortic hematoma

Martin L. Gunn

Imaging description

Blunt traumatic aortic injury is often, but not always, associated with mediastinal widening and aortic contour abnormalities on chest radiography (Cases 40 and 41) Most, but not all, patients with blunt thoracic aortic injury (BTAI) and an aortic pseudoaneurysm will have periaortic hematoma on the abdominal CT (Figures 42.1 and 42.2). In one retrospective study, 11/14 CTs of the abdomen in patients with known BTAI demonstrated periaortic hematoma in the retrocrural region [1]. In another study, 14/20 patients with confirmed BTAI had periaortic hematoma [2]. Overall, the sensitivity for the detection of BTAI in these studies was 70% and 88% [1, 2].

Importance

The importance of periaortic hematoma is that it is a marker of potential aortic injury. Patients with a blunt trauma mechanism that is compatible with traumatic aortic injury should undergo CT angiography (CTA) of the thoracic aorta to exclude BTAI.

Typical clinical scenario

Blunt thoracic aortic injury is a highly lethal injury. Up to 80% of patients with BTAI die at the scene. It is diagnosed in less than 0.5–2% of non-lethal motor vehicle collisions. Based on the historic work of Parmley, we know that unrecognized and untreated aortic injury has a high mortality, with a 1% mortality rate per hour for the first 48 hours after admission [3]. Fortunately, CTA of the chest, and whole body CT are now commonly performed to evaluate patients at risk of BTAI and survival is greatly improved [4]. Signs of chest injury (e.g., thoracic cage fractures, pneumothorax, pulmonary contusions, or diaphragmatic injury) are present in most patients with BTAI who undergo chest radiography as the initial evaluation following major trauma. However, these signs can be subtle, or even absent, particularly in young patients who have a more pliable chest wall [5]. A recent study found that extrathoracic injuries are present in 96% of patients with BTAI. The most common were intra-abdominal injuries and pelvic fractures [6]. Hence, it is possible that CTA of the chest might be omitted from the initial evaluation of some major trauma patients.

Differential diagnosis

Retrocrural periaortic fluid is not specific for BTAI. It may also be due to spinal injury, posterior rib fracture, or it may be idiopathic.

Teaching point

Retrocrural periaortic hematoma should be considered as a marker for aortic injury. If it is identified on an abdominal CT scan, and no CTA of the chest is available, a CTA of the chest should be performed.

REFERENCES

1. Curry JD, Recine CA, Snavely E, Orr M, Fildes JJ. Periaortic hematoma on abdominal computed tomographic scanning as an indicator of thoracic aortic rupture in blunt trauma. *J Trauma*. 2002;**52**(4):699–702.
2. Wong H, Gotway MB, Sasson AD, Jeffrey RB. Periaortic hematoma at diaphragmatic crura at helical CT: sign of blunt aortic injury in patients with mediastinal hematoma. *Radiology*. 2004;**231**(1):185–9.
3. Parmley LF, Mattingly TW, Manion WC, Jahnke EJ, Jr. Nonpenetrating traumatic injury of the aorta. *Circulation*. 1958; **17**(6):1086–101.
4. Demetriades D, Velmahos GC, Scalea TM, *et al.* Diagnosis and treatment of blunt thoracic aortic injuries: changing perspectives. *J Trauma*. 2008;**64**(6):1415–18; discussion 1418–19.
5. Gunn ML. Imaging of aortic and branch vessel trauma. *Radiol Clin North Am*. 2012;**50**(1):85–103.
6. Teixeira PG, Inaba K, Barmparas G, *et al.* Blunt thoracic aortic injuries: an autopsy study. *J Trauma*. 2011;**70**(1):197–202.

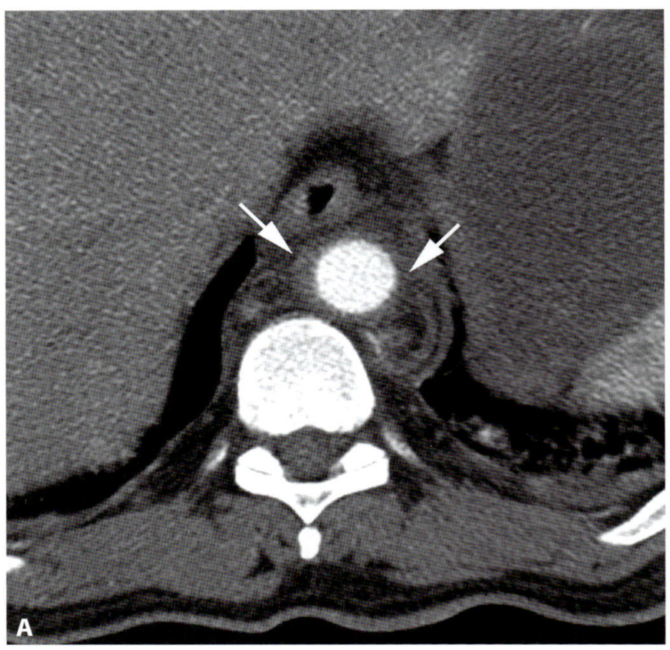

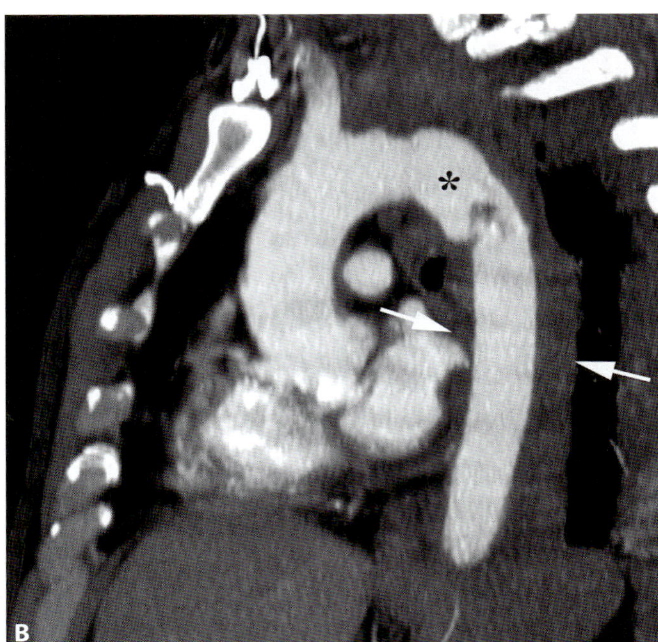

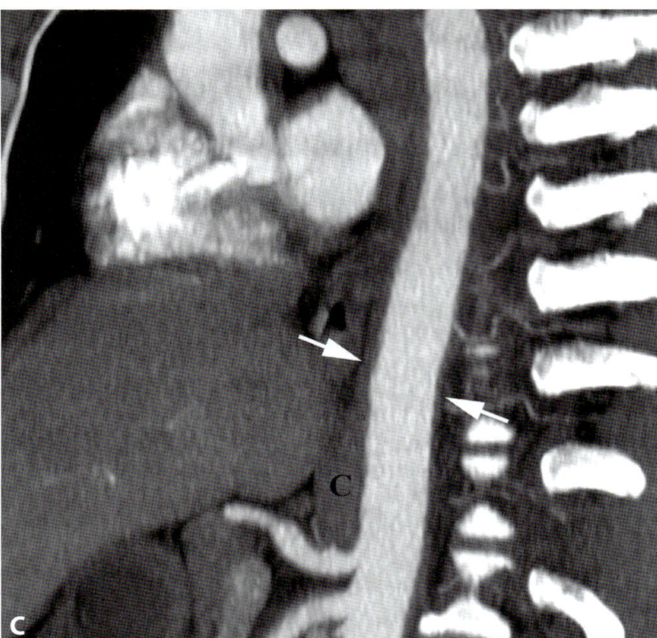

Figure 42.1 A 41-year-old male pedestrian who was hit by a car. **A.** Axial image from a CTA of the abdomen reveals periaortic hematoma in the retrocrural region, surrounding the aorta (arrows). This is likely due to subadventitial hematoma. In addition, there is some hemorrhage visible in the surrounding retrocrural fat. **B.** Sagittal reformation of the upper chest shows extensive periaortic hematoma (arrows) and an aortic transection in the region of the aortic isthmus (asterisk). **C.** However, the sagittal reformation of the retrocrural location indicates that the quantity of blood extending to this area is small (arrows). "C" indicates the crus of the hemidiaphragm.

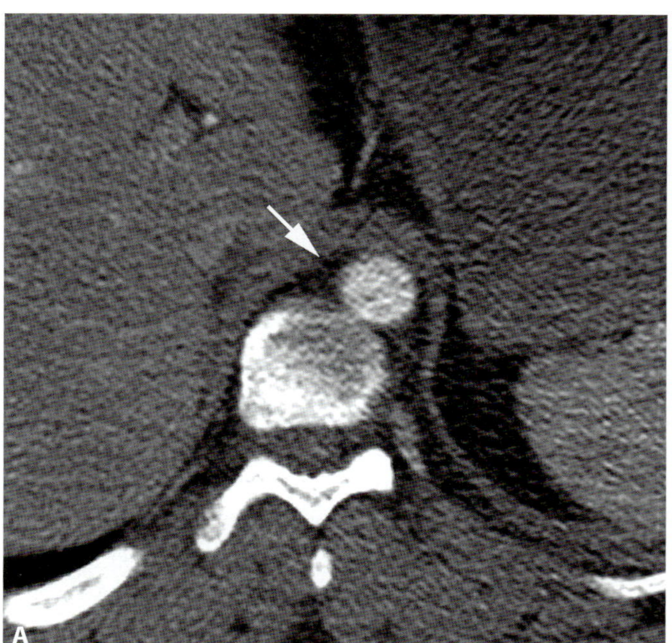

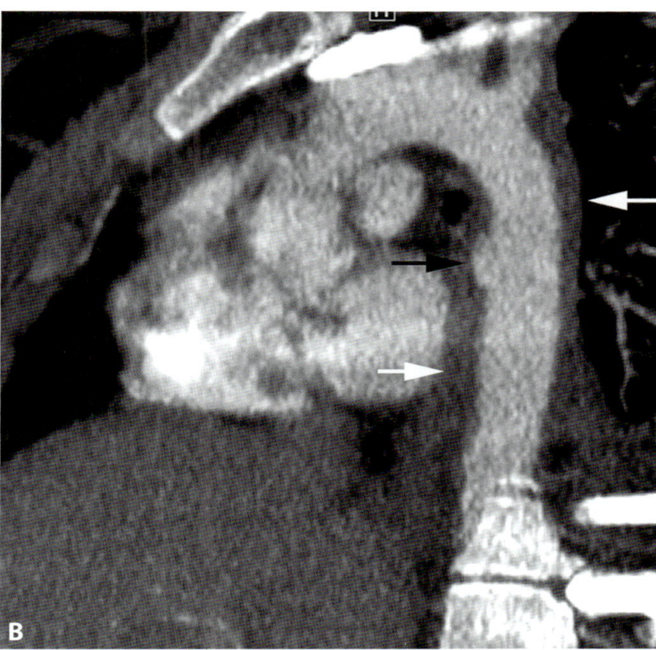

Figure 42.2 An 18-year-old man who was involved in a motorcycle collision. **A.** On axial CT of the upper abdomen, there was no periaortic hematoma (arrow). **B.** Sagittal reformation of the chest reveals extensive periaortic hematoma more cranially (white arrows), and a pseudoaneurysm of the descending thoracic aorta (black arrow). The absence of periaortic hematoma in the upper slices in an abdominal CT should not be used to exclude blunt thoracic aortic injury.

Mimicks of hemopericardium on FAST

Matthew H. Nett

Imaging description

During the focused assessment with sonography for trauma (FAST) scan, the epicardial fat pad can be mistaken for hemopericardium or a pericardial effusion [1]. This is more often seen in obese patients and can appear sonographically anechoic (Figure 43.1). To avoid this pitfall, Blaivas *et al.* suggest using a modified sub-xiphoid view [1]. In this technique the probe is angled perpendicular to the skin, visualizing the inferior vena cava (IVC) entering the right atrium, which allows visualization of the right side of the heart next to the diaphragm. Any amount of pericardial fluid that is not loculated will be seen in this location. Also, in this view the IVC should collapse at least 50% with a sudden inspiration. Collapse of less than 50% indicates increased central pressure, possibly due to tamponade. Imaging the IVC as it enters the heart can be performed quickly and does not add significant time to the FAST examination.

If differentiating between a hemopericardium and an epicardial fat pad remains difficult despite this technique, and the patient is stable, further imaging with computed tomography provides rapid differentiation (Figure 43.2).

Another potential pitfall with pericardial effusion on FAST occurs with penetrating precordial injuries. These patients may have concurrent lacerations to the heart and pericardial sac. Blood might not accumulate in the pericardial sac in these patients, as it can escape into the pleural space [2].

Importance

FAST is a means of rapidly triaging patients to the operating room, and is becoming more popular among emergency physicians and radiologists [3]. However, unnecessary thoracotomy can result from the false-positive diagnosis of hemopericardium [4].

Typical clinical scenario

This diagnostic dilemma may occur in an unstable blunt or penetrating trauma patient. Additionally, patients may undergo emergent pericardial ultrasound when they present in hemodynamic shock with pulseless electrical activity, as pericardial tamponade is a readily treatable cause. The use of emergent bedside echocardiography is continuing to expand [5].

FAST is advantageous because it can be performed during resuscitation, and in experienced hands can be performed in a few minutes. Although there is great variability in the reported sensitivity and specificity of the FAST exam, it is useful and accurate to assess for pericardial or intraperitoneal fluid in patients with hemodynamic instability from blunt trauma [6]. In patients in whom no abnormality is identified on FAST, further diagnostic assessment is required as significant injuries may be missed.

Differential diagnosis

When space occupying hypoechogenic material is identified in the pericardial space, the differential diagnosis includes hemopericardium, pericardial effusion from other causes, or an epicardial fat pad, which is a normal finding.

Teaching point

The epicardial fat pad can mimic hemopericardium on FAST scan. Awareness of this mimicker may help prevent unnecessary interventions in critically ill patients. If the FAST scan is negative or equivocal and there are clinical signs of abdominal trauma, further imaging with CT is recommended.

REFERENCES

1. Blaivas M, DeBehnke D, Phelan MB. Potential errors in the diagnosis of pericardial effusion on trauma ultrasound for penetrating injuries. *Acad Emerg Med.* 2000;**7**(11):1261–6.
2. Ball CG, Williams BH, Wyrzykowski AD, *et al.* A caveat to the performance of pericardial ultrasound in patients with penetrating cardiac wounds. *J Trauma.* 2009;**67**(5):1123–4.
3. Tayal VS, Beatty MA, Marx JA, Tomaszewski CA, Thomason MH. FAST (focused assessment with sonography in trauma) accurate for cardiac and intraperitoneal injury in penetrating anterior chest trauma. *J Ultrasound Med.* 2004;**23**(4):467–72.
4. Rozycki GS, Feliciano DV, Ochsner MG, *et al.* The role of ultrasound in patients with possible penetrating cardiac wounds: a prospective multicenter study. *J Trauma.* 1999;**46**(4):543–51; discussion 551–2.
5. Beaulieu Y. Bedside echocardiography in the assessment of the critically ill. *Crit Care Med.* 2007;**35**(5 Suppl):S235–49.
6. Patel NY, Riherd JM. Focused assessment with sonography for trauma: methods, accuracy, and indications. *Surg Clin North Am.* 2011; **91**(1):195–207.

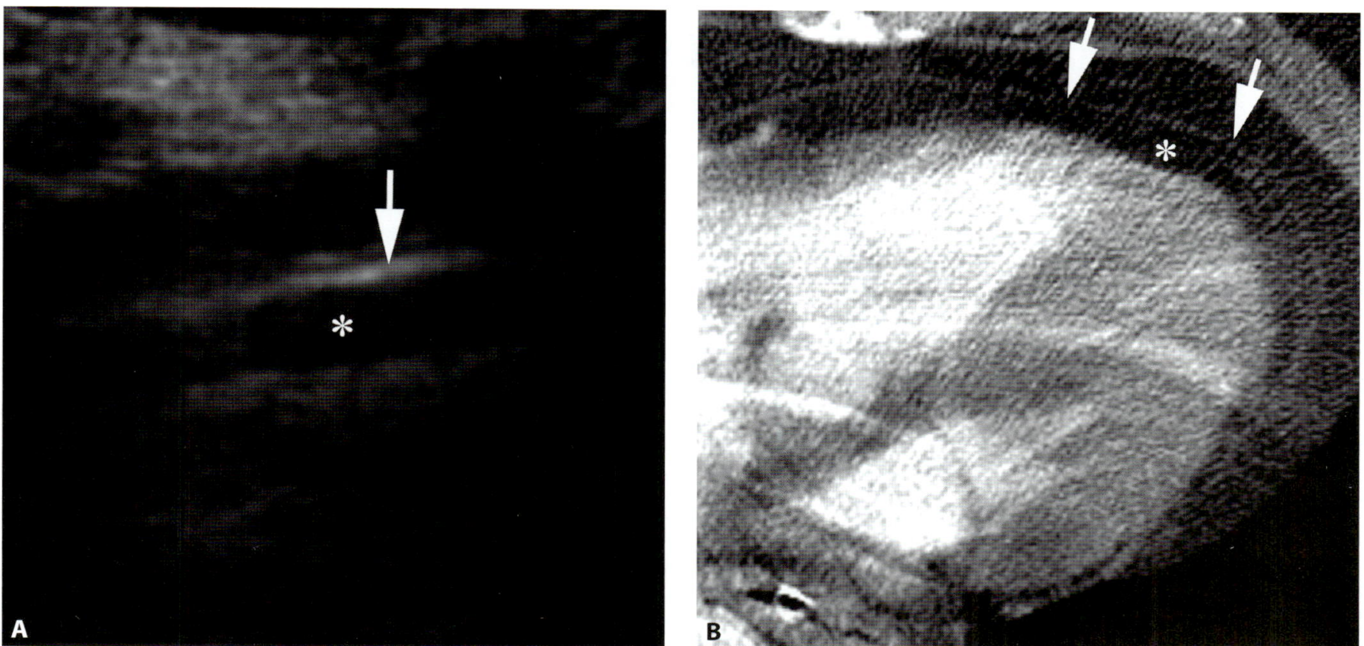

Figure 43.1 **A.** Subxiphoid ultrasound view from a FAST in a 54-year-old man involved in blunt trauma demonstrates apparent anechoic fluid in the pericardial space (asterisk) between the right ventricle wall and pericardial line (arrow), suggestive of a traumatic hemopericardium. **B.** Corresponding contrast-enhanced CT shows an abundance of pericardial fat (asterisk) outlined by the pericardium (arrows), mimicking pericardial effusion on the prior ultrasound.

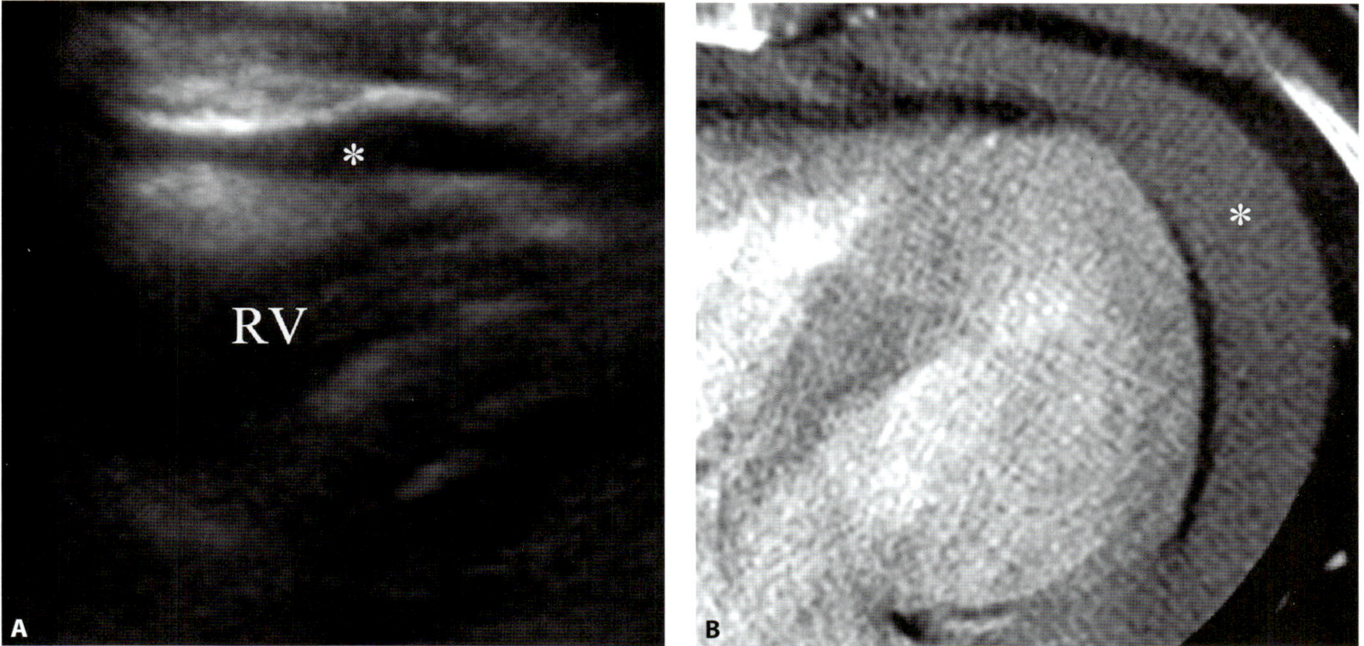

Figure 43.2 **A.** Subxiphoid sonographic view from a FAST scan in a 28-year-old man involved in a penetrating trauma demonstrates pericardial fluid (asterisk) between the right ventricular (RV) myocardium and pericardial white line. **B.** Contrast-enhanced chest CT confirms the hemopericardium (asterisk).

Mimicks of acute thoracic aortic syndromes: aortic dissection, intramural hematoma, and penetrating aortic ulcer

Martin L. Gunn

Imaging description

Acute thoracic aortic dissection is the most common aortic catastrophe. It is caused by a tear along the laminar planes of the media layer of the aortic wall, with formation of a blood-filled channel. The diagnosis of aortic dissection is usually quite simple using contrast-enhanced multi-detector CT angiography (CTA). However, several artifacts can simulate aortic dissection.

Most artifacts lead to a false-positive diagnosis.

Pulsation artifact mimics aortic dissection, particularly Stanford type A aortic dissection. Within the ascending aorta, pulsation artifact occurs principally in the left anterior and right posterior wall of the ascending aorta. To discriminate, look for the artifact extending into the adjacent mediastinal fat, and similar "pseudoflaps" in the adjacent superior vena cava and main pulmonary artery. This can be particularly valuable to discriminate between type A and type B aortic dissection, and can even be performed using prospective gating without beta blockade in the acutely unwell patient (Figure 44.1). Aortic pulsation artifact can obscure subtle intimal irregularities, and impairs assessment of aortic valve and coronary involvement for type A dissections. These relationships are well assessed with gating (Figure 44.2) [1].

Aortic pulsation artifact, which causes pseudointimal flaps and can simulate traumatic aortic injury, is covered in Case 39.

Another potential mimicker of aortic dissection is atelectasis. Atelectatic lung usually enhances vigorously, and may simulate a thin false lumen surrounding the aorta. Clues to the presence of atelectasis include a limited extent, and careful inspection of lung windows (Figure 44.3).

Acute intramural hematoma (IMH) is the acute appearance of a hematoma in the aortic wall without a demonstrable intimal flap or ulceration. The importance of performing non-contrast CT to detect acute IMH is discussed in Case 45. On non-contrast CT, IMH typically appears as an eccentric, hyperdense area of thickening of the aortic wall. Adjacent structures, such as atelectasis and pleural effusions may obscure or simulate acute IMH.

Plaques most commonly ulcerate in patients with advanced atherosclerosis. Early ulceration commonly occurs in the descending aorta and is initially asymptomatic and confined to the intimal layer. Penetrating aortic ulcers (PAUs) are ulcers with deeper penetration, extending through the internal elastic lamina of the aorta and burrowing deeply through the intima and into the aortic media [2]. These patients often present with pain, and have a higher rate of progression to saccular aneurysm, IMH, dissection, or rupture. Penetrating aortic ulcers appear as focal outpouchings of contrast through an area of intimal calcification, ulceration through the expected location of the intima of the aorta in the absence of calcification, or a focal outpouching of contrast with associated IMH (Figure 44.4). This distinction between a penetrating ulcer and an ulcerating plaque is important.

Importance

Misinterpretation of acute thoracic aortic disease has been identified as a cause of unnecessary hospital transfer, and unnecessary surgery [3].

The use of ECG-gated CTA in the emergency department improves definition of the walls of the ascending aorta, and aids the evaluation of the presence and extent of intimomedial flaps, and clarifies coronary artery and aortic valve involvement.

Acute IMH is best identified with non-contrast CT. However, as this disease is uncommon, performing a single phase gated post-contrast CT could be considered as the initial examination in the emergency department setting. A delayed non-contrast examination 10–15 minutes after the initial examination could be performed if there is any suspicion of aortic wall thickening to suggest an acute IMH [4].

Patients presenting with an acute aortic syndrome who are diagnosed with a PAU have a high rupture rate, 38% (10/26) [5]. These patients require intervention or surveillance imaging [6]. Patients with asymptomatic non-penetrating ulcers have a more benign course with low rates of disease progression, and likely do not require imaging surveillance.

Typical clinical scenario

Acute aortic syndrome is a term that includes acute aortic dissection, IMH, and PAU. Aortic dissection has a prevalence of 2.6 to 3.5 cases per 100 000 patient years, with approximately two-thirds of patients being male, and a mean age of all patients of 63 years [7]. Males tend to be affected earlier than females. Whereas hypertension, Marfan's syndrome, Ehlers–Danlos syndrome, and annulo-aortic ectasia are risk factors for aortic dissection, most patients who present with PAU tend to be elderly and have advanced atherosclerotic disease.

The presence of refractory pain in the presence of either acute aortic dissection or PAU is associated with a worse outcome than asymptomatic disease [6].

Differential diagnosis

For non-gated examinations, the presence of similar adjacent "pseudoflap" in the main pulmonary artery and superior vena cava should allow for discrimination between pulsation artifact and aortic dissection in the ascending aorta. Before diagnosing

PAU, consider whether the plaque penetrates into the wall of the aorta, or simply represents ulceration of atheromatous plaque.

Teaching point

Pulsation artifact is very common in the ascending aorta. Always consider gated CTA of the aorta (even without heart rate control) in the emergent setting in questionable cases of ascending aortic dissection. Using prospectively ECG-gated CTA, motion-free images of the thoracic aorta can be obtained at a low radiation dose.

REFERENCES

1. Fleischmann D, Mitchell RS, Miller DC. Acute aortic syndromes: new insights from electrocardiographically gated computed tomography. *Semin Thorac Cardiovasc Surg*. 2008; **20**(4):340–7.
2. Hayashi H, Matsuoka Y, Sakamoto I, *et al*. Penetrating atherosclerotic ulcer of the aorta: imaging features and disease concept. *Radiographics*. 2000;**20**(4):995–1005.
3. Beaver TM, Herrbold FN, Hess PJ, Jr., Klodell CT, Martin TD. Transferring diagnosis versus actual diagnosis at a center for thoracic aortic disease. *Ann Thorac Surg*. 2005;**79** (6):1957–60.
4. Sodickson A. Strategies for reducing radiation exposure in multi-detector row CT. *Radiol Clin North Am*. 2012;**50**(1):1–14.
5. Tittle SL, Lynch RJ, Cole PE, *et al*. Midterm follow-up of penetrating ulcer and intramural hematoma of the aorta. *J Thorac Cardiovasc Surg*. 2002;**123**(6):1051–9.
6. Nathan DP, Boonn W, Lai E, *et al*. Presentation, complications, and natural history of penetrating atherosclerotic ulcer disease. *J Vasc Surg*. 2012;**55**(1):10–15.
7. Tsai TT, Nienaber CA, Eagle KA. Acute aortic syndromes. *Circulation*. 2005;**112**(24):3802–13.

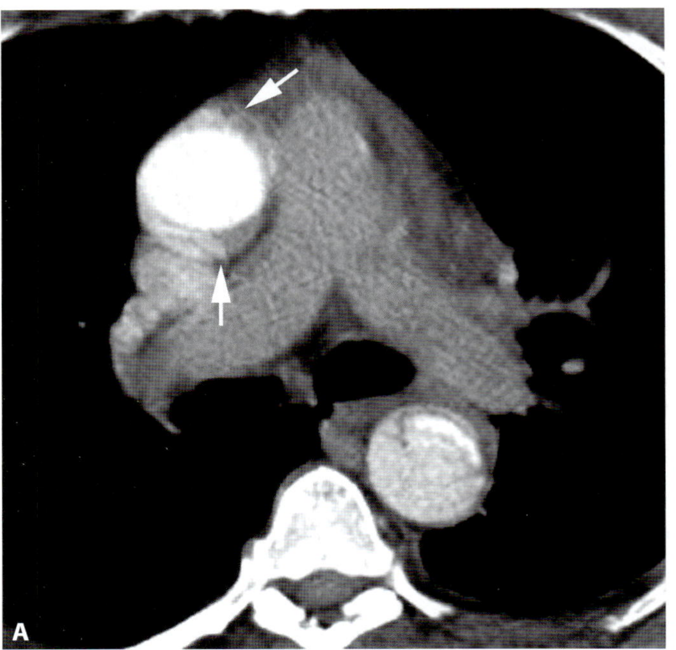

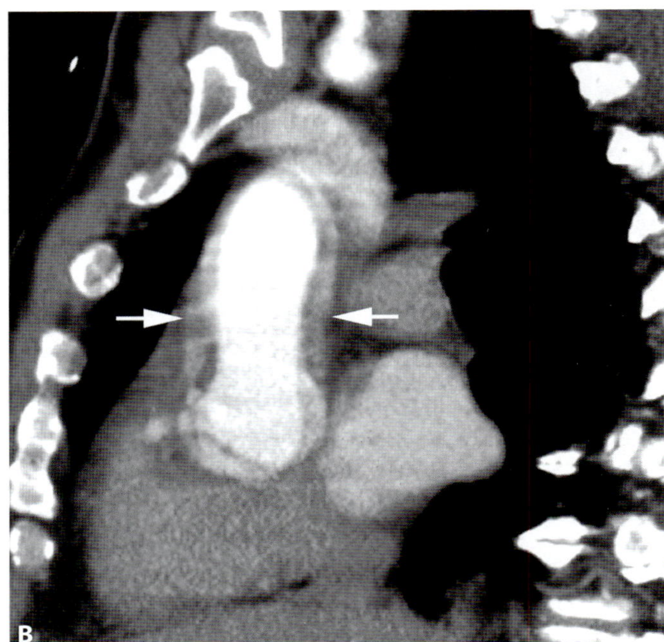

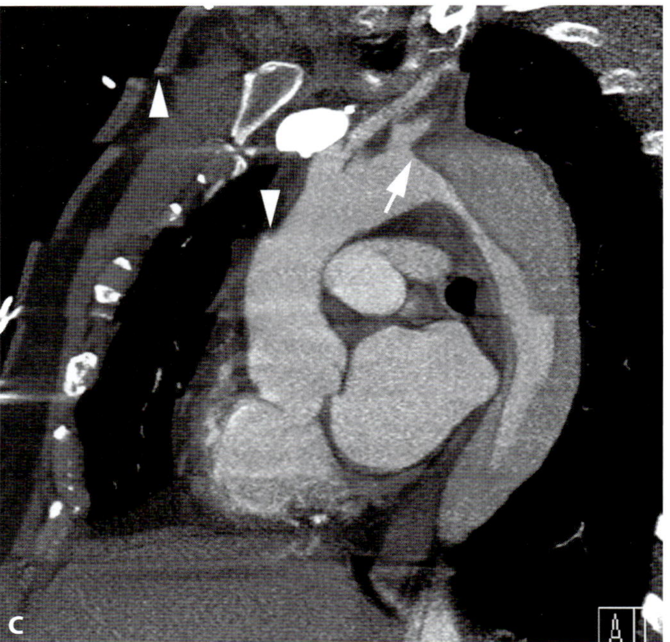

Figure 44.1 A 55-year-old man with acute sharp chest and back pain, transferred for management of aortic dissection, suspected as a type A dissection on outside imaging. The patient was mildly tachycardic, and could not hold his breath. Axial (**A**) and sagittal (**B**) reconstruction of the non-ECG-gated CTA from the referring facility demonstrates a double contour to the ascending aorta (arrows), suggestive of dissection at that level. **C.** Oblique sagittal reformation of a follow-up prospectively ECG-gated CTA performed in the emergency department confidently images the proximal extent of the intimomedial flap (arrow), indicating a Stanford type B aortic dissection. It showed no flap in the ascending aorta. Note the misregistration artifact in the chest due to breathing when prospective ECG-gating was used (arrowheads).

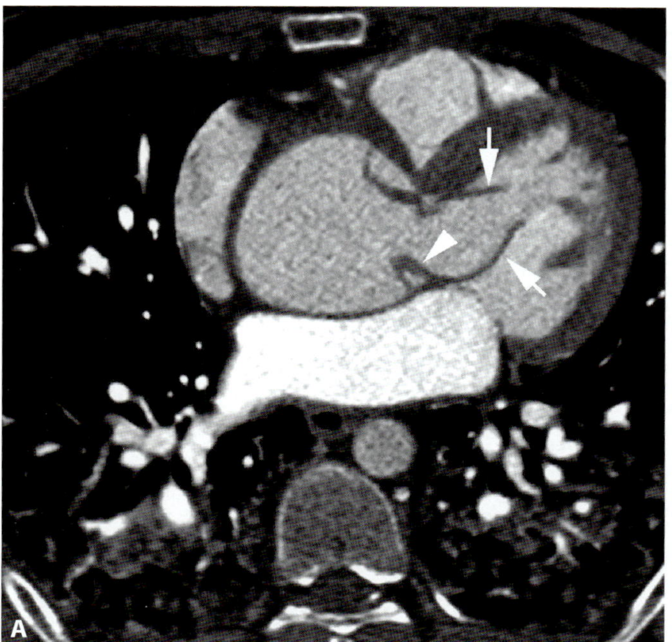

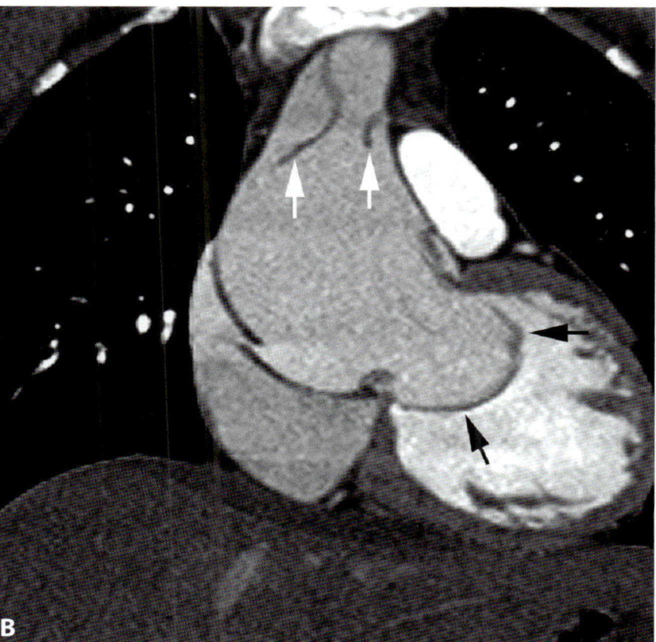

Figure 44.2 Gated CTA demonstrating an aortic dissection extending through the aortic valve in a 34-year-old man who developed sudden severe substernal chest pain and hypotension. **A.** Axial image at the level of the aortic valve shows prolapse of the intimomedial flaps (arrows) through the aortic valve (arrowhead) and into the left ventricle. **B.** Coronal images demonstrate the type A dissection flaps (white arrows) in the ascending aorta. The prolapsed flaps (black arrows) were causing outflow obstruction during ventricular systole.

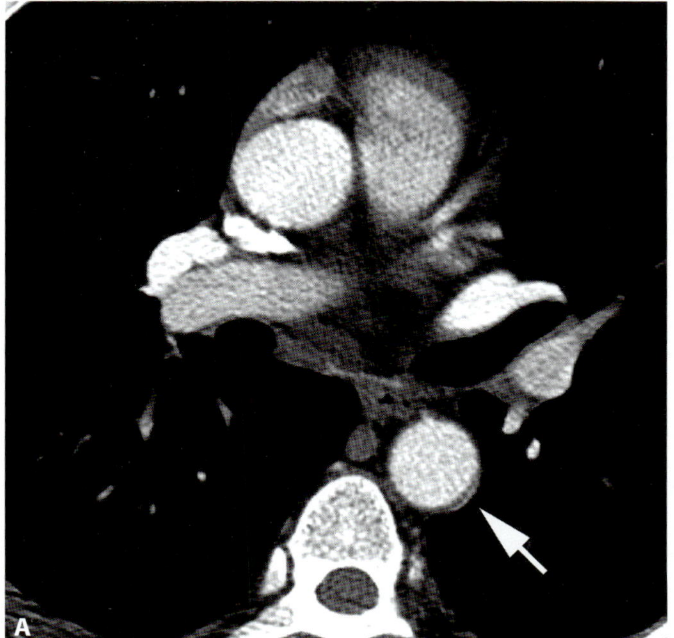

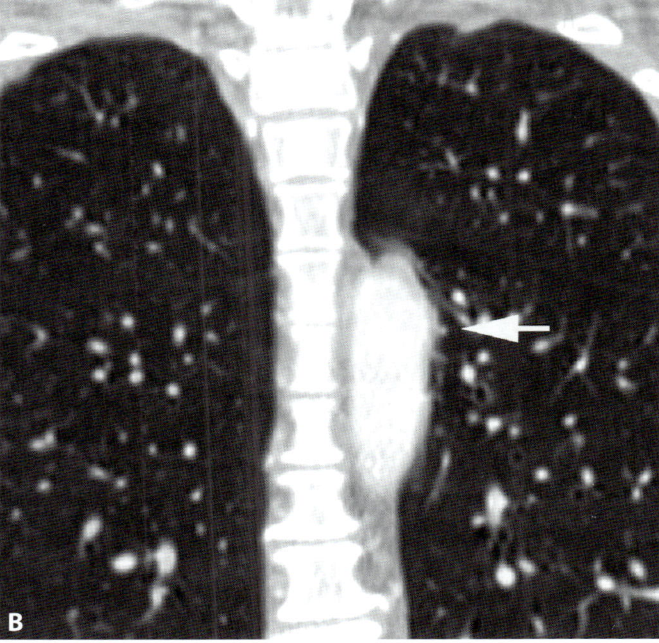

Figure 44.3 Pulmonary atelectasis misinterpreted as a small aortic dissection in a 47-year-old man with chest pain and hypertension. Follow-up study performed prone at full inspiration two months later showed no dissection. **A.** Axial CTA demonstrates a thin crescentic ruin of enhancement similar in density to the aortic lumen in the left posterior-lateral edge of the descending thoracic aorta (arrow). **B.** Coronal lung windows of the same CTA shows localized lung volume loss with pulmonary vessels (arrow) croweded against the aortic lumen.

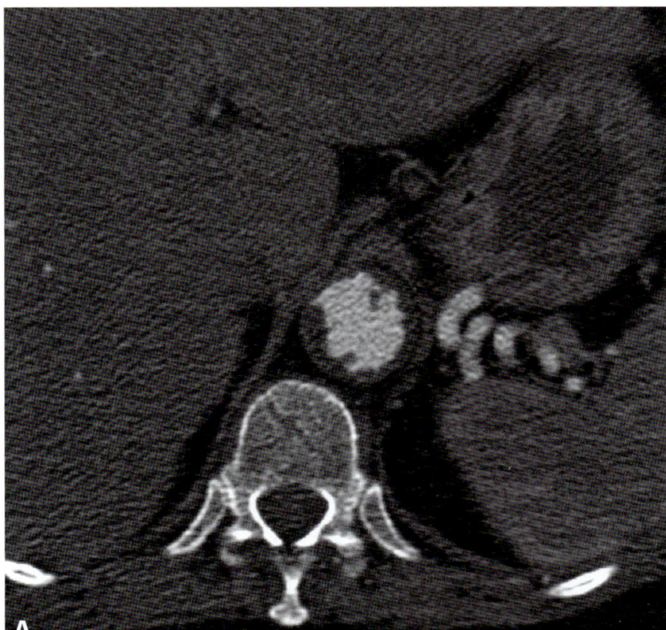

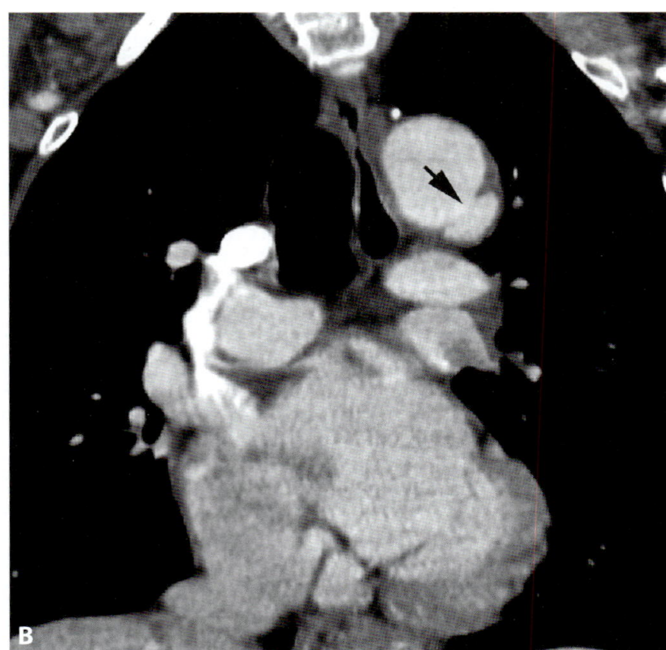

Figure 44.4 Ulcerating plaque and penetrating aortic ulcer. **A.** Axial CTA of a 73-year-old man, undergoing outpatient evaluation of an abdominal aortic aneurysm (not shown) demonstrates nearly concentric fibrofatty plaque projecting within the lumen of the lower thoracic aorta. Multiple ulcerations are present within this plaque, however, none project into, or beyond, the wall of the aorta. **B.** Coronal reformat of a CTA in a 67-year-old man with chest pain radiating to the back demonstrates a focal outpouching extending through the expected contour of the normal wall of the aortic arch (arrow). This penetrating aortic ulcer enlarged on serial CTAs.

Aortic intramural hematoma

Matthew H. Nett

Imaging description

Aortic intramural hematoma (IMH) results from rupture of the vasa vasorum and hemorrhage into the arterial media, which leads to weakening of the aortic wall, but absence of the intimal disruption that characterizes an aortic dissection [1]. Intramural hematoma of the thoracic aorta can be classified with the Stanford system similar to aortic dissections, with a Stanford type A involving the ascending aorta, and a Stanford type B involving the descending aorta.

Intramural hematoma can be diagnosed with echocardiography, CT, and MRI. Transesophageal echocardiography has been reported to have a sensitivity of up to 100% and specificity of 91%, and transthoracic echocardiography has a reported sensitivity range of 77–80%, although both of these modalities are operator dependent. CT and MR also have a sensitivity and negative predictive value which approach 100% [1, 2], but because it is less time-consuming and typically more readily available than MR, CT has become the diagnostic test of choice in suspected acute aortic syndrome (AAS).

The typical appearance of IMH on CT is a hyperdense, typically crescentic collection located eccentrically in the wall of the aorta (Figure 45.1A). Because intravascular contrast may obscure an IMH and therefore prevent early detection (Figure 45.1B), a non-contrast phase performed prior to CT angiography (CTA) is generally considered a necessary component of an AAS protocol [1, 3, 4]. Narrow window settings when reviewing these images can also increase the conspicuity of subtle areas of hyperdensity and focal aortic wall thickening.

However, a prospective study evaluating the benefit of this additional non-contrast phase is yet to be performed. As IMH is uncommon, performing a single phase gated post-contrast CTA could be considered an optional screening technique in the emergency department setting, similar to protocols used for gated "multi-rule out" CTA. A delayed non-contrast examination 10–15 minutes after the initial examination could be performed if there is any suspicion of aortic wall thickening to suggest an acute IMH [5].

The differentiation of IMH from either atherosclerotic thickening of the aorta, thrombus, or thrombosed dissection may infrequently be difficult on CT, and in such settings, MRI can be a valuable problem-solving tool, especially when dynamic cine gradient-echo sequences are applied [3]. MRI may also help to determine the age of a hematoma based on the signal characteristics of hemoglobin degradation products.

Importance

Intramural hematoma is a potentially lethal syndrome which can progress to aortic rupture, dissection, or aneurysm [6]. Because IMH may lie on a pathologic continuum that precedes aortic dissection, early detection could mean earlier intervention and improved clinical outcomes [4].

Typical clinical scenario

Aortic IMH is part of AAS, along with aortic dissection and penetrating atherosclerotic ulcer. The typical presenting clinical scenario is chest pain radiating to the back, with hypertension being the most common predisposing factor. In a retrospective evaluation of 373 cases of suspected AAS syndrome in patients with these symptoms, 67 (18%) of patients had an aortic dissection on CT and a further 14 (3.8%) were positive for an IMH [2]. Very rarely, IMH may be due to trauma [7].

Options for treatment of IMH include surgery, or medical management. Morbidity and mortality studies have shown type A and type B IMH led to aortic dissection in 25% and 13%, to aortic rupture in 28% and 9%, and did not progress in 28% and 76%, respectively. There is a 30-day mortality with surgical repair of 18% for type A IMH, and 33% for type B IMH, compared to 60% and 8% 30-day mortality with medical treatment, respectively [6]. Management of type B IMHs is therefore typically conservative. Treatment of type A IMHs is less well established and requires consideration of the imaging characteristics, patient demographics, surgical and perioperative risks, persistent or recurrent episodes of pain, and the risk of progression to type A dissection. All patients, regardless of the type of IMH, require surveillance imaging [1].

Differential diagnosis

The three causes of AAS are clinically indistinguishable, but have characteristic imaging findings. A thrombosed false lumen of aortic dissection can mimic IMH. An intimal flap will be seen on contrast-enhanced imaging with aortic dissection. If an IMH is seen adjacent to a contrast-filled luminal outpouching, this is typical of a penetrating aortic ulcer. The ulcers may vary in size and depth, but are usually best visualized on axial images and most commonly occur in the mid to distal descending aorta.

> To evaluate for hyperdense IMH on CTA, either include a non-contrast CT phase as a component of an AAS CT protocol, or perform a delayed phase 10–15 minutes post-contrast CT in all patients with mural thickening.

REFERENCES

1. Chao CP, Walker TG, Kalva SP. Natural history and CT appearances of aortic intramural hematoma. *Radiographics*. 2009;**29**(3):791–804.
2. Hayter RG, Rhea JT, Small A, Tafazoli FS, Novelline RA. Suspected aortic dissection and other aortic disorders: multi-detector row CT in 373 cases in the emergency setting. *Radiology*. 2006;**238**(3):841–52.

3. Litmanovich D, Bankier AA, Cantin L, Raptopoulos V, Boiselle PM. CT and MRI in diseases of the aorta. *AJR Am J Roentgenol.* 2009; **193**(4):928–40.

4. Ledbetter S, Stuk JL, Kaufman JA. Helical (spiral) CT in the evaluation of emergent thoracic aortic syndromes. Traumatic aortic rupture, aortic aneurysm, aortic dissection, intramural hematoma, and penetrating atherosclerotic ulcer. *Radiol Clin North Am.* 1999; **37**(3):575–89.

5. Sodickson A. Strategies for reducing radiation exposure in multi-detector row CT. *Radiol Clin North Am.* 2012;**50**(1):1–14.

6. Baikoussis NG, Apostolakis EE, Siminelakis SN, Papadopoulos GS, Goudevenos J. Intramural haematoma of the thoracic aorta: who's to be alerted the cardiologist or the cardiac surgeon? *J Cardiothorac Surg.* 2009;**4**:54.

7. Gunn ML. Imaging of aortic and branch vessel trauma. *Radiol Clin North Am.* 2012;**50**(1):85–103.

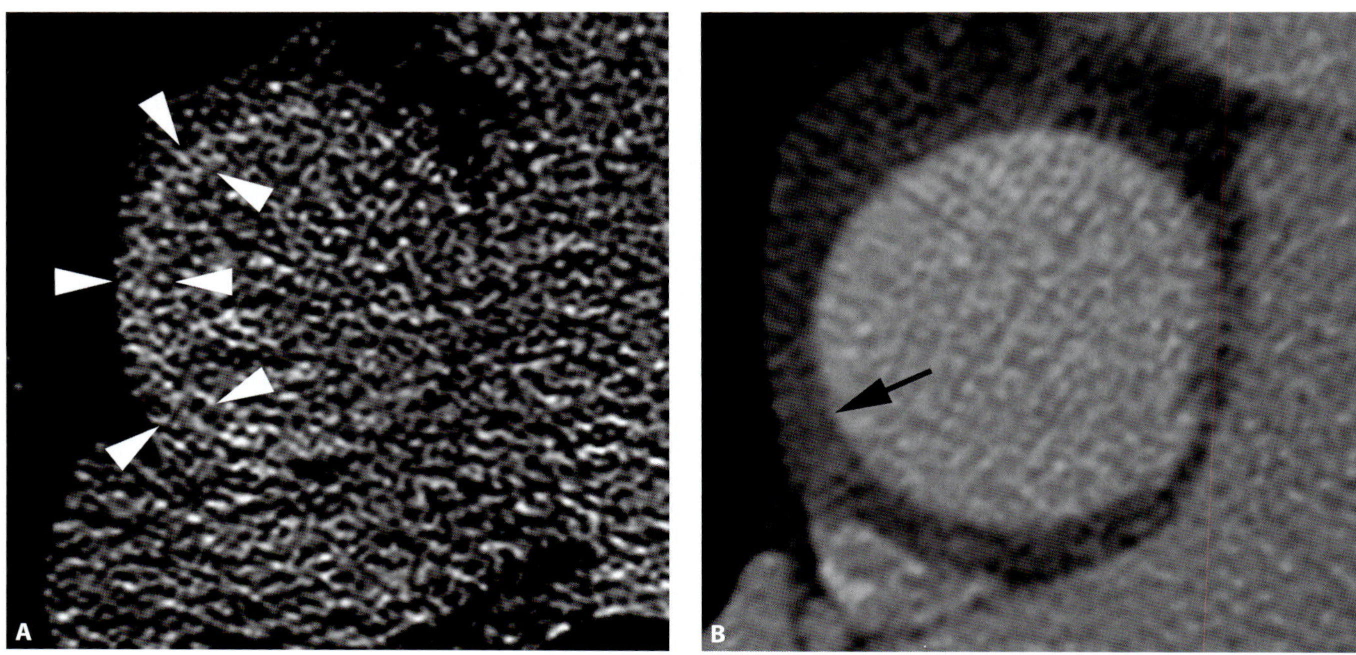

Figure 45.1 **A.** Axial low-dose non-contrast CT from a 63-year-old man with chest pain radiating to the back demonstrates crescentic hyperdensity in the wall of the ascending aorta (arrowheads) diagnostic of an intramural hematoma, Stanford type A. **B.** Axial contrast-enhanced CT at the same level again shows hyperdensity in the ascending aortic wall (arrow) although it is much less conspicuous and could be easily overlooked. Note the intima is intact.

Pitfalls in peripheral CT angiography

Martin L. Gunn

Imaging description

On fast multichannel CT scanners (16 channel and greater), "out running" the contrast is quite easy to do. You can calculate your scanner table speed if you know your detector configuration, pitch, and rotation time, using the formula:

$$\text{Speed (mm per second)} = \frac{\text{Rows (n)} \times \text{Collimation (mm)} \times \text{Pitch Factor}}{\text{Rotation Time (second)}}.$$

For example, the table speed for a 16-row CT scanner with a 16×1.25 millimeter (mm) detector configuration, a pitch of 1:1.375, and a rotation time of 0.5 seconds (s), is approximately 55 mm/s, and will cover 1 meter in 18 seconds. A dual-source scanner using a pitch of 3.4, can operate at a table speed of up to 450 mm/s, covering 1 meter in 2.2 seconds. This is unnecessarily fast for lower extremity CT angiography (CTA).

Blood flow velocity in the lower extremities in the normal resting individual ranges from approximately 100 mm/s in the common femoral artery to 65 mm/s in the dorsalis pedis artery [1]. The time for contrast to travel from the aorta to the popliteal artery ranges from 4 to 24 seconds, with an average of approximately 10 seconds [2]. However, flow velocity and rate vary considerably between individuals. In the emergency patient, the contrast transit time through the arteries might be longer, especially in patients who are hypovolemic or have arterial occlusive disease. Consequently, if peripheral CTA is triggered by contrast arrival in the aorta and there is an insufficient pre-scan delay, it is easy to "out run" the contrast in the lower extremities and obtain a non-diagnostic scan. A table speed of 60 mm/s will result in the contrast being outrun in approximately 30% of patients with peripheral arterial occlusive disease (Figure 46.1) [2].

To overcome this problem, various techniques have been proposed to ensure adequate synchronization of contrast delivery and CT scanning. One is to reduce the table speed by decreasing the pitch and/or slowing the gantry rotation speed to ensure a long enough scan duration (e.g., 25 to 30 seconds). A table speed of 40 mm/s provides the best synchronization of CT data acquisition and contrast bolus propagation overall [3], but this can be adjusted according to patient age (e.g., 25.0 to 53.8 mm/s), with slower speeds used for older patients [4]. To achieve adequate arterial enhancement with a longer acquisition time, a higher volume of contrast, high contrast concentration (e.g., 370–400 mg iodine per mL), and a biphasic injection protocol should be considered. Another technique is to use a timing bolus, with a delay determined by the scan duration displayed on the CT console, subtracted from the time that the contrast takes to get to the feet in a patient with advanced occlusive disease [2].

$$\text{Delay (seconds)} = t_{CMT} + 40 \text{ seconds} - t_{scan}$$

where t_{CMT} is the contrast transit time to the aorta (determined by a test bolus) in seconds, 40 seconds is an assumed average aorta to foot contrast transit time, and t_{scan} is the scan duration in seconds. This technique works well with fast scanners, as the scan delay increases with increasing scan speed. However, these single phase acquisition techniques do not account for asymmetric blood flow in the lower extremities, which may be present in patients with advanced peripheral vascular disease. If a second phase is acquired through the calf immediately after the initial helical phase is complete, excellent arterial opacification may be obtained when both acquisitions are examined (Figure 46.2).

The use of lower tube potentials (e.g., a peak tube voltage of 100 keV) generally achieves superior iodine attenuation for CTA. However, at lower tube potentials, beam-hardening artifacts from metal-ware and bullet fragments will be accentuated, potentially obscuring vessels (Figure 46.3).

Importance

Inadequate synchronization between scan acquisition time and contrast delivery can result in non-diagnostic CT scans.

Typical clinical scenario

The risk of "outrunning" the contrast bolus is greatest with faster CT scanners (16 channel and above) and in patients with cardiac disease.

Differential diagnosis

The differential diagnosis is vascular occlusion from embolism, dissection, injury, compartment syndrome, or spasm.

Teaching point

Modern CT scanners, which have high table speeds, can easily "out run" the contrast bolus, particularly in patients with advanced peripheral arterial occlusive disease. Consider increasing the scan delay and performing a second helical acquisition below the knees.

REFERENCES

1. Holland CK, Brown JM, Scoutt LM, Taylor KJW. Lower extremity volumetric arterial blood flow in normal subjects. *Ultrasound Med Biol.* 1998;**24**(8):1079–86.
2. Fleischmann D. A practical approach to contrast delivery for peripheral CTA. *Appl Radiol.* 2004;Suppl:61–70.
3. Meyer BC, Oldenburg A, Frericks BB, *et al.* Quantitative and qualitative evaluation of the influence of different table feeds on visualization of peripheral arteries in CT angiography of aortoiliac and lower extremity arteries. *Eur Radiol.* 2008;**18**(8):1546–55.
4. Siriapisith T, Wasinrat J, Mutirangura P, Ruangsetakit C, Wongwanit C. Optimization of the table speed of lower extremity CT angiography protocols in different patient age groups. *J Cardiovasc Comput Tomogr.* 2010;**4**(3):173–83.

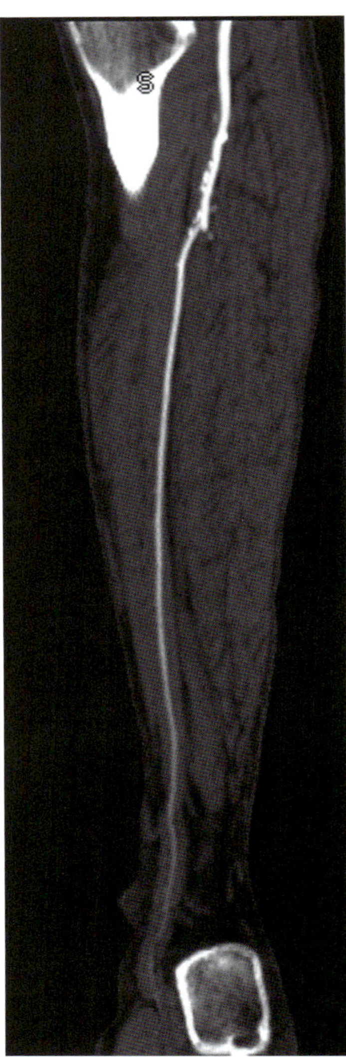

Figure 46.1 Curved planar reformat through the left posterior tibial artery of a peripheral CTA in a 44-year-old man with peripheral vascular disease and a right leg amputation. The scan was performed on a 16-channel CT scanner with a table speed of 55 mm/s. Note the fading contrast in the distal artery, typical in appearance for contrast that has been "outrun."

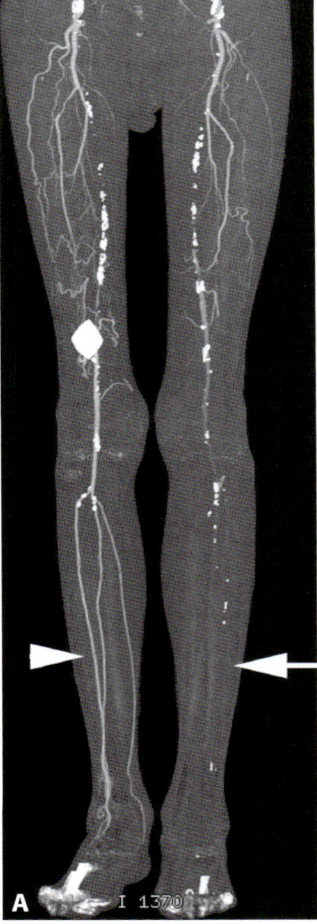

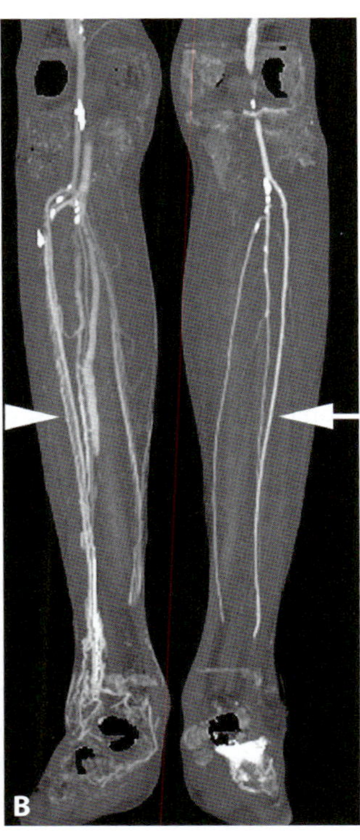

Figure 46.2 **A.** Coronal volume maximum intensity projection (MIP) through the lower extremities in a 67-year-old man with advanced occlusive peripheral arterial disease. This scan was performed on a 64-channel CT scanner, using the equation described in the text to calculate the scan delay. The scan through the left calf is too early with insufficient contrast enhancement (arrow), whereas timing is optimal for the right calf (arrowhead). This is a typical pattern in some patients with advanced occlusive disease and asymmetric flow to the lower extremities. **B.** An immediate second phase through the calf demonstrates excellent enhancement of the arteries in the left calf but venous contamination on the right (arrowhead). No single phase acquisition would have been ideal for this patient, but a combination of the two phases provides a complete assessment of both calves.

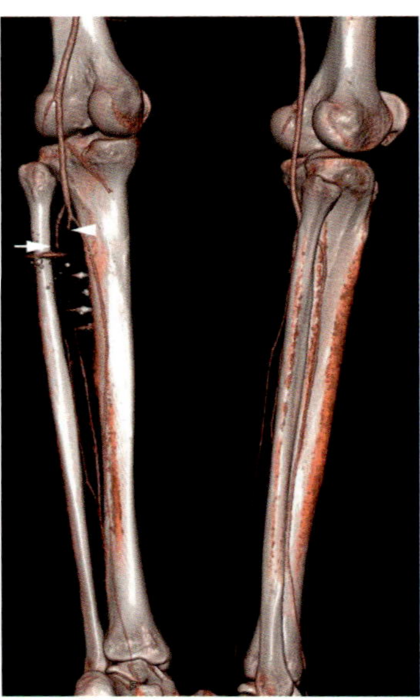

Figure 46.3 Posterior 3D volume rendered view of the calf in a 34-year-old man who was shot multiple times in his lower extremities. Note the occlusion of the anterior tibial (arrowhead) and peroneal arteries (arrow) for several centimeters due to injury related to the gunshot wounds. Also note the streak-like beam-hardening artifact arising from bullet fragments. A kVp of 120 was used for scan. The artifact would have been worse if a lower kVp was used.

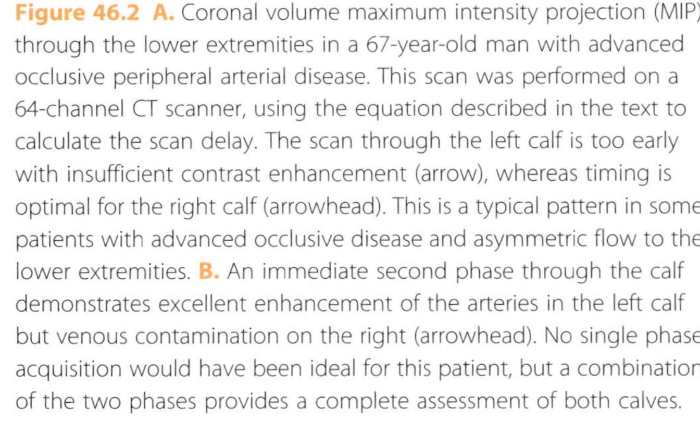

Breathing artifact simulating pulmonary embolism

Martin L. Gunn

Imaging description

Although CT pulmonary angiography (CTPA) is regarded by many as the reference standard for the diagnosis of pulmonary embolism in the acute setting, several studies have reported a high rate of non-diagnostic examinations due to poor image quality. In the PIOPED II study, which utilized 4-, 8-, and 16-channel CT scanners, nearly 10% of patients had an inconclusive scan. This was within the 6–11% range reported in other studies [1, 2]. The most common causes for an indeterminate CTPA are motion artifact and poor contrast enhancement [3]. Breathing artifact can be the cause of both, but rapid advances in CT technology have resulted in shorter scan times, reducing the frequency of these artifacts [4].

Despite optimal contrast timing, some patients still demonstrate suboptimal pulmonary arterial opacification, which can be either generalized or localized [5]. It is believed that this artifact is due to suspended deep inspiration with a Valsalva maneuver, which causes unopacified blood from the inferior vena cava to pass preferentially into the right heart and pulmonary arteries [5]. In patients with a patent foramen ovale the Valsalva maneuver may cause blood to pass preferentially from the right to the left heart [6]. This phenomenon has been termed "transient contrast bolus interruption."

Transient contrast bolus interruption can lead to a non-diagnostic CT scan, or if it is localized, it may simulate a pulmonary embolism (Figure 47.1A and B).

When this artifact is recognized, repeating the scan at end expiration may result in improved contrast opacification (Figure 47.1C) [5–7].

Breathing artifact is also associated with decreased vascular opacification of the smaller pulmonary arteries as the motion causes partial volume artifact, averaging the density of the enhanced artery with the surrounding black lung [8]. On CT images, this may also lead to a "seagull wing" appearance to the pulmonary arteries, apparent paring of vessels and fissures, hypoattenuation beside intrapulmonary vessels, and stair-step artifacts on multiplanar reformations and volume rendered reconstructions [3, 9]. These signs should be readily identified using lung windows (Figure 47.2).

Breathing artifact should also be considered when multiple pulmonary arteries appear poorly opacified at the same cranio-caudal scan level (Figure 47.3).

Importance

Breathing artifacts simulate pulmonary emboli. Recognition of transient contrast bolus interruption should prompt the radiologist to repeat the CT scan in the expiratory phase.

Typical clinical scenario

Patients with suspected pulmonary embolism are commonly short of breath, and this reduces the achievable breath-hold duration in these patients. If the preprogrammed "hold your breath" instructions on the CT scanner are given to the patient, this might encourage the compliant patient to take a deep inspiration followed by a Valsalva maneuver, and lead to transient contrast-bolus interruption. Pre-oxygenation and practice breath-hold techniques (mid inspiration without Valsalva maneuver, or end expiration) will reduce the risk of a non-diagnostic scan that needs to be repeated. Although a caudocranial scan direction might reduce the risk of motion artifacts with slower CT scanners in the lower lobe pulmonary arteries, where most pulmonary emboli occur, scan direction has not been found to influence diagnostic accuracy on 64-channel or higher multidetector CT (MDCT) [4].

Differential diagnosis

The differential diagnosis is pulmonary embolism.

Teaching point

Breathing artifacts are a common cause of inconclusive CTPA. Examine the lung windows for signs of breathing artifact at the level of questionable pulmonary emboli. If there is good contrast enhancement of the superior vena cava and aorta, but not the pulmonary arteries, or if multiple pulmonary arteries are poorly enhanced for a short section of the scan, consider transient contrast bolus interruption. In this setting, try repeating the CTPA while the patient holds their breath in expiration.

REFERENCES

1. Stein PD, Fowler SE, Goodman LR, *et al.* Multidetector computed tomography for acute pulmonary embolism. *N Engl J Med.* 2006; **354**(22):2317–27.
2. Revel MP, Cohen S, Sanchez O, *et al.* Pulmonary embolism during pregnancy: diagnosis with lung scintigraphy or CT angiography? *Radiology.* 2011;**258**(2):590–8.
3. Jones SE, Wittram C. The indeterminate CT pulmonary angiogram: imaging characteristics and patient clinical outcome. *Radiology.* 2005; **237**(1):329–37.
4. Wu C, Sodickson A, Cai T, *et al.* Comparison of respiratory motion artifact from craniocaudal versus caudocranial scanning with 64-MDCT pulmonary angiography. *AJR Am J Roentgenol.* 2010;**195** (1):155–9.

5. Mortimer AM, Singh RK, Hughes J, Greenwood R, Hamilton MC. Use of expiratory CT pulmonary angiography to reduce inspiration and breath-hold associated artefact: contrast dynamics and implications for scan protocol. *Clin Radiol.* 2011;**66**(12):1159–66.
6. Henk CB, Grampp S, Linnau KF, *et al.* Suspected pulmonary embolism: enhancement of pulmonary arteries at deep-inspiration CT angiography – influence of patent foramen ovale and atrial-septal defect. *Radiology.* 2003;**226**(3):749–55.
7. Chen YH, Velayudhan V, Weltman DI, *et al.* Waiting to exhale: salvaging the nondiagnostic CT pulmonary angiogram by using expiratory imaging to improve contrast dynamics. *Emerg Radiol.* 2008;**15**(3):161–9.
8. Wittram C, Maher MM, Yoo AJ, *et al.* CT angiography of pulmonary embolism: diagnostic criteria and causes of misdiagnosis. *Radiographics.* 2004;**24**(5):1219–38.
9. Schoepf UJ, Costello P. CT angiography for diagnosis of pulmonary embolism: state of the art. *Radiology.* 2004;**230**(2):329–37.

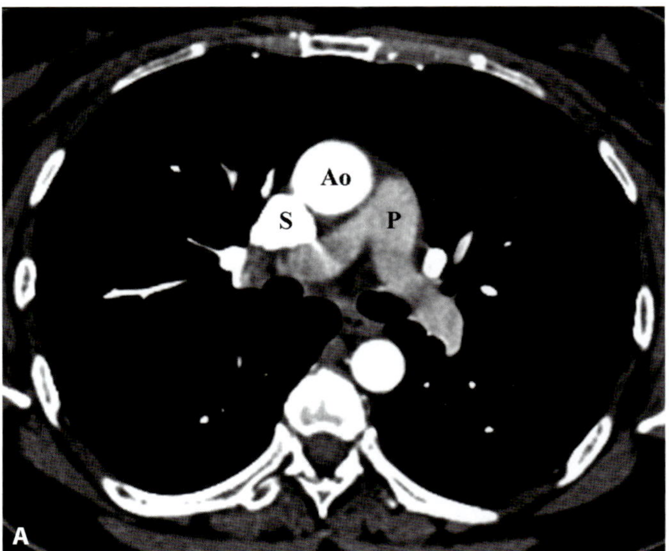

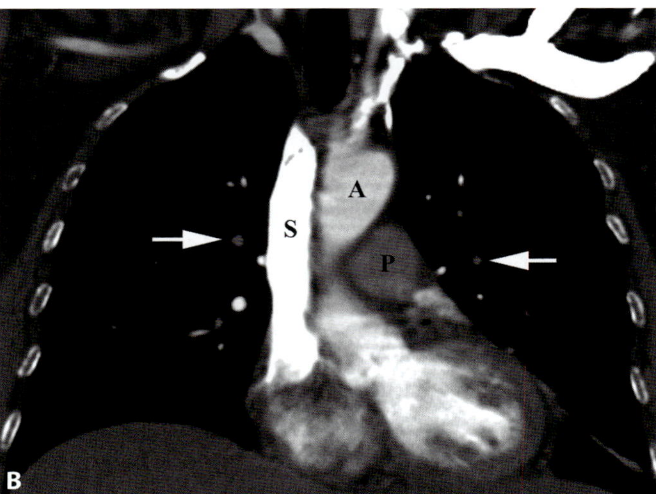

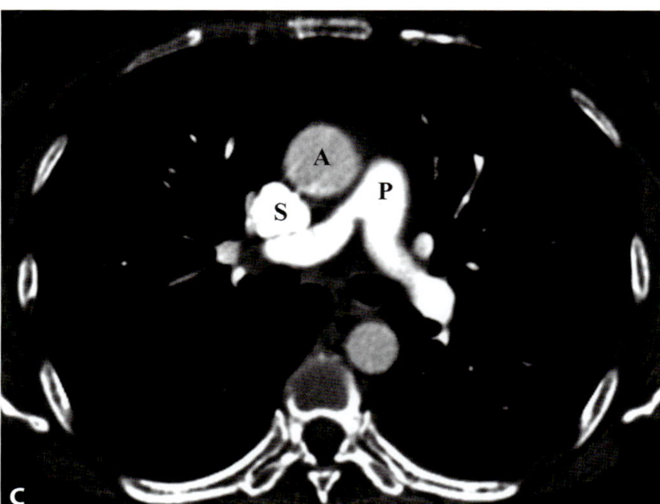

Figure 47.1 Transient interruption of contrast medium column during CTPA. Axial (**A**) and coronal (**B**) image demonstrates bright contrast in the aorta (Ao) and superior vena cava (S), but poor contrast opacification of the main pulmonary arteries (P) and branch pulmonary arteries (arrows), resulting in an inconclusive scan. A repeat CTPA (**C**) with the same contrast volume and flow rate, but performed at end expiration, demonstrates excellent enhancement of the main pulmonary artery (P), with less enhancement of the aorta (A). Images provided courtesy of Dr. Aaron Sodickson, Brigham and Women's Hospital, Boston, MA, USA.

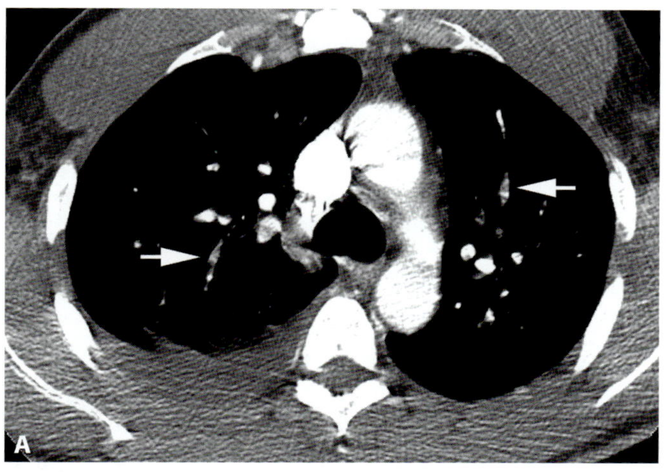

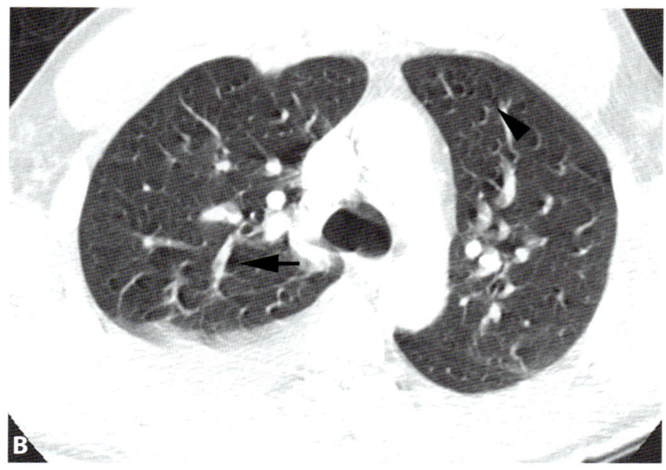

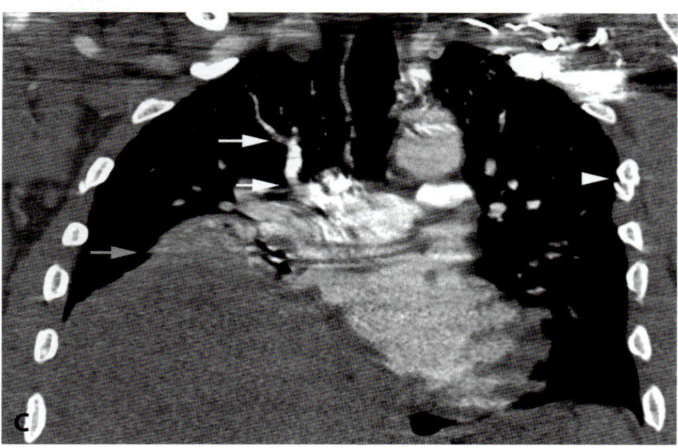

Figure 47.2 A 56-year-old man with known hepatocellular carcinoma presents with tachypnea. **A.** Axial reconstruction from a CTPA shows two areas within the pulmonary arteries that simulate pulmonary emboli (arrows.) Note the noise of the study, principally due to using a low kVp technique to optimize iodine attenuation, 100 kVp in this patient. **B.** Lung windows at the same level demonstrate dark regions adjacent to the arteries, due to motion artifact (arrow). In many places the small arteries appear archiform, and in some places the "seagull sign" is visible (arrowhead). Also note the double contour to the trachea caused by respiratory misregistration. **C.** Coronal multiplanar reconstruction of the same study demonstrates the areas of artifactual decreased attenuation in the pulmonary arteries from the same artifact (white arrows). Also present are other signs of breathing artifact, including double contours to a rib (arrowhead) and stair-step artifact in the right hemidiaphragm (gray arrow). The examination was reported as indeterminate for pulmonary embolism distal to the lobar pulmonary arteries.

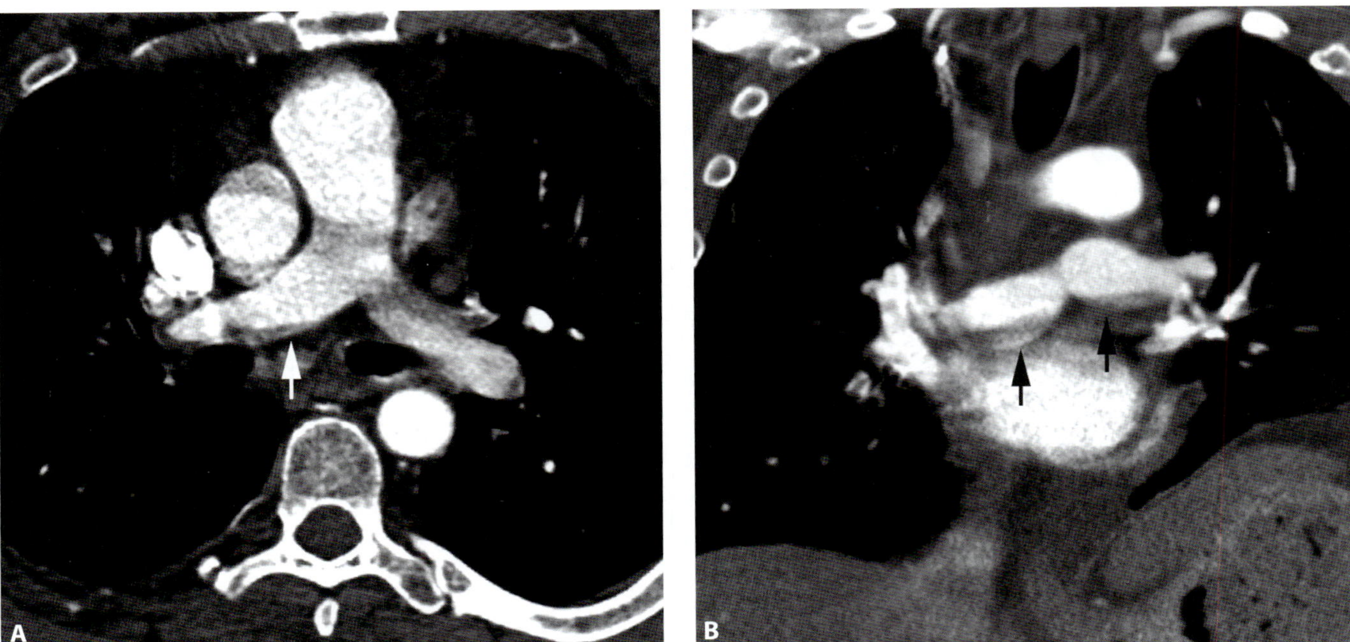

Figure 47.3 A 48-year-old man with dyspnea, reduced blood oxygen saturation, and tachycardia. **A.** Axial reconstruction from a CTPA demonstrates an apparent filling defect in the right main pulmonary artery (arrow). **B.** Coronal reformation demonstrates symmetric bilateral apparent filling defects in the lower edge of both the main pulmonary arteries (arrows), consistent with breathing artifact.

Acute versus chronic pulmonary thromboembolism

Randy K. Lau

Imaging description

Pulmonary emboli, both acute and chronic, appear on CT as intraluminal filling defects that have sharp interfaces [1]. Signs of acute pulmonary embolism include complete occlusion of a vessel with enlargement compared to adjacent vessels, a partial filling defect surrounded by contrast, and a peripheral intraluminal filling defect that forms acute angles with the arterial wall (Figures 48.1 and 48.2) [1, 2]. Occasionally, peripheral wedge-shaped areas, the so-called Hampton's hump, that may represent infarcts are identified. In contrast, chronic pulmonary emboli are characterized by complete occlusion in a vessel smaller than adjacent vessels, a peripheral crescent-shaped filling defect that forms obtuse angles with the arterial wall, and a web or flap (Figure 48.1) [1]. Other direct signs of chronic pulmonary artery emboli include eccentric thrombus or calcified thrombus (Figure 48.3) [3]. Extensive bronchial or systemic collateral vessels, mosaic perfusion, or calcifications within eccentric vessel thickening are secondary signs of chronic pulmonary emboli (Figure 48.3) [1]. When pulmonary emboli are identified, signs of right heart strain including septal bowing convex toward the left ventricle and reflux of contrast into the inferior vena cava with dilated hepatic veins should be sought [4].

Importance

Acute pulmonary embolism is a major cause of morbidity and mortality associated with surgery, injury, and medical illness. Annual incidence ranges between 23 and 69 per 100000 [5, 6]. Mortality rates approach 20–30% in those presenting with hemodynamic impairment [7, 8]. Chronic pulmonary embolism may result in pulmonary hypertension. Early recognition of chronic pulmonary emboli may lead to appropriate management, such as pulmonary artery thromboendarterectomy, and may improve outcome.

Typical clinical scenario

There are innumerable risk factors for pulmonary emboli, all of which are related to Virchow's triad: hypercoagulability, hemodynamic changes including stasis or turbulent flow, and endothelial injury or dysfunction. Signs and symptoms of pulmonary emboli are non-specific and include unexplained dyspnea, tachypnea, palpitations, tachycardia, and pleuritic chest pain with or without signs and symptoms of deep venous thrombosis.

Differential diagnosis

Causes of false-positive diagnoses of pulmonary emboli include flow-related artifacts related to poor pulmonary artery opacification, breathing artifact, beam-hardening artifacts and misidentification of pulmonary veins for pulmonary arteries.

Breathing artifacts causing a false-positive diagnosis of pulmonary embolism are described in Case 47 [1, 9].

Beam-hardening and streak artifact is common and can be due to dense contrast within the superior vena cava, catheters, or other medical devices. Artifact is easily recognized by identifying radiating nature of abnormality that extends outside the pulmonary artery on bone windows.

Vascular bifurcations can mimic linear filling defects characteristic of chronic pulmonary emboli on axial images. However, evaluation of coronal images and sagittal images will reveal the site of abnormality is located at the bifurcation of the pulmonary arteries.

Non-thrombotic pulmonary emboli from embolized polymethylmethacrylate, mercury, and ethiodol can mimic calcified, chronic pulmonary emboli. The appropriate history of vertebroplasty, mercury injection, hepatic artery chemoembolization with ethiodol, or hysterosalpingogram with oil-based contrast are critical to diagnosis.

Non-thrombotic emboli related to tumor may also mimic filling defects of thrombotic pulmonary emboli.

Teaching point

Imaging findings suggesting chronic nature of pulmonary emboli include eccentric filling defect forming obtuse angles with the arterial wall, linear webs, calcified thrombus, extensive bronchial collaterals, and mosaic perfusion. These should be sought during evaluation of CT pulmonary angiography for acute pulmonary embolism.

REFERENCES

1. Wittram C, Maher MM, Yoo AJ, *et al.* CT angiography of pulmonary embolism: diagnostic criteria and causes of misdiagnosis. *Radiographics.* 2004;**24**(5):1219–38.
2. Wittram C. How I do it: CT pulmonary angiography. *AJR Am J Roentgenol.* 2007;**188**(5):1255–61.
3. Castaner E, Gallardo X, Ballesteros E, *et al.* CT diagnosis of chronic pulmonary thromboembolism. *Radiographics.* 2009;**29**(1):31–50; discussion 50–3.
4. Kang DK, Ramos-Duran L, Schoepf UJ, *et al.* Reproducibility of CT signs of right ventricular dysfunction in acute pulmonary embolism. *AJR Am J Roentgenol.* 2010;**194**(6):1500–6.
5. Silverstein MD, Heit JA, Mohr DN, *et al.* Trends in the incidence of deep vein thrombosis and pulmonary embolism: a 25-year population-based study. *Arch Intern Med.* 1998;**158**(6):585–93.

6. Anderson FA, Jr., Wheeler HB, Goldberg RJ, *et al.* A population-based perspective of the hospital incidence and case-fatality rates of deep vein thrombosis and pulmonary embolism. The Worcester DVT Study. *Arch Intern Med.* 1991;**151**(5):933–8.

7. Carson JL, Kelley MA, Duff A, *et al.* The clinical course of pulmonary embolism. *N Engl J Med.* 1992;**326**(19):1240–5.

8. Alpert JS, Smith RE, Ockene IS, *et al.* Treatment of massive pulmonary embolism: the role of pulmonary embolectomy. *Am Heart J.* 1975;**89**(4):413–18.

9. Jones SE, Wittram C. The indeterminate CT pulmonary angiogram: imaging characteristics and patient clinical outcome. *Radiology.* 2005;**237**(1):329–37.

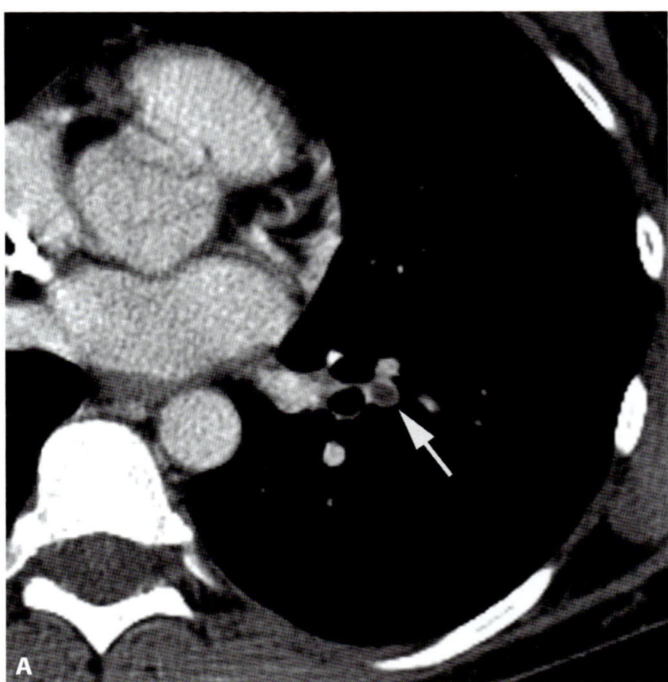

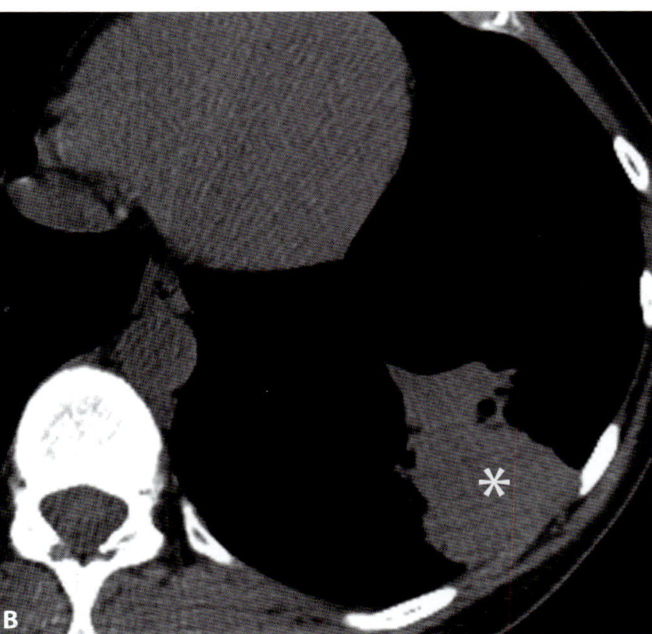

Figure 48.1 A. Axial contrast CT demonstrating acute pulmonary embolus to a pulmonary artery branch of the left lower lobe (white arrow). Notice the acute angle formed between the thrombus and pulmonary arterial wall, typical for acute thrombus. **B.** Noncontrast CT performed in the same patient 14 days later demonstrating peripheral wedge-shaped hypodensity in the left lower lobe representing an evolving pulmonary infarct, the so-called Hampton's hump (white asterisk).

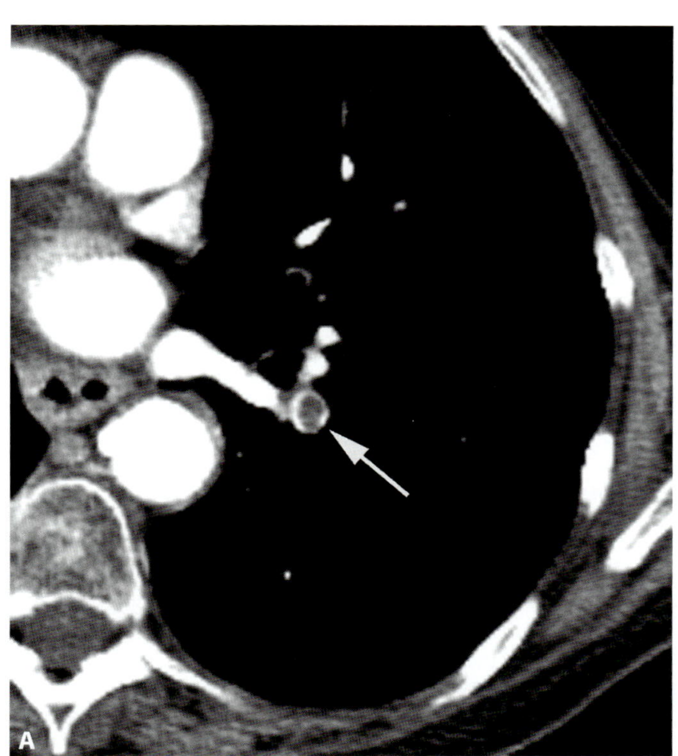

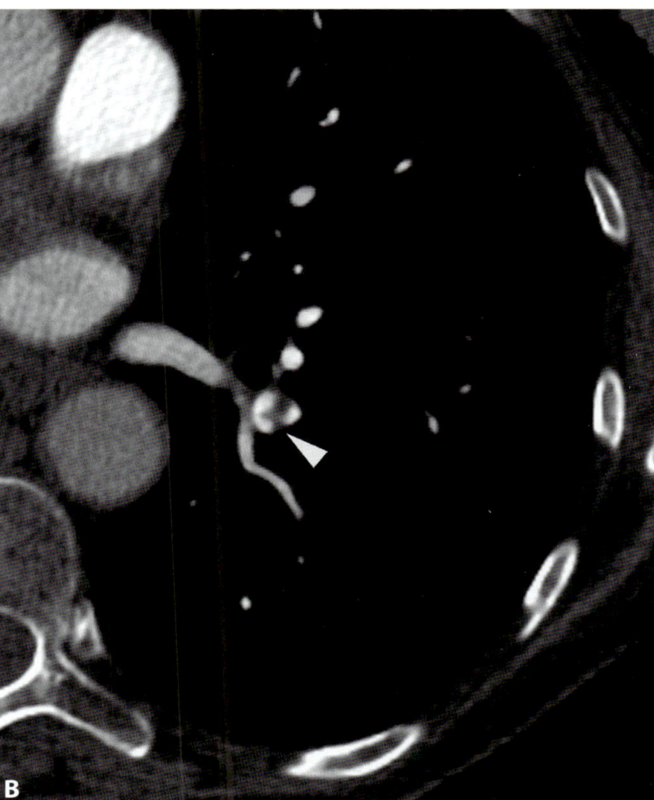

Figure 48.2 A. Axial contrast CT from an 80-year-old woman with non-small cell lung cancer and remote history of left mastectomy for breast cancer with incidental, acute left lower lobe pulmonary embolus (PE) (arrow) on staging CT. Again noted is the acute angle formed by the thrombus and vessel lumen. **B.** Axial contrast CT for follow-up performed 36 days later, demonstrating partial recanalization with web- or flap-like defect (arrowhead) typical of chronic pulmonary thromboemboli.

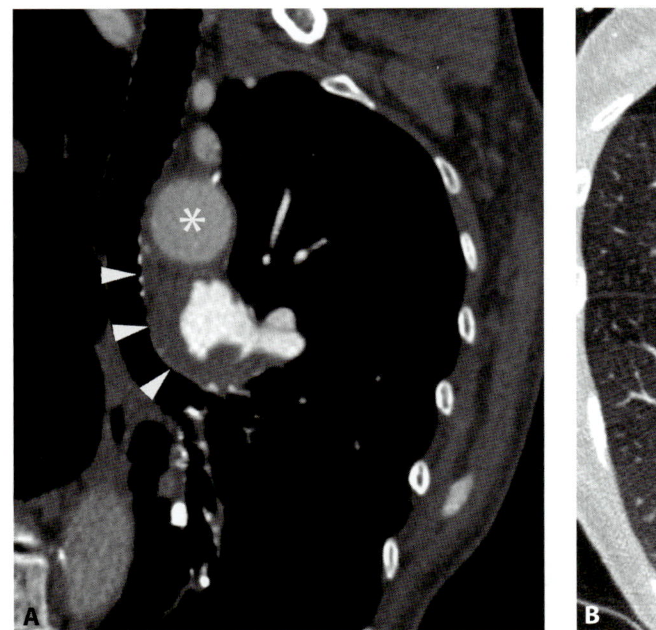

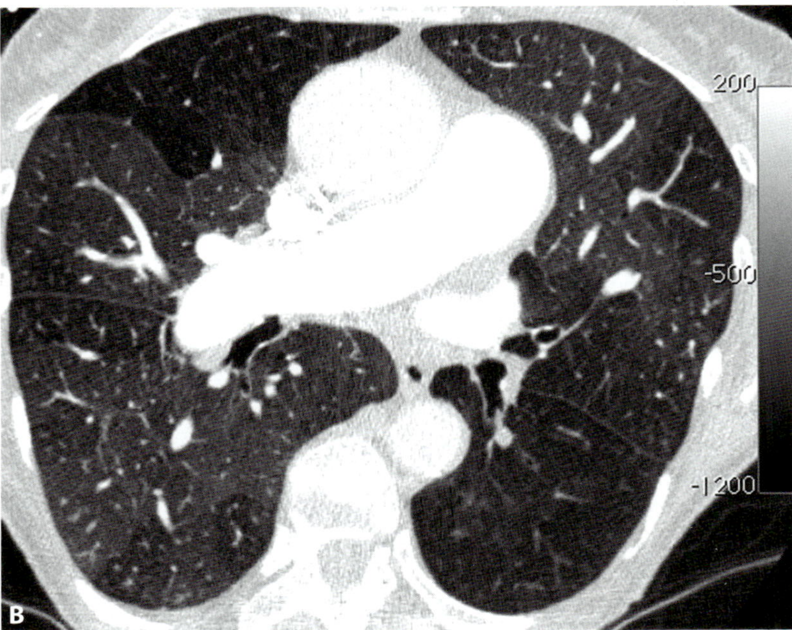

Figure 48.3 A. Contrast CT from a 76-year-old woman with past medical history significant for essential thrombocytosis with deep vein thrombosis and PE following spine surgery nine years prior to current CT. Double oblique reformat through the left pulmonary artery demonstrates peripheral crescent shaped filling defect (arrowhead) typical for chronic pulmonary thromboembolism. Note that the size of the left pulmonary artery is equal to that of the aortic arch (asterisk) consistent with pulmonary arterial hypertension. **B.** Coronal reformat demonstrates mosaic perfusion, a secondary sign of chronic pulmonary thromboemboli.

Vascular embolization of foreign body

Martin L. Gunn

Imaging description

Approximately 14% of aortic injuries by autopsy are caused by penetrating injuries [1]. Over 95% of these injuries are caused by knives or guns, and 90–100% of patients will die before arriving in hospital [1, 2]. In survivors, vascular embolization of a foreign body is rare, but can cause life-threatening ischemia. For a projectile to embolize in a vessel, it must be small enough and travel at sufficiently low velocity to dissipate most of its energy in soft tissues before it reaches the aorta, and penetrate just one wall of the vessel [3]. If the wall of the vessel is elastic enough, the vascular defect may close without significant exsanguination [4]. Shotgun and bullet emboli can be clinically elusive as symptoms from the embolized fragment may be absent or remote from the site of primary injury [3]. Needle embolization can also complicate intravenous drug use (Figure 49.1) [5].

Most foreign body emboli follow the direction of blood flow. They can be classified as arterial, venous, or paradoxical [6]. Systemic arterial emboli are at least twice as common as systemic venous emboli [7]. Bullets entering the ascending aorta may embolize to the great vessels supplying the head and upper extremities, whereas bullets entering the descending thoracic aorta and abdominal aorta will embolize to the abdominal viscera or lower extremities. Because the left common iliac artery has a less acute angle of origin from the abdominal aorta than the right, it has been speculated that arterial emboli are more likely to travel to the left lower extremity (Figure 49.2) [4].

Arterial emboli tend to travel in the direction of blood flow, and lodge at the branch points in vessels. They usually become symptomatic early on.

Venous emboli may either travel in the direction of blood flow, or may travel retrograde, due to gravity [6]. About two-thirds of patients with venous bullet emboli are asymptomatic [7]. Usually the emboli lodge in to the right side of the heart or pulmonary arteries, potentially causing arrhythmias, valve injury, hemoptysis, or chest pain.

Paradoxical emboli travel from the right heart or systemic venous circulation to the systemic arterial circulation, typically through a patent foramen ovale [8].

Importance

Due to the potential for arterial occlusion and ischemia, arterial and paradoxical emboli should be retrieved in most circumstances, using endovascular techniques where possible [6]. However, the appropriate treatment of intracerebral arterial embolism remains controversial [3].

There are no standard indications for the removal of intracardiac foreign body emboli, although intracardiac emboli that cause valvular dysfunction or dysrhythmias should be removed [6, 7]. Increasingly, percutaneous transvenous techniques are used. If foreign body emboli are not removed, surveillance imaging should be performed.

Pulmonary arterial emboli can be safely observed in most cases, especially if they are likely to be difficult to retrieve [6].

Typical clinical scenario

One of three clues may prompt the radiologist and clinician to the possibility of a bullet embolus [6].

1. An unequal number of entry and exit wounds, not accounted for by imaging or surgical confirmation of the bullet in the expected body region.
2. A radiograph that shows a bullet in a body region inconsistent with the presumed trajectory of the bullet.
3. Serial imaging that shows the foreign body changing location.

Projectiles can also migrate along other anatomic pathways, in particular the spinal canal, airways, and gastrointestinal tract [9].

Differential diagnosis

Discrimination between an embolized projectile and a metallic foreign body from direct penetrating injury may be difficult in the setting of prior injury, shotgun injuries, and multiple gunshot injury. Surgical vascular clips and angioembolization coils may be confused with embolized bullet fragments. Examination of multiplanar reformats and radiographs may assist with discrimination.

When the primary injury is below the diaphragm and the embolus lodges in the upper extremity or intracranial arteries, consider paradoxical embolism through a patent foramen ovale, atrial septal defect, or ventricular septal defect.

Teaching point

Whenever there is an unequal number of entry and exit gunshot wounds, not accounted for by the identification of the projectiles by imaging or surgery, embolization should be suspected.

REFERENCES

1. Dosios TJ, Salemis N, Angouras D, Nonas E. Blunt and penetrating trauma of the thoracic aorta and aortic arch branches: an autopsy study. *J Trauma*. 2000;**49**(4):696–703.

2. Baldwin ZK, Phillips LJ, Bullard MK, Schneider DB. Endovascular stent graft repair of a thoracic aortic gunshot injury. *Ann Vasc Surg*. 2008;**22**(5):692–6.

3. da Costa LB, Wallace MC, Montanera W. Shotgun pellet embolization to the posterior cerebral circulation. *AJNR Am J Neuroradiol*. 2006; **27**(2):261–3.

4. Klein CP. Gunshot wounds of the aorta with peripheral arterial bullet embolism. Report of 2 cases. *Am J Roentgenol Radium Ther Nucl Med.* 1973;**119**(3):547–50.

5. Lewis TD, Henry DA. Needle embolus: a unique complication of intravenous drug abuse. *Ann Emerg Med.* 1985;**14**(9):906–8.

6. Miller KR, Benns MV, Sciarretta JD, *et al.* The evolving management of venous bullet emboli: a case series and literature review. *Injury.* 2011; **42**(5):441–6.

7. Chen JJ, Mirvis SE, Shanmuganathan K. MDCT diagnosis and endovascular management of bullet embolization to the heart. *Emerg Radiol.* 2007;**14**(2):127–30.

8. Corbett H, Paulsen EK, Smith RS, Carman CG. Paradoxical bullet embolus from the vena cava: a case report. *J Trauma.* 2003; **55**(5):979–81.

9. Rawlinson S, Capper S. Descent of a bullet in the spinal canal. *Emerg Med J.* 2007;**24**(7):519.

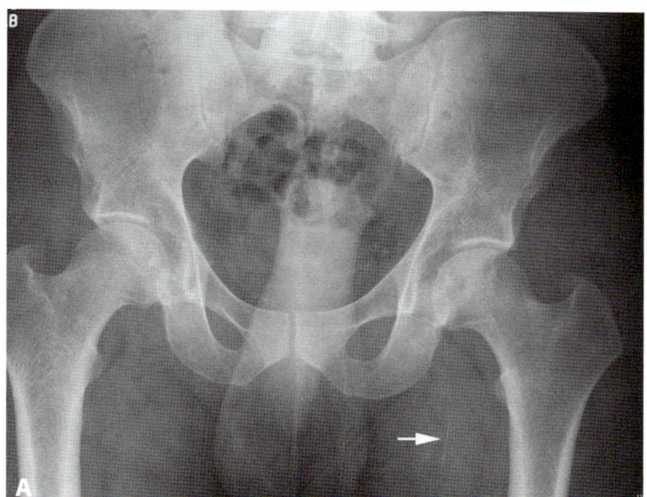

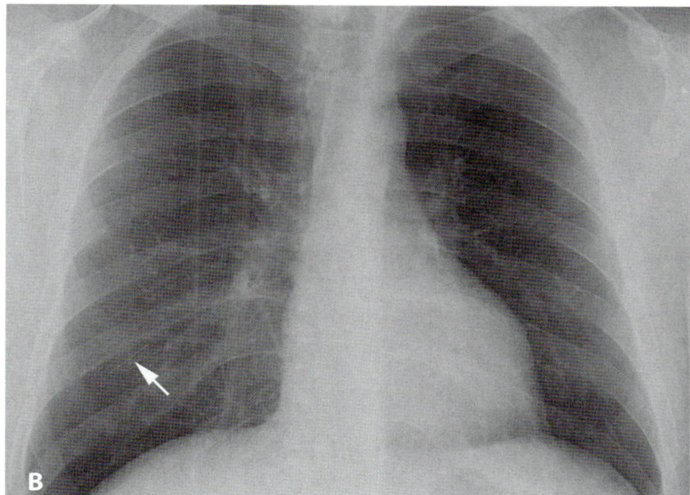

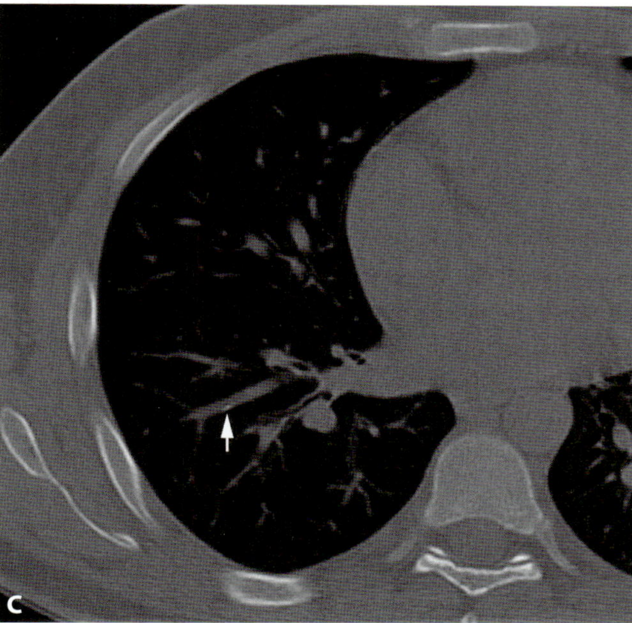

Figure 49.1 A. Anterior-posterior radiograph of the pelvis in a 28-year-old man with a history of intravenous drug use and acute chest pain. Previously two broken needles had been identified in the left groin. A radiograph of the pelvis taken after the onset of chest pain revealed only one needle (arrow). **B.** Posterior-anterior chest radiograph showed that the other needle had embolized to the right lung (arrow). **C.** Axial view from a non-contrast CT of the chest shows the needle within a small pulmonary artery (arrow). There was no visible pneumothorax or pulmonary hemorrhage. The patient was treated conservatively.

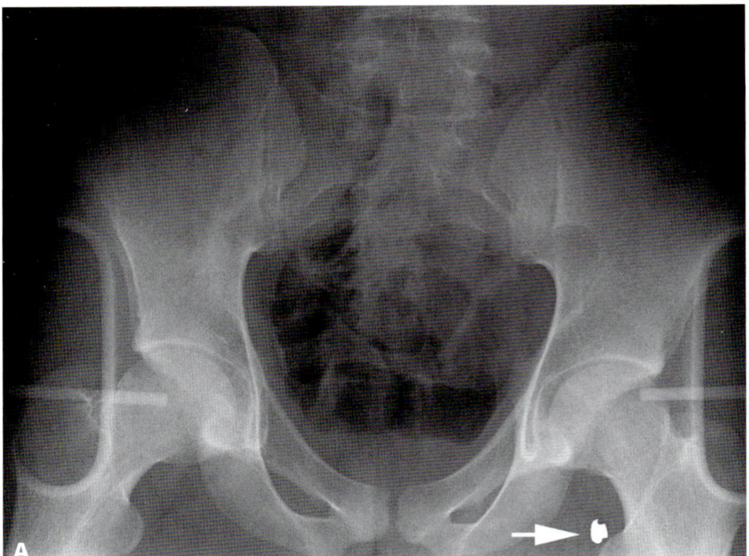

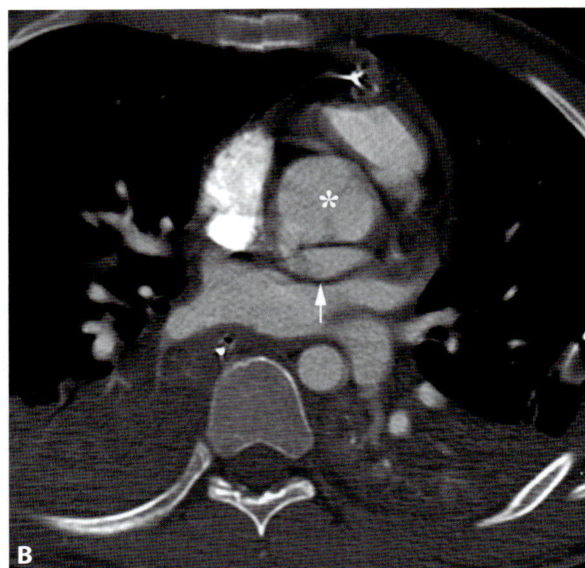

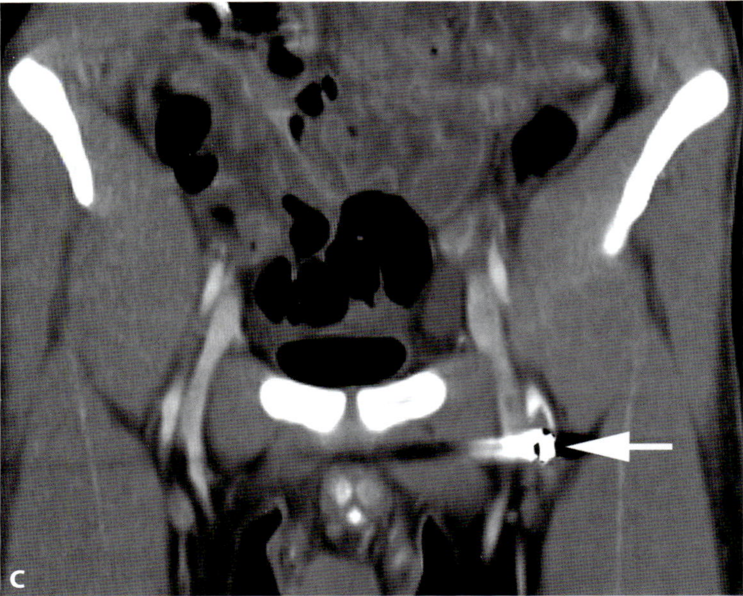

Figure 49.2 A 19-year-old man who was shot in the back of his chest. There was no exit wound, and a portable chest radiograph did not identify the bullet. **A.** Pelvic radiograph reveals a bullet in the region of the left groin (arrow). Due to hemodynamic instability, the patient was taken to the operating room, and a thoracotomy was performed. **B.** Immediate postoperative chest CT angiogram (CTA) identifies a pseudoaneurysm (arrow) between the aortic root (asterisk) and the left atrium, in the expected trajectory of the bullet. **C.** Coronal multiplanar reformat from a CTA of the lower extremity shows the embolized bullet at the bifurcation of the left common femoral artery (arrow) with a small area of adherent thrombus. The bullet was surgically retrieved, and the aortic pseudoaneurysm repaired.

Simulated active bleeding

Joel A. Gross

Imaging description

On contrast-enhanced images, foci of high density that do not conform to the shape, location, and enhancement of normal parenchyma usually represent active bleeding. Occasionally, a similar appearance can be seen with islands of perfused parenchyma surrounded by hematoma (Figure 50.1).

Active arterial or venous extravasation presents with foci of high density on portal venous phase images, corresponding to contrast-enhanced blood that has extravasated from a disrupted blood vessel. If this material is significantly denser than parenchyma, then contrast extravasation can be diagnosed at this time. However, if this material is similar in density to enhancing parenchyma, delayed images are necessary to make the diagnosis. Delayed imaging findings diagnostic of active extravasation include persistence or additional accumulation of high-density contrast and/or diffusion of contrast into the surrounding spaces.

Islands of perfused parenchyma surrounded by hematoma will generally follow the enhancement pattern of normal parenchyma. They enhance on portal venous phase images, and wash out on delayed phase images. If there is contusion of these fragments, the enhancement pattern may be different, and can match that of contused parenchyma within an organ. Whether contused or not, their appearance will be different from that seen with significant extravasations.

If the enhancement pattern follows the arterial blood pool (i.e., aorta), consider a pseudoaneurysm, and perform multiphase CT of the organ if an arterial phase was not already performed.

Importance

The spleen is the most common organ to demonstrate arterial extravasation following blunt abdominal trauma [1]. While criteria are still being debated [2], patients with active extravasation frequently undergo more aggressive treatment than those without extravasation. Treatment for splenic extravasation may include splenectomy or embolization [3, 4] although not all patients require intervention [5]. In contrast, stable patients without active extravasation are usually closely observed, avoiding embolization and splenectomy.

Typical clinical scenario

Blunt or penetrating abdominal trauma with injuries to the liver, spleen, or kidney.

Differential diagnosis

The differential diagnosis of hyperdense areas of contrast visible only on a venous phase CT of the abdomen includes active vascular extravasation, a small portion of perfused parenchyma, pseudoaneurysm, an arteriovenous fistula, and when seen in the kidneys, urinary extravasation.

> **Teaching point**
>
> Delayed phase CT performed at 5–10 minutes after contrast administration may be necessary to differentiate islands of perfused parenchyma from active vascular extravasation.

REFERENCES

1. Yao DC, Jeffrey RB, Jr., Mirvis SE, *et al.* Using contrast-enhanced helical CT to visualize arterial extravasation after blunt abdominal trauma: incidence and organ distribution. *AJR Am J Roentgenol.* 2002; **178**(1):17–20.
2. Rhodes CA, Dinan D, Jafri SZ, Howells G, McCarroll K. Clinical outcome of active extravasation in splenic trauma. *Emerg Radiol.* 2005; **11**(6):348–52.
3. Shanmuganathan K, Mirvis SE, Boyd-Kranis R, Takada T, Scalea TM. Nonsurgical management of blunt splenic injury: use of CT criteria to select patients for splenic arteriography and potential endovascular therapy. *Radiology.* 2000;**217**(1):75–82.
4. Franklin GA, Casos SR. Current advances in the surgical approach to abdominal trauma. *Injury.* 2006;**37**(12):1143–56.
5. Burlew CC, Kornblith LZ, Moore EE, Johnson JL, Biff WL. Blunt trauma induced splenic blushes are not created equal. *World J Emerg Surg* 2012;7:8.

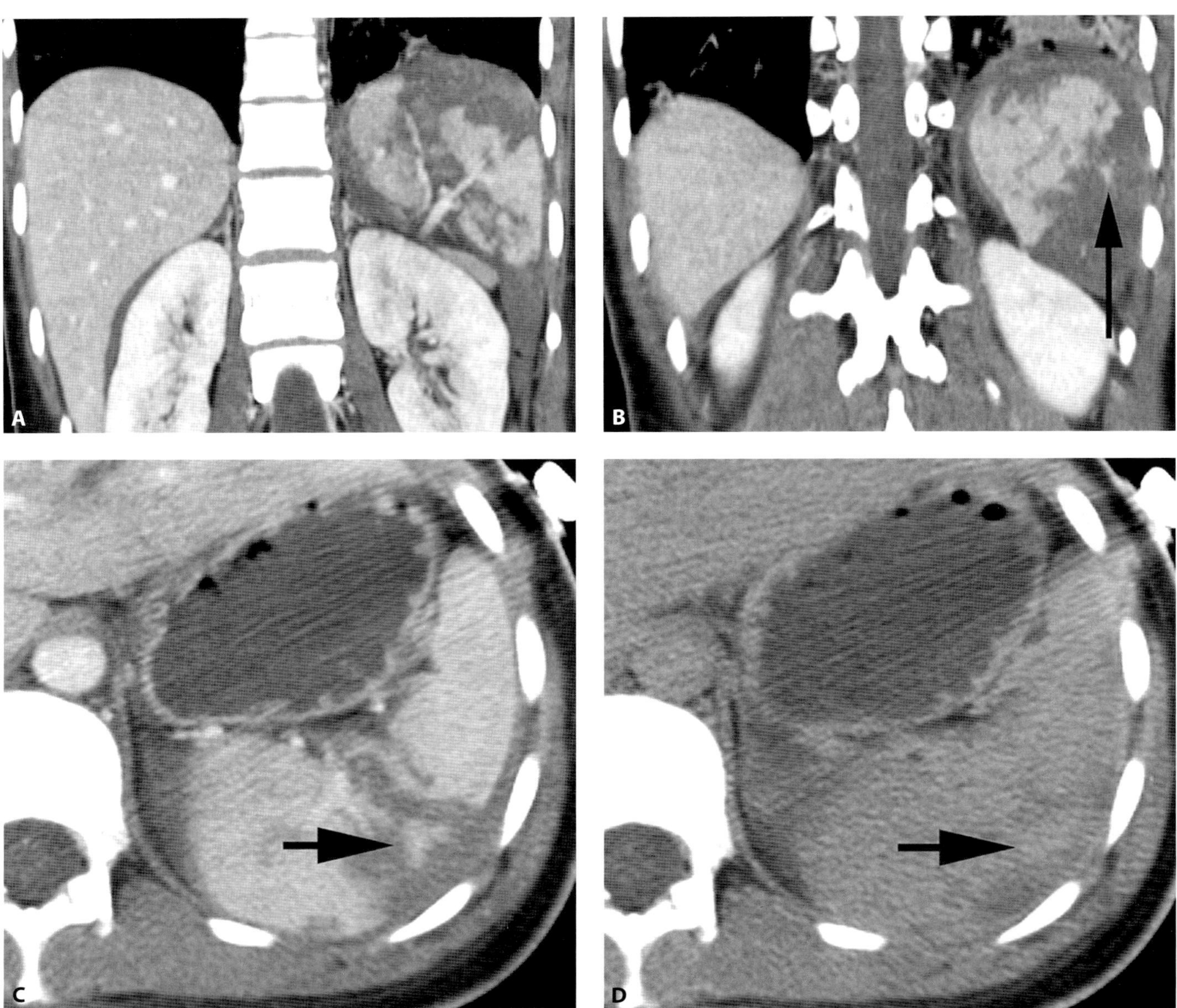

Figure 50.1 Contrast-enhanced CT from an 18-year-old woman following a motor vehicle crash. **A.** and **B.** Coronal reformations during the portal venous phase demonstrate a shattered spleen with a focal area of relatively high density (arrow) surrounded by hematoma, suggestive of active extravasation of vascular contrast. **C.** Axial image from the portal venous phase demonstrates the same focal density (arrow) suggestive of active extravasation. **D.** Axial image at the same level obtained 10 minutes later demonstrates similar density in the region of suspicion (arrow) compared with the remainder of the spleen. No persistent foci of high density were present to suggest accumulation or spread of extravasated vascular contrast, and the area of suspicion on portal venous phase images represented an island of perfused parenchyma surrounded by hematoma, rather than active hemorrhage with vascular extravasation.

CASE 51

Pseudopneumoperitoneum

Martin L. Gunn

Imaging description

In the emergency setting, free intraperitoneal gas, or pneumoperitoneum, usually represents perforated bowel. Unless the patient has a history of recent abdominal surgery or penetrating abdominal injury, pneumoperitoneum usually indicates an emergent laparotomy.

However, accumulation of gas in the extraperitoneal spaces which lie adjacent to the peritoneal space may simulate pneumoperitoneum. Moreover, gas within bowel loops on plain radiography may simulate subdiaphragmatic air.

Gas beneath the right hemidiaphragm can be simulated by colonic or small bowel interposition between the liver and hemidiaphragm (Figure 51.1), and was first described by Chilaiditi in 1910 [1, 2].

Gas within the subperitoneal space, which is extraperitoneal, can originate within the mediastinum or rectum and spread throughout the abdomen, and simulate intraperitoneal air [3]. The subperitoneal space lies subjacent to the parietal peritoneum, whereas the peritoneal space lies between the visceral and parietal peritoneum [3].

There are areas of continuity between the extrapleural space in the chest and the subperitoneal space in the abdomen. The principal communications lie posteriorly, and are formed by the esophageal and aortic hiatus of the diaphragm. Additionally, two areas of continuity lie between the anterior attachments of the diaphragm to the lower sternum and ribs. These are termed the sternocostal triangle (or foramina of Morgagni) and contain connective tissues, the superior epigastric continuations of the internal thoracic (mammary) arteries, and lymphatics.

When extraperitoneal spread of gas from the thorax to the abdomen is suspected, usually the quantity of gas in the thorax is disproportionately greater than the quantity in the abdomen. Because the gas lies within connective tissue pathways, small webs can often be identified within the gas (Figure 51.2). Moreover, the gas usually does not conform to the expected contour of the visceral peritoneum covering the bowel.

Importance

In patients with pneumomediastinum due to thoracic injuries (e.g., aerodigestive injury) or from non-traumatic pneumomediastinum (e.g., the Macklin effect caused by an acute asthma attack), subperitoneal gas in the abdomen must not be mistaken for abdominal bowel perforation [4]. A similar process can occur with extraperitoneal bladder and rectal injury, where gas can track throughout the subperitoneal space.

Typical clinical scenario

Patients with extensive pneumomediastinum, sometimes due to positive pressure ventilation, may appear to have pneumoperitoneum, which is not suspected clinically.

Differential diagnosis

Following trauma, not all pneumoperitoneum is caused by a ruptured hollow viscus. Consider gas introduced during diagnostic peritoneal lavage, gas traveling from the bladder in the setting of Foley catheterization and an intraperitoneal bladder rupture, and stab wounds that enter the peritoneal space but do not penetrate the bowel. Gelfoam angioembolization of solid organ injuries often leads to branching gas patterns within solid organs [5]. Several other rare causes of non-traumatic pneumoperitoneum have been reported [6].

Teaching point

In patients with pneumomediastinum, apparent pneumoperitoneum on CT may be due to subperitoneal gas.

REFERENCES

1. Saber AA, Boros MJ. Chilaiditi's syndrome: what should every surgeon know? *Am Surg.* 2005;**71**(3):261–3.
2. Chilaiditi D. Zur Frage der Hepatoptose und Ptose im allgemeinen im Anschluss an drei Fälle von temporärer, partieller Leberverlagerung. *Fortschr Geb Rontgenstr.* 1910;**16**:173–208.
3. Meyers MA, Charnsangavej C, Oliphant M. *Meyers' Dynamic Radiology of the Abdomen: Normal and Pathologic Anatomy*, 6th edn. New York: Springer, 2010.
4. Sakai M, Murayama S, Gibo M, Akamine T, Nagata O. Frequent cause of the Macklin effect in spontaneous pneumomediastinum: demonstration by multidetector-row computed tomography. *J Comput Assist Tomogr.* 2006;**30**(1):92–4.
5. Killeen KL, Shanmuganathan K, Boyd-Kranis R, Scalea TM, Mirvis SE. CT findings after embolization for blunt splenic trauma. *J Vasc Interv Radiol.* 2001;**12**(2):209–14.
6. Mularski RA, Sippel JM, Osborne ML. Pneumoperitoneum: a review of nonsurgical causes. *Crit Care Med.* 2000;**28**(7):2638–44.

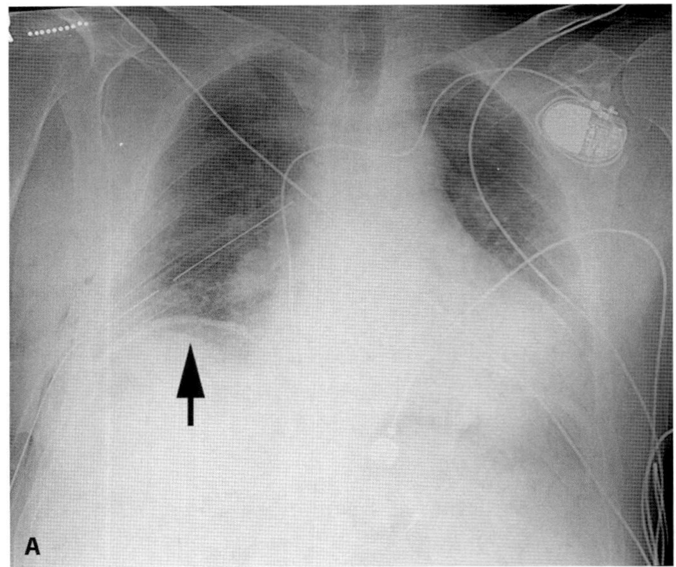

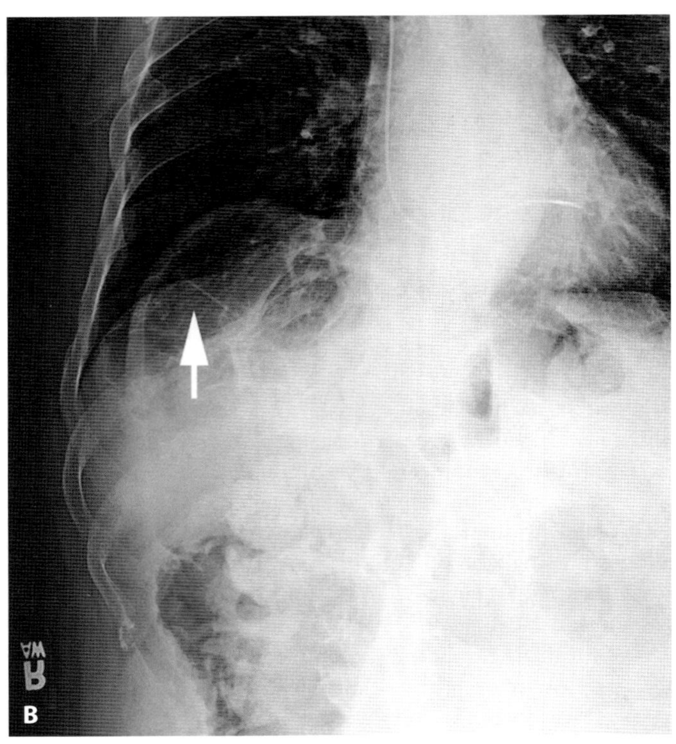

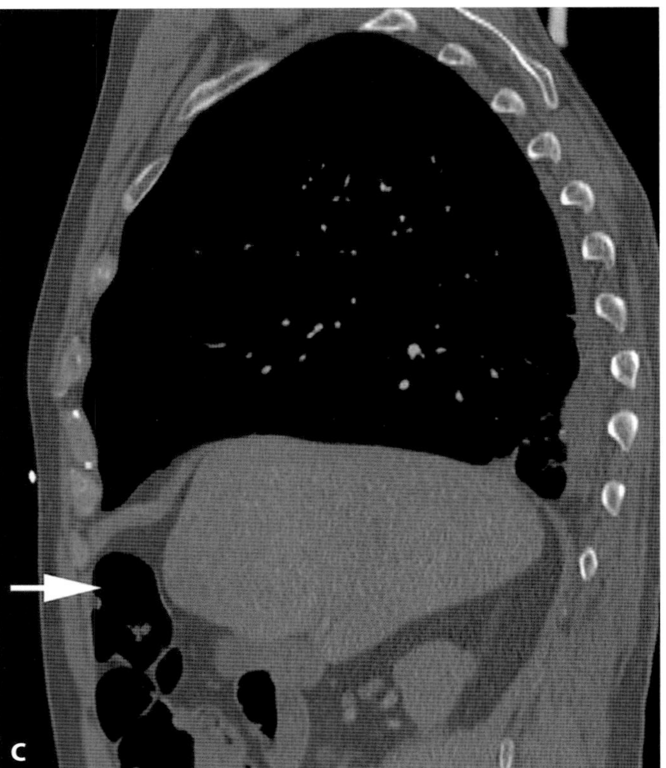

Figure 51.1 A 64-year-old man who suffered a ground-level fall and right pneumothorax. **A.** Chest radiograph following intercostal tube placement demonstrates a lucency beneath the right hemidiaphragm, which resembles intraperitoneal air (arrow). **B.** Decubitus radiograph clearly shows the hepatic flexure of the colon beneath the right hemidiaphragm (arrow). **C.** Sagittal reformation from a chest CT done hours before at an outside facility shows the colon loop (arrow) anterior to the liver and beneath the right hemidiaphragm.

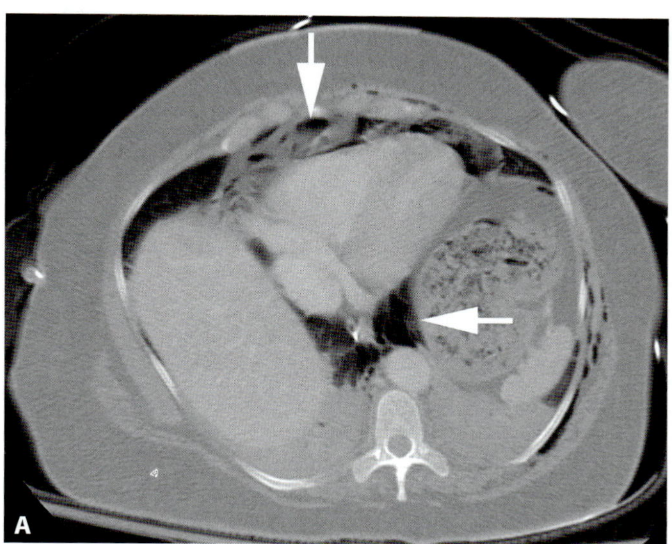

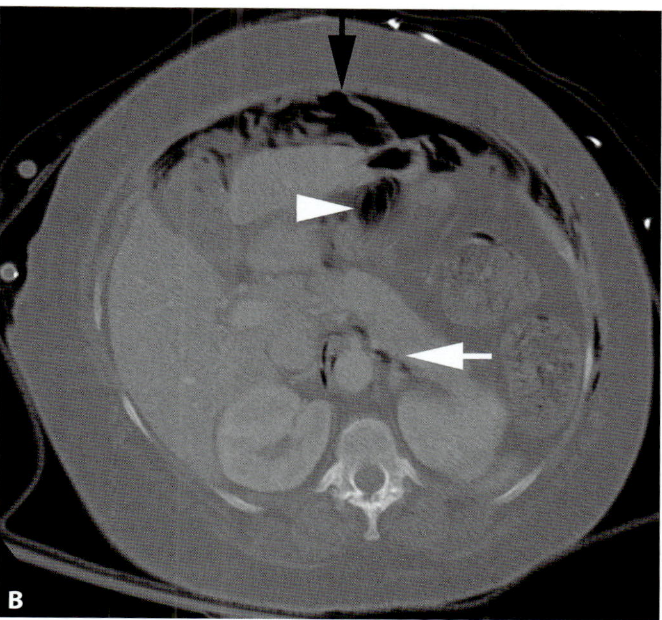

Figure 51.2 Axial enhanced CT images from the lower chest and upper abdomen in a 54-year-old woman who had a cardiac arrest, necessitating cardiopulmonary resuscitation which included chest compressions, defibrillation, intubation, and positive pressure ventilation. **A.** Extensive pneumomediastinum, with gas anterior to the heart, adjacent to the sternocostal triangle, and surrounding the esophagus (arrows). **B.** Axial image from the upper abdomen demonstrates retroperitoneal gas anterior to the crura of the diaphragm (white arrow), as well as gas within the gastrohepatic ligament, which connects to the retroperitoneal space (arrowhead). Gas could be seen to surround the mesenteric vessels, a space which is continuous with the retroperitoneal space. Gas within the anterior abdomen (black arrow) is located just below the sternocostal triangle, communicating with the anterior mediastinum. Note the strands of connective tissue within the gas, something not identified in pneumoperitoneum.

CASE

52 Intra-abdominal focal fat infarction: epiploic appendagitis and omental infarction

Claire K. Sandstrom

Imaging description

Inflammatory changes in the fat generally provide reliable evidence of an underlying acute intra-abdominal process. Disproportionate fat stranding in the pericolic region with mild colonic wall thickening suggests a pericolonic inflammatory process, including diverticulitis, appendicitis, epiploic appendagitis, and omental infarction, as opposed to colonic wall thickening centered on the colon, typical of infection, ischemia, and inflammatory bowel disease [1]. Occasionally, the underlying pathologic process is in the fat itself, involving the omentum, mesentery, or epiploic appendages.

Appendices epiploicae arise in two rows from the serosal surface of the colon, from the cecum to the rectosigmoid junction (Figure 52.1). Pedunculated and with tenuous blood supply, they are prone to torsion, infarction, and subsequent inflammation.

Imaging of epiploic appendagitis reveals an ovoid or lobular lesion of fat density less than 5 cm in size (usually 1–4 cm) adjacent to the anterior colonic wall (Figures 52.2 and 52.3) [2]. There is typically a well-defined hyperattenuating rim and surrounding inflammatory change. As with other pericolic processes, wall thickening of adjacent colon is mild compared with the degree of fat stranding.

Epiploic appendagitis can occur anywhere along the colon, but is most common adjacent to the sigmoid [3]. The "central dot sign" has been described as a specific sign for epiploic appendagitis and reflects thrombosed or obstructed central vessels within the torsed appendage.

Detachment of infarcted epiploic appendages is likely to be the cause of calcified fatty bodies that are incidentally identified on CT, a finding that has been termed "peritoneal mice" (Figure 52.4) [3].

When inflammatory changes are centered in the greater omentum, deep to the abdominal wall, the diagnosis of omental infarction should be considered. Omental infarction is typically seen in the right abdomen, medial to the cecum or ascending colon, possibly because of anomalous venous anatomy [4]. Classic imaging characteristics are of a solitary, heterogeneous, and high-attenuating fatty mass, larger than epiploic appendagitis with an average diameter of 7 cm (Figures 52.5 and 52.6). Lack of a hyperattenuating ring or central dot sign further help distinguish omental infarction from epiploic appendagitis.

While imaging characteristics overlap, discrimination between epiploic appendagitis and omental infarction is probably unnecessary as both are managed conservatively. Some suggest grouping them under the term, "intra-abdominal focal fat infarction" (IFFI) [5].

Importance

Epiploic appendagitis and omental infarction are self-limiting causes of abdominal pain and do not require emergent surgery [5]. Though immediate surgery can be curative, most now advise conservative management of omental infarction with surgery considered only if symptoms persist or an abscess develops. Patients with epiploic appendagitis are treated as outpatients, with anti-inflammatory medications.

Typical clinical scenario

The diagnosis of epiploic appendagitis is typically made in young adults presenting with acute abdominal pain, often in the lower abdomen. Omental infarction has been described in a broader population of patients, including the pediatric and elderly. Clinical presentation mimics that of appendicitis or diverticulitis with acute onset of abdominal pain and possibly nausea, vomiting, and anorexia. Fever, diarrhea, or constipation are rare, and leukocyte count is normal to mildly elevated [5]. Factors that predispose to intra-abdominal fat infarction include obesity, vascular compromise from strenuous activity, congestive heart failure, digitalis treatment, and recent abdominal surgery or trauma.

Differential diagnosis

The clinical and imaging differential diagnosis depends on the site of the infarction. However, the most important diagnosis to exclude is typically acute appendicitis. An appendix measuring less than 6 mm in diameter and containing air or oral contrast virtually excludes a diagnosis of acute appendicitis.

If the infarction is on the left, the main consideration is acute diverticulitis. Patients are usually older and will often have diverticulosis. The inflammation involves a segment of colonic wall, and there may be a fluid collection or gas, reflecting an abscess in the pericolic fat. Most patients with acute diverticulitis do not require emergent surgery but should receive antibiotics.

Other differential considerations include mesenteric panniculitis, which is usually seen in the root of the small bowel mesentery and does not abut the colonic wall [2]. "Omental caking" from peritoneal carcinomatosis can mimic omental infarction, but multiple lesions and peritoneal nodules in the setting of known malignancy point to the correct diagnosis. Fat-containing masses, including liposarcoma, lipoma, exophytic renal angiomyolipoma, and teratoma, may occasionally present diagnostic dilemmas.

Teaching point

Epiploic appendagitis and omental infarction are self-limited causes of abdominal pain and do not require emergent surgery. They can be recognized by distinctive inflammatory changes centered in the intra-abdominal fat.

REFERENCES

1. Pereira JM, Sirlin CB, Pinto PS, *et al.* Disproportionate fat stranding: a helpful CT sign in patients with acute abdominal pain. *Radiographics.* 2004;**24**:703–15.

2. Singh AK, Gervais DA, Hahn PF, *et al.* Acute epiploic appendagitis and its mimics. *Radiographics.* 2005;**25**:1521–34.

3. Almeida AT, Melao L, Viamonte B, Cunha R, Pereira JM. Epiploic appendagitis: an entity frequently unknown to clinicians – diagnostic imaging, pitfalls, and look-alikes. *AJR Am J Roentgenol* 2009; **193**(5):1243–51.

4. Epstein LI, Lempke RE. Primary idiopathic segmental infarction of the greater omentum: case report and collective review of the literature. *Ann Surg.* 1968;**167**(3):437–43.

5. van Breda Vriesman AC, Lohle PNM, Coerkamp EG, Puylaert JBCM. Infarction of omentum and epiploic appendage: diagnosis, epidemiology and natural history. *Eur Radiol.* 1999;**9**:1886–92.

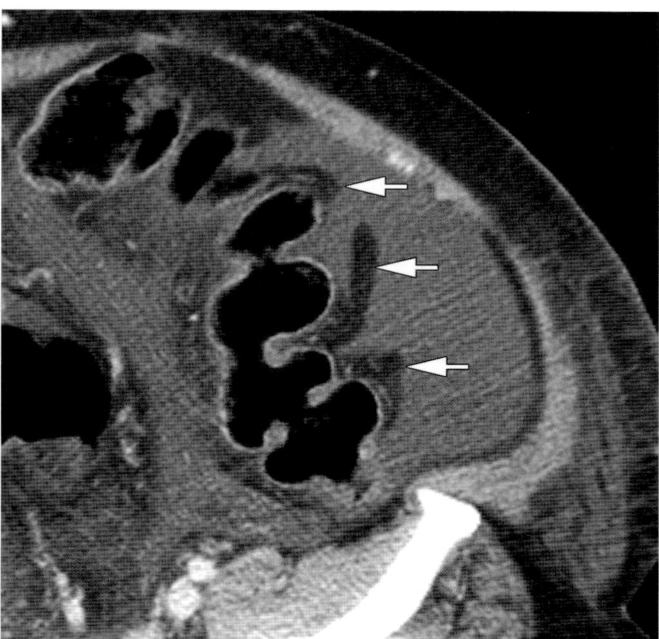

Figure 52.1 Axial contrast-enhanced CT of the abdomen in a 48-year-old man with cirrhosis and right lower quadrant pain shows ascites outlining normal epiploic appendages of the sigmoid colon (arrows).

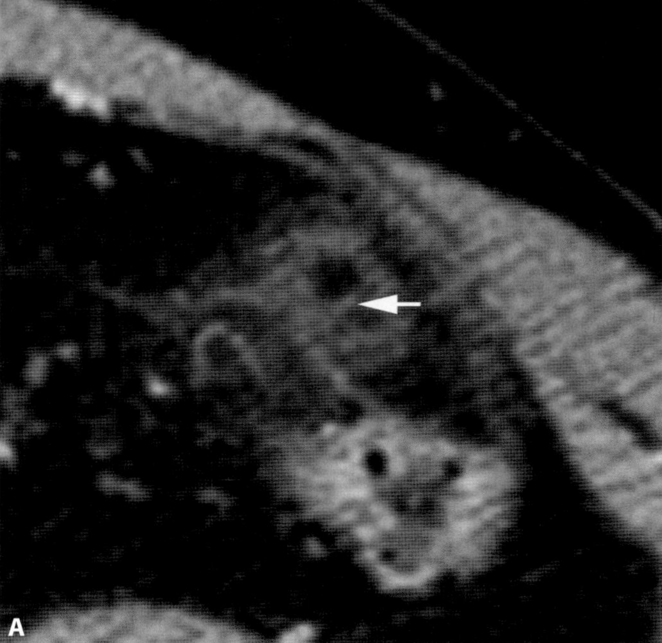

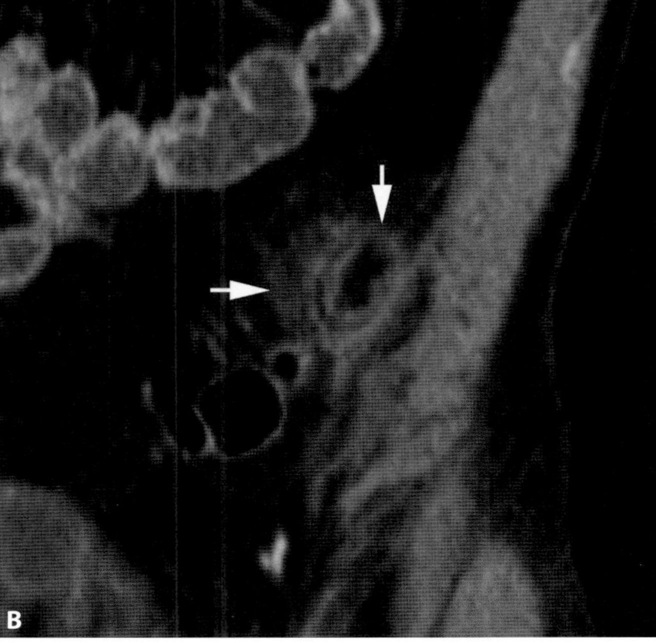

Figure 52.2 **A.** Axial contrast-enhanced CT of the abdomen in a 43-year-old man with left lower quadrant pain and diarrhea shows an inflammatory mass along the anterior border of the descending colon, with central fat density. The "central dot sign" is noted as a punctate hyperdensity within the fat (arrow). **B.** Similar appearance of fat with surrounding inflammation (arrows) on coronal reformation, consistent with acute epiploic appendagitis.

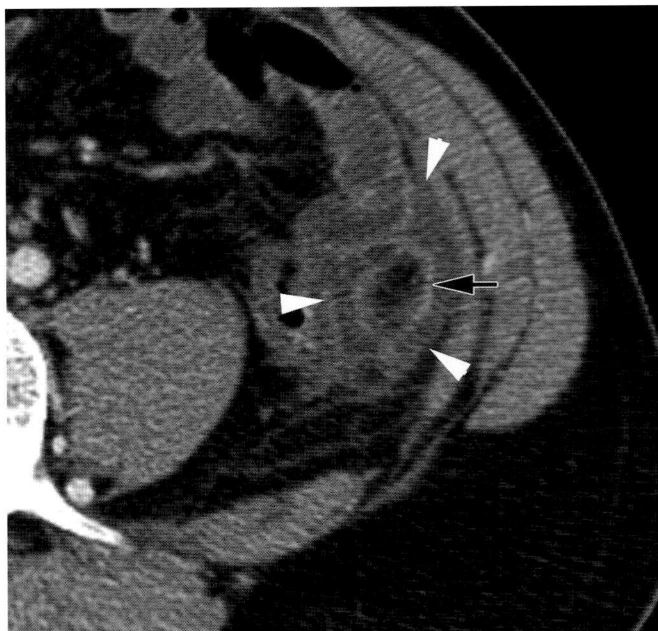

Figure 52.3 Axial contrast-enhanced CT of the abdomen in a 43-year-old man with three days of left lower quadrant pain shows an ovoid fatty lesion with enhancing rim of inflammation (arrow), compatible with acute epiploic appendagitis. In addition, there is adjacent free fluid (arrowheads).

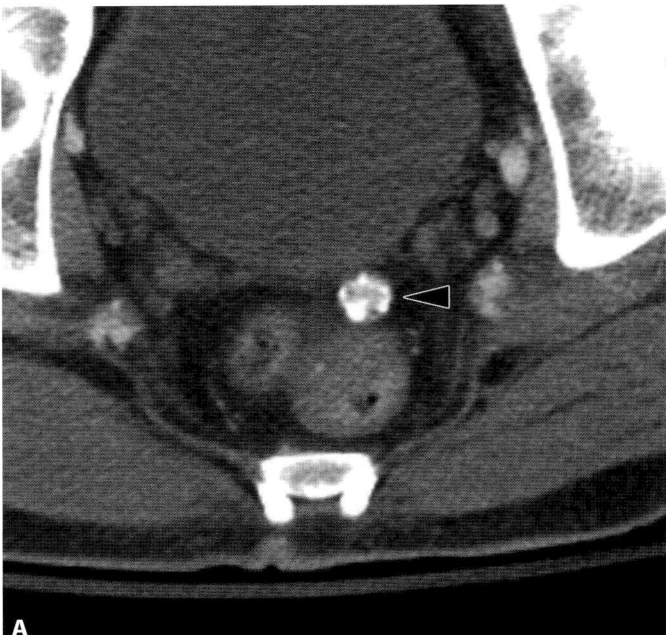

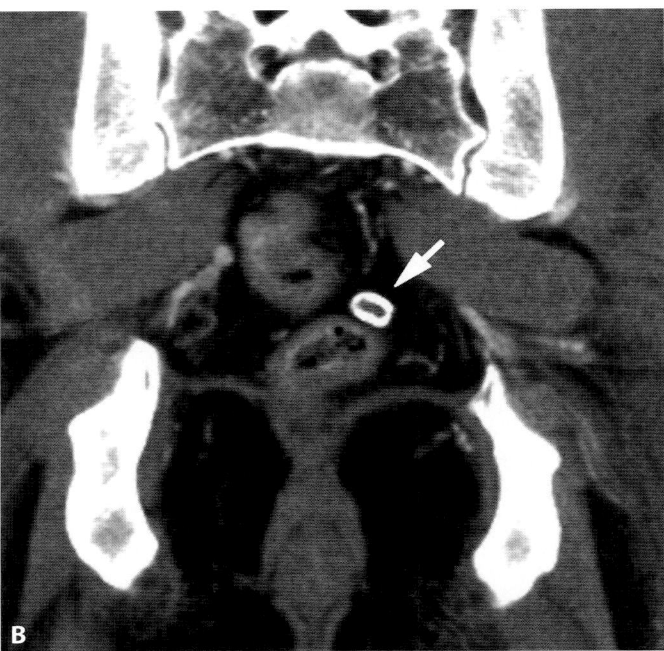

Figure 52.4 **A.** Axial contrast-enhanced CT of the abdomen and pelvis in a 39-year-old man following a bicycle crash shows a densely rim-calcified lesion (arrowhead) in the rectovesical fat. This most likely represents residua from remote torsion of an epiploic appendage. **B.** Coronal reformation better depicts the rim calcification of this lesion (arrow).

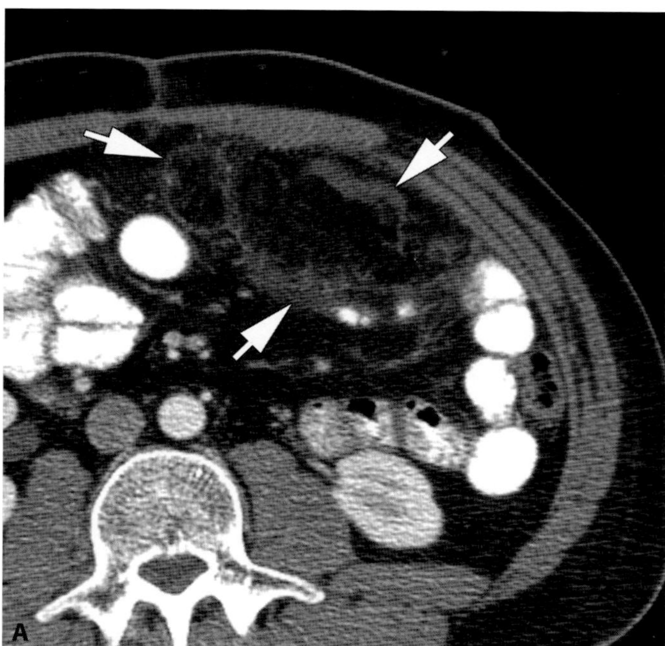

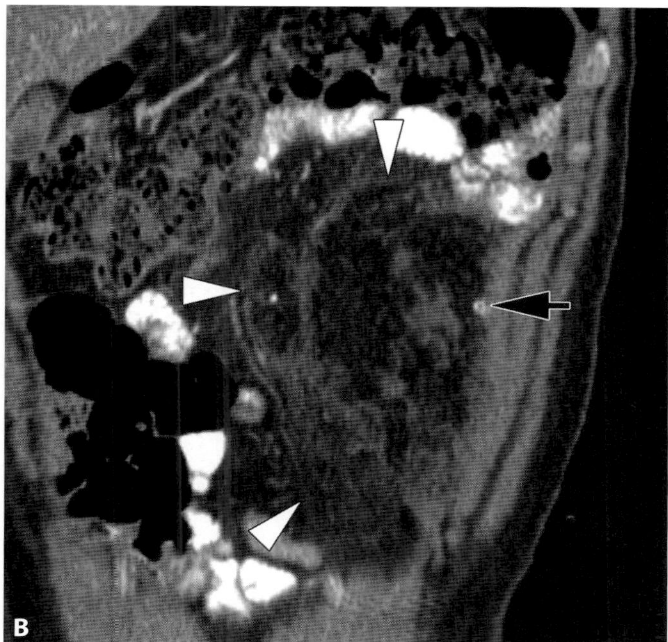

Figure 52.5 A. Axial contrast-enhanced CT of the abdomen in a 24-year-old man status post partial gastrectomy, splenectomy, and pancreatectomy shows a large ovoid area of inflammation in the left anterior midabdomen, consistent with omental infarct (arrows). **B.** Coronal reformation shows the large ovoid inflammatory mass (arrowheads) adjacent to the surgical drain (arrow).

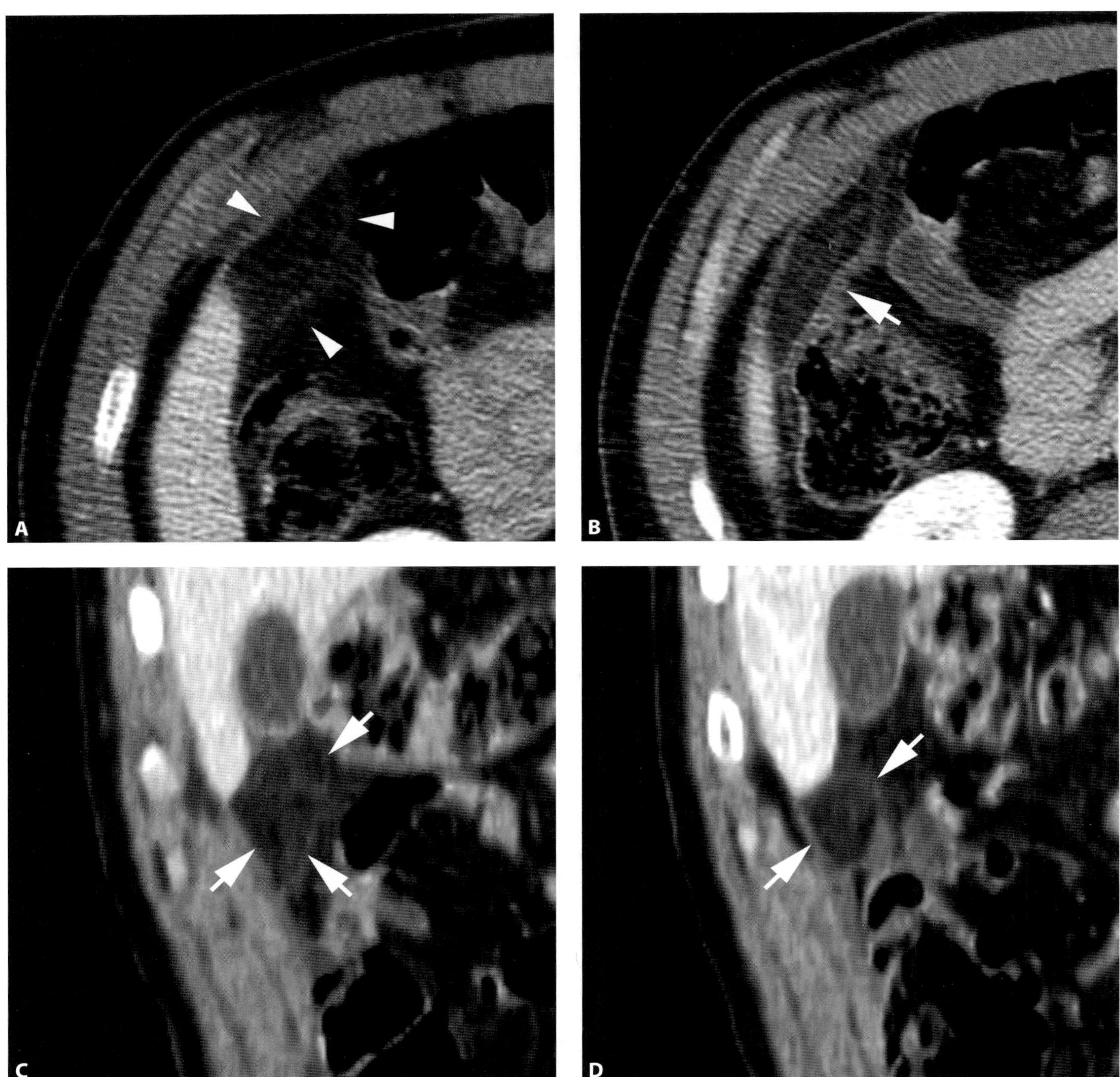

Figure 52.6 A. Axial contrast-enhanced CT of the abdomen in a 30-year-old man with abdominal pain shows hazy fat stranding (white arrowheads) in the right upper quadrant, interposed between the liver and hepatic flexure. **B.** On an axial image more inferiorly, the margins become more well defined (white arrow), but central fat density remains apparent. Size of the lesion favors an omental infarct. **C.** and **D.** Selected coronal reformations again show central fat stranding with rim-like thickening (arrows).

False-negative and False-positive FAST

Martin L. Gunn

Imaging description

Focused assessment with sonography for trauma (FAST) is an adjunct to the American College of Surgeons' Advanced Trauma Life Support (ATLS) primary survey.

Unfortunately, there are very few prospective randomized trials that examine the efficacy and effectiveness of FAST on patients with blunt abdominal trauma [1]. Multiple retrospective studies have generally demonstrated a high specificity ($\geq$95%), but widely ranging sensitivity (43–89%) for the detection on free intraperitoneal fluid [2–5]. Consequently, a negative FAST examination should not be considered as a means of excluding significant abdominal injury, especially in the hemodynamically stable patient. Moreover, FAST has not been shown to conclusively reduce the rate of trauma laparotomy or diagnostic peritoneal lavage (DPL), although it might lead to a slight reduction in the rate of CT scanning [1]. Limited data suggest a follow-up FAST scan might increase sensitivity [6, 7]. However, if there is a high pre-test probability of an abdominal injury, contrast-enhanced CT should be considered [1] following a negative FAST.

The intra-abdominal component of the FAST exam includes at least three views (Figure 53.1): (a) around the liver (perihepatic, hepato-renal recess, and subphrenic space), (b) around the spleen (perisplenic, splenorenal recess, and subphrenic), and (c) pelvis. The minimum volume of intraperitoneal fluid detectable with FAST is thought to range between 100 and 620mL [3].

In addition to an evaluation for intraperitoneal fluid, the indications for FAST have been increased to include sonography of the pericardial space and heart, especially in the setting of penetrating thoracic injury. Moreover, an extended FAST (eFAST) can be performed which also evaluates for pneumothorax and hemothorax [8].

False-negative and false-positive FAST scans can cause significant morbidity [9].

False-positive examinations are uncommon, but can occur when fluid-filled bowel adjacent to a solid organ mimics intraperitoneal fluid (Figure 53.2), due to pre-existing ascites or physiologic fluid in young patients (especially females), or by mistaking the seminal vesicles for rectovesical fluid. False-negative examinations can occur due to an empty bladder obscuring visualization of the rectovesical or rectouterine pouch, subcutaneous emphysema, obesity, and areas of non-visualized fluid. A FAST examination is a real-time evaluation for intraperitoneal blood. It is important to move and angle the transducer to fully evaluate for fluid in the upper abdomen, pleural space, and pelvis, rather than simply documenting static views of the upper quadrants and pelvis.

Importance

FAST should be considered a "rule in" and not a "rule out" investigation. A positive FAST in a hemodynamically unstable blunt abdominal trauma patient may expedite operative intervention. However, FAST should only be performed with caution, if at all, in hemodynamically stable patients with blunt abdominal trauma. In these patients, a negative study should be considered with suspicion as significant solid organ injuries may be missed. In these patients, if there are signs of intro-abdominal injury, consideration should be given to CT scanning [1, 2, 9].

A negative FAST after penetrating abdominal trauma should be followed by other diagnostic studies such as CT, diagnostic peritoneal lavage [10–12], or operative intervention.

An International Consensus Panel statement emphasized the importance of considering the quantity of free fluid observed before taking the patient to the operating room for a laparotomy as a small quantity of intraperitoneal fluid can be observed in the setting of important extraperitoneal injuries (e.g., pelvic fracture requiring angioembolization) [13].

Typical clinical scenario

Trauma imaging algorithms vary greatly between institutions, depending on local resources, patient demographics, and expertise. Important clinical factors including the hemodynamic stability, injury mechanism, availability of contrast-enhanced CT, or a surgeon experienced in DPL, should be considered before performing FAST. Hemodynamic instability is generally considered as a systolic blood pressure of $<$90mm Hg or pulse $>$100 beats per minute. These factors should also be considered when interpreting the significance of either a positive, or particularly a negative, FAST scan.

Differential diagnosis

In the hypotensive patient with a negative FAST, firstly consider a bleeding from other sites, such as the chest, retroperitoneum, pelvis, or lower extremity. Consider non-hemorrhagic causes such as tension pneumothorax, diaphragmatic rupture, cardiac ischemia, or spinal injury. Also, always consider a false-negative FAST examination.

Teaching point

A frankly positive FAST in the setting of blunt abdominal trauma can reduce the time from hospital arrival to the operating room. However, a negative FAST scan should be treated with suspicion, in both the hemodynamically stable and unstable trauma patient.

REFERENCES

1. Stengel D, Bauwens K, Sehouli J, *et al.* Emergency ultrasound-based algorithms for diagnosing blunt abdominal trauma. *Cochrane Database Syst Rev.* 2005;(2):CD004446.

2. Natarajan B, Gupta PK, Cemaj S, *et al.* FAST scan: is it worth doing in hemodynamically stable blunt trauma patients? *Surgery.* 2010; **148**(4):695–700; discussion 701.

3. Patel NY, Riherd JM. Focused assessment with sonography for trauma: methods, accuracy, and indications. *Surg Clin North Am.* 2011; **91**(1):195–207.

4. Gaarder C, Kroepelien CF, Loekke R, *et al.* Ultrasound performed by radiologists-confirming the truth about FAST in trauma. *J Trauma.* 2009;**67**(2):323–7; discussion 328–9.

5. McGahan JP, Rose J, Coates TL, Wisner DH, Newberry P. Use of ultrasonography in the patient with acute abdominal trauma. *J Ultrasound Med.* 1997;**16**(10):653–62; quiz 663–4.

6. Blackbourne LH, Soffer D, McKenney M, *et al.* Secondary ultrasound examination increases the sensitivity of the FAST exam in blunt trauma. *J Trauma.* 2004;**57**(5):934–8.

7. Nunes LW, Simmons S, Hallowell MJ, *et al.* Diagnostic performance of trauma US in identifying abdominal or pelvic free fluid and serious abdominal or pelvic injury. *Acad Radiol.* 2001; **8**(2):128–36.

8. Gillman LM, Ball CG, Panebianco N, Al-Kadi A, Kirkpatrick AW. Clinician performed resuscitative ultrasonography for the initial evaluation and resuscitation of trauma. *Scand J Trauma Resusc Emerg Med.* 2009;**17**:34.

9. Miller MT, Pasquale MD, Bromberg WJ, Wasser TE, Cox J. Not so FAST. *J Trauma.* 2003;**54**(1):52–9; discussion 59–60.

10. Udobi KF, Rodriguez A, Chiu WC, Scalea TM. Role of ultrasonography in penetrating abdominal trauma: a prospective clinical study. *J Trauma.* 2001;**50**(3):475–9.

11. Chiu WC, Shanmuganathan K, Mirvis SE, Scalea TM. Determining the need for laparotomy in penetrating torso trauma: a prospective study using triple-contrast enhanced abdominopelvic computed tomography. *J Trauma.* 2001;**51**(5):860–8; discussion 868–9.

12. Quinn AC, Sinert R. What is the utility of the Focused Assessment with Sonography in Trauma (FAST) exam in penetrating torso trauma? *Injury.* 2011;**42**(5):482–7.

13. Scalea TM, Rodriguez A, Chiu WC, *et al.* Focused Assessment with Sonography for Trauma (FAST): results from an international consensus conference. *J Trauma.* 1999;**46**(3):466–72.

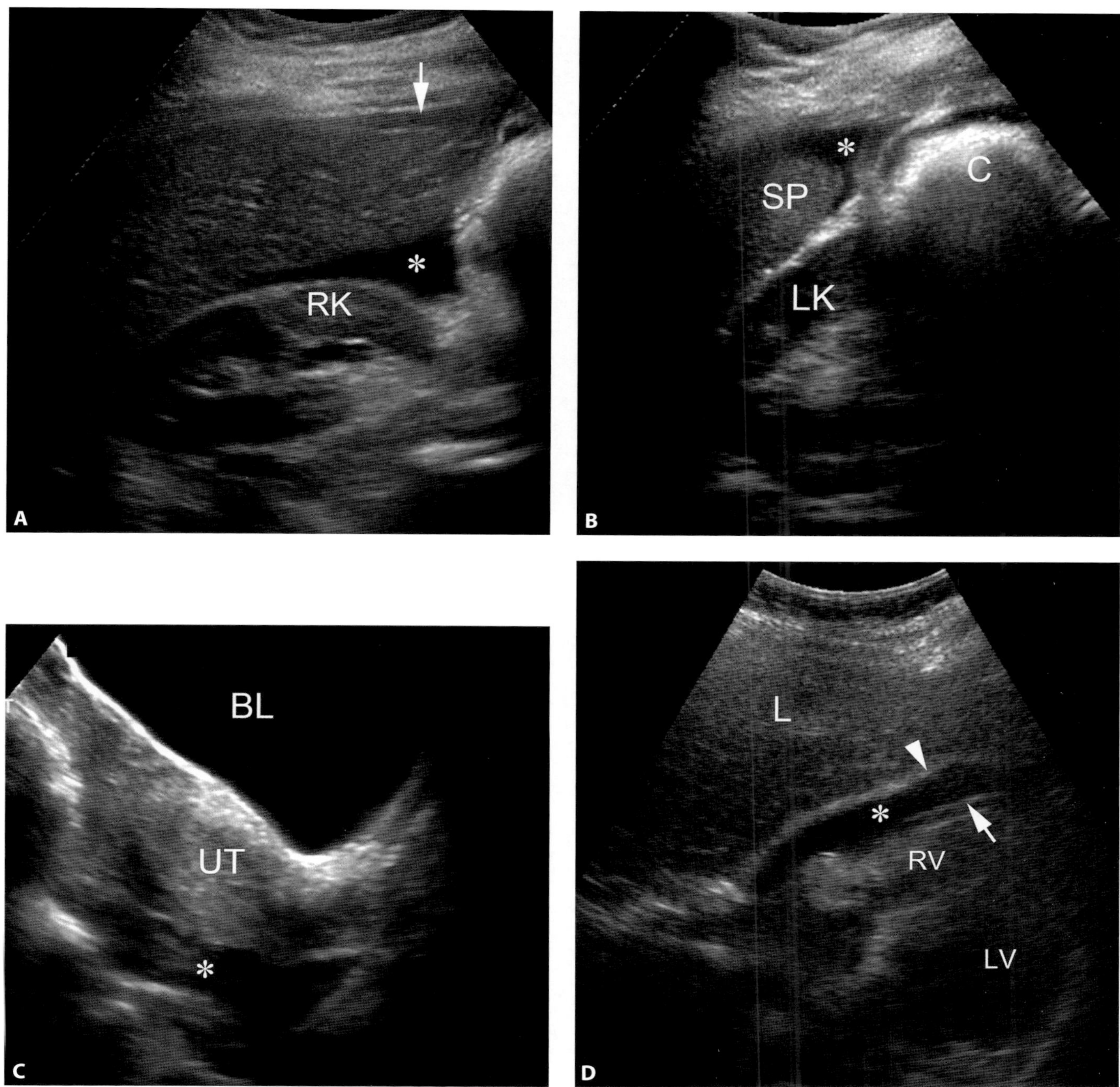

Figure 53.1 Views from true positive FAST scans. **A.** Subcostal oblique view of the right upper quadrant shows fluid containing debris (asterisk) in the hepato-renal pouch, anterior to the right kidney (RK) and superior to the splenic flexure of the colon (C). Also note the fluid anterior to the right lobe of the liver (arrow). From this position, angling the transducer will assist with evaluation for right pleural fluid, and right subdiaphragmatic fluid. **B.** Coronal view of the left upper quadrant. For a good acoustic window, the transducer is best placed almost in the coronal plane posterior in the left flank to avoid the gas-filled stomach. Note the fluid (asterisk) surrounding the interior pole of the spleen (SP). No fluid is noted in the splenorenal pouch, between the spleen and the left kidney (LK). **C.** Sagittal view of the pelvis through a distended bladder (BL) demonstrates the heterogeneous appearance of partially clotted blood (asterisk) posterior to the uterus (UT). **D.** Subcostal view of the pericardium through the liver (L) shows a pericardial effusion (asterisk) between the fibrous pericardium (arrowhead) and visceral pericardial line (arrow). Note the relative compression of the right ventricle (RV) compared to the left (LV), a sign of cardiac tamponade physiology.

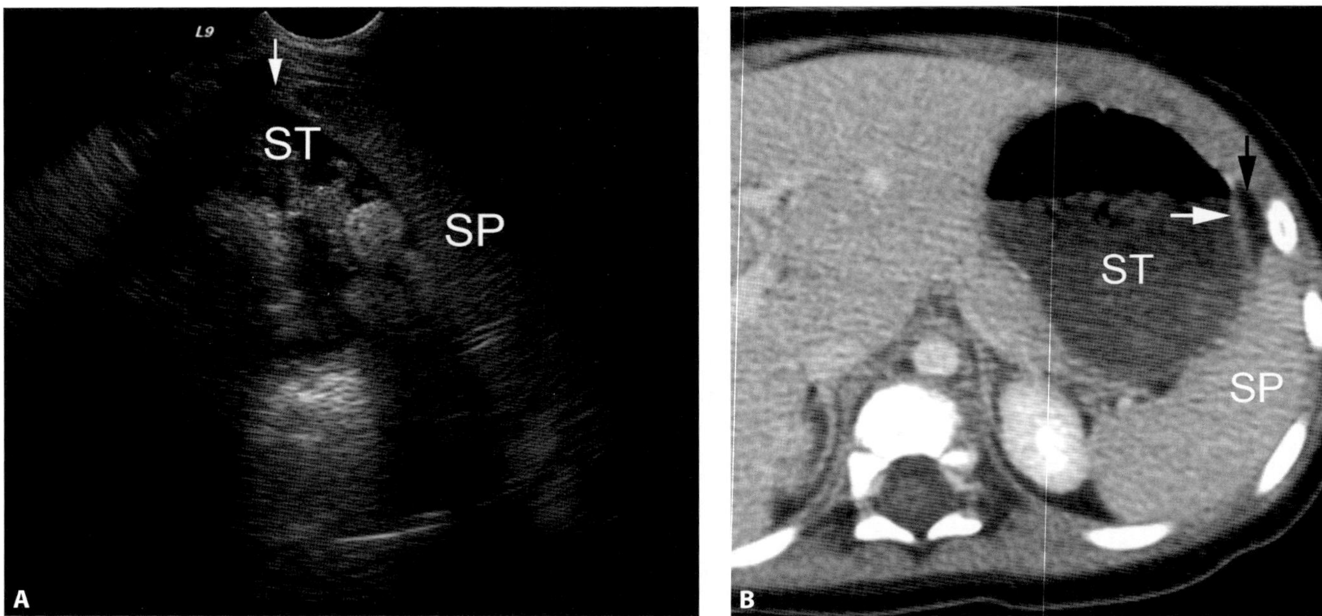

Figure 53.2 A. Single image from an outside hospital FAST scan in a one-year-old girl who was involved in a high-speed motor vehicle collision, sustaining a femoral shaft fracture. This FAST scan was interpreted as positive for perisplenic fluid at the outside institution. Note that the hypoechoic fluid in the stomach (ST) does not wrap around the tip (arrow) of the spleen (SP). This is a clue that the fluid is in the bowel rather than the spleen. A repeat sonogram at the receiving hospital was negative for free fluid. **B.** Due to the presence of the femur fracture and the outside sonogram, a contrast-enhanced CT was ordered, which shows the wall of the stomach (arrow) separating the fluid from the spleen. The black arrow depicts intra-abdominal fat.

Diaphragmatic slip simulating liver laceration

Michael J. Modica

Imaging description

A diaphragmatic slip is a muscular bundle or flap projecting from the inferior surface of the diaphragm (Figure 54.1A). They are most commonly seen incidentally on CT and are usually found in the superior right hepatic lobe [1]. On CT, diaphragmatic slips appear as wedge-shaped, round, or oval structures of muscle attenuation that range in size from 1 to 3.5 cm, and are surrounded by thin strips of fat attenuation (Figures 54.1B and 54.2). In many patients, multiple slips present at regular intervals on the liver surface of the right hepatic lobe. Occasionally, slips can course obliquely through the liver parenchyma, an appearance that can mimic a laceration (Figure 54.3). Sagittal and coronal reformations can better characterize their smooth course along the liver edge or within the liver parenchyma. Difficult cases may be resolved by following the slip along its long axis with ultrasound, decubitus CT, or CT in deep inspiration. The CT appearance of the slips corresponds well with their appearance on ultrasound where the slip appears as a highly echogenic structure indenting the liver edge [2].

Importance

A slip indenting the liver surface or projecting through the liver parenchyma may be confused with a laceration, mass, or focal contusion.

Typical clinical scenario

Diaphragmatic slips are present in 25% of CT scans, and their frequency increases with age [1].

Differential diagnosis

The liver is the second most commonly injured intra-abdominal solid organ. Diaphragmatic slips can mimic traumatic liver laceration or hematoma, peritoneal implants, or parenchymal liver masses. A laceration should be suspected based on its irregular linear or branching low-attenuation area on contrast-enhanced CT (Figure 54.4).

Parenchymal and subcapsular hematomas usually have well-defined elliptical margins that indent the liver parenchyma. The density can vary depending on whether the hematoma is unclotted (30–45 Hounsfield Units [HU]) or clotted (60–90 HU) blood.

Peritoneal implants on the liver are small nodular masses generally found along the right lateral peritoneal surface [3]. They are randomly distributed and usually accompanied by ascites.

Decubitus CT might help discriminate between a diaphragmatic slip and true lesions [4].

Teaching point

Diaphragmatic slips are smooth, linear or wedge-shaped, and low attenuation. Identification of fat on either side of the slip or contiguity with the inferior aspect of the diaphragm helps distinguish this entity from a laceration or hematoma. Multiplanar reformations, ultrasound, or decubitus CT can be useful in difficult cases.

REFERENCES

1. Auh YH, Rubenstein WA, Zirinsky K, *et al.* Accessory fissures of the liver: CT and sonographic appearance. *AJR Am J Roentgenol.* 1984; **143**(3):565–72.
2. Yeh HC, Halton KP, Gray CE. Anatomic variations and abnormalities in the diaphragm seen with US. *Radiographics.* 1990;**10**(6):1019–30.
3. Jeffrey RB, Jr. CT demonstration of peritoneal implants. *AJR Am J Roentgenol.* 1980;**135**(2):323–6.
4. Rosen A, Auh YH, Rubenstein WA, *et al.* CT appearance of diaphragmatic pseudotumors. *J Comput Assist Tomogr.* 1983;**7**(6):995–9.

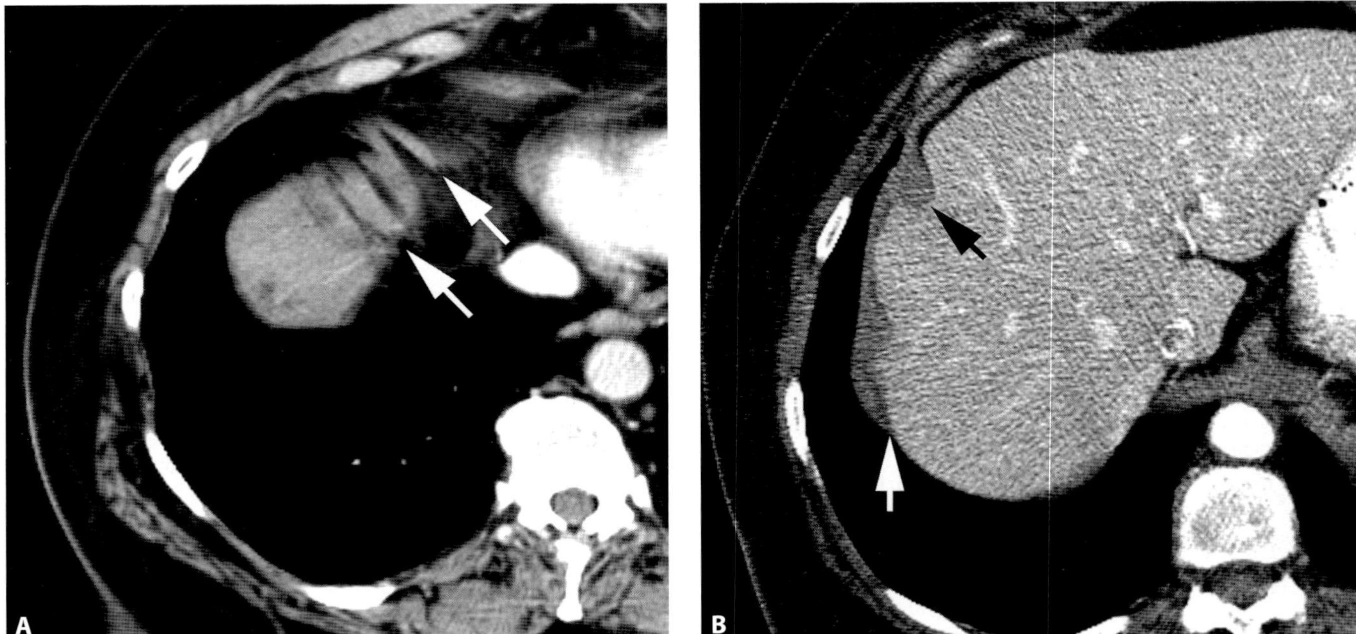

Figure 54.1 A. Axial contrast-enhanced CT of the abdomen in a 29-year-old man involved in a high-speed motor vehicle collision. Soft tissue density bundles are seen coursing over the right hepatic lobe (arrows). These are muscular diaphragmatic slips projecting from the inferior surface of the diaphragm. **B.** More caudal images through the liver demonstrate the smooth indentation (black arrow) these slips have on the liver surface. Notice the fat (white arrow) invaginating along the liver edge, separating it from the slip.

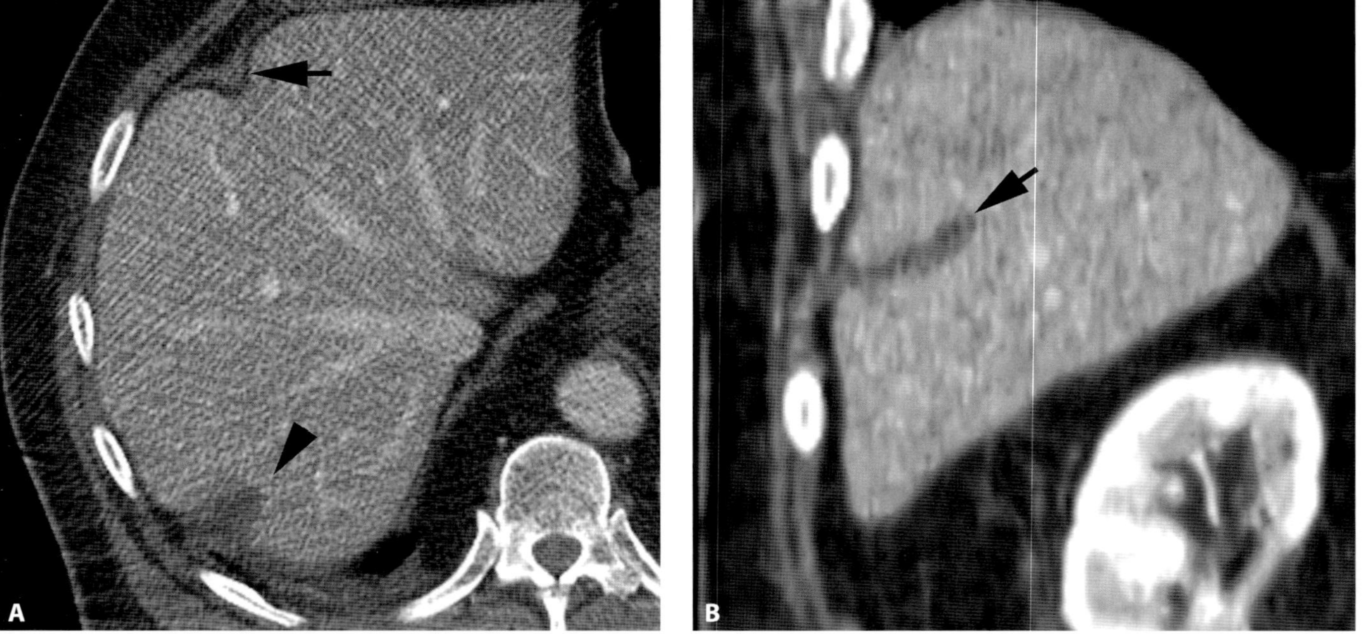

Figure 54.2 A. Axial contrast-enhanced CT of the abdomen in a 52-year-old blunt trauma patient. Notice the soft tissue and fat indentation (arrow) indenting the anterior right hepatic lobe. This is characteristic of a diaphragmatic slip. The linear low-density focus in the posterior right lobe (arrowhead) was thought to represent a focal subcaspular contusion. **B.** Coronal reformat from the same patient demonstrates a smooth linear soft tissue density structure extending from the chest wall coursing through the liver parenchyma (arrow). This is a characteristic appearance of a diaphragmatic slip.

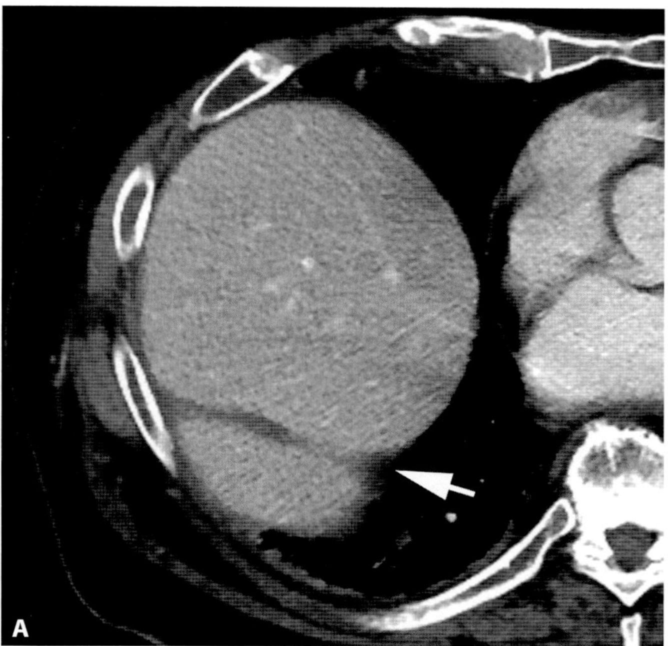

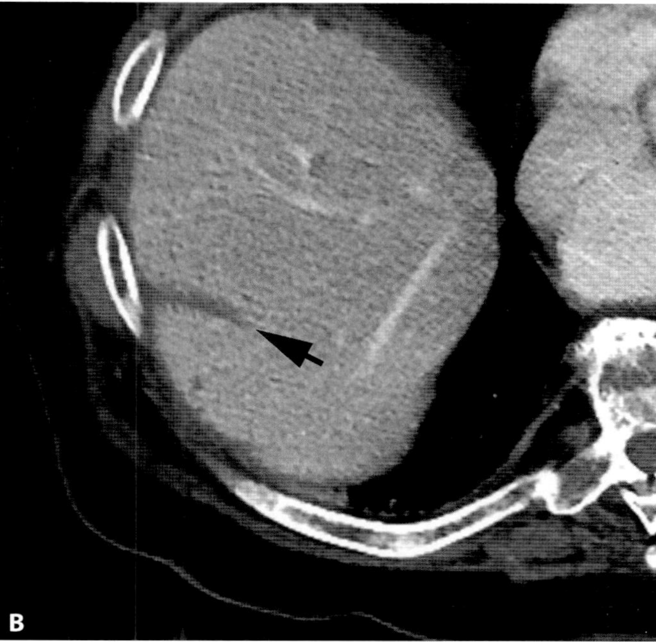

Figure 54.3 A. Axial contrast-enhanced CT of the abdomen in a 22-year-old man involved in a high-speed motor vehicle collision shows a muscular slip extending from the inferior surface of the diaphragm indenting the liver dome (arrow). **B.** Caudal images from the same patient show the slip coursing through the right hepatic lobe (arrow).

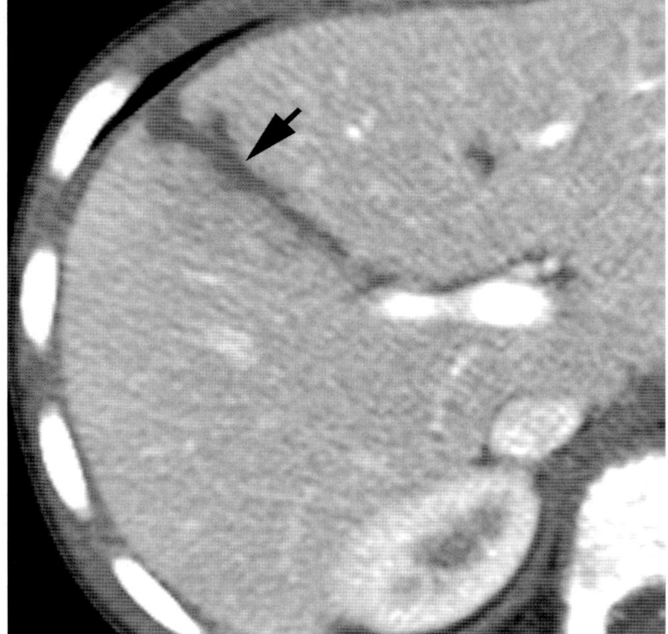

Figure 54.4 Axial contrast-enhanced CT of the abdomen in a 41-year-old man involved in a high-speed motor vehicle collision shows an irregular branching hypodensity extending from the capsule of the right hepatic lobe towards the right portal vein representing a liver traumatic laceration (arrow).

Gallbladder wall thickening due to non-biliary causes

Claire K. Sandstrom

Imaging description

By ultrasound, gallbladder wall edema is present when the gallbladder mural thickness is 3 mm or more (Figure 55.1). The wall may also appear striated, with alternating hypoechoic and hyperechoic layers (Figure 55.2). These imaging features are non-specific in isolation. Additional sonographic signs of acute cholecystitis include the presence of mobile stones or sludge, a non-mobile stone in the gallbladder neck, gallbladder luminal distension, pericholecystic fluid, and extrahepatic and intrahepatic biliary dilation. A sonographic Murphy's sign (SMS) is positive when there is maximal tenderness over the sonographically localized gallbladder, and negative if the pain is diffuse or localized to a site distant from the gallbladder [1]. If both gallstones and a positive SMS are present with gallbladder wall thickening, the positive predictive value is greater than 90%. Gallbladder wall thickening found in the absence of stones or a SMS has a reported negative predictive value for acute cholecystitis of approximately 95% [1].

Importance

Gallbladder wall edema can be found in patients with both biliary and non-biliary causes of right upper quadrant pain. Additional sonographic features that favor a non-biliary source of the gallbladder wall thickening include sonographic signs of cirrhosis, a hypoechoic liver, and decompressed gallbladder lumen.

Typical clinical scenario

Patients will often present with right upper quadrant abdominal pain and tenderness on clinical exam, prompting ultrasound or CT.

Differential diagnosis

The differential diagnosis for gallbladder wall edema remains broad until additional clinical and imaging findings are considered.

During the routine right upper quadrant ultrasound, hepatic size, morphology, and echogenicity should be inspected, with careful attention to hepatic enlargement and hypoechogenicity that may reflect acute hepatitis, or a shrunken, nodular hepatic contour suggestive of cirrhosis. The surface contour of the liver and the echotexture of the superficial liver parenchyma can be assessed using a high-resolution linear transducer when cirrhosis is suspected.

Often, the gallbladder lumen will be decompressed in the setting of a non-biliary cause of gallbladder wall edema. This is in contradistinction to acute cholecystitis, in which gallbladder distension suggests the underlying obstruction. Stones, sludge, pericholecystic fluid, and SMS may or may not accompany edema.

Non-biliary conditions causing gallbladder wall thickening include chronic or acute liver diseases, hypoproteinemia, right heart failure, and generalized edema such as in an acute allergic reaction (Figures 55.3–55.6) [2, 3]. The striated gallbladder wall is similarly non-specific. It has been described in the setting of congestive heart failure, renal failure, hepatic failure, hepatitis, ascites, hypoalbuminemia, pancreatitis, lymphatic or venous obstruction of the gallbladder, and prominent Rokitansky–Aschoff sinuses [4].

In children, gallbladder wall thickening has been found in the setting of hypoalbuminemia, ascites, partial wall contraction, and systemic venous hypertension [5]. Interestingly, in this series of 793 children, gallbladder wall thickening was not a feature of the five cases of acute cholecystitis.

Teaching point

Gallbladder wall edema on right upper quadrant ultrasound is a non-specific finding that can be associated with both biliary and non-biliary causes.

REFERENCES

1. Ralls PW, Colletti PM, Chandrasoma P, *et al.* Real-time sonography in suspected acute cholecystitis. Prospective evaluation of primary and secondary signs. *Radiology.* 1985;**155**:767–71.
2. Brook OR, Kane RA, Tyagi G, Siewert B, Kruskal JB. Lessons learned from quality assurance: errors in the diagnosis of acute cholecystitis on ultrasound and CT. *AJR Am J Roentgenol.* 2011; **196**(3):597–604.
3. Shlaer WJ, Leopold GR, Scheible FW. Sonography of the thickened gallbladder wall: a non-specific finding. *AJR Am J Roentgenol.* 1981;**136**:337–9.
4. Teefey SA, Baron RL, Bigler SA. Sonography of the gallbladder: significance of striated (layered) thickening of the gallbladder wall. *AJR Am J Roentgenol.* 1991;**156**:945–7.
5. Patriquin HB, DiPietro M, Barber FE, Teele RL. Sonography of thickened gallbladder wall: causes in children. *AJR Am J Roentgenol.* 1983;**141**:57–60.

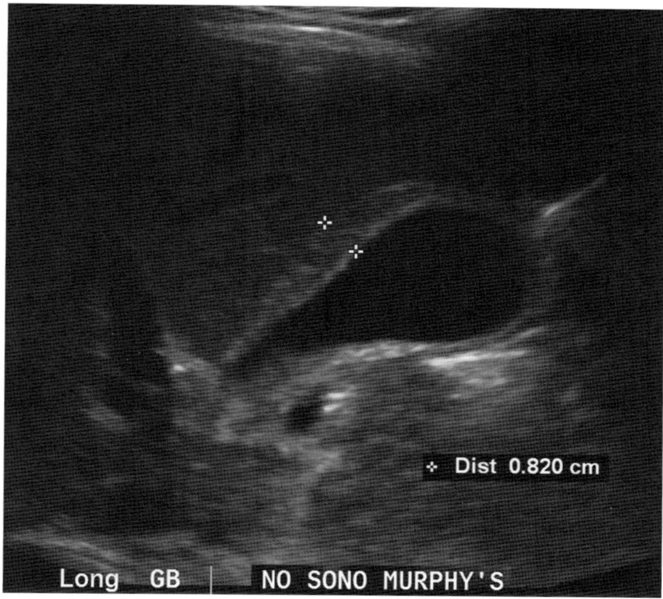

Figure 55.1 U trasound image of the gallbladder in a 38-year-old man with fulminant hepatitis attributed to isoniazid shows gallbladder wall thickening (calipers). Sonographic Murphy's sign was negative.

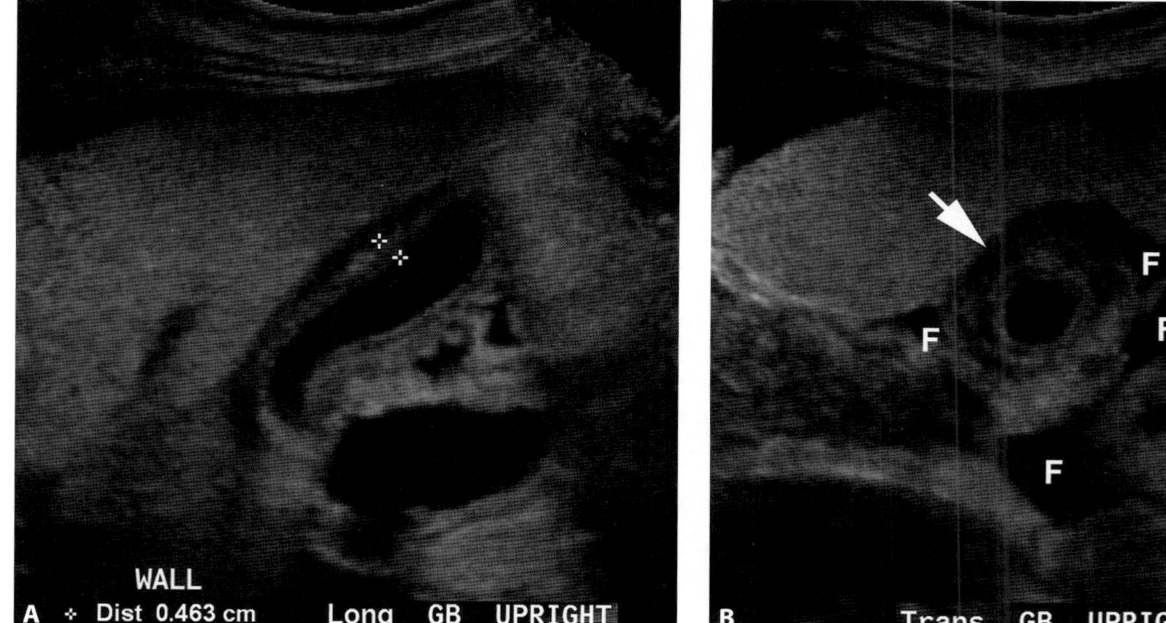

Figure 55.2 A. Ultrasound image of the gallbladder in a 54-year-old woman with history of dilated cardiomyopathy, suffering from acute cardiogenic and septic shock as well as severe hypoalbuminemia, shows gallbladder wall thickening (calipers). Note the striated pattern to the gallbladder wall. **B.** Transverse ultrasound image of the gallbladder again shows the wall thickening (arrow), as well as small amounts of free fluid (F) throughout the right upper quadrant. Sonographic Murphy's sign was negative.

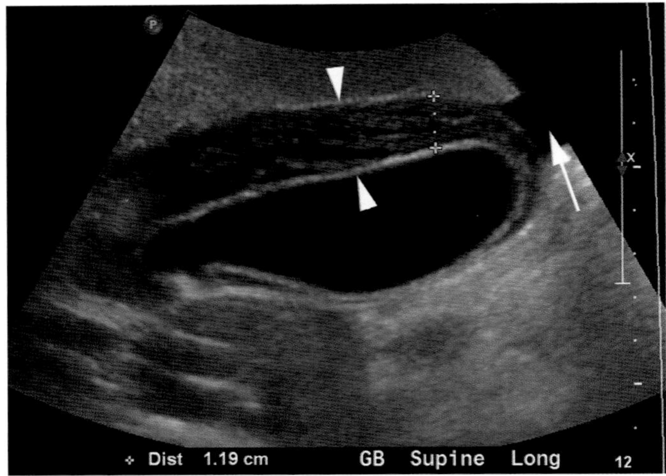

Figure 55.3 Ultrasound image of the gallbladder in a 54-year-old woman with history of alcoholism, elevated liver enzymes, hypoalbuminemia, and acute gastrointestinal bleeding shows marked laminated thickening of the gallbladder wall (arrowheads). There is also a small amount of ascites (arrow). No gallstones or sonographic Murphy's sign were detected.

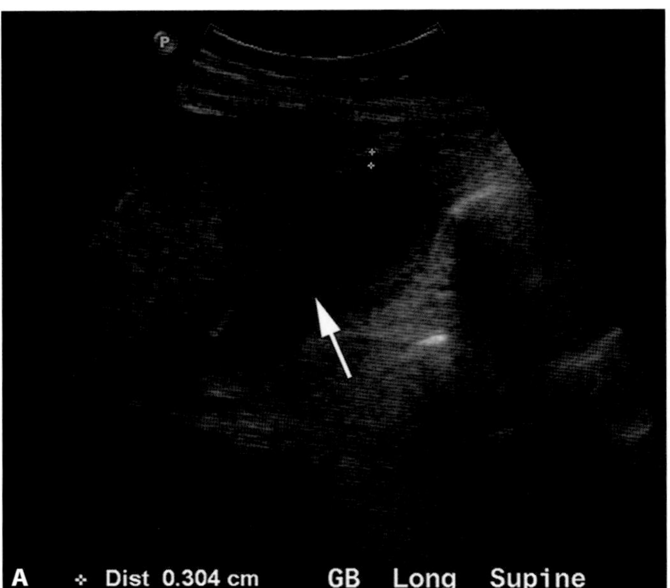

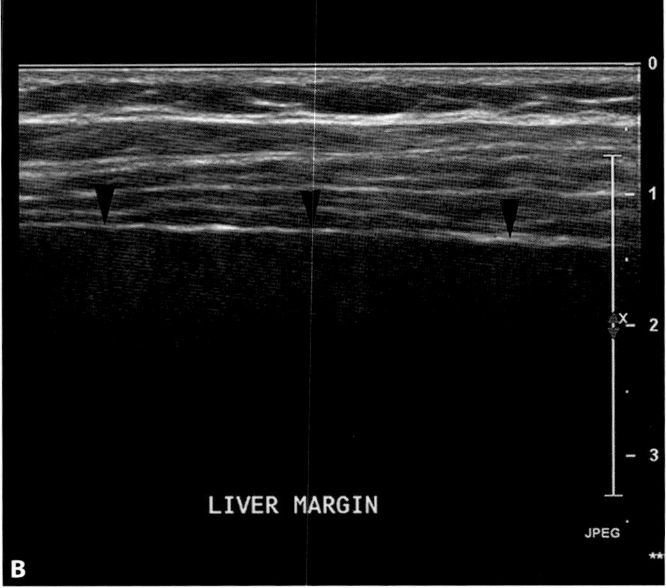

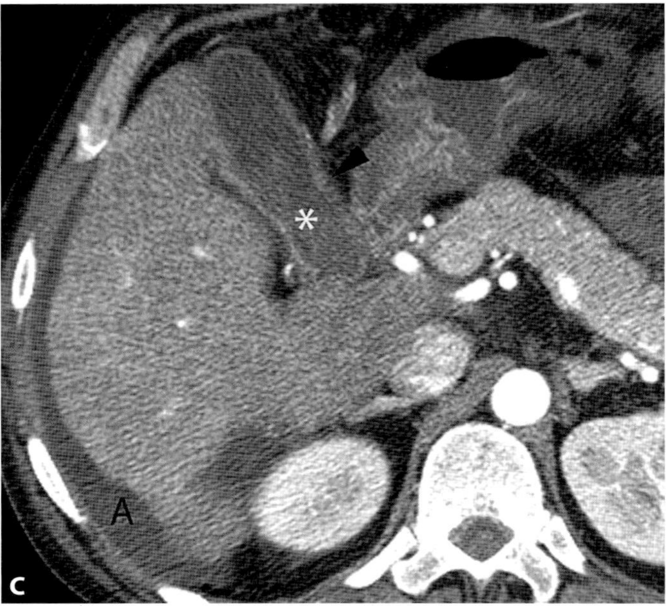

Figure 55.4 **A.** Ultrasound image of the gallbladder in a 49-year-old man with elevated hepatic and pancreatic enzymes shows sludge layering dependently within the gallbladder (arrow) and mild wall thickening (calibers). **B.** With the 12-MHz linear transducer, ultrasound image of the liver margin shows nodularity (arrowheads) consistent with underlying cirrhosis. **C.** CT with intravenous contrast obtained three days later shows shrunken liver, compatible with cirrhosis, and small perihepatic ascites (*A*). The gallbladder wall remains mildly thick (arrowhead), and hyperdense sludge is seen within the gallbladder lumen (asterisk). The patient was treated for acute hepatitis superimposed on alcoholic cirrhosis.

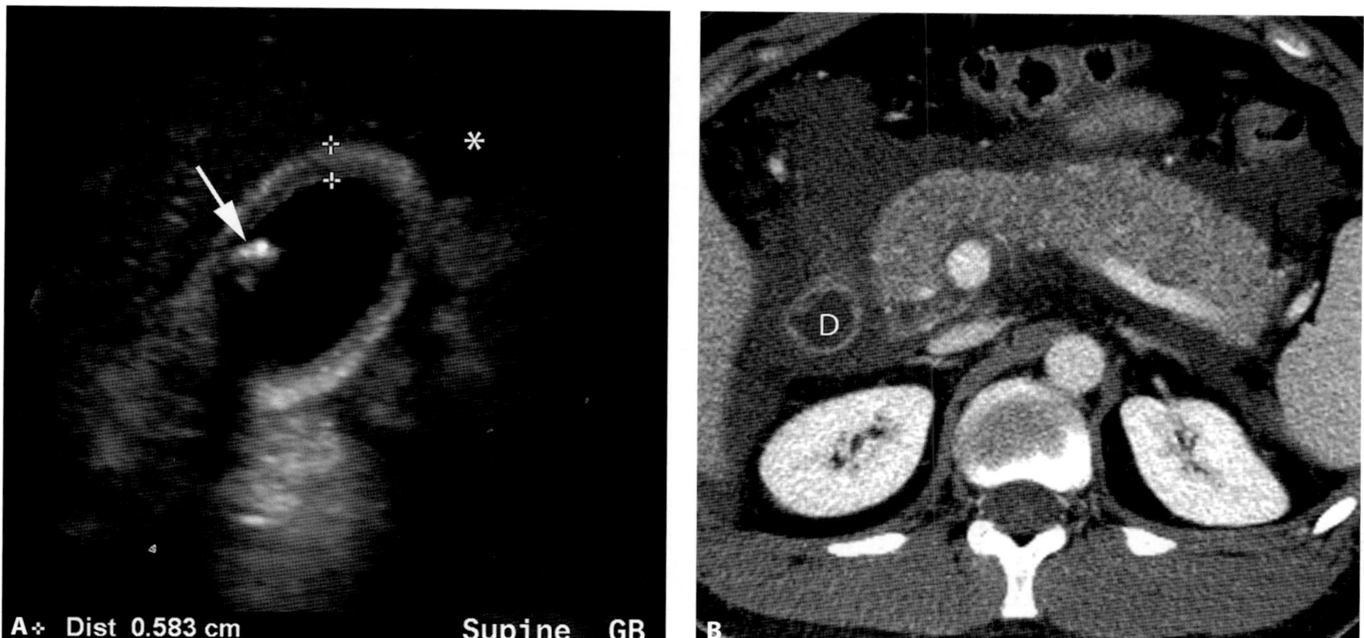

Figure 55.5 **A.** Ultrasound image of the gallbladder in a 38-year-old man with acute pancreatitis shows gallbladder wall thickening (calipers), a small amount of free fluid (asterisk), and a shadowing gallstone (arrow). **B.** Abdominal CT with intravenous contrast shows the enlarged, edematous pancreas, compatible with acute pancreatitis. There is free fluid surrounding the pancreas and descending duodenum (D). The gallbladder (not shown) was decompressed and therefore difficult to assess.

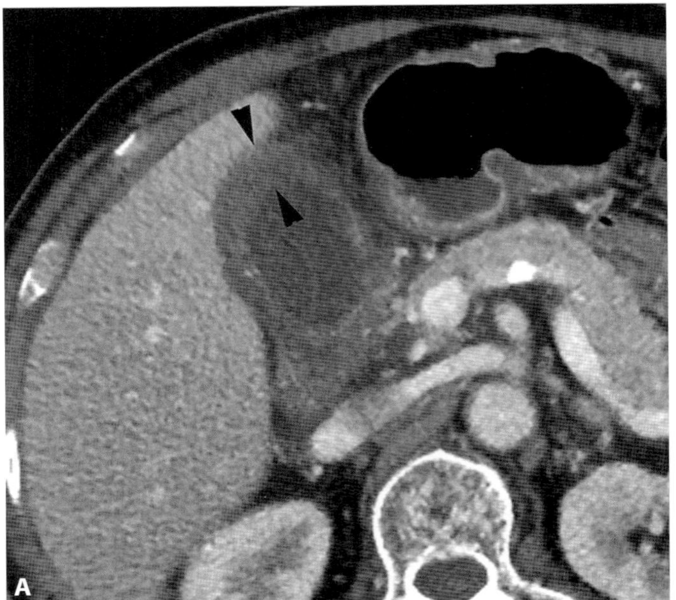

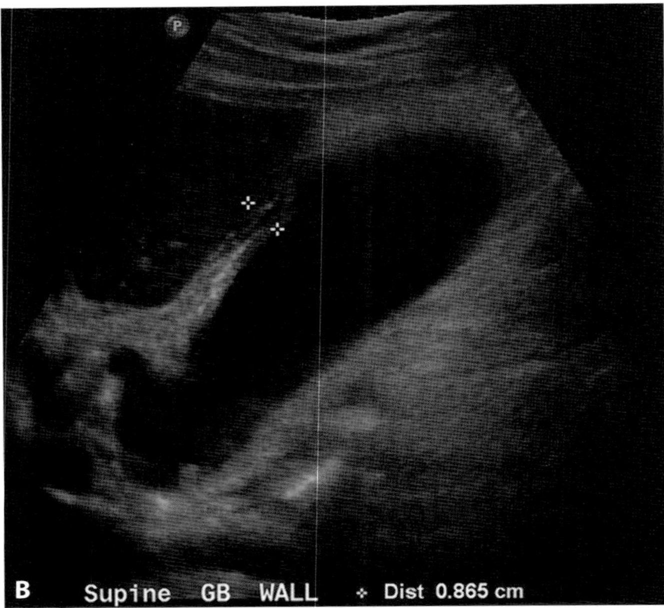

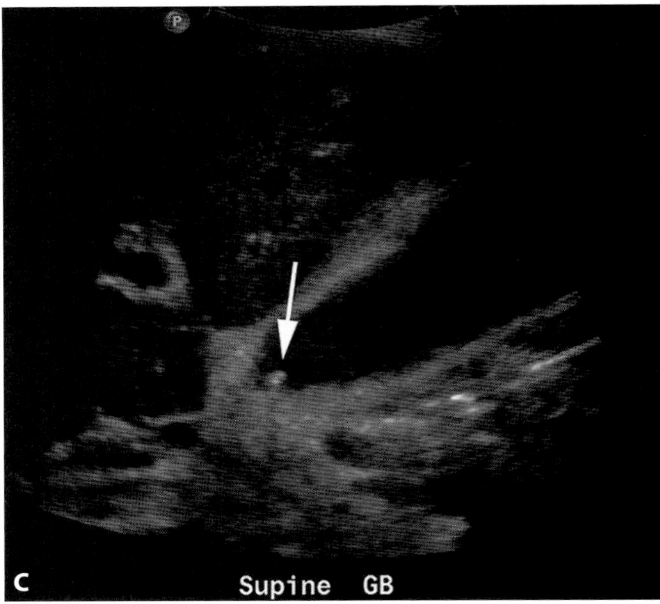

Figure 55.6 A. Axial contrast-enhanced CT from an 83-year-old man with right upper quadrant pain shows marked gallbladder wall thickening (arrowheads) and pericholecystic fat stranding. Note that the mucosa of the gallbladder enhances, but the remainder of the edematous wall does not. This appearance sometimes resembles pericholecystic fluid. **B.** Ultrasound of the gallbladder confirms a 9-mm thick gallbladder wall (calipers), which also appears laminated. **C.** Additional ultrasound image shows a small shadowing gallstone (arrow) in the gallbladder neck. Sonographic Murphy's sign was positive, and acute calculous cholecystitis was confirmed at surgery.

CASE

56 Splenic clefts

Michael J. Modica

Imaging description

The fetal spleen is characterized by numerous lobulations that may persist into adulthood [1]. Persistent splenic lobulations are most common along the medial border [1]. A separation between adjacent lobulations is known as a cleft, and may be mistaken for a laceration. Differentiating a cleft from a laceration on contrast-enhanced CT is facilitated by recognition of its characteristic appearance along with additional findings. Splenic clefts typically have smooth rounded margins and are not associated with perisplenic or subcapsular hematoma (Figures 56.1 and 56.2). A cleft may be quite large, measuring up to 3 cm in length. Larger clefts will usually contain fat. Typical imaging appearance with lack of surrounding perisplenic fluid or hematoma favors a cleft.

Importance

Splenic clefts are normal anatomic variants and have no clinical significance. When seen on contrast-enhanced CT, they may be mistaken for lacerations in abdominal trauma patients. Misdiagnosis may lead to unnecessary admission, observation, or further diagnostic workup, but is unlikely to lead to laparotomy now that conservative management of splenic lacerations is so strongly favored.

Typical clinical scenario

Incidentally found on abdominal CT done for any indication. Clefts are potentially problematic when present in the acute trauma patient [2].

Differential diagnosis

In contrast to clefts, splenic lacerations are often irregular hypodensities, with sharp cortical contours, and are most prominent on venous phase CT imaging (Figure 56.3) [3]. Associated signs of injury include perisplenic fluid or hematoma, left rib fractures, and other solid organ injury.

Teaching point

Splenic clefts typically have rounded external margins, may contain fat, and have sharp external margins. Lack of perisplenic fluid, hemorrhage, or other signs of abdominal trauma strongly supports the diagnosis of splenic cleft.

REFERENCES

1. Freeman JL, Jafri SZ, Roberts JL, Mezwa DG, Shirkhoda A. CT of congenital and acquired abnormalities of the spleen. *Radiographics.* 1993;**13**(3):597–610.
2. Hewett JJ, Freed KS, Sheafor DH, Vaslef SN, Kliewer MA. The spectrum of abdominal venous CT findings in blunt trauma. *AJR Am J Roentgenol.* 2001;**176**(4):955–8.
3. Roberts JL, Dalen K, Bosanko CM, Jafir SZ. CT in abdominal and pelvic trauma. *Radiographics.* 1993;**13**(4):735–52.

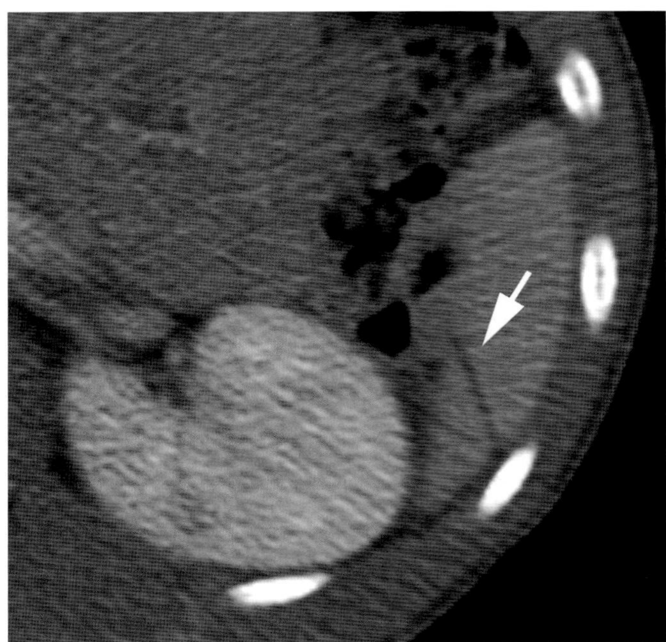

Figure 56.1 Axial contrast-enhanced CT from a 38-year-old man involved in a high-speed motor vehicle collision shows a well-defined linear parenchymal hypodensity extending from the splenic capsule (arrow). There is no surrounding perisplenic fluid. A follow-up scan for another indication did not reveal any change in the cleft. This is the typical appearance for a long splenic cleft.

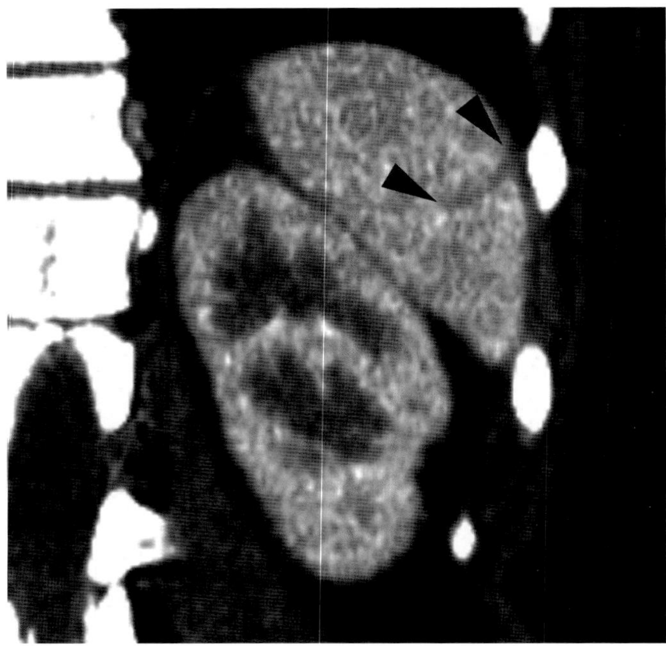

Figure 56.2 Coronal reformation from a contrast-enhanced CT scan in a 19-year-old man after a two-story fall shows a linear parenchymal hypodensity extending from the splenic capsule (arrowheads), typical of a splenic cleft. Note the rounded edges. There was no surrounding fluid.

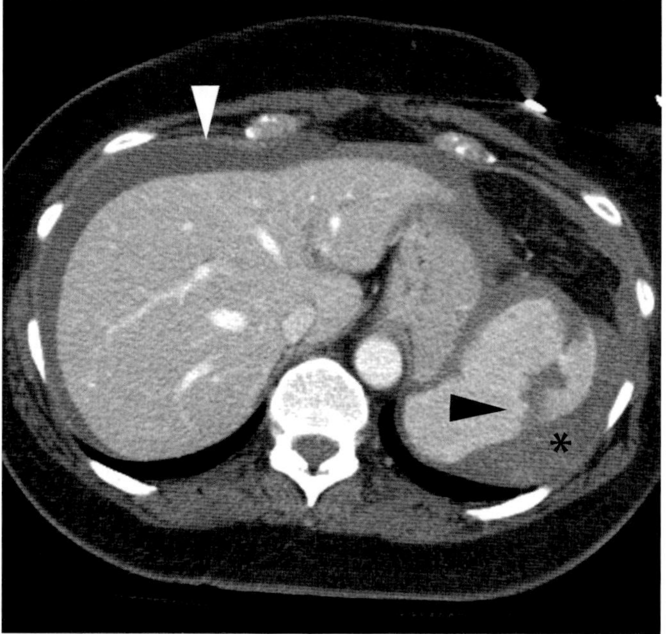

Figure 56.3 Axial contrast-enhanced CT image from a 22-year-old man involved in a high-speed motor vehicle collision shows an irregular hypodensity extending through the splenic parenchyma (black arrowhead) consistent with a splenic laceration. There is surrounding perisplenic hemorrhage (asterisk) and hemoperitoneum (white arrowhead).

Inhomogeneous splenic enhancement

Michael J. Modica

Imaging description

The spleen is primarily composed of red and white pulp, separated by a marginal zone of reticular cells (Figure 57.1) [1]. The red pulp consists of large numbers of blood-filled sinuses and sinusoids and is responsible for splenic filtration – filtering foreign material and damaged red blood cells. The white pulp is composed of aggregates of lymphoid tissue responsible for the immunologic function of the spleen [1]. The enhancement dynamics of the spleen are largely attributed to the different blood flow rates through these two tissue structures [2]. If the spleen is imaged in an early arterial phase, the parenchyma can appear heterogeneous as patches of unenhanced white pulp contrast with normally enhanced red pulp.

Three principal patterns of splenic enhancement have been described – archiform, focal, and diffuse. Archiform patterns typically appear as alternating bands of high and low attenuation, a ring-like pattern, or a zebra-stripe pattern (Figure 57.2) [3]. Focal lesions appear as a single area of low attenuation (Figure 57.3). Diffuse enhancement appears as a mottled pattern throughout the splenic parenchyma.

Importance

Inhomogeneous splenic enhancement can mimic splenic laceration, contusion, or infarct in the acute trauma patient. Familiarity with the typical appearance of archiform, focal, and diffuse enhancement patterns is critical to prevent overcalls. Difficult cases can often be resolved with delayed CT images through the spleen, which can be performed 5–10 minutes after the initial contrast infusion, using a low radiation dose technique [4].

Typical clinical scenario

Inhomogeneous splenic enhancement is typically an incidental CT finding and can present in any clinical context.

Differential diagnosis

Inhomogeneous splenic enhancement may be mistaken for traumatic splenic injury on contrast-enhanced CT. Types of splenic injury include perisplenic or subcapsular hematoma, laceration, infarct, active hemorrhage, or vascular injury. Perisplenic hematomas imply capsular injury or rupture, or represent hemorrhage from nearby injury. The sentinal-clot sign – high-attenuation clot around the spleen – is more specific for splenic origin of the bleed. Subcapsular hematomas are crescentic peripheral collections contained by the capsule, compressing the underlying splenic parenchyma (Figure 57.4). Splenic lacerations typically present as irregular, linear, branching low attenuation lesions within the splenic parenchyma that persist on all phases of contrast imaging (Figure 57.5). Splenic infarction usually presents as a wedge-shaped area of hypoattenuation extending from the capsule (Figure 57.6).

Active hemorrhage is identified when active extravasation of contrast (300–400 Hounsfield Units, [HU]) is visualized within the splenic parenchyma or in an extracapsular location [5]. On delayed imaging, the focus of active extravasation remains high attenuation and often increases in size, indicating active bleeding.

Post-traumatic vascular injuries include intrasplenic pseudoaneurysms and arteriovenous fistulas [6]. These lesions typically appear as focal high-attenuation regions within the splenic parenchyma, with attenuation values similar to adjacent enhanced arteries. Attenuation decreases with other arterial structures on delayed imaging and the lesions blend into the background splenic parenchyma [6].

Teaching point

Archiform and focal inhomogeneous splenic enhancement patterns can be differentiated from splenic injury by recognition of their typical appearance and lack of perisplenic fluid or hemorrhage. Difficult cases can often be resolved with delayed CT images.

REFERENCES

1. Cesta MF. Normal structure, function, and histology of the spleen. *Toxicol Pathol.* 2006;**34**(5):455–65.
2. Glazer GM, Axel L, Goldberg HI, Moss AA. Dynamic CT of the normal spleen. *AJR Am J Roentgenol.* 1981 Aug;**137**(2):343–6.
3. Donnelly LF, Foss JN, Frush DP, Bisset GS, 3rd. Heterogeneous splenic enhancement patterns on spiral CT images in children: minimizing misinterpretation. *Radiology.* 1999;**210**(2):493–7.
4. Anderson SW, Varghese JC, Lucey BC, *et al.* Blunt splenic trauma: delayed-phase CT for differentiation of active hemorrhage from contained vascular injury in patients. *Radiology.* 2007;**243**(1):88–95.
5. Federle MP, Courcoulas AP, Powell M, Ferris JV, Peitzman AB. Blunt splenic injury in adults: clinical and CT criteria for management, with emphasis on active extravasation. *Radiology.* 1998;**206**(1):137–42.
6. Marmery H, Shanmuganathan K, Mirvis SE, *et al.* Correlation of multidetector CT findings with splenic arteriography and surgery: prospective study in 392 patients. *J Am Coll Surg.* 2008;**206**(4):685–93.

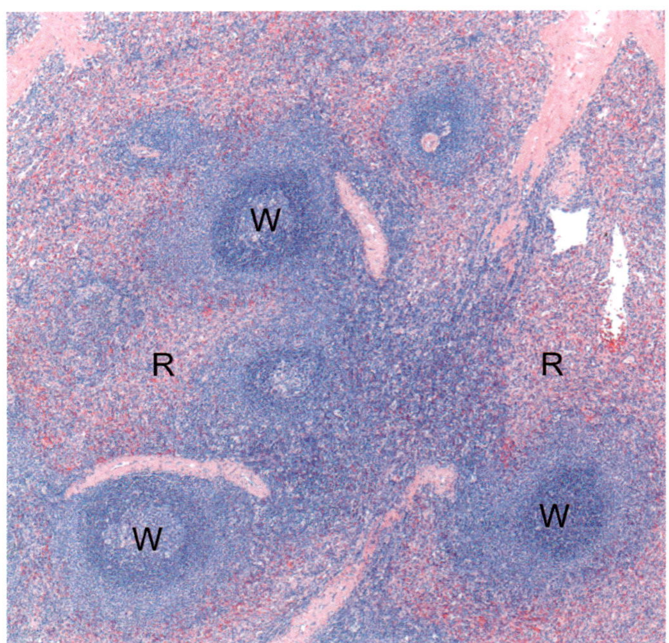

Figure 57.1 Histologic sample of the splenic parenchyma with hematoxylin and eosin staining (H&E) shows the typical relationship of red (R) and white (W) pulp. Notice the aggregates of white pulp have an oval or circular arrangement within the background of the red pulp. Image provided by Heike Deubner, MD, Department of Pathology, University of Washington, Seattle, WA, USA.

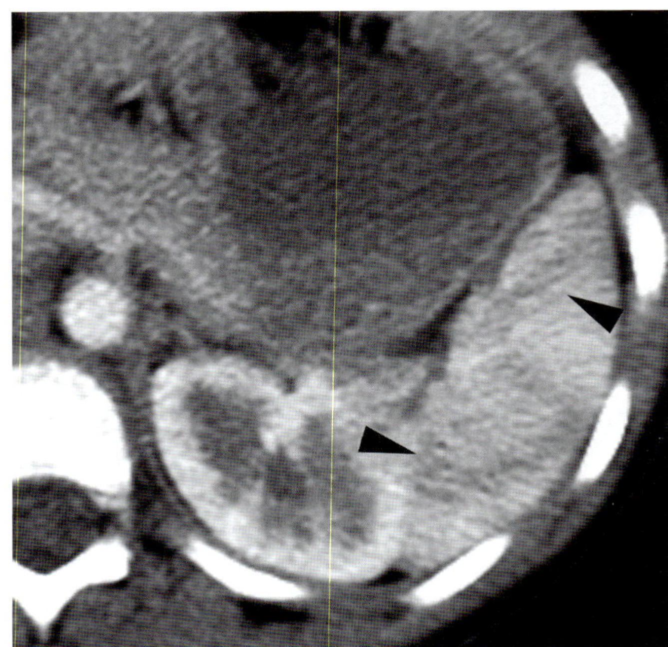

Figure 57.2 Axial contrast-enhanced CT image of the spleen in the early arterial phase in a 22-year-old woman involved in a high-speed motor vehicle collision shows a smooth, well-defined "S-shaped" stripe of hypodensity extending through the splenic parenchyma (arrowheads). There is no perisplenic fluid and no traumatic injury was found in the abdomen or pelvis. This is the typical appearance for archiform splenic enhancement and should not be confused with a laceration or contusion.

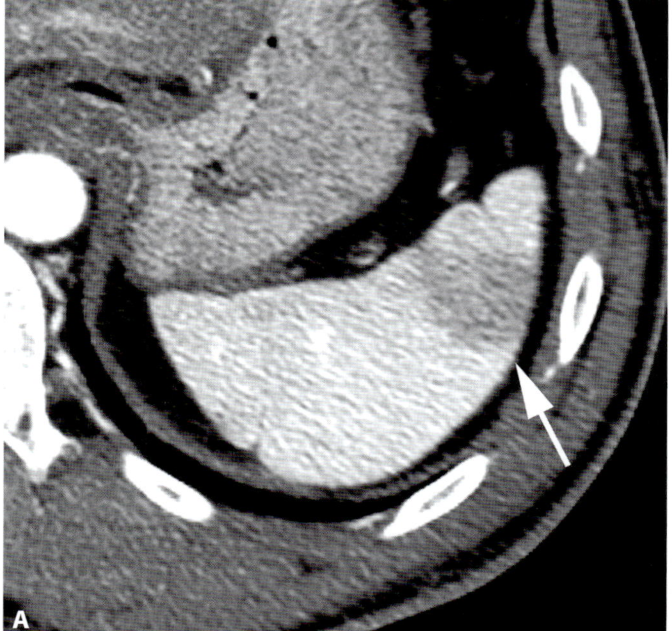

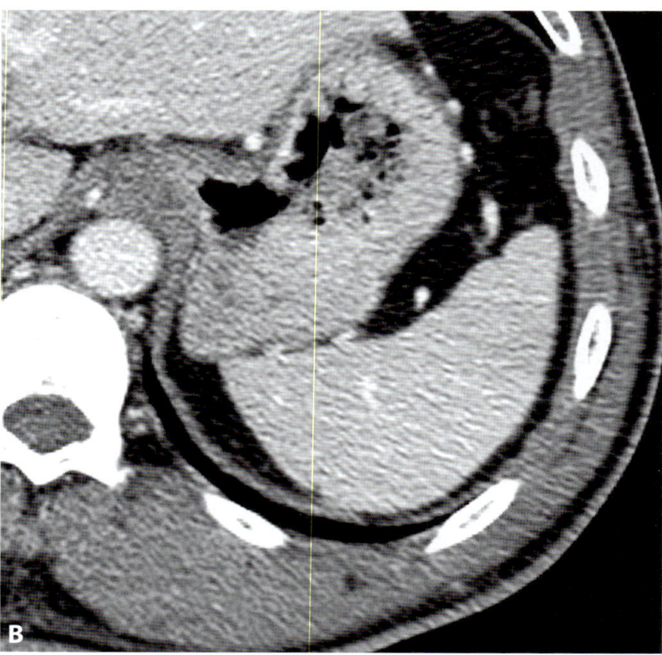

Figure 57.3 **A.** Axial contrast-enhanced CT image of the spleen in the early arterial phase in a 38-year-old man after a 10-foot fall shows a focal hypodensity in the splenic parenchyma (arrow). This lesion was thought to represent a contusion. No perisplenic fluid or other traumatic injury was found in the abdomen or pelvis. **B.** Axial contrast-enhanced CT image from the same patient taken 60 seconds after the initial image shows near complete resolution of the lesion, consistent with focal inhomogeneous splenic enhancement.

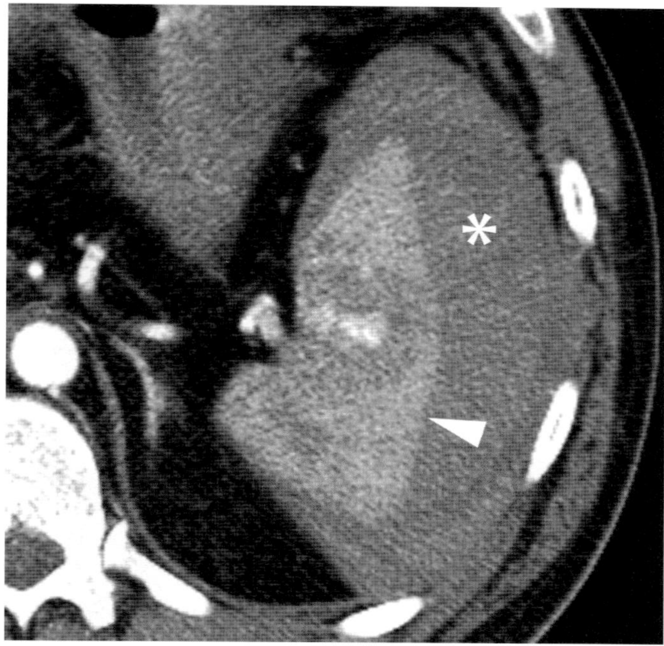

Figure 57.4 Axial contrast-enhanced CT image of the spleen in a 40-year-old man involved in a high-speed motor vehicle collision shows a large subcapsular hematoma (asterisk) compressing the underlying splenic parenchyma (arrowhead). Note the active vascular extravasation within the splenic parenchyma.

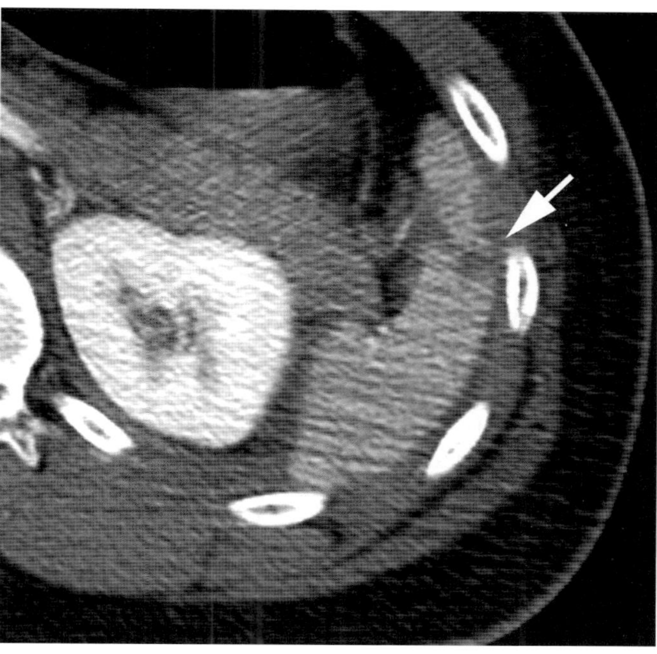

Figure 57.5 Axial contrast-enhanced CT image of the spleen in a 26-year-old man involved in a motorcycle accident shows an irregular hypodensity extending through the splenic parenchyma (arrow) consistent with a laceration.

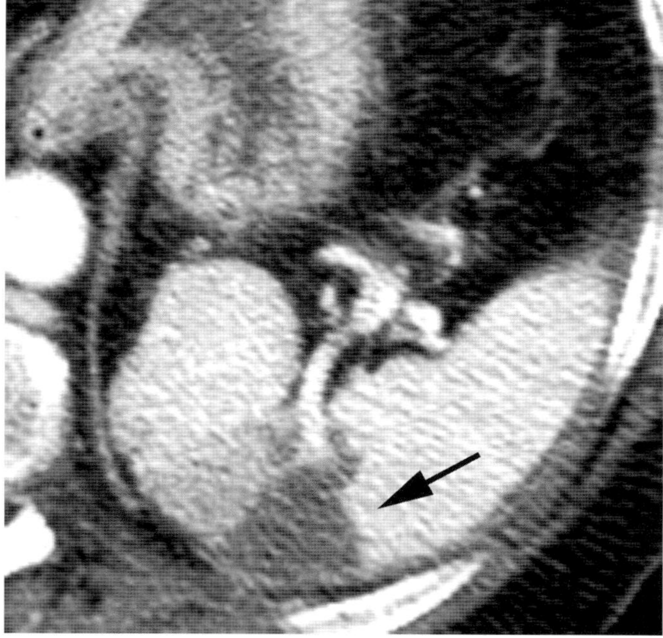

Figure 57.6 Axial contrast-enhanced CT image of the spleen in a 30-year-old woman shows a wedge-shaped parenchymal hypodensity (arrow) representing a splenic infarct.

Pseudosubcapsular splenic hematoma

Michael J. Modica

Imaging description

The left lobe of the liver can extend across the midline and hug the splenic capsule. When imaged on ultrasound, this can appear as a heterogeneously echogenic or hypoechoic mass surrounding the spleen. This can be mistaken for a perisplenic hemorrhage, subcapsular splenic hematoma, or mass [1].

Importance

The spleen is the most frequently injured solid organ in the abdomen. Frequently, focused assessment with sonography for trauma (FAST) is used, particularly in hemodynamically unstable patients, to rapidly assess for intraperitoneal and pericardial hemorrhage [2]. An elongated left hepatic lobe can be confused with subcapsular hematoma and potentially result in unnescessary laparotomy, further imaging, or tertiary care transfer (Figure 58.1).

Typical clinical scenario

An elongated left hepatic lobe can be an anatomic variant, most commonly in women, or it can be due to compensatory left lobe hypertrophy secondary to cirrhosis or right hepatic lobe atrophy or surgery [3].

Differential diagnosis

The differential diagnosis includes perisplenic hemorrhage or subcapsular splenic hematoma. Distinguishing an elongated left lobe of the liver is possible using ultrasound. Following the abnormality across the midline to the right upper quadrant will often show contiguity with the liver [1]. Color Doppler imaging over the region of interest should demonstrate the normal peripheral arterial and venous blood flow found in the liver parenchyma, separate from splenic flow [4]. In difficult cases, abdominal CT typically demonstrates this variant without ambiguity [1, 4].

Teaching point

Demonstrating contiguity with the liver by scanning across the midline and utilizing color Doppler can distinguish an elongated left lobe of the liver from a pseudocapsular splenic hematoma.

REFERENCES

1. Crivello MS, Peterson IM, Austin RM. Left lobe of the liver mimicking perisplenic collections. *J Clin Ultrasound.* 1986;**14**(9):697–701.
2. Jehle D, Guarino J, Karamanoukian H. Emergency department ultrasound in the evaluation of blunt abdominal trauma. *Am J Emerg Med.* 1993;**11**(4):342–6.
3. Hammond LJ. III. Congenital elongation of the left lobe of the liver. *Ann Surg.* 1905;**41**(1):31–5.
4. Li DK, Cooperberg PL, Graham MF, Callen P. Pseudo perisplenic "fluid collections": a clue to normal liver and spleen echogenic texture. *J Ultrasound Med.* 1986;**5**(7):397–400.

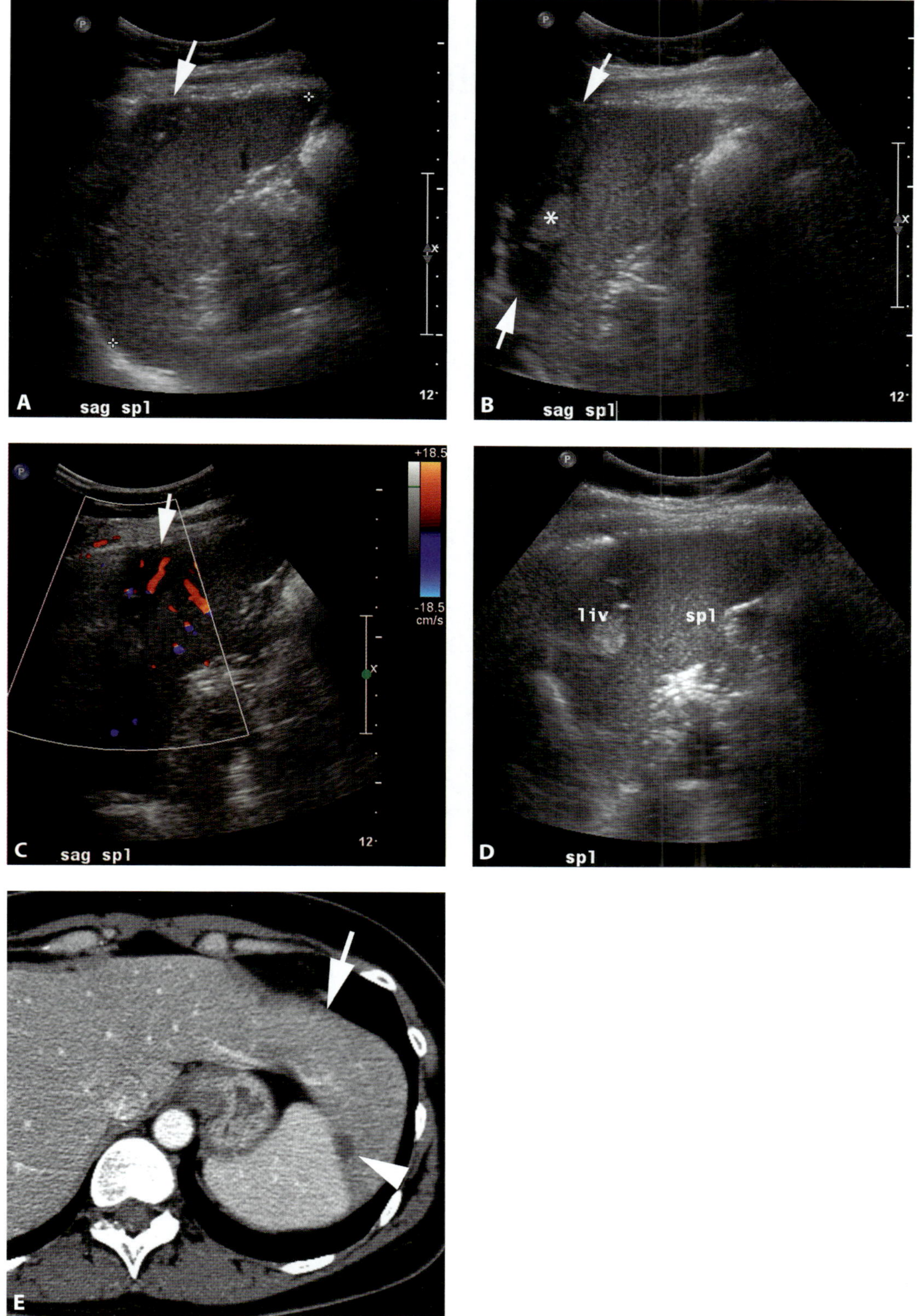

Figure 58.1 **A.** Sagittal ultrasound image of the spleen in a 25-year-old man involved in a high-speed motor vehicle collision shows an apparent mass (arrow) compressing the splenic capsule, initially thought to represent a subcapsular hematoma. Notice that the apparent mass is hypoechoic relative to splenic parenchyma. **B.** Ultrasound image taken more medially shows this hypoechoic mass with a focus of hyperechogenicity (asterisk). Arrows indicate the liver parenchyma. **C.** Color Doppler of the presumed subcapsular hematoma shows arterial and venous flow (arrow) consistent with liver parenchyma. **D.** Sagittal ultrasound image labeling the liver (liv) and spleen (spl). **E.** Axial contrast-enhanced CT from the same patient shows an elongated left lobe of the liver draped over the spleen (arrow). Notice the hypodense lesion (arrowhead) closely approximated to the splenic capsule. This corresponds to the hyperechogenic lesion seen in **B**.

Pseudopancreatitis following trauma

Martin L. Gunn

Imaging description

Missed pancreatic injuries result in considerable morbidity and increased mortality [1]. Unfortunately, evidence suggests that multidetector CT (MDCT) is of suboptimal sensitivity for the detection of pancreatic parenchymal and duct injury [1, 2]. One of the signs of pancreatic injury is peripancreatic fluid. In particular, fluid between the pancreas and splenic vein is suggestive of injury [3]. However, intrapancreatic and peripancreatic fluid has been observed in trauma patients who have no other signs or symptoms of pancreatic trauma at all [4].

Pseudopancreatitis appears as peripancreatic and intrapancreatic low density, resembling fluid (Figure 59.1). The pancreas may appear swollen. Other findings that support pseudopancreatitis include periportal low density (edema) within the liver, distension of the inferior vena cava (IVC), and small bowel edema.

Direct signs of pancreatic injury, such as lacerations or areas of hypoperfusion of the pancreas, are absent. Lacerations can be simulated by pancreatic clefts, which tend to be smooth and linear, and have rounded margins. Clefts may contain small penetrating vessels, and they do not traverse the full width of the glands.

Because the peripancreatic space is continuous with the extraperitoneal space in the pelvis, fluid can surround the pancreas as an extension of extraperitoneal pelvic fluid arising from pelvic fractures, extraperitoneal bladder rupture, or pelvic arterial injury.

Importance

Peripancreatic fluid should not be mistaken as a definitive sign of pancreatic injury in the trauma patient, leading to unnecessary investigations. If there are no direct signs of pancreatic injury, and there are supportive signs of pseudopancreatitis, such as periportal edema and small bowel edema, a dedicated pancreatic CT should be considered to evaluate for pancreatic injury about 24 hours after the injury. Where there is concern for pancreatic duct injury then endoscopic retrograde cholangiopancreatography (ERCP) or magnetic resonance cholangiopancreatograph (MRCP) to clarify the diagnosis, or surgery, should be performed [5].

Typical clinical scenario

Peripancreatic edema is usually seen in patients with severe trauma. One of the postulated causes of peripancreatic fluid in these patients is acute hypovolemia and hypotensive shock, which may cause pancreatic ischemia, followed by a short period of massive fluid resuscitation. Fluid volume loading of the ischemic pancreas may cause fluid to leak into the intrapancreatic and peripancreatic tissues.

Over-hydration is another potential cause of pseudopancreatitis. This would explain associated signs such as periportal edema and distension of the IVC that are usually seen in these patients [6].

In most cases of pseudopancreatitis, the serum amylase level will be normal [4]. However, this is not always the case. Serum amylase levels in pancreatic trauma can be misleading. Ischemic injury to the pancreas from hypovolemic shock may elevate the serum amylase level [4]. Moreover, the serum amylase level may be normal in the first three hours following pancreatic injury or may be elevated due to other injuries. The degree of hyperamylasemia does not correlate with the severity of pancreatic injury [7].

Differential diagnosis

The differential diagnosis includes traumatic pancreatitis, fluid tracking within the retroperitoneal space from other retroperitoneal injuries, and chronic pancreatitis.

Teaching point

Peripancreatic and intrapancreatic fluid is occasionally identified on MDCT when there are no other direct signs of pancreatic injury. Other signs of shock and elevated venous pressure such as periportal low density, a distended IVC, and small bowel edema support the diagnosis of traumatic pseudopancreatitis.

REFERENCES

1. Phelan HA, Velmahos GC, Jurkovich GJ, *et al.* An evaluation of multidetector computed tomography in detecting pancreatic injury: results of a multicenter AAST study. *J Trauma.* 2009;**66**(3):641–6; discussion 646–7.
2. Ilahi O, Bochicchio GV, Scalea TM. Efficacy of computed tomography in the diagnosis of pancreatic injury in adult blunt trauma patients: a single-institutional study. *Am Surg.* 2002;**68**(8):704–7; discussion 707–8.
3. Lane MJ, Mindelzun RE, Sandhu JS, McCormick VD, Jeffrey RB. CT diagnosis of blunt pancreatic trauma: importance of detecting fluid between the pancreas and the splenic vein. *AJR Am J Roentgenol.* 1994;**163**(4):833–5.
4. Brook OR, Fischer D, Militianu D, *et al.* Pseudopancreatitis in trauma patients. *AJR Am J Roentgenol.* 2009;**193**(3):W193–6.
5. Rekhi S, Anderson SW, Rhea JT, Soto JA. Imaging of blunt pancreatic trauma. *Emerg Radiol.* 2010;**17**(1):13–19.
6. Shanmuganathan K, Mirvis SE, Amoroso M. Periportal low density on CT in patients with blunt trauma: association with elevated venous pressure. *AJR Am J Roentgenol.* 1993;**160**(2):279–83.
7. Degiannis E, Glapa M, Loukogeorgakis SP, Smith MD. Management of pancreatic trauma. *Injury.* 2008;**39**(1):21–9.

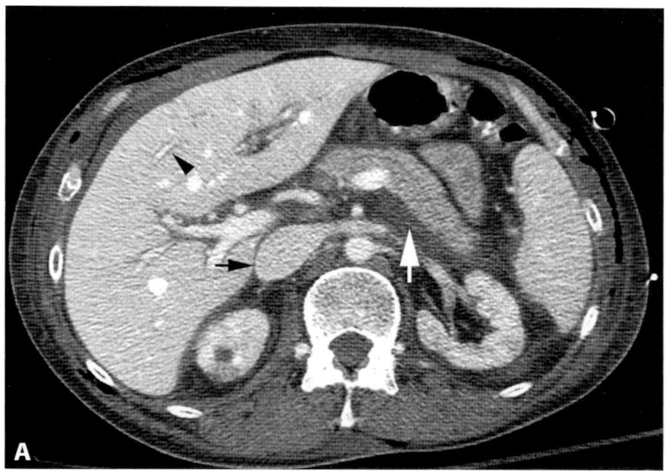

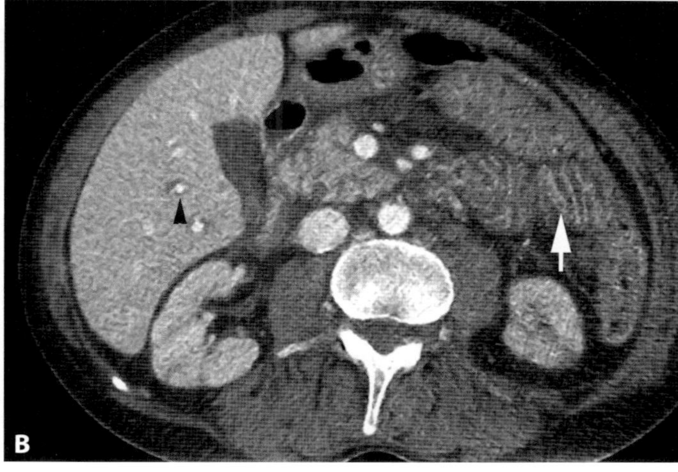

Figure 59.1 A. Follow-up axial contrast-enhanced CT image at the level of the pancreatic body in a 41-year-old man who was involved in a high-speed motor vehicle collision requiring a prolonged extrication at the scene, taken one day after injury. Peripancreatic fluid is present in the retropancreatic soft tissue and surrounding the tail of the pancreas (white arrow). In addition, there is a distended inferior vena cava (black arrow) and periportal low density within the liver (arrowhead). The main pancreatic duct is visible on this dedicated pancreas CT, and is normal. **B.** Contrast-enhanced CT on the preceding day showed peripancreatic low fluid as well as small bowel edema (arrow) and periportal edema (arrowhead).

CASE 60 Pancreatic clefts

Michael J. Modica

Imaging description

Pancreatic clefts are usually found at the junction of the neck and body. Histologically, they represent peripancreatic fat trapped within normal tissue, and they are found where vessels penetrate the pancreatic parenchyma [1]. Clefts typically appear smooth and linear with well-defined margins. On CT fat is typically visible, and they do not completely traverse the full width of the gland (Figures 60.1 and 60.2). The presence of fat surrounding penetrating vessels arising from the pancreatic arteries is indicative of a cleft [1]. Pancreatic clefts can be confused with a pancreatic laceration or transection on contrast-enhanced CT.

The lateral margin of the head and neck of the pancreas is normally convex. Lobular contour abnormalities, reported in up to 35% of normal subjects, are most common near the junction of the head and neck [2]. A deep fissure separating these lobules can be mistaken for laceration while the lobulations themselves can be misinterpreted as a pancreatic mass. Pancreatic lobulations, and the clefts that separate them, increase with age and they should not be misinterpreted as lacerations (Figure 60.3).

When an apparent laceration could be a cleft consider thin (1–2 mm) image reconstructions, pancreatic parenchymal phase scanning, coronal and sagittal multiplanar reformations, and MRI to identify the penetrating vessels within the parenchymal cleft.

Importance

Misdiagnosis of pancreatic injury can lead to unnecessary investigation with endoscopic retrograde cholangiopancreatography (ERCP) or magnetic resonance choloangiopancreatography (MRCP), or unnecessary surgery. Trauma surgeons investigate and treat pancreatic injuries aggressively as missed pancreatic injuries that result in delayed management can lead to considerable morbidity and mortality [3].

Typical clinical scenario

Pancreatic clefts can be found incidentally on contrast-enhanced CT in any clinical scenario. Typically, they are found in the acute abdominal trauma patient and mistaken for injury. Unfortunately, the use of the serum amylase levels to diagnose pancreatic injury in the acute phase is neither sufficiently sensitive nor specific [4].

Differential diagnosis

The differential diagnosis of pancreatic cleft includes pancreatic contusion, laceration and transection. Peripancreatic fat stranding, hemorrhage, and fluid between the splenic vein and pancreas are secondary signs of pancreatic injury and should be sought [4]. A single study found 90% of patients with pancreatic injury had fluid between the splenic vein and pancreas [4]. However, caution should be used if this is the only sign of pancreatic injury, as described in Case 59 on pseudopancreatitis.

A contusion appears as a larger area of hypoattenuation within the gland (Figure 60.4). A hematoma is usually isodense or hyperdense to parenchyma and often extends from the gland into the surrounding retroperitoneal tissues; often tracking between the pancreas and splenic vein. A laceration often involves the pancreatic body and appears as a hypodense linear defect extending through the parenchyma. Lacerations injuring the pancreatic duct are not always evident on CT; however, injury to the duct can be suggested by the degree of parenchymal injury present. Superficial lacerations (<50% pancreatic thickness) are unlikely to have duct injury, whereas patients with pancreatic head or tail transection are almost certain to have duct injury at surgery. In patients with pancreatic transection or deep lacerations (>50% pancreatic thickness) urgent evaluation may be facilitated by ERCP (Figure 60.5) or MRCP to guide surgical intervention.

Pancreatic lobulations can also mimic neoplasms [2].

> **Teaching point**
>
> Pancreatic clefts are normal structures most commonly found near the junction of the head and body. They are typically smooth, linear fat-containing structures with well-defined margins. They may contain a penetrating vessel.

REFERENCES

1. Cirillo RL, Jr., Koniaris LG. Detecting blunt pancreatic injuries. *J Gastrointest Surg.* 2002;**6**(4):587–98.
2. Brandon JC, Izenberg SD, Fields PA, *et al.* Pancreatic clefts caused by penetrating vessels: a potential diagnostic pitfall for pancreatic fracture on CT. *Emerg Radiol.* 2000;**7**(5):283–6.
3. Ross BA, Jeffrey RB, Jr., Mindelzun RE. Normal variations in the lateral contour of the head and neck of the pancreas mimicking neoplasm: evaluation with dual-phase helical CT. *AJR Am J Roentgenol.* 1996; **166**(4):799–801.
4. Lane MJ, Mindelzun RE, Sandhu JS, McCormick VD, Jeffrey RB. CT diagnosis of blunt pancreatic trauma: importance of detecting fluid between the pancreas and the splenic vein. *AJR Am J Roentgenol.* 1994;**163**(4):833–5.

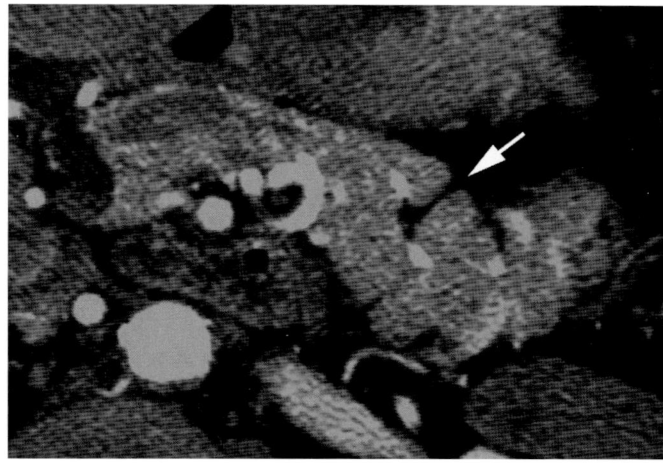

Figure 60.1 Axial contrast-enhanced CT image of the pancreas in a 40-year-old woman involved in an assault shows a smooth fat-containing cleft within the body of the pancreas (arrow). Notice that the margins are well defined and there is no peripancreatic fluid.

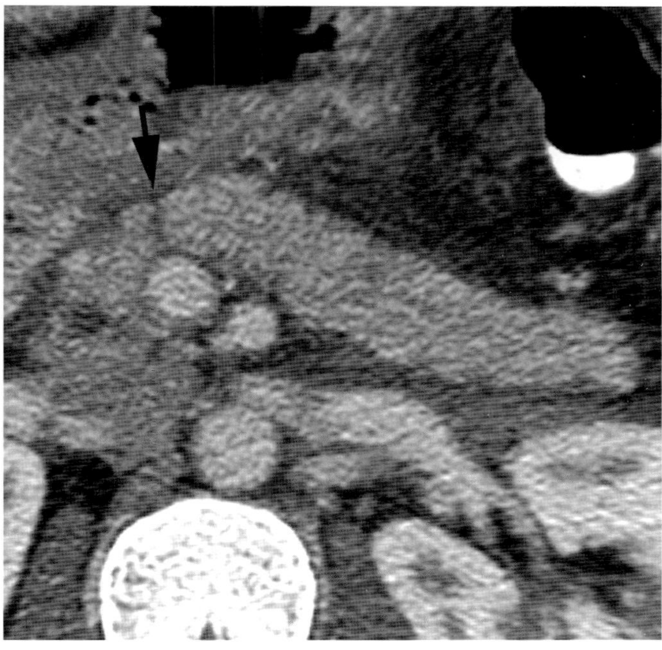

Figure 60.2 Axial contrast-enhanced CT image of the pancreas in a 29-year-old man shows a small, linear cleft near the junction of the pancreatic head and neck (arrow).

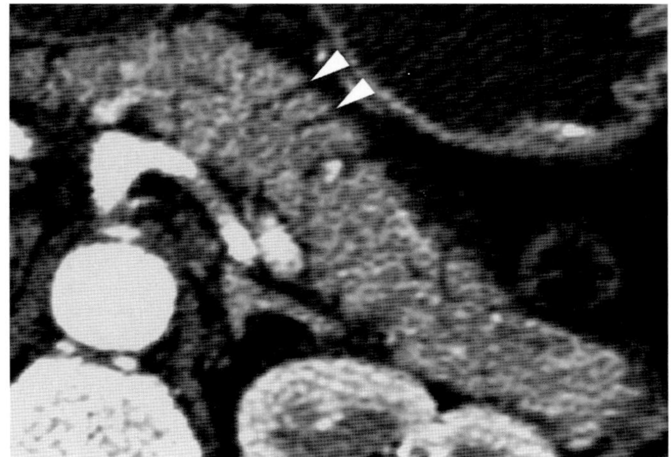

Figure 60.3 Axial contrast-enhanced CT image of the pancreas in a 71-year-old woman with abdominal pain shows smooth, linear, fat-containing clefts near the junction of the pancreatic head and neck (arrowheads). These are typical clefts thought to be separating pancreatic lobules.

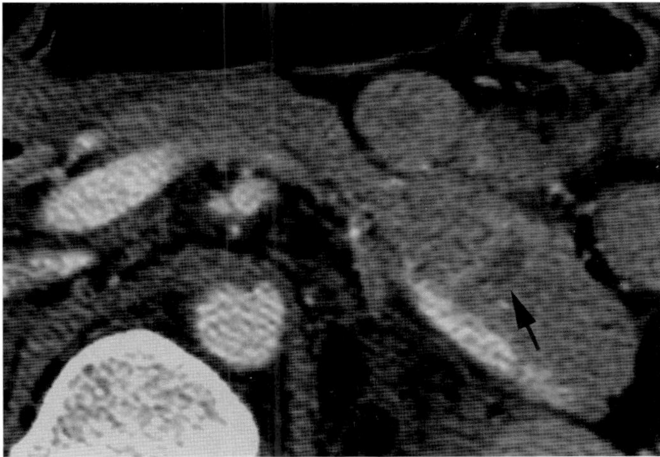

Figure 60.4 Axial contrast-enhanced CT image of the pancreas in a 30-year-old woman involved in a high-speed motor vehicle collision shows a hypodense lesion in the pancreatic tail (arrow). The parenchyma is swollen and edematous. Findings are most consistent with a pancreatic contusion.

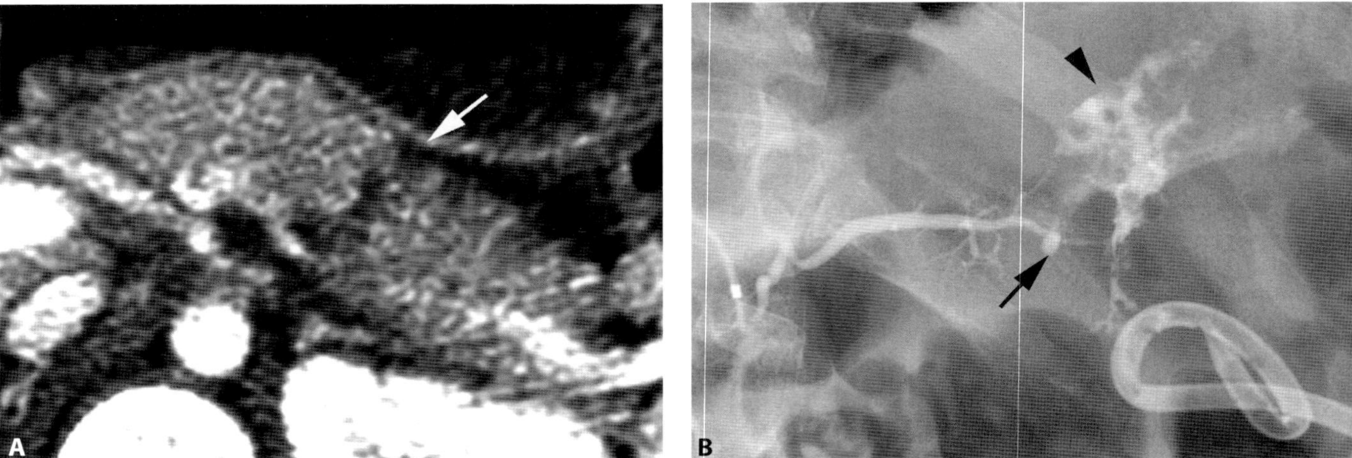

Figure 60.5 A. Axial contrast-enhanced CT image of the pancreas in a 44-year-old man involved in a motorcycle collision shows an irregular hypodensity traversing the pancreatic body (arrow). This lesion traverses greater than 50% of parenchymal thickness, suggestive of complete transection. Due to the likelihood of pancreatic duct injury, urgent surgical exploration, ERCP, or MRCP was recommended. **B.** Single image from an ERCP shows duct disruption (arrow) and contrast extravasation (arrowhead) consistent with main duct injury.

CASE **61**

Pseudothickening of the bowel wall

Martin L. Gunn

Imaging description

Pseudothickening can affect any segment of the bowel.

Two locations in the stomach commonly mimic bowel wall thickening. More proximally, in the region of the gastroesophageal junction and gastric cardia, there may be a transient area of apparent thickening (Figure 61.1). This is usually seen at the level of the ligamentum venosum [1]. A second common location of gastric pseudothickening is in the gastric antrum. This is present in the majority of patients undergoing CT [2]. Antral contractions increase the wall thickness to 5 to 10 mm in most patients, but in 5% of normal patients, it will exceed 1 cm [2]. Gastric wall thickening of 1 cm or more has been found to be sensitive but not specific for the diagnosis of malignant lesions (Figure 61.2) [3]. Moreover, localized antral wall thickening has been found to be a poor predictor of subclinical *Helicobacter pylori* infection [4]. However, if there are inflammatory changes in the surrounding fat, or heterogeneous enhancement, underlying disease such as gastritis or malignancy should be suspected.

When the lumen of the small bowel is distended, the wall should be imperceptible. In partially collapsed bowel, the wall should measure approximately 2–3 mm [1]. However, the circular folds of the jejunum (Kerckring's folds), which commence in the duodenum and predominate in the jejunum, can measure up to 8 mm in thickness, and lead to a false diagnosis of jejunal wall thickening (Figure 61.3) [5]. Moreover, when the jejunum contains water and oral contrast, it can appear quite thick. In cases where there is concern about pseudothickening, a small bowel study should be performed [6].

Distended normal colon should measure no more than 3 mm in thickness [7]. However, usually the colon is not distended during CT, and the normal wall may measure 6–8 mm in contracted segments [8]. Axial CT across haustral contractions may simulate focal wall thickening or a pseudomass. Due to the three longitudinal muscle layers, the pseudomass caused by haustral contractions often surrounds a triangular central lumen. Using multiplanar reformats in the coronal or sagittal plane usually clarifies the etiology (Figure 61.4).

Other forms of benign, but real, bowel wall thickening should not be mistaken as an indicator of significant pathology. Traditionally, submucosal intestinal fat has been considered to be a marker for inflammatory bowel disease. However, more recent evidence suggests that this sign is identified in up to 21% of patients with no clinical history of inflammatory bowel disease, and it is possibly related to obesity (Figures 61.5 and 61.6) [9]. Colonic wall thickening in the absence of bowel symptoms is observed in more than one-third of patients with cirrhosis, and in most of these cases it is primarily located in the right side of the colon [10].

Importance

Overcalling the presence of bowel wall thickening may lead to either unnecessary testing and treatment, or distract the clinician from the true diagnosis.

Typical clinical scenario

These errors are most commonly made in patients presenting with abdominal pain, or following abdominal trauma.

Differential diagnosis

As differentiation between pseudothickening and true thickening can be difficult, further imaging is usually necessary. Distension of the stomach with water or gas before performing a limited CT scan may be helpful. To evaluate the jejunum, a fluoroscopic contrast examination provides greater mucosal detail, and a dynamic evaluation of peristalsis. Colonic wall thickening can be reassessed after the insufflation of rectal gas or contrast agents. This may be supplemented with an intravenous spasmolytic agent.

Teaching point

Bowel wall pseudothickening is common in the gastroesophageal junction, gastric antrum, jejunum, and colon. In these regions exercise caution, and when necessary, repeat the examination following administration of oral or rectal contrast.

REFERENCES

1. Shirkhoda A. Diagnostic pitfalls in abdominal CT. *Radiographics.* 1991;**11**(6):969–1002.
2. Pickhardt PJ, Asher DB. Wall thickening of the gastric antrum as a normal finding: multidetector CT with cadaveric comparison. *AJR Am J Roentgenol.* 2003;**181**(4):973–9.
3. Insko EK, Levine MS, Birnbaum BA, Jacobs JE. Benign and malignant lesions of the stomach: evaluation of CT criteria for differentiation. *Radiology.* 2003;**228**(1):166–71.
4. Kul S, Sert B, Sari A, *et al.* Effect of subclinical *Helicobacter pylori* infection on gastric wall thickness: multislice CT evaluation. *Diagn Interv Radiol.* 2008;**14**(3):138–42.
5. Zalcman M, Sy M, Donckier V, Closset J, Gansbeke DV. Helical CT signs in the diagnosis of intestinal ischemia in small-bowel obstruction. *AJR Am J Roentgenol.* 2000;**175**(6):1601–7.

6. Macari M, Balthazar EJ. CT of bowel wall thickening: significance and pitfalls of interpretation. *AJR Am J Roentgenol.* 2001;**176**(5):1105–16.

7. Fisher JK. Normal colon wall thickness on CT. *Radiology.* 1982; **145**(2):415–18.

8. Wiesner W, Mortele KJ, Ji H, Ros PR. Normal colonic wall thickness at CT and its relation to colonic distension. *J Comput Assist Tomogr.* 2002;**26**(1):102–6.

9. Harisinghani MG, Wittenberg J, Lee W, *et al.* Bowel wall fat halo sign in patients without intestinal disease. *AJR Am J Roentgenol.* 2003;**181** (3):781–4.

10. Guingrich JA, Kuhlman JE. Colonic wall thickening in patients with cirrhosis: CT findings and clinical implications. *AJR Am J Roentgenol.* 1999;**172**(4):919–24.

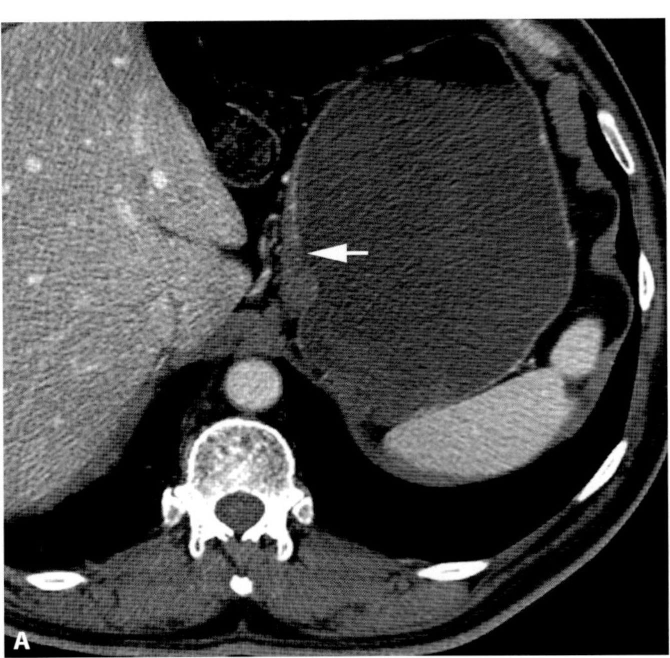

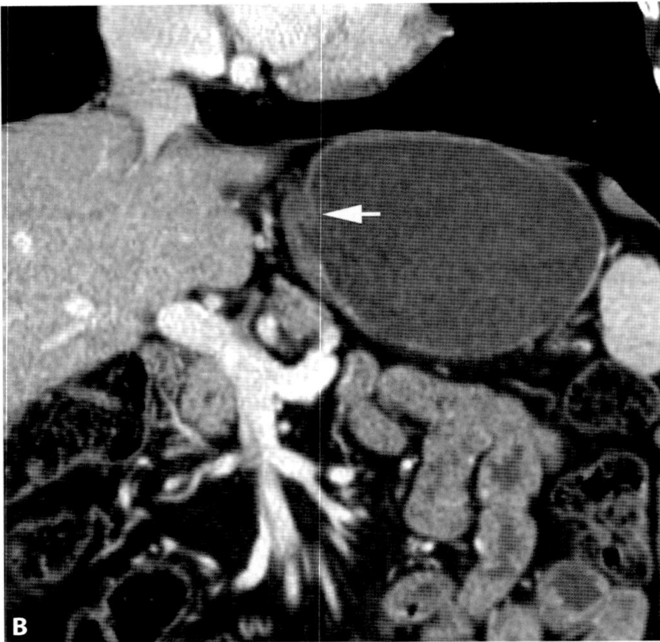

Figure 61.1 Axial (**A**) and coronal (**B**) contrast-enhanced CT of the upper abdomen in a 73-year-old man with iron deficiency anemia and a normal gastroscopy demonstrates apparent wall thickening of the gastric cardia (arrows).

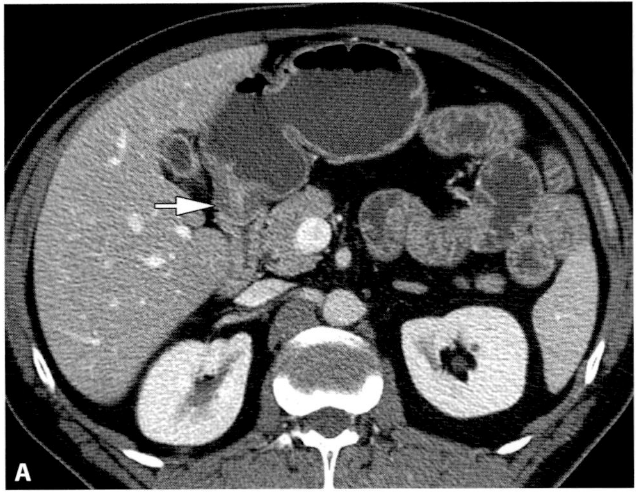

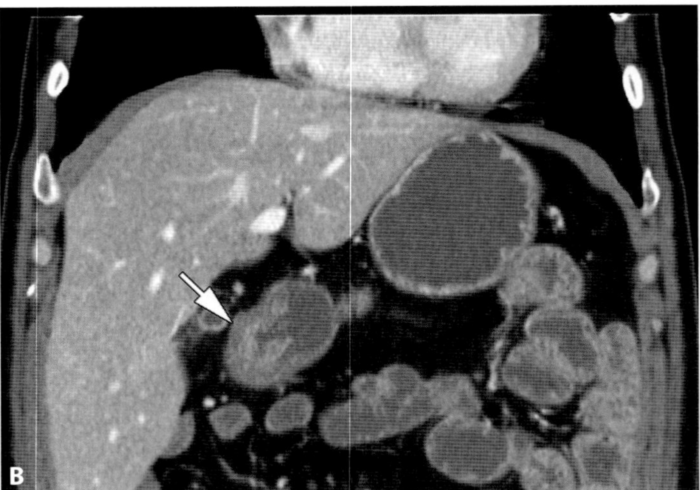

Figure 61.2 Axial (**A**) and coronal (**B**) contrast-enhanced CT of the upper abdomen in a 49-year-old man with abdominal pain demonstrates thickening in the gastric antrum and first part of the duodenum (arrow), without surrounding inflammatory change. The patient underwent a normal gastroscopy.

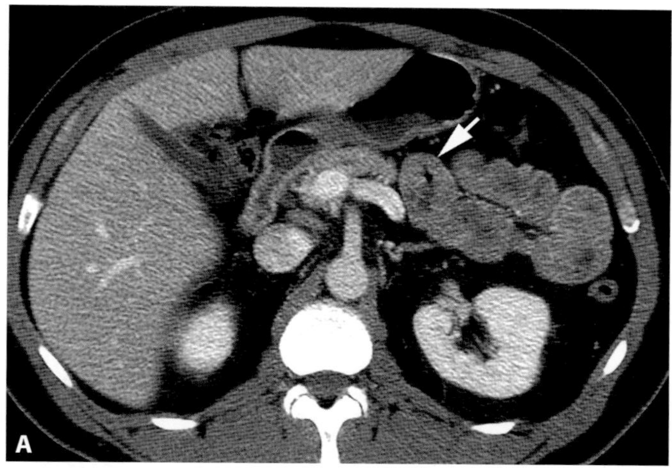

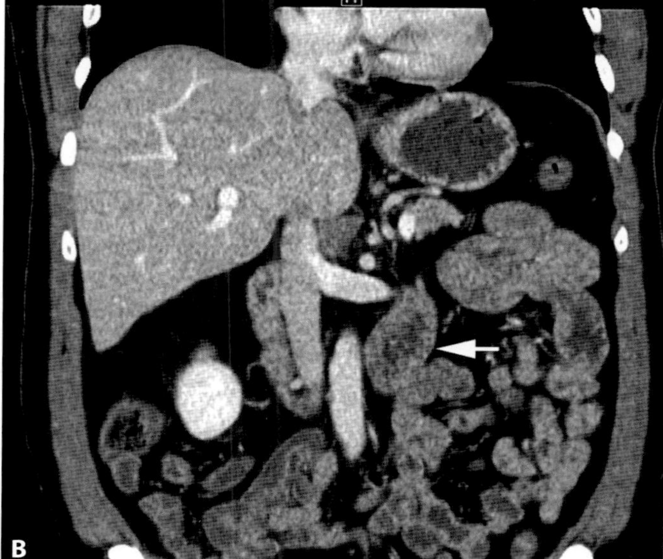

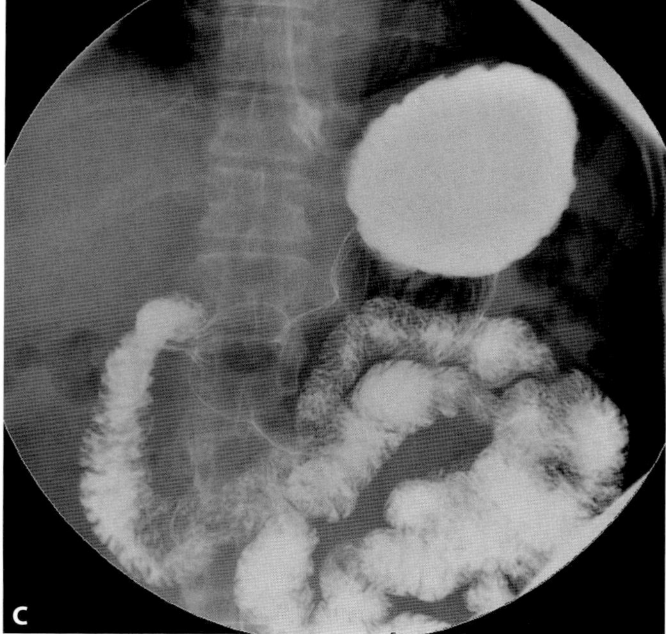

Figure 61.3 Axial (**A**) and coronal (**B**) contrast-enhanced CT of the abdomen in a 48-year-old man with a history of ulcerative colitis, and sudden onset of acute severe abdominal pain, nausea, and vomiting reveals apparent thickening of the jejunum (arrows). This is the typical pattern for the Kerckring's folds in non-distended jejunum. **C.** An upper gastrointestinal barium examination performed the following day reveals no mucosal abnormality.

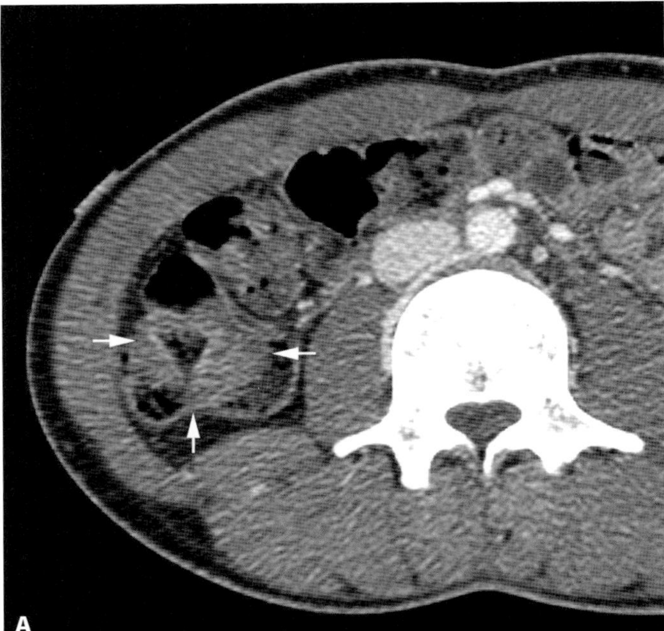

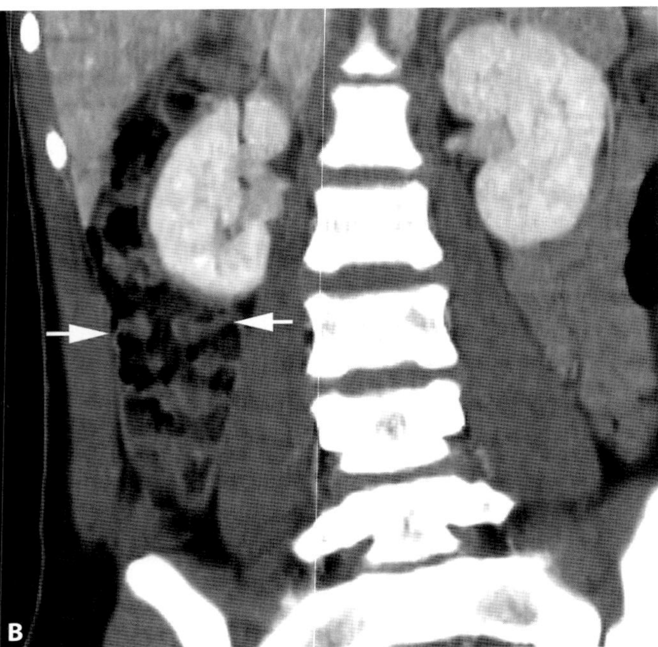

Figure 61.4 Axial (**A**) and coronal (**B**) contrast-enhanced CT of the abdomen and pelvis in a 20-year-old man who was involved in a low-speed motor vehicle collision demonstrates pseudothickening of the wall of the ascending colon (arrows). Note the triangular shape of the lumen, which is caused by the three longitudinal muscle layers.

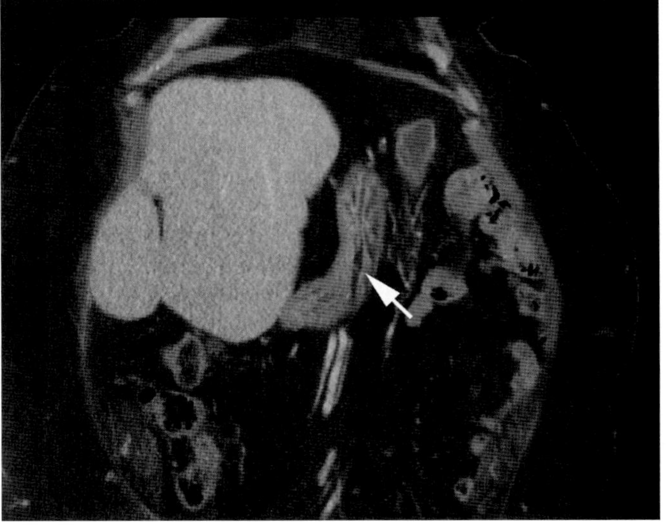

Figure 61.5 Coronal contrast-enhanced CT of the abdomen in a 64-year-old woman with morbid obesity and type II diabetes mellitus demonstrates gastric submusocal fat (arrow). The patient had no known history of inflammatory bowel disease.

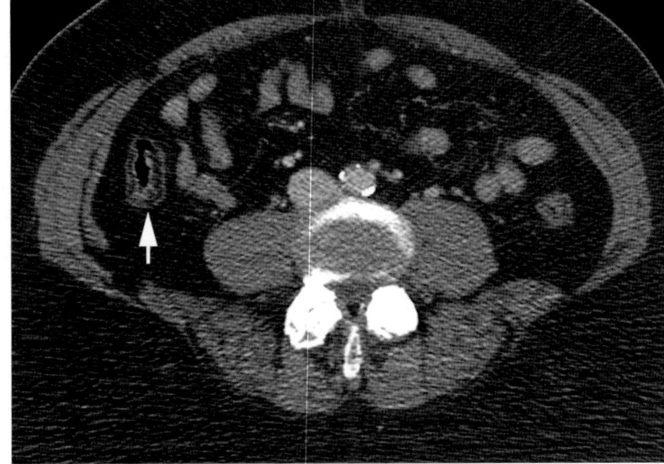

Figure 61.6 Axial contrast-enhanced CT of the abdomen and pelvis in an obese 52-year-old woman who was involved in a motor vehicle collision demonstrates submucosal fat in the ascending colon (arrow). The patient had no known or suspected inflammatory bowel disease.

CASE **62**

Small bowel transient intussusception

Joel A. Gross

Imaging description

The widespread use of CT has led to increased detection of small bowel intussusceptions (SBI) [1], which are identified in 0.5% of non-selective abdominal-pelvic CTs [2].

An intussusception occurs when a segment of bowel and its mesentery (the intussusceptum) invaginates or telescopes into a contiguous segment of bowel (the intussuscipiens) [1]. An intussusception may be identified on CT, MRI, ultrasound, or fluoroscopy.

On contrast-enhanced CT, a bowel within bowel appearance is noted. On transverse short axis images through the bowel, a classic "target" appearance results if mesenteric fat is visible between the two segments of bowel in the intussusceptum, and/or fluid is present between the intussusceptum and intussuscipiens (Figure 62.1). On a longitudinal view, similar structures may be identified in a "sausage-shaped" mass.

Similar corresponding findings are noted on ultrasound, where they may be referred to as the "doughnut sign" on a transverse image or the "sandwich sign" on a longitudinal image [3].

Importance

Small bowel intussusception accounts for 5–16% of adult intussusceptions [4]. In the past, the standard of care for intussusception in adults was surgical resection to treat symptoms and for pathologic evaluation [5], as malignancy was reported in 24% of enteric intussusceptions and 54% of colonic intussusceptions [6]. These data were generally based on surgical series in patients with obstructive symptoms and intraoperative diagnoses [2]. These studies clearly would not have included many of the transient SBI currently detected on imaging, which explains the different rates of pathology associated with SBI in more recent studies.

With the increased utilization of abdominal CT imaging, incidental SBI are more frequently identified in patients with no imaging evidence for, or clinical suspicion of bowel obstruction. Sending all of these patients for surgery would be inappropriate and distinguishing patients requiring additional workup or surgery from those whose findings are incidental is critical.

A lead point is a mass or other abnormality of the bowel or mesentery that results in the intussusception. Intussusception without a lead point tends to be transient and generally does not cause proximal bowel obstruction, while intussusception with a lead point is usually persistent or recurrent, and often causes proximal bowel obstruction [1].

Lead point colonic intussusception in an adult is often related to a primary or secondary malignant neoplasm. In contrast, lead point SBI is usually due to a benign lesion and less often to a malignancy (usually a metastatic lesion) [1].

SBI more commonly occurs without a lead point, and it is important to distinguish those cases that are transient and incidental from those that require further evaluation or treatment.

When a SBI is identified on ultrasound, its transient nature (and subsequent resolution) may be identified during the initial ultrasound exam or a repeat ultrasound exam performed a short time later. Without exposing a patient to ionizing radiation repeatedly, the transient nature of the SBI is not directly observable on CT, but several characteristics of SBI noted on ultrasound and CT studies permit identification of transient SBI.

Incidentally detected SBI on ultrasound that met the following criteria, were demonstrated to reduce spontaneously and to be of no clinical significance: absence of lead point, normal wall thickness, length <3.5 cm, normal vascularity, and no evidence of dilated bowel [3]. CT also demonstrated that SBI ≤ 3.5 cm in length were transient and inconsequential [2].

In another study of patients with acute abdominal pain and SBI diagnosed on CT, the only statistically significant predictors of the need for surgery were evidence of small bowel obstruction or a lead point [5].

Despite these studies, several authors are not yet ready to ignore incidentally detected intussusceptions as benign transient findings that need no further evaluation, although appropriate follow-up has not been clearly defined [1, 5].

In our practice, short segment SBI without evidence of a lead point or bowel obstruction, and with normal wall enhancement and bowel wall thickness, are reported as incidental findings, and no further evaluation is recommended in the absence of recurrent symptoms of obstruction.

Typical clinical scenario

Identification of SBI should prompt careful evaluation for a lead point mass and for evidence of proximal bowel obstruction. If these findings are absent in short segment intussusception, and the bowel otherwise appears normal, findings are consistent with a transient intussusception, which accounts for a majority of SBI (84%, [2])

Differential diagnosis

When the typical bowel within bowel appearance is present on CT (target, doughnut or sandwich sign), findings are pathognomonic for SBI [1]. When bowel wall edema or other abnormalities obscure these findings, it may be difficult to differentiate between inflammation, bowel wall mass, and intussusception.

Teaching point

Incidentally detected uncomplicated short SBIs without a lead point reduce spontaneously and do not require surgical evaluation or treatment.

Large bowel intussusceptions are often caused by neoplasm, and warrant further evaluation.

REFERENCES

1. Kim YH, Blake MA, Harisinghani MG, *et al.* Adult intestinal intussusception: CT appearances and identification of a causative lead point. *Radiographics.* 2006;**26**(3):733–44.
2. Lvoff N, Breiman RS, Coakley FV, Lu Y, Warren RS. Distinguishing features of self-limiting adult small-bowel intussusception identified at CT. *Radiology.* 2003;**227**(1):68–72.
3. Mateen MA, Saleem S, Rao PC, Gangadhar V, Reddy DN. Transient small bowel intussusceptions: ultrasound findings and clinical significance. *Abdom Imaging.* 2006;**31**(4):410–16.
4. Maconi G, Radice E, Greco S, *et al.* Transient small-bowel intussusceptions in adults: significance of ultrasonographic detection. *Clin Radiol.* 2007;**62**(8):792–7.
5. Olasky J, Moazzez A, Barrera K, *et al.* In the era of routine use of CT scan for acute abdominal pain, should all adults with small bowel intussusception undergo surgery? *Am Surg.* 2009;**75**(10): 958–61.
6. Weilbaecher D, Bolin JA, Hearn D, Ogden W, 2nd. Intussusception in adults. Review of 160 cases. *Am J Surg.* 1971;**121**(5):531–5.

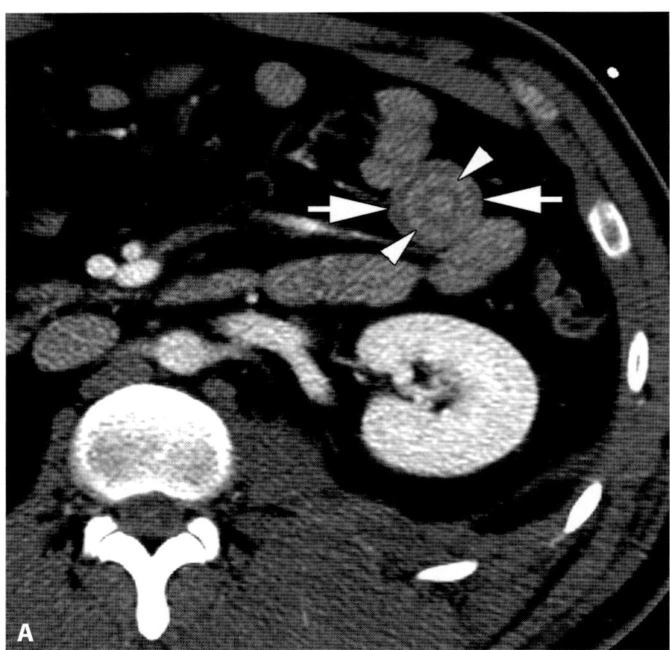

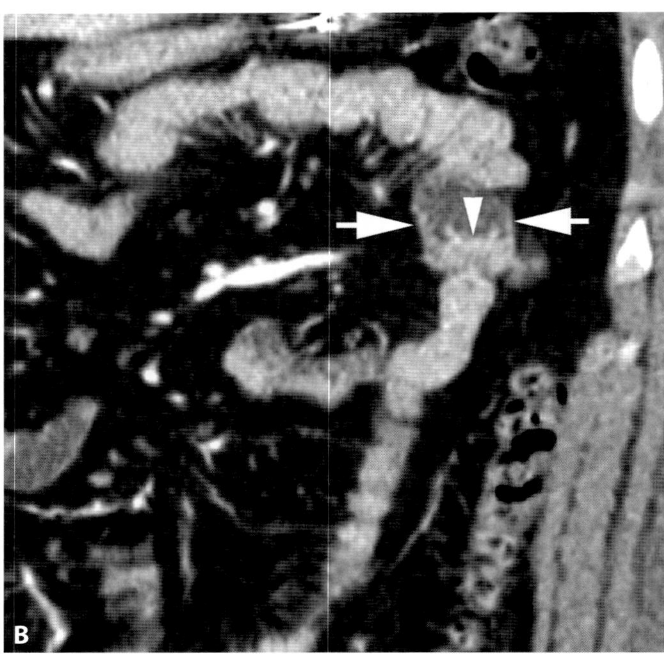

Figure 62.1 Contrast-enhanced abdomen and pelvis CT of a 28-year-old man following fall from a ladder. **A.** Axial CT image demonstrates transverse short axis view of a small bowel intussusception, showing the target sign manifested by the intussusceptum (between arrowheads) within the intussuscipiens (arrows). The low density between the two structures represents a small amount of fluid in the bowel. Mesenteric fat is not well visualized between the two layers of bowel in the intussusceptum. **B.** Coronal CT image demonstrates the short segment of intussusceptum (arrowhead) invaginating into the intussuscipiens (arrows) which is mildly distended compared to proximal bowel below and distal bowel above. No lead point or bowel distension was present on the study consistent with a transient intussusception.

CASE 63 Duodenal diverticulum

Martin L. Gunn

Imaging description

Duodenal diverticula are common, affecting up to 22% of the population [1]. They can be diagnostically challenging for two reasons. Firstly, uncomplicated diverticula can be mistaken for another emergent pathology requiring surgery, and secondly they can cause complications.

Almost all (95%) duodenal diverticula arise from the medial wall of the duodenum, mostly in the second and third parts, and they can be either congenital diverticula (so-called "intraluminal diverticula") or, more commonly, pulsion type diverticula [1]. The most common location is adjacent to the entry point of the common bile duct into the duodenum, where they are termed "periampullary diverticula."

The typical imaging appearance on CT or MRI is of a saccular outpouching from the medial wall of the duodenum, which may contain gas, a gas–fluid or gas-contrast level, or debris (Figures 63.1 and 63.2).

Duodenal diverticula rarely arise from the bulb of the duodenum; most that do are probably healed duodenal ulcers.

Duodenal diverticula may mimic a duodenal ulcer, paraduodenal abscess, or duodenal injury with extraluminal gas (Figure 63.3). Complications of duodenal diverticula include bleeding, diverticulitis, perforation, and biliary duct obstruction (Figure 63.4) [1].

If a small cystic lesion such as a pseudocyst in the head of the pancreas is encountered that lies adjacent to the ampulla of Vater, consider a duodenal diverticulum. A repeat study may demonstrate gas, indicating its communication with the bowel lumen. The absence of surrounding inflammatory change will usually assist to differentiate between abscess, diverticulitis, and pseudocyst.

Importance

Duodenal diverticula are commonly encountered, and should not be mistaken for duodenal perforation in blunt or penetrating trauma, or for a duodenal ulcer in patients with upper abdominal pain.

Typical clinical scenario

For penetrating abdominal trauma, a "triple contrast" technique, which employs oral, rectal, and intravenous contrast, yields the greatest sensitivity for bowel injury [2, 3]. Transmural duodenal injury may be associated with extraluminal gas. In blunt trauma, a duodenal diverticulum will typically appear as a small air–fluid level medial to the second portion of the duodenum [4].

Differential diagnosis

Differential considerations include a pancreatic pseudocyst, cystic neoplasm of the pancreas, duodenal ulcer, contained duodenal rupture, transmural duodenal injury, abscess, and duodenal diverticulitis.

Teaching point

Duodenal diverticula are common, and can mimic both cystic paraduodenal lesions and duodenal perforation.

REFERENCES

1. Bittle MM, Gunn ML, Gross JA, Rohrmann CA. Imaging of duodenal diverticula and their complications. *Curr Probl Diagn Radiol*. 2012; **41**(1):20–9.
2. Shanmuganathan K, Mirvis SE, Chiu WC, *et al.* Penetrating torso trauma: triple-contrast helical CT in peritoneal violation and organ injury – a prospective study in 200 patients. *Radiology*. 2004;**231**(3): 775–84.
3. Shanmuganathan K, Mirvis SE, Chiu WC, Killeen KL, Scalea TM. Triple-contrast helical CT in penetrating torso trauma: a prospective study to determine peritoneal violation and the need for laparotomy. *AJR Am J Roentgenol*. 2001;**177**(6):1247–56.
4. Brofman N, Atri M, Hanson JM, *et al.* Evaluation of bowel and mesenteric blunt trauma with multidetector CT. *Radiographics*. 2006; **26**(4):1119–31.

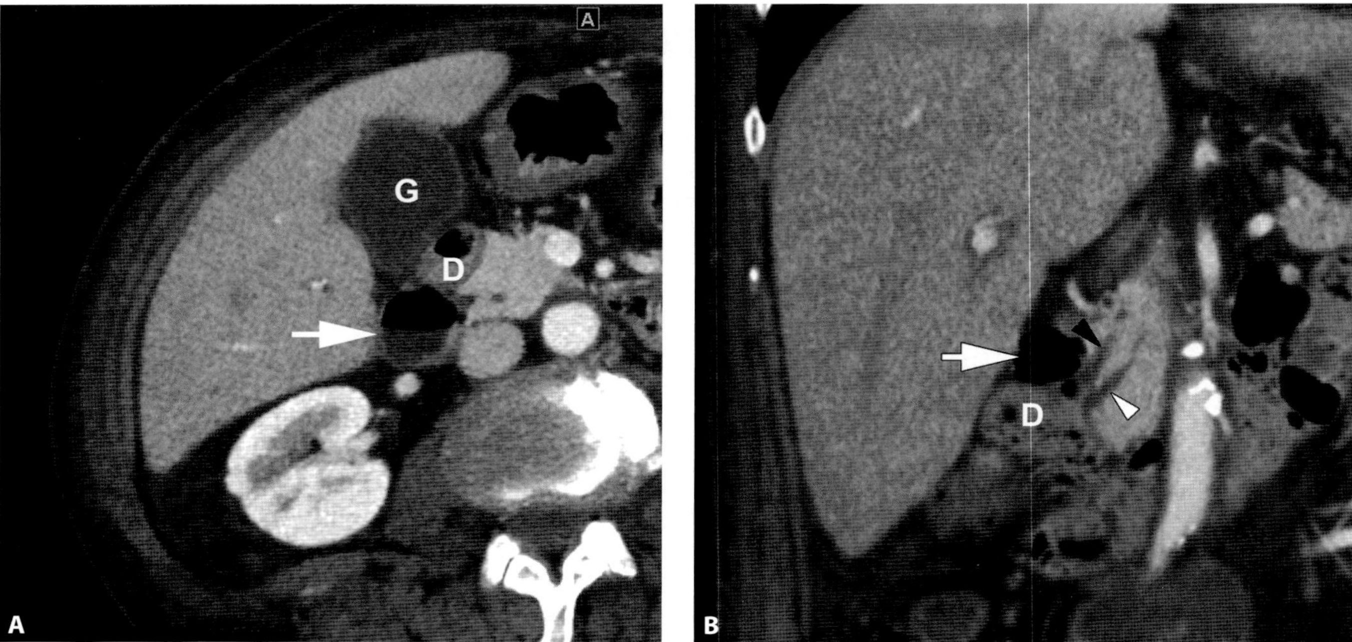

Figure 63.1 Axial (**A**) and coronal (**B**) intravenous contrast-enhanced CT of the abdomen in a 59-year-old woman with nausea, vomiting, and right lower quadrant pain reveals the typical appearance of an incidental peri-ampullary duodenal diverticulum (white arrow). This one contains an air–fluid level. The duodenum (D) and gallbladder (G) appear discrete from this. Note the relationship to the distal common bile duct (black arrowhead) and the pancreatic duct (white arrowhead) on the coronal image.

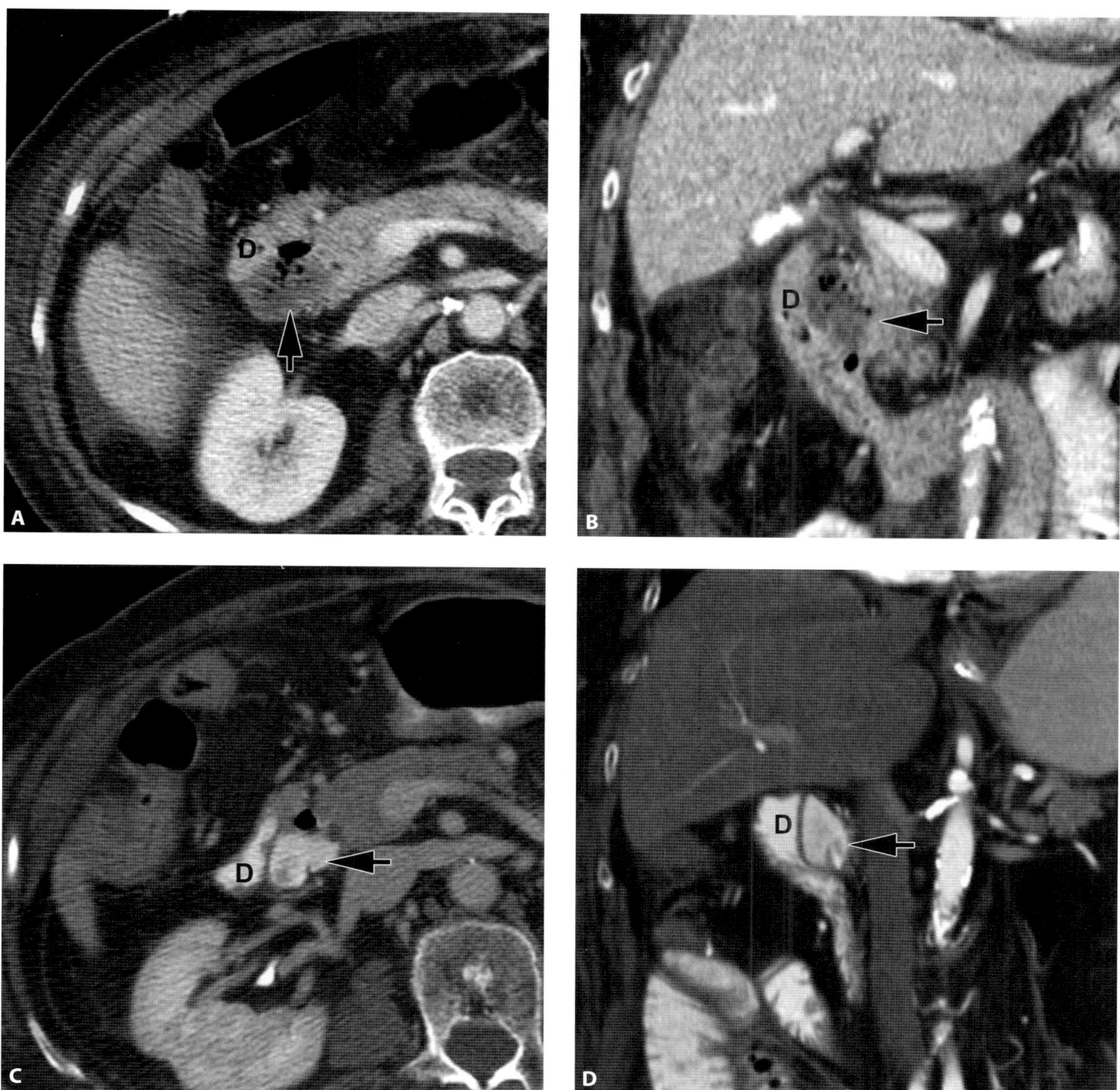

Figure 63.2 Axial (**A**) and coronal (**B**) intravenous contrast-enhanced CT of the abdomen in an 85-year-old woman with abdominal pain due to a closed-loop small bowel obstruction. Note the debris in the duodenal diverticulum (arrow), a finding commonly encountered. Duodenum is denoted by D. **C.** and **D.** CT following oral contrast administration demonstrates filling of the diverticulum, with only a small quantity of debris remaining.

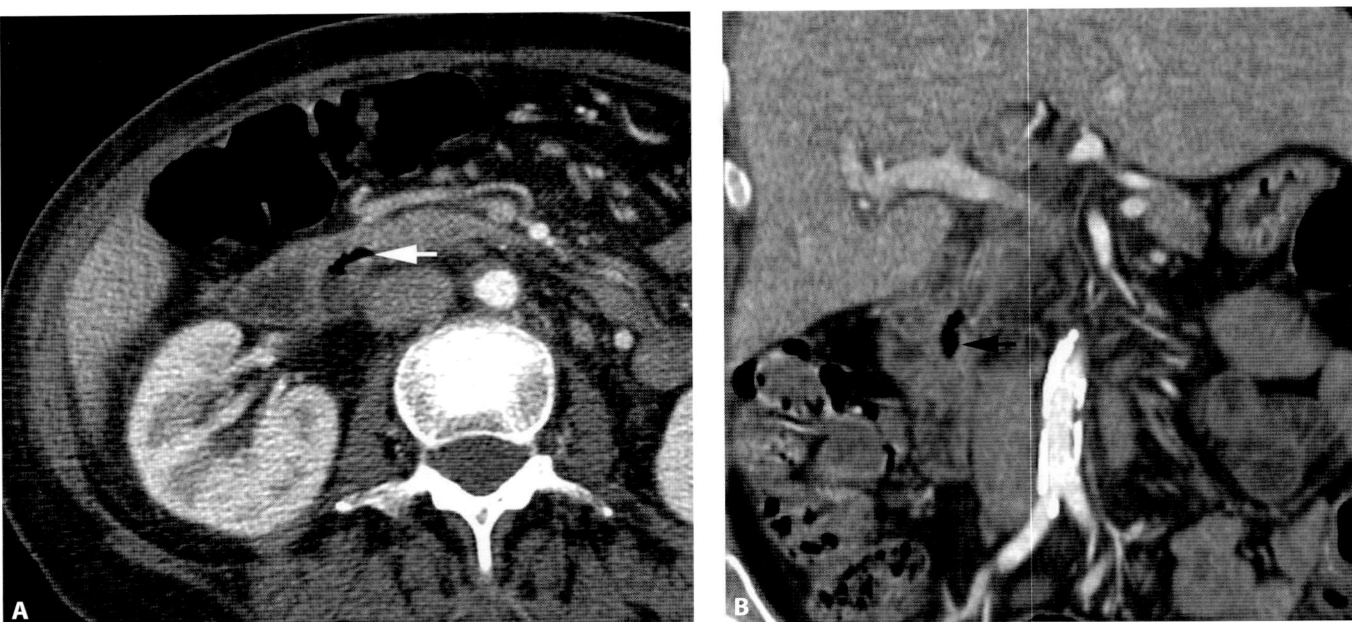

Figure 63.3 Axial (**A**) and coronal (**B**) contrast-enhanced CT of the abdomen in a 62-year-old woman with abdominal pain and elevated liver enzymes following a cholecystectomy demonstrates a small amount of gas initially interpreted as gas within the distal common bile duct (pneumobilia) (arrow). This is gas within a small periampullary duodenal diverticulum, confirmed by endoscopic retrograde cholangiopancreatography.

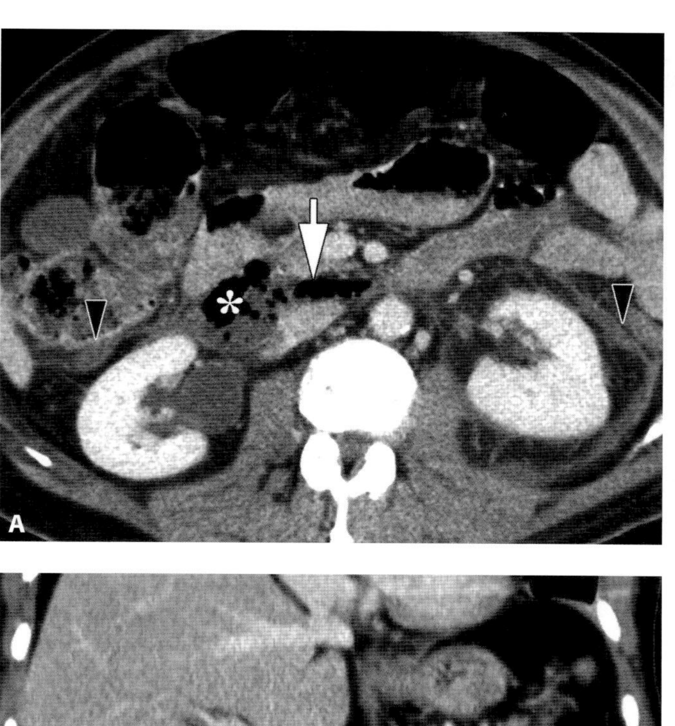

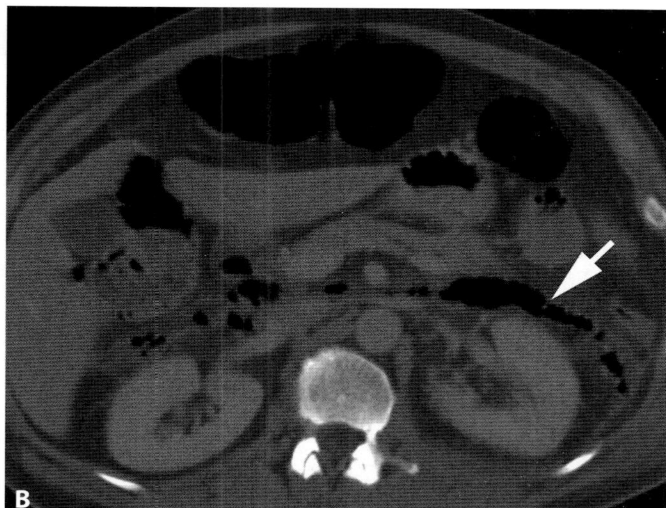

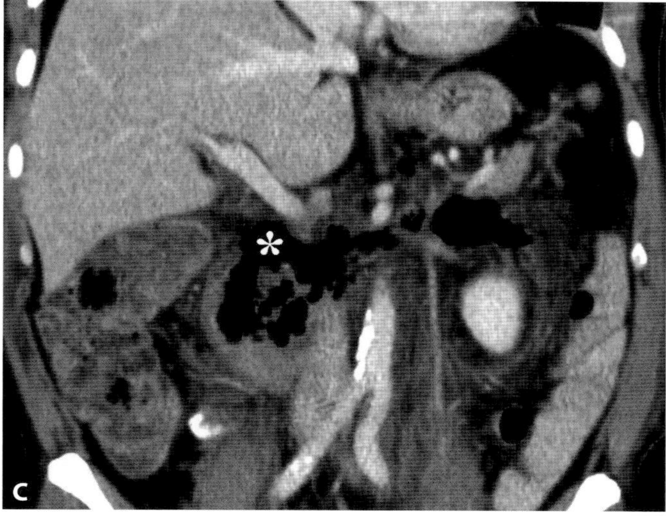

Figure 63.4 Axial (**A** and **B**) and coronal (**C**) intravenous and oral contrast-enhanced CT of the abdomen in a 55-year-old man with fever and abdominal pain. A perforated duodenal diverticulum (asterisks) is present with retroperitoneal gas (arrow) and fluid (arrowheads).

Pseudopneumatosis

Martin L. Gunn

Imaging description

Pneumatosis intestinalis (PI) refers to the presence of gas within the wall of the bowel. Pneumatosis intestinalis can be "benign" or life-threatening (Figure 64.1). CT is more sensitive than radiographs for the detection of PI, although the identification of this finding does not mandate surgery [1].

Pseudopneumatosis may resemble PI, but it occurs when intraluminal gas is trapped against the bowel mucosa not within the bowel wall (Figure 64.2). This may occur between mucosal folds, between the mucosa and viscous bowel contents (Figure 64.3), or when bubbles adhere to the mucosa of the bowel. Pseudopneumatosis most commonly occurs in the cecum and ascending colon (Figure 64.4) [2].

Importance

The importance of not mistaking pseudopneumatosis for PI is that life-threatening causes of PI, such as mesenteric ischemia, toxic megacolon, acute graft-versus-host diseases, bowel obstruction, and cecal ileus should be considered whenever pneumatosis is identified [3]. The presence of portomesenteric gas and PI is suggestive but not diagnostic of transmural bowel infarction [1, 4]. However, a very large number of "benign" causes have been identified [3]. These are covered more completely in Case 65 and include pulmonary diseases such as obstructive airway disease, cystic fibrosis, and lung transplantation, iatrogenic causes such as placement of percutaneous feeding tubes (Figure 64.5), endoscopic procedures, and corticosteroid administration [3]. Clinical predictors, such as the presence of abdominal distension, peritonitis, and lactic acidemia are most predicative of positive intraoperative findings [5].

A false-positive diagnosis of PI may lead to unnecessary laparotomy, especially when the patient has corroborative clinical or laboratory signs suggestive of bowel ischemia.

Typical clinical scenario

Pseudopneumatosis is common. The incorrect diagnosis of PI may be made when bowel ischemia is questioned clinically due to abdominal pain, or abnormal clinical findings.

Differential diagnosis

Differentiating pseudopneumatosis from PI is difficult. The presence of portal venous gas is strongly suggestive of PI. The pattern of intramural gas may be unhelpful; for example, the presence of a bubbly versus a linear pattern of PI has not been found to be discriminative between a clinically worrisome or benign cause in children [6]. The presence of soft tissue thickening of the bowel, stranding, and intraperitoneal fluid are more predictive [6]. Although it has been suggested that the presence of bubbles in the dependent portion of the bowel is more suggestive of PI than pseudopneumatosis, this sign has not been verified and its finding is frequently identified in clinical practice [7]. If there is a question, rescanning the patient in a prone or decubitus position may provide a definitive answer. Another feature related to the distribution is gas not extending anterior to a gas–fluid interface. Gas extending along the bowel wall above a gas–fluid interface is more consistent with true pneumatosis intestinalis than pseudopneumatosis [8].

Teaching point

Gas trapped between bowel contents and the bowel mucosa can mimic pneumatosis intestinalis, even when the gas lies in the dependent portion of the bowel. Porto-mesenteric gas strongly suggests true pneumatosis. Even when true pneumatosis is identified, clinical signs of bowel ischemia are more predictive of the need for emergent laparotomy than the pattern or extent of pneumatosis.

REFERENCES

1. Wayne E, Ough M, Wu A, *et al.* Management algorithm for pneumatosis intestinalis and portal venous gas: treatment and outcome of 88 consecutive cases. *J Gastrointest Surg.* 2010;**14**(3):437–48.
2. Simon AM, Birnbaum BA, Jacobs JE. Isolated infarction of the cecum: CT findings in two patients. *Radiology.* 2000;**214**(2):513–16.
3. Ho LM, Paulson EK, Thompson WM. Pneumatosis intestinalis in the adult: benign to life-threatening causes. *AJR Am J Roentgenol.* 2007; **188**(6):1604–13.
4. Kernagis LY, Levine MS, Jacobs JE. Pneumatosis intestinalis in patients with ischemia: correlation of CT findings with viability of the bowel. *AJR Am J Roentgenol.* 2003;**180**(3):733–6.
5. Duron VP, Rutigliano S, Machan JT, Dupuy DE, Mazzaglia PJ. Computed tomographic diagnosis of pneumatosis intestinalis: clinical measures predictive of the need for surgical intervention. *Arch Surg.* 2011;**146**(5):506–10.
6. Olson DE, Kim YW, Ying J, Donnelly LF. CT predictors for differentiating benign and clinically worrisome pneumatosis intestinalis in children beyond the neonatal period. *Radiology.* 2009;**253**(2):513–19.
7. Taourel P, Garibaldi F, Arrigoni J, *et al.* Cecal pneumatosis in patients with obstructive colon cancer: correlation of CT findings with bowel viability. *AJR Am J Roentgenol.* 2004;**183**(6):1667–71.
8. Wang JH, Furlan A, Kaya D, *et al.* Pneumatosis intestinalis versus pseudo-pneumatosis: review of CT findings and differentiation. *Insights Imaging.* 2011;**2**(1):85–92.

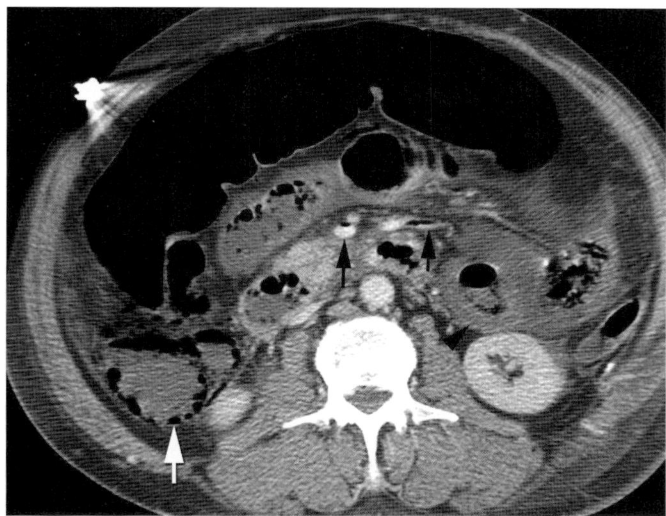

Figure 64.1 Contrast-enhanced CT of the abdomen demonstrating pneumatosis intestinalis in a 54-year-old man with multi-organ failure, hypotension, and an elevated serum lactate level. There is pneumatosis intestinalis in the cecum (white arrow) accompanied by portal venous gas (small black arrows), and hypoenhancement of the jejunum (black arrowhead). Free intraperitoneal gas was also visible. The patient died soon after the CT scan, and the findings were confirmed post-mortem.

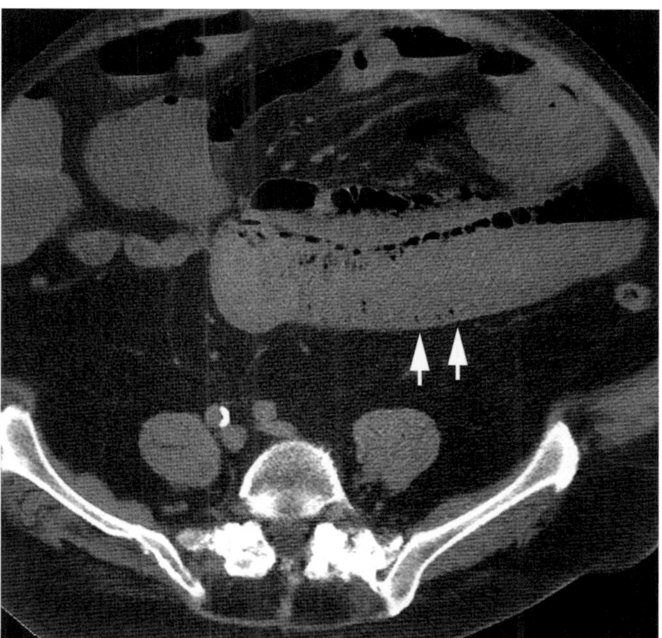

Figure 64.2 Non-contrast CT of the abdomen and pelvis in a 68-year-old man with abdominal pain due to small bowel obstruction demonstrates gas in the dependent part of the small bowel. Note the vertically oriented pattern of the gas. The circular folds of the jejunum are clearly visible anteriorly within the jejunum. The dependent gas (arrows) is almost certainly due to gas trapped beneath folds. The patient was treated without surgery and made a full recovery.

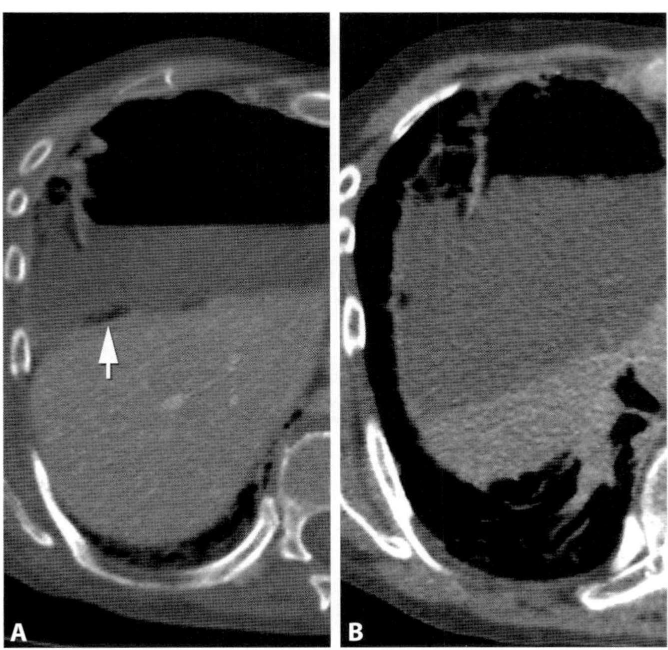

Figure 64.3 Suspected pneumatosis coli in a 72-year-old man with cecal volvulus. **A.** Axial venous phase contrast-enhanced CT demonstrates gas posteriorly, apparently within the colon wall (arrow). **B.** Repeat CT scan 30 minutes later demonstrates that the gas is no longer present in the same location, confirming that the gas was not intramural. Another technique that may be used is rolling the patient, and performing an immediate repeat CT.

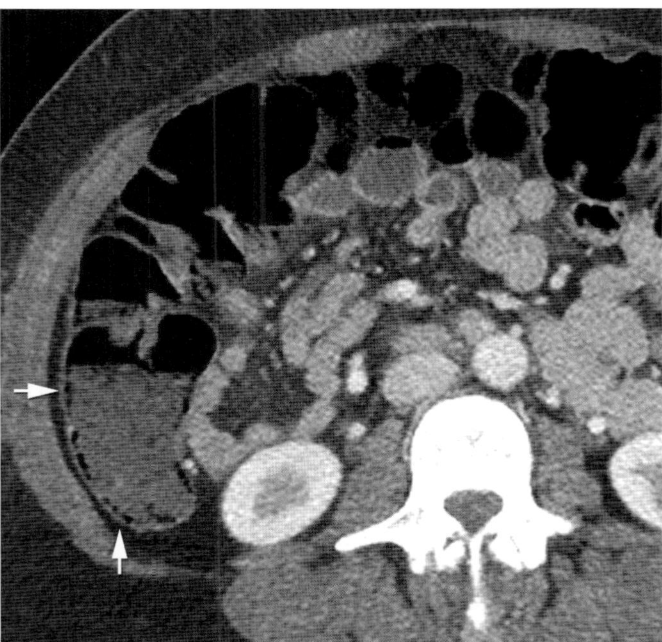

Figure 64.4 Pseudopneumatosis due to gas trapped between bowel mucosa and fecal material in a 44 year old woman with left lower quadrant pain who was discharged well from the emergency department. Note how the gas only extends anteriorly as far as the gas-fluid level, a clue suggestive of pseudopneumatosis. This pattern is more commonly encountered in the cecum and ascending colon because of an admixture of liquid stool and gas. When gas extends beyond the gas-fluid level, pneumatosis intestinalis should be considered.

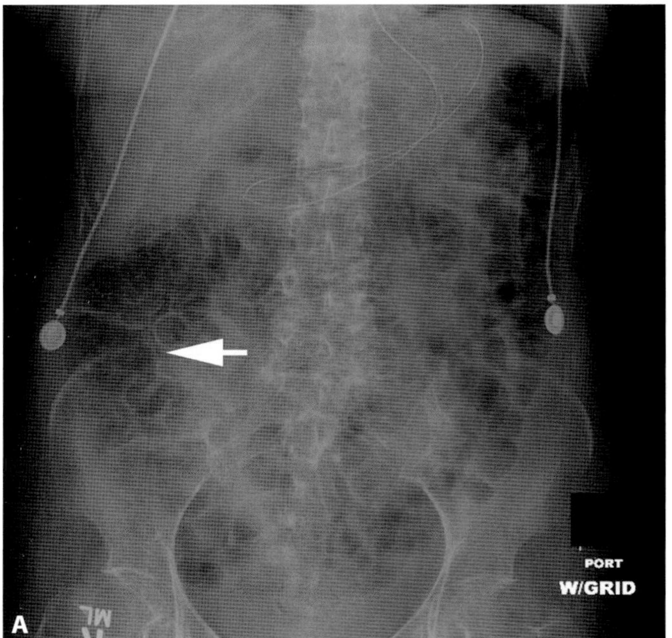

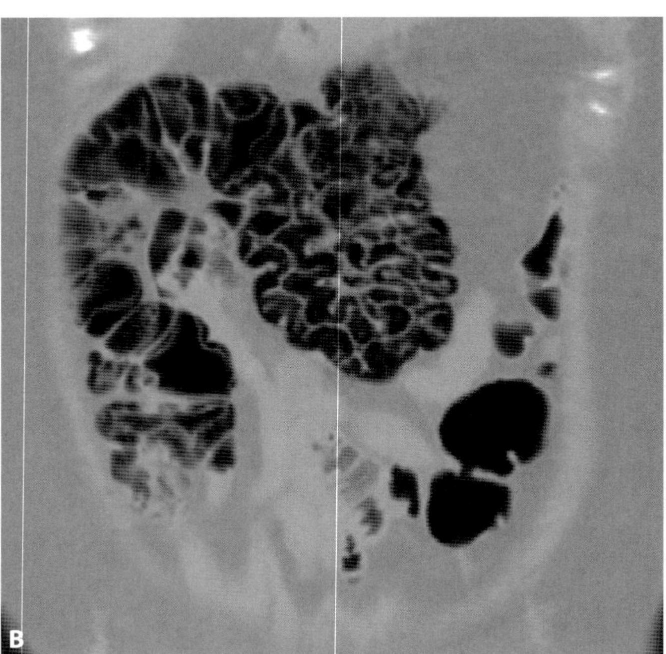

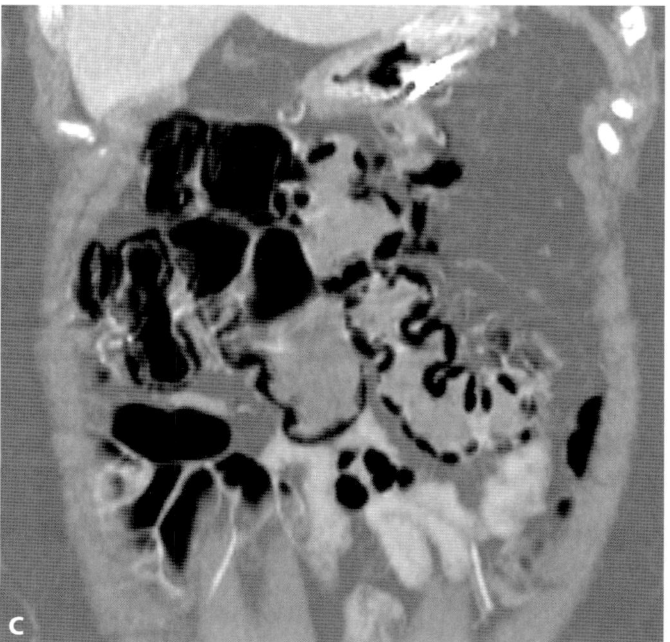

Figure 64.5 A. Anterior-posterior supine radiograph of the abdomen in a 61-year-old woman with systemic candiasis demonstrates pneumatosis coli of the right hemicolon (arrow). The patient had recently undergone attempted percutaneous gastrostomy tube placement. She did not have clinical signs of bowel ischemia. **B.** Follow-up non-contrast CT scan performed 10 days later demonstrates widespread pneumatosis. She remained asymptomatic from this. **C.** Coronal image from an abdominal CT performed 11 days later shows a slight reduction of the pneumatosis. The patient was discharged well 7 days later.

CASE

65 Pneumatosis intestinalis

Ken F. Linnau

Imaging description

Pneumatosis intestinalis (PI) is defined as gas within the bowel wall. It is not a disease, but a radiographic finding that may be secondary to any one of many pathologic processes [1].

In adults, PI is most commonly identified on radiography or CT, but can be detected on MRI or sonography. Sonography is more commonly utilized in pediatric patients. CT is most sensitive for the detection of PI owing to its high contrast resolution and ability to directly depict gas [2]. Particularly in the colon, viewing of CT images in lung windows tends to accentuate PI.

The imaging hallmarks of PI are circumferential cystic (bubbly) or linear radiolucent clusters of gas in the bowel wall. Gas collections within the bowel wall that join together result in a more circular appearance (Figures 65.1 and 65.2). As described in Case 64 intraluminal bowel gas trapped between mucosal folds or mixed with fecal material can mimic PI, particularly in the dependent portion of bowel loops.

Depiction of associated radiographic findings may improve identification of life-threatening causes of PI and distinguish them from benign causes. However, there is substantial overlap between the imaging findings of life-threatening PI with benign PI. The strongest CT predictors for abnormal surgical findings (positive laparotomy) are the presence of (1) dilated bowel, (2) portal venous gas, (3) atherosclerosis and vascular occlusion, and (4) ascites [3]. However, the presence of portal venous gas and PI does not predict partial versus transmural bowel infarction [4]. Bowel wall thickening, absent or intense mucosal enhancement, pneumoperitoneum, pneumoretroperitoneum, and short segment small bowel distribution of PI have also been associated with life-threatening etiologies of PI [1].

The presence of pneumomediastinum with PI is an imaging finding which favors a benign pulmonary etiology [1].

Importance

Historically, PI was considered a sign of bowel ischemia requiring swift intervention to avoid bowel infarction, particularly if it was associated with gas in the portal venous system. However, there are several non-life-threatening causes of PI and the most common are listed in Table 65.1. While the prevalence of PI in the general population is unknown because many patients are asymptomatic and never come to clinical attention, it is likely that increased use of CT has resulted in more common detection [1, 3, 5]. Although this finding was rarely identified on autopsy in the past, it may now be much more commonly detected in clinical practice [5]. This observation is supported by falling mortality rates from PI when historical series (65–85%) are compared to a recent retrospective series (22%) [3].

Table 65.1. Non-life-threatening causes of pneumatosis intestinalis

Infection
Inflammatory bowel disease
Immunosuppression
Corticosteroid use Connective tissue disease Transplantation Chemotherapy
Pulmonary disease Chronic obstructive pulmonary disease Cystic fibrosis Asthma
Iatrogenic causes Enteric tube placement (including percutaneous tubes) CT colonography Endoscopy Post-surgical anastomosis
Primary pneumatosis Idiopathic Pneumatosis cystoides intestinalis

The critical treatment decision is whether to proceed to emergency exploratory laparotomy for salvage of an intra-abdominal catastrophe, usually bowel ischemia, or to manage the patient conservatively. Non-operative treatment options are successful in about half of the cases of PI and include diet, antibiotics, and inhalation oxygen therapy [3, 5]. A recent non-validated retrospective multivariate regression analysis of 30 radiographic, clinical, and laboratory parameters suggests that dilated bowel loops on CT (with or without abdominal distension on physical exam), peritonitis (rebound tenderness or guarding), and lactic acidemia (>2.5 mg/dL or 0.28 mmol/L) predict the presence of life-threatening intra-abdominal findings requiring surgical intervention [3].

Typical clinical scenario

The detection of PI should always prompt careful evaluation for associated radiographic findings which are predictors of bowel ischemia at surgery. Correlation with clinical signs and serum lactate is indispensable in the setting of PI to avoid unnecessary laparotomy. Asymptomatic PI can sometimes be an incidental self-limiting finding of questionable clinical relevance.

Differential diagnosis

More than 50 causes for PI have been described in adults, which can be benign or life threatening. A listing of the most common causes is provided in Table 65.1. For a comprehensive listing, the reader is referred to a recent review article by Ho *et al.* [1].

Teaching point

Pneumatosis intestinalis may be either idiopathic, benign, or life-threatening. Correlation of imaging findings with clinical and laboratory parameters, such as abdominal distension, peritonitis, and serum lactate concentration, is imperative to avoid unnecessary laparotomy.

REFERENCES

1. Ho LM, Paulson EK, Thompson WM. Pneumatosis intestinalis in the adult: benign to life-threatening causes. *AJR Am J Roentgenol.* 2007; **188**(6):1604–13.
2. Lund EC, Han SY, Holley HC, Berland LL. Intestinal ischemia: comparison of plain radiographic and computed tomographic findings. *Radiographics.* 1988;**8**(6):1083–108.
3. Duron VP, Rutigliano S, Machan JT, Dupuy DE, Mazzaglia PJ. Computed tomographic diagnosis of pneumatosis intestinalis: clinical measures predictive of the need for surgical intervention. *Arch Surg.* 2011;**146**(5):506–10.
4. Wiesner W, Mortele KJ, Glickman JN, Ji H, Ros PR. Pneumatosis intestinalis and portomesenteric venous gas in intestinal ischemia: correlation of CT findings with severity of ischemia and clinical outcome. *AJR Am J Roentgenol.* 2001;**177**(6):1319–23.
5. Heng Y, Schuffler MD, Haggitt RC, Rohrmann CA. Pneumatosis intestinalis: a review. *Am J Gastroenterol.* 1995;**90**(10):1747–58.

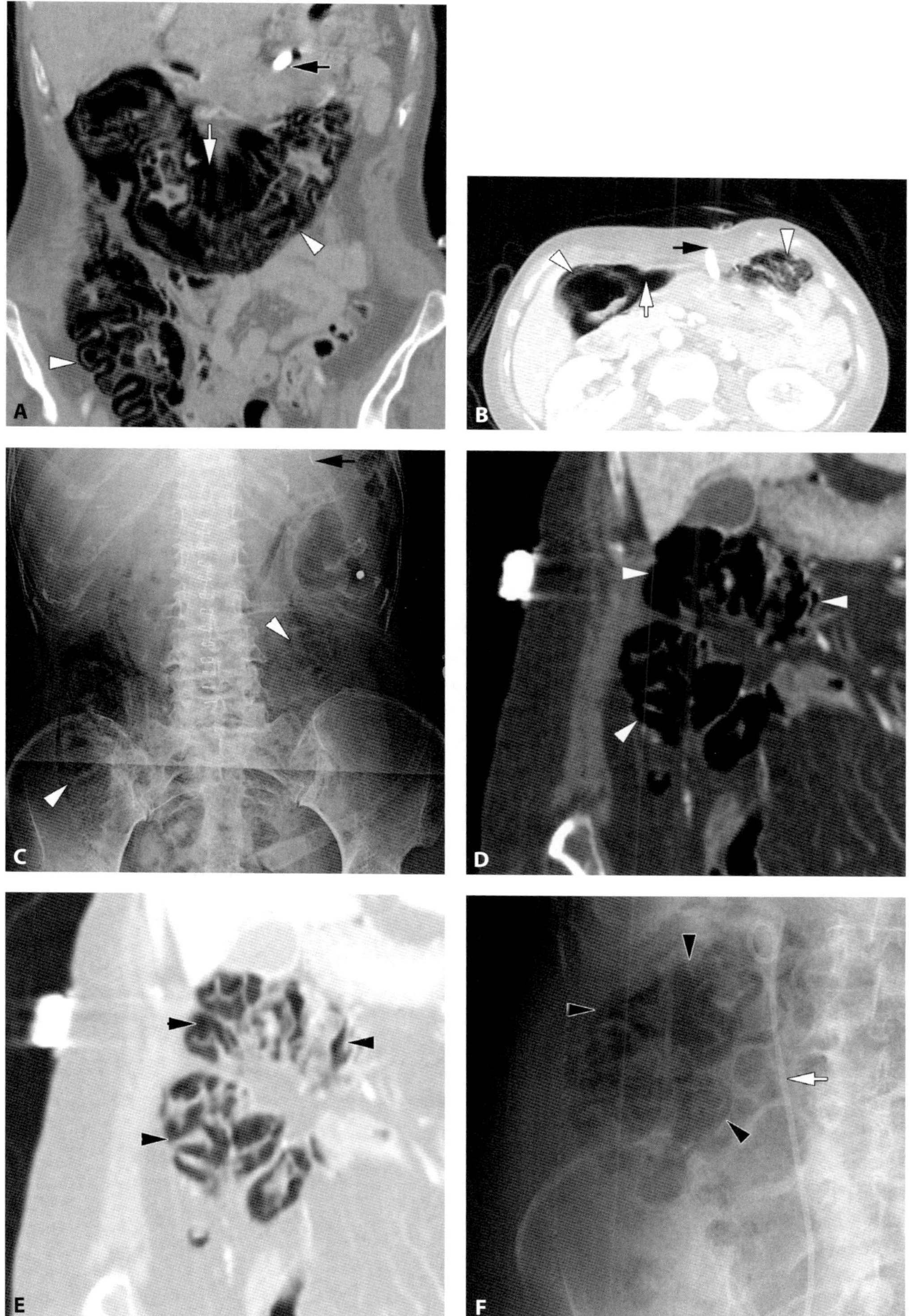

Figure 65.1 Contrast-enhanced abdomen and pelvis CT from a 61-year-old woman with abdominal pain 10 days after percutaneous gastrostomy tube placement for ischemic stroke. The patient was also on long-term steroid therapy for idiopathic thrombocytopenia. Serum lactate levels were normal (<2.5 mg/dL). Coronal (**A**) and axial (**B**) CT images in lung windows show diffuse linear PI (arrowheads) of the ascending and transverse colon. There is gas in the transverse mesocolon (white arrow). The gastrostomy tube is noted superior to the transverse colon (black arrow). It was felt that PI was a complication of gastrostomy tube placement yet a laparotomy was performed, but no bowel ischemia was identified. **C.** Postoperative radiograph shows residual linear PI (arrowheads). The gastrostomy has been removed with the tube fixation tacks left behind (arrow). The patient was then discharged to a nursing home. **D.** Coronal CT image in soft tissue window from the same patient 13 months after initial presentation, when she complained of mild right-sided abdominal discomfort (with normal serum lactate), shows persistent linear PI at the hepatic flexure (arrowheads), which is much more accentuated in lung window (**E**), consistent with benign PI. Abdominal pain was due to an obstructing right ureteric stone. **F.** Abdominal radiograph after ureter stent placement (arrow) shows persistent benign PI (arrow heads) 15 months after initial presentation.

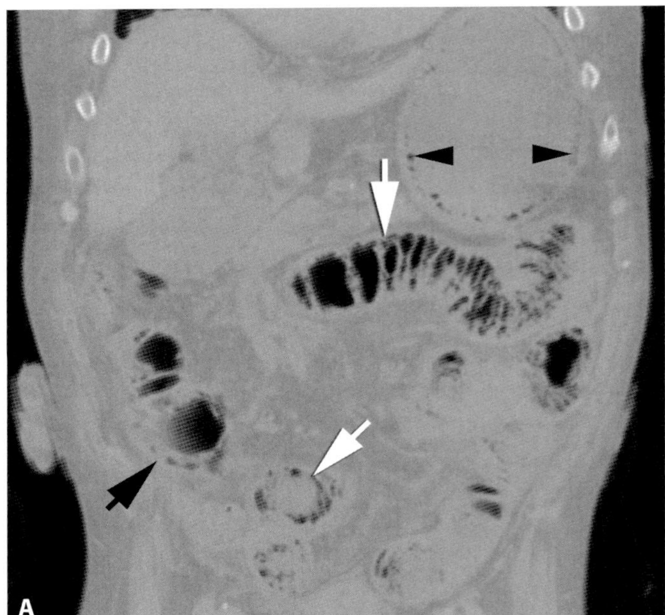

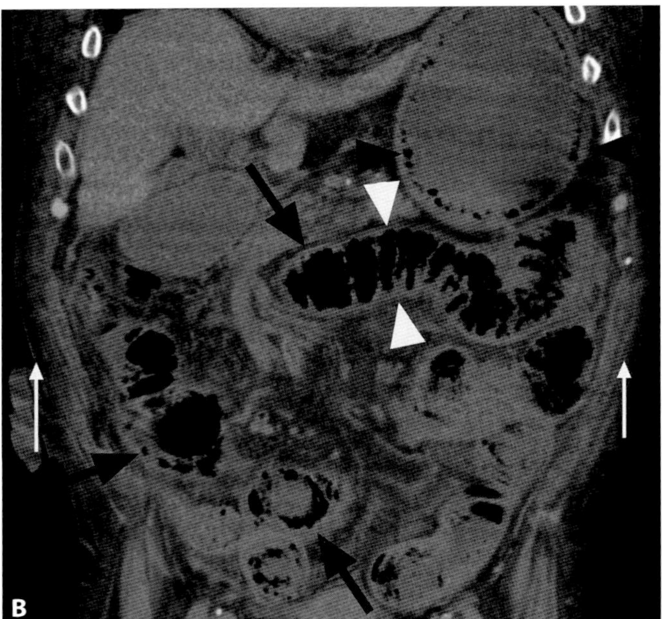

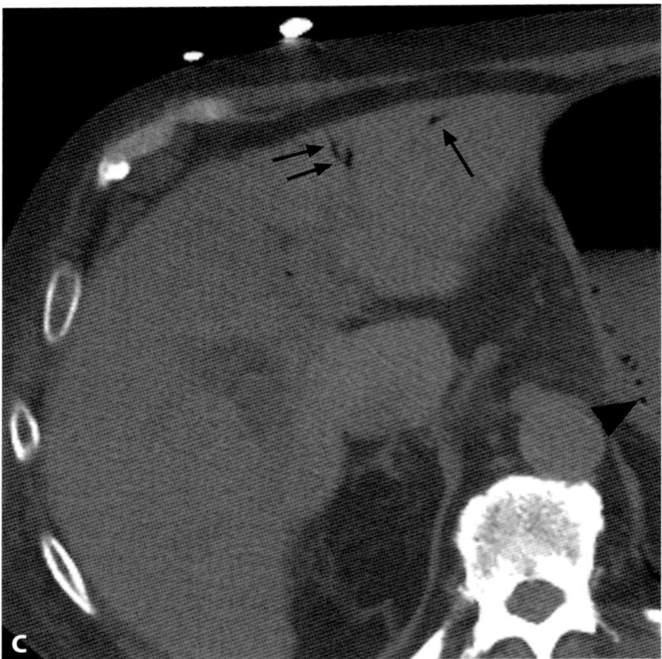

Figure 65.2 Non-contrast coronal (**A** and **B**) and axial (**C**) CT images from a 79-year-old man who presents with severe abdominal pain with distension and blood in his enterostomy bag (colectomy was performed several years prior). Peritonitis was suspected clinically. His serum lactate level at presentation was 13 mg/dL. Bubbly collections of gas are present in the wall of the stomach (black arrowheads), small bowel (white arrows), colon (black arrows), and esophagus (not shown) extending circumferentially along the bowel wall, consistent with PI. There is small bowel dilation (white arrowheads) and bulging of the flanks (small white arrows), indicative of abdominal distension (predictors of abnormal surgical findings). Linear branching gas collections in the periphery of the liver (**C**) are characteristic of portal venous gas (small black arrows), common with life-threatening PI. The extent of disease was classified as unsurvivable and the patient died hours later after comfort care was instituted. Diffuse bowel ischemia and necrosis due to small bowel infection with *Clostridium difficile* was confirmed at autopsy.

Pseudoappendicitis

Martin L. Gunn

Imaging description

CT is a highly accurate test for the diagnosis of acute appendicitis, with a sensitivity and specificity of 94% or greater [1–4]. Appendiceal enlargement and periappendiceal inflammation are the most sensitive signs of acute appendicitis.

The maximum diameter of the normal appendix varies widely among patients and ranges from 3 to 11 mm [5]. Various maximum diameter thresholds for the diagnosis of appendicitis with CT have been suggested, the most common being 6 mm. However, 24–45% of the population has an appendiceal diameter greater than 6 mm, making this threshold sensitive, but not specific [5, 6]. Hence, one should not diagnose acute appendicitis based on outer diameter alone. In particular, many normal appendices containing air will have an increased outer–outer wall diameter but non-thickened wall (Figure 66.1).

Airless fluid within an appendix of greater than 6 mm outer wall diameter is rarely identified in normal patients, and although this does occur in 1–4% of normal appendices (Figure 66.2), it should raise the suspicion for appendicitis in the symptomatic patient [6, 7]. A "depth" of fluid within the appendix of greater than 2.6 mm has been suggested as more sensitive and specific for acute appendicitis by one institution [8, 9].

As the outer diameter of the appendix includes a variable quantity of luminal contents, many have suggested evaluating the mural thickness of the appendix. The mural thickness can be assessed either by measuring it directly in patients with luminal contents or, in patients without appendiceal luminal contents, by measuring the outer wall diameter of the appendix and dividing by two. The average single-wall thickness of the normal appendix on CT is approximately 1.5 mm (range, 0.25–3.1 mm), and only 0.9–2% of normal patients will have an appendiceal wall thickness of 3 mm or more [7, 10].

The absence of oral contrast filling the appendix is not helpful in the diagnosis of acute appendicitis (Figure 66.3). Ingested oral contrast will opacify the lumen in only approximately 50% of patients without appendicitis [7]. Moreover, recent evidence suggests that the addition of oral contrast does not increase sensitivity for the diagnosis acute appendicitis compared to intravenous contrast alone [4].

Multiplanar reformations and thin axial reconstruction thickness images (1 to 2.5 mm) have been found to increase appendix identification and reader confidence when CT is performed for acute appendicitis [11, 12].

Importance

The use of CT has been shown to reduce the negative appendectomy rate in patients presenting to the emergency department with symptoms suggestive of acute appendicitis [13]. However, a false-positive diagnosis of acute appendicitis may lead to unnecessary appendectomy.

Typical clinical scenario

Patients who are evaluated for right lower quadrant pain who do not have CT evidence of periappendiceal fat stranding or fluid, mural enhancement, or definitive mural thickening.

Differential diagnosis

The primary differential diagnosis is a normal appendix. However, thickening of the appendix can be due to a number of causes other than acute appendicitis. These include inflammatory conditions such as Crohn's disease (Figure 66.4), chronic appendicitis (Figure 66.5) [14], granulomatous appendicitis, lymphoid hyperplasia, tuberculosis, mucocele of the appendix, as well as malignancies, especially colon cancer, peritoneal metastases, and carcinoid.

Teaching point

The maximum diameter of the outer wall of the appendix measures more than 6 mm on CT in almost half of the population without acute appendicitis.

REFERENCES

1. Rao PM, Rhea JT, Novelline RA, *et al.* Helical CT technique for the diagnosis of appendicitis: prospective evaluation of a focused appendix CT examination. *Radiology*. 1997;**202**(1):139–44.
2. Balthazar EJ, Megibow AJ, Siegel SE, Birnbaum BA. Appendicitis: prospective evaluation with high-resolution CT. *Radiology*. 1991; **180**(1):21–4.
3. Rao PM, Rhea JT, Novelline RA, *et al.* Helical CT combined with contrast material administered only through the colon for imaging of suspected appendicitis. *AJR Am J Roentgenol*. 1997; **169**(5):1275–80.
4. Anderson SW, Soto JA, Lucey BC, *et al.* Abdominal 64-MDCT for suspected appendicitis: the use of oral and IV contrast material versus IV contrast material only. *AJR Am J Roentgenol*. 2009;**193**(5):1282–8.
5. Kim HC, Yang DM, Jin W, Park SJ. Added diagnostic value of multiplanar reformation of multidetector CT data in patients with suspected appendicitis. *Radiographics*. 2008;**28**(2):393–405; discussion 406.
6. Webb EM, Wang ZJ, Coakley FV, *et al.* The equivocal appendix at CT: prevalence in a control population. *Emerg Radiol*. 2010;**17**(1):57–61.
7. Tamburrini S, Brunetti A, Brown M, Sirlin CB, Casola G. CT appearance of the normal appendix in adults. *Eur Radiol*. 2005; **15**(10):2096–103.

8. Moteki T, Horikoshi H. New CT criterion for acute appendicitis: maximum depth of intraluminal appendiceal fluid. *AJR Am J Roentgenol.* 2007;**188**(5):1313–19.
9. Moteki T, Ohya N, Horikoshi H. Evaluation of the maximum depth of intraluminal appendiceal fluid to diagnose appendicitis with a 64-detector row CT scanner. *J Comput Assist Tomogr.* 2011;**35**(6):703–10.
10. Webb EM, Joe BN, Coakley FV, *et al.* Appendiceal wall thickening at CT in asymptomatic patients with extraintestinal malignancy may mimic appendicitis. *Clin Imaging.* 2009;**33**(3):200–3.
11. Johnson PT, Horton KM, Kawamoto S, *et al.* MDCT for suspected appendicitis: effect of reconstruction section thickness on diagnostic accuracy, rate of appendiceal visualization, and reader confidence using axial images. *AJR Am J Roentgenol.* 2009;**192**(4):893–901.
12. Paulson EK, Harris JP, Jaffe TA, Haugan PA, Nelson RC. Acute appendicitis: added diagnostic value of coronal reformations from isotropic voxels at multi-detector row CT. *Radiology.* 2005; **235**(3):879–85.
13. Raja AS, Wright C, Sodickson AD, *et al.* Negative appendectomy rate in the era of CT: an 18-year perspective. *Radiology.* 2010;**256**(2):460–5.
14. Checkoff JL, Wechsler RJ, Nazarian LN. Chronic inflammatory appendiceal conditions that mimic acute appendicitis on helical CT. *AJR Am J Roentgenol.* 2002;**179**(3):731–4.

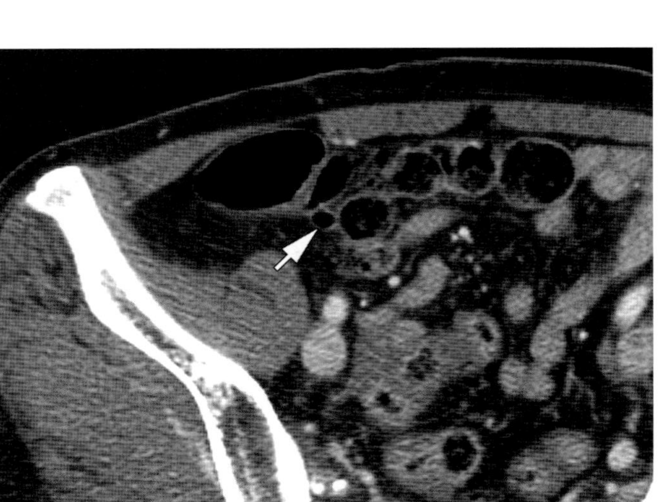

Figure 66.1 Gas filled appendix in a 69-year-old pedestrian who was hit by a car. CT of the pelvis with intravenous contrast demonstrates gaseous distension of the appendix, increasing its outer diameter to 8mm, without wall thickening (arrow).

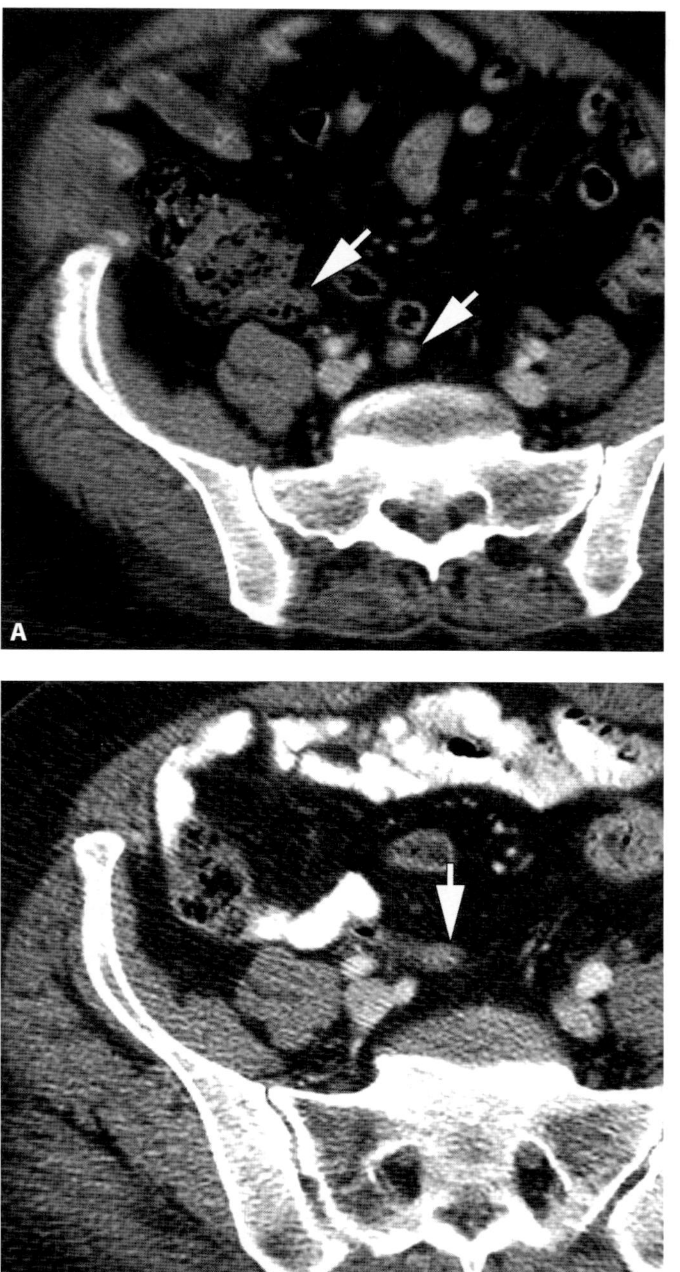

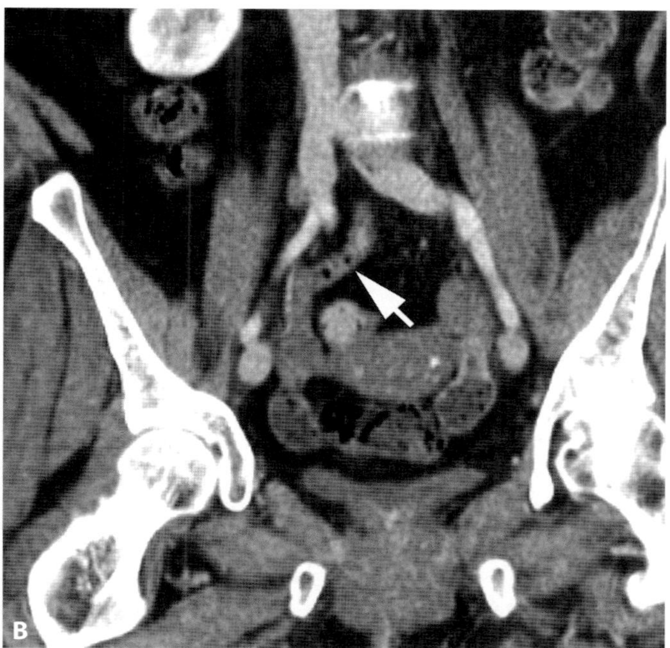

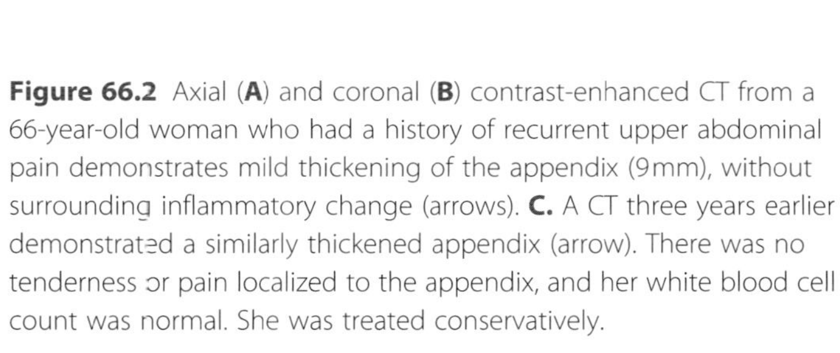

Figure 66.2 Axial (**A**) and coronal (**B**) contrast-enhanced CT from a 66-year-old woman who had a history of recurrent upper abdominal pain demonstrates mild thickening of the appendix (9mm), without surrounding inflammatory change (arrows). **C.** A CT three years earlier demonstrated a similarly thickened appendix (arrow). There was no tenderness or pain localized to the appendix, and her white blood cell count was normal. She was treated conservatively.

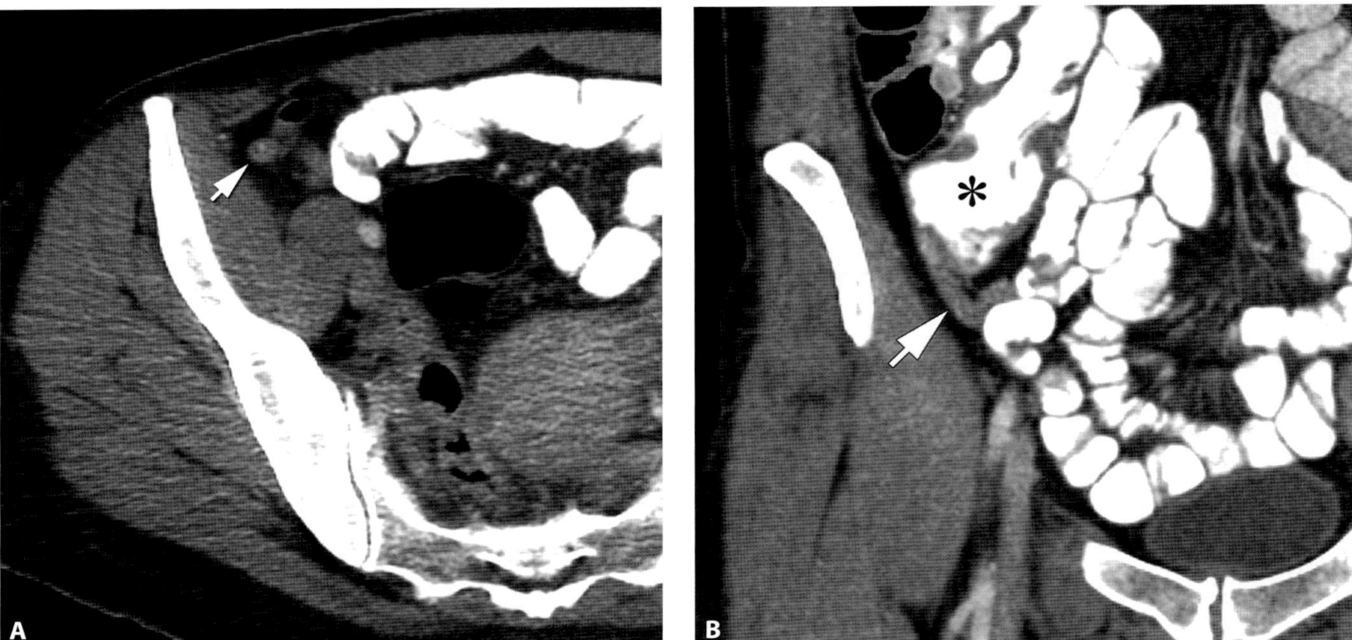

Figure 66.3 Axial (**A**) and coronal (**B**) contrast-enhanced CT from a 19-year-old woman with acute myeloid leukemia undergoing a surveillance CT. Her abdomen was asymptomatic at the time. Note orally administered contrast filling the cecum (asterisk) but not the appendix, which contains some fluid (arrow). Her appendix measured 8mm, with each wall measuring 3mm, and the lumen measuring 2mm. Absence of oral contrast filling of the appendix is not a reliable sign of acute appendicitis.

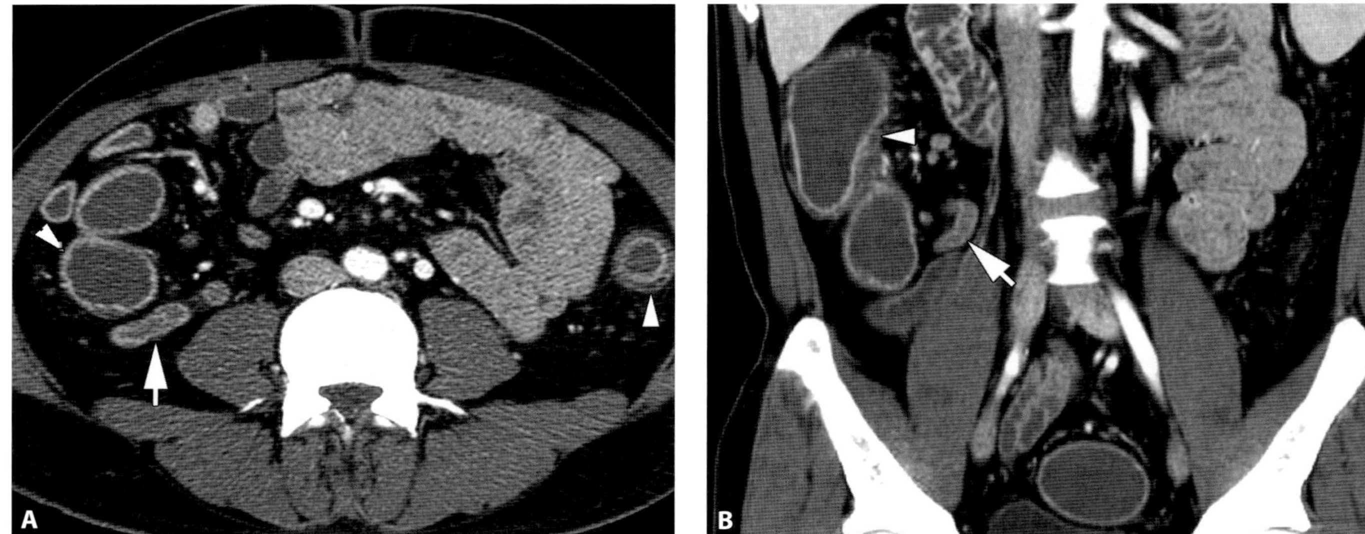

Figure 66.4 Axial (**A**) and coronal (**B**) contrast-enhanced CT from a 35-year-old man with Crohn's disease, with pancolitis and ileitis. Note the mural hyper-enhancement of the fluid-filled appendix (arrow). There is also colonic wall thickening (arrowhead).

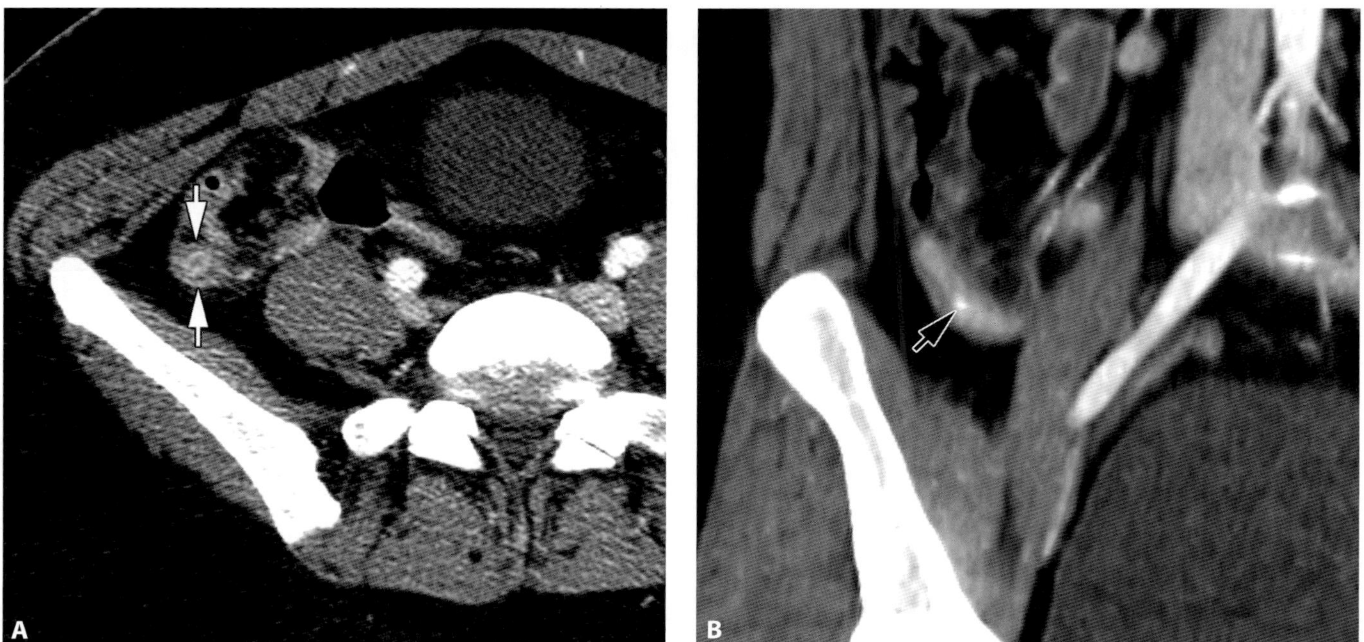

Figure 66.5 Axial (**A**) and coronal (**B**) contrast-enhanced CT from a 47-year-old man who was involved in a motor vehicle collision. The appendix (arrows) measures 11 mm in diameter, and demonstrates serosal and mucosal enhancement, and mild periappendiceal stranding. A tiny fecolith is also present (small black arrow). The patient was asymptomatic and an appendectomy was not performed. Although not pathologically proven, this is likely a case of chronic appendicitis.

Missed renal collecting system injury

Joel A. Gross

Imaging description

Renal collecting system injury is diagnosed when dense contrast-enhanced urine leaks out of the collecting system (Figure 67.1). It is most commonly identified within the perinephric space, but sometimes is seen traversing a laceration within the kidney, within the renal sinus, or adjacent to the ureter.

The renal collecting system injury may be associated with a renal injury, as seen when a renal laceration extends into the collecting system. It may also occur as an isolated collecting system injury without an associated renal injury, such as rupture of a dilated obstructed renal pelvis, or laceration of a ureter.

Fluid adjacent to an organ following trauma commonly represents hematoma secondary to organ injury. The kidney is more complex to evaluate, since it produces urine. Fluid around the kidney may simply represent hematoma or it could represent urine extravasation from a collecting system injury, or both.

Parenchymal phase images (usually portal venous phase) do not adequately evaluate for urinary extravasation, as minimal, if any contrast is excreted by the time images are obtained.

Unexplained fluid adjacent to the kidney or collecting system in a patient at risk for collecting system injury requires an additional delayed imaging phase. A 5–10 minute post-contrast delay allows excreted high-density contrast to leak from a collecting system injury into the adjacent soft tissues, thus permitting the differentiation of a urinary leak from a simple hematoma [1–3].

The delayed phase images are performed with a low dose, to minimize patient radiation. Although images are noisier than the parenchymal phase images, they are adequate for the identification of high-density contrast external to the collecting system. If no findings suggestive of urinary extravasation are present on the parenchymal phase images, no delayed phase images are obtained.

Importance

Delayed diagnosis of renal collecting system injury will lead to persistent urinoma formation in the minority of patients. Delayed diagnosis of ureteral injuries is associated with prolonged hospital stays and an increased rate of nephrectomy [4, 5].

Typical clinical scenario

Patients with major renal injuries or with unexplained fluid associated with the kidney or collecting system require delayed CT images for further evaluation for urinary extravasation. If a urinoma is identified, a non-operative approach is usually favored, with bladder decompression, broad-spectrum antiobiotics, bed rest and low-dose repeat CT 3–7 days after injury [6].

Differential diagnosis

Mild perinephric stranding is often visualized bilaterally in older patients. If findings are fairly symmetric and typical for the patient's age, and no renal injury (such as an adjacent laceration) is identified, then findings are extremely unlikely to represent urinary extravasation, and delayed images are not usually obtained.

Many perinephric neoplasms and inflammatory processes including lymphoma, leukemia, sarcomas, and fibrosis have been reported [7], but these are rare and often present with additional findings to suggest their true diagnosis. If there is any suspicion that imaging findings may represent urinary extravasation on parenchymal phase imaging, delayed images will confirm the presence or absence of extravasation.

Perinephric fluid following trauma most commonly represents perinephric hematoma. While hyperdense fluid may represent hematoma, and hypodense fluid may represent urine, the findings overlap considerably and this differentiation cannot reliably be made on parenchymal phase images. Delayed images at 5–10 minutes are necessary for adequate evaluation.

Arterial extravasation may also result in high-density contrast external to the collecting system. Unless the extravasation is extremely small or slow, it should be evident on the parenchymal phase images.

Teaching point

In patients with unexplained fluid adjacent to the kidney or collecting system following trauma, delayed phase images are necessary to evaluate for urinary extravasation.

Parenchymal phase images alone will result in missed collecting system injuries.

REFERENCES

1. Alonso RC, Nacenta SB, Martinez PD, Guerrero AS, Fuentes CG. Kidney in danger: CT findings of blunt and penetrating renal trauma. *Radiographics*. 2009;**29**(7):2033–53.

2. Lee YJ, Oh SN, Rha SE, Byun JY. Renal trauma. *Radiol Clin North Am*. 2007;**45**(3):581–92, ix.

3. Kawashima A, Sandler CM, Corl FM, *et al*. Imaging of renal trauma: a comprehensive review. *Radiographics*. 2001;**21**(3):557–74.

4. Kunkle DA, Kansas BT, Pathak A, Goldberg AJ, Mydlo JH. Delayed diagnosis of traumatic ureteral injuries. *J Urol.* 2006; **176**(6 Pt 1):2503–7.
5. Mulligan JM, Cagiannos I, Collins JP, Millward SF. Ureteropelvic junction disruption secondary to blunt trauma: excretory phase imaging (delayed films) should help prevent a missed diagnosis. *J Urol.* 1998; **159**(1):67–70.
6. Alsikafi NF, McAninch JW, Elliott SP, Garcia M. Nonoperative management outcomes of isolated urinary extravasation following renal lacerations due to external trauma. *J Urol.* 2006; **176**(6 Pt 1):2494–7.
7. Surabhi VR, Menias C, Prasad SR, *et al.* Neoplastic and non-neoplastic proliferative disorders of the perirenal space: cross-sectional imaging findings. *Radiographics.* 2008;**28**(4):1005–17.

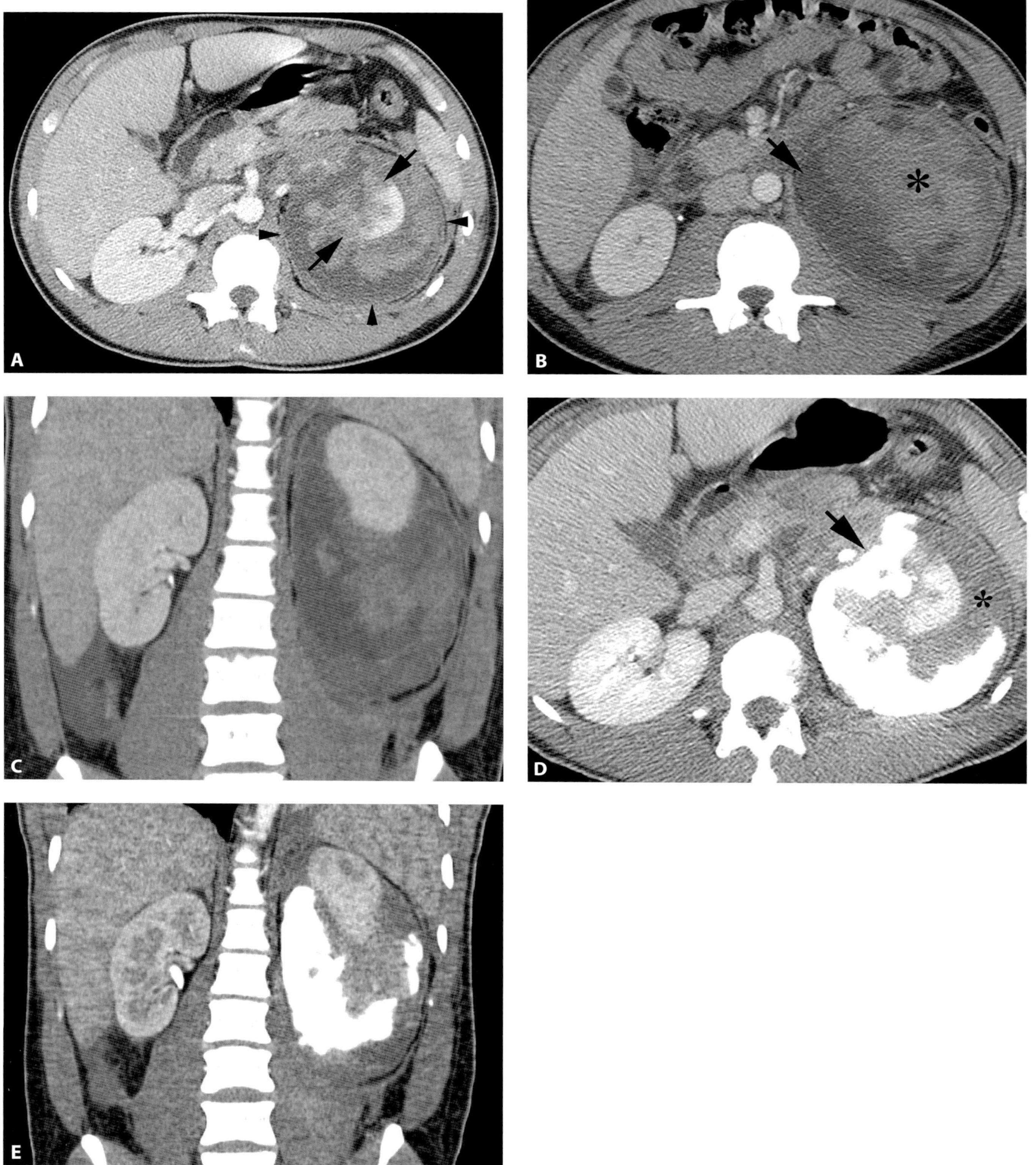

Figure 67.1 Portal venous phase contrast-enhanced abdominal CT from an 18-year-old man following trauma. Initial study (A–C) performed at an outside hospital. Follow-up study (D and E) performed at our facility 4 hours later. **A.** Axial CT demonstrates deep lacerations (arrows) of the inferior left kidney, with extensive heterogeneous density material within the left perinephric space (demarcated by arrowheads). **B.** Axial CT more inferiorly demonstrates a large amount of relatively high- (asterisk) and low-density (arrow) material within the perinephric space. **C.** Coronal image demonstrates normally enhancing kidneys bilaterally, with expansion of the left perinephric space, as shown on axial images. No delayed images were obtained, and findings were interpreted as a large perinephric hematoma. **D.** The patient was reimaged at our facility 4 hours later. If contrast had not been required for other purposes, non-contrast images through the kidneys would have been adequate to evaluate for urinary extravasation. Image obtained at the same level as A demonstrates a large urinoma arising from the renal collecting system (arrow) as well as a hematoma (asterisk) within the perinephric space. **E.** Similar findings are demonstrated on coronal images.

Pseudohydronephrosis

Ken F. Linnau

Imaging description

CT imaging characteristics of hydronephrosis due to renal calculus disease partially overlap with alternate etiologies which may be mistaken for hydronephrosis.

Evaluation of CT images usually starts with side-to-side comparison of both kidneys to evaluate for secondary signs of urinary tract obstruction (Figure 68.1A, 68.2A). Thereby the clinical history is tremendously helpful to correctly identify the affected side.

In pseudohydronephrosis, a portion of the urinary tract is enlarged, or appears dilated, by fluid density material (Figure 68.1). However, on CT asymmetric stranding in the perinephric fat, unilateral ureteric dilation and unilateral renal enlargement of the affected side are usually absent [1].

Renal and ureter stones of clinical relevance are virtually all visible on non-contrast CT [2, 3]. Pseudohydronephrosis will not be associated with an obstructing calculus in the urinary tract.

At times, a very full urinary bladder can prevent unobstructed flow of urine and lead to dilation of the renal collecting system, mimicking hydronephrosis (Figure 68.3).

Importance

Pseudohydronephrosis can be mistaken for urinary tract obstruction from renal stone disease. Renal and ureter stones of clinical relevance are virtually all visible on non-contrast CT [3]. Even on contrast-enhanced CT, urinary tract obstruction can usually be differentiated from pseudohydronephrosis (Figure 68.4).

Typical clinical scenario

Patients with acute onset of flank pain and suspected renal calculi usually undergo CT of the abdomen and pelvis without intravenous contrast. No renal calculus is detected, but a portion of the renal collecting system appears dilated.

Differential diagnosis

- *Parapelvic cysts* (Figure 68.1) [4]: parapelvic cysts (also known as peripelvic cysts) are cysts that are confined to the renal sinus. They are present in up to 1.5% of the population [5], are usually multiple, and probably arise from the lymphatics in the renal sinus. On CT and ultrasound they do not appear to be interconnected, and cannot be traced to the renal calyces, renal pelvis, or other parts of the collecting system. They may be separated from each other by fat or apparent thick septa, and are more spherical, rather than the typical "cauliflower" appearance to a dilated calyx. Occasionally a single large parapelvic cyst may mimic a dilated renal pelvis. Once again, there will be no calyceal dilation, and the cyst cannot be shown to communicate with the renal collecting system.
- *Bladder distention* (Figure 68.3): one series that examined 72 healthy volunteers with distended bladders using ultrasound showed 13.9% had mild and 2.8% had moderate hydronephrosis which disappeared when the bladder was emptied [6].
- *Hydronephrosis of pregnancy*: hydronephrosis occurs secondary to physiologic dilation of the maternal renal collecting system during pregnancy, and is more frequently identified on the right side. Distinguishing between physiologic hydronephrosis of pregnancy, and ureteric obstruction can be difficult. Although it is likely due to a form of ureteric obstruction, it has been found that hydronephrosis of pregnancy is not associated with an elevation of the renal resistive index (RI) [7]. A renal RI of >0.7 during pregnancy should be considered to be an indicator of ureteric obstruction. The rise in RI tends to occur within six hours of acute obstruction [8].
- *Renal sinus varices*: dilated renal sinus vessels can mimic hydronephrosis. Color Doppler imaging, or post-contrast CT will usually easily discriminate.

> **Teaching point**
>
> Pseudohydronephrosis may be due to parapelvic cysts, bladder distension, pregnancy, or renal sinus varices. Parapelvic cysts will demonstrate no macroscopic communication with the renal collecting system. Repeat imaging following bladder emptying should assist in cases of pseudohydronephrosis due to bladder distension.

REFERENCES

1. Dalrymple NC, Casford B, Raiken DP, Elsass KD, Pagan RA. Pearls and pitfalls in the diagnosis of ureterolithiasis with unenhanced helical CT. *Radiographics*. 2000;**20**(2):439–47; quiz 527–8, 532.

2. ACR Appropriateness Criteria: Acute Onset Flank Pain – Suspicion of Stone Disease. *Am Coll Radiol*;2011. Available from: http://www.acr.org/SecondaryMainMenuCategories/quality_safety/app_criteria/pdf/ExpertPanelonUrologicImaging/AcuteOnsetFlankPainSuspicionofStoneDiseaseDoc1.aspx (accessed March 14, 2012).

3. Tublin ME, Murphy ME, Delong DM, Tessler FN, Kliewer MA. Conspicuity of renal calculi at unenhanced CT: effects of calculus composition and size and CT technique. *Radiology*. 2002;**225**(1):91–6.

4. Cronan JJ, Amis ES, Jr., Yoder IC, *et al.* Peripelvic cysts: an impostor of sonographic hydronephrosis. *J Ultrasound Med*. 1982;**1**(6):229–36.

5. Rha SE, Byun JY, Jung SE, *et al.* The renal sinus: pathologic spectrum and multimodality imaging approach. *Radiographics.* 2004;24 Suppl 1: S117–31.

6. Mann MJ. Hydronephrosis secondary to bladder distension. *J DMS.* 1990;**6**(2):87–91.

7. Hertzberg BS, Carroll BA, Bowie JD, *et al.* Doppler US assessment of maternal kidneys: analysis of intrarenal resistivity indexes in normal pregnancy and physiologic pelvicaliectasis. *Radiology.* 1993; **186**(3):689–92.

8. Patel SJ, Reede DL, Katz DS, Subramaniam R, Amorosa JK. Imaging the pregnant patient for nonobstetric conditions: algorithms and radiation dose considerations. *Radiographics.* 2007; **27**(6):1705–22.

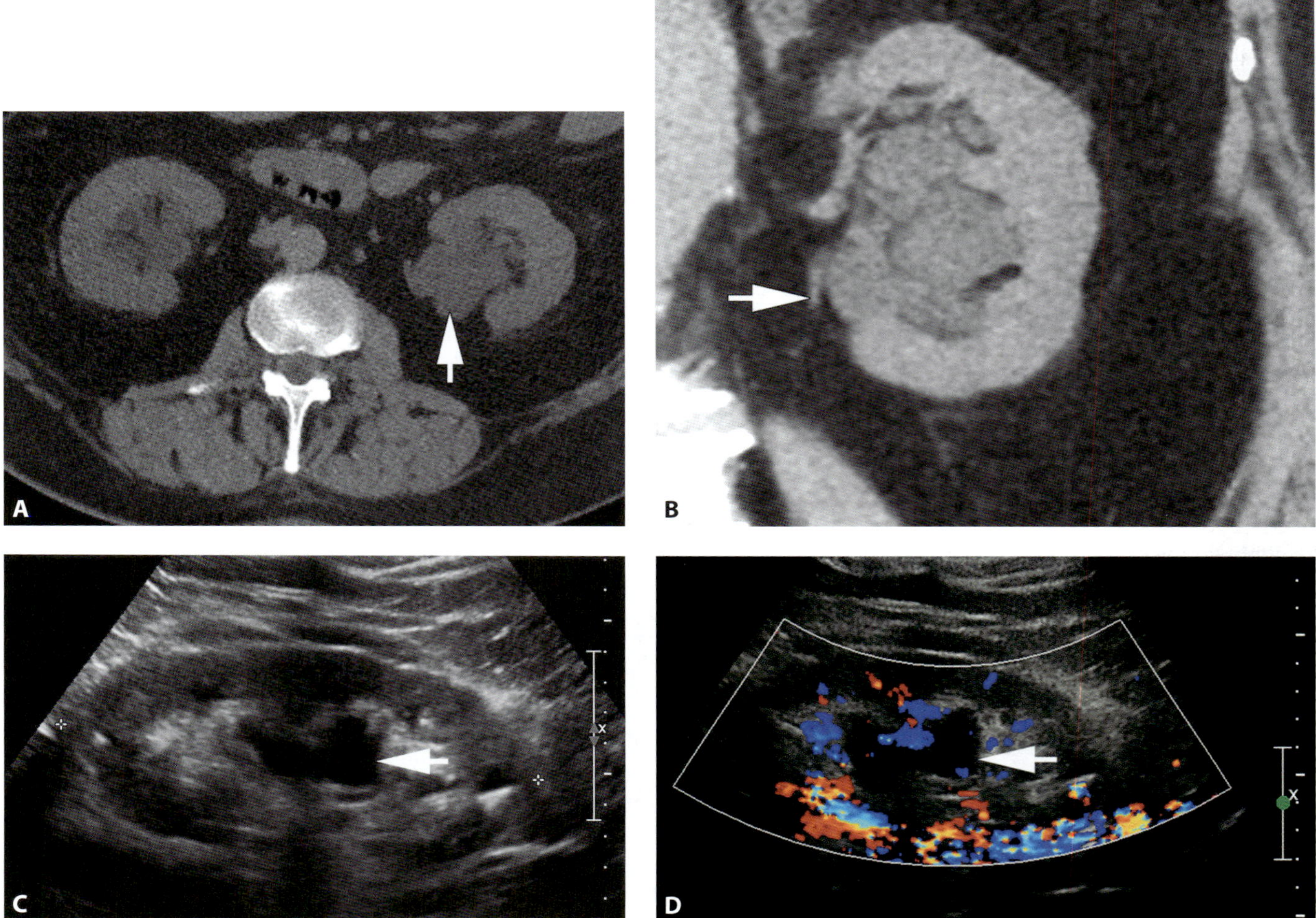

Figure 68.1 A. Axial non-contrast CT of the kidneys in a 61-year-old woman with acute left flank pain and a history of kidney stones shows asymmetric fluid-density attenuation in the left renal pelvis (arrow), suggestive of left hydronephrosis. **B.** Coronal reformation shows that the left ureter is not dilated (arrow). Note how the cysts appear to be rounded and separated by a thin layer of fat. No calculus was identified. **C.** Long axis gray-scale ultrasound image of the left kidney shows the echo-free structures with through-transmission that do not communicate with the collecting system (arrow), confirming the diagnosis of a parapelvic cyst and excluding hydronephrosis. **D.** The absence of flow on color Dopper ultrasound within the cyst (white arrow) excludes a vascular etiology, such as a varix.

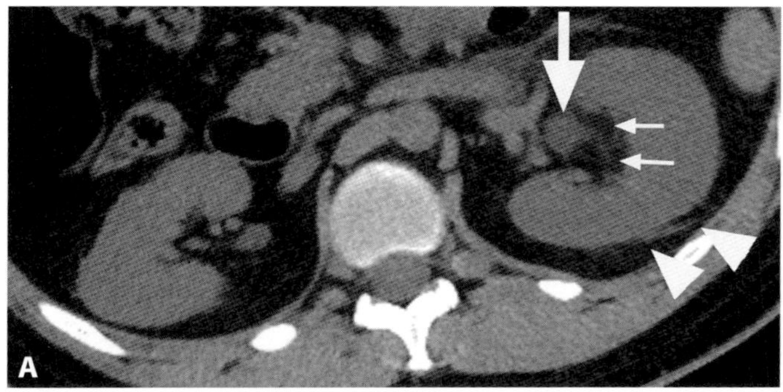

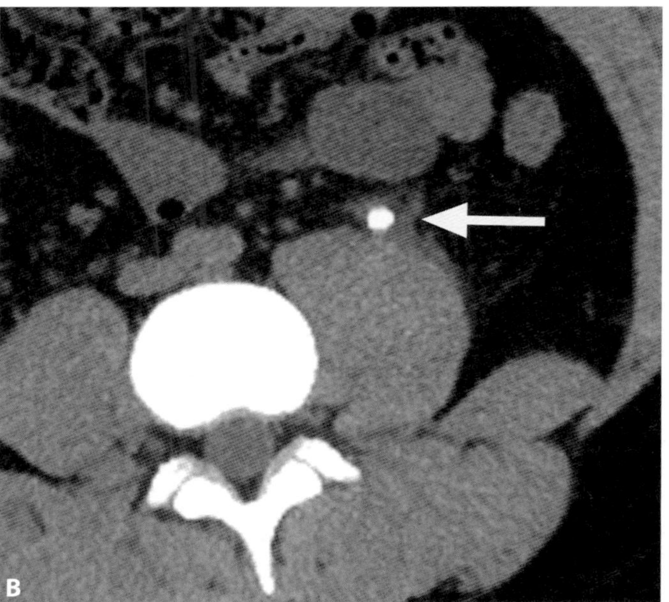

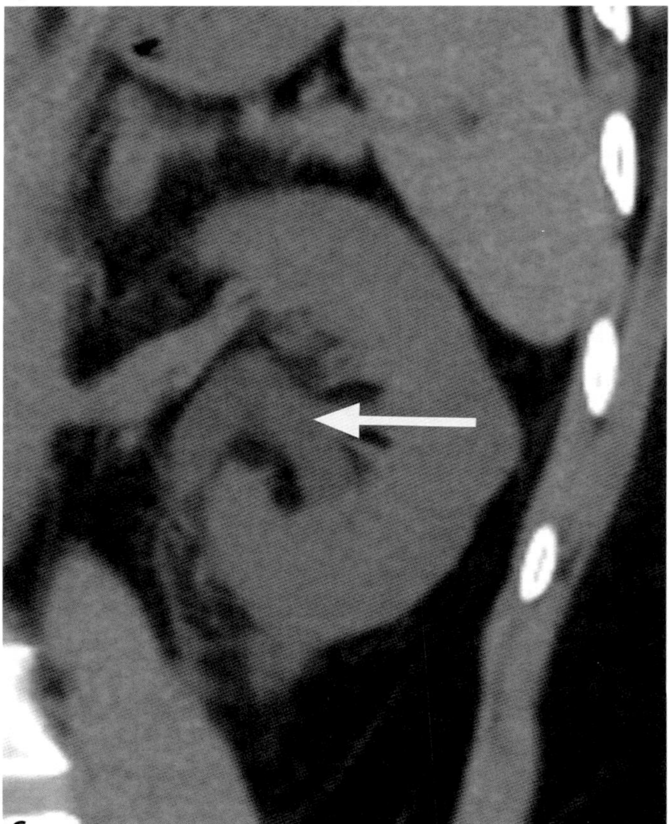

Figure 68.2 A. Axial non-contrast CT from a 26-year-old man with acute onset of left flank pain and microhematuria shows fluid attenuation in a mildly dilated left renal pelvis (large white arrow) and secondary signs of acute urinary tract obstruction: an enlarged left kidney with stranding of the fat in the renal sinus (small white arrows) and perinephric region (arrowheads). Secondary signs of obstruction are absent on the normal right side. **B.** Axial non-contrast CT image at the level of the mid-ureter shows a 6mm left ureteric calculus (arrow) with surrounding fat stranding, confirming mild true hydronephrosis. **C.** Coronal reformation of the left kidneys shows hydronephrosis with dilation of the lower pole calyx (arrow) with some perinephric and peri-ureteric stranding.

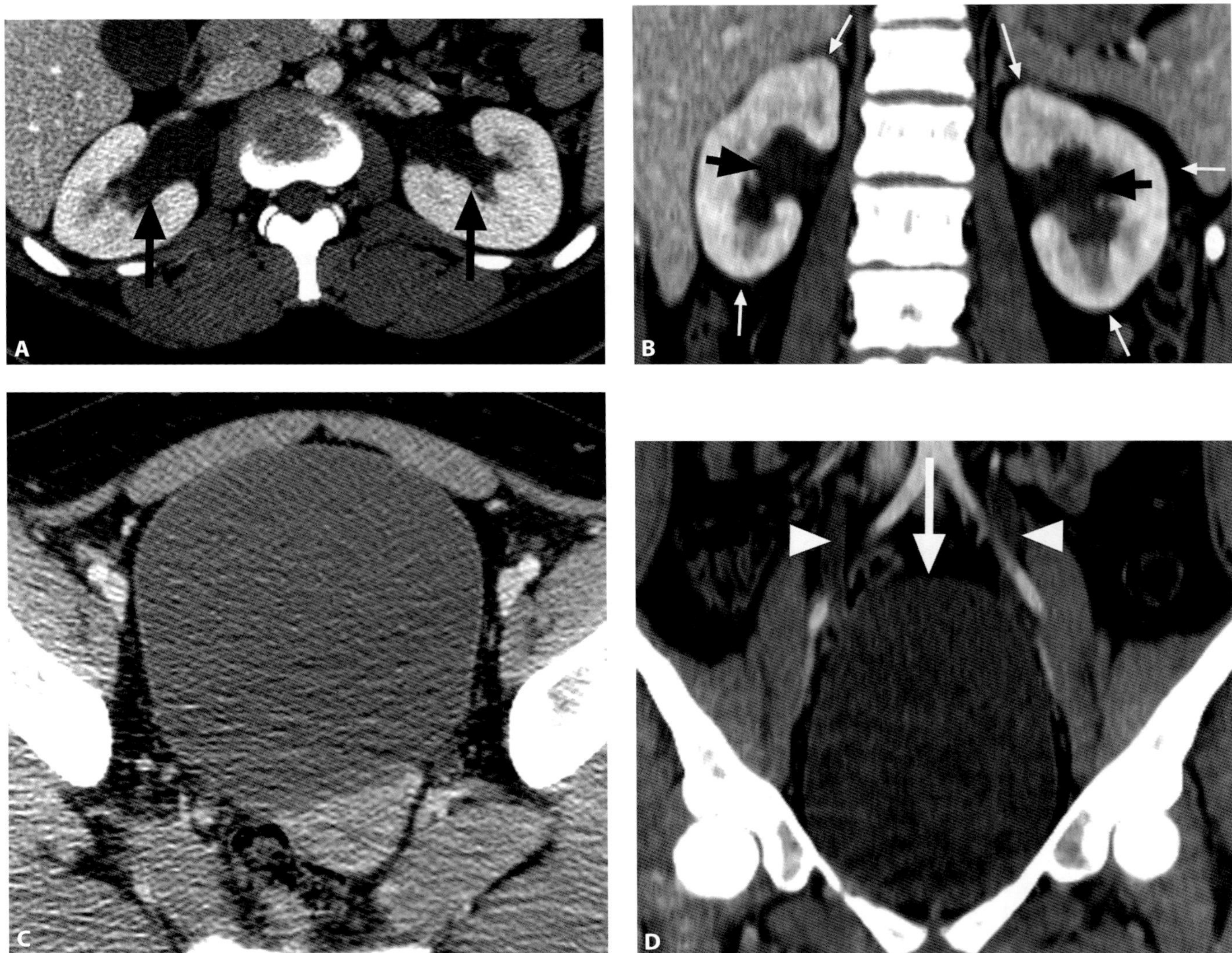

Figure 68.3 A. Axial contrast-enhanced CT image of the kidneys demonstrates bilateral symmetric dilation of the renal pelvis and effacement of the renal infundibula (arrows) in a 19-year-old woman after a motor vehicle collsion, which could be mistaken for hydronephrosis. **B.** Coronal image confirms bilateral symmetric dilation of the renal collecting system (black arrows). The kidneys are symmetric and normal in size and there is no perinephric fluid (white arrows). **C.** Axial CT image at the level of the urinary bladder shows gross distension, in keeping with pseudohydronephrosis. **D.** Coronal CT image shows the urinary bladder extending beyond the iliac crests (arrow) for a urine volume of about 1000 mL. The distal ureters are bilaterally dilated without obstucting stones (arrow heads), confirming pseudohydronephrosis, which resolved after the bladder was emptied.

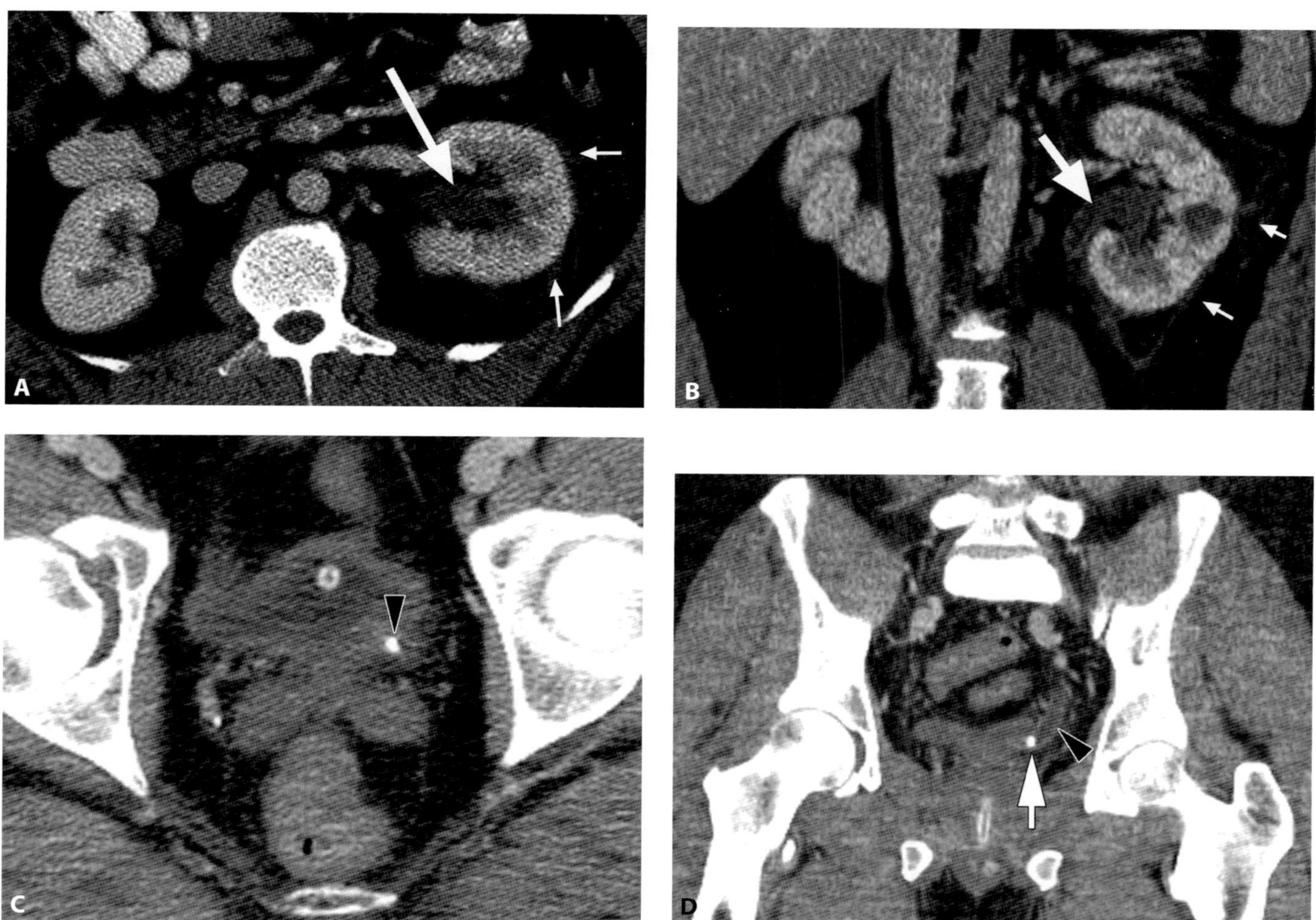

Figure 68.4 Axial (**A**) and coronal (**B**) contrast-enhanced CT images from a 43-year-olc man with colicky left lower quadrant pain show signs of acute renal collecting system obstruction including an enlarged left kidney with mild to moderate hydronephrosis (large arrow) and perinephric fat stranding (small arrows). **C.** Despite the administration of intravenous contrast, an obstructing calculus at the ureterovesical junction (UVJ) (arrowhead) is readily appreciated on axial contrast-enhanced CT in this case. **D.** Left hydroureter (arrowhead) is present proximal to the left UVJ calculus (arrow).

Physiologic pelvic intraperitoneal fluid

Martin L. Gunn

Imaging description

Given the frequency that CT of the abdomen and pelvis is performed to evaluate the trauma patient, intraperitoneal "free" fluid is encountered quite often. Although a small quantity of isolated free intraperitoneal fluid is usually attributed to physiologic fluid in women of reproductive age, it is also seen in males.

Whilst in the past it was believed that free fluid identified on CT was an indicator of occult bowel, mesenteric, or solid organ injury, and it mandated laparotomy [1], it is now generally accepted that conservative management with a non-operative approach for hemodynamically stable patients who have a small quantity of free pelvic fluid is safe (Figure 69.1). These patients should have no clinical or imaging signs of bowel injury [2, 3]. The change in approach to small quantities of free pelvic fluid is likely related to the evolution of CT itself. Decreasing reconstruction slice thickness, and the introduction of routine multiplanar reformations likely increases the sensitivity of CT for the detection of small quantities of free intraperitoneal fluid [4].

Clues that the free fluid will not require surgery include a cranial extent below the third sacral body, low density (< 15 Hounsfield Units [HU]), and small volume (i.e., $\leq$ five 5 mm thick axial slices) (Figure 69.2) [2, 5]. However, isolated intraperitoneal fluid located outside the deep pelvis, such as between mesenteric leaves or paracolic gutters, should be treated with suspicion [3–6].

Importance

The importance of free intraperitoneal fluid is twofold. Firstly, it is critical not to underestimate the fact that free fluid may be the only marker of bowel or mesenteric injury. Many studies emphasize that free intraperitoneal fluid is the most common finding of bowel or mesenteric injury on CT [6]. Conversely, the absence of any free intraperitoneal fluid almost excludes the presence of bowel or mesenteric injury [7]. Secondly, the importance of a small quantity of isolated free pelvic fluid on high-quality multidetector CT (MDCT) should not prompt further unnecessary investigations or laparotomy following a careful search for a bowel, mesenteric, or small bowel injury.

The most common traumatic injury missed by CT in the abdomen and pelvis remains bowel injury [8]. Avoiding delays in diagnosis of bowel or mesenteric injuries is important, as they can lead to significant morbidity and mortality from peritonitis, sepsis, and hemorrhage [9].

Typical clinical scenario

A small quantity of free intraperitoneal fluid is identified in approximately 3–5% of trauma CT scans [3–5].

Differential diagnosis

The potential causes of intraperitoneal fluid include physiologic fluid from follicular rupture in women of reproductive age; fluid related to shock and/or rapid intravascular volume resuscitation; occult bladder, bowel, mesenteric, solid organ, or vascular injury. The most common iatrogenic cause is diagnostic peritoneal lavage.

Teaching point

On MDCT, the presence of a small quantity of isolated low-density intraperitoneal pelvic fluid should prompt a careful evaluation of the abdomen and pelvis for sites of bleeding, bowel, and mesenteric injury. If the patient does not have clinical signs of bowel injury, conservative management (which may include observation) appears safe.

REFERENCES

1. Cunningham MA, Tyroch AH, Kaups KL, Davis JW. Does free fluid on abdominal computed tomographic scan after blunt trauma require laparotomy? *J Trauma.* 1998;**44**(4):599–602; discussion 603.
2. Brofman N, Atri M, Hanson JM, *et al.* Evaluation of bowel and mesenteric blunt trauma with multidetector CT. *Radiographics.* 2006; **26**(4):1119–31.
3. Rodriguez C, Barone JE, Wilbanks TO, Rha CK, Miller K. Isolated free fluid on computed tomographic scan in blunt abdominal trauma: a systematic review of incidence and management. *J Trauma.* 2002; **53**(1):79–85.
4. Yu J, Fulcher AS, Wang DB, *et al.* Frequency and importance of small amount of isolated pelvic free fluid detected with multidetector CT in male patients with blunt trauma. *Radiology.* 2010; **256**(3):799–805.
5. Drasin TE, Anderson SW, Asandra A, Rhea JT, Soto JA. MDCT evaluation of blunt abdominal trauma: clinical significance of free intraperitoneal fluid in males with absence of identifiable injury. *AJR Am J Roentgenol.* 2008;**191**(6):1821–6.
6. LeBedis CA, Anderson SW, Soto JA. CT imaging of blunt traumatic bowel and mesenteric injuries. *Radiol Clin North Am.* 2012; **50**(1):123–36.
7. Atri M, Hanson JM, Grinblat L, *et al.* Surgically important bowel and/or mesenteric injury in blunt trauma: accuracy of multidetector CT for evaluation. *Radiology.* 2008;**249**(2):524–33.
8. Lawson CM, Daley BJ, Ormsby CB, Enderson B. Missed injuries in the era of the trauma scan. *J Trauma.* 2011;**70**(2):452–6; discussion 456–8.
9. Killeen KL, Shanmuganathan K, Poletti PA, Cooper C, Mirvis SE. Helical computed tomography of bowel and mesenteric injuries. *J Trauma.* 2001;**51**(1):26–36.

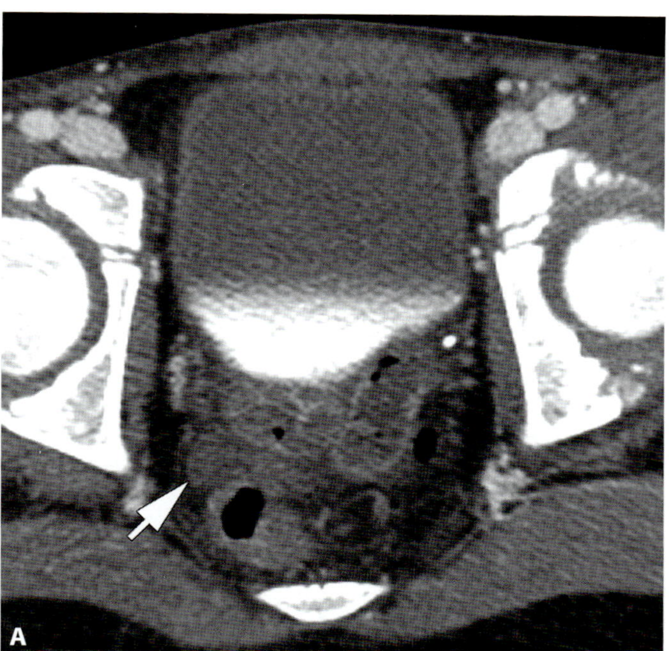

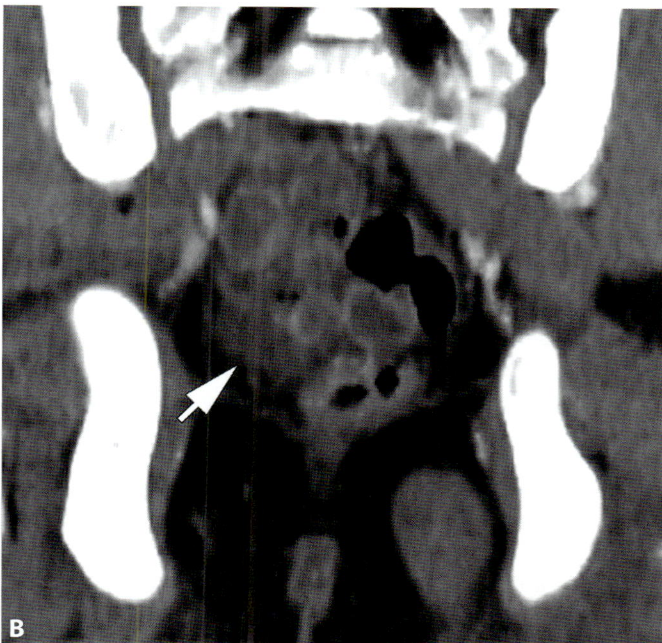

Figure 69.1 Axial (**A**) and coronal (**B**) contrast-enhanced CT of the pelvis in an 8-year-old boy, who was involved in a bicycle accident, demonstrates a trace of low-density intraperitoneal fluid (white arrows). The remainder of the abdomen and pelvis were normal, and no intervention was necessary. The boy made a full recovery.

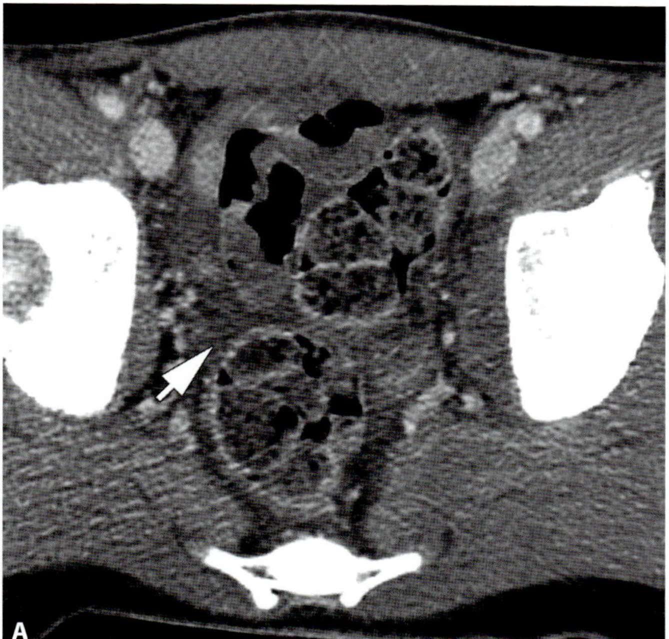

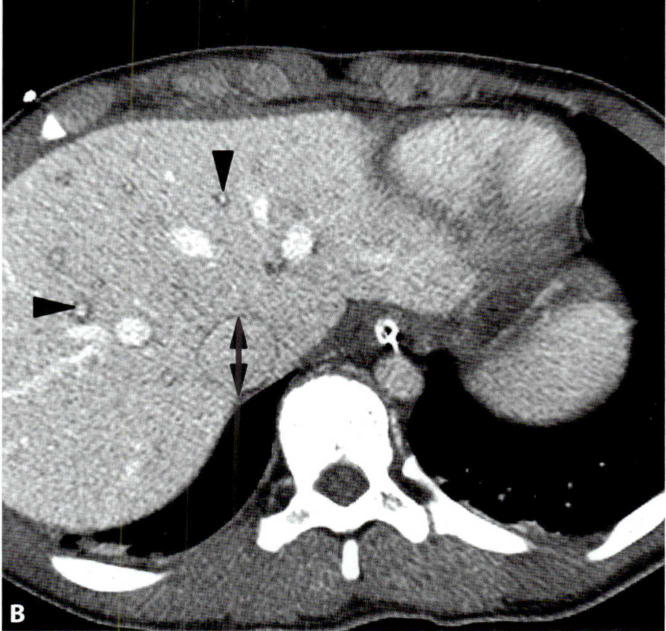

Figure 69.2 **A.** Axial view of a contrast-enhanced CT of the abdomen and pelvis in a 13-year-old male pedestrian who suffered severe head injuries, and was intubated and resuscitated by ambulance officers at the roadside, shows a trace of low-density (5 HU) intraperitoneal fluid anterior to the rectum (arrow). **B.** Axial image at the level of the liver demonstrates periportal edema (arrowheads) and distension of the inferior vena cava (double-arrow) consistent with vigorous fluid resuscitation. No abdominal or pelvic injuries were identified, and there were no abdominal complications.

Avoiding missed injuries to the bowel and mesentery: the importance of intraperitoneal fluid

Martin L. Gunn

Imaging description

All intraperitoneal fluid identified by a CT of the abdomen and pelvis in a trauma patient should prompt a careful re-examination of the bowel, mesentery, bladder, and solid organs to identify the potential cause.

Fluid that is located outside the deep pelvis should be treated with great suspicion. Traumatic injury to the solid organs usually leads to accumulation of blood in Morison's pouch, the paracolic gutters, or the subphrenic spaces. Moreover, intraperitoneal blood originating from lacerations of solid organs tends to be of higher density ("sentinel clot") than physiologic fluid or fluid resulting from bladder or biliary injuries. Free intraperitoneal fluid arising from solid organ lacerations and bowel injuries usually has a density of greater than 30–40 Hounsfield Units (HU) (Figure 70.1), whereas the density of physiologic fluid tends to be lower, averaging 15 HU or less.

Free intraperitoneal fluid from bowel or mesenteric injuries tends to lie between the leaves of the small bowel mesentery, and may form a triangular shape (Figure 70.2). Other signs of bowel injury should be sought (Table 70.1) whenever this configuration of fluid is identified.

If the density of free intraperitoneal fluid is low but the volume of fluid is sufficient to extend beyond the deep pelvis, consideration should also be given to intraperitoneal bladder rupture and biliary injury.

In the setting of penetrating abdominal trauma, significant bowel injuries can occur in the absence of a significant quantity of free intraperitoneal fluid. At our institution, we perform CT of the abdomen and pelvis with oral, rectal, and intravenous contrast, and perform portal venous and delayed phase CT whenever there has been penetrating abdominal trauma with potential violation of the peritoneum, retroperitoneum, or bowel.

Although there are CT signs that are highly specific for bowel or mesenteric injuries (Table 70.1), these are present in such a small percentage of patients who have bowel injuries that the absence of these signs should not be especially reassuring when more than a small volume of low-density isolated pelvic fluid is encountered. Moreover, false-positive diagnoses of bowel perforation can occur when pneumoperitoneum is encountered in the setting of penetrating abdominal trauma (where gas is introduced at the time of peritoneal violation), intraperitoneal bladder rupture with urinary catherization, pseudopneumoperitoneum (see Case 51), and following diagnostic peritoneal lavage.

In a retrospective study that examined the preoperative CT with oral and intravenous contrast performed on 54 consecutive patients who had surgically confirmed bowel or mesenteric injury, the following signs of bowel or mesenteric injuries were identified [1]:

Table 70.1. CT signs of bowel and mesenteric injury

CT Signs of blunt bowel injury [1]	Frequency
DEFINITIVE SIGNS	
Visible bowel wall discontinuity	7%
Extraluminal oral contrast (used for penetrating injury)	6%
Pneumoperitoneum or pneumoretroperitoneum	20%
SUGGESTIVE SIGNS	
Bowel wall thickening	55%
Decreased bowel wall enhancement	13%
CT Signs of mesenteric injury	**Frequency**
DEFINITIVE SIGNS	
Mesenteric active vascular extravasation	17%
Mesenteric vascular beading	39%
Termination of mesenteric vessels	35%
SUGGESTIVE SIGNS	
Mesenteric infiltration	69%
Mesenteric hematoma	39%
Bowel wall thickening	49%
Common signs of bowel and mesenteric injuries	**Frequency**
Intraperitoneal fluid	93%
Abdominal wall injury (hematoma, "seat-belt sign")	17%

Importance

Discriminating pathologic intraperitoneal fluid from a trace of "physiologic" pelvic fluid is important, and should have a direct impact on management. Unfortunately, the most common traumatic injury in the abdomen and pelvis missed by CT remains bowel injury [2]. Avoiding delays in diagnosis of bowel or mesenteric injuries is critical, as delayed repair can lead to significant morbidity and mortality from peritonitis, sepsis, and hemorrhage [3].

Typical clinical scenario

Although they can lead to significant morbidity, blunt bowel and mesenteric injuries are quite uncommon. They are identified in approximately 5% of cases of intra-abdominal injury requiring surgery [4]. While patients with bowel injuries usually have abdominal pain and tenderness, it may take several hours before they develop clear clinical signs of peritonitis. In patients with head injuries or distracting injuries, the diagnosis may remain clinically ambiguous for days. Consequently, detecting an abnormal location, quantity, or density of free intraperitoneal fluid, and considering bowel, mesenteric, bladder, or biliary injury early in the diagnostic workup is imperative.

Differential diagnosis

The differential diagnosis of unexplained moderate or larger quantities of intraperitoneal fluid includes occult injuries to the solid viscera, bowel, mesentery, biliary system, and vasculature; intraperitoneal bladder rupture (Figure 70.3); and diagnostic peritoneal lavage. Small quantities of low-density "physiologic" fluid may also be encountered, as discussed in Case 69.

Teaching point

Intraperitoneal fluid that extends beyond the deep pelvis, has a HU density of >40, or lies within the leaves of the mesentery or paracolic gutters should be considered suspicious for an occult intra-abdominal injury, especially an injury to the bowel, mesentery, or bladder.

REFERENCES

1. Brofman N, Atri M, Hanson JM, *et al.* Evaluation of bowel and mesenteric blunt trauma with multidetector CT. *Radiographics.* 2006; **26**(4):1119–31.
2. Lawson CM, Daley BJ, Ormsby CB, Enderson B. Missed injuries in the era of the trauma scan. *J Trauma.* 2011;**70**(2):452–6; discussion 456–8.
3. Killeen KL, Shanmuganathan K, Poletti PA, Cooper C, Mirvis SE. Helical computed tomography of bowel and mesenteric injuries. *J Trauma.* 2001;**51**(1):26–36.
4. LeBedis CA, Anderson SW, Soto JA. CT imaging of blunt traumatic bowel and mesenteric injuries. *Radid Clin North Am.* 2012;**50**(1):123–36.

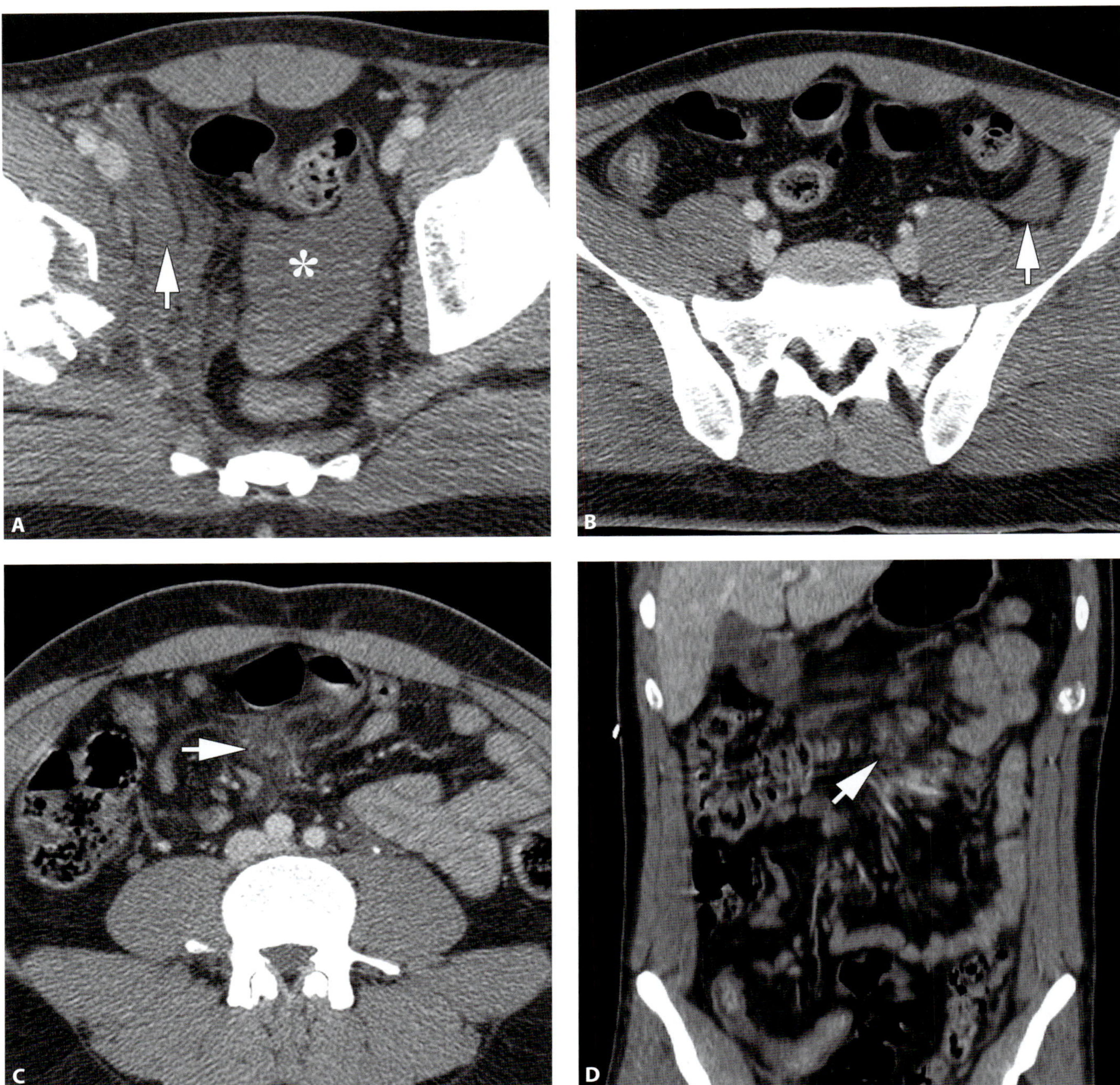

Figure 70.1 Contrast-enhanced CT of the abdomen and pelvis in a 27-year-old male who was involved in a high-speed motor vehicle collision. **A.** Axial image through the pelvis demonstrates high-density fluid (asterisk) with some layering debris. The density was 40 HU. Also note the extraperitoneal pelvic fluid (arrow) adjacent to the right acetabular fracture. **B.** Note that the fluid is not just in the deep pelvis, but also in the left paracolic gutter (arrow). **C.** Axial image from the mid-abdomen demonstrates a contusion in the small bowel mesentery (arrow), which was also visible on a coronal image (**D**). No solid organ injury was identified on the CT. The patient was taken to the operating room, and a mesenteric injury just distal to the ligament of Treitz was identified and repaired. A small perforation that required a resection was also found in the distal jejunum.

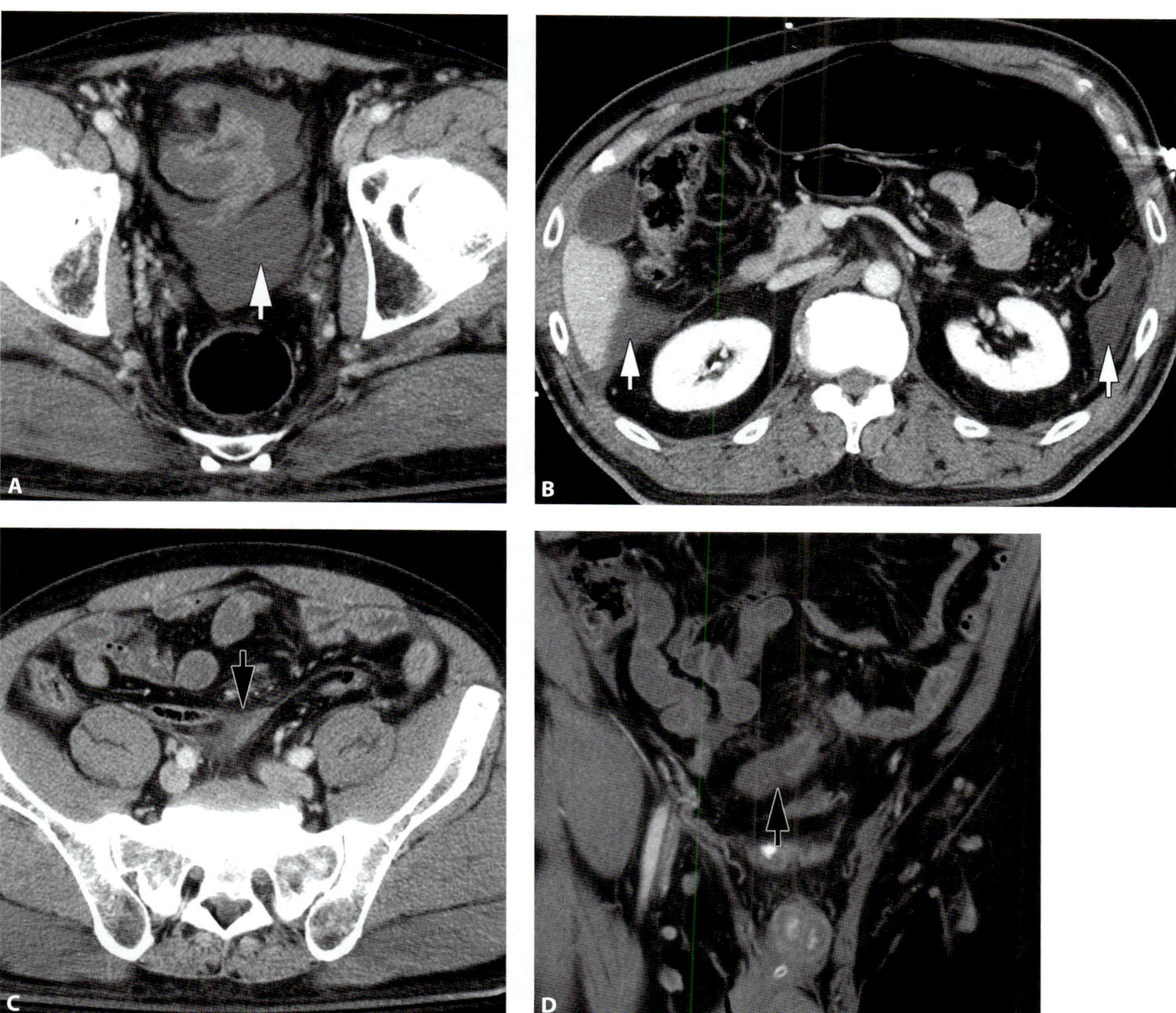

Figure 70.2 Contrast-enhanced CT of the abdomen and pelvis in a 65-year-old man who was involved in a high-speed motor vehicle collision. **A.** Axial image of the pelvis demonstrates a large quantity of intraperitoneal pelvic fluid (arrow) with a density of 15 HU. **B.** Intraperitoneal fluid was also present within the upper abdomen (white arrow), and within the leaves of the distal small bowel mesentery (black arrow) (**C**). Close inspection of a coronal image (**D**) revealed a short section of ileum that was devascularized (arrow), highly suspicious for bowel injury. At surgery, two small bowel injuries were discovered, including a perforation in the proximal ileum that required resection.

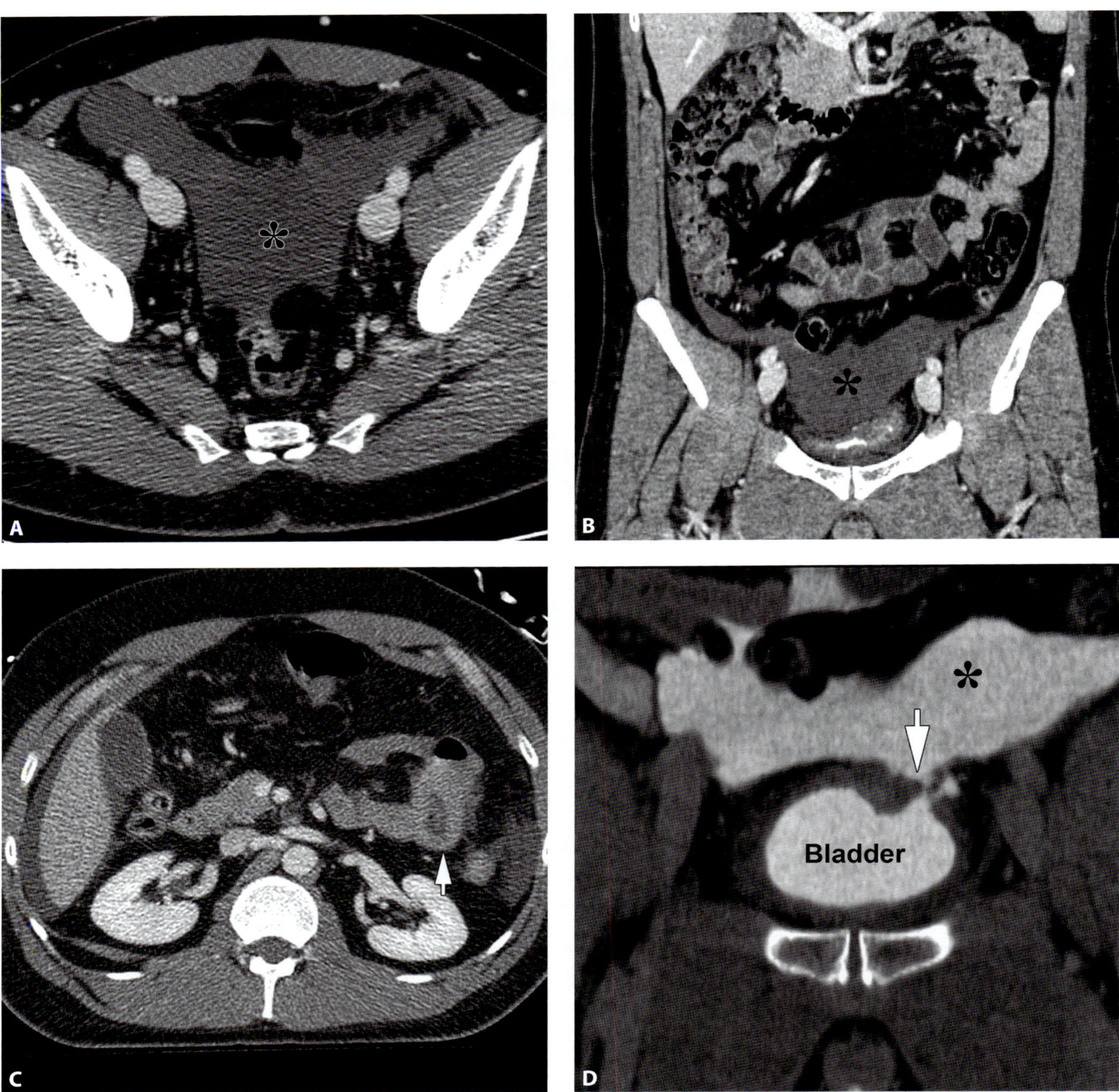

Figure 70.3 Axial (**A**) and coronal (**B**) contrast-enhanced CT from a 32-year-old man who was involved in a high-speed motor vehicle collision and transferred to our hospital demonstrates a large volume of unexplained low-density (HU = 7) intraperitoneal fluid in the pelvis (asterisk). No solid organ injury or pelvic fracture was identified. A prior CT performed at an outside institution (**C**) suggested jejunal thickening (arrow), but this was thought to be within the spectrum of normal by our radiologist in the Emergency Department. A CT cystogram was recommended (**D**). Coronal image from the CT cystogram demonstrates a defect in the dome of the bladder (arrow) with intravesical contrast ("Bladder") flowing into the peritoneal space (asterisk), indicating an intraperitoneal bladder rupture.

CASE **71**

Endometrial hypodensity simulating fluid

Martin L. Gunn

Imaging description

Although the imaging reference standard for the assessment of the endometrium is transvaginal sonography [1], CT is increasingly performed in patients for non-gynecologic reasons, and may reveal several endometrial abnormalities, including endometrial thickening, endometrial fluid, distortion of the endometrium by uterine masses, and intrauterine pregnancy.

In both premenopausal and postmenopausal women, the normal endometrium is hypodense compared to the enhancing myometrium, qualitatively resembling fluid ("pseudofluid"), but density measurements are often significantly greater than water [2]. The myometrium enhances rapidly in both the arterial and venous phases, in comparison to the endometrium, which enhances much more slowly and less vigorously (Figure 71.1) [3, 4]. Four types of normal myometrial enhancement have been described, with subendometrial uterine enhancement predominating in the premenopausal age group [3, 4]. The cervix also often demonstrates delayed and reduced enhancement in comparison with the body of the uterus and may have a low-attenuation appearance, simulating cervical carcinoma [4].

One should not diagnose endometrial fluid on CT based solely on the presence of low-attenuation endometrium in comparison with the adjacent myometrium (Figures 71.1 and 71.2). Unfortunately, neither the abnormal nor normal endometrial appearances have been extensively studied using CT. In one study of 54 postmenopausal women without known endometrial disease who underwent CT, the endometrium was visible in 26 women (48%), and in those it had a mean attenuation of 63 Hounsfield Units (HU) on 5–7 mm axial CT reconstructions, and an average short axis thickness of 7.5 mm [2]. Due to the poor quality of sagittal reformats available at the time (1999), and the axial reconstruction slice thickness, the endometrial thickness measurement was not precise, and the endometrial attenuation was affected by partial volume averaging artifact. Moreover, the thickness of central uterine hypodensity visible on CT in postmenopausal women, due to lack of central uterine and endometrial contrast enhancement, likely does not correlate with the endometrial "stripe" visible on sonography [2].

Grossman and others examined the endometrium on CT in pre- and postmenopausal woman and compared it to transvaginal sonography [1]. They noted that the outer wall of the endometrium is not usually sharply demarcated on CT, and found that CT overall has a low sensitivity for the detection of an abnormally thickened endometrium, only 53% in pre- and postmenopausal women. However, they found that when the endometrium was thickened on CT, it was thick in two-thirds of cases on sonography, and when there was gross thickening demonstrated on CT, this finding was also present on ultrasound.

In cases where the uterus is ante- or retroverted, axial images traverse the endometrium in a plane coronal to the uterine body, and this may lead to overestimation of endometrial measurements [1]. A clue to this is the identification of a triangular endometrial shape on axial CT images (Figure 71.3). Sagittal or curved planar reformats of the endometrium are likely to be the best way to assess endometrial contour and thickness on CT (Figure 71.4) [3].

Early gestational sacs can also be identified on CT (Figure 71.5). Usually these appear as fluid-filled cyst-like structures in the endometrium [5]. There may be surrounding ring enhancement. Fetal structures are usually not visible until late in the first trimester. A corpus luteum (which is a small [≤3 cm] cystic ovarian lesion that often has vigorous peripheral contrast enhancement) within either ovary is supportive, but not diagnostic, of a pregnancy.

Importance

Identification of a thickened endometrium on CT, rather than endometrial fluid, should prompt ultrasonography in postmenopausal patients. Fluid in the endometrial cavity is an unusual finding on transvaginal sonography in non-menstruating women, and consideration should be given to endometrial thickening when an apparently thickened, hypodense endometrium is identified.

Typical clinical scenario

Endometrial thickening, "pseudofluid," and intrauterine gestational sacs are usually encountered incidentally on abdominal CT scans performed for other indications.

Differential diagnosis

The differential diagnosis of endometrial hypodensity includes fluid, physiologic endometrial enlargement, decidualization due to pregnancy, intrauterine pregnancy, endometrial polyp, hyperplasia or carcinoma, and a degenerating submucosal fibroid.

Teaching point

Whenever endometrial hypodensity is identified on CT, consider endometrial thickening or intrauterine pregnancy, and measure the thickness using multiplanar reformations. In post-menopausal women with endometrial thickening (>5 mm), correlate with transvaginal sonography.

REFERENCES

1. Grossman J, Ricci ZJ, Rozenblit A, *et al.* Efficacy of contrast-enhanced CT in assessing the endometrium. *AJR Am J Roentgenol.* 2008;**191**(3):664–9.

2. Lim PS, Nazarian LN, Wechsler RJ, Kurtz AB, Parker L. The endometrium on routine contrast-enhanced CT in asymptomatic postmenopausal women: avoiding errors in interpretation. *Clin Imaging.* 2002;**26**(5):325–9.

3. Yitta S, Hecht EM, Mausner EV, Bennett GL. Normal or abnormal? Demystifying uterine and cervical contrast enhancement at multidetector CT. *Radiographics.* 2011;**31**(3):647–61.

4. Kaur H, Loyer EM, Minami M, Charnsangavej C. Patterns of uterine enhancement with helical CT. *Eur J Radiol.* 1998;**28**(3):250–5.

5. Shin DS, Poder L, Courtier J, *et al.* CT and MRI of early intrauterine pregnancy. *AJR Am J Roentgenol.* 2011;**196**(2):325–30.

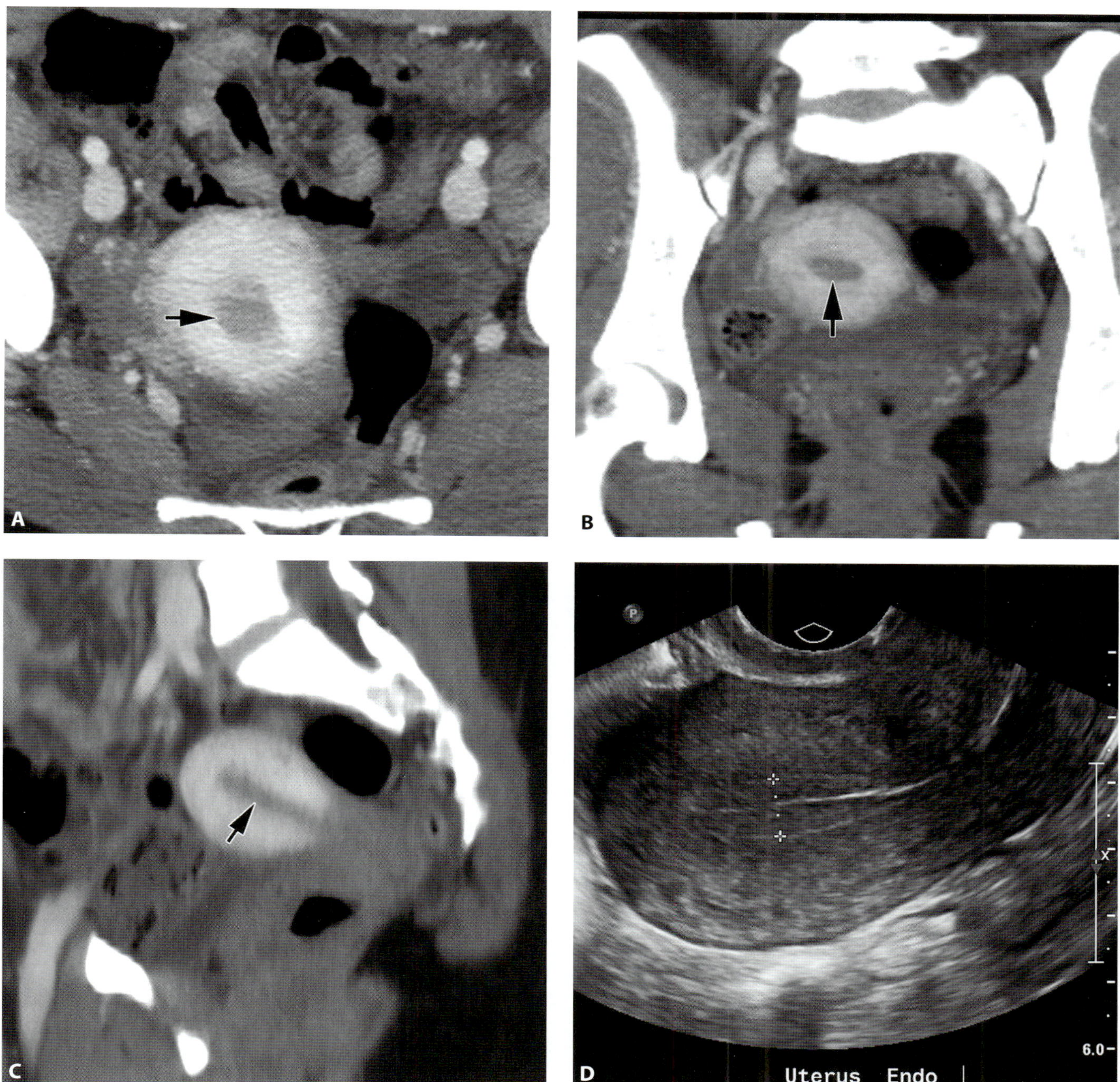

Figure 71.1 Normal endometrial enhancement described as endometrial fluid by the interpreting radiologist. Contrast-enhanced CT of the pelvis in a 28-year-old woman with right lower quadrant pain who was being evaluated for acute appendicitis. Axial (**A**), coronal (**B**), and double-oblique sagittal (**C**) demonstrate a hypodense endometrium (arrows) that measures 9 mm (normal). Note the clearly demarcated relative hypo-enhancement of the uterine cervix. This is a normal phenomenon, likely related to increased fibrous tissue in the cervical stroma. **D.** Sagittal image from a transvaginal sonogram performed 45 minutes later reveals a normal-appearing endometrium measuring 8 mm without evidence of endometrial fluid.

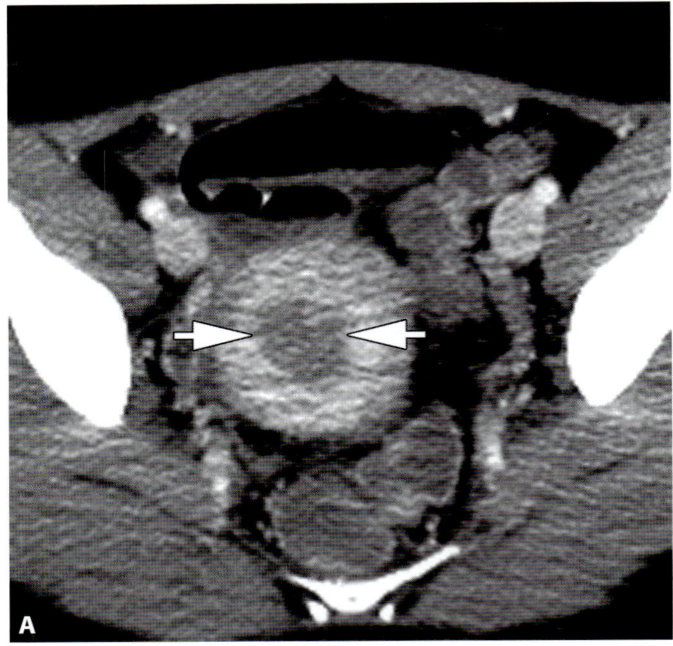

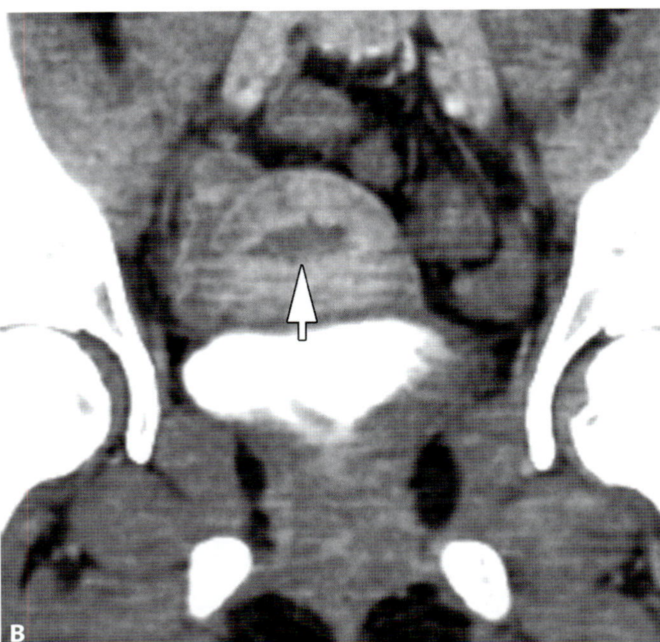

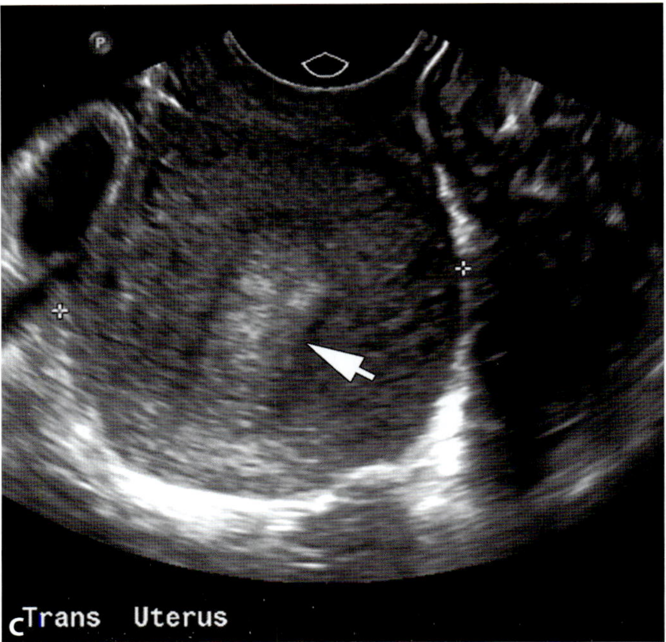

Figure 71.2 Normal endometrium in a 19-year-old woman with right lower quadrant pain. **A.** Axial contrast-enhanced CT of the pelvis. Note that the density of the endometrium (arrows) appears qualitatively only slightly denser than the fluid in the rectum, a phenomenon accentuated by the myometrial enhancement. The endometrial density measures 63 HU. Due to the anteversion of the uterus into the axial plane, the endometrium may appear thick. **B.** However, a coronal reformation shows that there is no thickening (arrow). **C.** Transvaginal sonogram demonstrates that the endometrium (arrow) is echogenic and of normal thickness (8 mm), and that the endometrial cavity does not contain fluid.

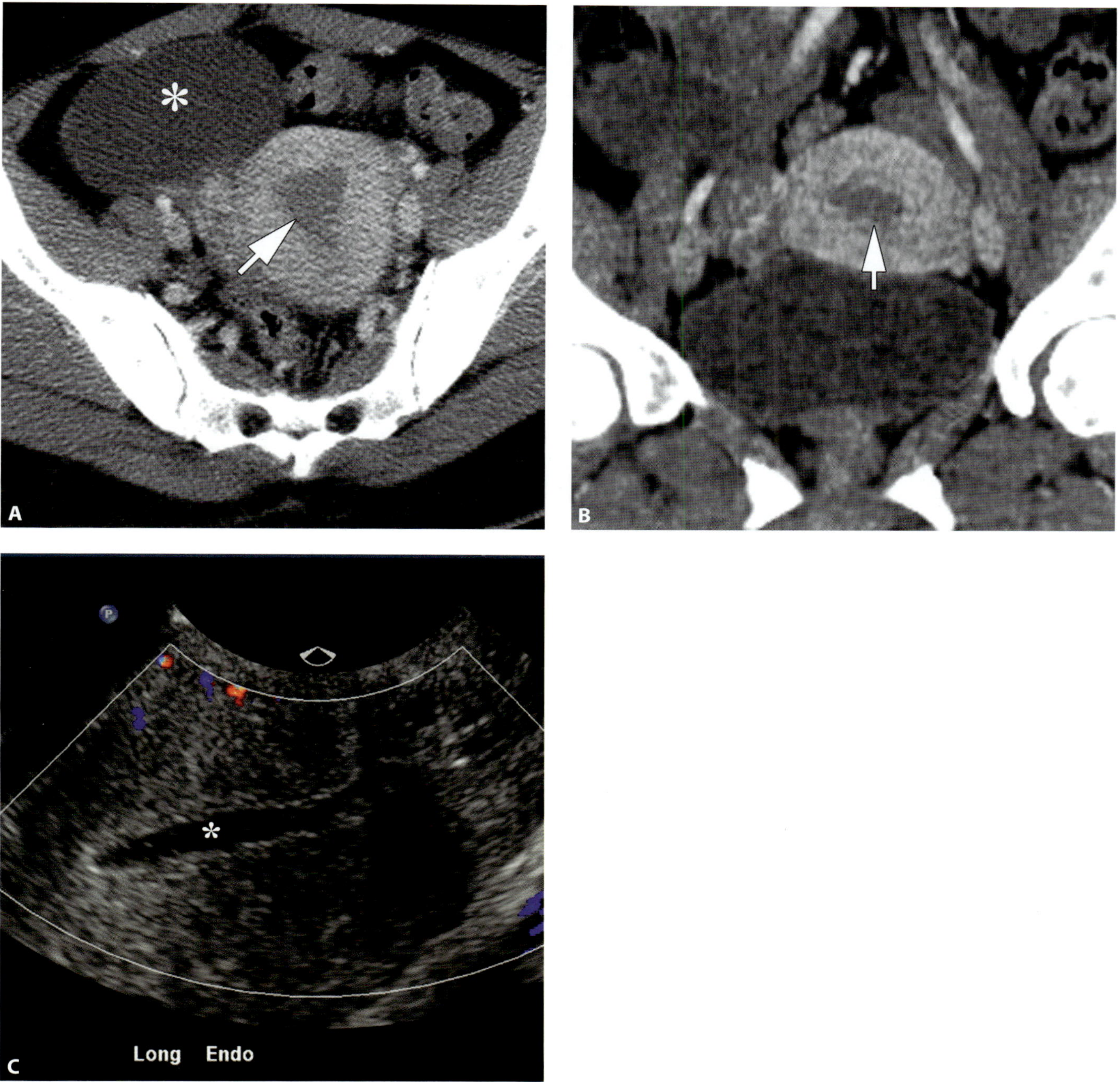

Figure 71.3 Axial (**A**) and (**B**) coronal contrast-enhanced CT of the pelvis in a 35-year-old woman with right lower quadrant pain shows hypodensity in a normal triangular-shaped endometrial cavity (arrow). Often the endometrium will appear triangular when the uterus is ante- or retroverted. The endometrial density measured 68 HU. The patient also had a large right ovarian cyst (asterisk). **C.** Sagittal transvaginal sonogram of the uterus performed approximately 90 minutes later shows fluid in the endometrial canal (asterisk) without endometrial thickening.

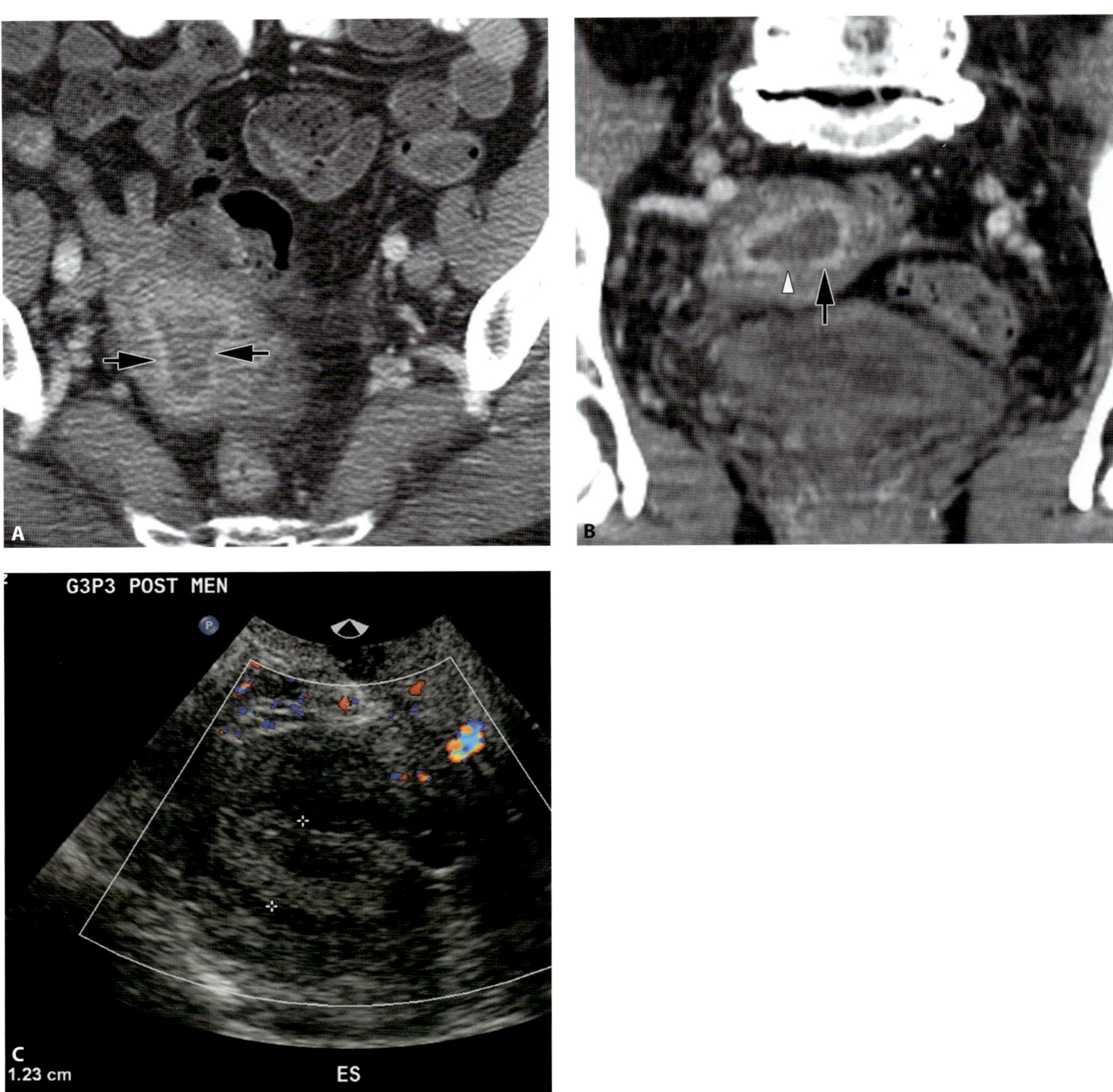

Figure 71.4 Axial (A) and (B) coronal contrast-enhanced CT of the pelvis in a 65-year-old post-menopausal woman with bloating and increasing abdominal girth demonstrates endometrial thickening (13 mm on the coronal image), and subendometrial enhancement (arrows). Note that the subendometrial enhancement is interrupted (white arrowhead). C. Transvaginal sonogram confirms thickening of the endometrium (12 mm). Endometrial biopsy revealed endometrial carcinoma.

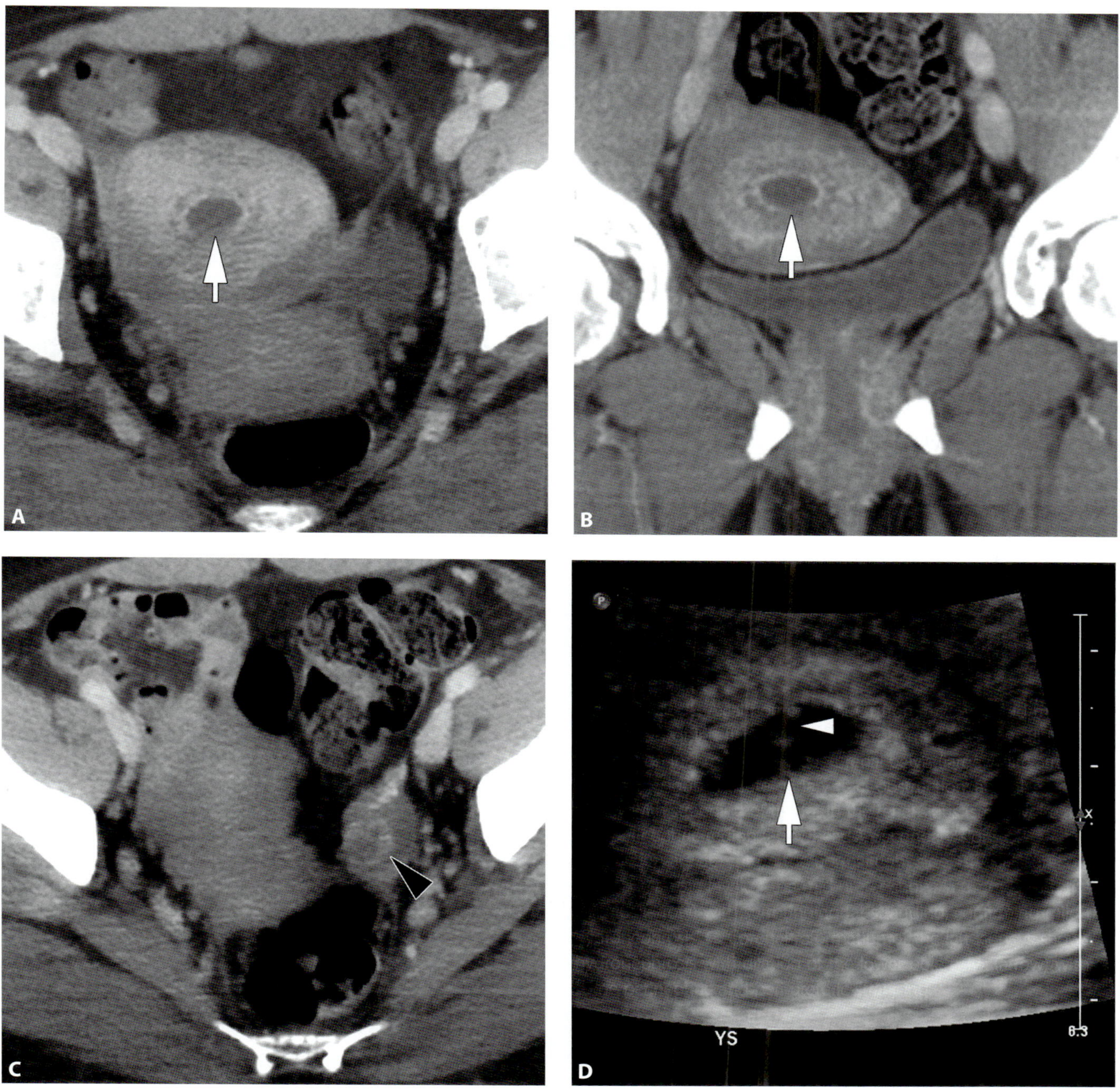

Figure 71.5 Axial (A) and coronal (B) contrast-enhanced CT of the abdomen and pelvis in a 29-year-old woman with severe head injuries following a high-speed motor vehicle collision. A rounded cystic area is present with n the fundal endometrium (arrow) with a ring of surrounding enhancing decidualized tissue, suspicious for an unexpected intrauterine gestational sac. **C.** Note the crenulated corpus luteum cyst in the left ovary (arrowhead). **D.** Emergency Department transabdominal sonogram confirmed the presence of a 6 week 0 day gestation sac (arrow) containing a yolk sac (arrowhead).

CASE **72**

Pseudogestational sac

Joel A. Gross

Imaging description

A pseudogestational sac (also known as a pseudosac) presents as an oval fluid collection centrally located within the uterine cavity of a pregnant patient.

The earliest sign of an intrauterine pregnancy (IUP) is the intradecidual sign, a small fluid collection surrounded by an echogenic ring, eccentrically located within the endometrium (Figure 72.1) [1].

Soon after, a double decidual sac sign may be identified, which is a small eccentric fluid collection within the endometrium, partially surrounded by two layers of decidua.

In contrast, a pseudosac consists of fluid located centrally within the uterine cavity (not within the endometrium), and surrounded by only one layer of decidualized endometrium (Figure 72.2). A double decidual sac sign is absent [2, 3]. "Beaking" of the fluid collection is suggestive of a pseudosac, but a leiomyoma or focal uterine contraction can cause a similar appearance in an IUP [4].

Decidual cysts are thin-walled cysts, usually at the junction of the endometrium and myometrium. They are non-specific findings and are identified in non-pregnant patients, normal IUPs, and ectopic pregnancies [1].

Although the intradecidual and double decidual signs have been reported as accurate, if there is any question about the location of the fluid, it is prudent to take a more conservative approach and specify that findings could represent an early IUP or a pseudosac associated with ectopic pregnancy, to ensure that an ectopic pregnancy diagnosis is not missed, resulting in a potentially fatal outcome.

Importance

Ectopic pregnancy is the leading cause of death during the first trimester of pregnancy, with a 9–14% mortality rate [5]. Pseudosacs occur in 10–20% of ectopic pregnancies [6], and may be mistaken for an early IUP, resulting in missed detection of an ectopic pregnancy.

Typical clinical scenario

In a pregnant patient presenting with pain and/or bleeding, it is critical to differentiate between an IUP and an ectopic pregnancy, and transvaginal ultrasound is the study of choice.

The most specific finding of an ectopic pregnancy is visualization of an extrauterine pregnancy, but this is not always visualized, and 15–35% of ectopic pregnancies will not demonstrate any identifiable extrauterine mass on transvaginal ultrasound [1].

In the absence of high clinical suspicion or imaging findings for a heterotopic pregnancy (a simultaneous intrauterine and extrauterine pregnancy), identification of an IUP is presumed to exclude the presence of an ectopic pregnancy. Incorrectly diagnosing a pseudosac as an IUP can result in missing the diagnosis of an ectopic pregnancy, a potentially fatal error.

An IUP is clearly present when a yolk sac or fetal pole is identified within the intrauterine gestational sac. An intrauterine fluid collection without a yolk sac or fetal pole, presents a more challenging diagnosis.

Differential diagnosis

Fluid in the uterus may represent a normal IUP, a pseudo sac, an anembryonic pregnancy, blood products, incomplete miscarriage, or decidual cysts.

Teaching point

A fluid collection within the uterus may represent a pseudogestational sac associated with ectopic pregnancy rather than an early IUP.

REFERENCES

1. Levine D. Ectopic pregnancy. *Radiology*. 2007;**245**(2):385–97.
2. Dogra V, Paspulati RM, Bhatt S. First trimester bleeding evaluation. *Ultrasound Q*. 2005;**21**(2):69–85; quiz 149–50, 53–4.
3. Bhatt S, Ghazale H, Dogra VS. Sonographic evaluation of ectopic pregnancy. *Radiol Clin North Am*. 2007;**45**(3):549–60, ix.
4. Yeh HC. Efficacy of the intradecidual sign and fallacy of the double decidual sac sign in the diagnosis of early intrauterine pregnancy. *Radiology*. 1999;**210**(2):579–82.
5. Lin EP, Bhatt S, Dogra VS. Diagnostic clues to ectopic pregnancy. *Radiographics*. 2008;**28**(6):1661–71.
6. Nyberg DA, Laing FC, Filly RA, Uri-Simmons M, Jeffrey RB, Jr. Ultrasonographic differentiation of the gestational sac of early intrauterine pregnancy from the pseudogestational sac of ectopic pregnancy. *Radiology*. 1983;**146**(3):755–9.

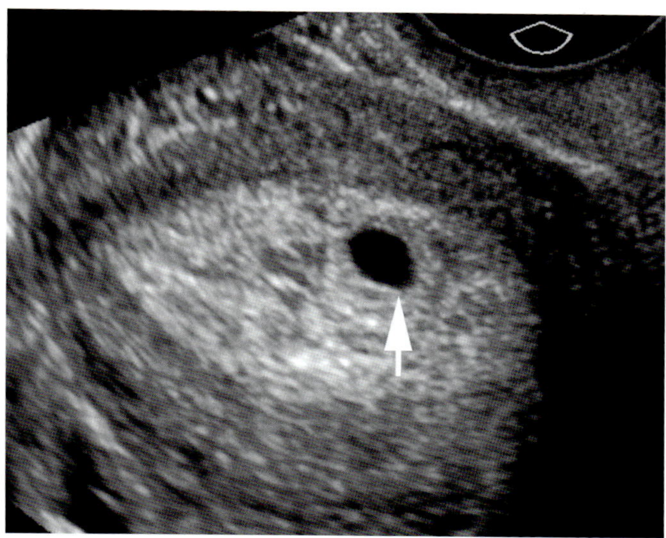

Figure 72.1 A 23-year-old woman with positive beta-hCG and pelvic pain. A small (<5 mm) fluid collection is identified within the uterus (arrow). The fluid collection is eccentrically located within the echogenic decidualized endometrium suggesting an IUP rather than a pseudosac. The absence of a yolk sac and fetal pole are consistent with an early IUP, and a normal IUP was confirmed on follow-up. If the location of fluid within the endometrium, rather than within the uterine cavity, could not confidently be confirmed, then the differential diagnosis should include early IUP versus pseudosac associated with ectopic pregnancy.

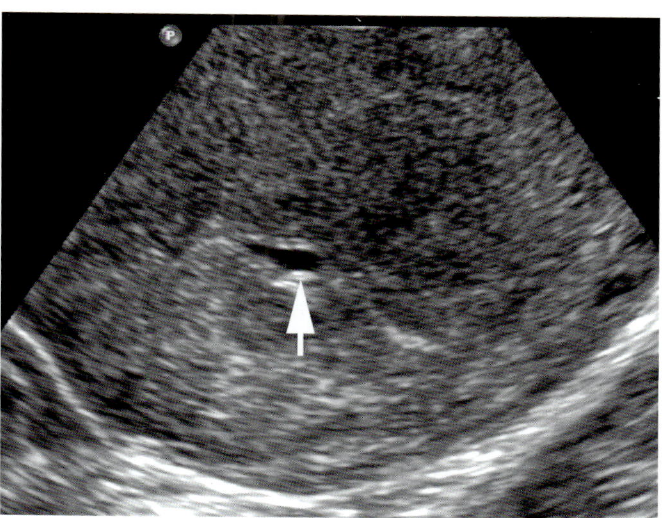

Figure 72.2 Transvaginal ultrasound in a 35-year-old woman with abnormal beta-hCG and history of prior ectopic pregnancies. A small oval fluid collection (arrow) is identified within the endometrial cavity, with a small amount of surrounding decidualized echogenic endometrium. Findings represented a pseudosac in a patient with an ectopic pregnancy.

CASE

73 Cystic pelvic mass simulating the bladder

Joel A. Gross

Imaging description

A dominant pelvic cystic structure or fluid collection may be mistaken for the urinary bladder, as it can present with an identical appearance to the bladder on multiple images (Figure 73.1), on ultrasound, CT, or MRI. This apparent bladder is sometimes referred to as a "pseudobladder."

Careful evaluation of the entire structure, and/or a high level of clinical suspicion leading to additional reformations or imaging, will usually help avoid this pitfall.

Importance

Mistaking a cystic structure for a distended urinary bladder can result in an erroneous diagnosis of bladder pathology, such as increased post-void residual secondary to bladder outlet obstruction [1]. This could result in inappropriate interventions such as Foley catheter or suprapubic catheter placement.

Moreover, a cystic mass mistaken for the urinary bladder can result in delayed identification of a pelvic cystic mass, such as a cystic ovarian carcinoma.

Typical clinical scenario

This potential pitfall arises when a large cystic mass is present in or near the pelvic midline, presenting with a similar or identical appearance to the bladder on many images. This may occur in studies performed specifically to evaluate the bladder, to evaluate for a pelvic mass, or for non-specific symptoms unrelated to the pelvis.

The lesions are often of adnexal origin resulting in the pseudobladder being more commonly diagnosed in women; however, other cystic lesions and fluid collections can occur in both sexes, and a pseudobladder may also be identified in men. Ovarian cystadenomas and lymphoceles were the most common etiology for pseudobladders in one ultrasound study [2].

Differential diagnosis

While a pseudobladder may look like a normal urinary bladder on some or most images, careful evaluation usually confirms that it is not in fact the urinary bladder. Carefully review the full extent of the cystic structure to ensure that it extends to the urethra (prostate or pelvic floor), and that it is appropriately positioned (anterior to the uterus in women). Small cystic structures may represent the non-distended urinary bladder.

Coronal reformations are usually diagnostic when the diagnosis is in question on axial CT.

If necessary, delayed images on contrast-enhanced CT or MRI can distend the bladder and highlight its location with higher density/intensity intraluminal contents. Even if contrast is not utilized, delayed images are often adequate to visualize the bladder on CT and MRI, due to its increased size and conspicuity. Retrograde filling of the bladder with contrast could also be performed.

On ultrasound, pre- and post-void images, or identification of urinary jets, may differentiate the bladder from a cystic mass. A functioning Foley catheter also helps in identification of the bladder, and can be utilized to drain or fill the bladder to assist in evaluation.

Lesions that may present as a pseudobladder include [2, 3]:

- Benign or malignant cystic ovarian neoplasm
- Benign ovarian cyst
- Paraovarian cyst
- Postoperative seroma
- Mesenteric or peritoneal cyst
- Abscess or hematoma
- Intestinal duplication cyst [4].

Teaching point

A cystic structure in the pelvic midline usually represents the bladder, but cystic pelvic masses can have a similar appearance.

If the cystic structure is suspicious in appearance or location, or if other structures are present that could represent a non-distended bladder, careful review of multiplanar reformats or additional targeted imaging can usually clarify the findings.

REFERENCES

1. Cooperberg MR, Chambers SK, Rutherford TJ, Foster HE, Jr. Cystic pelvic pathology presenting as falsely elevated post-void residual urine measured by portable ultrasound bladder scanning: report of 3 cases and review of the literature. *Urology.* 2000;**55**(4):590.

2. Vick CW, Viscomi GN, Mannes E, Taylor KJ. Pitfalls related to the urinary bladder in pelvic sonography: a review. *Urol Radiol.* 1983;**5**(4):253–9.

3. Fiske CE, Callen PW. Fluid collections ultrasonically simulating urinary bladder. *J Can Assoc Radiol.* 1980;**31**(4):254–5.

4. Loff S, Jaeger TM, Lorenz C, Waag KL. Large, septated ileal duplication cyst in a 4-year old, simulating the urinary bladder. *Pediatr Surg Int.* 1998;**13**(5–6):433–4.

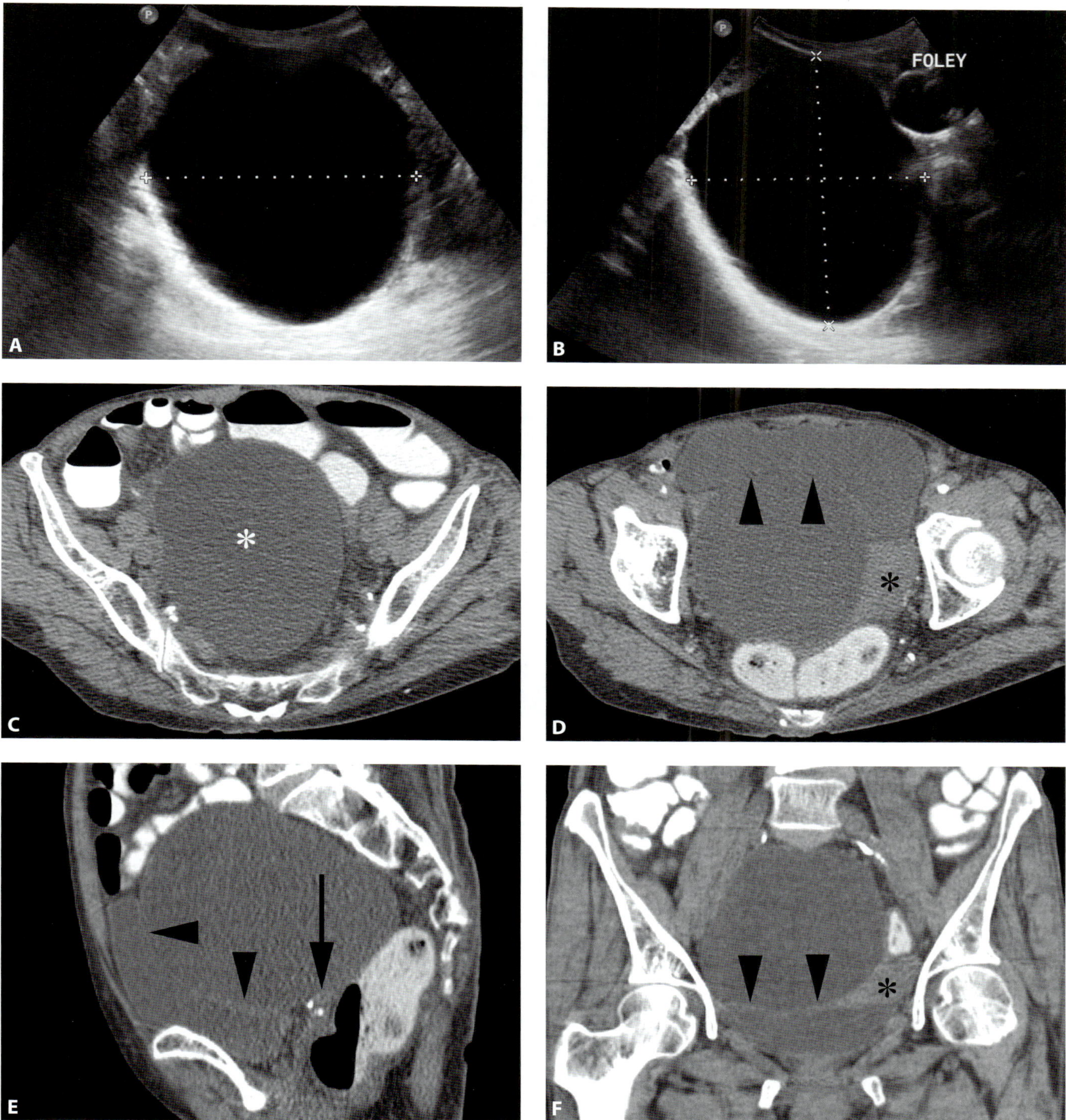

Figure 73.1 A 73-year-old woman with acute renal insufficiency. **A.** Transverse ultrasound through the pelvis demonstrates an oval cystic structure (between measurement cursors) with anechoic contents and prominent through transmission posteriorly, believed to be a distended urinary bladder. **B.** Sagittal ultrasound through the pelvis demonstrates the same cystic structure (between measurement cursors), also consistent with a distended urinary bladder. A Foley catheter (FOLEY) is identified inferior to the cystic structure, and was reported to be extravesical and within the urethra. **C.–F.** Prior pelvic CT performed with oral contrast but without intravenous contrast. **C.** Axial image through the mid pelvis demonstrates cystic structure in the midline (asterisk), consistent with distended urinary bladder. There are no findings on this image to suggest that the structure is not the bladder. **D.** Axial image more inferiorly demonstrates two cystic structures. The more anterior one is the bladder, and is draped over the oval posterior cystic structure (which is continuous with the structure in C). A thin bladder and cyst wall (arrowheads) separates the two cystic structures. The uterus (asterisk) is also visualized. **E.** Sagittal image again demonstrates two cystic masses. The anterior structure is the bladder, separated from the posterior cystic mass by a thin wall (arrowheads). Uterus (arrow) is visualized. **F.** Similar findings in a coronal image demonstrate the anterior inferior bladder separated from the cystic mass by a thin wall (arrowheads). The asterisk indicates the uterus. Additional imaging (not shown) demonstrated the cystic mass to arise from the right ovary.

CASE 74

Ovarian torsion

Joel A. Gross

Imaging description

Ultrasound is the initial study of choice for the evaluation of suspected ovarian torsion, due to its ability to evaluate the structure and size of the adnexa without ionizing radiation, and its ability to evaluate vascular flow.

It is erroneous to assume that the presence of Doppler flow excludes ovarian torsion and the absence of flow confirms torsion; evaluation is more complex and subtle than this simple binary approach would suggest.

Overall, there are quite variable data supporting the accuracy of ultrasound for the preoperative diagnosis of ovarian torsion: correct preoperative diagnostic rates have been reported to be as low as 23–66% of cases [1]. Absence of both arterial and venous flow in an enlarged ovary is highly suggestive of ovarian torsion, with a sensitivity approaching 100% and a specificity of 97% [2], while normal flow in a normal-sized and normal-appearing ovary is extremely unlikely to represent torsion. However, arterial or venous flow may occur in an ovary that has undergone torsion. In one series of 199 patients who presented with adnexal pain, 29 of whom had surgically proven ovarian torsion, the absence of Doppler flow in the ovarian artery was noted in only 22 (76%), although there was abnormal (non-continuous) or absent flow in the ovarian vein in 29 (100%) [2]. Our local experience demonstrates the presence of normal arterial and continuous venous flow does not entirely exclude torsion. An enlarged ovary with clinical symptoms of torsion may in fact have undergone torsion, either currently, or intermittently in the past, even if Doppler flow is demonstrated (Figure 74.1) [1, 3]. Other signs such as visible coiling of the ovarian pedicle (the "whirlpool sign") have been reported to reliably predict ovarian torsion [4].

Large non-vascular ovarian lesions may displace and obscure the smaller normal remaining ovarian parenchymal component. In one series, a hemorrhagic corpus luteum was reported as the most common laparoscopic finding following a false-positive diagnosis of ovarian torsion [1]. If an adnexal lesion is mistaken for ovarian parenchyma, Doppler interrogation will fail to demonstrate normal arterial and venous flow, and an erroneous diagnosis of ovarian torsion may be made (Figure 74.2).

Other findings supportive, but not diagnostic, of ovarian torsion include ovarian edema (usually seen as hypoechoic peripheral foci), relative enlargement of the ovary on the symptomatic side, and an ipsilateral ovarian cyst or mass.

Importance

Delayed diagnosis of ovarian torsion can lead to infertility, ovarian necrosis, peritonitis, or even death [5].

Typical clinical scenario

Ovarian torsion typically occurs in the first three decades of life. Approximately 17–20% of cases occur in pregnant women [6]. Torsion is believed to be due to increased mobility of the ovary in young women, with twisting of the ovary on its ligamentous supports, resulting in compromise of the venous and lymphatic drainage, followed by the arterial supply [6]. Women with enlarged ovaries due to ovarian masses such as teratomas, hemorrhagic cysts, cystadenomas, or ovarian hyperstimulation are predisposed to torsion. Although torsion can occur, it is unusual in normal-sized ovaries or in ovaries containing cysts less than 5 cm [6]. Torsion of a normal-sized ovary is most commonly seen in adolescents. Overall, it is an uncommon diagnosis, representing the fifth most common gynecologic emergency [1].

Although women present with pelvic or abdominal pain localized to a lower quadrant, there are no specific clinical signs or manifestations, nor any sensitive biochemical markers to assist with diagnosing ovarian torsion [3]. Consequently the diagnosis is challenging, and requires a high level of clinical suspicion.

Differential diagnosis

Mimics (and predisposing causes) of ovarian torsion include: hemorrhagic ovarian cysts, cystadenoma, ovarian hyperstimulation syndrome (OHHS) [6], and endometriomas.

Teaching point

Ovarian torsion may be present despite the presence of arterial and venous flow. Moreover, a non-vascular lesion in an ovary may be mistaken for ovarian parenchyma, resulting in an erroneous diagnosis of ovarian torsion.

REFERENCES

1. Mashiach R, Melamed N, Gilad N, Ben-Shitrit G, Meizner I. Sonographic diagnosis of ovarian torsion: accuracy and predictive factors. *J Ultrasound Med.* 2011;**30**(9):1205–10.
2. Nizar K, Deutsch M, Filmer S, *et al.* Doppler studies of the ovarian venous blood flow in the diagnosis of adnexal torsion. *J Clin Ultrasound.* 2009;**37**(8):436–9.
3. Shadinger LL, Andreotti RF, Kurian RL. Preoperative sonographic and clinical characteristics as predictors of ovarian torsion. *J Ultrasound Med.* 2008;**27**(1):7–13.
4. Auslender R, Shen O, Kaufman Y, *et al.* Doppler and gray-scale sonographic classification of adnexal torsion. *Ultrasound Obstet Gynecol.* 2009;**34**(2):208–11.
5. Schultz LR, Newton WA, Jr., Clatworthy HW, Jr. Torsion of previously normal tube and ovary in children. *N Engl J Med.* 1963;**268**:343–6.
6. Chang HC, Bhatt S, Dogra VS. Pearls and pitfalls in diagnosis of ovarian torsion. *Radiographics.* 2008;**28**(5):1355–68.

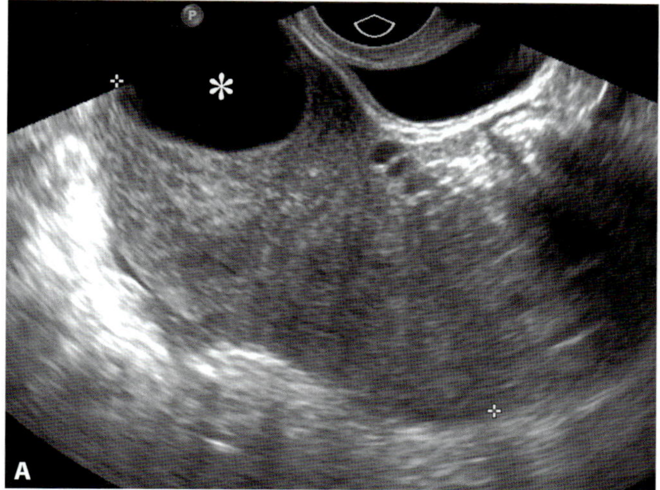

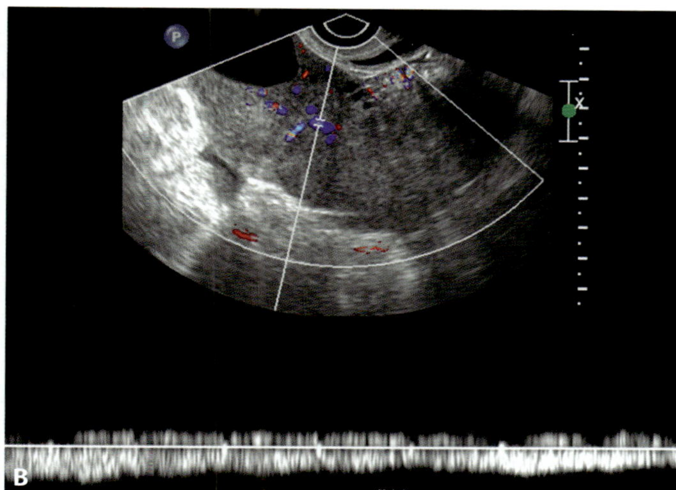

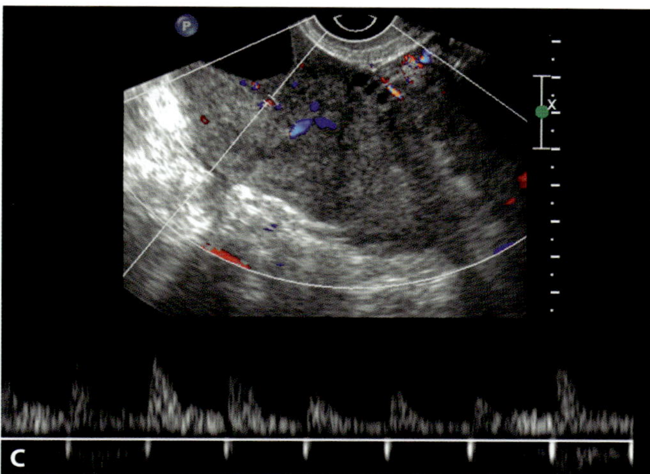

Figure 74.1 Transvaginal ultrasound of a 38-year-old woman with right lower quadrant pain, clinically suspicious for ovarian torsion. **A.** An enlarged right ovary is present, measuring 9 cm in length (between cursors), with a volume of 70 cubic centimeters. A simple 3.5 cm cyst is present (asterisk). **B.** and **C.** Spectral Doppler demonstrates normal continuous venous (**B**) and arterial (**C**) flow in the ovary. Despite the presence of normal flow, the patient was taken to surgery for suspicion of intermittent torsion, due to the typical clinical presentation and an enlarged ovary. An enlarged and hemorrhagic ovary was found, twisted three times on its pedicle. The ovary was untwisted and considered viable.

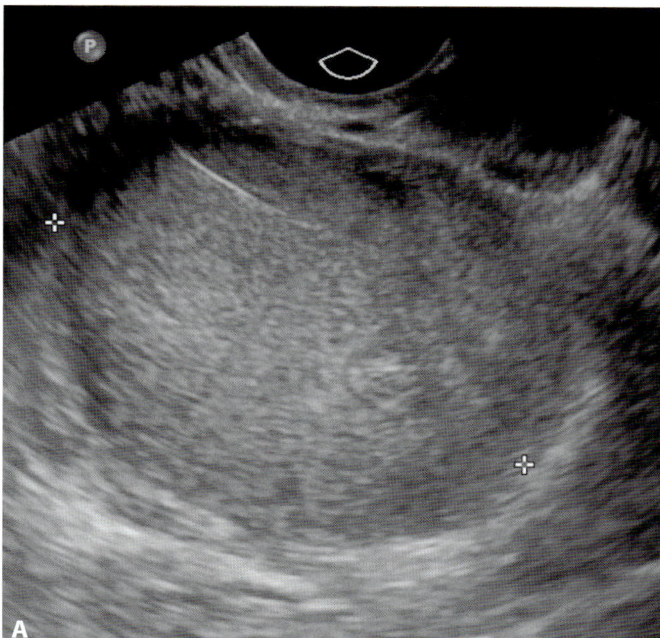

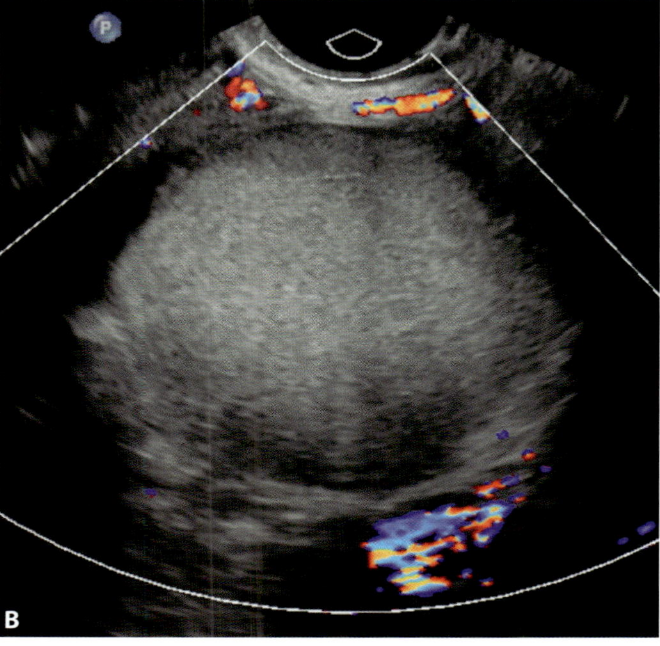

Figure 74.2 Transvaginal ultrasound of a 37-year-old woman with left lower quadrant pain. **A.** The left ovary is enlarged, measuring 6.6 cm in length (between cursors) and 114 cc in volume. **B.** No color Doppler flow is identified within the ovary, although flow is clearly identified anterior and posterior to the ovary. At surgery, a large left ovarian endometrioma was discovered, which was treated with cystectomy. The remainder of the left ovary was unremarkable without evidence of torsion. On ultrasound, the endometrioma obscured visualization of the residual normal ovarian tissue, and was mistaken for ovarian parenchyma. Lack of flow within the endometrioma resulted in an erroneous diagnosis of torsion.

CASE 75

Urine jets simulating a bladder mass

Ken F. Linnau

Imaging description

When performing contrast-enhanced CT of the pelvis, physiologic excretion of urine into the urinary bladder can cause the spurious appearance of a filling defect which should not be confused with a bladder mass.

The pseudomass appearance in the bladder due to urinary jets can be caused by either an ureteral contrast jet entering the bladder that contains mostly hypodense non-opacified urine (Figure 75.1) or a non-opacified urine jet entering the contrast-filled bladder (Figure 75.2). The latter can mimic a hypodense mass.

Since most CT of the abdomen and pelvis scans in emergency departments are performed with intravenous contrast, the more commonly observed pattern is that of a radiodense jet (Figure 75.1). Depending on the timing of the scan, variable mixtures of excreted contrast and non-opacified urine may result, which can be mistaken for bladder masses. Recognition of pseudomass artifact is important and can easily be obviated by obtaining delayed scans through the bladder or ultrasonography of the bladder.

Importance

The ureteric jet phenomenon occurs through the peristaltic contraction of the ureters at the ureterovesical junction. The frequency of these jets depends on the degree of patient hydration, and ranges from one jet per minute to continuous flow. The ureteric jet phenomenon is well recognized in ultrasonography where it is usually evaluated with color Doppler and may aid in the evaluation of ureteric obstruction and reflux [1–3].

This phenomenon on CT is less well documented, but observed almost daily in clinical practice. Incomplete mixing of urine and contrast can at times mimic filling defects or masses in the bladder.

Typical clinical scenario

The finding is usually identified incidentally, but may be seen in patients being investigated for hematuria.

Differential diagnosis

The principal differential diagnoses include intravesical blood clot, transitional carcinoma of the bladder, prostatic carcinoma invading the bladder, and hypertrophy of the median lobe of the prostate.

Teaching point

Incomplete mixing of contrast with non-opacified urine can mimic tumors in the bladder.

REFERENCES

1. Burge HJ, Middleton WD, McClennan BL, Hildebolt CF. Ureteral jets in healthy subjects and in patients with unilateral ureteral calculi: comparison with color Doppler US. *Radiology*. 1991; **180**(2):437–42.
2. Cox IH, Erickson SJ, Foley WD, Dewire DM. Ureteric jets: evaluation of normal flow dynamics with color Doppler sonography. *AJR Am J Roentgenol*. 1992;**158**(5):1051–5.
3. Leung VY, Chu WC, Yeung CK, Metreweli C. Doppler waveforms of the ureteric jet: an overview and implications for the presence of a functional sphincter at the vesicoureteric junction. *Pediatr Radiol*. 2007;**37**(5):417–25.

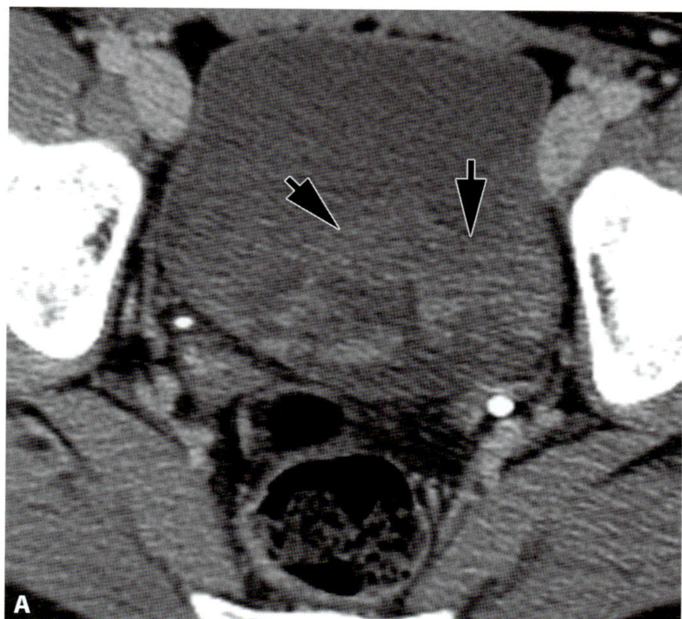

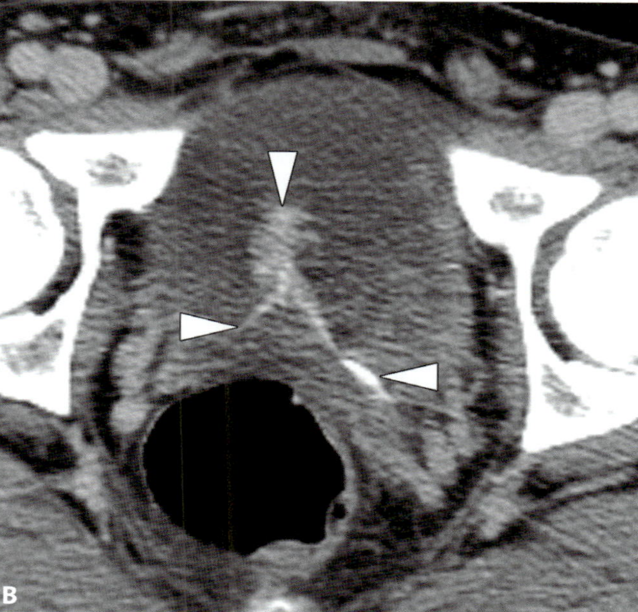

Figure 75.1 **A.** Axial contrast-enhanced CT from a 19-year-old man who was assaulted shows a hyperdense fungating pseudomass in the left posterolateral aspect of the urinary bladder (arrows), simulating a hematoma. **B.** A more inferior axial CT image shows bilateral jets of contrast extending from the ureteral orifices into the bladder with inhomogeneous mixing of urine and contrast (white arrowheads). No filling defect was present on delayed phase images.

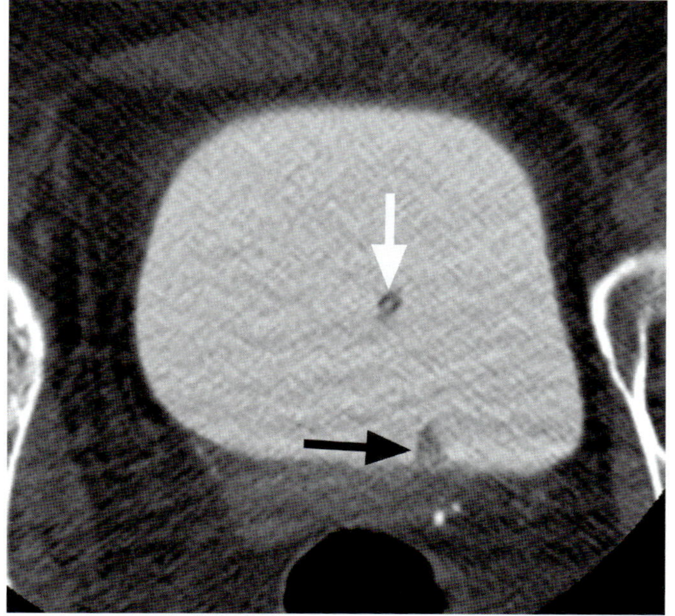

Figure 75.2 Axial CT cystography image of a 69-year-old woman who sustained a pelvic ring disruption during a motor vehicle collision shows a hypodense filling defect in the posterior aspect of the bladder (black arrow) in the area of the expected left ureteral orifice. This pseudomass mimics the appearance of a bladder tumor. However, it is due to non-opacified urine jet entering the contrast filled bladder following retrograde instillation of contrast through a Foley catheter (white arrow). The bladder wall was normal on subsequent cystoscopy.

CASE **76**

Extraluminal bladder Foley catheter

Joel A. Gross

Imaging description

A Foley bladder catheter without visualized surrounding bladder is usually due to bladder rupture and extraluminal location of the catheter. However, a similar appearance can occur with a collapsed bladder (Figure 76.1).

On a pelvic CT, distal portions of a urinary catheter may be visualized in the pelvis without evidence of surrounding urine or bladder. Findings may be suggestive of an extraluminal catheter location, usually secondary to bladder rupture.

While initial review of images may suggest an extraluminal catheter location, careful evaluation of the bladder and catheter may confirm an intraluminal location. Multiplanar reformations are helpful and sometimes essential to adequately evaluate the catheter tip position. Moreover, if the patient is being evaluated following trauma, they are also useful to confirm a normal bladder or diagnose bladder injuries [1]. If the catheter position cannot be resolved following careful review of all images, additional evaluation may be necessary.

If the patient is not at risk for bladder injury and CT cystography (CTC) is not indicated for other reasons, low-dose delayed images with a clamped and adequately distended bladder will usually clearly demonstrate the catheter location within the bladder.

If the patient is at risk for bladder injury or a bladder wall defect, or if there is intraperitoneal or extraperitoneal fluid that might have arisen from a bladder injury, then further imaging is indicated. In most cases, CTC with retrograde distension of the bladder with contrast is advised. While post-contrast delayed CT images with partial filling and distention of the bladder with contrast may provide the correct diagnosis, this approach is unreliable and CTC should be considered to optimize the likelihood of making the correct diagnosis [2–4]. Other simple techniques to determine the catheter position include clamping the catheter and performing transabdominal ultrasound, or instilling iodinated contrast through the catheter and performing fluoroscopy or plain radiography.

Importance

A urinary catheter in a collapsed bladder may be mistakenly diagnosed as representing a bladder rupture, resulting in unnecessary urgent surgery.

Typical clinical scenario

This is usually an incidental finding. However, a high degree of suspicion is necessary in patients who are being evaluated for trauma or who have undergone recent surgery to the lower urinary tract.

Differential diagnosis

The primary differential diagnosis is a urinary catheter in a collapsed bladder with an imperceptible bladder wall; a catheter balloon in a bladder diverticulum [5], a urethra [6], ureter [7]; or an extraluminal position of the catheter if the patient has a bladder wall defect.

Teaching point

Apparent extraluminal location of a urinary catheter in a non-distended bladder should prompt careful reevaluation of images and consideration of repeat imaging with a distended bladder.

REFERENCES

1. Chan DP, Abujudeh HH, Cushing GL, Jr., Novelline RA. CT cystography with multiplanar reformation for suspected bladder rupture: experience in 234 cases. *AJR Am J Roentgenol.* 2006;**187**(5):1296–302.
2. Haas CA, Brown SL, Spirnak JP. Limitations of routine spiral computerized tomography in the evaluation of bladder trauma. *J Urol.* 1999;**162**(1):51–2.
3. Quagliano PV, Delair SM, Malhotra AK. Diagnosis of blunt bladder injury: a prospective comparative study of computed tomography cystography and conventional retrograde cystography. *J Trauma.* 2006; **61**(2):410–21; discussion 421–2.
4. Power N, Ryan S, Hamilton P. Computed tomographic cystography in bladder trauma: pictorial essay. *Can Assoc Radiol J.* 2004;**55**(5):304–8.
5. Abadi S, Brook OR, Solomonov E, Fischer D. Misleading positioning of a Foley catheter balloon. *Br J Radiol.* 2006;**79**(938):175–6.
6. Vaidyanathan S, Hughes PL, Soni BM. A simple radiological technique for demonstration of incorrect positioning of a foley catheter with balloon inflated in the urethra of a male spinal cord injury patient. *ScientificWorldJournal.* 2006;**6**:2445–9.
7. George J, Tharion G. Transient hydroureteronephrosis caused by a Foley's catheter tip in the right ureter. *Scientific World Journal.* 2005;**5**:367–9.

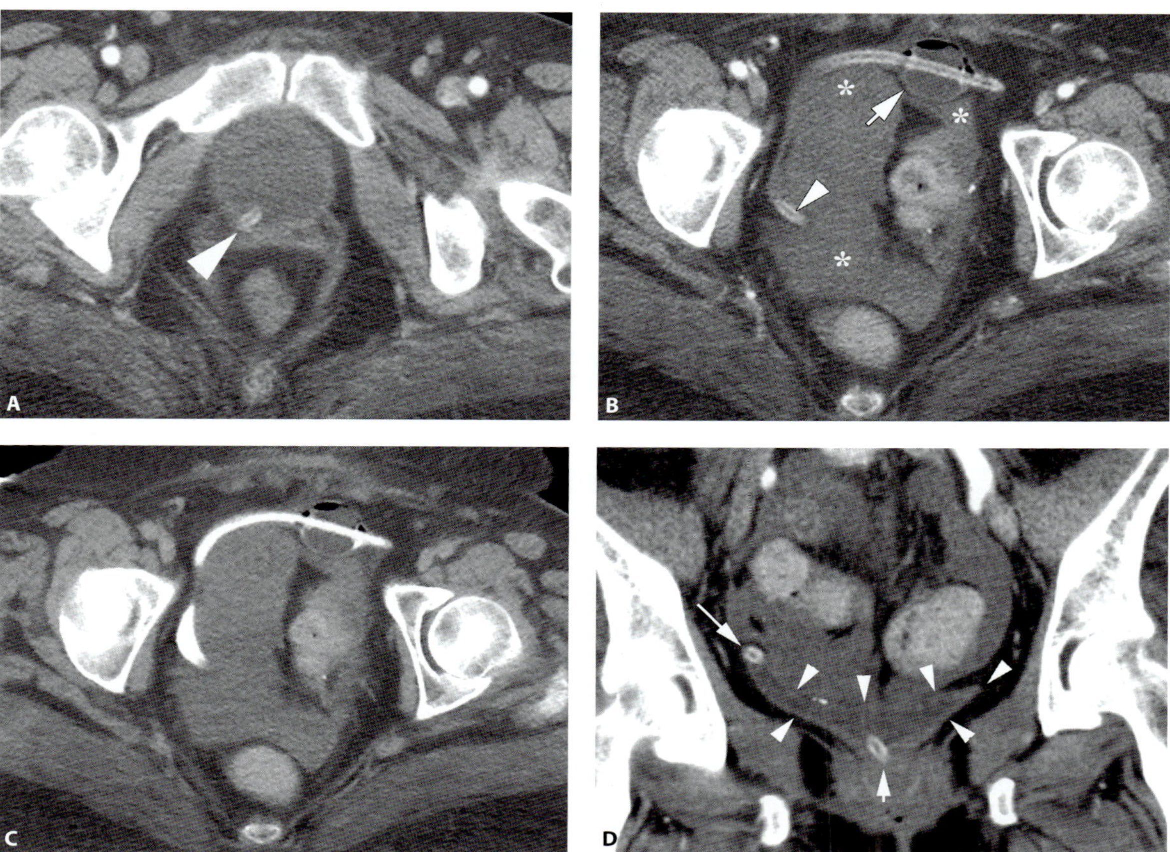

Figure 76.1 A 50-year-old woman following a motor vehicle crash, with hemoperitoneum. **A.** Axial intravenous contrast-enhanced CT demonstrates the tube of a Foley catheter (arrowhead) within the posterior lumen of the bladder. **B.** More cranial image demonstrates the catheter balloon (arrow) in the anterior pelvis. No bladder or urine is visualized around the catheter. Fluid in the pelvis (asterisks) represents intraperitoneal hemoperitoneum from a splenic injury. The arrowhead identifies the catheter lumen as it passes to the right lateral pelvis. **C.** Delayed images with partially distended bladder demonstrate a similar appearance, and a bladder rupture was reported, with an extraluminal location of the catheter balloon. However, no bladder injury was identified at surgery. **D.** Careful evaluation of a posterior coronal reformation from the venous phase of the study demonstrates a poorly distended bladder (arrowheads), the catheter entering the bladder inferiorly (short arrow) and extending into the right superior recess of the bladder (long arrow).

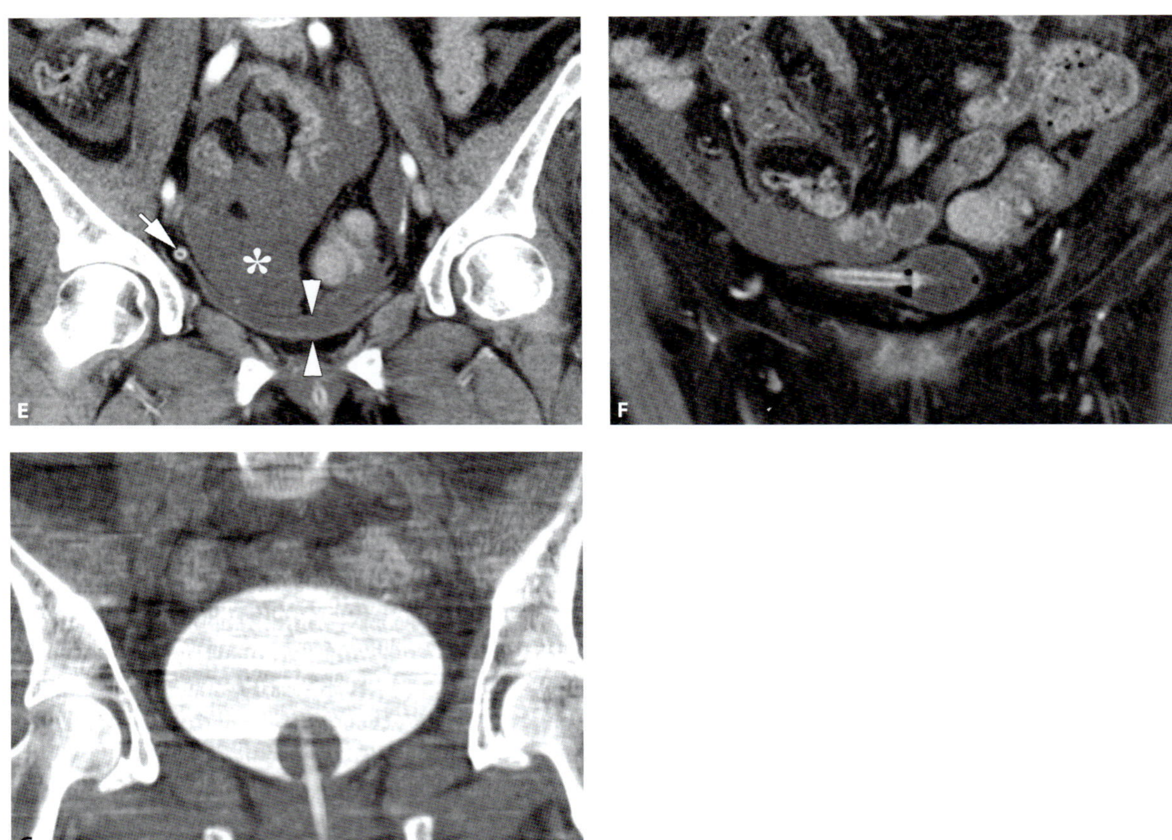

Figure 76.1 (cont.) **E.** Coronal image obtained more anteriorly better delineates the enhancing wall of the non-distended bladder (arrowheads), inferior to hemoperitoneum (asterisk). Foley catheter in the collapsed right superior recess of the bladder is again noted (arrow). **F.** Anterior coronal image of the catheter balloon. Note the very subtle finding of the bladder tapering smoothly around the "neck" of the catheter, between the balloon and the tubing. Moreover, the previous coronal images demonstrated the bladder to be collapsed and barely perceptible around the catheter in the right superior recess, and a similar finding is likely present here as well, especially given the absence of streaky adjacent extraperitoneal fluid and gas. **G.** Coronal image from CT cystogram obtained following surgery demonstrates the catheter balloon in the bladder and no evidence of injury. Note that the delayed images demonstrated in (**C**), did not adequately distend the bladder to permit the correct diagnosis.

CASE **77**

Missed bladder rupture

Joel A. Gross

Imaging description

CT cystography (CTC) has replaced conventional fluoroscopic cystography for the evaluation of bladder rupture in trauma patients. The bladder is filled with diluted contrast using the Foley catheter to generate adequate distension. At our institution, we usually perform CTC immediately after venous phase CT scan by emptying the bladder, mixing 30mL of iohexhol 350mg/mL in a 500cc bag of normal saline warmed to body temperature, connecting this to the Foley catheter using an intravenous "drip" set, and hanging the bag 40cm above the symphysis pubis. We perform a low-dose CT scan through the pelvis after 350mL of contrast has been administered, or contrast stops dripping, or when the patient cannot tolerate bladder distention. We routinely perform multiplanar reformations.

A markedly distended bladder (such as in a patient with chronic bladder outlet obstruction) may require considerably more than 350cc of contrast to fill the bladder. If images do not demonstrate a distended oval-shaped bladder (e.g., a floppy bladder draping around adjacent structures) then the bladder is inadequately distended and has not been "stressed" adequately to evaluate for rupture.

If the bladder was not adequately drained prior to filling via the urinary catheter, it may contain the instilled contrast-enhanced fluid in its dependent portions, but unenhanced fluid more anteriorly. The presence of the unenhanced fluid limits evaluation, as unenhanced fluid that leaks through the anterior bladder may not be distinguishable from other unenhanced fluid in the pelvis from other causes (such as hematoma), and the diagnosis of a bladder rupture may be missed.

Contrast leakage into the intraperitoneal or extraperitoneal spaces diagnoses intraperitoneal (IPBR), extraperitoneal (EPBR), or combined (CBR) bladder ruptures.

When an EPBR is present, it may be difficult to adequately distend the bladder to exclude a concomitant IPBR. Figure 77.1 illustrates a patient with a large EPBR with a poorly distended bladder during CTC. An IPBR was also present at surgery, but was missed on imaging.

When delayed phase CT images following intravenous contrast are used instead of a dedicated CTC, a bladder rupture may be missed, as demonstrated in Figure 77.2.

Sagittal and coronal reformations improve detection of the location of bladder rupture, especially for ruptures occurring at the dome of the bladder, which can be difficult to visualize on axial images [1].

Importance

Previous studies demonstrated that antegrade filling of the bladder with contrast-enhanced urine excreted by the kidneys does not adequately distend the bladder to reliably diagnose bladder ruptures [2, 3]. This is consistent with experience using conventional cystography in which bladder ruptures were not visible with partially distended bladders, but only became visible as the bladder was well distended.

There are rare situations in which CTC might fail to identify a bladder injury.

A large EPBR may provide a low resistance pathway for the contrast to rapidly leak out of the bladder, preventing adequate distension of the bladder, and limiting evaluation for a concomitant IPBR [4]. This pitfall is demonstrated in Figure 77.1.

Typical clinical scenario

CTC is performed for patients at risk of bladder injury following trauma.

The situations in which the standard CTC technique may fail to detect bladder rupture are rare, but knowledge of these potential pitfalls will decrease the risk of these preventable errors.

Differential diagnosis

A large EPBR with inadequate distension of the bladder may simply be due to the visualized EPBR; however, additional sites of EPBR and the presence of a concomitant IPBR should also be considered.

Extravesical contrast may arise from other sources and falsely suggest a bladder rupture. Carefully review all imaging to evaluate for extravasated contrast from vascular, bowel, ureteric, or urethral injury.

Teaching point

Inadequate distension or opacification of the bladder limits evaluation for bladder rupture. Delayed phase CT following intravenous contrast is unreliable for the identification of bladder rupture.

REFERENCES

1. Chan DP, Abujudeh HH, Cushing GL, Jr., Novelline RA. CT cystography with multiplanar reformation for suspected bladder rupture: experience in 234 cases. *AJR Am J Roentgenol.* 2006;**187**(5):1296–302.
2. Haas CA, Brown SL, Spirnak JP. Limitations of routine spiral computerized tomography in the evaluation of bladder trauma. *J Urol.* 1999;**162**(1):51–2.
3. Quagliano PV, Delair SM, Malhotra AK. Diagnosis of blunt bladder injury: a prospective comparative study of computed tomography cystography and conventional retrograde cystography. *J Trauma.* 2006;**61**(2):410–21; discussion 421–2.
4. Power N, Ryan S, Hamilton P. Computed tomographic cystography in bladder trauma: pictorial essay. *Can Assoc Radiol J.* 2004;**55**(5):304–8.

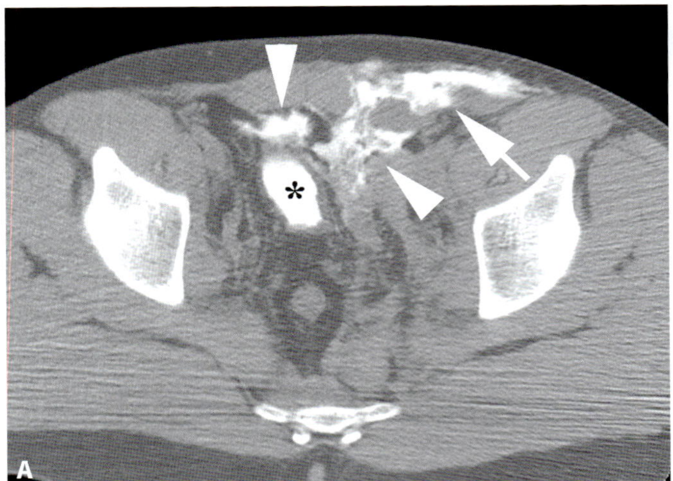

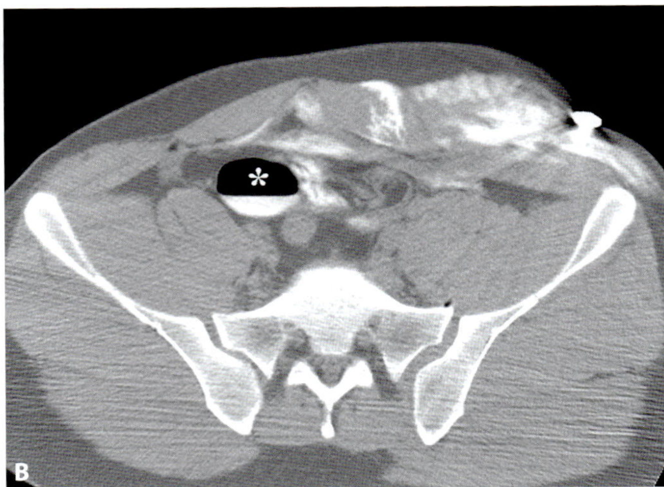

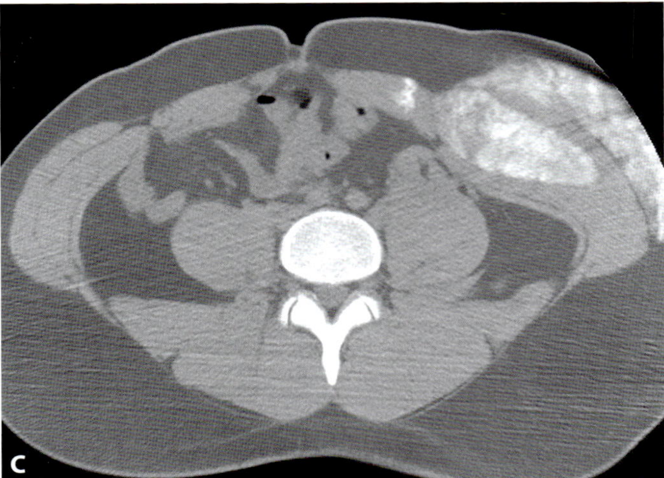

Figure 77.1 Axial images from CTC obtained for a 26-year-old man following a motorcycle crash and pelvic fracture. **A.** Inferior most image demonstrates a poorly distended bladder (asterisk) with adjacent extraperitoneal contrast (arrowheads), identifying an EPBR. Contrast also extends into and through the abdominal wall (arrow). **B.** Image obtained more superiorly demonstrates the maximum distension of the bladder (asterisk) on this study. Extraperitoneal and abdominal wall contrast is again noted. **C.** Superior most image demonstrates extensive abdominal wall contrast, without evidence of intraperitoneal contrast. Patient was diagnosed with a large EPBR, but at surgery an IPBR was also identified. Absence of intraperitoneal contrast was believed due to inadequate distension of the bladder, secondary to contrast rapidly leaking into the extraperitoneal space.

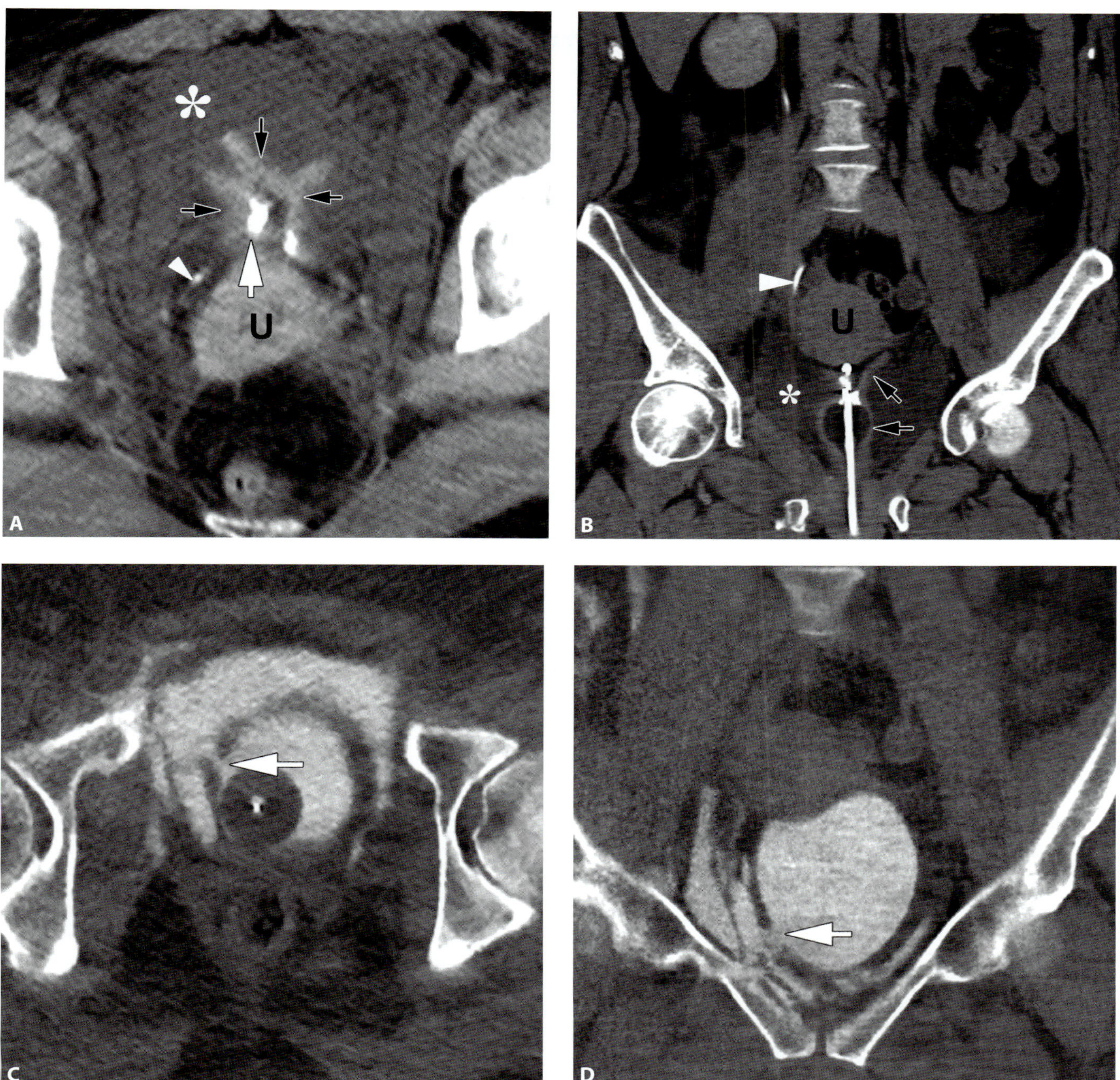

Figure 77.2 Axial (**A**) and coronal (**B**) 10 minute delayed phase intravenous post-contrast CT images in a 37-year-old woman who was involved in a high-speed motor vehicle collision, fracturing her pelvis. Fluid (asterisk) surrounds a collapsed bladder (black arrows). On this delayed phase, contrast has passed through the ureters (arrowhead), and into the bladder (white arrow). The uterus is identified by the letter "U." The vertical white line in (B) is a Foley catheter with a balloon near its tip. This CT was performed at an outside institution prior to transfer. Owing to the perivesical fluid, and the unreliability of delayed phase CT for the identification of bladder ruptures, a dedicated low-radiation dose CT cystogram was performed after transfer to our institution. Axial (**C**) and coronal (**D**) images from the CTC clearly demonstrate a defect in the right wall of the bladder (arrow), with contrast leaking into the extraperitoneal pelvic spaces, confirming an EPBR.

CASE **78**

Pseudofracture from motion artifact

Claire K. Sandstrom

Imaging description

Misregistration is a CT artifact caused by both gross patient or physiologic motion during helical or axial acquisition [1, 2]. Motion artifacts can cause a variety of appearances on CT, including shading, streaking and double contours [1]. The appearance and severity of CT motion artifacts varies depending on the magnitude, speed, and direction of patient movement, as well as on the speed of the CT scanner itself. Slight patient motion, such as from cardiac motion, peristalsis, or tremor, can cause misregistration during image reconstruction, and is detected as bands and streaks on the axial image at the level of motion [1]. This motion is most problematic for evaluation of soft tissues and rarely causes significant diagnostic dilemmas in the skeleton. Gross patient motion, which occurs in intoxicated patients who cannot lie still, or can be due to respiratory motion, can cause step-off between axial images that may be confused for fracture or dislocation (Figures 78.1–78.3). In this case, the step-off will involve not only the cortex but also overlying soft tissue planes, such as the posterior wall of the pharynx in the neck or the skin overlying the sternum (Figure 78.4). Blurring may also be seen in the adjacent soft tissues, confirming the presence of motion (Figure 78.5).

Importance

Patient motion is less likely and less problematic with the modern CT scanners than with prior generations of scanners. Nevertheless, motion occurring with significant enough magnitude or velocity cannot be overcome with present-day techniques [2]. Instead, it is incumbent upon the radiologist to recognize the resulting appearance as artifact rather than true pathology.

Typical clinical scenario

Patients evaluated in the emergency department are frequently uncooperative due to medical or surgical disease, substance abuse, or injuries, and may have tachypnea and restlessness due to pain or hypoxia. Motion artifacts are therefore not uncommon and are sometimes unavoidable (Figures 78.6 and 78.7).

Differential diagnosis

The primary differential diagnosis of motion artifact in the skeleton is a true fracture (Figure 78.8). Fractures should not have obvious motion within the adjacent soft tissues, will usually not be directly in the same plane as the image was acquired, and may have adjacent hematoma. Step-offs in the soft tissues may also represent lacerations overlying the fractures.

Multiplanar reformations are very helpful to discriminate between motion artifact and true fracture [3]. The scout (or topogram) image may show evidence of a step-off, confirming a fracture. In our practice, we check with the patient or ordering clinician directly, to determine whether there are focal confirmatory physical exam findings [4], or if the history adequately explains a fracture in that location. If question remains, additional radiographs or repeat CT through the level of the possible injury can be considered but do involve additional radiation [3].

Teaching point

Motion artifact may cause pseudofractures in any osseous structure in the body. Clues are often apparent in multiplanar reformats, which should be examined carefully. Look for step artifacts in the soft tissues (such as the skin, pharynx, or airway) or similar pseudofractures in the lines and tubes that cross the same axial location as the fracture. If the skin is not included in the reconstruction field of view (FOV), consider retrospectively reconstructing CT images with a wider FOV.

REFERENCES

1. Barrett JF, Keat N. Artifacts in CT: recognition and avoidance. *Radiographics.* 2004;**24**:1679–91.
2. McCullough CH, Bruesewitz MR, Daly TR, Zink FE. Motion artifacts in subsecond conventional CT and electron-beam CT: pictorial demonstration of temporal resolution. *Radiographics.* 2000;**20**:1675–81.
3. Kim EY, Yang HJ, Sung YM, *et al.* Sternal fracture in the emergency department: diagnostic value of multidetector CT with sagittal and coronal reconstruction images. *Eur J Radiol.* 2012;**81**: e708–11.
4. Courter BJ. Pseudofractures of the mandible secondary to motion artifact. *Am J Emerg Med.* 1994;**12**:88–9.

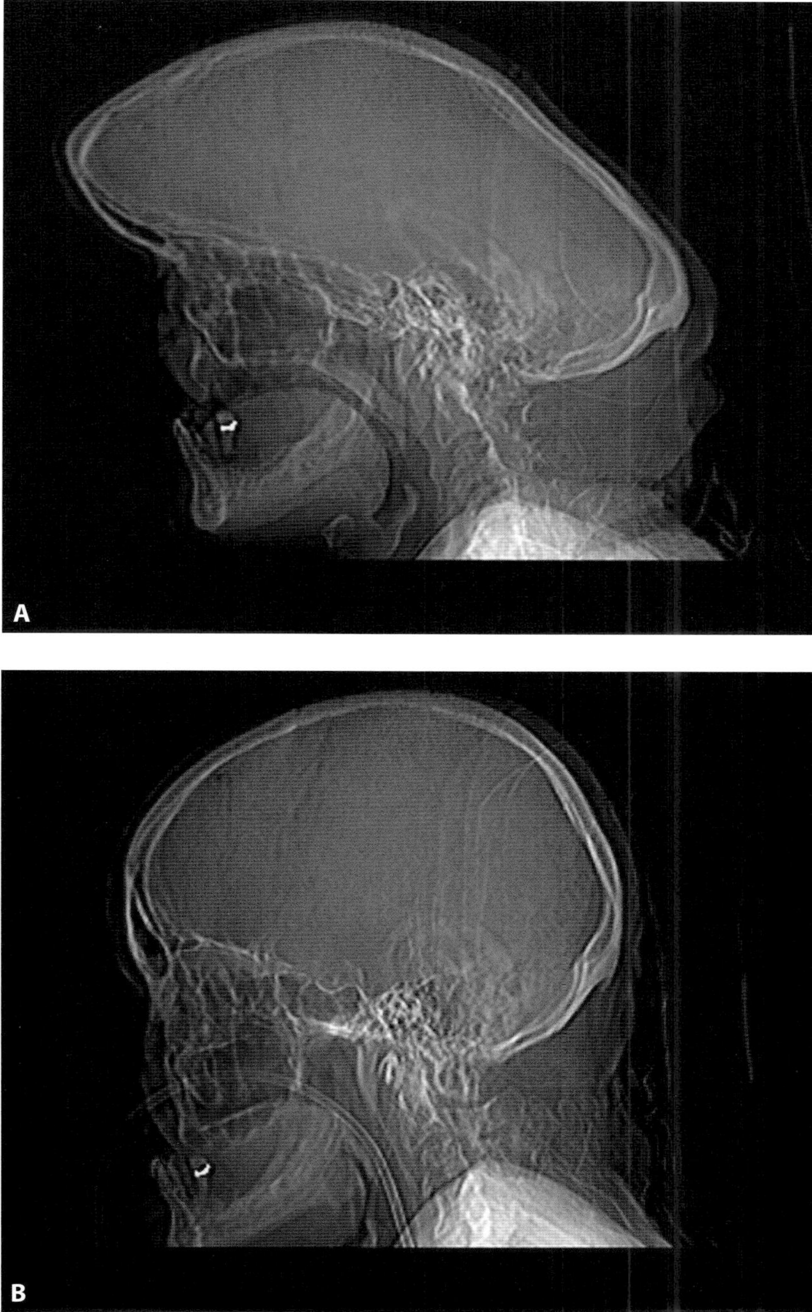

Figure 78.1 A. Lateral topogram image from head CT in a 70-year-old man with multiple recent falls shows an abnormally elongated appearance of the skull. **B.** Comparison with the lateral scout image from a CT two days prior confirms that the patient has normal skull shape. Figure **A** was acquired while the patient was trying to sit up. While unlikely to lead to misdiagnosis, motion during scout image acquisition may affect tube current modulation and thus image quality and radiation dose in automated CT systems.

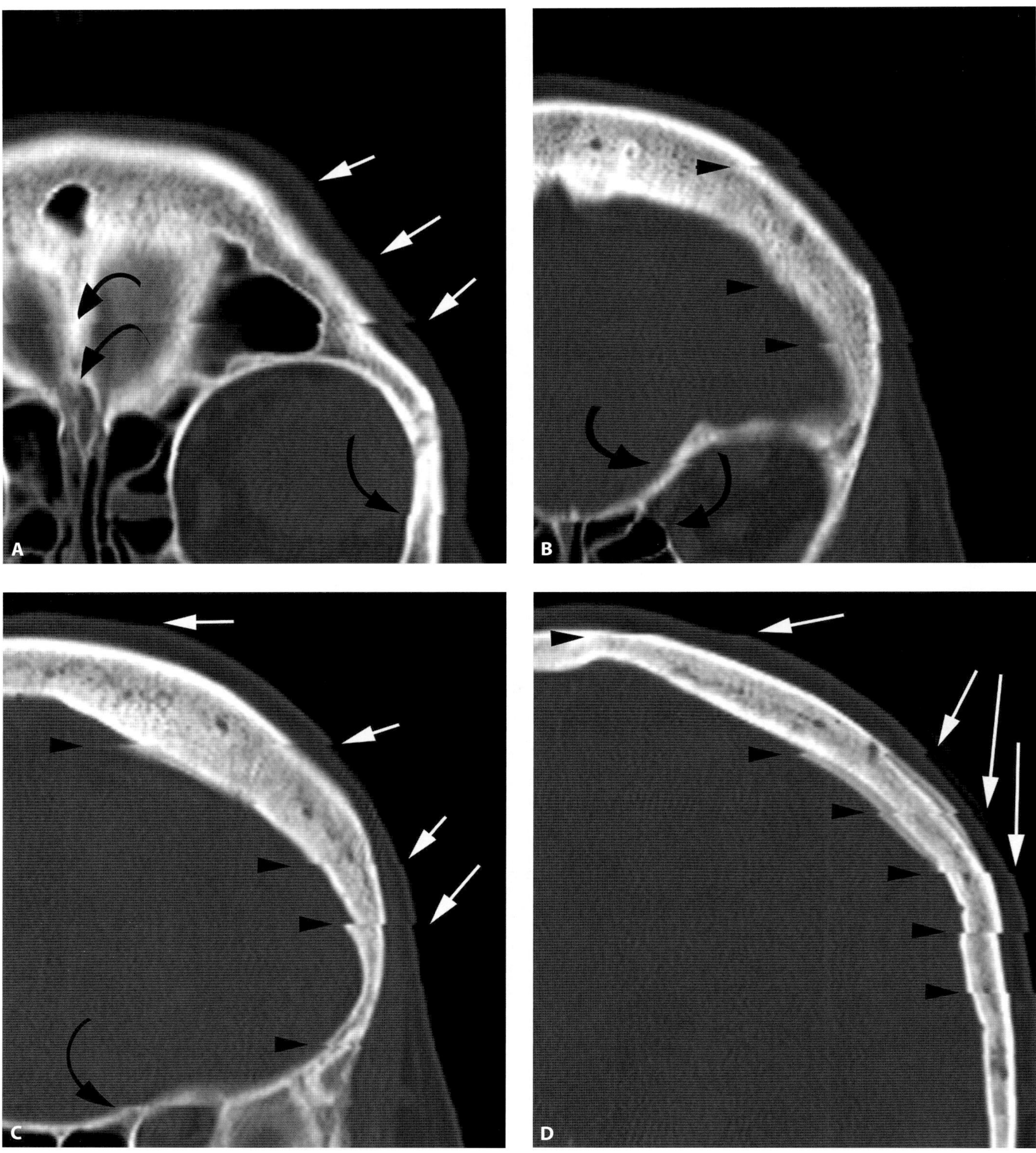

Figure 78.2 A.–D. Selected coronal images, from anterior (**A**) to posterior (**D**) through the skull, in a 59-year-old woman with epistaxis after facial trauma show multiple step-offs through the sinuses and orbits (curved black arrows), skull (arrowheads), and overlying scalp (white arrows). These are pseudofractures from motion and do not represent acute fractures.

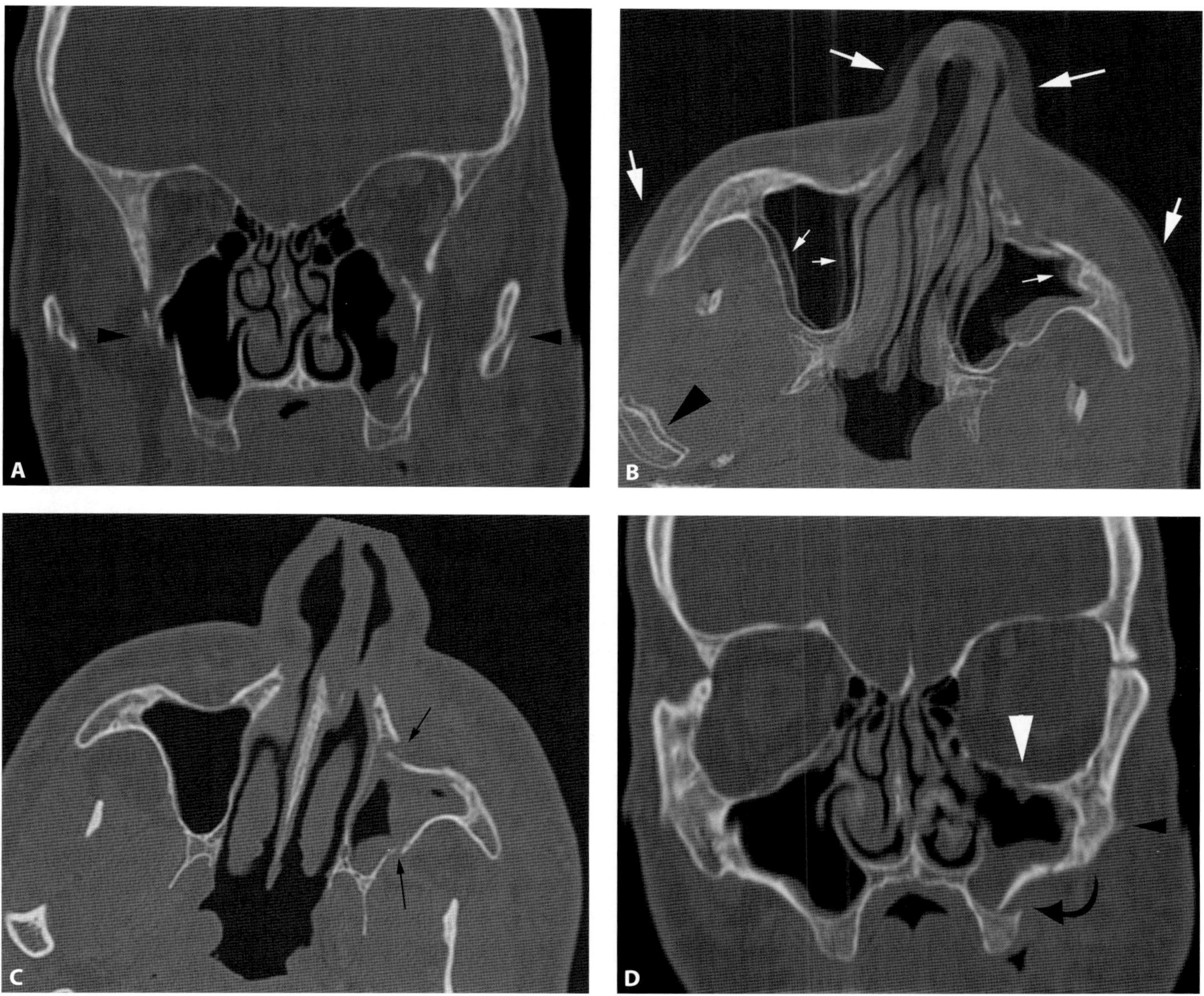

Figure 78.3 **A.** Coronal reformation image in bone windows from maxillofacial CT in a 45-year-old man who developed cellulitis not long after left zygomaticomaxillary complex fractures shows step-off (black arrowheads) through the nasal septum, medial and lateral walls of the maxillary sinuses, mandibles, and skin surface bilaterally due to motion artifact. **B.** Axial image through the level of the step-offs confirms the presence of motion artifact on the image, including ghosting of the soft tissues (large white arrows), maxilla (small white arrows), and right mandible (arrowhead). **C.** More inferior axial image without degradation by motion artifact clearly shows the subacute fractures of the anterior and lateral walls of the left maxillary sinus (arrows). **D.** Coronal reformation image more anterior than in **A** shows the maxillary sinus (curved black arrow) and orbital floor fractures (white arrowhead), clearly separate from the level of motion artifact (black arrowhead).

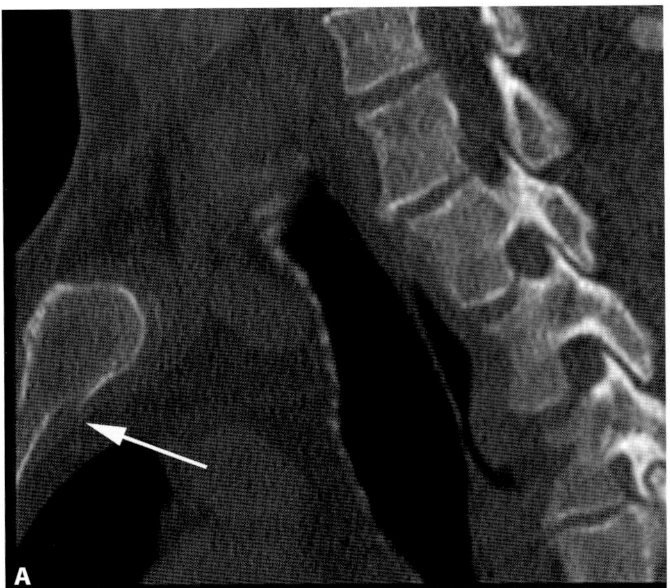

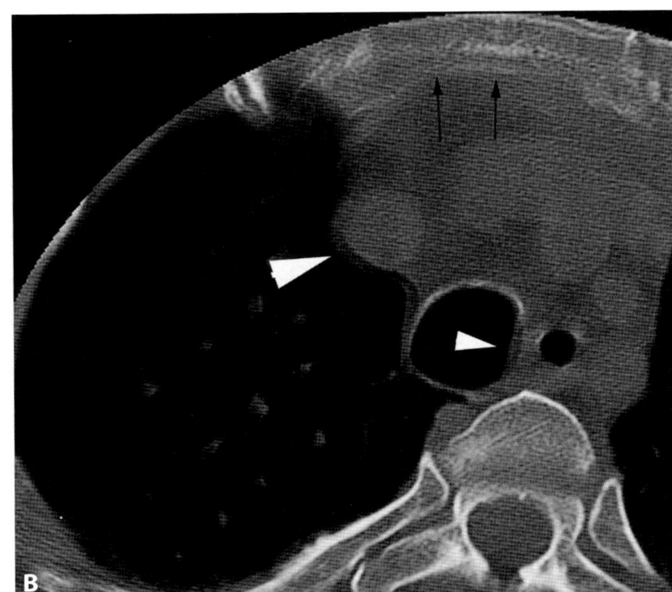

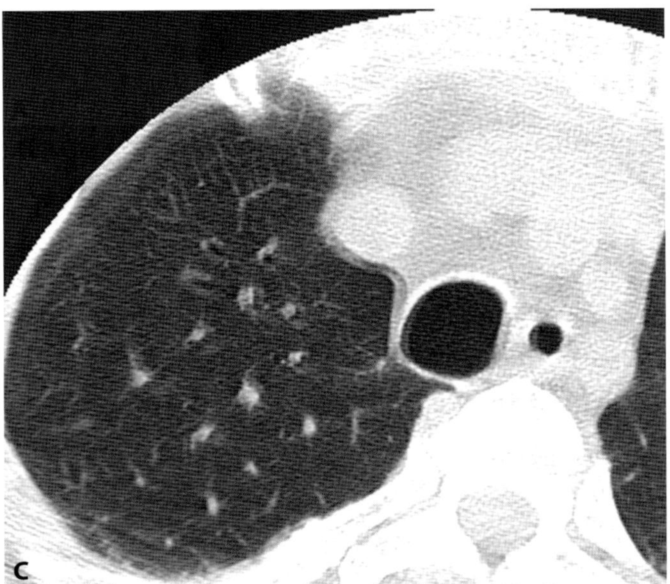

Figure 78.4 A. Sagittal reformation from cervical spine CT in bone windows in a 61-year-old man after a fall from a ladder shows an edge-of-the-scan finding. There is cortical step-off of the posterior margin of the manubrium (white arrow). The overlying skin surface was not included in the reconstruction field of view (FOV). It is often helpful to retrospectively reconstruct images such as this with a wider FOV, but this was not felt necessary in this case. **B.** Axial image in bone window through the level of the step-off shows ghosting of the posterior margin of the sternum (small black arrows), the right mediastinal border (large white arrowhead), and trachea (small white arrowhead), all signs indicating motion. **C.** Same axial image in lung windows confirms motion artifact in the right upper lung.

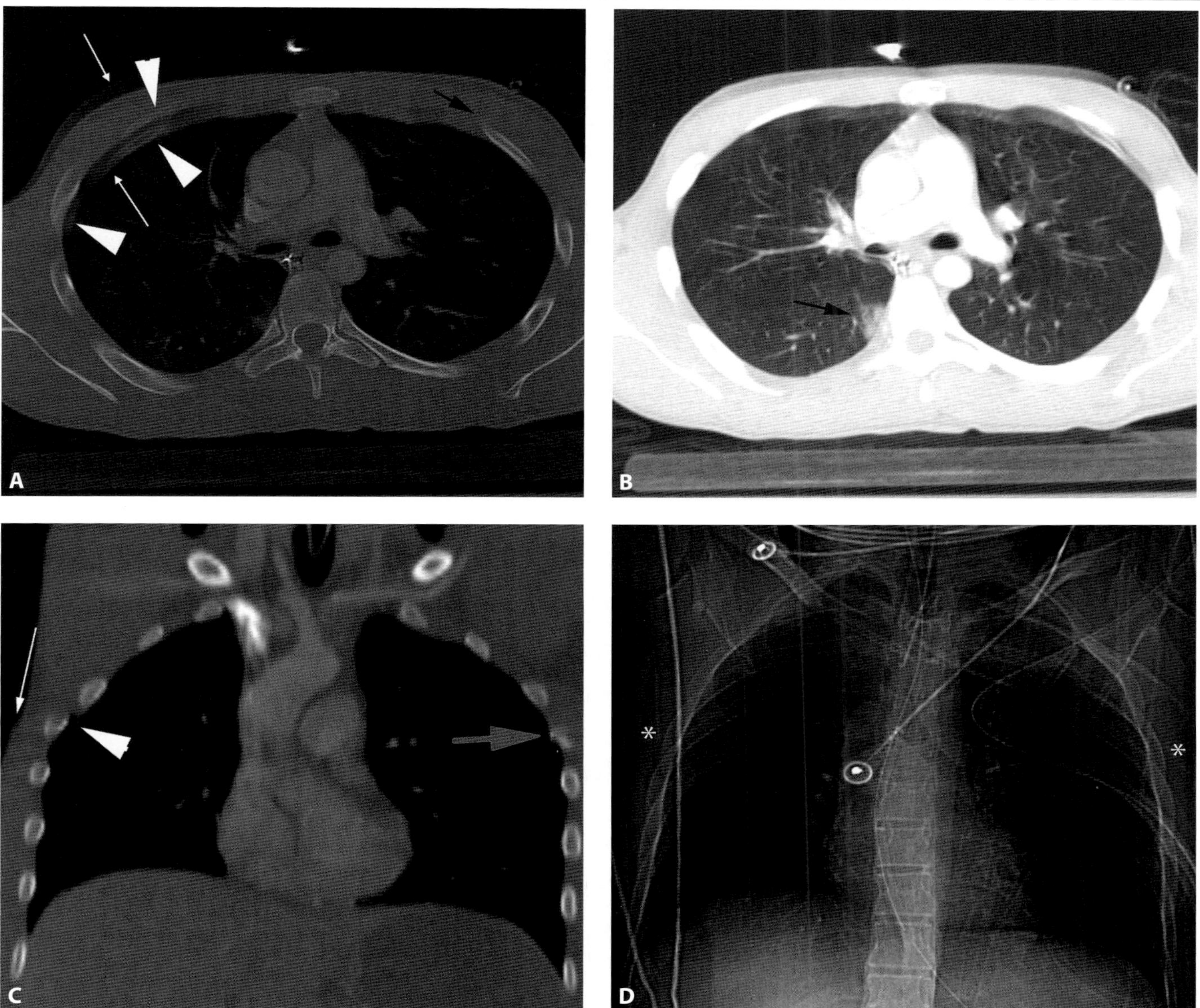

Figure 78.5 A. Axial image from contrast-enhanced chest CT in bone windows in a 21-year-old man after a motor vehicle collision shows step-off of the right anterior ribs (arrowheads) that may initially suggest displaced and overriding fracture fragments. However, ghosting of the chest wall (thin white arrows) indicates chest wall motion. Blurring of the anterior margin of the left-sided rib (black arrow) is also a sign of motion. **B.** Additional signs of motion are seen on an axial image at a similar level in lung windows, including blurring of pulmonary vessels and anterior pleural surface. There is also a pulmonary contusion (arrow) along the right aspect of the spine. **C.** Coronal reformation from chest CT shows the effects of respiratory motion at a single level, with step-off of the left fourth rib (gray arrow), loss of the superior cortex of the right fourth rib (arrowhead), and step-off of the right lateral chest wall contour (thin white arrow). The mediastinal structures are relatively spared by respiratory motion. **D.** Frontal scout image confirms that the bilateral fourth ribs are intact (asterisks).

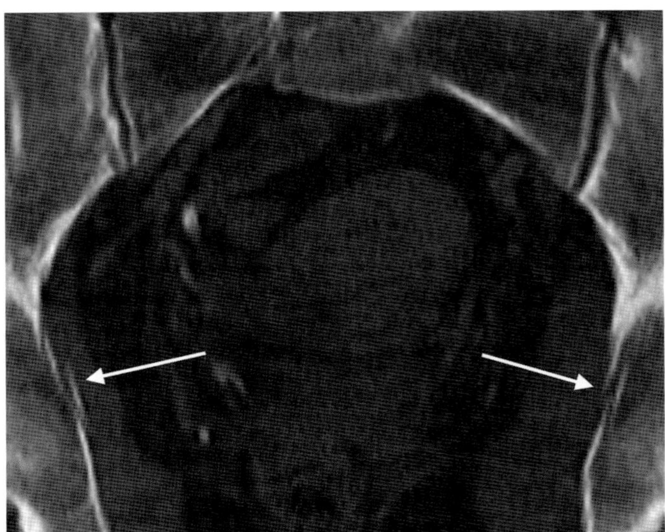

Figure 78.6 Bone window coronal reformation from contrast-enhanced abdominopelvic CT in a 55-year-old woman involved in a high-speed motor vehicle accident shows pseudofractures of the bilateral acetabulae (white arrows) due to motion artifact.

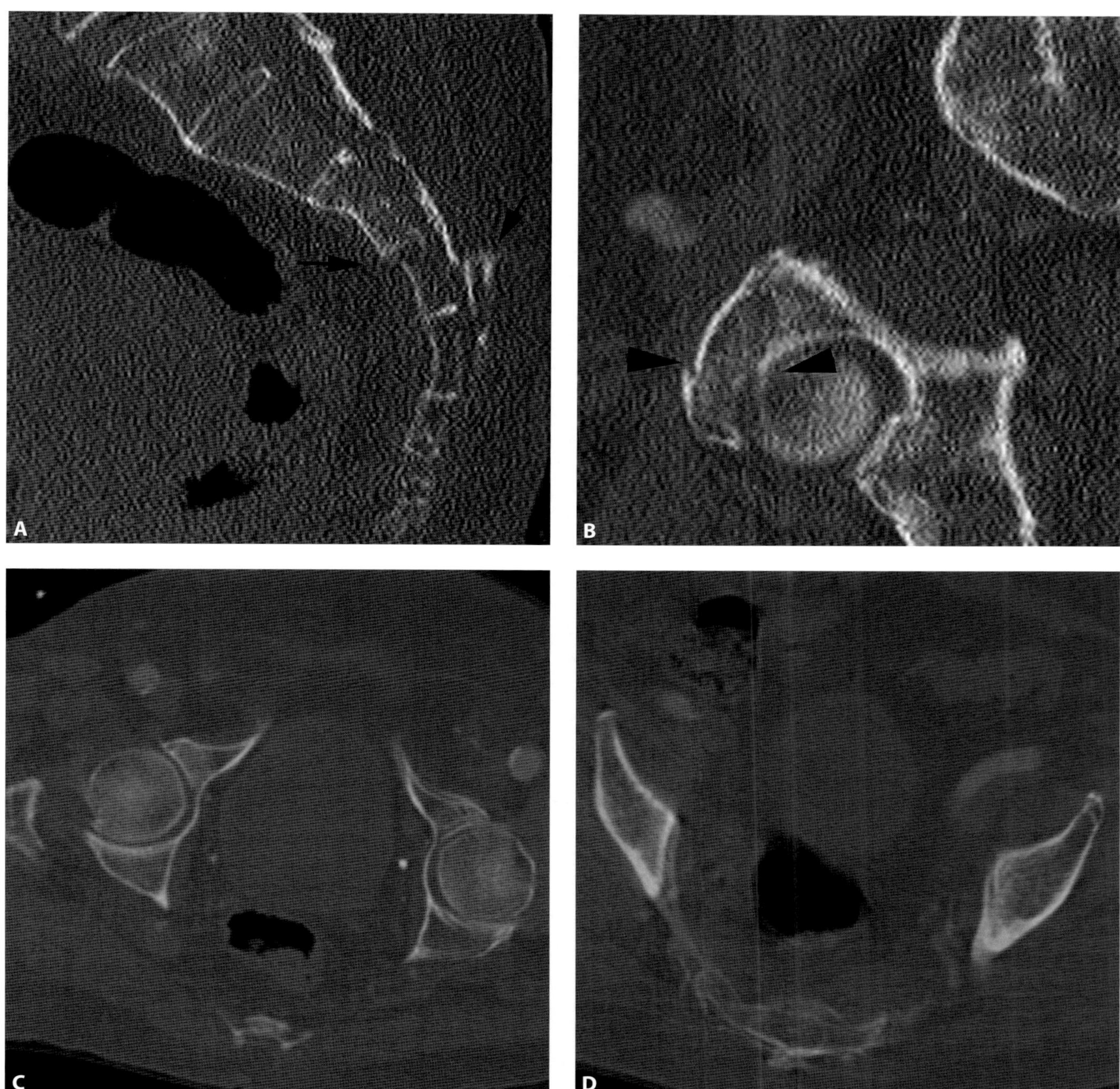

Figure 78.7 A. and **B.** Selected sagittal images from reformatted lumbar spine CT through the sacrum (**A**) and left acetabulum (**B**) in an 89-year-old man with back pain after a fall two days prior show step-off of the sacrum (arrows) and anterior left acetabulum (arrowheads) initially concerning for acute fractures. **C.** and **D.** However, correlation with axial images from CT of the abdomen and pelvis from which the lumbar spine had been reconstructed shows motion at the same levels as the suspected fractures through the sacrum (**C**) and acetabulum (**D**), confirming artifact.

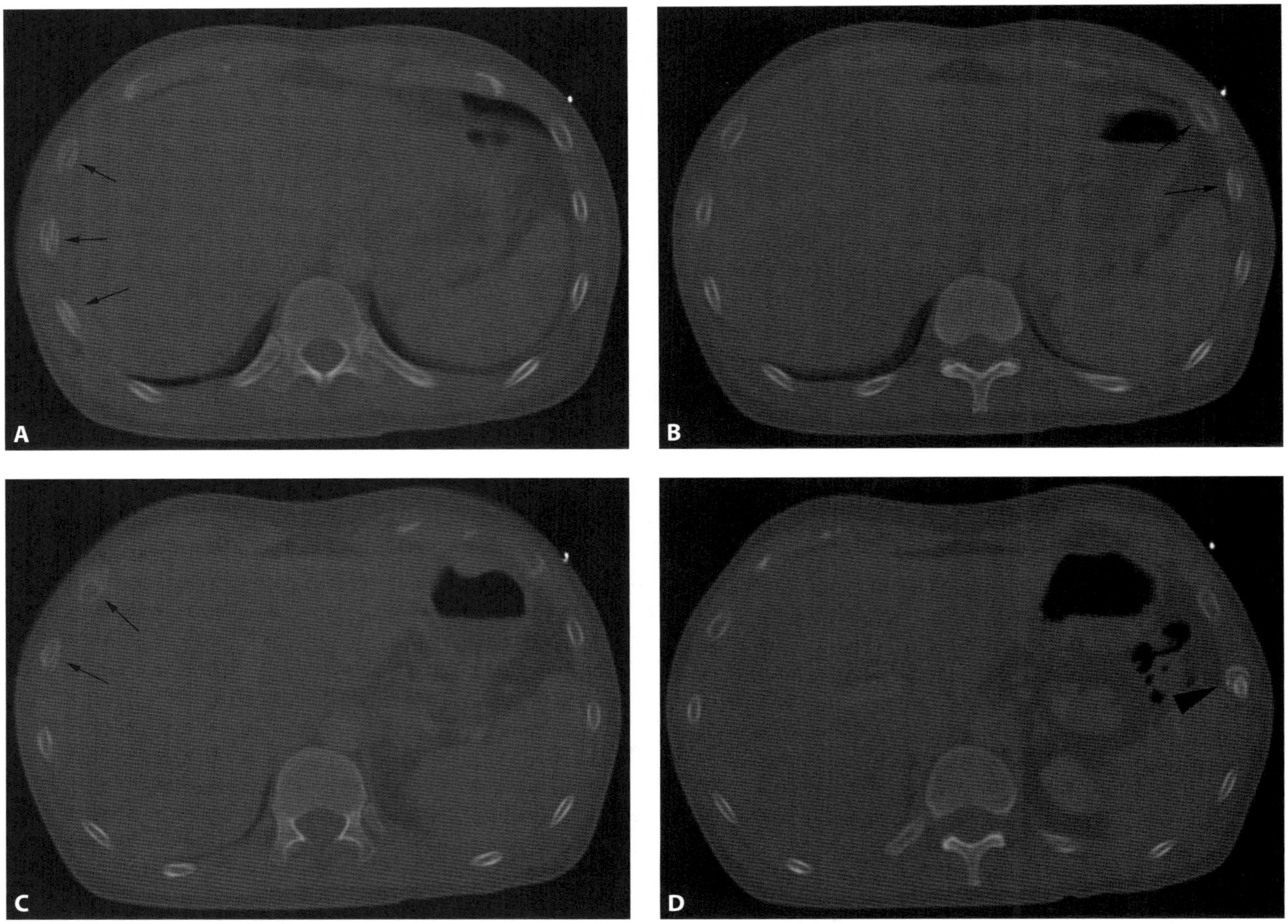

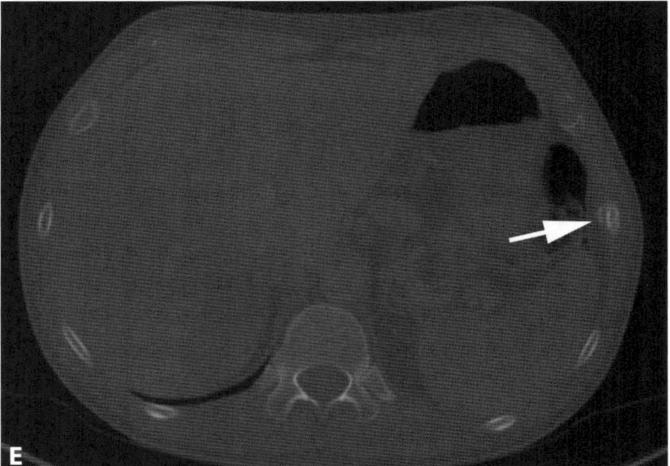

Figure 78.8 A–C. Selected axial images in bone windows from contrast-enhanced abdominal CT in a 42-year-old man with abdominal pain but no recent history of trauma shows multiple pseudofractures of the bilateral ribs (arrows) due to motion artifact. **D.** More inferior axial CT image shows another left rib deformity (arrowhead) that may initially be ignored because of the above motion artifacts. However, close inspection reveals sclerotic margins and lack of blurring expected of motion. **E.** Review of an axial CT image obtained four months earlier indicates at that time a subacute fracture with callus formation (arrow). The lower left rib fracture is therefore real and not artifactual, but could have been overlooked because of the motion artifact at other levels.

CASE **79**

Mach effect

Claire K. Sandstrom

Imaging description

When multiple structures overlap or abut on a radiograph, an optical illusion known as the "Mach effect" may simulate light and dark lines. This enhances edges and in some cases makes the structures easier to differentiate. However, the Mach effect may also simulate a fracture. The Mach effect is a normal phenomenon that results from the physiologic process called "lateral inhibition" in the radiologist's retina [1]. As shown on Figure 79.1, each gray rectangle appears shaded, light at the top and darker at the bottom. Now use a sheet of paper to cover all but one of the gray rectangles – the illusion of a gradient of gray within the remaining rectangle disappears, revealing that each rectangle is, in fact, homogeneous in color.

When the Mach effect is being considered, careful inspection is necessary to avoid misdiagnosis. Additional views may remove the superimposition of structures creating the effect. In other cases, a CT may be necessary to exclude underlying pathology.

Importance

The radiologist must be cognizant of the possibility of the Mach effect and mindful of how to differentiate this from pathology such as a fracture.

Typical clinical scenario

There is possibility for Mach effect on almost any radiograph, so this is an important concept for many different clinical scenarios. In the setting of trauma, the Mach band is most likely to be mistaken for an acute fracture.

Differential diagnosis

Most commonly, the Mach effect, or Mach band, is encountered due to bony overlap at a number of anatomic sites, particularly those with multiple small, closely opposed bones, such as in the spine, wrist, and foot (Figures 79.2–79.6). A fracture should be suspected if there is cortical disruption in addition to the linear lucency (Figure 79.7). Mach effect is particularly common in the posterior-inferior aspect of the cervical vertebral bodies (Figure 79.6), most frequently due to either degenerative osteophytes or hypertrophied transverse processes, and it occurrs in up to 27% of normal adults [2]. Differentiation is usually possible by considering the normal location of cervical spine fractures (anterior vertebral body) and lack of secondary signs of a cervical spine injury.

A Mach band may also create the illusion of pneumomediastinum, when a lucency is seen along the lateral margin of the heart on a chest radiograph (Figure 79.8) [3]. The absence of an opaque line representing the elevated pleura may help differentiate artifact from pneumomediastinum. This is described in Case 30. Nevertheless, a CT may be needed to exclude the presence of gas outlining the mediastinum.

Teaching point

Mach effect can cause a potentially distracting or confusing lucency on any radiograph obtained in the emergency setting. Careful evaluation of the radiograph may be sufficient to allow correct diagnosis, while CT can be used to solve indeterminate cases.

REFERENCES

1. Daffner RH. Visual illusions in the interpretation of the radiographic image. *Curr Probl Diagn Radiol.* 1989;**18**(2):62–87.
2. Daffner RH, Deeb ZL, Rothfus WE. Pseudofractures of the cervical vertebral body. *Skeletal Radiol.* 1986;**15**(4):295–8.
3. Zylak CM, Standen JR, Barnes GR, Zylak CJ. Pneumomediastinum revisited. *Radiographics.* 2000;**20**(4):1043–57.
4. Ciures A. The Mach bands: the visualization of Gibbs phenomenon in the space domain. *J Theor Biol.* 1977;**66**:195–7.

Figure 79.1 These solid gray blocks, when opposed, appear lighter or darker near the edges due to the Mach effect. However, if all but one of the blocks is covered up, this grading effect is removed. This figure is based on a diagram by Alexandru Ciures [4].

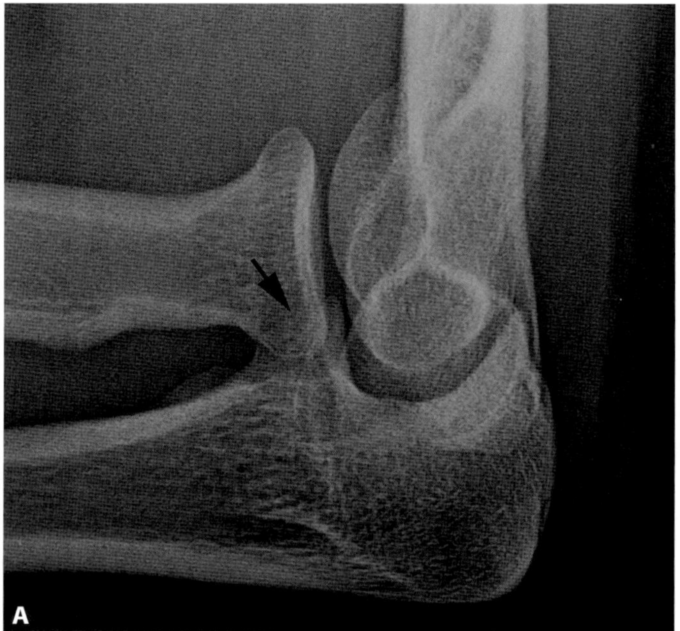

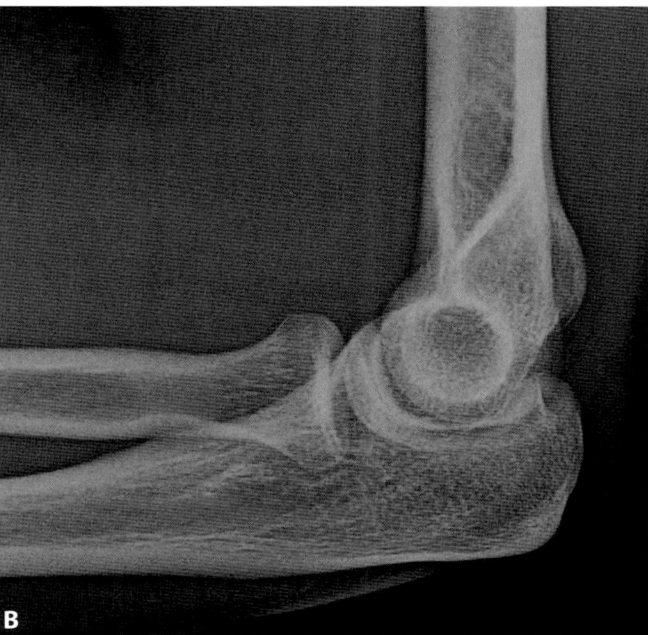

Figure 79.2 **A.** Oblique radiograph of the left elbow in a 28-year-old woman involved in a bicycle accident shows a linear lucency through the radial head due to superimposition on the coronoid process (arrow). **B.** Lateral radiograph of the elbow shows no effusion or other evidence of intra-articular fracture.

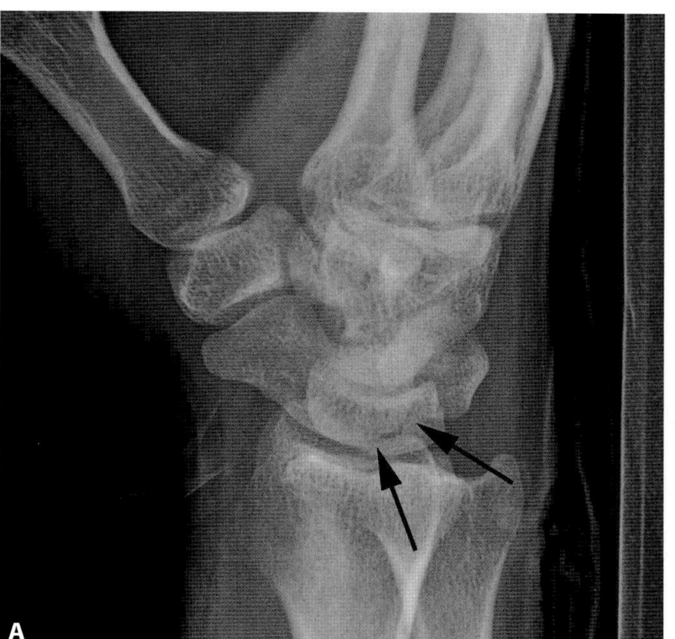

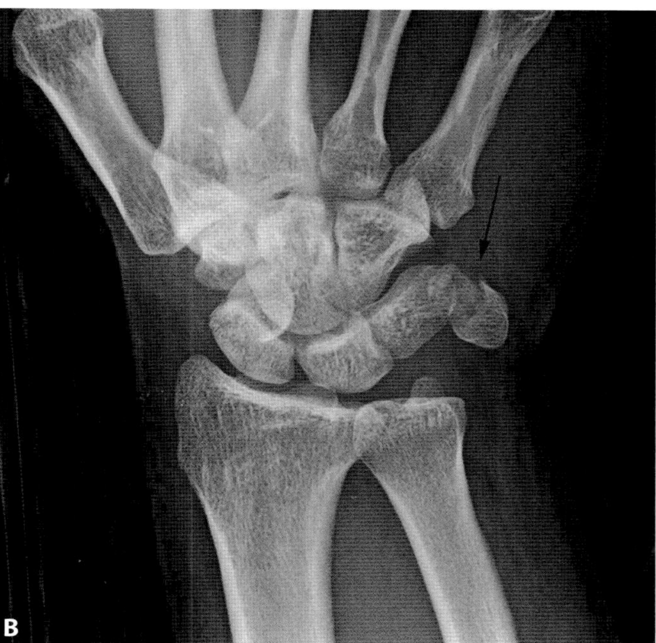

Figure 79.3 **A.** Lateral radiograph of the right hand in a 25-year-old man, obtained at slight obliquity due to difficult positioning, demonstrates a curvilinear lucency in the lunate (arrows) related to Mach effect from overlying scaphoid. **B.** Anterior-posterior radiograph of the hand shows a fracture of the pisiform (arrow), with overlying soft tissue swelling, but confirms the absence of a lunate injury.

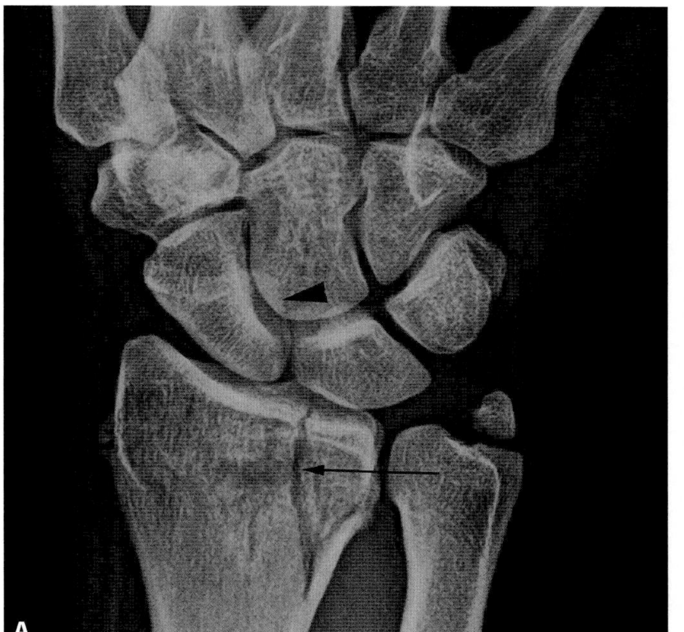

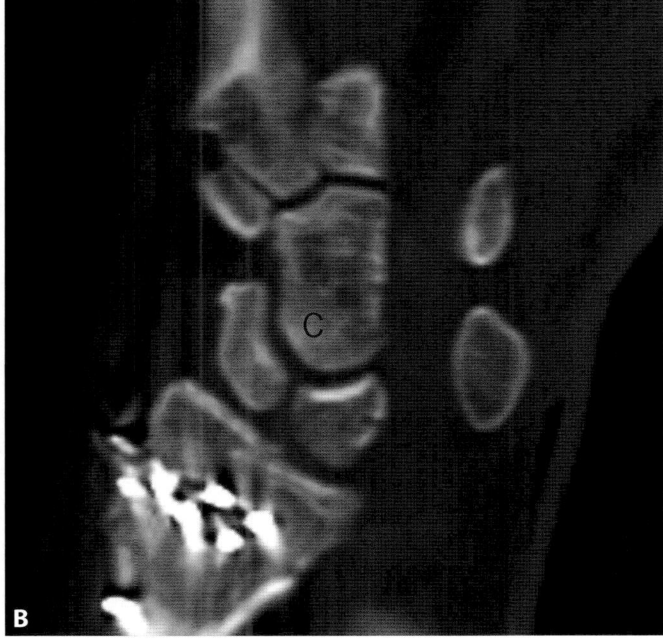

Figure 79.4 **A.** Frontal radiograph of the right wrist in a 53-year-old man following a fall demonstrates a comminuted, intra-articular fracture of the distal radius (arrow) and an ulnar styloid avulsion. There is also a linear lucency in the base of the capitate (arrowhead), which is likely due to Mach effect from the superimposed scaphoid. **B.** Coronal CT of the wrist performed after open reduction and internal fixation of the distal radius fracture shows no fracture of the capitate (*C*). Lucencies in the cortices of the lunate and capitate reflect vascular channels.

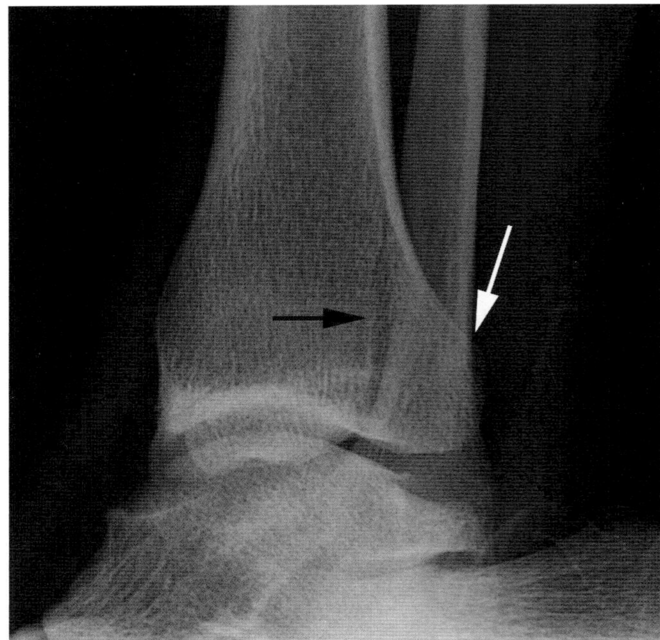

Figure 79.5 Slightly oblique, lateral radiograph of the right tibia in a 34-year-old man following a bicycle accident demonstrates two linear lucencies in the distal tibia, one from superimposition of the distal tibiofibular joint (black arrow) and the second arising from Mach effect of the posterior fibular cortex (white arrow). There is no cortical step-off related to either lucency, correctly identifying them as artifact.

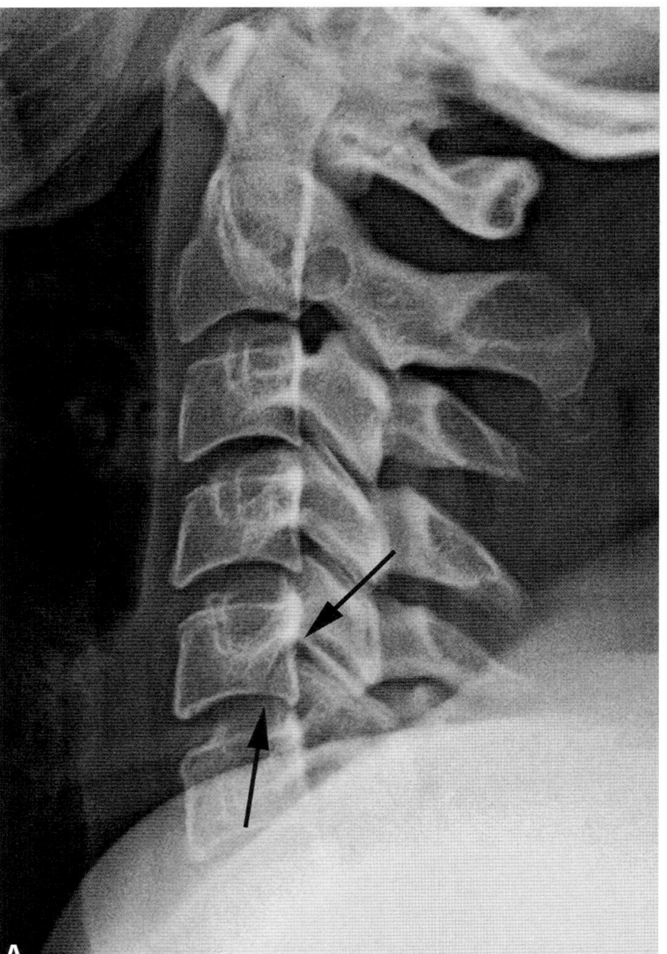

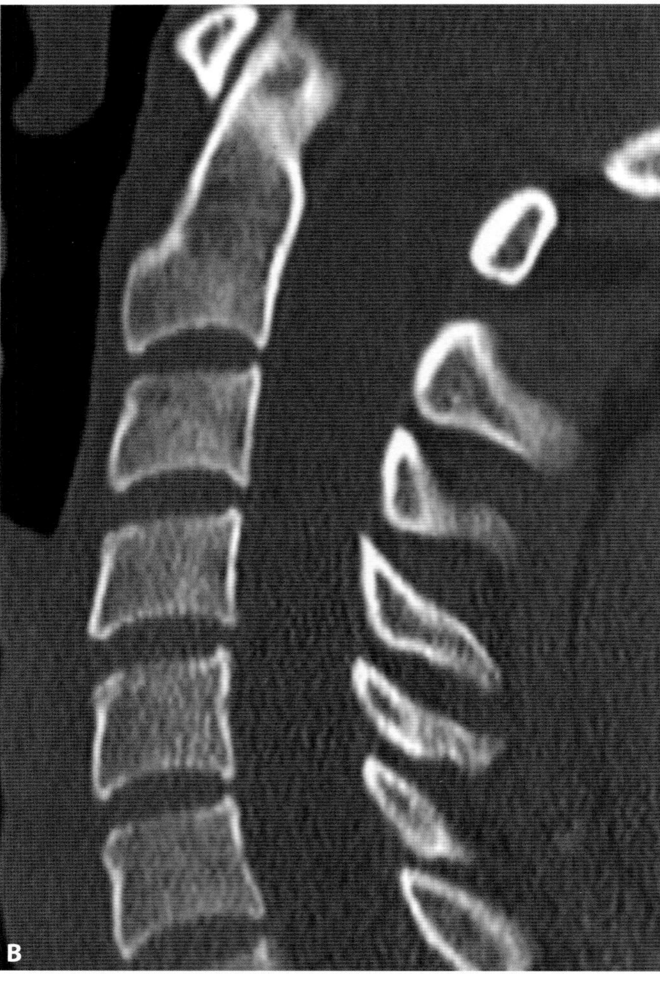

Figure 79.6 A. Lateral radiograph of the cervical spine in a 33-year-old woman involved in a motor vehicle collision demonstrates a linear lucency in the posterior vertebral body of C5 (arrows), presumably related to Mach effect from overlapping osseous structures. **B.** Subsequent cervical spine CT confirms the absence of a vertebral body fracture.

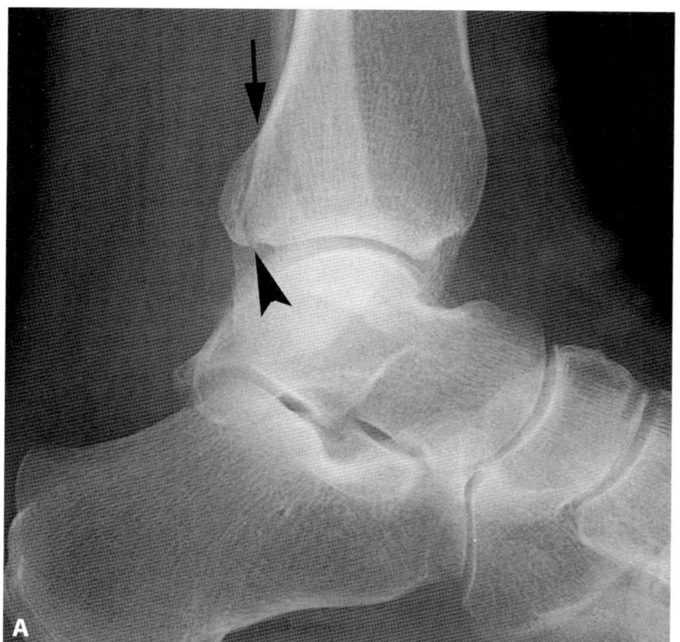

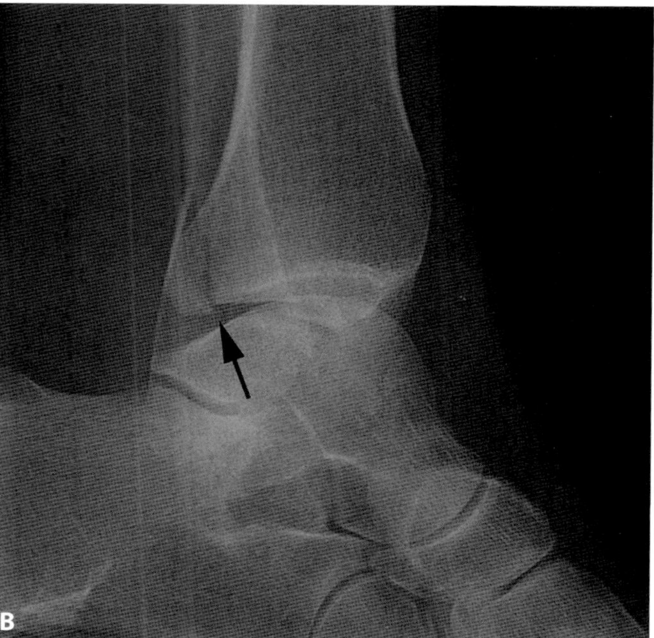

Figure 79.7 A. Lateral radiograph of the left ankle in a 49-year-old woman after twisting her ankle shows a linear lucency in the posterior malleolus which parallels the dorsal cortex of the fibula (arrow) and simulates the Mach line seen in Figure 79.5. However, a fracture was suspected given the irregularity of the articular cortex of the posterior malleolus (arrowhead). **B.** The posterior malleolar fracture is better appreciated on accompanying lateral radiograph of the left tibia (arrow), which was obtained at a slightly different obliquity.

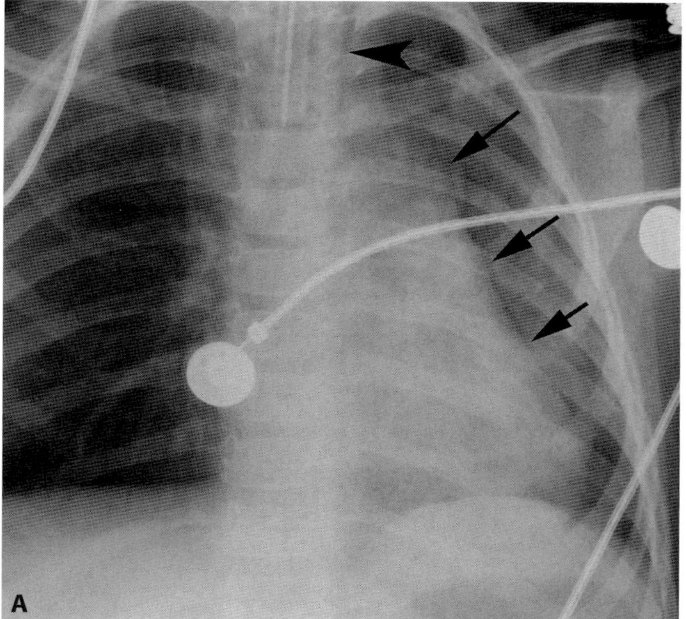

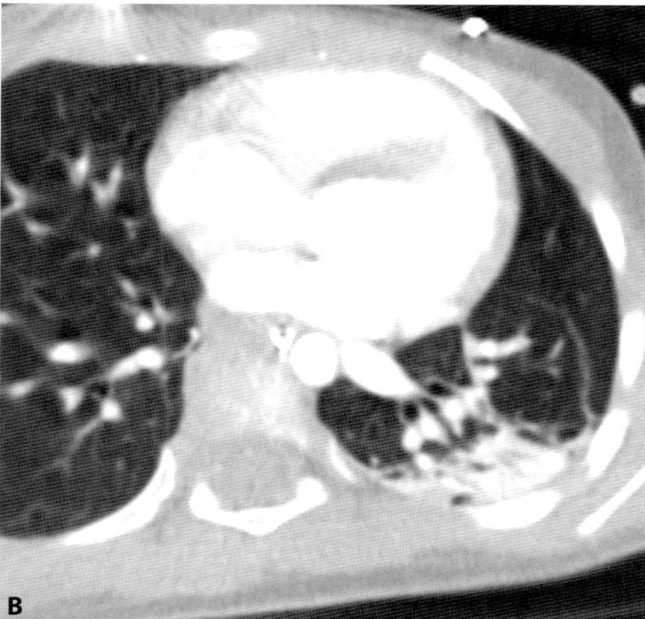

Figure 79.8 A. Frontal radiograph of the chest in a 4-year-old girl who was a victim in an automobile accident. There is a linear lucency bordering the left lateral margin of the heart (arrows), which was interpreted as "suggestive of pneumomediastinum." There was also a linear gas collection in the left superior mediastinum (arrowhead), which could represent gas in the esophagus or pneumomediastinum. Clinical suspicion for a chest or airway injury was low. **B.** A slice through the chest from a subsequent abdominopelvic CT, confirming that the finding on radiograph was pseudopneumomediastinum from Mach effect.

CASE

80 Foreign bodies not visible on radiographs

Ken F. Linnau

Imaging description

When foreign bodies are retained in the subcutaneous or deep soft tissue of the appendicular skeleton, accurate imaging interpretation aims to precisely describe the location and composition of the object, as well as its relationship to the adjacent anatomic structures.

Radiographs are usually obtained if a retained foreign body is suspected, because most retained foreign material is readily visualized on plain films. Radiography is also inexpensive and readily available. At least two orthogonal views are obtained to identify the location of the foreign body.

Unfortunately, some organic foreign bodies such as small slivers of wood (Figure 80.1) and some types of plastic are not clearly visible radiographically [1].

If the location of the foreign body is known, sonography is a sensitive and specific method for the detection of almost all superficial foreign bodies (Figure 80.2). Additionally, sonography may guide foreign body removal [2]. If a foreign body is detected using sonography, a good practice is to mark the skin directly over the location of the foreign body to guide surgical removal.

For small deep organic foreign bodies, CT scanning is a valuable imaging alternative, which allows depiction of wood that is occult radiographically. If a wooden foreign body is suspected, view images with a wide window (between 500 and 1000 Hounsfield Units [HU]) setting [1].

Table 80.1 summarizes the appearances of the most common foreign bodies on CT, radiography, ultrasound, and MRI.

Importance

Failure to identify and remove foreign bodies can cause acute or late complications, which range from inflammation, infection, tumors (rarely) and long-term retention [1–3].

MRI provides excellent soft tissue contrast, assisting with the detection of small foreign bodies [3]. In our experience, MRI is not helpful for the detection of some retained foreign material. Particularly, glass from car windshields can cause substantial susceptibility artifact on MR, likely due to metallic components in the glass, such as tinting. MR is contraindicated for the detection of ferromagnetic foreign bodies as they can move during MR evaluation.

Typical clinical scenario

A patient presents with a soft tissue laceration of an extremity and a retained foreign body is suspected.

Table 80.1 Common foreign bodies: visibility on imaging modalities

Material	CT (windows)	Radiography	Ultrasound	MRI
WOOD				
Tongue depressor	Lung	Not visible	Hyperechoic	Not visible
Toothpick	Lung	Not visible	Hyperechoic	Not visible
Dry bamboo	Bone, soft tissue	Visible	Hypoechoic	Not visible
ROAD GRAVEL	Bone, lung, soft tissue	Visible	Shadowing	Not visible
ALUMINUM	Bone, soft tissues	Visible	Shadowing	T1,T2, and STIR artifact
GLASS				
Window glass (non-tinted)	Bone, lung, soft tissue	Visible	Hyperechoic	T1 hypointense, T2 and STIR artifact
Windshield glass (tinted)	Bone, lung, soft tissue	Visible	Hyperechoic	T1, T2, and STIR artifact
Bottle glass	Bone, lung, soft tissue	Visible	Hyperechoic	T1, T2, and STIR artifact
PLASTIC				
#1 plastic	Not visible	Visible	Hyperechoic	Not visible
#2 plastic	Not visible	Not visible	Hyperechoic	Not visible
#3 plastic	Soft tissue	Faintly visible	Hyperechoic	Not visible
#5 plastic	Bone, lung, soft tissue	Visible	Hyperechoic	Not visible
Compostable fork	Soft tissue	Faintly visible	Hyperechoic	Not visible
FISH BONE	Bone, lung, soft tissue	Visible	Hyperecholic	Not visible

Note: The conspicuity of any foreign body depends on its size, shape, and superimposition of other body structures.

Differential diagnosis

None.

> **Teaching point**
>
> Wood and some types of plastic are radiolucent. CT is particularly useful at localizing foreign bodies and determining their relationship to surrounding structures. Ultrasound can identify all types of foreign bodies if the location is known and the foreign bodies is in the plane of the transducer. MRI is usually not helpful in detecting foreign bodies and many types of glass will cause significant artifact.

REFERENCES

1. Peterson JJ, Bancroft LW, Kransdorf MJ. Wooden foreign bodies: imaging appearance. *AJR Am J Roentgenol.* 2002;**178**(3): 557–62.
2. Callegari L, Leonardi A, Bini A, *et al.* Ultrasound-guided removal of foreign bodies: personal experience. *Eur Radiol.* 2009; **19**(5):1273–9.
3. Specht CS, Varga JH, Jalali MM, Edelstein JP. Orbitocranial wooden foreign body diagnosed by magnetic resonance imaging. Dry wood can be isodense with air and orbital fat by computed tomography. *Surv Ophthalmol.* 1992;**36**(5):341–4.

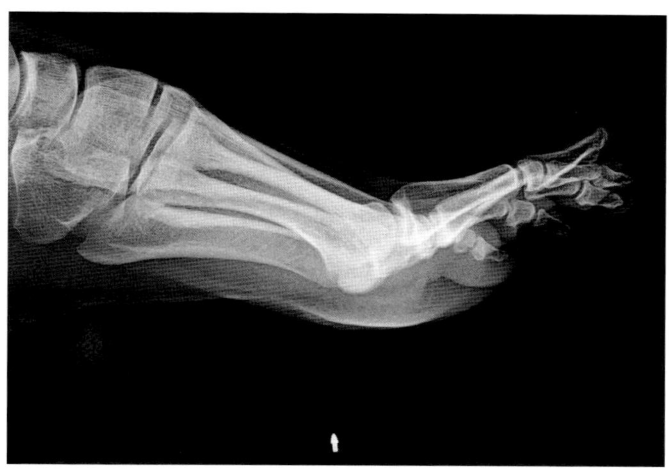

Figure 80.1 Lateral radiograph of the foot of a 16-year-old woman who stepped onto a deck and "felt a wood splinter go through her foot." The ballpoint pen tip at the inferior margin of the radiograph indicates a small skin laceration. No foreign body is detected radiographically.

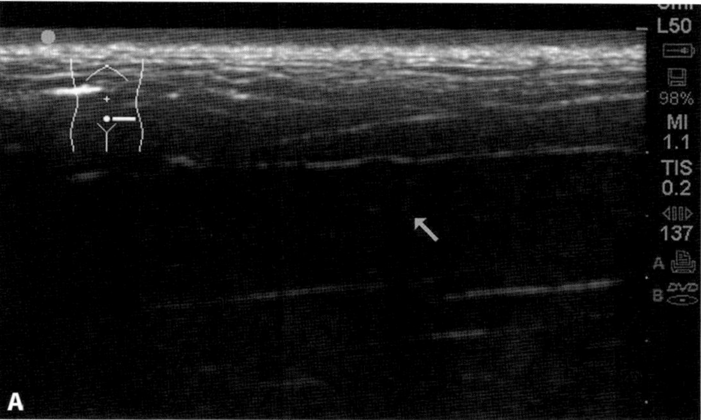

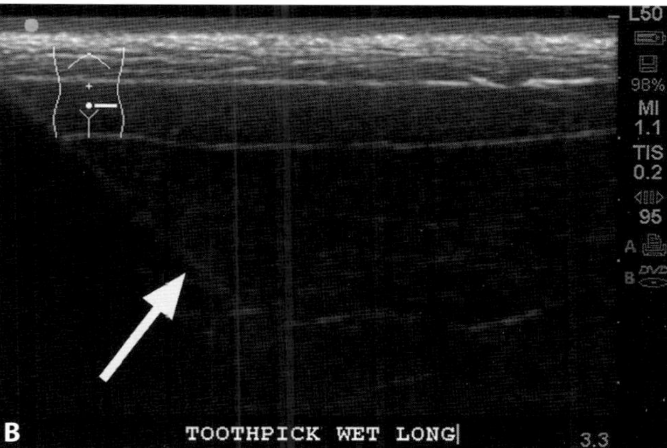

Figure 80.2 A. Transverse sonographic image of a tooth pick inserted into a ham phantom shows the wooden foreign body as a mildly hyperechoic focus (arrow) with acoustic shadowing. **B.** In the longitudinal plane, the full length of the tooth pick is easily appreciated (arrow).

CASE **81**

Accessory ossicles

Claire K. Sandstrom

Imaging description

A radiograph obtained for acute trauma or musculoskeletal extremity pain reveals an extraosseous structure. Close inspection reveals that the bone is round and well corticated on all sides, as opposed to the appearance expected for an acutely avulsed fragment of bone. Perhaps correlation with physical exam reveals no focal tenderness at the site of the accessory ossicle, effectively excluding acute fracture. Alternatively, if the physical exam is unreliable or if clinical suspicion persists, a CT might show the complete cortication and lack of osseous donor site. An MRI may be obtained if there is concern that the accessory ossicle is itself the source of ongoing subacute or chronic pain, in which case bone marrow edema would be revealed on both sides of the synchrondrosis between the accessory ossicle and parent bone.

Importance

The number of described accessory ossicles is very large. Their clinical significance lies in the fact that they can simulate acute avulsions radiographically and they can be symptomatic themselves.

Typical clinical scenario

Accessory ossicles are small bones that result from failure of fusion of the ossification centers of the parent bone, and represent anatomic variants. They are most commonly found in the ankle and foot. Twenty-four ossicles are depicted in the ankle and foot section of Keats and Anderson's *Atlas of Normal Roentgen Variants That May Simulate Disease* [1], but more variants in this region exist. The most common accessory ossicles seen in the foot include the os trigonum, accessory navicular, and os vesalianum [2]. The os peroneum is also relatively common but is actually a sesamoid, discussed in greater detail in Case 85.

Differential diagnosis

Figure 81.1 illustrates the location of many pertinent accessory ossicles in the foot and the fractures that may mimic them.

The os trigonum (Figure 81.2) is a large accessory ossicle that forms posterior to the talus and can become symptomatic from forced plantar flexion of the foot, for example in acute traumatic fracture or chronic microtrauma in dancers. The latter, so-called the os trigonum syndrome, can be diagnosed by MRI, while radiographs and CT may be sufficient in the setting of acute trauma [3].

There are several subtypes of accessory navicular (also called os tibiale externum), including a small sesamoid (type I) within the distal tibialis posterior tendon (Figure 81.3), a larger ossicle (type II) joined to the navicular by a cartilaginous synchrondrosis (Figures 81.4 and 81.5), and a cornuate navicular (type III), essentially an ossicle fused to the navicular by an osseous bridge to form a prominent navicular tuberosity [3, 4]. Medial foot pain can occur as a result of the accessory navicular, again due to chronic microtrauma or acute fracture. In either case, MRI may show bone marrow edema.

An os calcaneus secundarius (Figure 81.6) is an accessory ossicle near the anterior process of the calcaneus and can therefore mimic an acute fracture of the anterior process [5]. While seen best radiographically on the medial oblique view, a CT is often helpful for characterization.

The os intermetatarseum (Figure 81.7) is not particularly common in population studies but because of its location it can be confused radiograpically with a Lisfranc fracture-dislocation [4]. MRI or stress views may be helpful when trying to exclude the latter clinically significant injury.

A number of other accessory ossicles occur in the ankle and foot (Figures 81.8–81.11).

At the shoulder, the acromion is normally formed by four ossification centers (pre-acromion, meso-acromion, and meta-acromion, all fusing to the basiacromion). Hence there are several potential variants of os acromiale (Figures 81.12 and 81.13), the most common of which is the meso-acromion [3]. The normal acromial apophysis can persist into the late second and early third decades of life [6], so care must be taken when evaluating the shoulder in children and teenagers. While best seen on the axillary radiographic view of the shoulder, os acromiale may appear as a double density sign on the anterior-posterior (AP) view, as shown in Figure 13 [7]. In adults, an os acromiale is most commonly confused for an acute acromial fracture (Figure 81.14). The os acromiale may predispose to rotator cuff tears from downward displacement by the deltoid but this is controversial [6].

Teaching point

Accessory ossicles may mimic avulsion fractures, or may cause pain, and should be considered when small ossified structures are seen in characteristic locations.

REFERENCES

1. Keats TE, Anderson MW. *Atlas of Normal Roentgen Variants That May Simulate Disease*, 8th edn. Philadelphia: Mosby Elsevier; 2007.
2. Coskun N, Yuksel M, Cevener M, *et al.* Incidence of accessory ossicles and sesamoid bones in the feet: a radiographic study of the Turkish subjects. *Surg Radiol Anat.* 2009;**31**(1):19–24.

3. Kalantari BN, Seeger LL, Motamedi K, Chow K. Accessory ossicles and sesamoid bones: spectrum of pathology and imaging evaluation. *Appl Radiol.* 2007;**36**(10):28–32.
4. Mellado JM, Ramos A, Salvado E, *et al.* Accessory ossicles and sesamoid bones of the ankle and foot: imaging findings, clinical significance and differential diagnosis. *Eur Radiol.* 2003;**13** Suppl 6:L164–77.
5. Hodge JC. Anterior process fracture or calcaneus secundarius: a case report. *J Emerg Med.* 1999;**17**(2):305–9.
6. Sammarco VJ. Os acromiale: frequency, anatomy, and clinical implications. *J Bone Joint Surg.* 2000;**82**A(3):394–400.
7. Lee DH, Lee KH, Lopez-Ben R, Bradley EL. The double-density sign: a radiographic finding suggestive of an os acromiale. *J Bone Joint Surg.* 2004;**86**A(12):2666–70.
8. Zatzkin HR. Trauma to the foot. *Semin Roentgenol.* 1970; **5**(4):419–35.

Table 81.1 Accessory ossicles of the ankle and foot

#	Accessory ossicle	Avulsion mimicker
1	Intercalary bones (sesamoids) between talus and malleoli	Medial or lateral cortex of talus
2	Os subfibulare	Fibular tip
3	Os subtibiale	Medial malleolar tip
4	Talus secundarius	Lateral process of the talus
5	Talus accessorius	Medial talus
6	Os trigonum	Posterior talar process or posterior malleolus
7	Accessory navicular (os tibiale externum)	Medial border of navicular
8	Os supratalare	Superior head of talus
9	Os supranaviculare	Superior cortex of navicular
10	Os infranaviculare	Inferior cortex of navicular or superior cuneiform
11	Calcaneus secundarius	Anterior process of calcaneus
12	Os peroneum	Proximal lateral cuboid
13	Os vesalianum	Base of fifth metatarsal
14	Os intermetatarseum	Lisfranc fracture-dislocation

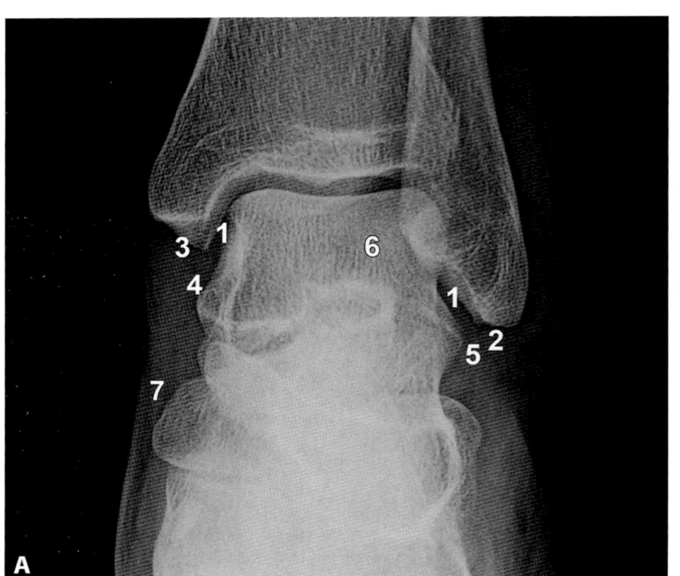

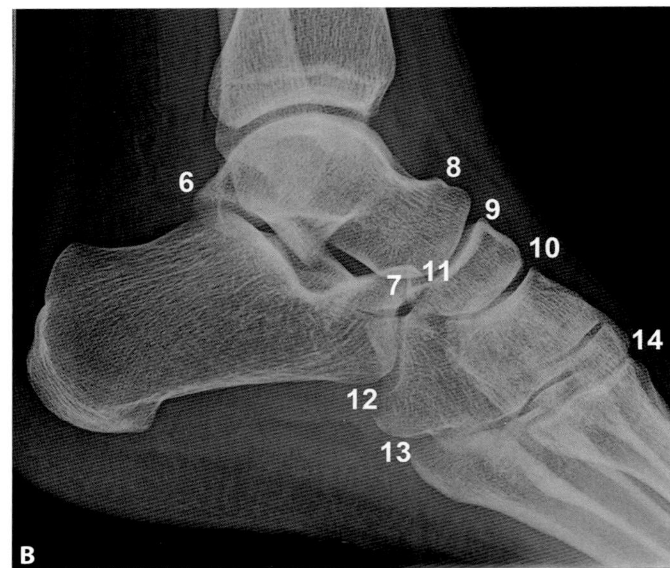

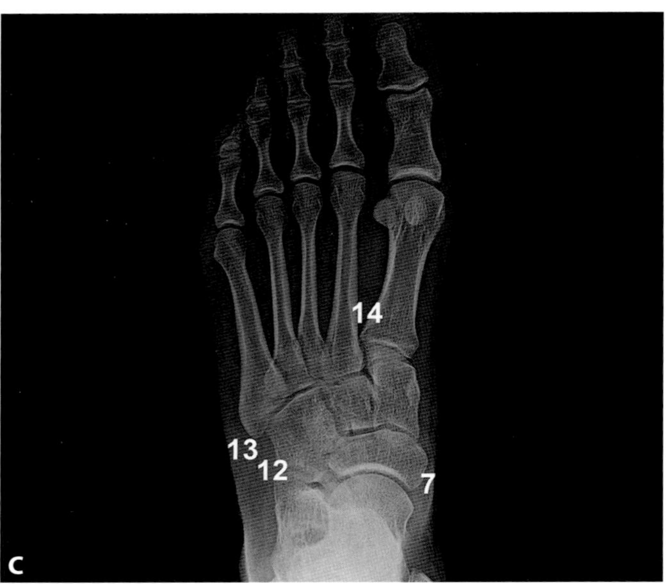

Figure 81.1 Anterior-posterior (AP) (**A**) and lateral (**B**) views of the ankle and AP view (**C**) of the foot in a 33-year-old woman with ankle pain shows soft tissue swelling along the lateral ankle but normal alignment of osseous structures and no acute fracture. Either accessory ossicles or avulsion fractures may be seen at the numbered sites listed in Table 81.1 (Adapted from [1] and [8]).

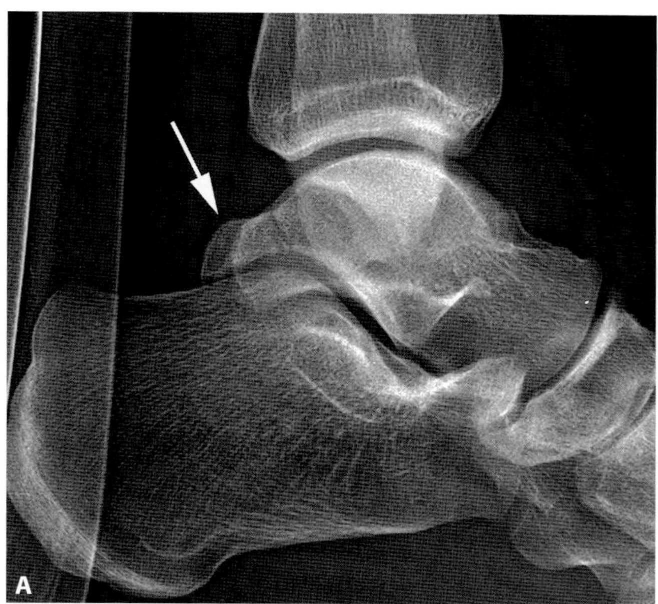

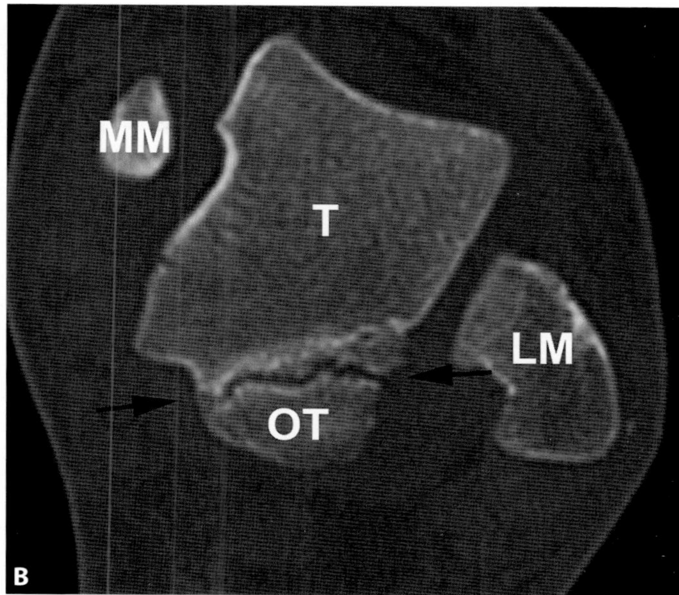

Figure 81.2 **A.** Lateral ankle radiograph from a 24-year-old woman complaining of bilateral ankle pain shows an os trigonum (white arrow), as well as a joint effusion (black arrowheads). **B.** Axial image from CT of the ankle in bone windows shows the pseudoarthrosis (black arrows) between the os trigonum (OT) and talus (T), at the level of the medial (MM) and lateral malleoli (LM).

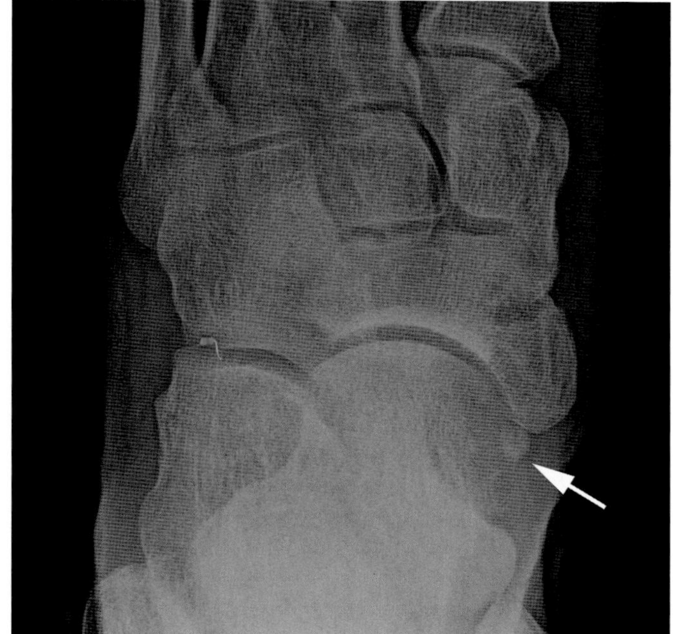

Figure 81.3 Frontal view of the foot in a 41-year-old man complaining of atraumatic foot pain shows a type I os tibiale externum (arrow).

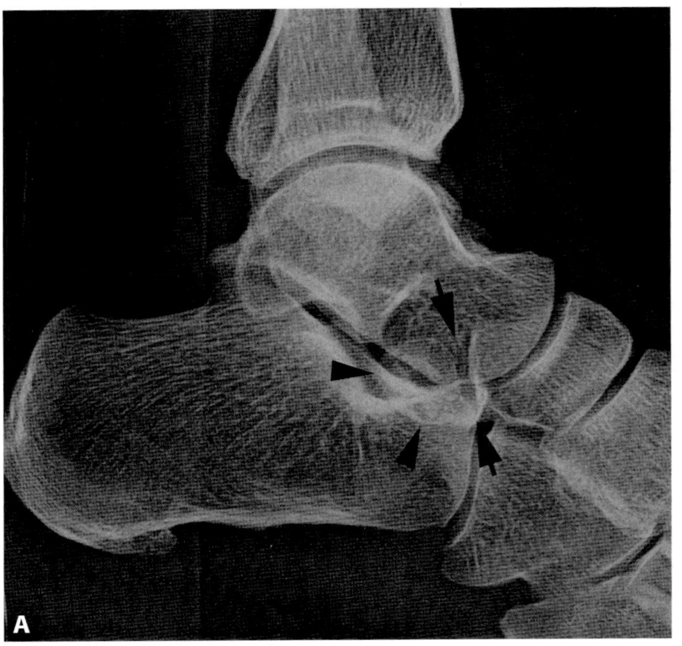

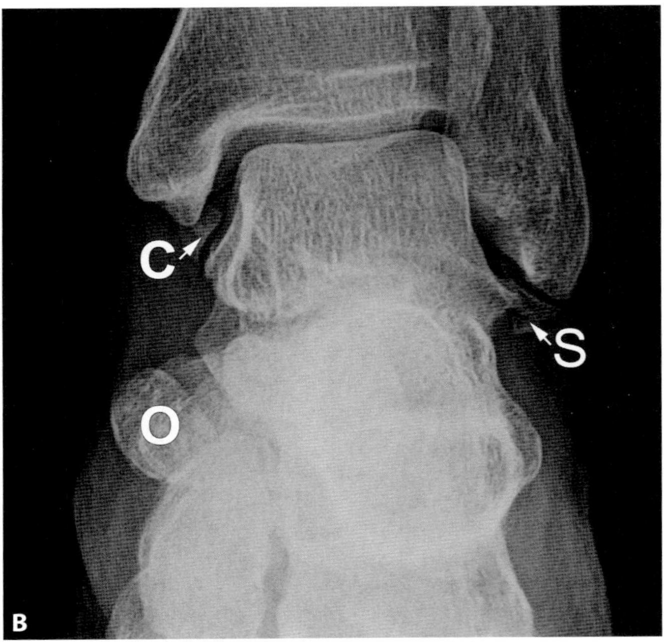

Figure 81.4 A. Lateral view of the ankle in a 55-year-old woman with chronic left ankle pain demonstrates an accessory navicula (black arrowheads) articulating via synchondrosis (thin black arrows) with the navicular. Given the size and cartilaginous articulation with the navicular, this is a type II os navicular. **B.** On the AP view, the accessory navicular is seen as a large round osseous structure (O) along the medial aspect of the foot. Incidentally, a medial intercalary bone (C) and an os subfibiale (S) are also seen at the ankle.

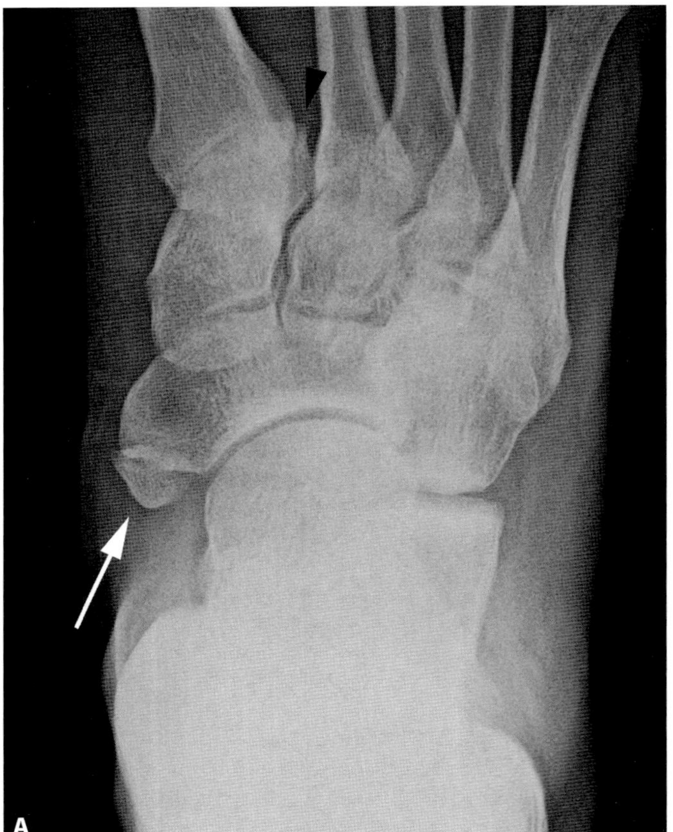

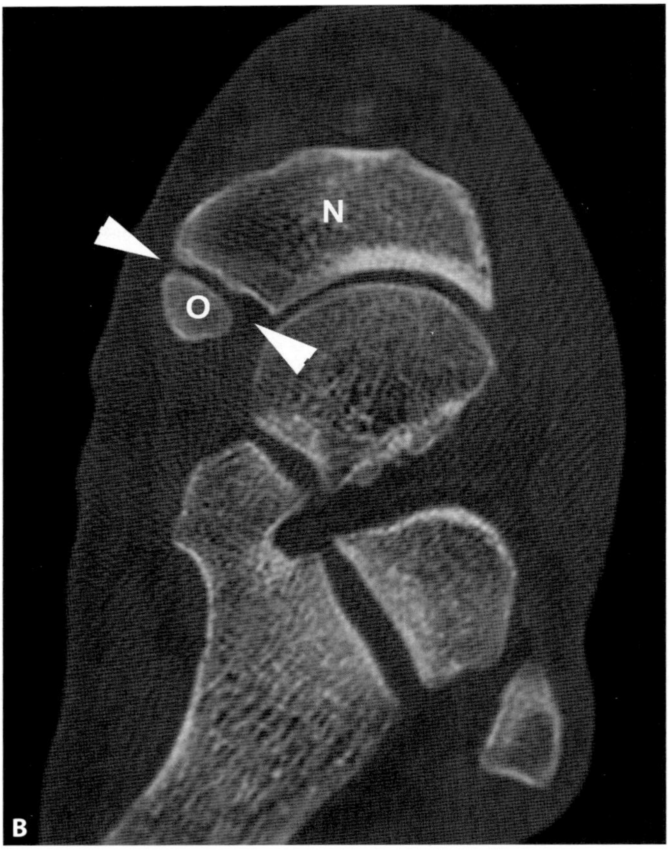

Figure 81.5 A. AP view of the foot in a 17-year-old man who was run over by a truck demonstrates a bony structure along the medial aspect of the navicular (arrow). The margin is not adequately visualized to allow differentiation of an accessory navicular from an acute fracture. A Lisfranc injury was diagnosed based on the subtle malalignment of the first metatarsal and medial cuneiform (arrowhead). **B.** Subsequent CT of the foot shows complete cortication of the accessory navicular (O) and the synchondrosis (arrowheads) with the navicular (N).

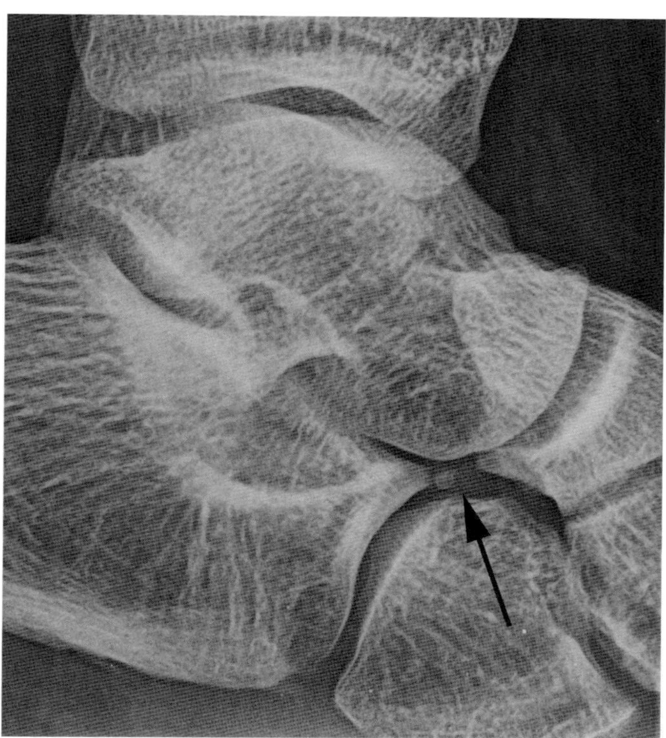

Figure 81.6 Lateral ankle radiograph from a 56-year-old man with foot pain following a motor vehicle collision shows a small calcaneus secundarius (black arrow) adjacent to the anterior process of the calcaneus.

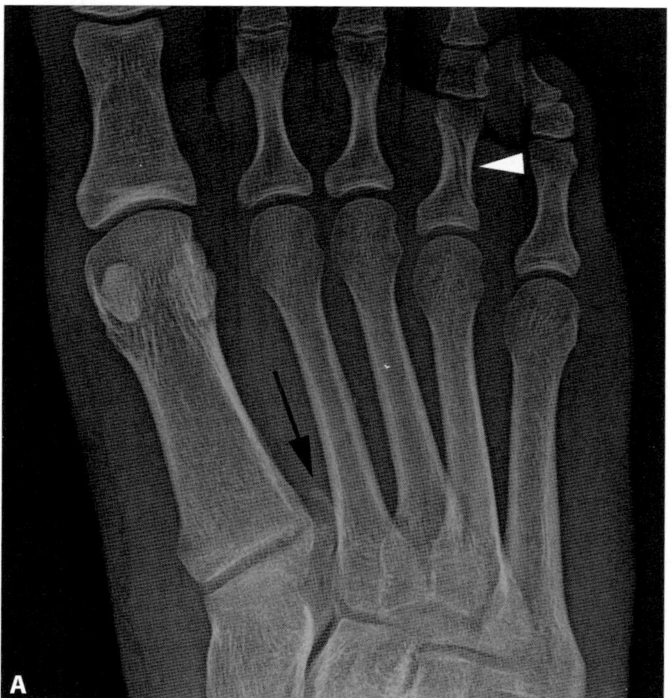

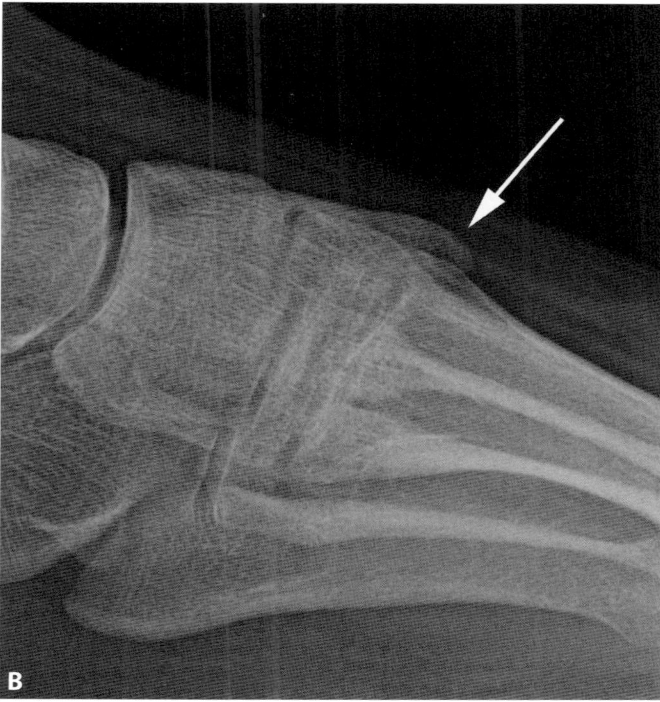

Figure 81.7 A. AP view of the right foot in a 23-year-old woman with fourth digit pain and a non-displaced spiral fracture (arrowhead) of the fourth proximal phalanx also shows a well-corticated osseous structure between the bases of the first and second metatarsals (arrow). **B.** The lateral view of the foot shows the os intermetatarseum (arrow) projecting superior to the tarsometatarsal joints.

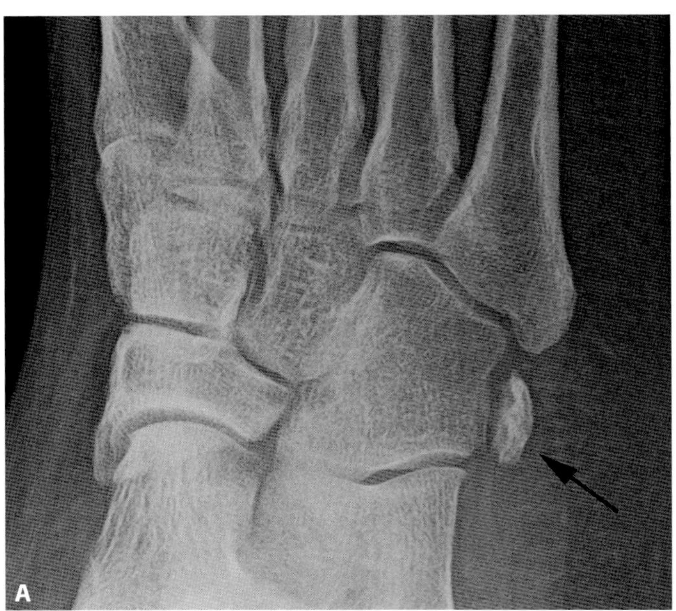

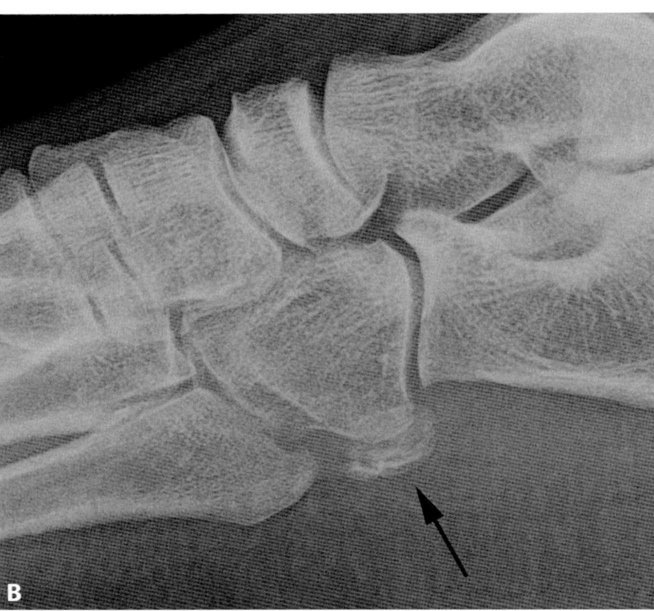

Figure 81.8 Oblique (**A**) and lateral (**B**) views of the foot in a 55-year-old woman with foot pain and erythema show an os vesalianum (arrow) in expected position.

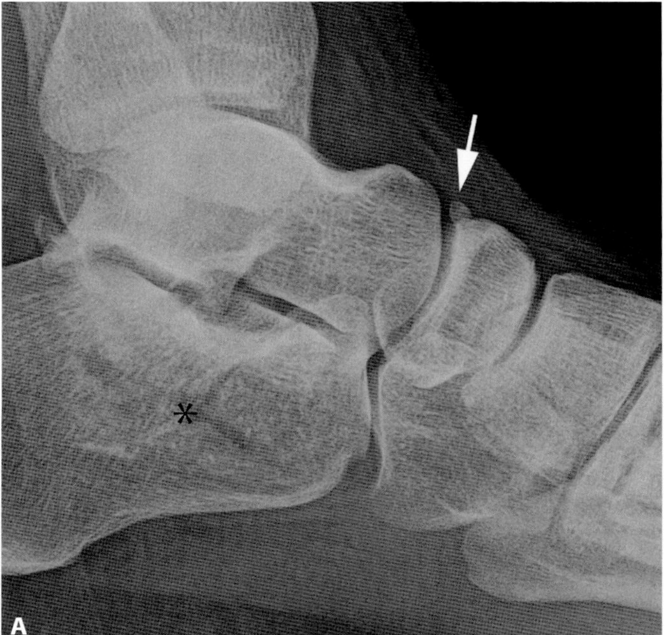

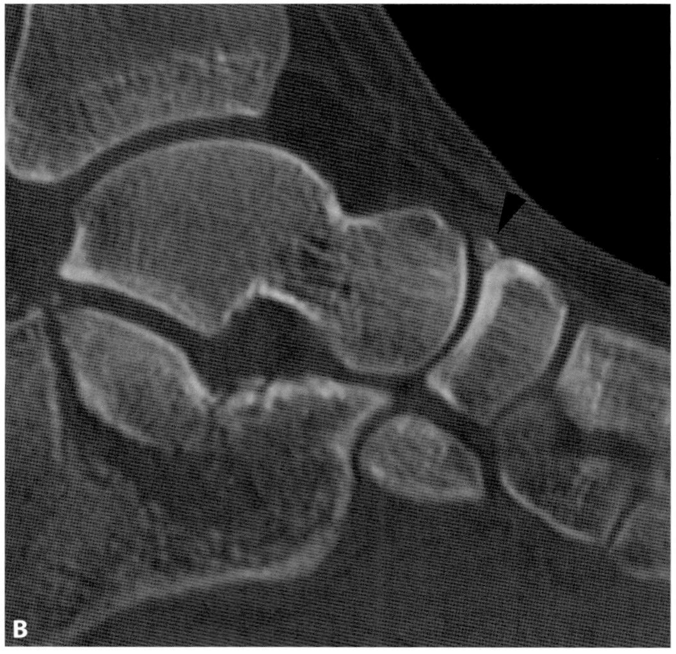

Figure 81.9 **A.** Lateral radiograph of the foot in a 26-year-old woman involved in a high-speed motor vehicle collision with obvious calcaneal fracture (asterisk) shows an os supranaviculare (arrow). **B.** CT for surgical planning of the calcaneal fracture confirms the well-corticated margins of the os supranaviculare (arrowhead).

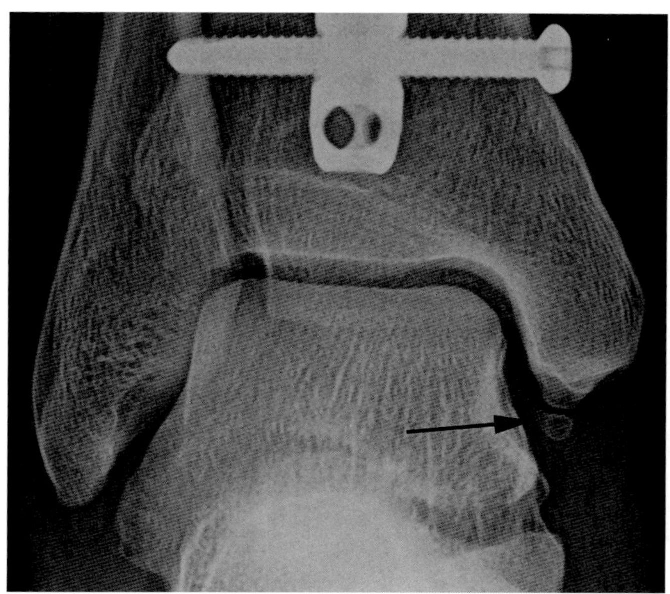

Figure 81.10 Frontal ankle radiograph from a 49-year-old man who recently had intramedullary nailing of a tibial shaft fracture secondary to a motorcycle crash shows an os subtibiale (thin black arrow) adjacent to the tip of the medial malleolus.

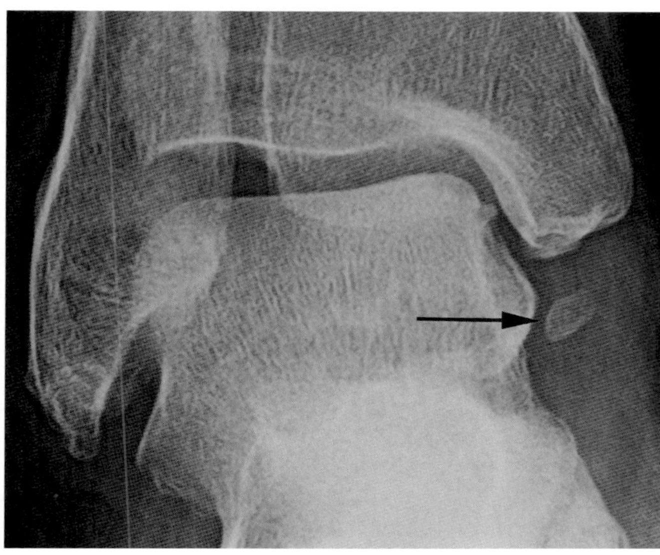

Figure 81.11 Frontal ankle radiograph from a 47-year-old woman who twisted her ankle two days earlier shows an accessory talus (arrow) along the medial aspect of the talus, slightly more distal than the medial intercalary bone seen in Figure 81.4B and the os subtibiale seen in Figure 81.10.

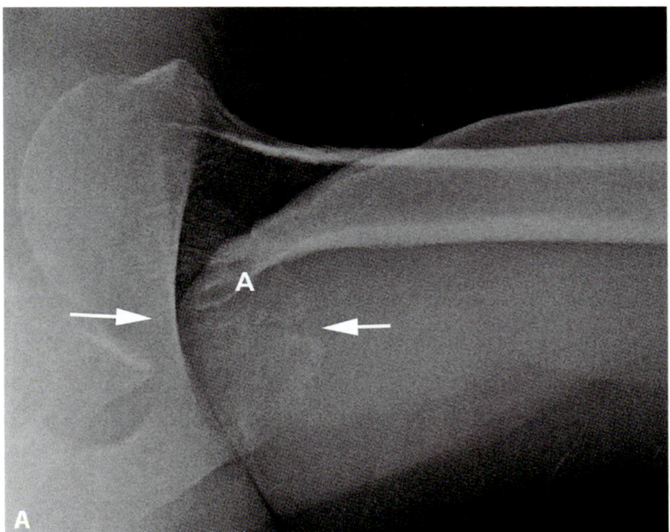

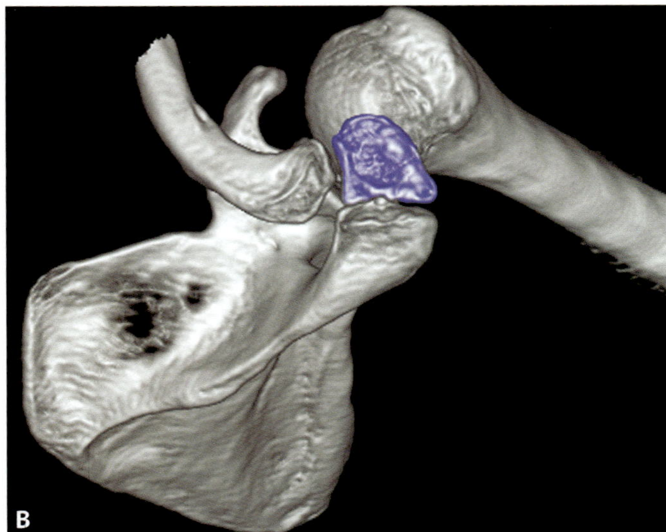

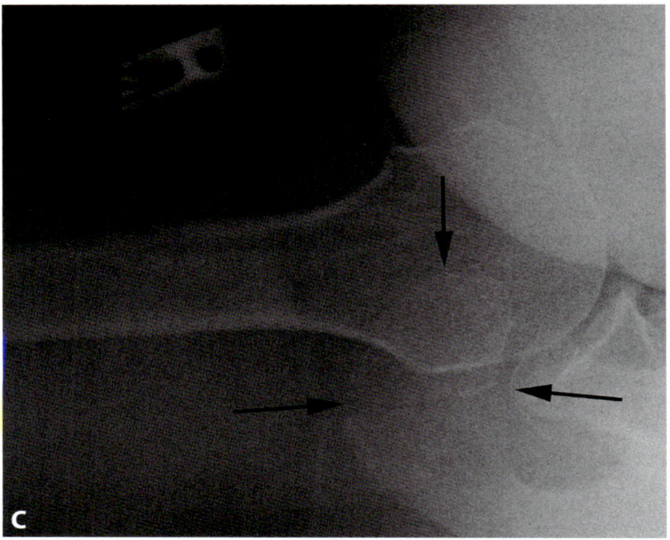

Figure 81.12 **A.** Axillary view in a 67-year-old woman with left shoulder pain following a ground-level fall shows an os acromiale (A) separated from the base of the acromion by a synchondrosis (arrows). A glenoid fracture was also detected (not shown), prompting CT. **B.** At CT, the rotated 3D reconstruction shows the accessory ossicle (blue shading) articulating with the acromium and distal clavicle. **C.** Axillary view of the right shoulder three years prior shows os acromiale (arrows) on the contralateral side.

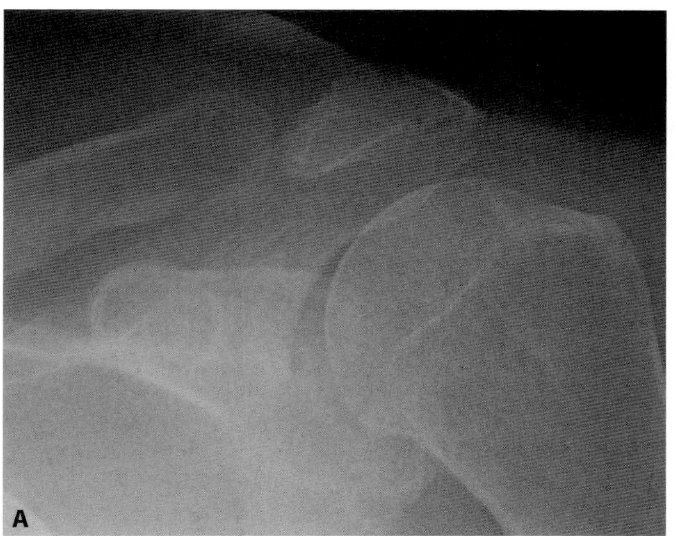

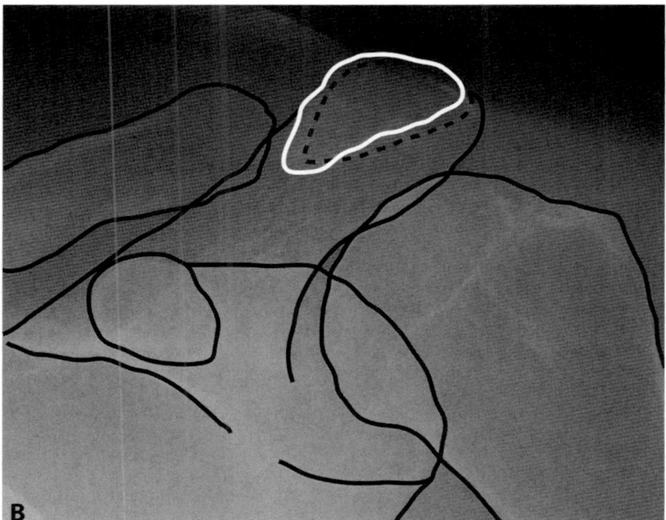

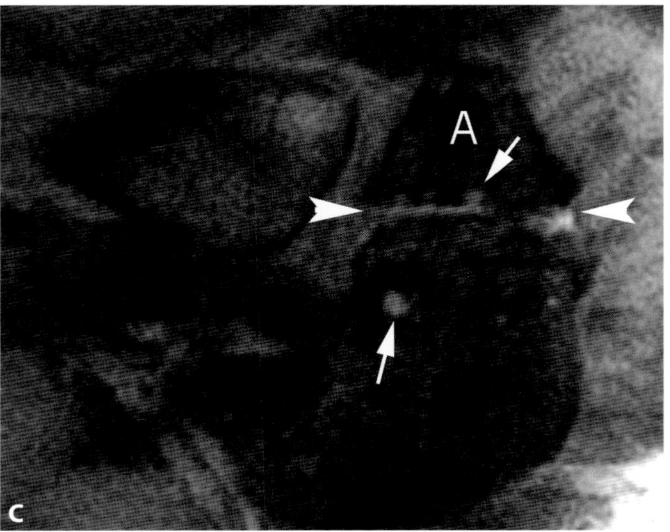

Figure 81.13 **A.** AP view in a 54-year-old man with left shoulder pain and limited adduction shows the double density sign [7]. **B.** Same image as A with line-drawing overlay. White line is the os acromiale, and black dashed line is the base of the acromion. **C.** Axial short tau inversion recovery (STIR) image from MRI shows the os acromiale (*A*) with degenerative changes, including subchondral cysts (arrows) at the synchondrosis (arrowheads).

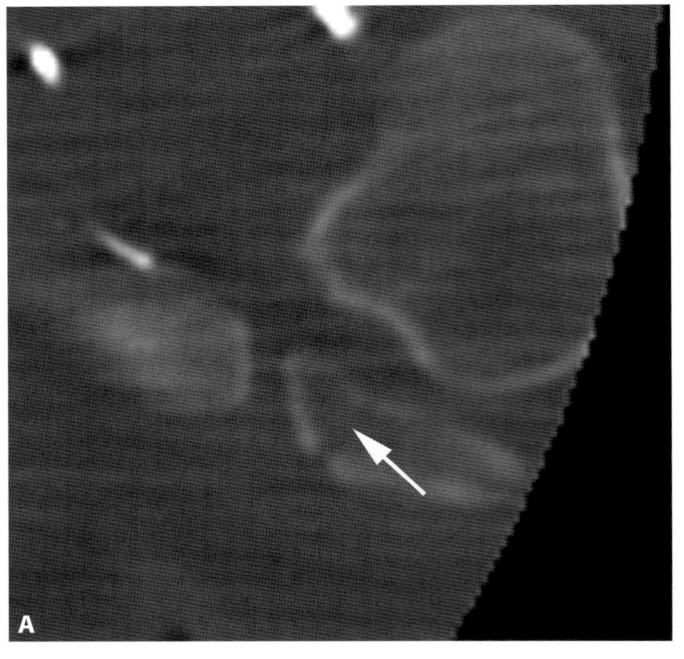

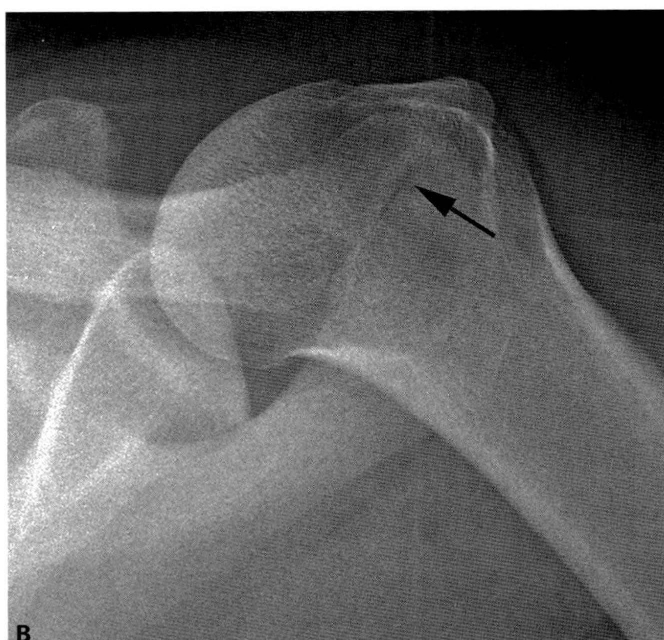

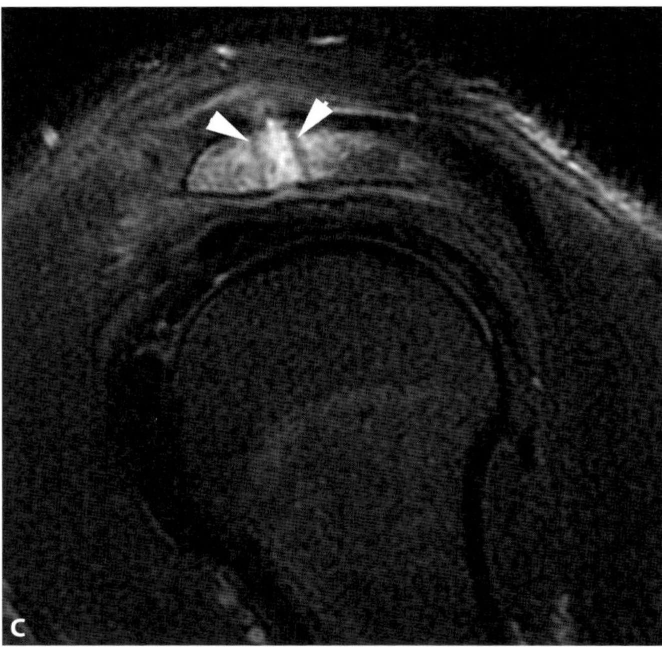

Figure 81.14 A. CT angiography of the chest was performed in a 32-year-old man following a three-story fall. There was a lucency through the left acromion (arrow) which was suggested to represent an os acromiale. **B.** An axillary view of the shoulder was obtained, again showing the acromial luncency (arrow) and again described as an os acromiale. **C.** However, MRI was obtained for continued pain. Sagittal STIR image shows an acute fracture through the acromion (arrowheads) and no accessory ossicle. This case illustrates the importance of seeing a fully corticated osseous structure for the diagnosis of an accessory ossicle.

CASE **82**

Fat pad interpretation

Claire K. Sandstrom

Imaging description

Fat planes are often present on radiographs but may be displaced or obliterated by soft tissue swelling and hemorrhage after acute trauma. Several of these fat pads have been described [1], but by far the most useful are the intracapsular fat pads of the elbow and suprapatellar fat pads of the knee.

In the absence of an effusion, the posterior fat pad of the elbow usually rests within the olecranon fossa, hidden from view by the humeral condyles on the lateral radiograph of the elbow in 90 degrees of flexion (Figure 82.1). If the elbow is extended, the posterior fat pad may be seen even in the absence of an effusion [2, 3]. There are also two normal anterior fat pads, which lie along the anterior aspect of the distal humerus; one in the coronoid fossa, and the second in the radial fossa. The anterior humeral fat pads are normally visible in adults without fractures. On the lateral elbow radiograph, they are superimposed and appear flat or triangular in shape (Figure 82.2).

Any process that distends the joint capsule of the elbow will elevate the fat pads. Identification of the posterior fat pad on a technically adequate radiograph virtually assures the presence of effusion [4]. Displacement of the anterior fat pad creates a sail shape, though this may be difficult to see if the paired fat pads are no longer superimposed [5]. Following trauma, the most likely reason for an effusion is an intracapsular fracture (Figures 82.3–82.5), most commonly a radial head fracture in adults (Figure 82.3) [5], and a supracondylar fracture in children [6]. However, 6–29% of children with an isolated posterior fat pad sign will not have an intracapsular fracture on follow-up imaging and clinical assessment [6].

The suprapatellar recess of the joint capsule of the knee lies between the triangular anterior suprapatellar fat pad and convex posterior suprapatellar fat pad (Figure 82.6). An effusion is diagnosed by separation of the anterior and posterior suprapatellar fat pads to greater than 4–7 mm on a cross-table lateral knee radiograph (Figures 82.7–82.9) [7, 8].

Importance

About three-quarters of patients with a post-traumatic elbow effusion will have an occult fracture diagnosed either by follow-up radiograph or by MR [6, 9]. Post-traumatic knee lipohemarthrosis always indicates intracapsular fracture, while simple effusions without visible fracture may be due to internal derangement such as ligamentous injury or pre-existing joint disease.

Typical clinical scenario

Patients with joint pain and swelling following trauma usually undergo radiographs of the joint and adjacent bones. Effusions in the setting of extremity trauma should prompt close inspection of the intracapsular osseous structures for subtle fractures. If none is found initially, clinical exam may help reveal ligamentous injury that would prompt MRI. Otherwise, most post-traumatic effusions are treated conservatively with splinting and follow-up.

Differential diagnosis

Not all effusions detected after acute trauma are due to fractures. Effusions from other causes, including inflammatory and crystal-induced arthritides, hemarthrosis from hemophilia and other bleeding diatheses, septic arthritis, and neoplasms, appear radiographically similar, though history and clinical presentation are usually revealing [3].

The absence of a fat pad sign does not exclude the presence of fracture (Figures 82.1 and 82.5). If the integrity of the joint capsule is disrupted, the effusion may decompress into the subcutaneous tissues or environment, or the fat pads themselves may be dislocated from their expected locations [5] (Figure 82.5). In either case, no effusion may be seen despite severe fractures and dislocations. These, however, are not usually a diagnostic dilemma. Other reasons for failing to visualize displacement of the posterior elbow fat pad when there is a fracture include poor radiographic technique, severe soft tissue swelling, and extracapsular fracture [3].

Teaching point

At the elbow, an effusion is diagnosed by visualization of the posterior fat pad and sail-shaped anterior fat pad. Visualization of a posterior fat pad strongly suggests an occult intracapsular fracture, but in a quarter of patients, none will be identified at follow-up.

REFERENCES

1. Zimmers TE. Fat plane radiological signs in wrist and elbow trauma. *Am J Emer Med.* 1984;**2**(6):526–32.
2. De Maeseneer M, Jacobson JA, Jaovisidha S, *et al.* Elbow effusions: distribution of joint fluid with flexion and extension and imaging implications. *Invest Radiol.* 1998;**33**:117–25.
3. Murphy WA, Siegel MJ. Elbow fat pads with new signs and extended differential diagnosis. *Radiology.* 1977;**124**:659–65.
4. Norell HG. Roentgenologic visualization of the extracapsular fat, its importance in the diagnosis of traumatic injuries to the elbow. *Acta Radiol.* 1954;**42**(3):205–10.
5. Smith DN, Lee JR. The radiological diagnosis of post-traumatic effusion of the elbow joint and its clinical significance: the 'displaced fat pad' sign. *Injury.* 1977;**10**:115–19.

6. Skaggs DL, Mirzayan R. The posterior fat pad sign in association with occult fracture of the elbow in children. *J Bone Joint Surg*. 1999; **81**-A:1429–33.

7. Tai AW, Alparslan HL, Townsend BA, *et al.* Accuracy of cross-table lateral knee radiography for evaluation of joint effusions. *AJR Am J Roentgenol*. 2009;**193**(4):W339–44.

8. Hall FM. Radiographic diagnosis and accuracy in knee joint effusions. *Radiology*. 1975;**115**(1):49–54.

9. O'Dwyer HO, O'Sullivan P, Fitzgerald D, *et al.* The fat pad sign following elbow trauma in adults. *J Comput Assist Tomogr*. 2004;**28**:562–5.

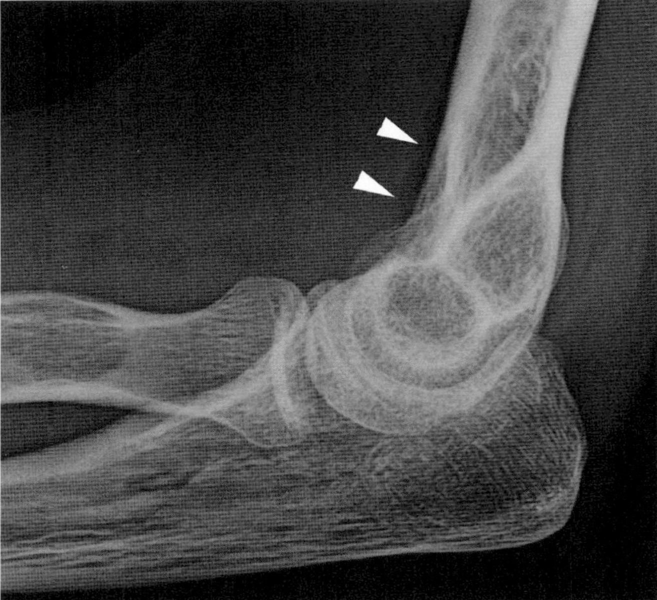

Figure 82.1 Lateral radiograph of the left elbow in a 42-year-old woman found to have a distal radial fracture after falling shows a normal, flat anterior fat pad (arrowheads). The posterior fat pad is not visualized. Extracapsular fractures of the elbow will not cause an effusion or displace the fat pads.

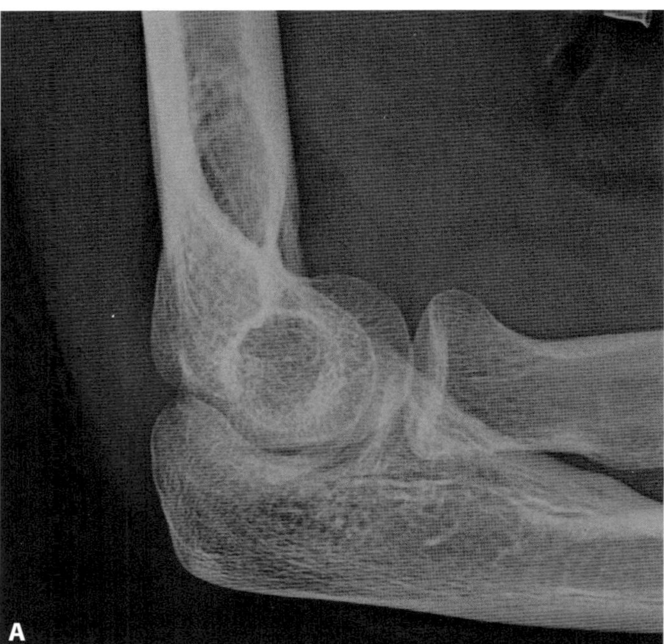

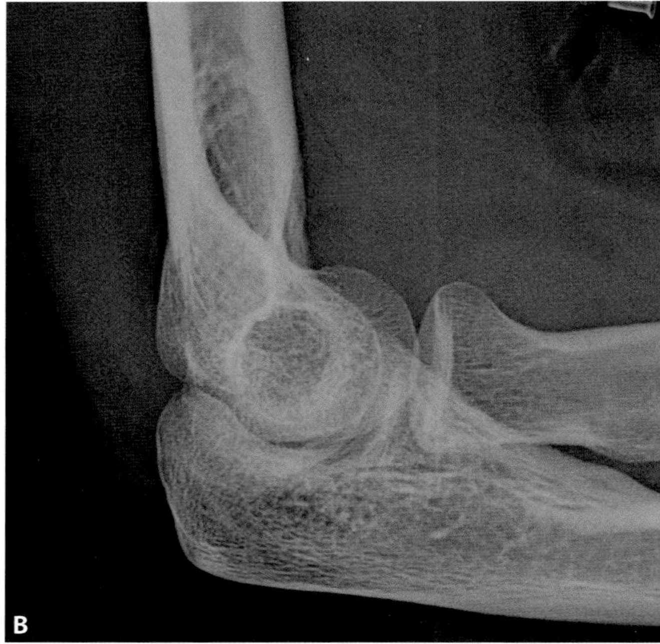

Figure 82.2 A. Lateral radiograph of the right elbow in a 22-year-old woman who fell shows elevation of the anterior and posterior fat pads, which are highlighted in blue in Figure **B.** No fracture was detected.

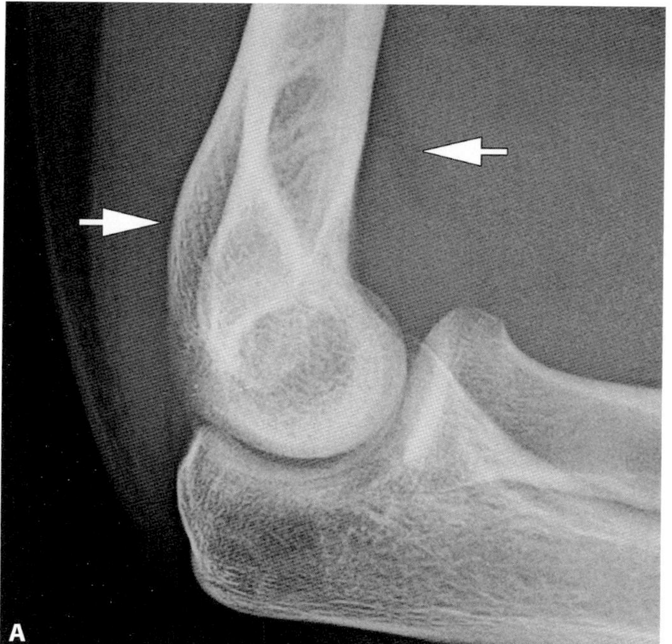

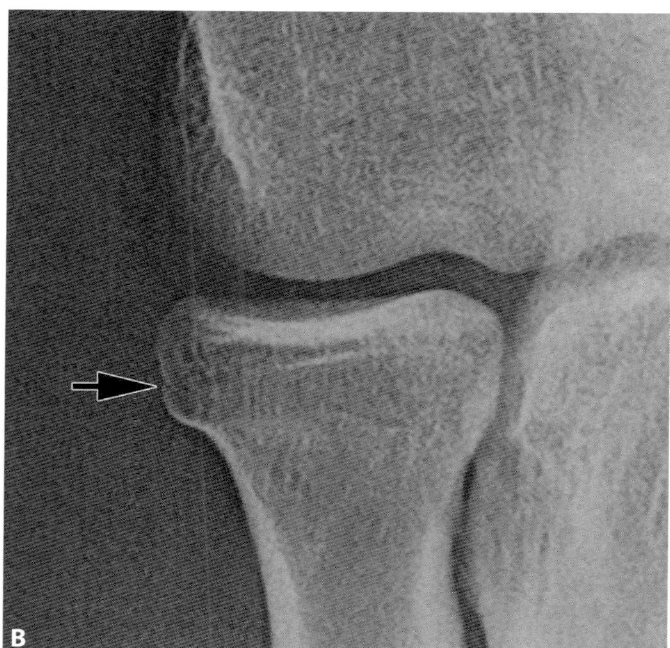

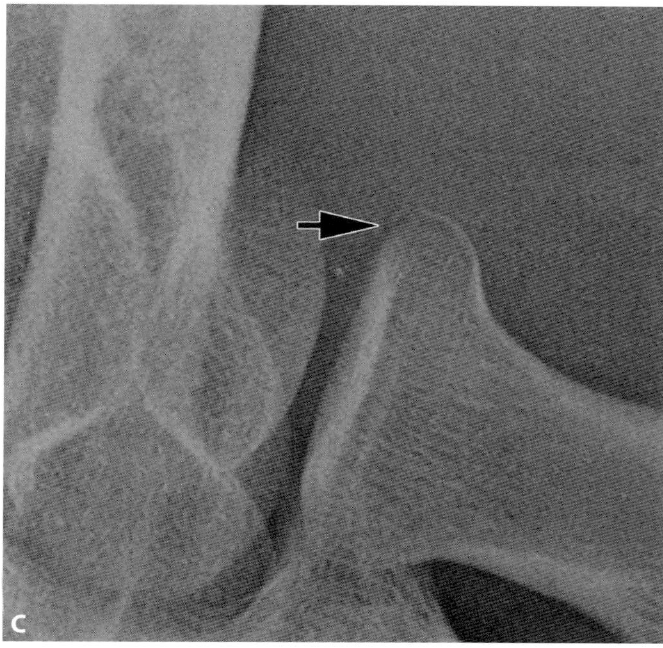

Figure 82.3 A. Lateral radiograph of the right elbow in a 20-year-old man who fell onto his right elbow shows elevation of the anterior and posterior fat pads (arrows). **B.** There is very minimal irregularity (arrow) of the lateral radial head on magnified frontal view of the elbow. **C.** The oblique elbow radiograph more clearly shows the non-displaced racial head fracture (arrow).

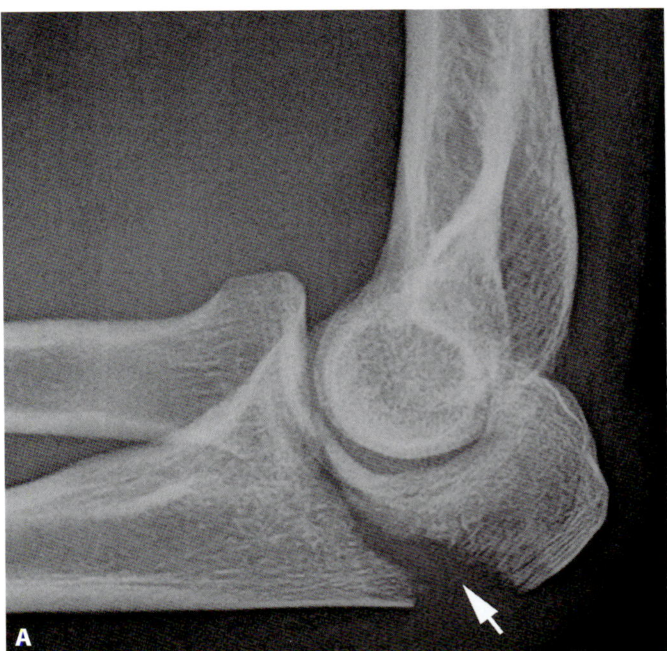

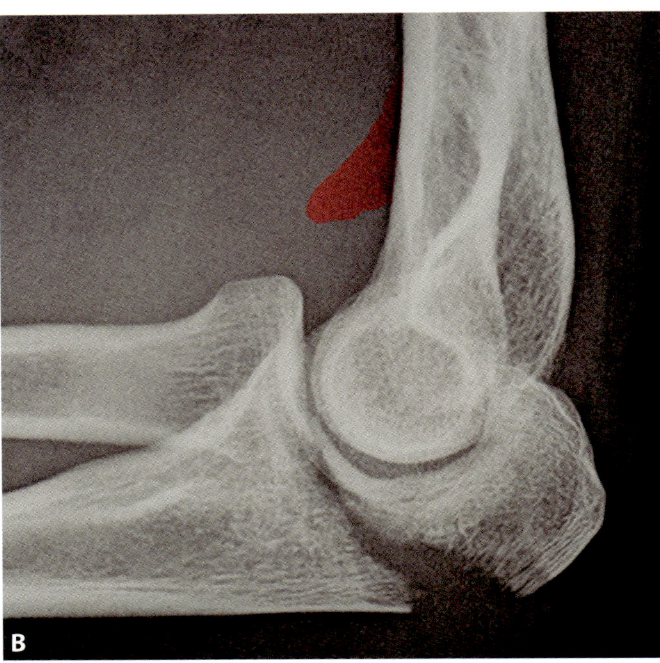

Figure 82.4 A. Lateral radiograph of the left elbow in a 23-year-old woman with elbow pain after a motor vehicle accident shows transverse olecranon fracture with mild displacement (arrow). There is elevation of the anterior fat pad, which is highlighted in red in Figure **B.** The posterior fat pad is not visible.

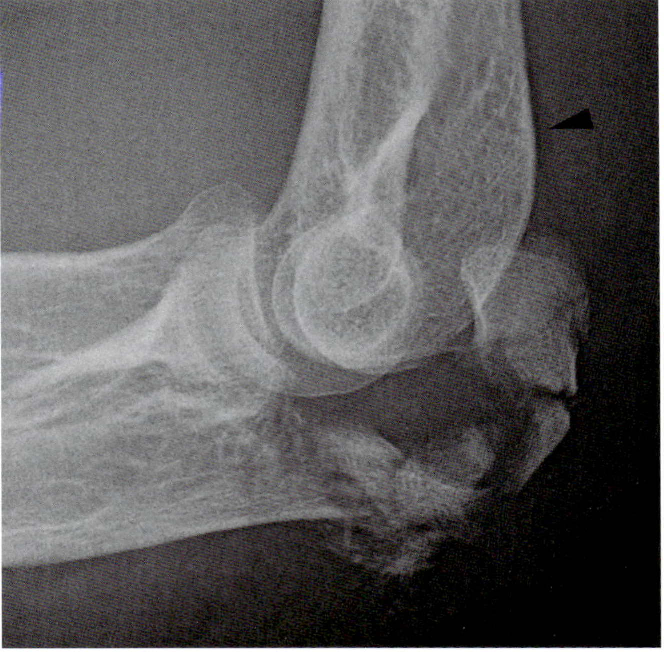

Figure 82.5 Lateral radiograph of the left elbow in a 44-year-old man who fell nine feet shows a markedly comminuted olecranon fracture. The posterior fat pad may be slightly elevated (arrowhead), but the anterior fat pads are not visualized. This is likely due to disruption of the joint capsule with dispersion of the hemarthrosis into the subcutaneous soft tissues.

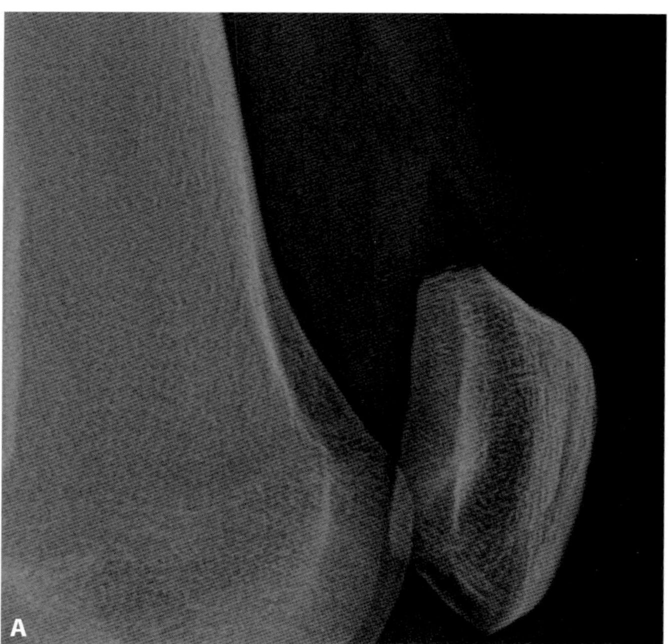

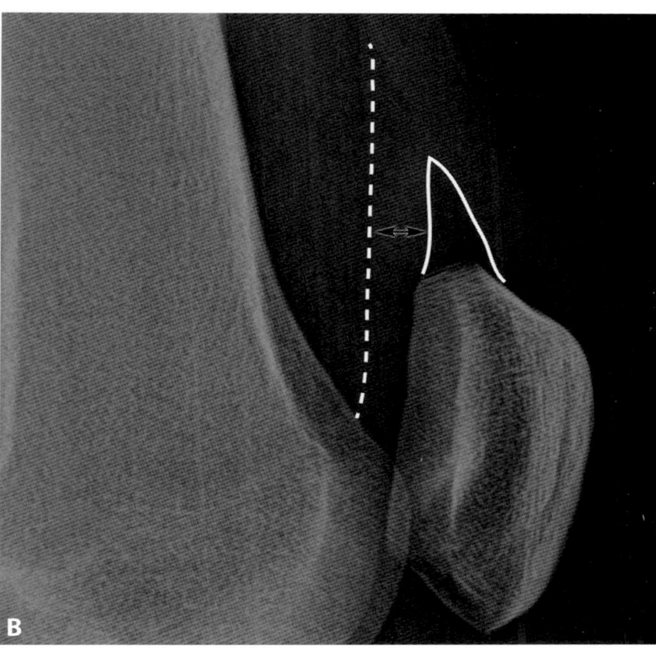

Figure 82.6 A. Lateral radiograph of the left knee in an 18-year-old woman who fell off a bicycle shows normal alignment of the knee. There is no joint effusion, which, as highlighted in **B**, is the distance (double-arrow) between the anterior suprapatellar fat pad (white solid line) and posterior suprapatellar fat pad (white dashed line).

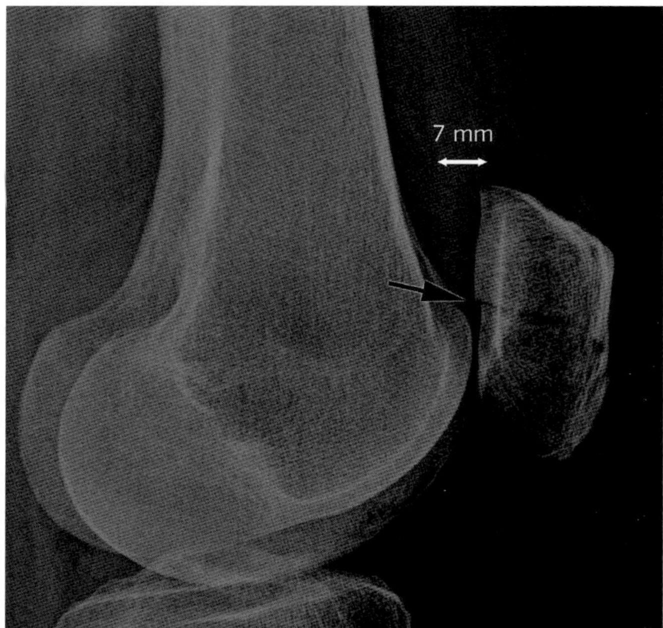

Figure 82.7 Lateral radiograph of the left knee in a 79-year-old woman who fell shows a small joint effusion (double-arrow) secondary to a minimally displaced transverse patellar fracture (arrow).

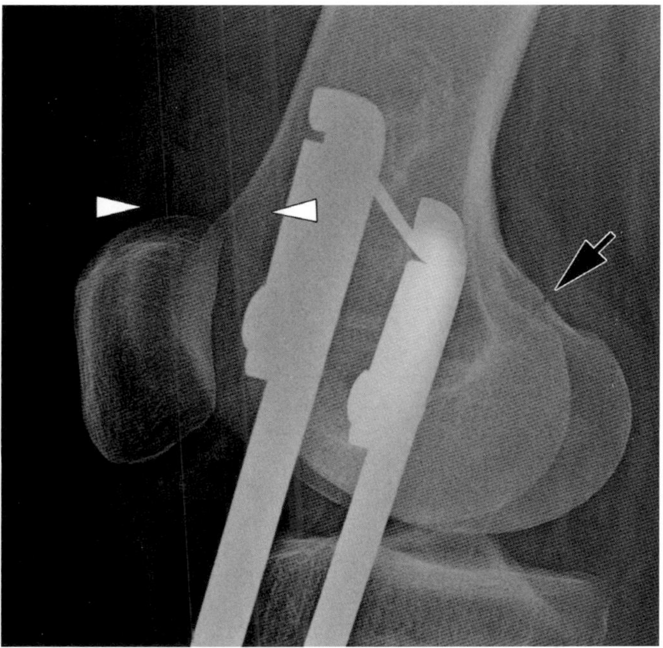

Figure 82.8 Lateral radiograph of the right knee in a 33-year-old man injured in a motor vehicle collision shows a lipohemarthrosis (arrowheads) and a non-displaced Hoffa fracture of the posterior femoral condyle (arrow). There is also traction hardware placed for a comminuted fracture of the femoral shaft (not shown).

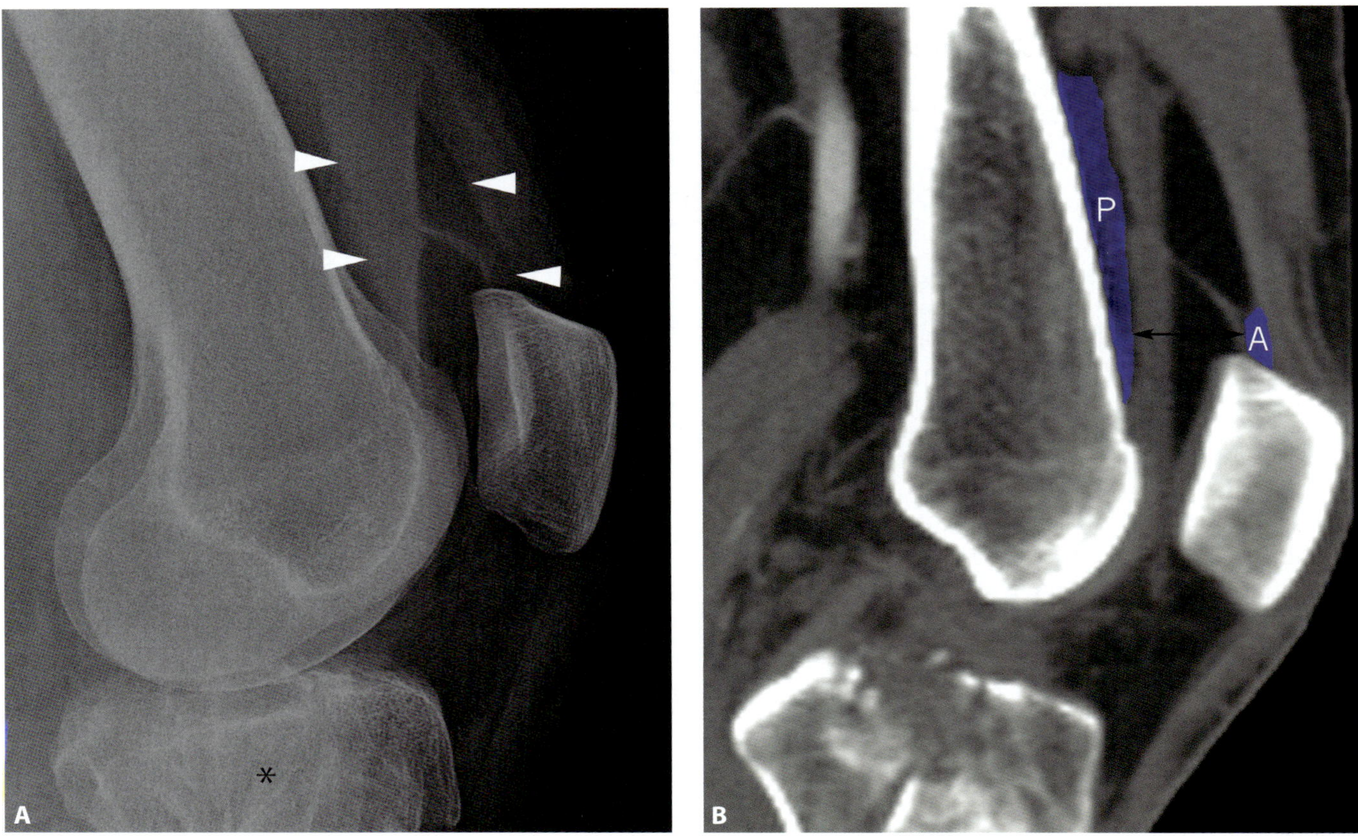

Figure 82.9 **A.** Lateral radiograph of the left knee in a 45-year-old man who fell shows a large lipohemarthrosis (arrowheads) secondary to a markedly comminuted tibial plateau fracture (asterisk). **B.** Sagittal CT image of the same knee shows the large lipohemarthrosis between the anterior (A) and posterior (P) suprapatellar fat pads, highlighted in blue. The preferred place to measure for suprapatellar recess widening is shown by the black double-arrow.

Posterior shoulder dislocation

Claire K. Sandstrom

Imaging description

As opposed to the relative ease of recognizing classic anterior subglenoid dislocation of the shoulder, the findings of posterior shoulder dislocation on the anterior-posterior (AP) view of the shoulder are subtle and require a high degree of suspicion to detect [1–3]. Nearly one-quarter of posterior shoulder dislocations are missed on initial radiographic assessment [4]. The normal anatomic appearance of the shoulder is reviewed in Figure 83.1. Signs to look for on the AP view include the "lightbulb" appearance of the humeral head, due to fixed internal rotation of the humerus (Figure 83.2), and a "vacant" glenoid fossa due to lateral displacement of the humeral head, creating the "rim sign" (Figure 83.3). The "rim sign" can be due to hemarthrosis or septic arthritis. There may be absence of normal half-moon overlap between the humeral head and glenoid. The "trough sign" is caused by impaction of the humeral head on the glenoid, and reflects the parallel lines created by the medial humeral head cortex and the reverse Hill–Sachs fracture fragment (Figure 83.4). More recently, the "Mouzopoulos" sign on AP radiographs has been described [2]. In posterior shoulder dislocation, projection of the greater and lesser tuberosities on the AP view of the internally rotated humeral head creates a capital "M" appearance (Figure 83.5). A false-positive Mouzopoulos sign may be seen when there is marked internal rotation of the humerus in the absence of dislocation [2].

Accurate radiographic diagnosis, though, usually requires an axillary view of the shoulder, on which the humeral head is clearly seen posterior to the glenoid fossa. The associated impaction fracture of the anterior humeral head, the so-called "reverse Hill–Sachs" injury may also be evident. Unfortunately, this view may be difficult to obtain because of pain associated with abduction of the dislocated arm [3].

CT of the shoulder will clearly demonstrate the dislocation, as well as any associated fractures (Figure 83.6). These include the "reverse Hill–Sachs" injury of the anterior humeral head and a "reverse Bankart" fracture of the posterior glenoid.

Importance

Posterior glenohumeral dislocation is much less common than anterior dislocation, with an oft-quoted incidence of 1–4% of dislocations. Unfortunately, this injury is initially missed in a high percentage of patients. Posterior dislocation of the glenohumeral joint can be difficult to diagnose without an axillary view. Missed fractures will likely present with chronic pain and restricted range of motion, or premature degeneration.

Typical clinical scenario

Seizure, electrocution, and falls on an outstretched hand are the most common causes of posterior glenohumeral dislocation [1]. The humeral head is locked in internal rotation and cannot be externally rotated. Patients may be unconscious or incapable of complaining of shoulder pain or restricted range of motion at the time of their presentation.

Differential diagnosis

Pseudosubluxation secondary to joint effusion or hemarthrosis may result in abnormal alignment of the glenohumeral joint on anterior views of the shoulder. However, the lack of dislocation will be evident on axillary views of the shoulder.

Teaching point

Posterior shoulder dislocations can be subtle and are often missed. A focused evaluation of the shoulder, examining for the specific radiographic signs of posterior dislocation, is recommended. A complete shoulder radiographic series should also include an axillary view.

REFERENCES

1. Cicak N. Posterior dislocation of the shoulder. *J Bone Joint Surg.* 2004;**86**B:324–32.
2. Mouzopoulos G. The "Mouzopoulos" sign: a radiographic sign of posterior shoulder dislocation. *Emerg Radiol.* 2010;**17**(4):317–20.
3. Clough TM, Bale RS. Bilateral posterior shoulder dislocation: the importance of the axillary radiographic view. *Eur J Emerg Med.* 2001;**8**:161–3.
4. Rouleau DM, Hebert-Davies J. Incidence of associated injury in posterior shoulder dislocation: systematic review of the literature. *J Orthop Trauma.* 2012;**26**(4):246–51.

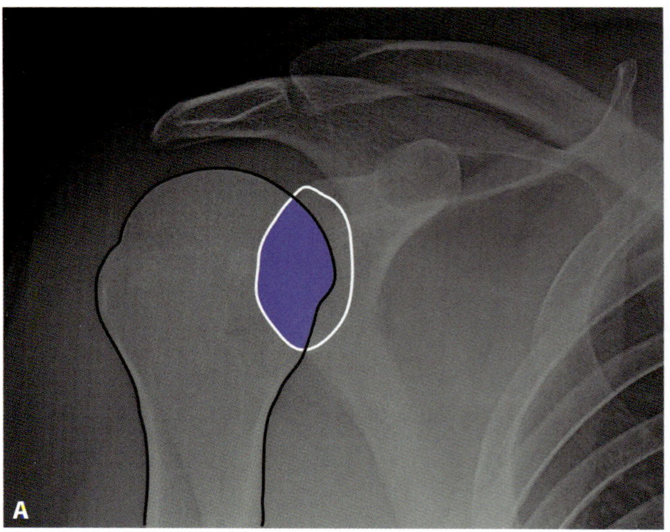

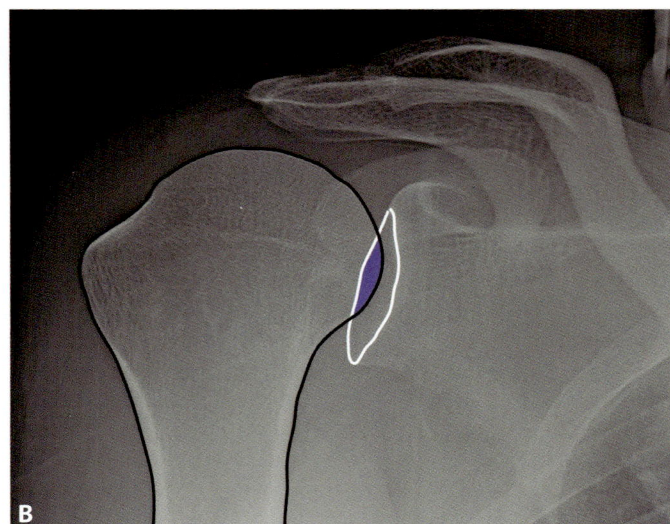

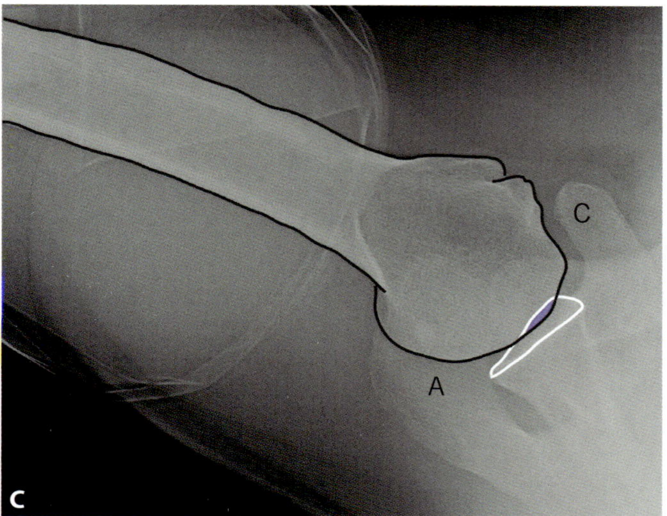

Figure 83.1 AP (**A**) and Grashey (**B**) views of the shoulder in a 39-year-old man with shoulder pain show normal alignment. There is normal overlap (blue shading) between the humeral head (black outline) and the glenoid (white outline). **C.** True axillary view of a normal shoulder shows the humeral head (black outline) normally articulating with the glenoid (white outline), with minimal overlap on this view (blue shading). The corocoid process (C) identifies the anterior aspect of the joint, while the acromion process (A) is posterior.

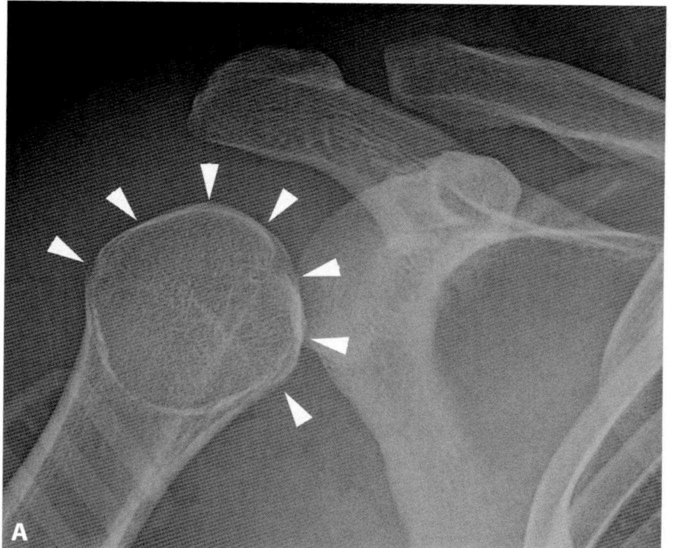

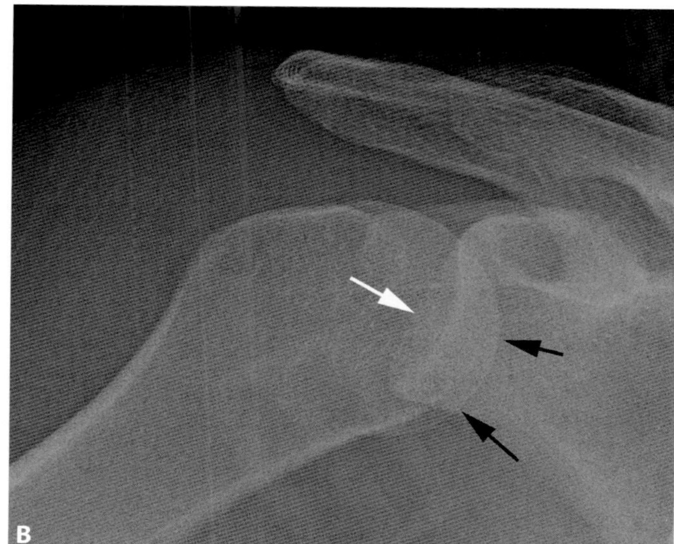

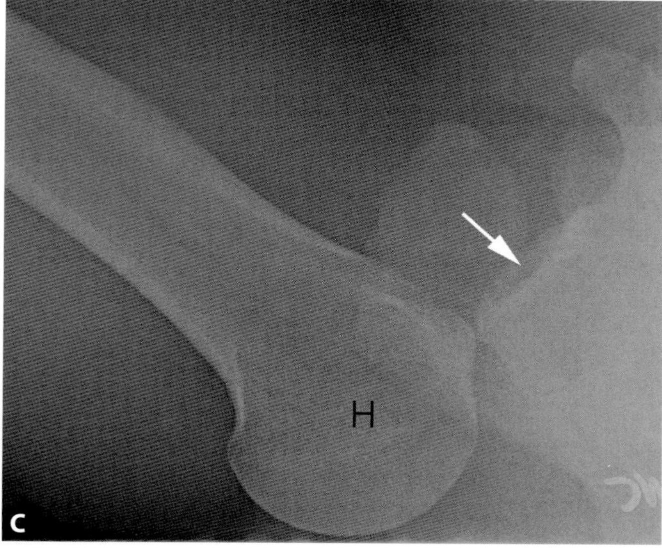

Figure 83.2 A. AP view of the shoulder in a 24-year-old man following a motorcycle crash demonstrates a rounded appearance of the humeral head (arrowheads) ("lightbulb" sign), which is due to internal rotation. There is also minimal overlap between the humeral head and glenoid. **B.** Grashey view of the same shoulder shows the abnormal overlap between the humeral head (black arrows) and glenoid (white arrow), indicating dislocation. **C.** True axillary view of the shoulder confirms posterior dislocation of the humeral head (H), relative to the glenoid (arrow).

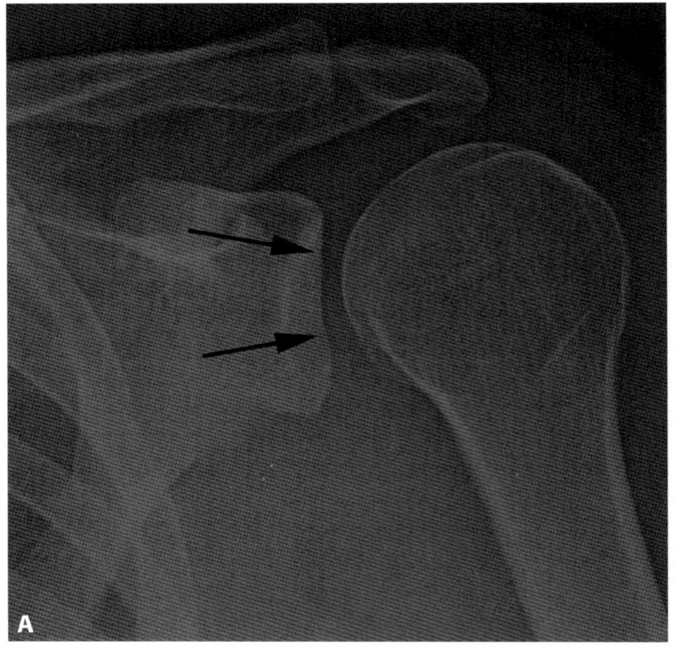

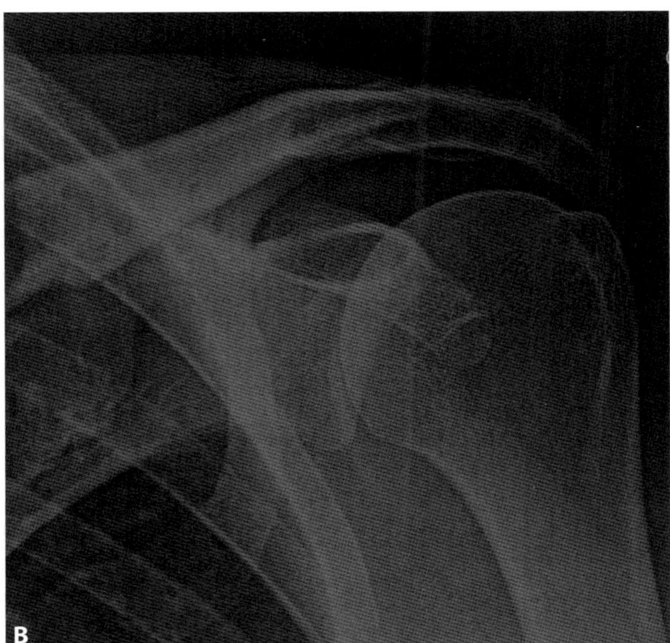

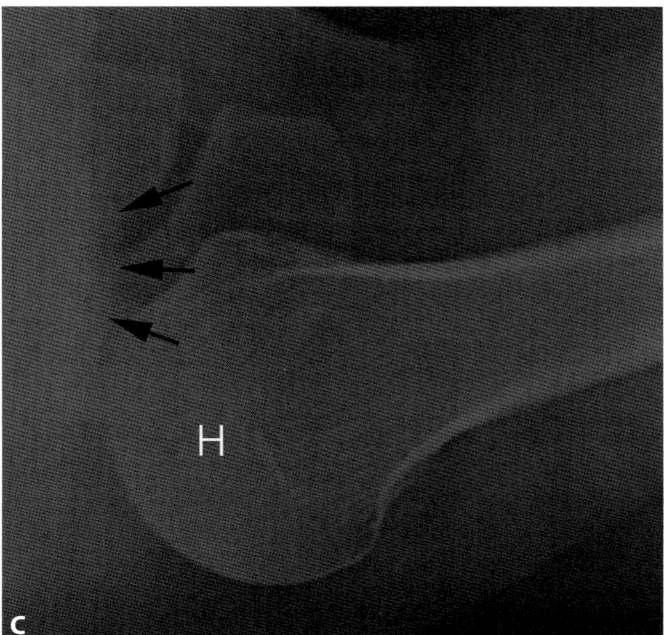

Figure 83.3 **A.** AP view of the shoulder in a 42-year-old man with left shoulder pain following a ground-level fall shows widening of the glenohumeral joint space, with empty appearance of the glenoid, the "rim sign" (arrows). **B.** Grashey AP view of the same shoulder fails to show the normal glenohumeral joint space, confirming the dislocation. **C.** True axillary view of the shoulder confirms posterior dislocation of the humeral head (H), even though the glenoid (arrows) is difficult to see well due to overlap with the chest wall.

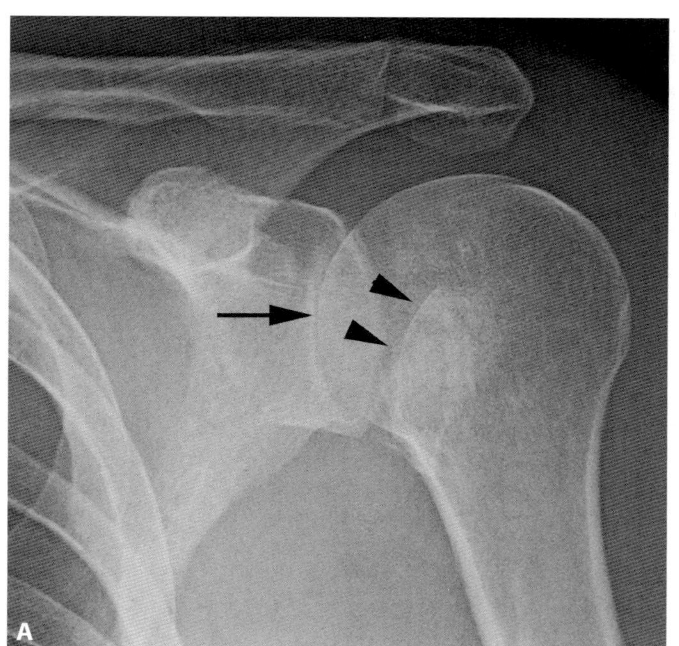

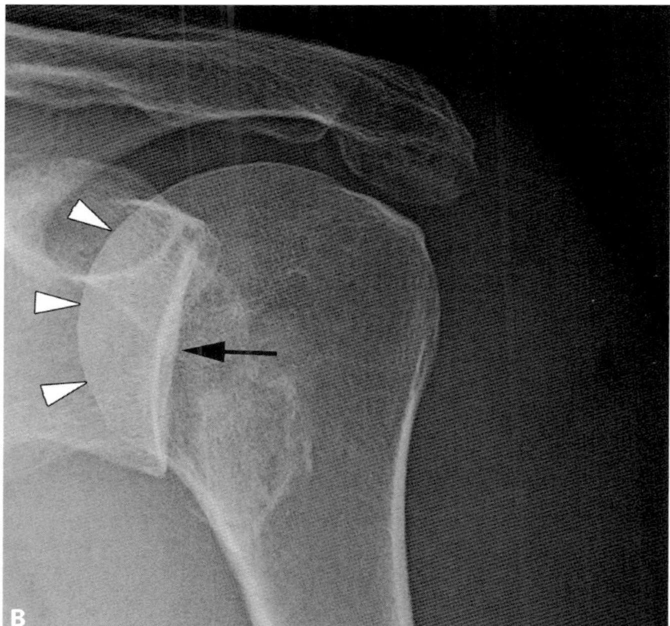

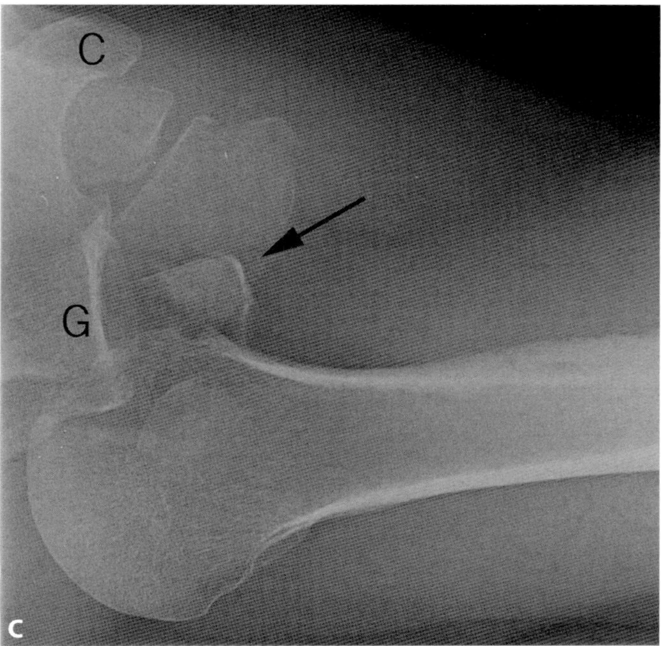

Figure 83.4 A. AP view of the shoulder in a 50-year-old man with shoulder pain initially suggested normal alignment, yet the glenohumeral joint space appeared too narrow (arrow). In addition, there was a "trough sign" in the humeral head (arrowheads). **B.** Grashey view of the same shoulder confirms dislocation, with the articular surface of the humeral head (arrowheads) projecting medial relative to the glenoid fossa (arrow). **C.** True axillary view of the shoulder confirms posterior dislocation of the humeral head relative to the glenoid (G) and corocoid (C). A large reverse Hill–Sachs fracture fragment is evident (arrow).

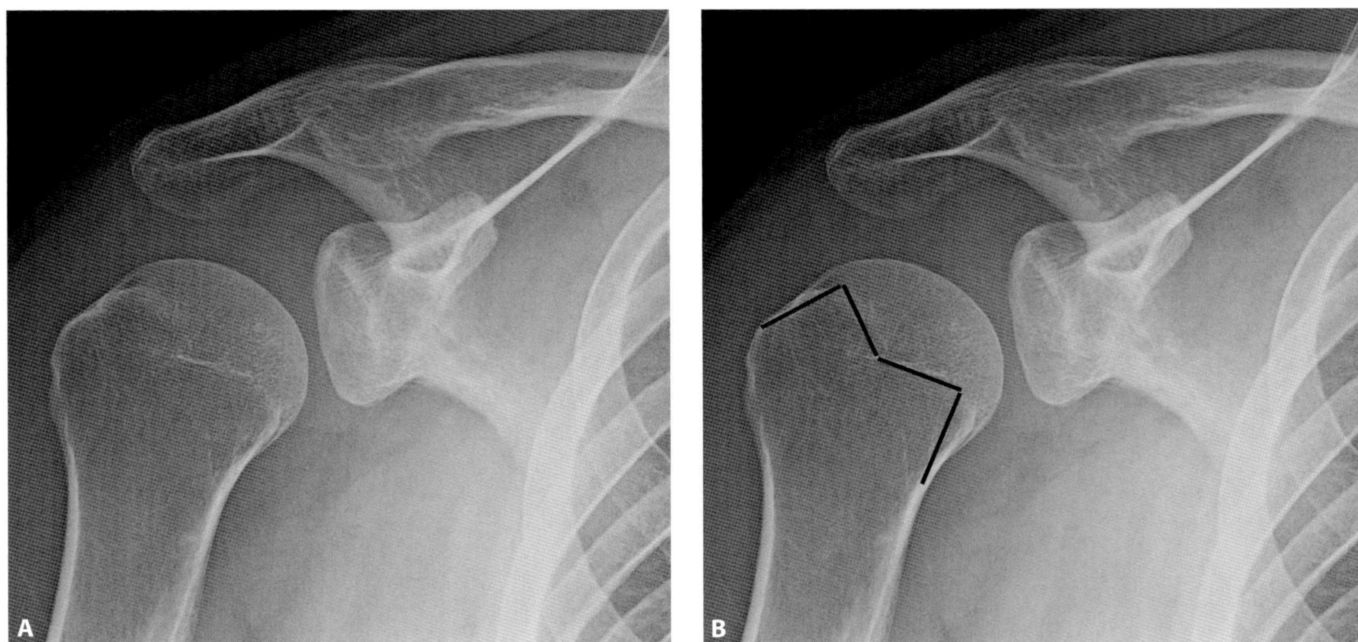

Figure 83.5 A. AP view of the shoulder in a 29-year-old man with shoulder pain and locking after a fall shows lack of the normal osseous overlap between the humeral head and glenoid. As shown by the black lines in **B,** there is internal rotation of the humerus with projection of the greater and lesser tuberosities creating the "Mouzopoulos M sign." Posterior shoulder dislocation was confirmed on the axillary view (not shown).

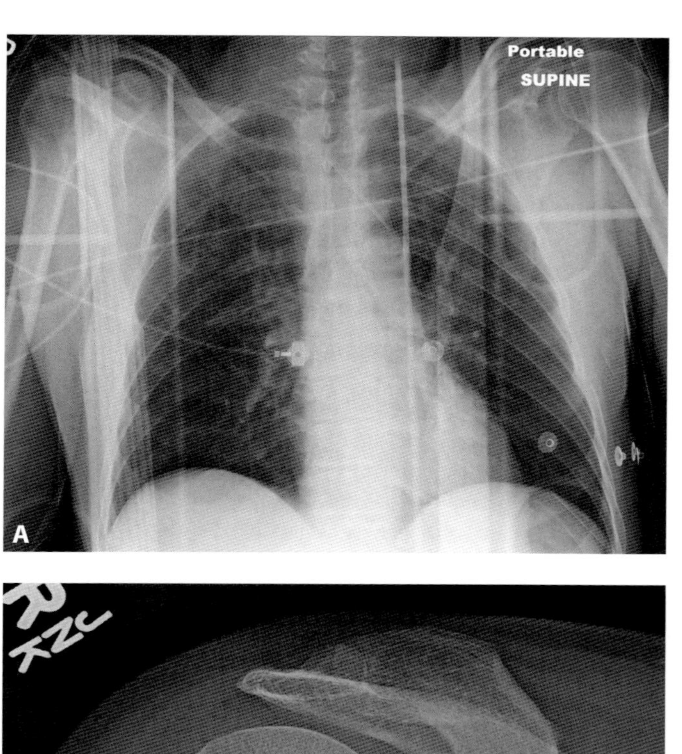

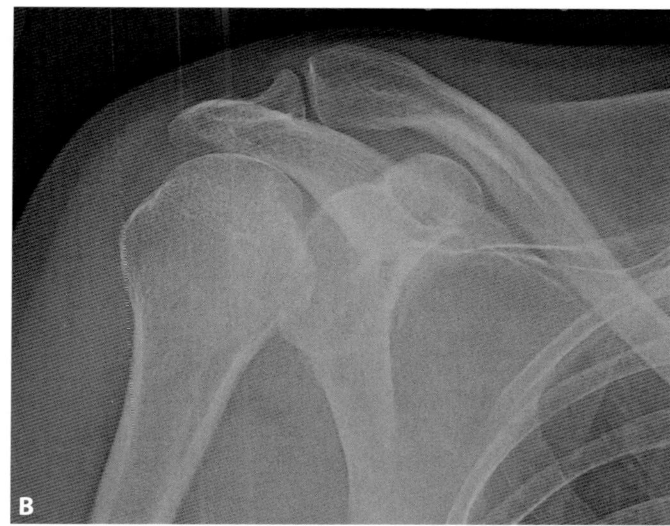

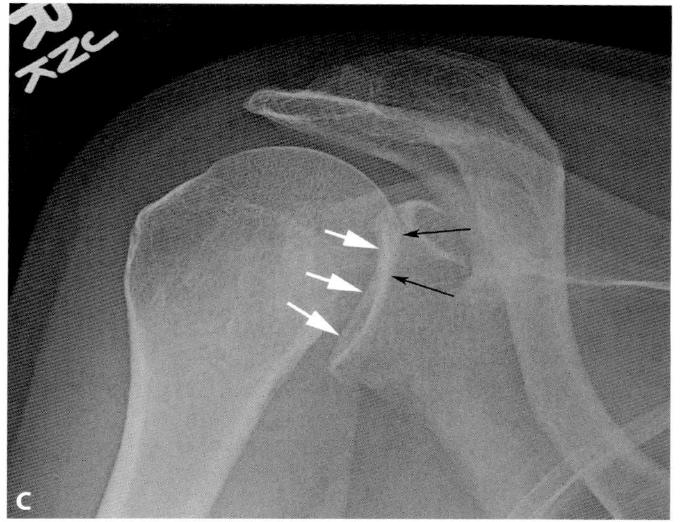

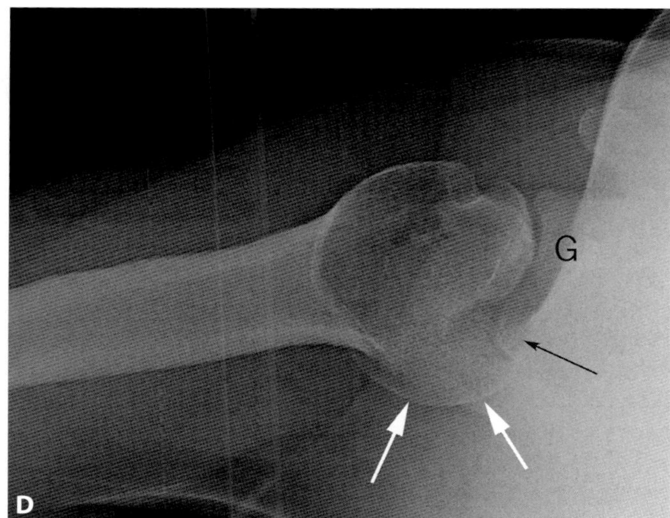

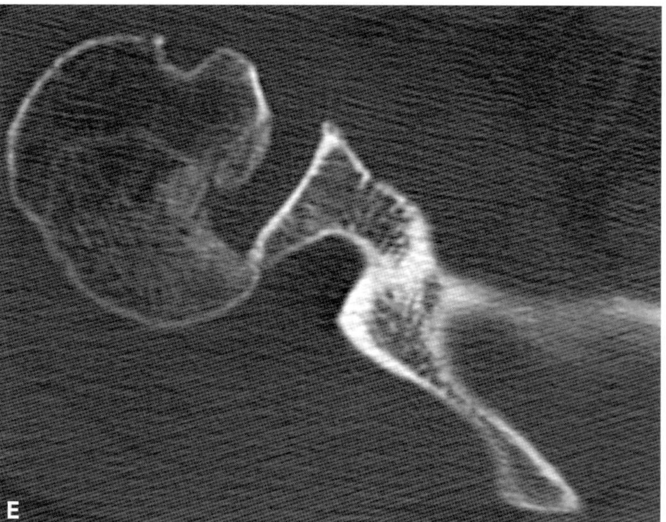

Figure 83.6 A. AP view of the chest in a 66-year-old man who was found down at home with a chronic subdural hematoma shows artifact from the trauma backboard obscuring the right glenohumeral joint. Unfortunately, distracting injuries prevented clinical or radiographic recognition of any abnormality in the right shoulder. **B.** AP view of the shoulder obtained five days later shows internal rotation of the humeral head. **C.** Grashey AP view of the same shoulder shows abnormal overlap of the humeral head (thin black arrows) and glenoid fossa (white arrows), and lack of visualization of a normal glenohumeral joint space. **D.** True axillary view of the shoulder confirms two large fracture fragments posteriorly, one from the humeral head (white arrows) and one from the posterior glenoid (G) (black arrow). **E.** Axial CT image of the shoulder without contrast after attempted closed reduction shows posterior subluxation of the humeral head and a large reverse Hill–Sachs fracture fragment. This patient later underwent humeral head arthroplasty because of the severe deformity of the humeral head.

CASE **84**

Easily missed fractures in thoracic trauma

Claire K. Sandstrom

Imaging description

In the setting of blunt trauma to the chest, injuries such as rib fractures, pneumothorax, hemothorax, and pulmonary contusions are relatively common and easily diagnosed by CT if not by radiograph. Fractures of the sternum and scapula and sternoclavicular joint dislocation are less common but reflect higher-energy trauma. Furthermore, they can be easily overlooked, even on CT.

Fractures of the sternum occur in up to 10% of polytrauma cases [1]. Substernal mediastinal hemorrhage should prompt close examination of the sternum, though hemorrhage is not universally present (Figure 84.1). The horizontal sternal fracture may be occult on axial CT images, and dedicated reformations in sagittal and coronal planes relative to the sternum may be necessary for diagnosis (Figure 84.2). An important caveat is that motion artifact may create artifactual step-off in the sternum, but close inspection for matching step-off in the overlying skin line usually helps differentiation. Because sternal fractures usually result from direct frontal injury, additional thoracic, cardiac, and spinal injuries should be excluded (Figure 84.3) [2].

Sternoclavicular dislocation is a rare injury. Anterior dislocations are more common than posterior and are usually clinically obvious. Posterior dislocations, conversely, are often clinically and radiographically occult. Furthermore, the posteriorly dislocated clavicle may be associated with vascular, nerve, or tracheal injuries, necessitating further evaluation. For greater detail, see Case 36.

Scapular fractures are easily overlooked [3]. Extra-articular fractures are usually treated conservatively, but fractures involving the glenoid or scapular neck require surgical fixation. Any suspicious lucencies evident on radiographs should be investigated by CT, with coronal and sagittal reformations in the plane of the scapula. Harris and Harris found a high percentage of patients with scapular fractures have other ipsilateral soft tissue and skeletal injuries, such as upper rib fractures, clavicular fracture or acromioclavicular joint disruption, or pulmonary contusion [3]. These findings should therefore prompt close evaluation of the scapula.

Another serious injury is scapulothoracic dissociation (Figure 84.4). Though rare, this closed amputation of the upper extremity is associated with extensive injury of the subclavian vessels and brachial plexus [1]. Lateral displacement of the scapula on a chest radiograph should prompt clinical and imaging evaluation for this potentially devastating injury.

Importance

Sternal and scapular fractures, scapulothoracic dissociation, and sternoclavicular joint dislocation are relatively rare complications of blunt trauma. However, they are easily missed and frequently associated with significant morbidity.

Typical clinical scenario

Rib fractures, hemothoraces, pneumothoraces, and lung contusions and lacerations are frequently encountered in the setting of severe blunt chest trauma. Less common but no less clinically important are injuries described above. Sternal fractures typically occur in motor vehicle collisions when the chest hits the steering wheel, though they can also be caused by the seat belt. When an isolated sternal fracture is encountered, if there are no ECG abnormalities, symptomatic treatment appears safe [4]. Scapular fractures often result from motor vehicle collisions, falls from height, or seizures [5], while scapulothoracic dissociation is most commonly from motorcycle accidents [1]. The clavicle may dislocate posteriorly from the sternoclavicular joint following a direct blow. Close inspection for these injuries should be prompted by history of significant blunt trauma to the chest or chest pain.

Differential diagnosis

Once recognized, these injuries rarely present a diagnostic dilemma. They are usually overlooked when multiple other distracting injuries, perhaps life-threatening, are also present.

Teaching point

In the setting of blunt chest trauma, targeted inspection may be necessary to exclude the presence of these potentially significant skeletal injuries involving the chest wall and shoulder girdle.

REFERENCES

1. Wicky S, Wintermark M, Schnyder P, Capasso P, Denys A. Imaging of blunt chest trauma. *Eur Radiol.* 2000;**10**:1524–38.
2. Peters S, Nicolas V, Heyer CM. Multidetector computed tomography-spectrum of blunt chest wall and lung injuries in polytraumatized patients. *Clin Radiol.* 2010;**65**(4):333–8.
3. Harris RD, Harris JH. The prevalence and significance of missed scapular fractures in blunt chest trauma. *AJR Am J Roentgenol.* 1988;**151**:747–50.
4. Hossain M, Ramavath A, Kulangara J, Andrew JG. Current management of isolated sternal fractures in the UK: time for evidence based practice? A cross-sectional survey and review of literature. *Injury.* 2010;**41**(5):495–8. Epub 2009 Aug 13.
5. Tadros AM, Lunsjo K, Czechowski J, Abu-Zidan FM. Causes of delayed diagnosis of scapular fractures. *Injury.* 2008;**39**(3):314–18.

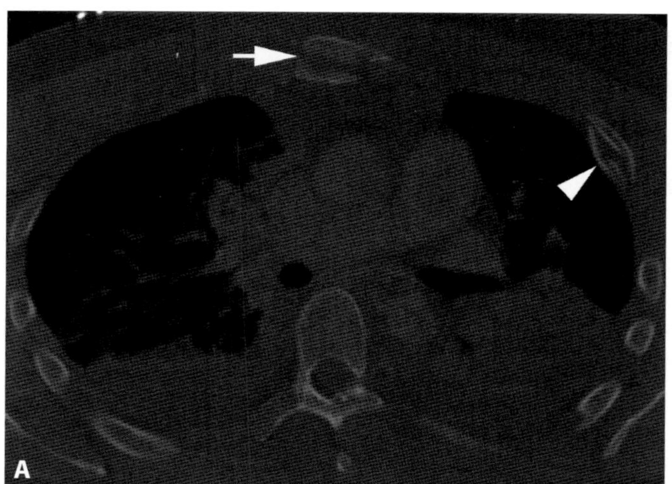

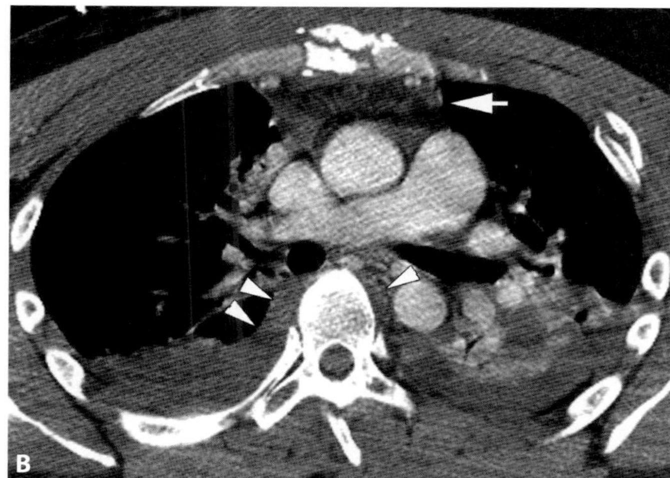

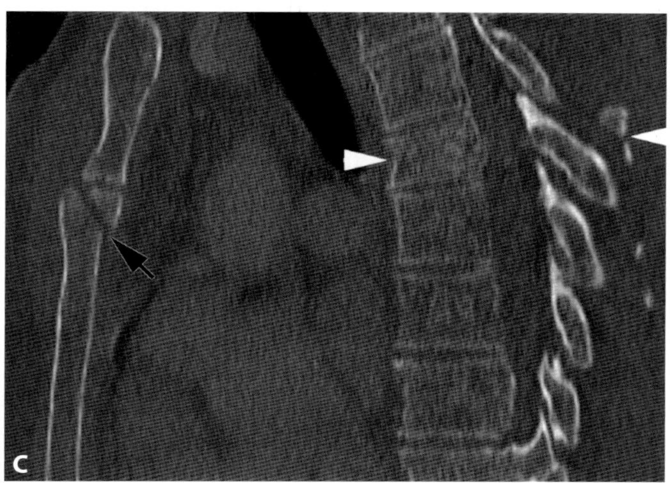

Figure 84.1 A. Axial contrast-enhanced CT angiogram of the chest in bone windows in a 42-year-old man after a motorcycle collision shows mildly displaced fracture in the sternum (arrow). Left rib fractures are also present (arrowhead). **B.** Axial image at the same level in soft tissue windows demonstrates a small retrosternal hematoma (arrow), as well as bilateral hemothoraces and posterior mediastinal hematoma (arrowheads). **C.** Sagittal reformation image better demonstrates the sternal fracture (arrow), as well as the T5 burst fracture (arrowheads) that was the source of the posterior mediastinal hematoma.

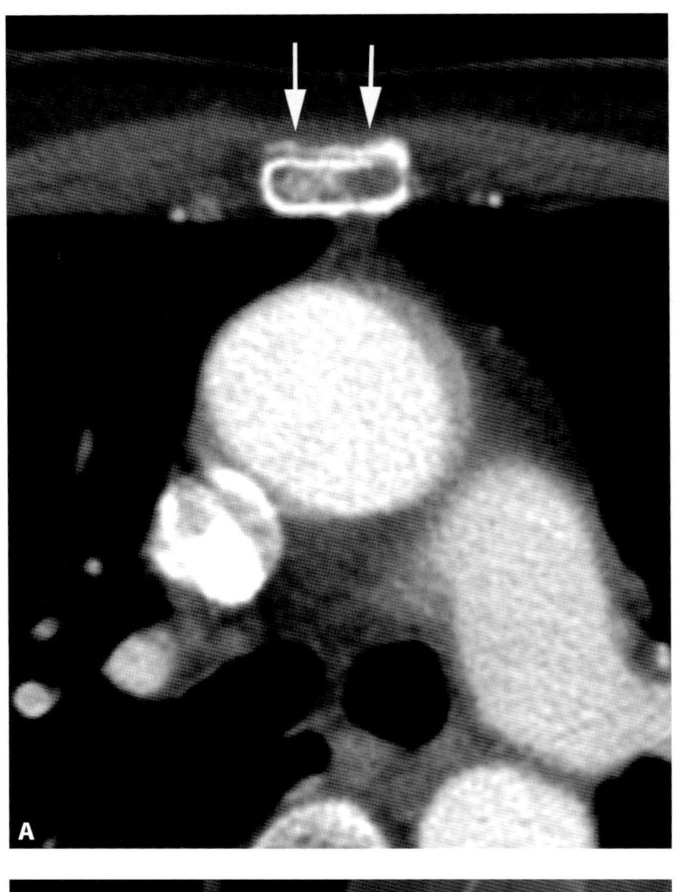

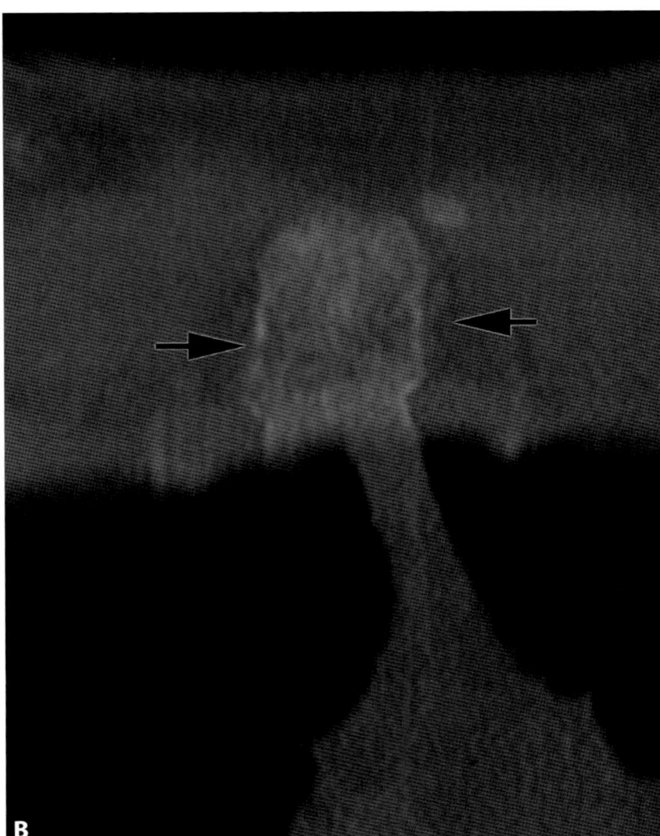

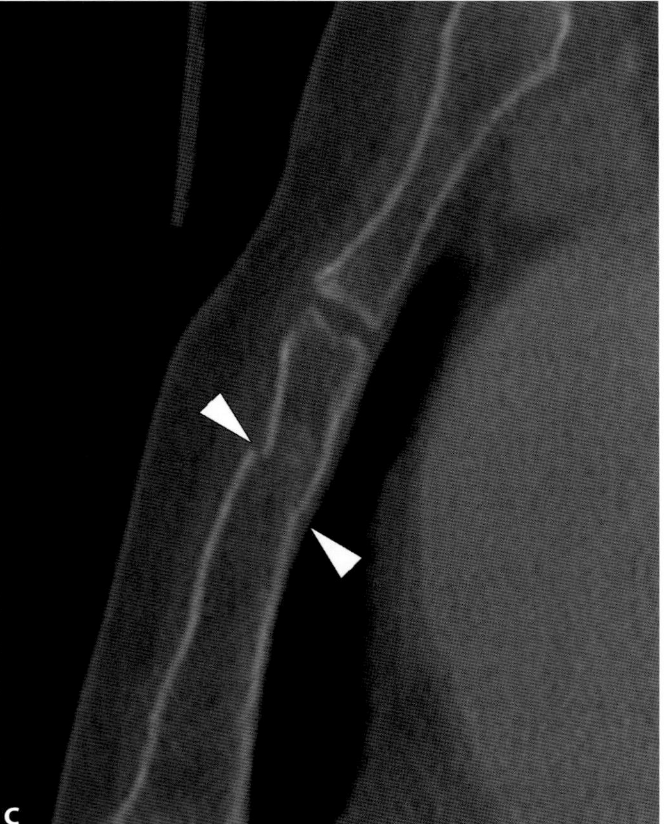

Figure 84.2 A. Axial contrast-enhanced CT of the chest in a 63-year-old woman with chest pain following a motor vehicle collision shows mild step-off in the sternum (arrows). Note the absence of retrosternal hematoma in this patient. **B.** The coronal image poorly demonstrates the minimally displaced fracture of the sternum (arrows). **C.** Sagittal reformations were performed, clearly demonstrating the minimally displaced fracture (arrowheads). Note that the overlying skin surface is smooth, excluding motion artifact as the source of the sternal step-off. Review of the remainder of the CT images, not shown, revealed no additional injuries.

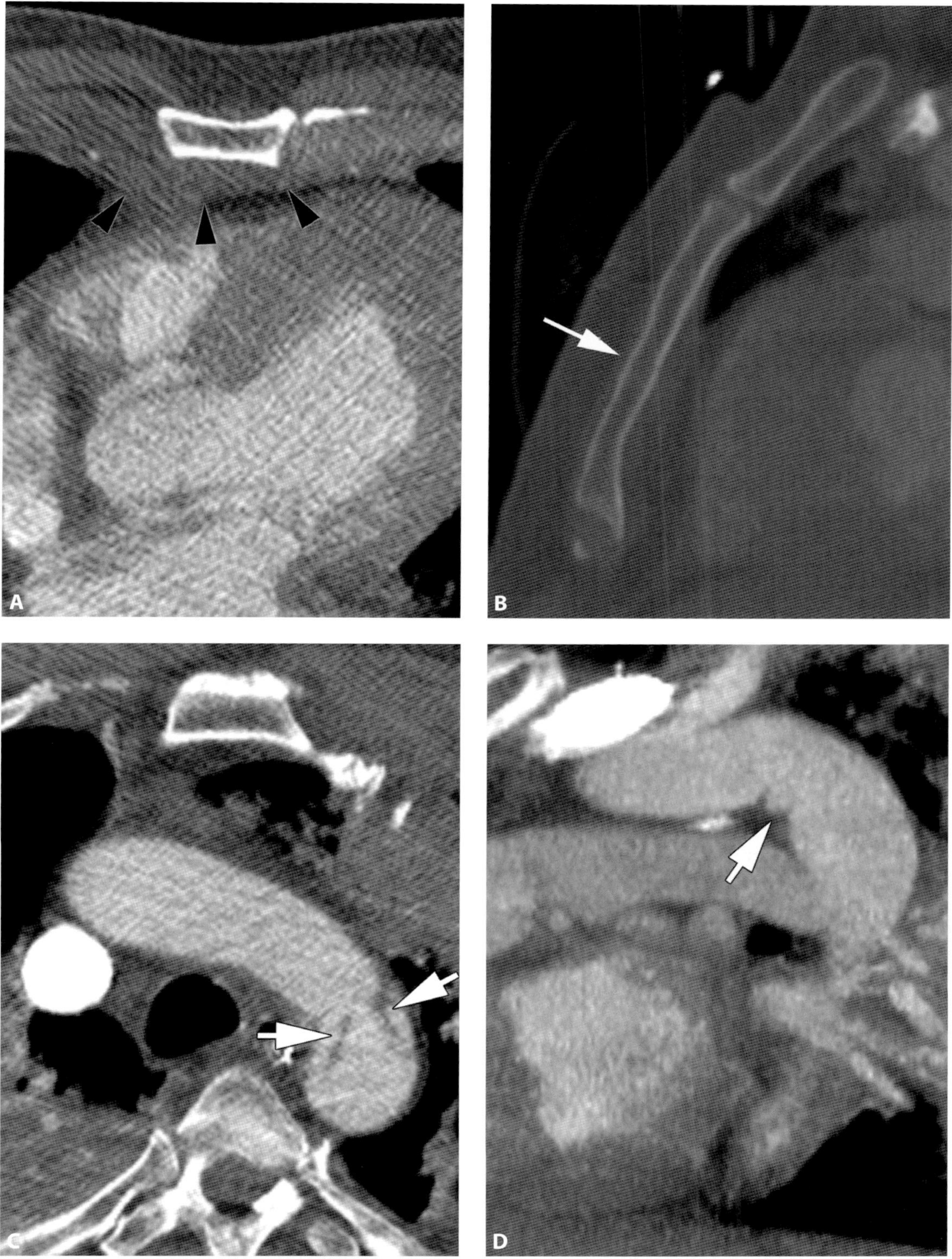

Figure 84.3 A. Axial contrast-enhanced CT angiogram of the chest in soft tissue windows in a 49-year-old man following a motorcycle collision shows a retrosternal hematoma (arrowheads), raising concern for sternal fracture. **B.** Sagittal reformation confirms a non-displaced sternal fracture (arrow). **C.** Axial image at a different level shows intimal tears (arrows) as a component of a blunt thoracic aortic injury. **D.** Oblique maximum intensity projection (MIP) image shows a small pseudoaneurysm of the acute aortic injury (arrow).

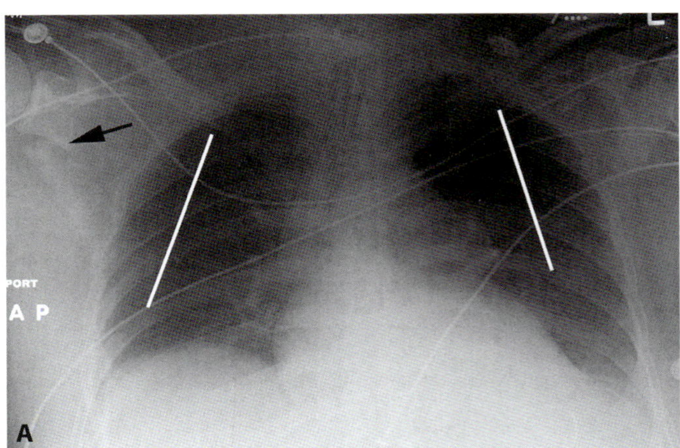

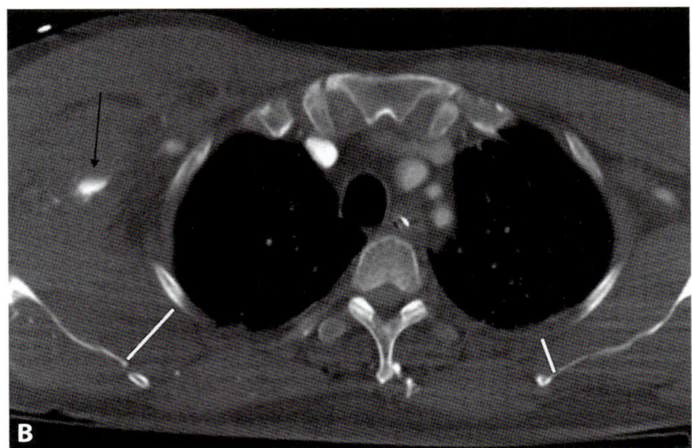

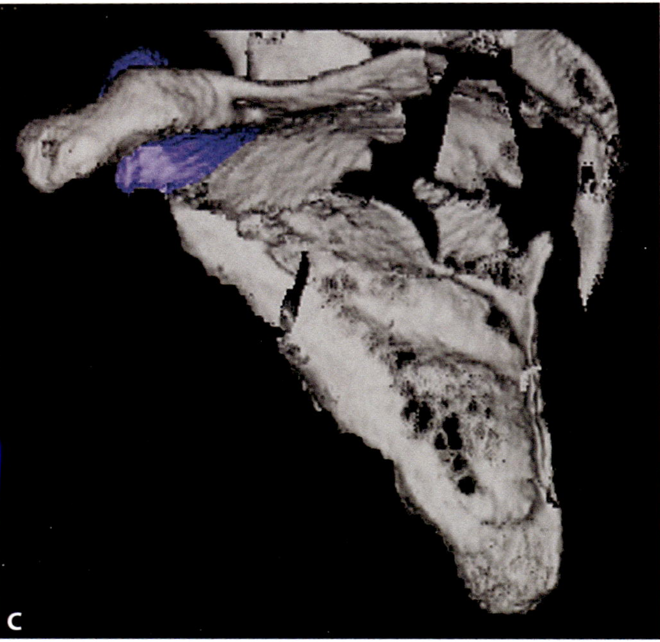

Figure 84.4 **A.** Portable chest radiograph from a 57-year-old man ejected during a motor vehicle collision shows a right scapular fracture (arrow). There is also lateral displacement of the right scapula, best appreciated by comparing the medial scapular margins (white lines). **B.** Axial contrast-enhanced CT of the chest demonstrates asymmetric elevation of the right scapula away from the ribs (white lines), suggestive of scapulothoracic dissociation. There is also generalized enlargement of the soft tissues of the right upper hemithorax, secondary to hematoma. Active contrast extravasation (arrow) is appreciated from the right axillary artery. **C.** 3D surface rendering from a dedicated CT of the right scapula shows severely comminuted fracture of the scapular body and neck. The fully avulsed glenoid fragment is highlighted in blue.

Sesamoids and bipartite patella

Claire K. Sandstrom

Imaging description

There are several sesamoid bones that we expect to see in most, if not all, patients, as well as other variants that may also be detected. The largest and best known sesamoid bone is the patella, which is part of the extensor mechanism of the knee (Figures 85.1–85.4).

In the foot, there are typically two first hallux sesamoid bones within the two heads of the flexor hallucis brevis tendon, forming part of the first metatarsophalangeal joint capsule along the plantar surface of the first metatarsal head (Figures 85.5–85.9) [1]. The tibial sesamoid is medial and the fibular sesamoid lies laterally. Additionally, the os peroneum may be seen along the lateral aspect of the midfoot within the distal peroneus longus tendon (Figures 85.10–85.12).

Each hand typically has five sesamoid bones: two at the first metacarpophalangeal (MCP) joint, one each at the second and fifth MCP joints, and one at the first interphalangeal joint (Figures 85.13 and 85.14).

Importance

Sesamoids are accessory ossific structures that are contained within a tendon or joint capsule and reduce friction during flexion and extension as they slide over adjacent structures. In distinction to accessory ossicles, sesamoids form from their own ossification center. Like accessory ossicles, the importance of sesamoids is recognizing their normal and variant appearances that maybe mimic or hide acute pathology.

Typical clinical scenario

Fractured or displaced sesamoid bones should raise concern for associated tendinous disruption and may prompt MRI. Partite sesamoids may be differentiated from acute fracture by radiograph or CT, or in some cases, MRI or bone scan.

Differential diagnosis

Patellar fractures are frequently transverse or longitudinal, secondary to forced rapid flexion against a fully contracted quadriceps or transient lateral patellar dislocation, respectively. Complex stellate fractures occur from severe direct trauma. They should always be differentiated from bipartite patella, which is the synchondrosis most commonly found at the superolateral margin of the patella [2] and much more commonly occurs in men than women. There is typically a smooth and clearly sclerotic margin between the components of the bipartite patella (Figures 85.1–85.4). The synchondrosis can become disrupted by direct trauma, in which case MRI may be required for diagnosis.

Either of the hallux sesamoids can be partite; a bipartite hallux sesamoid is present in approximately 19% of the population [3]. Acute sesamoid fractures do occur and, just like bipartite sesamoids, most commonly involve the tibial sesamoid [2]. To differentiate a bipartite sesamoid from an acute fracture, one should note that a bipartite sesamoid is usually larger and has smooth margins with well-corticated synchondrosis. However, since acute trauma can disrupt the synchrondrosis of a bipartite sesamoid, focal pain and tenderness overlying a bipartite patella should still prompt further evaluation with MRI or bone scan [1]. "Turf toe" refers to a hyperextension injury with acute rupture of the plantar capsule and injury to the flexor hallucis brevis tendon, with or without fracture or dislocation of the sesamoids [4, 5].

Also in the foot, the os peroneum should be differentiated from avulsion fractures of the base of the fifth metatarsal or base of the cuboid. Fractures of the os peroneum may be seen radiographically and are typically associated with disruption of the peroneus longus tendon [2].

> **Teaching point**
>
> Identification of a partite sesamoid in the setting of focal pain and tenderness following an appropriate history of direct trauma does not exclude acute injury to the tendon or synchondrosis. MRI is usually diagnostic in such cases.

REFERENCES

1. Taylor JAM, Sartoris DJ, Huang GS, Resnick DL. Painful conditions affecting the first metatarsal sesamoid bones. *Radiographics*. 1993;**13**:817–30.
2. Kalantari BN, Seeger LL, Motamedi K, Chow K. Accessory ossicles and sesamoid bones: spectrum of pathology and imaging evaluation. *Appl Radiol*. 2007;**36**(10):28–32.
3. Karadaglis D. Morphology of the hallux sesamoids. *Foot Ankle Surg*. 2003;**9**(3):165–7.
4. Bowers KD, Martin RB. Turf-toe: a shoe-surface related football injury. *Med Sci Sports*. 1976;**8**:81–3.
5. Clanton TO, Ford JJ. Turf toe injury. *Clin Sports Med*. 1994; **13**(4):731–41.

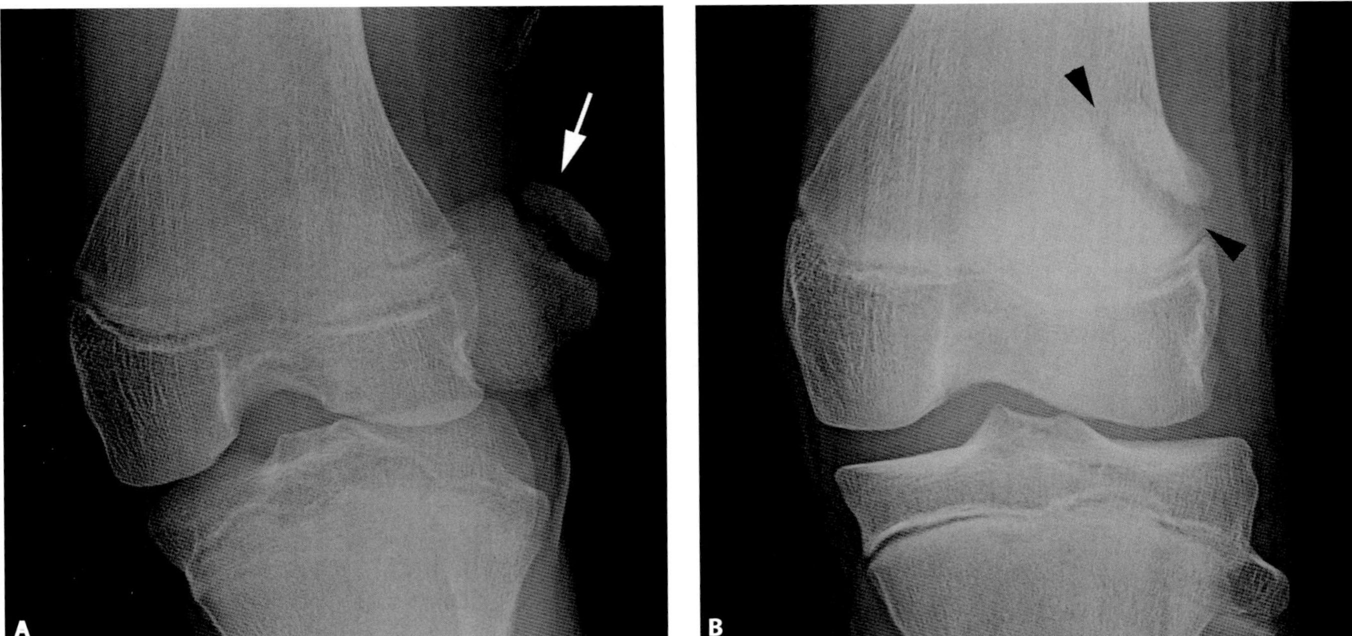

Figure 85.1 **A.** Anterior-posterior (AP) radiograph of the knee in a 15-year-old boy who fell awkwardly while playing ultimate Frisbee shows non-transient lateral patellar dislocation. A bipartite patella (arrow) is nicely shown as a result of the dislocation. **B.** After reduction, an AP radiograph of the same knee shows the more common imaging appearance of the bipartite patella superimposed over the distal femur. The cleft (arrowheads) is still visible but less well characterized.

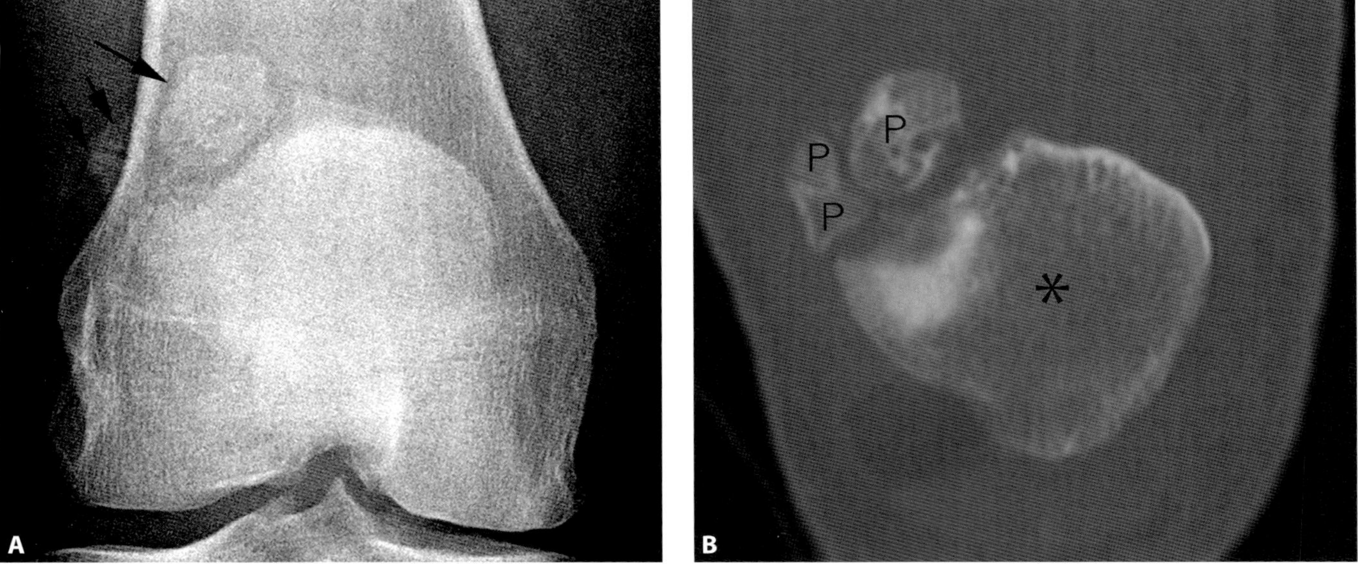

Figure 85.2 **A.** AP radiograph of the knee in a 50-year-old man with recent knee trauma and lipohemarthrosis shows multiple rounded osseous structures (arrows) at the superolateral aspect of the patella. **B.** Coronal reformations of the knee subsequently showed an osteochondral defect of the femur as the source of the lipohemarthrosis. There are at least three well-corticated bones (P) at the superolateral aspect of the patella (asterisk), consistent with multipartite patella.

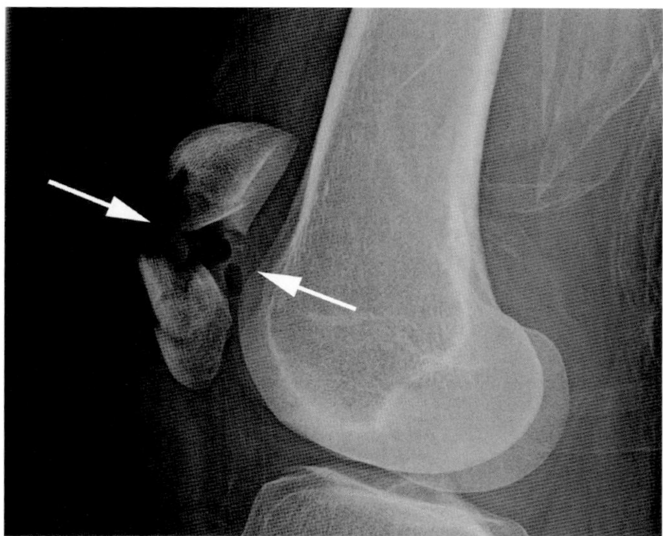

Figure 85.3 Lateral knee radiograph from a 52-year-old man involved in a motor vehicle collision shows an open transverse fracture (arrows) of the patella. Because the fracture is open and the joint capsule is disrupted, there is no joint effusion.

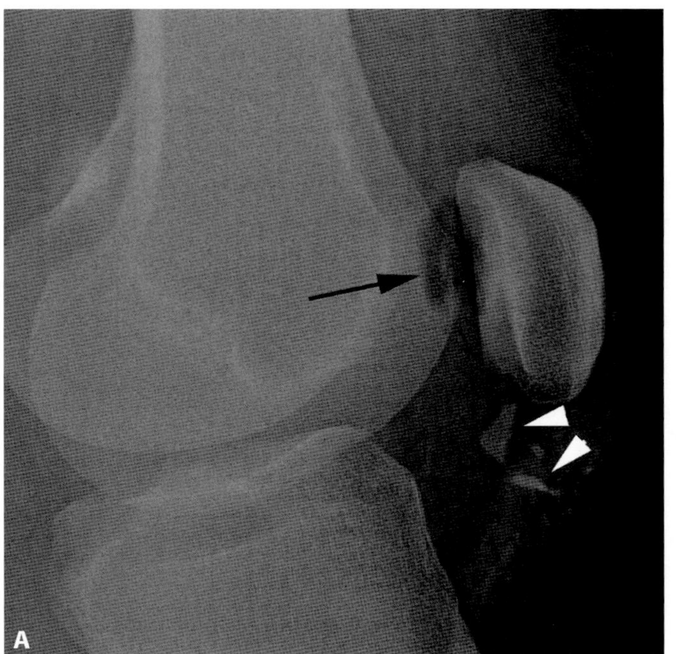

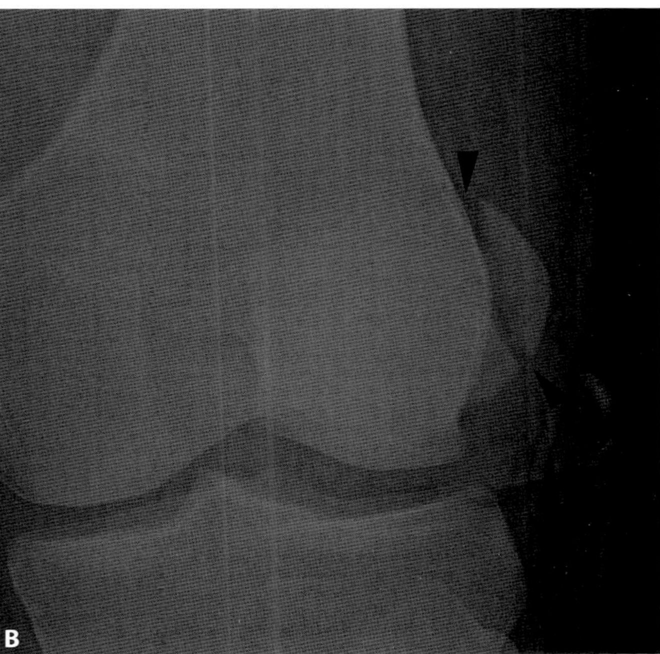

Figure 85.4 **A.** Lateral radiograph of the left knee in an 18-year-old man following a motor vehicle collision shows gas within the joint capsule (arrow) and several bone fragments (arrowheads) inferior to the patella. **B.** Lateral oblique radiograph shows a vertical fracture (arrowheads) through the lateral aspect of the patella. Note the sharp angles on the fracture fragment, compared with the rounded margins on the bipartite patella in Figure 85.1.

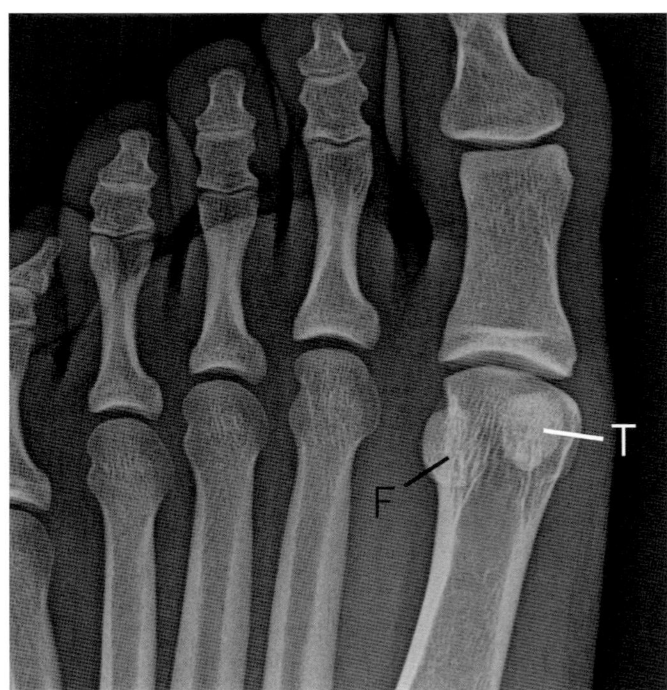

Figure 85.5 Frontal foot radiograph from a 21-year-old man with two days of foot pain while walking shows normal tibial (T) and fibular (F) hallux sesamoids.

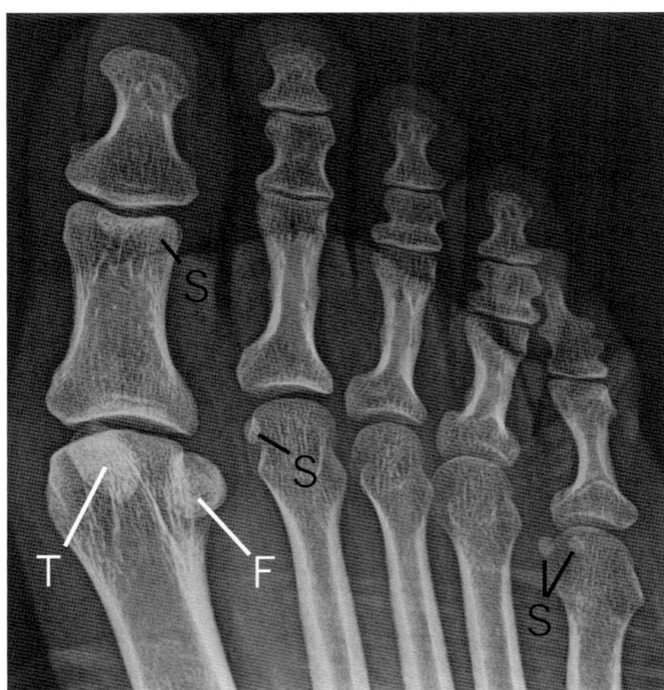

Figure 85.6 Frontal foot radiograph from a 38-year-old female after falling shows multiple unnamed sesamoid bones (S) in addition to the tibial (T) and fibular (F) hallux sesamoids.

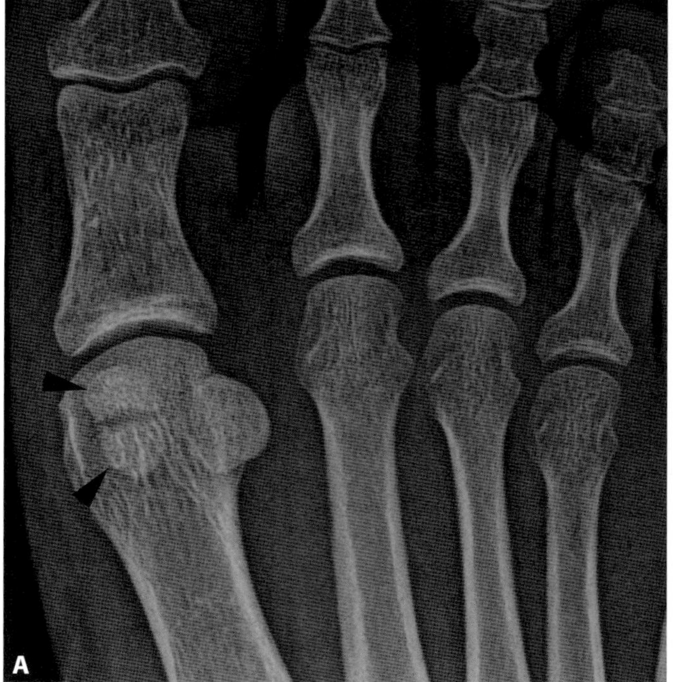

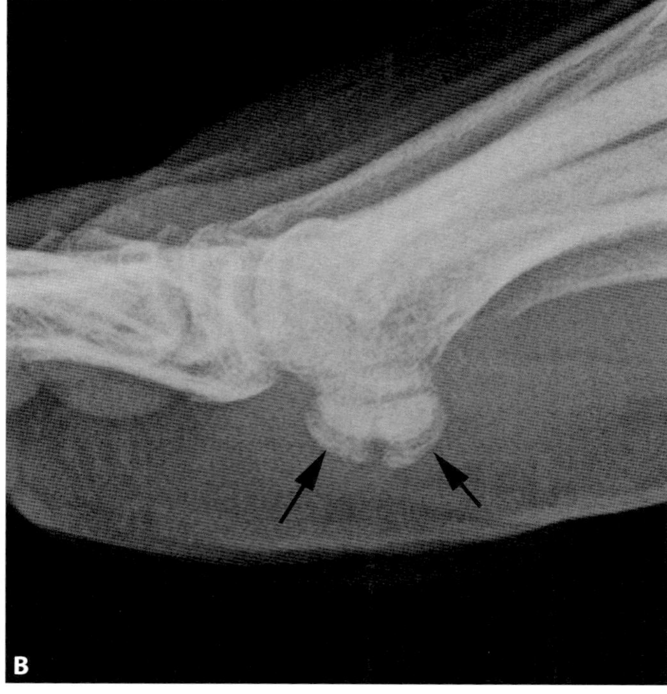

Figure 85.7 **A.** Frontal foot radiograph from a 24-year-old woman with a known calcaneal fracture (not shown) from a fall shows a bipartite tibial hallux sesamoid (arrowheads). **B.** Lateral foot radiograph shows the two parts of the sesamoid (arrows) and the intervening synchondrosis.

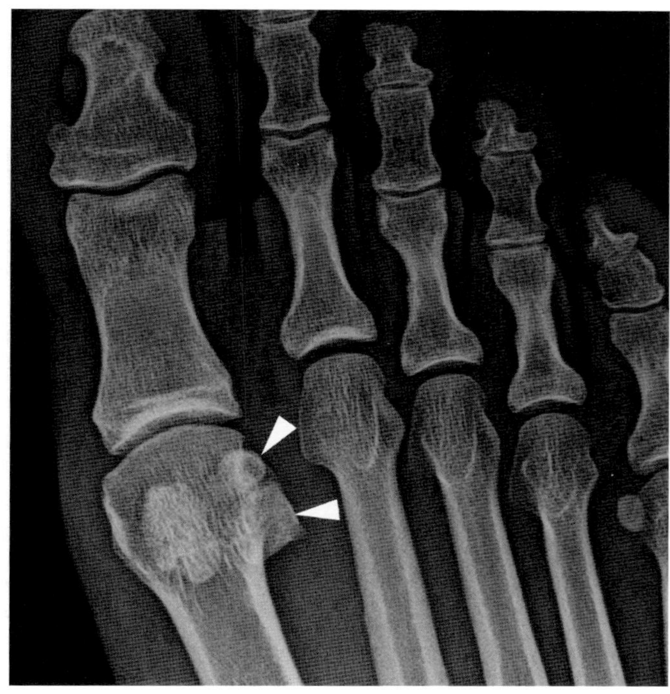

Figure 85.8 Frontal foot radiograph from a 43-year-old man with right foot pain shows a bipartite fibular hallux sesamoid (arrowheads).

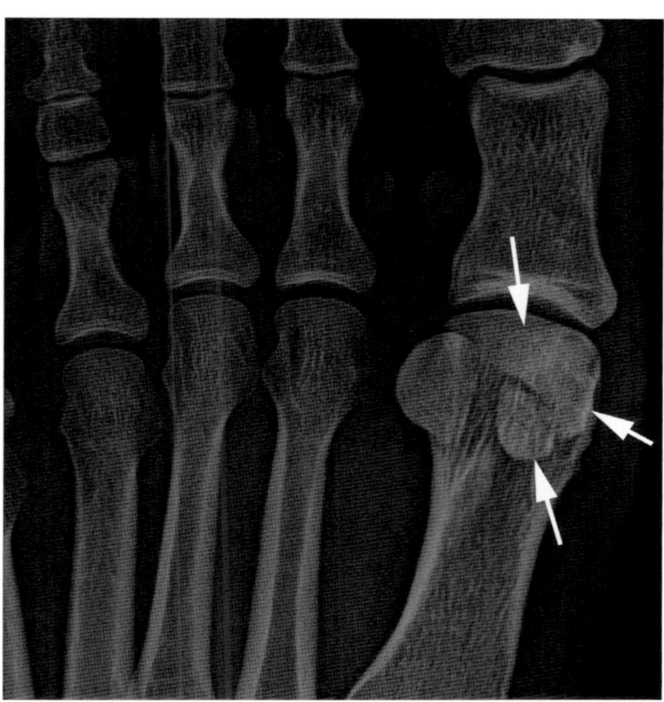

Figure 85.9 Frontal foot radiograph from a 44-year-old man suffering a ground-level fall shows a multipartite tibial hallux sesamoid (arrows).

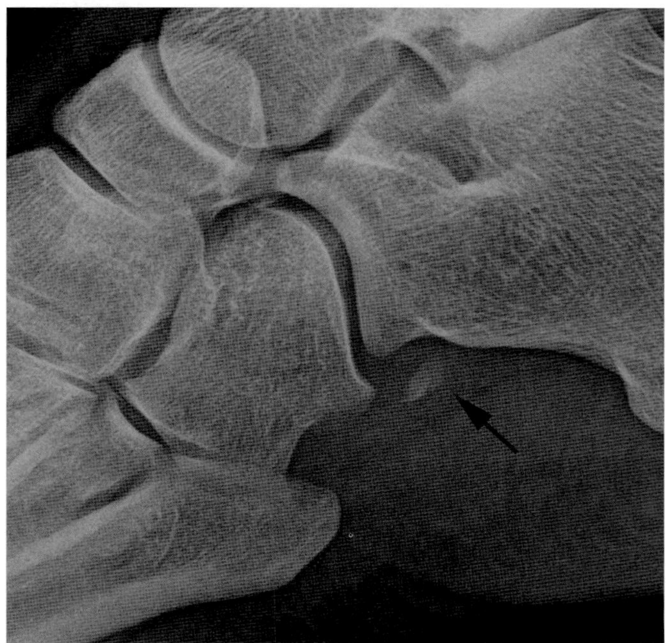

Figure 85.10 Lateral foot radiograph from a 24-year-old man suspected of having an ankle fracture after motor vehicle collision shows a normal os peroneum (arrow).

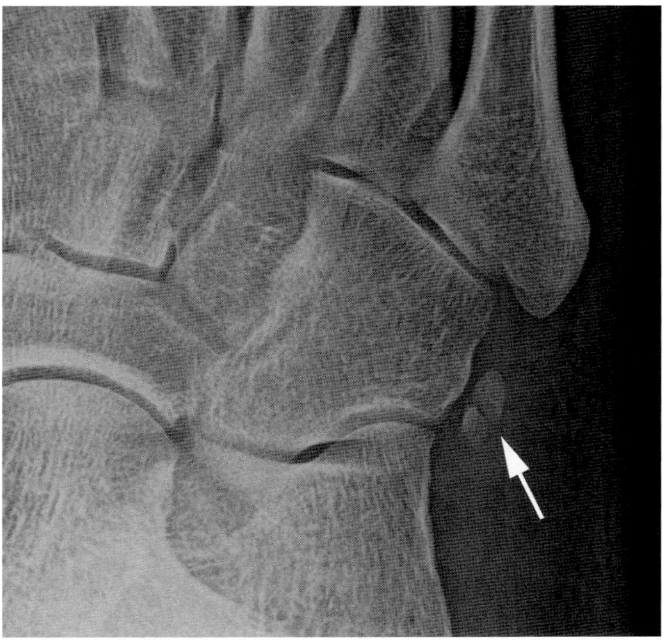

Figure 85.11 Oblique foot radiograph from a 40-year-old man with foot pain and swelling related to a third metatarsal fracture (not shown). There is also a bipartite os peroneum along the lateral aspect of the cuboid (arrow).

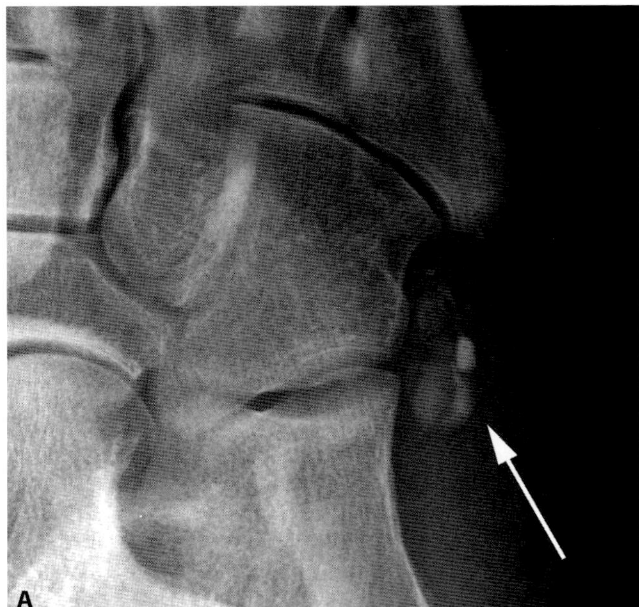

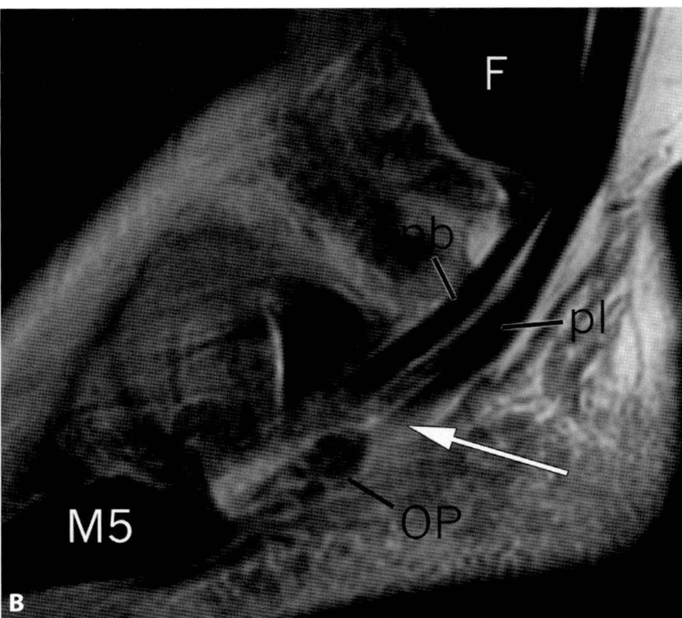

Figure 85.12 A. Oblique foot radiograph from a 47-year-old woman with acute lateral ankle pain and swelling shows a multipartite os peroneum (arrow) that was radiographically similar five years earlier. **B.** Sagittal proton density MR with fat saturation shows the multipartite os peroneum (OP) in relation to the fifth metatarsal (M5) and distal fibula (F). The sesamoids lie within the distal peroneus longus tendon (pl), inferior to the peroneus brevis tendon (pb). High signal within the distal peroneus longus tendon (arrow) suggests tendinosis without high-grade tear.

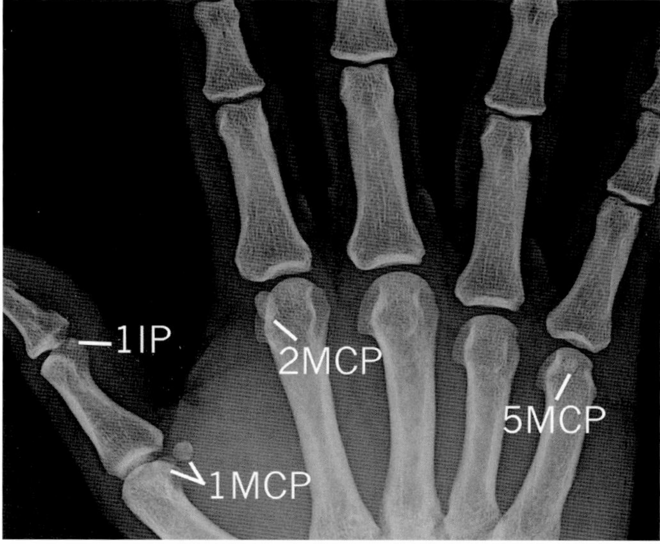

Figure 85.13 Frontal hand radiograph from a 31-year-old man following an assault shows all five of the common sesamoid bones that may be found in the hand, including the paired first MCP sesamoids (1MCP), first interphalangeal sesamoid (1IP), and sesamoids at the second and fifth MCP joints (2MCP and 5MCP respectively). Many patients will not have all five but some variable number.

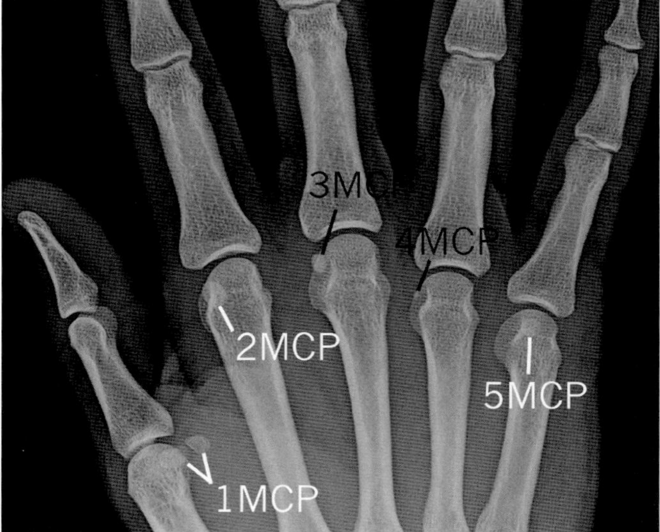

Figure 85.14 Frontal hand radiograph from a 22-year-old man following an assault shows four of the common sesamoid bones (labeled in white, 1MCP, 2MCP, and 5MCP), with the exception of the first interphalangeal sesamoid. In addition, there are two additional sesamoids that are uncommonly seen at the third and fourth MCP joints (labeled in black, 3MCP and 4MCP).

Subtle knee fractures

Claire K. Sandstrom

Imaging description

In addition to the standard anterior-posterior (AP) and lateral radiographs of the knee, medial and lateral oblique views, or tangential views, should be obtained as part of the standard radiographic assessment of the injured knee. While the lateral radiograph is sensitive for knee effusion and therefore has been suggested as a screening tool for intra-articular pathology [1], additional views are often needed to identify the fracture. Lateral and medial oblique views at 45 degrees were advocated by Daffner and Tabas to remove superimposition of the patella over the distal femur and to better show the medial and lateral tibial plateaus [2]. The combination of tangential, AP, and lateral radiographs of the knee has been reported to be more sensitive, at 85%, for acute fracture detection than AP and lateral radiographs alone (79% sensitive) [3]. Multidetector CT still plays a role in fracture characterization and preoperative planning.

Any cortical defect or unexplained sclerotic or lucent line should be viewed with suspicion, even if only seen on one projection. CT or MR should be used in confirming or excluding a fracture when radiographs are indeterminate or clinical suspicion is high despite negative radiographs.

Importance

Treatment of intra-articular fractures of the knee is frequently surgical, to restore joint stability and integrity of the articular surface and decrease the likelihood of post-traumatic osteoarthritis. Recognition of radiographically subtle fractures is important, as they can have significant clinical consequences if undiagnosed [4].

Typical clinical scenario

Following acute lower extremity trauma, knee radiographs may be ordered if the patient complains of pain, or if there is suspicious deformity, bruising, or swelling in an unconscious patient. While two views of the knee may be faster to obtain and more comfortable for the patient, the addition of the tangential views is important to detect and characterize fractures.

Differential diagnosis

The acute fractures that may be more apparent on tangential knee radiographs include tibial plateau injuries (Figures 86.1 and 86.2), patellar fractures (Figure 86.3), and proximal fibular fractures (Figure 86.4). A pre-existing arthropathy may have associated osteophytes or spurs that may be confused for marginal fractures. An effusion seen on lateral view may be due to acute non-osseous soft tissue internal derangement or an underlying arthropathy, hemarthrosis (in the coagulapathic patient), or septic arthritis. Careful inspection for a lipohemarthrosis limits the differential diagnosis to a fracture. Developmental and congenital abnormalities, such as physeal scar or bipartite patella, may also mimic fractures.

Teaching point

Medial and lateral oblique radiographs of the knee improve sensitivity for potentially subtle fractures.

REFERENCES

1. Verma A, Su A, Golin AM, O'Marrah B, Amorosa JK. A screening method for knee trauma. *Acad Radiol.* 2001;**8**(5):392–7.
2. Daffner RH, Tabas JH. Trauma oblique radiographs of the knee. *J Bone Joint Surg.* 1987;**69**A(4):568–72.
3. Gray SD, Kaplan PA, Dussault RG, *et al.* Acute knee trauma: how many plain film views are necessary for the initial examination? *Skeletal Radiol.* 1997;**26**(5):298–302.
4. Capps GW, Hayes CW. Easily missed injuries around the knee. *Radiographics.* 1994;**14**(6):1191–210.

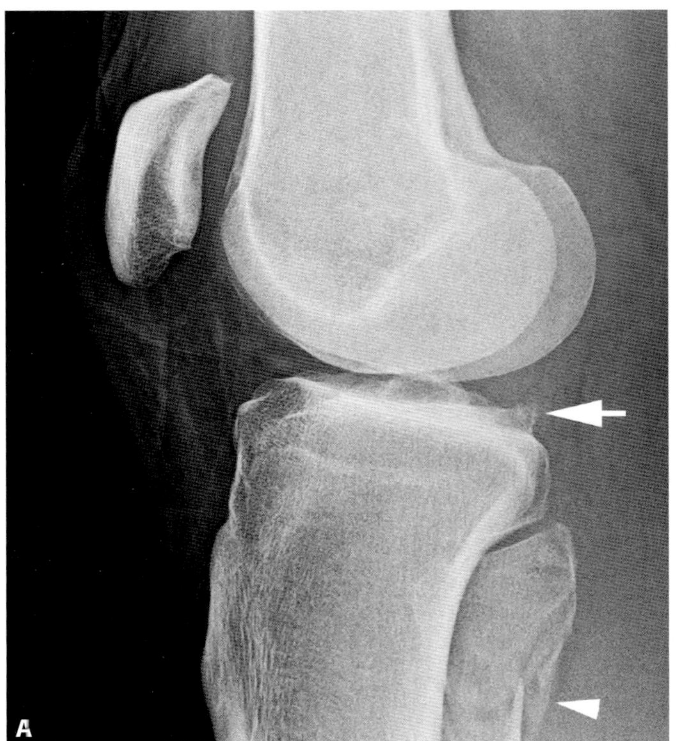

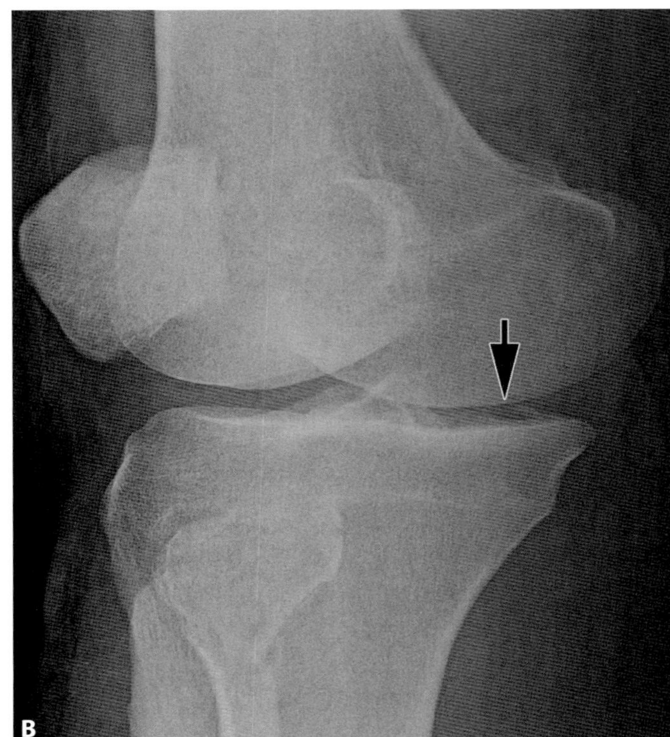

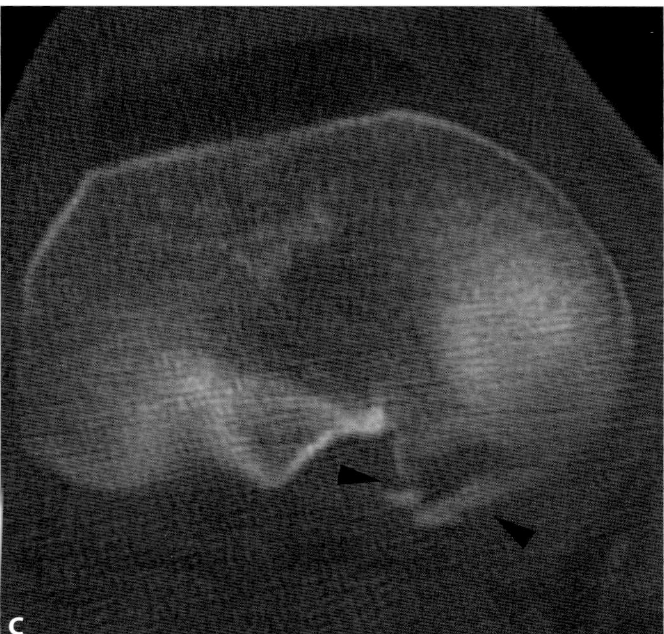

Figure 86.1 A. Lateral radiograph of the right knee in a 40-year-old man who fell 30 ft onto his lower extremities shows a fracture of the proximal fibula (arrowhead). An effusion was difficult to assess. There is also questionable irregularity of the posterior tibial plateau (arrow), of uncertain age. AP view, not shown, was unremarkable. **B.** Medial oblique view shows cortical step-off along the medial tibial plateau (arrow), confirming that the posterior plateau irregularity suspected on the lateral view is a fracture. **C.** Axial CT through the tibial plateau shows small fracture fragments from the posterior margin of the medial tibial plateau (arrowheads).

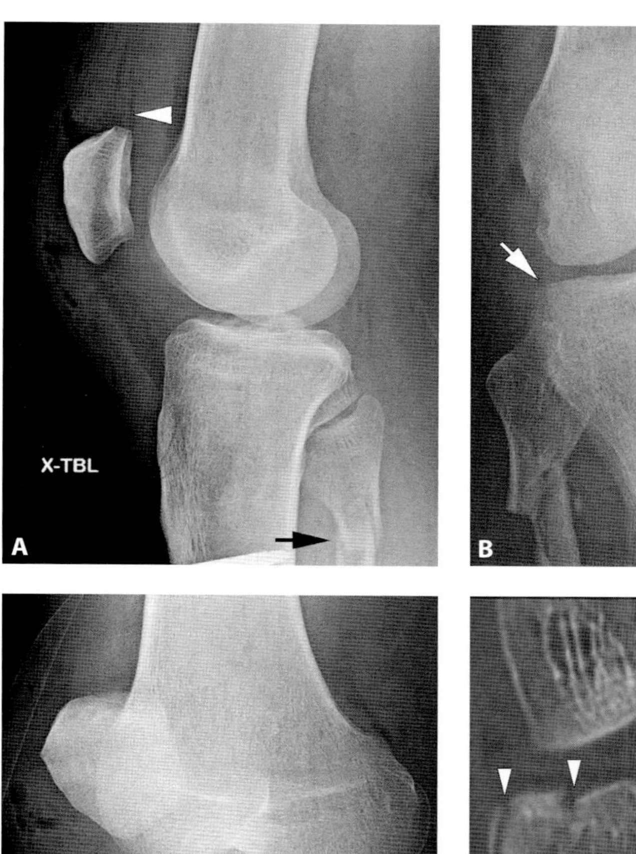

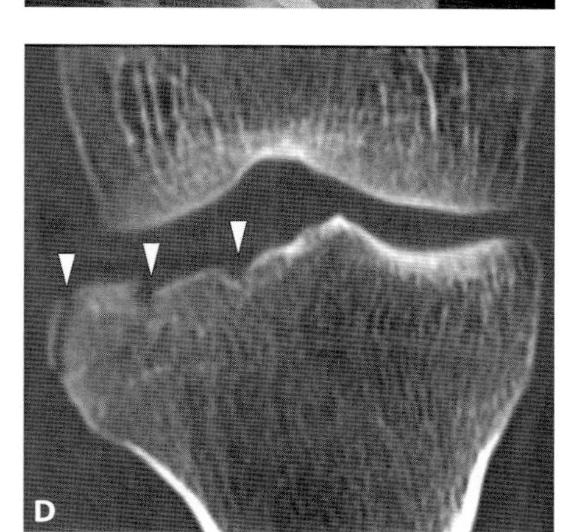

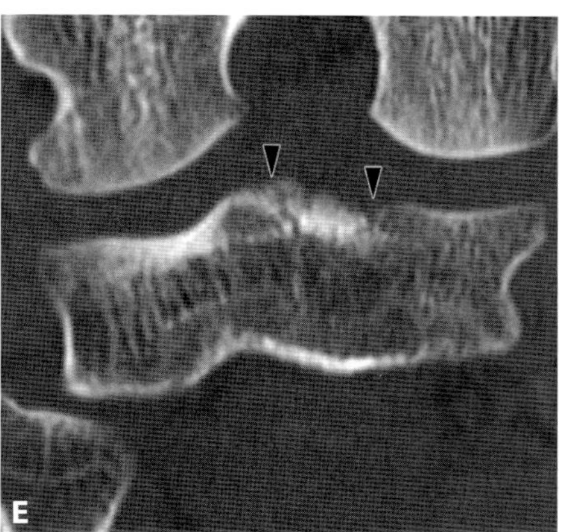

Figure 86.2 A. Cross-table lateral radiograph of the knee in a 40-year-old man following collision between a motor vehicle and his all-terrain vehicle shows a lipohemarthrosis (arrowhead). There is also a fracture of the proximal fibular shaft (arrow) and external fixation hardware in the proximal tibia. It is important to note that the fibular fracture is not intracapsular and therefore does not adequately explain the lipohemarthrosis. **B.** AP view of the same knee shows normal alignment. Questionable irregularity is seen along the lateral tibial plateau (arrow). **C.** Oblique radiograph of the knee better shows the fracture at the margin of the lateral tibial plateau (arrow). **D.** Coronal image from subsequent CT of the knee obtained for surgical planning purposes show the minimally displaced lateral tibial plateau fracture (arrowheads), which is more extensive than previously recognized. **E.** More posterior coronal image from the same CT shows a non-displaced, radiographically occult fracture of the medial tibial spine (arrowheads).

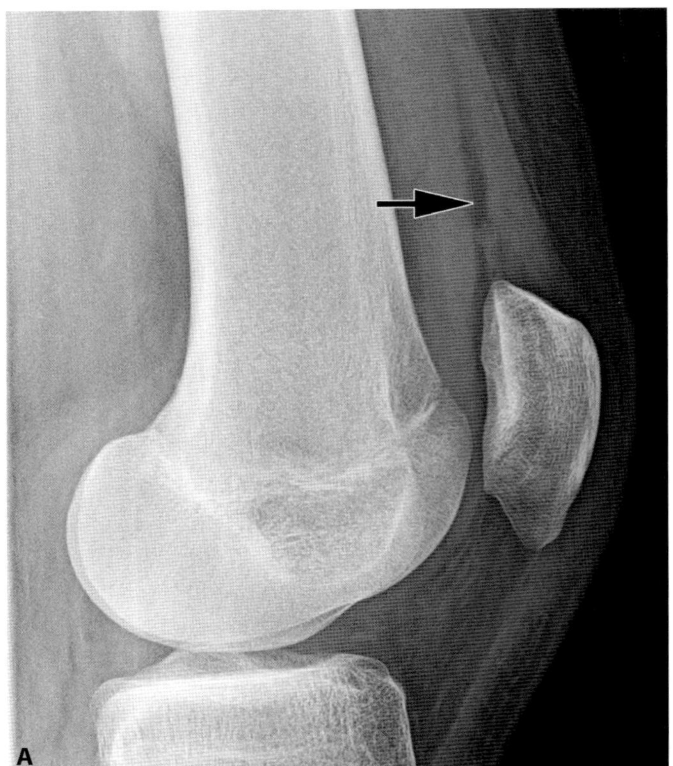

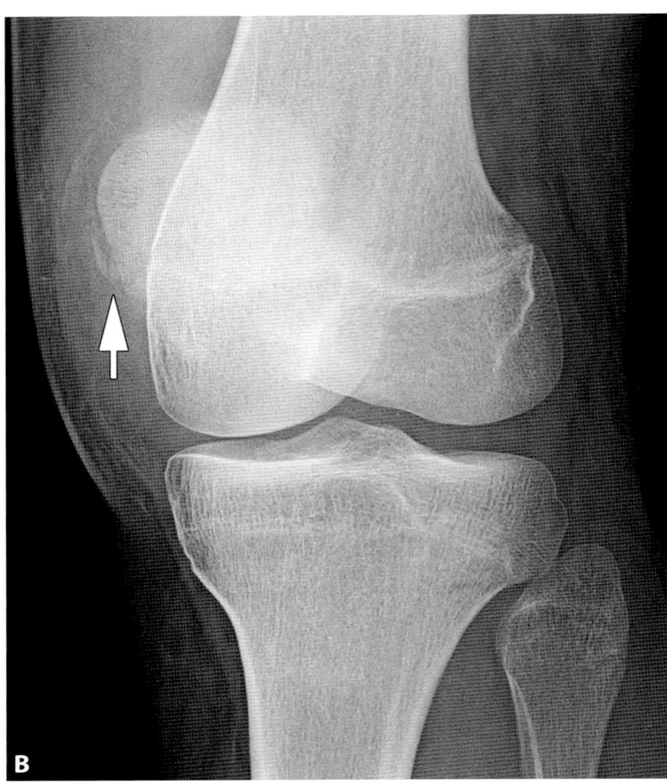

Figure 86.3 **A.** Cross-table lateral radiograph of the knee in a 16-year-old girl following transient lateral patellar dislocation shows a lipohemarthrosis (arrow). **B.** Oblique radiograph of the knee shows a small bone fragment medial to the patella (arrow), which is an osteochondral defect of the medial patellar facet resulting from transient lateral patellar dislocation. This finding was radiographically occult on the AP view (not shown). No other fractures were detected.

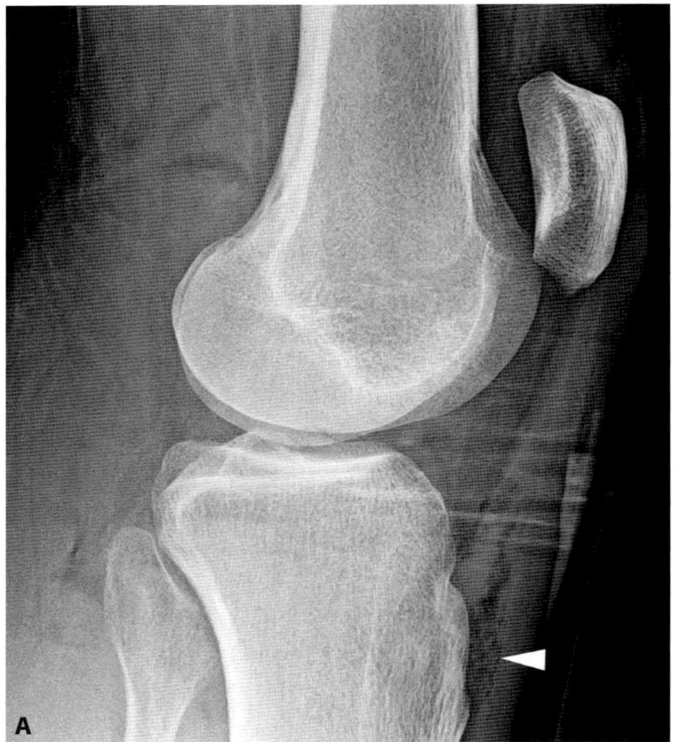

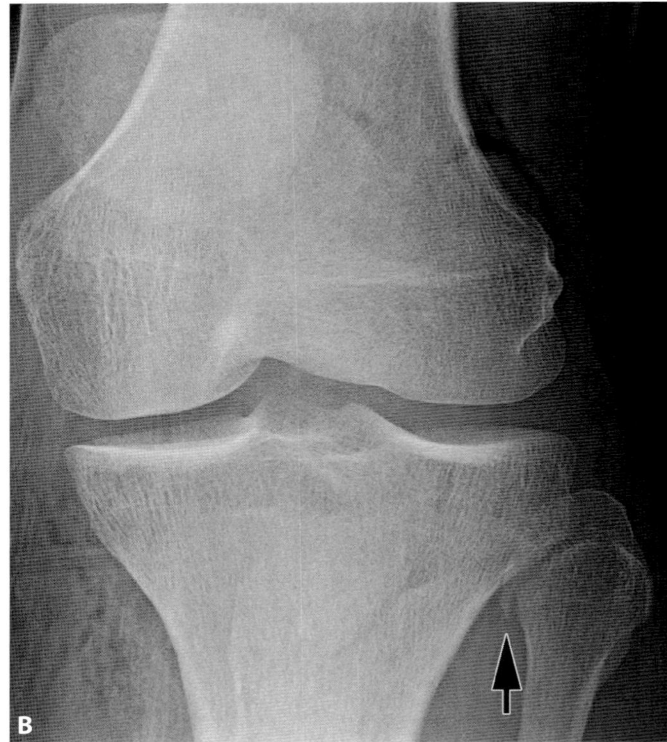

Figure 86.4 **A.** Lateral radiograph of the knee in a 30-year-old man with known open tibial shaft fracture (not shown) resulting from a motor vehicle collision shows gas within the anterior infrapatellar soft tissues (arrowhead). Other than patella alta, alignment at the knee appears intact, and there was no joint effusion. AP radiograph (not shown) was normal. **B.** However, a lateral oblique view shows a minimally displaced fracture of the proximal fibula (arrow) that would otherwise have been radiographically occult.

CASE 87 Lateral condylar notch sign

Claire K. Sandstrom

Imaging description

In the setting of knee pain following trauma, an effusion raises concern for an internal joint derangement. Close inspection of the radiographs may reveal subtle clues to the presence of acute or chronic anterior cruciate ligament (ACL) tear, such as a deep lateral condylar notch [1]. To assess the lateral condylar notch sign, draw a line tangential to the lower articular surface of the lateral femoral condyle. Measure the depth of the notch perpendicular to this line [2]. If the sulcus measures greater than 1.5 mm in depth, an ACL injury should be suspected (Figure 87.1) [3]. Additionally abnormal angularity of the notch should also raise concern for ACL injury. A sulcus shallower than 1.5 mm does not assure integrity of the ACL (Figure 87.2) [3, 4].

Pathologically, the deep lateral femoral sulcus reflects the presence of a transchondral fracture, presumed to result from impaction on the lateral or posterolateral tibia during the twisting injury that simultaneously tore the ACL. If the impaction injury results in only a bone bruise or isolated chondral injury, it will be occult on radiograph but not on MRI.

Importance

Significant soft tissue injuries, including tears of the ACL, are difficult to detect on radiographs and usually require high clinical suspicion and MRI for diagnosis. However, if a deep lateral femoral notch is present, an MRI should be performed to evaluate the extent of underlying soft tissue injuries. A deep lateral femoral notch can occur following acute rupture or chronic instability [2, 5].

Typical clinical scenario

Searching for subtle signs of internal derangement is particularly worthwhile in twisting injuries, such as those incurred during athletic events and falls. The lateral femoral notch sign with an ACL tear is more common in young men and in those with concomitant lateral meniscal injuries [5, 6].

Differential diagnosis

The differential diagnosis for a notch along the articular surface of the femoral condyle includes a normal sulcus, a transchondral fracture related to an ACL tear, an osteochondral defect or subchondral osteonecrosis from other causes. A normal sulcus of the lateral femoral condyle should be within 10 mm of Blumensaat's line (a line that corresponds to the roof of the intercondylar notch) on a true lateral radiograph and should always measure less than 1 mm in depth (Figure 87.3) [2].

If a concavity along the lateral femoral articular surface is more than 10 mm distal to Blumensaat's line, even less than 1 mm deep, a transchondral fracture should be suspected [5, 6].

Teaching point

Post-traumatic knee pain accompanied by an effusion should prompt close inspection of the radiograph for evidence of ACL tear. A lateral femoral condylar notch greater than 1.5 mm in depth, reflecting osteochondral injury, on the lateral knee radiograph is a specific sign of ACL tear.

REFERENCES

1. Pao DG. The lateral femoral notch sign. *Radiology*. 2001;**219**:800–1.
2. Warren RF, Kaplan N, Bach BR. The lateral notch sign of anterior cruciate ligament insufficiency. *Am J Knee Surg*. 1988;**1**:119–24.
3. Cobby MJ, Schweitzer ME, Resnick D. The deep lateral femoral notch: an indirect sign of a torn anterior cruciate ligament. *Radiology*. 1992;**184**:855–8.
4. Jones AR, Finlay DBL, Learmonth DJ. A deep lateral femoral notch as a sign of acutely torn anterior cruciate ligament. *Injury*. 1993;**24**(9):601–2.
5. Garth WP, Greco J, House MA. The lateral notch sign associated with acute anterior cruciate ligament disruption. *Am J Sports Med*. 2000; **28**(1):68–73.
6. Nakauchi M, Kurosawa H, Kawakami A. Abnormal lateral notch in knees with anterior cruciate ligament injury. *J Orthop Sci*. 2000;**5**(2):92–5.

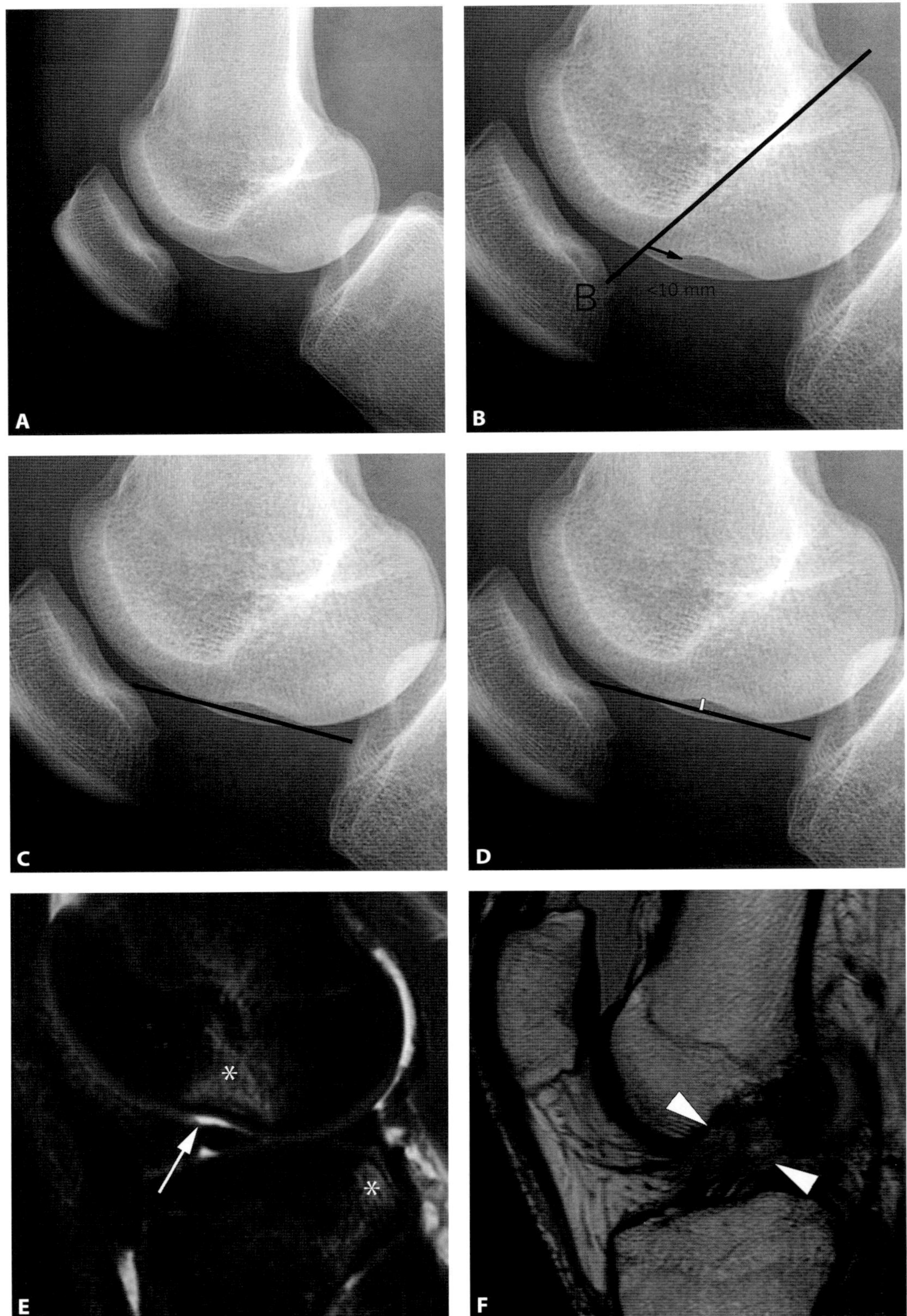

Figure 87.1 **A.** Lateral radiograph of the right knee in a 22-year-old male soccer player with pain after twisting his knee shows an effusion. **B.** Close-up of the same lateral view shows a femoral notch within 10 mm of Blumensaat's line (black line with B). **C.** To measure the depth based on the technique reported by Warren *et al.* [2], draw a line tangential to the lateral femoral condyle (black line). **D.** The notch depth is then measured (short white line) perpendicular to the black line. In this case, the sulcus was 1.7 mm deep. **E.** Sagittal T2 image of the knee with fat saturation shows bone marrow edema in the lateral femoral condyle and posterior tibial plateau (asterisks), as well as the lateral femoral sulcus and the more apparent overlying chondral defect (arrow). **F.** Sagittal oblique proton density image through the knee confirms mid-substance tear of the ACL (arrowheads).

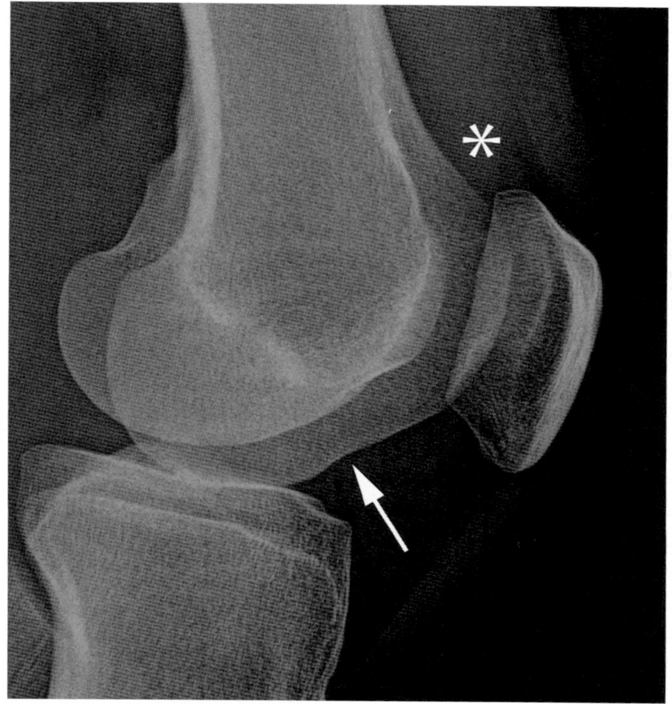

Figure 87.2 Lateral radiograph of the left knee in a 21-year-old woman after a fall shows a joint effusion (asterisk). No deep femoral notch sign (white arrow) is present, though the patient was later diagnosed with acute ACL tear by MRI.

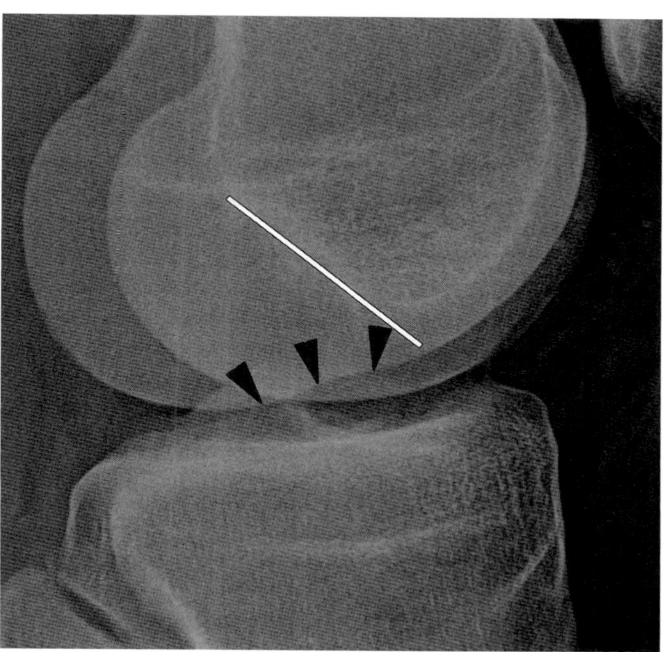

Figure 87.3 A lateral left knee radiograph from a 35-year-old man shows a normal lateral femoral condylar sulcus (arrowheads) relative to Blumensaat's line (white line).

Easily missed fractures of the foot and ankle

Ken F. Linnau

Imaging description

Metatarsal stress fractures

Metatarsal bones are the most common site of stress fracture. These fractures were originally recognized in military recruits after excessive marching, but can occur with any excessive activity including hiking, running, dancing, and marching. Metatarsal stress fractures can also occur after foot surgery, owing to the altered biomechanics of the foot. Stress fractures are often occult on initial radiographs.

If an occult stress fracture is suspected clinically, imaging options include repeat radiography in 7–10 days at which point periosteal bone formation and fracture healing will often make the injury apparent (Figure 88.1). Alternatively, bone scintigraphy or MRI can be utilized for diagnosis [1].

Lisfranc fracture dislocation

Tarsometatarsal (Lisfranc) fracture dislocations usually are caused by forced plantar flexion, twisting, or crush injury due to the unique osseous and ligamentous anatomy between the first and second metatarsal base. The bases of the second through fifth metatarsal bones are tightly connected through the transverse metatarsal ligaments. However, no such ligament exists between the first and second metatarsal. Therefore, the oblique Lisfranc ligament connects the medial cuneiform to the base of the second metatarsal. Furthermore, the base of the second metatarsal is locked in a mortise formed by the medial and lateral cuneiform due to the fact that the middle cuneiform is shorter than its neighbors.

While severe Lisfranc fracture dislocations do not pose a diagnostic challenge and are easily classified as homolateral or divergent, lower-energy injuries can easily be overlooked. Clues to subtle Lisfranc injuries include soft tissue swelling at the dorsum of the foot and mild widening of the joint space between first and second metatarsal (Figure 88.2).

Stress views can sometimes expedite diagnosis, and lead to improved outcomes. Stress is applied through forefoot abduction over a fulcrum at the calcaneo-cuboid joint (Figure 88.2). If the first tarsometatarsal joint displaces more than 2 mm under stress, surgical repair is indicated [2].

For complete characterization of the injury and surgical planning, CT of the foot is usually obtained.

Fractures of the lateral process of the talus

The lateral process of the talus is best visualized on the anteroposterior view of the ankle series (Figure 88.3). It should be carefully inspected every time that soft tissue swelling is present at or inferior to the lateral malleolus. The lateral process of the talus bears the lateral portion of the posterior articular facet. Fractures through the lateral process of the talus are also known as "snowboarder fracture" and predispose to premature osteoarthritis if overlooked and untreated [3].

Importance

Metatarsal stress fractures are often occult on initial radiographs. Continued untreated use of a metatarsal stress fractures may lead to fracture displacement.

Lisfranc fracture dislocations, no matter how subtle radiographically, always require surgical repair owing to the importance of the Lisfranc ligament to maintain the arch of the foot.

Missed fractures of the lateral process of the talus predispose to premature osteoarthritis.

Typical clinical scenario

Foot or ankle pain.

Differential diagnosis

None.

> **Teaching point**
>
> Scintigraphy or MRI may be necessary if metatarsal stress fracture is suspected and radiographs are normal. CT is indicated if Lisfranc fracture dislocation is suspected. The lateral process of the talus is best visualized on the anterior-posterior view of the ankle series.

REFERENCES

1. Rogers LF. *The Foot. Radiology of Skeletal Trauma.* Philadelphia: Churchill Livingston; 2002, 1319.
2. Digiovanni CW, Benirschke SK, Hansen ST. Foot injuries. In: Browner BD, Jupiter JB, Levine AM, eds. *Skeletal Trauma.* Philadelphia WB Saunders; 2003, 2466.
3. Yu JS, Cody ME. A template approach for detecting fractures in adults sustaining low-energy ankle trauma. *Emerg Radiol.* 2009;**16**(4):309–18.

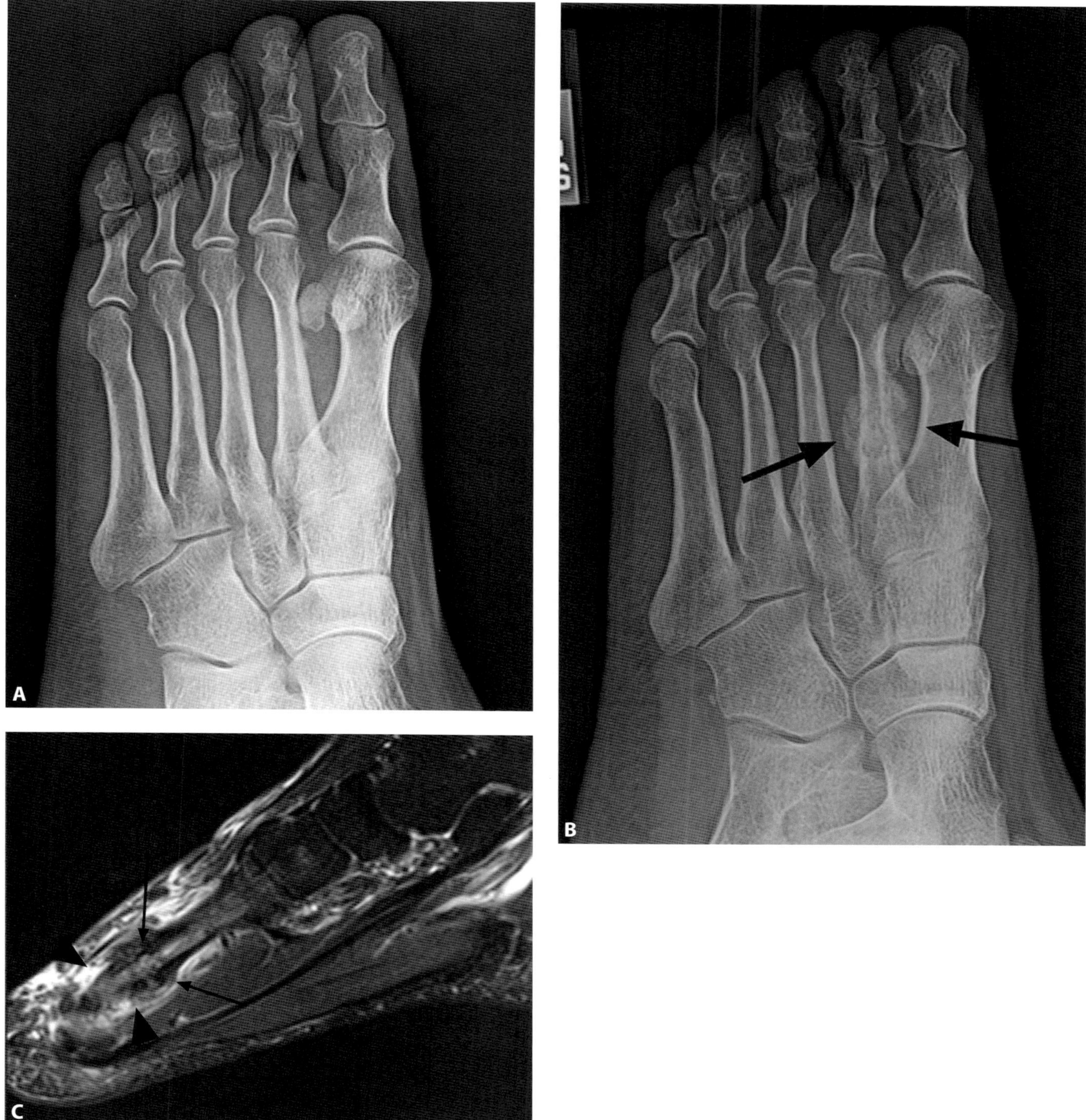

Figure 88.1 Oblique radiograph of the foot (**A**) of a 43-year-old runner with foot pain shows no fracture but subtle periosteal reaction. Repeat oblique radiograph one month after the original presentation (**B**) shows exuberant fluffy periosteal reaction in the mid-diaphysis of the second metatarsal (arrows) consistent with stress fracture. Sagittal short tau inversion recovery image of the same patient (**C**) shows the stress fracture with periosteal reaction (arrows) and soft tissue swelling (arrowheads).

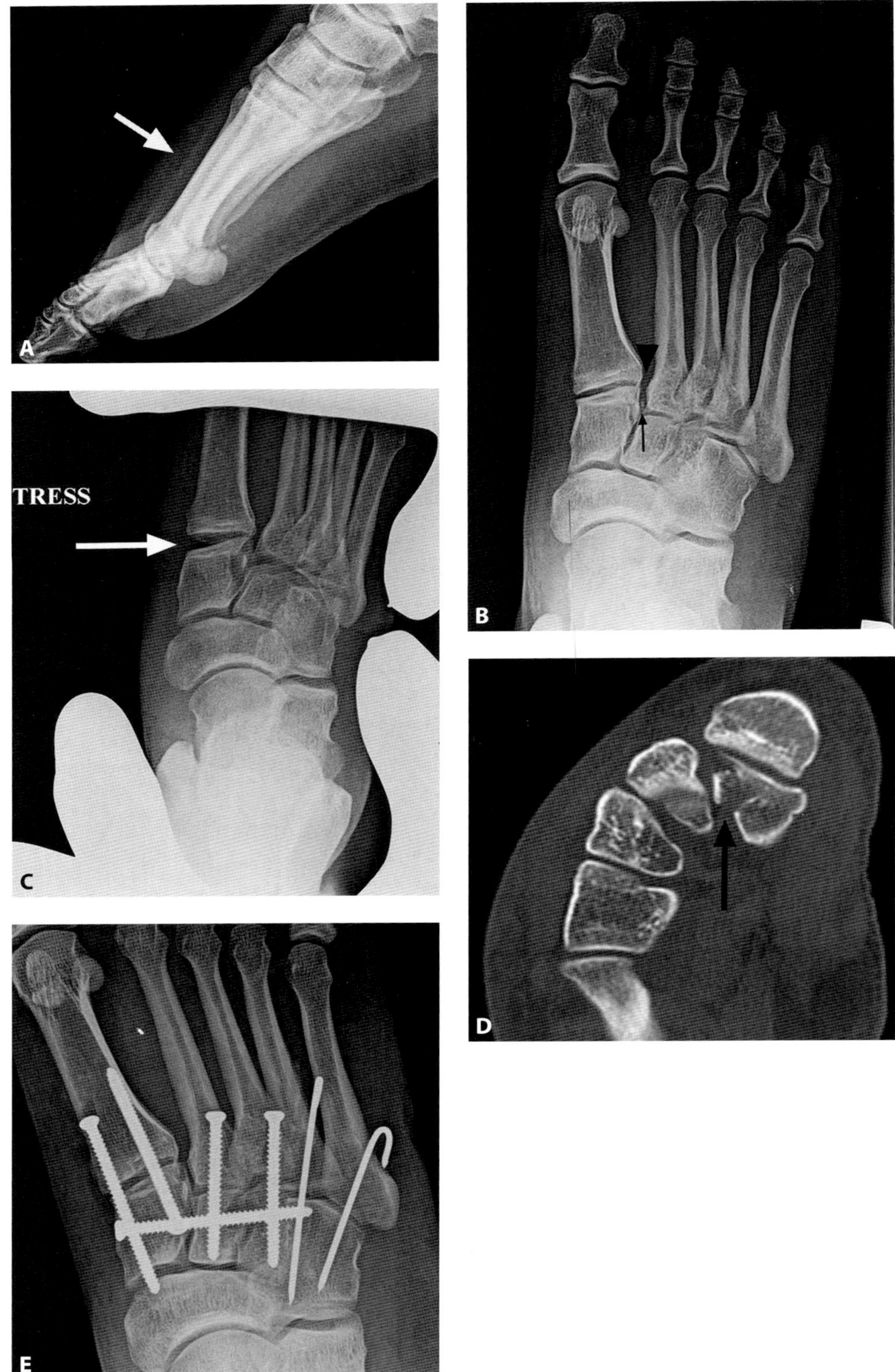

Figure 88.2 Lateral radiograph of the foot (**A**) in a 22-year-old woman who jumped from a 9 ft (3 m) wall, landing on concrete, shows substantial dorsal soft tissue swelling (arrow). On the frontal projection (**B**) there is questionable widening of the joint space between first and second metatarsal (arrowhead) and a possible subtle avulsion fragment (arrow). On stress view (**C**) there is widening of the first tarsometatarsal articulation (arrow). Axial CT image (**D**) confirms an avulsion fragment at the level of the Lisfranc ligament (arrow). **E.** Postoperative radiograph following open reduction and internal fixation.

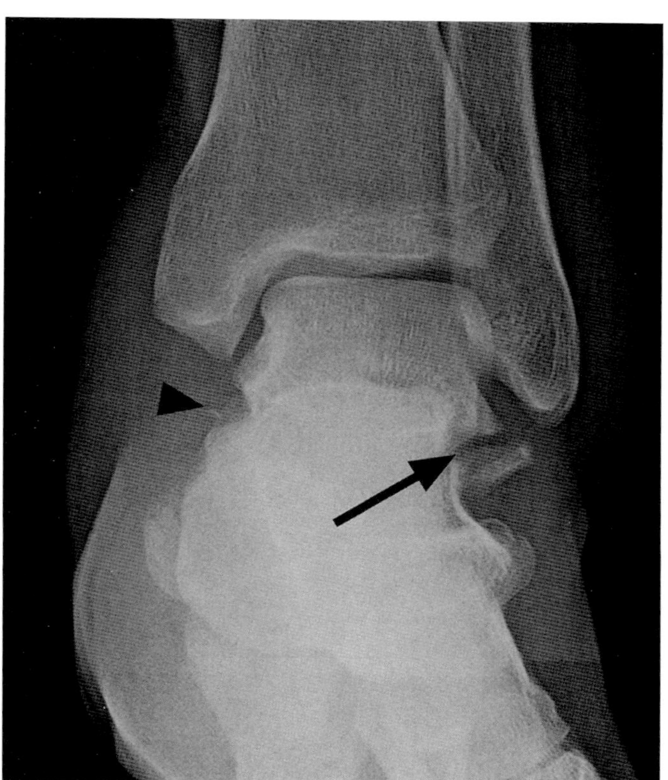

Figure 88.3 Anterior-posterior radiograph of the ankle in a 22-year-old bicyclist who crashed shows diffuse ankle swelling. There is a large fracture fragment of the lateral process of the talus (arrow), which may be obscured on oblique (mortise view). Additional small avulsion fragment is noted at the medial aspect of the talus (arrowhead).

Thymus simulating mediastinal hematoma

Ken F. Linnau

Imaging description

The thymus is the organ of T-cell maturation. The retrosternal gland increases in weight from birth to the age of about 12 years and subsequently involutes with gradual fatty replacement of cellular components. During infancy the ratio of thymus weight to body weight is highest, which can lead to its prominence on chest radiographs of small children. Gradual fat replacement of thymus tissue starts at puberty, which is why the residual thymus is typically detected on CT scans of young adults. MR data suggest that the thymus thickness itself does not actually change much with increasing age [1]. The thymic density on CT is highest in infancy where it measures about 80 Hounsfield Units (HU) on non-contrast CT of the chest [2], which is similar to the attenuation of acute hematoma. The thymus density in teenagers and young adults usually approximates muscle tissue. Above the age of 50 years, residual thymic tissue is uncommonly separated from surrounding mediastinal fat on CT. "Thymic rebound" occurs in some adults who have undergone chemotherapy [3].

Normally, the thymus fills the mediastinal perivascular space up to the age of 20 years [4]. The thymic borders are initially convex but become straight or concave as a child grows, assuming a triangular shape [4]. On radiographs, the thymic sail or notch sign, a triangular margin of the upper mediastinum, is specific for the normal thymus when it is present and should not be confused with the spinnaker sail sign, which is seen with pneumomediastinum [4]. More specific findings for aortic injury include abnormalities of the aortic arch or loss of concave margin seen normally at the aortopulmonary window [5]. One recent review of pediatric thoracic injuries found an indistinct aortic knob to be the most specific sign of blunt thoracic aortic injuries (BTAI) on chest radiographs [6]. Rightward tracheal or esophageal deviation, left mainstem bronchus depression, and a left apical cap are other corroborative radiographic signs of BTAI [6]. Imaging of BTAI is described in detail in Cases 39 to 41.

The most common CT appearance of the thymus in adults is triangular, pointing anterior, with borders that tend to be concave towards the midline. At times, small blood vessels can be seen traversing the thymus.

Usually the thymus extends from the sternal notch to about the fourth intercostal space, but the gland can extend into the neck and down to the diaphragm. It is due to this variable retrosternal extension and appearance in the anterior mediastinum that thymus tissue may be mistaken for mediastinal hematoma (Figure 89.1).

Importance

In the setting of trauma, the thymus should not be mistaken for mediastinal hematoma that can be due to blunt traumatic aortic injury, chest injury, sternal fractures, or spinal fractures [7].

Typical clinical scenario

Generally, one should not over-investigate children younger than 10 years of age for BTAI. Blunt thoracic aortic injury is uncommon in children <10 years of age, and vanishingly rare in children <5 years [8], the same age groups in which the thymus will be visible on radiography. Due to pulsation artifact, and the small size of the aorta, in young children interpretation may be more difficult. Chest CT may show a soft tissue density in the anterior mediastinum with density values similar to hematoma, misleading the radiologist into a false positive diagnosis of periaortic hematoma.

Differential diagnosis

Following blunt trauma, the differential diagnosis is usually between the thymus and mediastinal hematoma due to fractures of the rib cage and small mediastinal vessel injuries.

Teaching point

The thymus obscures the mediastinal structures in young children (<5 years of age). Due to the very low risk of blunt aortic injury in this age group, and concerns about radiation exposure, one should have a higher threshold for chest CT angiography than in an adult patient.

REFERENCES

1. de Geer G, Webb WR, Gamsu G. Normal thymus: assessment with MR and CT. *Radiology*. 1986;**158**(2):313–17.
2. Sklair-Levy M, Agid R, Sella T, Strauss-Liviatan N, Bar-Ziv J. Age-related changes in CT attenuation of the thymus in children. *Pediatr Radiol*. 2000;**30**(8):566–9.
3. Kissin CM, Husband JE, Nicholas D, Eversman W. Benign thymic enlargement in adults after chemotherapy: CT demonstration. *Radiology*. 1987;**163**(1):67–70.
4. Nasseri F, Eftekhari F. Clinical and radiologic review of the normal and abnormal thymus: pearls and pitfalls. *Radiographics*. 2010;**30**(2): 413–28.
5. Anderson SA, Day M, Chen MK, *et al.* Traumatic aortic injuries in the pediatric population. *J Pediatr Surg*. 2008;**43**(6):1077–81.
6. Pabon-Ramos WM, Williams DM, Strouse PJ. Radiologic evaluation of blunt thoracic aortic injury in pediatric patients. *AJR Am J Roentgenol*. 2010;**194**(5):1197–203.
7. Gunn ML. Imaging of aortic and branch vessel trauma. *Radiol. Clin North Am*. 2012;**50**(1):85–103.
8. Barmparas G, Inaba K, Talving P, *et al.* Pediatric vs adult vascular trauma: a National Trauma Databank review. *J Pediatr Surg*. 2010; **45**(7):1404–12.

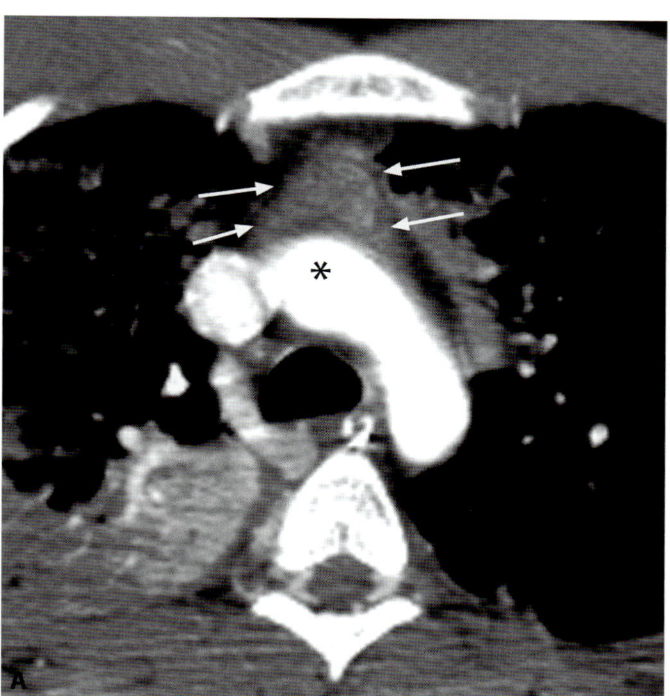

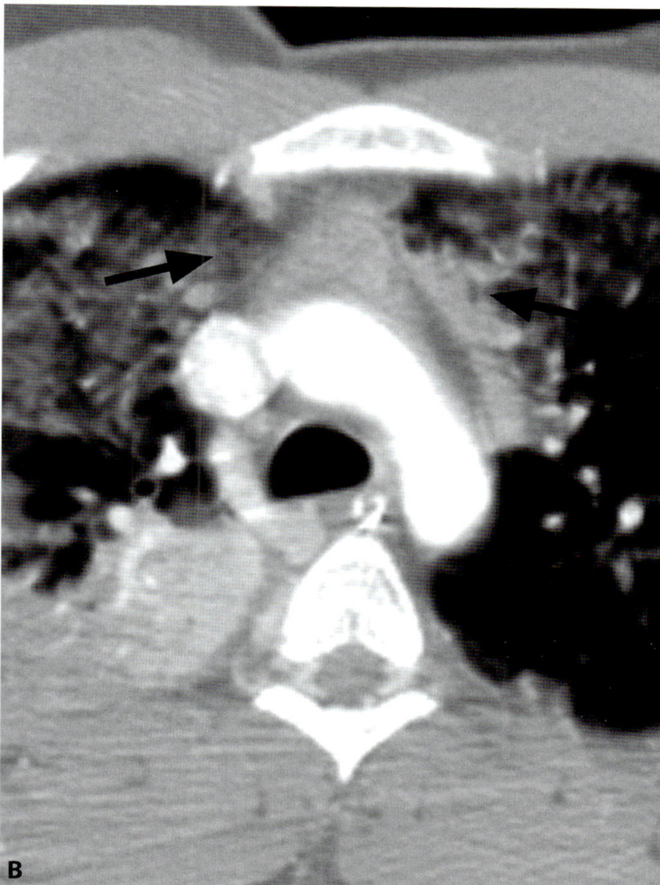

Figure 89.1 A. Axial contrast-enhanced CT of the chest of a 19-year-old man who was involved in a high-speed motor vehicle crash shows soft tissue attenuation with density value of 80 HU in the anterior mediastinum (arrows). The characteristic anterior pointing triangular shape with borders mildly concave towards the midline is consistent with thymus. The aorta (asterisk) shows no direct signs of traumatic injury and is separated from the thymus by a fat plane. **B.** On lung windows pulmonary contusions (arrows) adjacent to the thymus are shown, which are also separated from the gland by fat.

Foreign body aspiration

Ramesh S. Iyer

Imaging description

The imaging hallmark of foreign body aspiration is static lung volume that persists over the respiratory cycle [1]. A unilateral hyperlucent lung is often identified but not required (Figure 90.1). In most pediatric patients this is evaluated with bilateral decubitus views in addition to the conventional frontal and lateral projections. Normally the dependent lung should become more opaque with atelectasis on decubitus imaging. In the setting of aspiration, the affected lung either remains lucent on the ipsilateral decubitus view, or does not increase in opacity to the same extent as the contralateral lung due to partial bronchial obstruction.

In older and more compliant patients, expiratory views may be performed in lieu of decubitus projections. Expiration may reveal persistent lucency in the setting of aspiration, while the normal lung should become more opaque.

The majority of foreign bodies are radiolucent [2]. In some cases there is interruption of the air-filled bronchus at the level of the aspirated foreign body, known as the interrupted bronchus sign (Figure 90.2) [3].

Importance

Suspicion of foreign body aspiration on radiographs should prompt an urgent bronchoscopy for retrieval. Delay in diagnosis may result in bronchial rupture, bronchopulmonary fistula, prolonged hospitalization, and chronic or recurrent pulmonary infection [4].

Typical clinical scenario

Foreign body aspiration occurs most commonly in children aged 1–3 years. Typical symptoms include sudden-onset cough, wheezing, or stridor. However, often the choking episode goes unwitnessed by parents and may not be elicited when questioning the pediatric patient [5].

Differential diagnosis

When there is a high clinical suspicion for aspiration, the differential diagnosis of asymmetric pulmonary aeration is limited and should prompt bronchoscopy. Asthma and viral pulmonary infection may cause hyperinflation. However, both entities typically have symmetrically elevated lung volumes with peribronchial thickening. Adjacent masses such as bronchogenic cysts or lymphadenopathy may compress a bronchus and produce asymmetric aeration. In Swyer-James syndrome, or post-infectious obliterative bronchiolitis, one lung is small and hyperlucent from diminished arterial flow and alveolar overdistention. "Pulmonary sling" occurs when the left pulmonary artery originates from the right pulmonary artery and courses between the trachea and esophagus. This vascular segment may compress either the main bronchus or trachea to produce upper airway obstruction.

Teaching point

A unilateral hyperlucent lung is not universally identified in cases of foreign body aspiration. If a foreign body is suspected, decubitus or expiration radiographs should be performed in addition to frontal and lateral projections. If static lung volume or unilateral hyperlucency is observed, bronchoscopy is indicated.

REFERENCES

1. Donnelly LF, Frush DP, Bisset GS, 3rd. The multiple presentations of foreign bodies in children. *AJR Am J Roentgenol.* 1998;**170**:471–7.
2. Girardi G, Contador AM, Castro-Rodriguez JA. Two new radiological findings to improve the diagnosis of bronchial foreign-body aspiration in children. *Pediatr Pulmonol.* 2004;**38**(3):261–4.
3. Swischuk LE. *Imaging of the Newborn, Infant and Young Child*, 5th edn. Philadelphia, PA: Lippincott Williams & Wilkins, 2004:134.
4. Shilzerman L, Mazzawi S, Rakover Y, *et al.* Foreign body aspiration in children: the effects of delayed diagnosis. *Am J Otolaryngol.* 2010; **31**(5):320–4.
5. Bittencourt PF, Camargos PA, Scheinmann P, *et al.* Foreign body aspiration: clinical, radiological findings and factors associated with its late removal. *Int J Pediatr Otorhinolaryngol.* 2006; **70**(5):879–84.

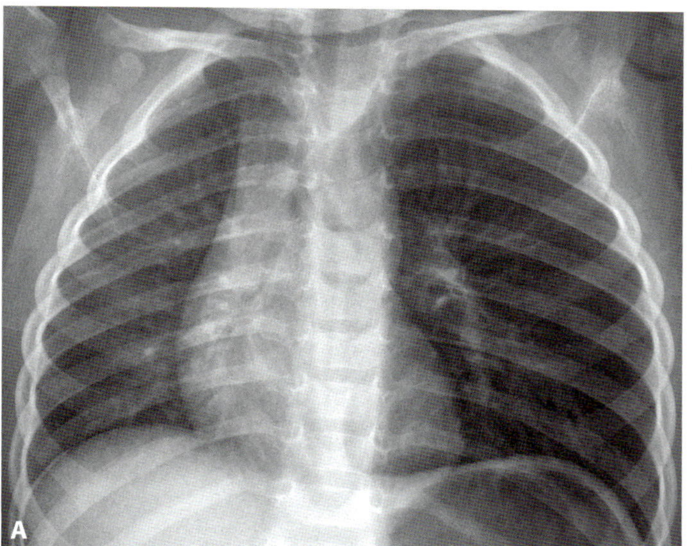

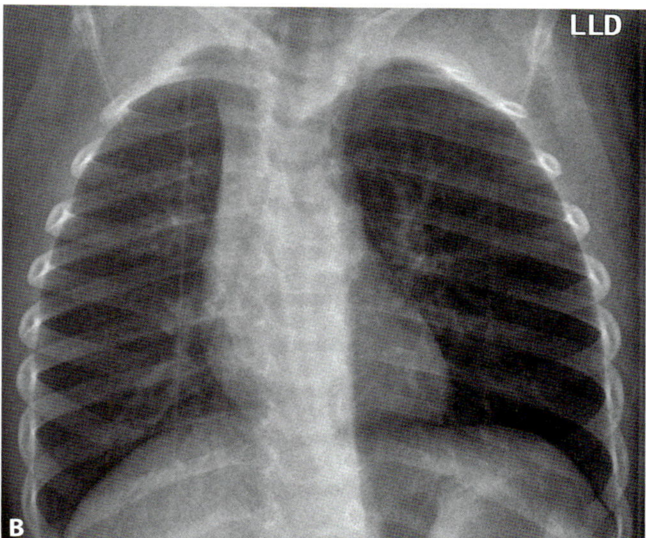

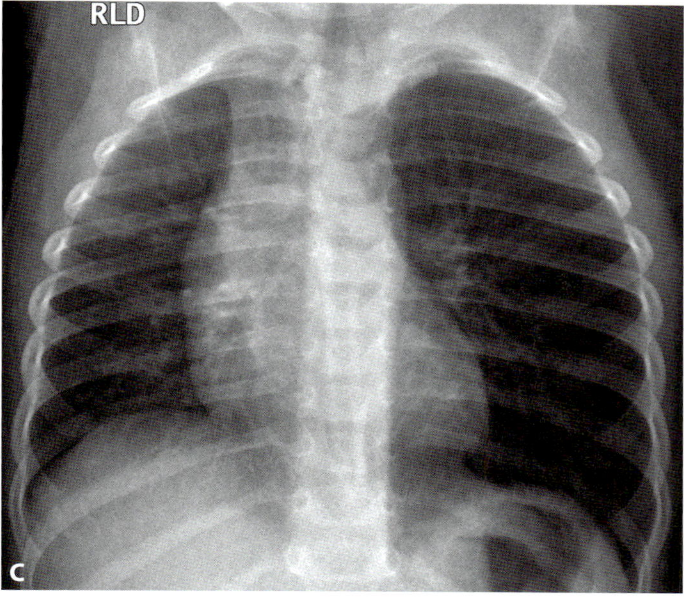

Figure 90.1 **A.** Frontal chest radiograph from a 2-year-old boy who aspirated a peanut into his left main bronchus demonstrates marked left lung hyperinflation. There is rightward mediastinal shift. **B.** Left lateral decubitus view corroborates air trapping in the left lung due to bronchial obstruction. **C.** Right lateral decubitus view exhibits expected ipsilateral atelectasis, shown for comparison.

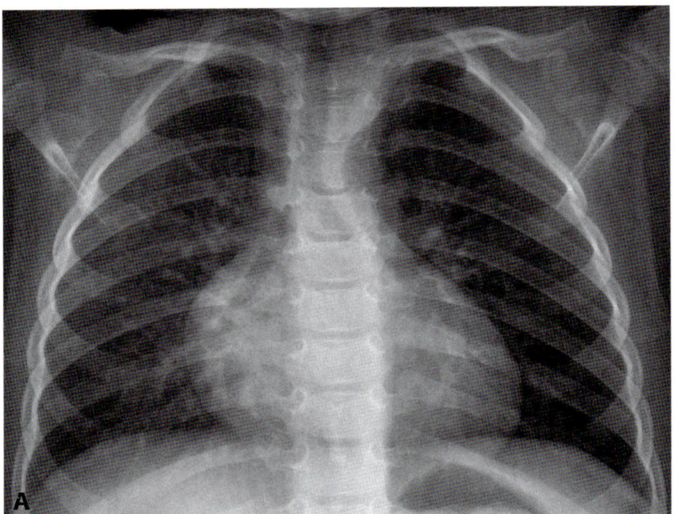

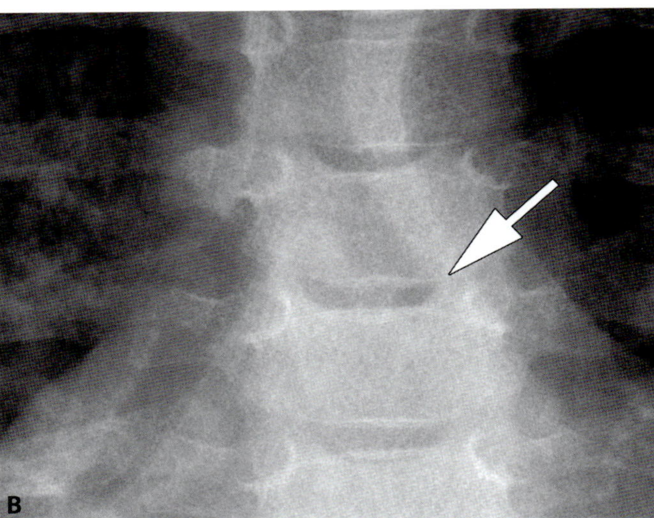

Figure 90.2 A. Frontal chest radiograph from a 3-year-old boy who aspirated carrots into his left main bronchus exhibits mildly increased left lung lucency that persisted on the left lateral decubitus view (not shown). **B.** Magnified view of the main bronchi demonstrates the interrupted bronchus sign (arrow) with cut-off of the left bronchial air column. Note the normal air column within the right main bronchus.

CASE **91**

Idiopathic ileocolic intussusception

Ramesh S. Iyer

Imaging description

"Idiopathic" intussusception refers to distal small bowel lymphoid hyperplasia causing invagination of the ileum into the proximal colon. Such ileocolic intussusceptions without a discrete lead point and comprise the vast majority (>90%) of all pediatric intussusceptions [1].

Ultrasound is the preferred diagnostic modality for intussusception evaluation. An intussusception's characteristic ultrasound appearance is a right abdominal mass exhibiting a swirled pattern of alternating hyperechogenicity and hypoechogenicity when the bowel is scanned transverse (Figure 91.1). This pattern of concentric rings represents alternating layers of mucosa, muscularis, and serosa. Names given to this appearance include "doughnut sign" and "pseudo-kidney" [2–4]. In longitudinal view, this bowel invagination and its various layers may resemble a sandwich ("sandwich sign") [5].

Several accompanying sonographic features have been assessed for their ability to predict reducibility and/or bowel necrosis. These include thickness of the peripheral hypoechoic ring, the presence of fluid either trapped in the intussusception or within the peritoneal cavity, and presence of blood flow in the intussusception on Doppler interrogation [6, 7–13]. In aggregate the literature has been controversial regarding their respective prognostic abilities, rendering it difficult to predict which intussusceptions will reduce with enema or progress to bowel inviability [1].

On radiography, findings suggestive of intussusception include the presence of a soft tissue mass creating a meniscus-like interface with the distal bowel, confirming its intraluminal position (Figure 91.2) [14, 15]. Visualization of a gas-filled cecum and ascending colon, such as on a left lateral decubitus view, renders intussusception highly unlikely [14]. Many authors have argued that abdominal radiography is unhelpful in evaluating for intussusception due to the high percentage of patients with a non-specific gas pattern [14–19]. There is also a potential pitfall of the gas-filled redundant sigmoid colon situated in the right lower abdomen and mimicking the cecum, which frequently occurs in young children [16].

Enema is the preferred therapy for intussusception. Air, barium, or water-soluble contrast may be utilized. After the rectum is catheterized and the colon is infused with contrast, the intussusceptum should be identified as an intraluminal soft tissue meniscus. Retrograde movement of the intussusceptum should occur over the course of the exam (Figure 91.2). The ileocecal valve is frequently a site where reduction is arrested. The intussusception can cause the ileocecal valve to become edematous and therefore restrict movement across it (Figure 91.3). Moreover, the swollen ileocecal valve may frequently mimic a small residual intussusception on fluoroscopy [20] (Figure 91.4). Both may appear as a round smoothly marginated filling defect along the medial aspect of the cecum. Ultrasound can facilitate overcoming this pitfall (Figure 91.5). The radiologist may sonographically differentiate residual intussusception from an edematous ileocecal valve [5]. The former will have the targetoid appearance. The edematous valve is best shown in longitudinal view with its echogenic leaflets protruding into the cecum. The terminal ileum may be thickened and feature hypoechoic mural edema. Seen in cross-section this should not be confused with the concentric ring pattern of intussusception. As with enemas, passage of fluid or gas between ileum and cecum would strongly suggest successful reduction. At the author's institution, ultrasound equipment may be brought into the fluoroscopy suite to make this distinction while the patient still rectally intubated.

Importance

Sonographic diagnosis of intussusception should prompt urgent enema for attempted reduction. Delay in diagnosis or unsuccessful enema necessitates surgical reduction with possible resection of non-viable bowel.

Typical clinical scenario

Idiopathic intussusception typically presents in children between three months and one year of age. Common presenting signs and symptoms include crampy abdominal pain, emesis, and alternating lethargy and irritability. Bloody diarrhea, or so-called "currant-jelly" stools, is a hallmark feature of intussusception but is relatively uncommon. Idiopathic intussusception is more common during winter and spring; this seasonal predilection is most likely due to viral infections producing lymphoid hyperplasia.

Differential diagnosis

The sonographic appearance of intussusception is highly specific in the hands of a skilled sonographer and interpreting radiologist. Occasional false-positive studies result from conditions that cause intramural edema or hemorrhage, such as Henoch–Schonlein purpura. As mentioned above, the sigmoid colon can be highly redundant in young children and may mimic a gas-filled proximal colon on frontal radiographs. Appendicitis may present with similar symptomatology, but typically afflicts older children.

Teaching point

Gas in a redundant sigmoid colon may resemble a normal cecum on radiography, and conversely, an edematous ileocecal valve may simulate a residual intussusception following enema reduction. If there is clinical suspicion for intussusception, ultrasound should be performed. If intussusception is diagnosed, a reduction enema should be attempted urgently.

REFERENCES

1. Daneman A, Navarro O. Intussusception. Part 1: a review of diagnostic approaches. *Pediatr Radiol.* 2003;**33**(2):79–85.
2. del-Pozo G, Albillos JC, Tejedor D. Intussusception: US findings with pathologic correlation – the crescent-in-doughnut sign. *Radiology.* 1996;**199**:688–92.
3. Holt S, Samuel E. Multiple concentric ring sign in the ultrasonographic diagnosis of intussusception. *Gastrointest Radiol.* 1978;**3**(3):307–9.
4. Swischuk LE, Hayden CK, Boulden T. Intussusception: indications for ultrasonography and an explanation of the doughnut and pseudokidney signs. *Pediatr Radiol.* 1985;**15**(6):388–91.
5. del-Pozo G, Albillos JC, Tejedor D, *et al.* Intussusception in children: current concepts in diagnosis and enema reduction. *Radiographics.* 1999;**19**:299–319.
6. Verschelden P, Filiatraut D, Garel L, *et al.* Intussusception in children: reliability of US in diagnosis – a prospective study. *Radiology.* 1992;**184**:741–4.
7. Britton I, Wilkinson AG. Ultrasound features of intussusception predicting outcome of air enema. *Pediatr Radiol.* 1999;**29**:705–10.
8. Mirilas P, Koumanidou C, Vakaki M, *et al.* Sonographic features indicative of hydrostatic reducibility of intestinal intussusception in infancy and early childhood. *Eur Radiol.* 2001;**11**(12):2576–80.
9. Feinstein KA, Myers M, Fernbach SK, *et al.* Peritoneal fluid in children with intussusception: its sonographic detection and relationship to successful reduction. *Abdom Imaging.* 1993;**18**:277–9.
10. del-Pozo G, Gonzalez-Spinola JG, Gomez-Anson B, *et al.* Intussusception: trapped peritoneal fluid detected with US – relationship to reducibility and ischemia. *Radiology.* 1996;**201**(2):379–83.
11. Lam AH, Firman K. Value of sonography including color Doppler in the diagnosis and management of long standing intussusception. *Pediatr Radiol.* 1992;**22**:112–14.
12. Lagalla R, Caruso G, Novara V, *et al.* Color Doppler ultrasonography in pediatric intussusception. *J Ultrasound Med.* 1994;**13**:171–4.
13. Kong MS, Wong HF, Lin SL, *et al.* Factors related to detection of blood flow by color Doppler ultrasonography in intussusception. *J Ultrasound Med.* 1997;**16**:141–4.
14. Sargent MA, Babyn P, Alton DJ. Plain abdominal radiography in suspected intussusception: a reassessment. *Pediatr Radiol.* 1994;**24**: 293–5.
15. Lee JM, Kim H, Byun JY, *et al.* Intussusception: characteristic radiolucencies on the abdominal radiograph. *Pediatr Radiol.* 1994;**24**:293–5.
16. Fiorella DJ, Donnelly LF. Frequency of right lower quadrant position of the sigmoid colon in infants and young children. *Radiology.* 2001;**219**:91–4.
17. Eklof O, Hartelius H. Reliability of the abdominal plain film diagnosis in pediatric patient with suspected intussusception. *Pediatr Radiol.* 1980;**9**:199–206.
18. White SJ, Blane CE. Intussusception: additional observations on the plain radiograph. *AJR Am J Roentgenol.* 1982;**139**:511–13.
19. Kirks DR. Diagnosis and treatment of pediatric intussusception: how far should we push our radiologic techniques? *Radiology.* 1994;**191**:622–3.
20. Crystal P, Barki Y. Using color Doppler sonography-guided reduction of intussusception to differentiate edematous ileocecal valve from residual intussusception. *AJR Am J Roentgenol.* 2004;**182**(5):1345.

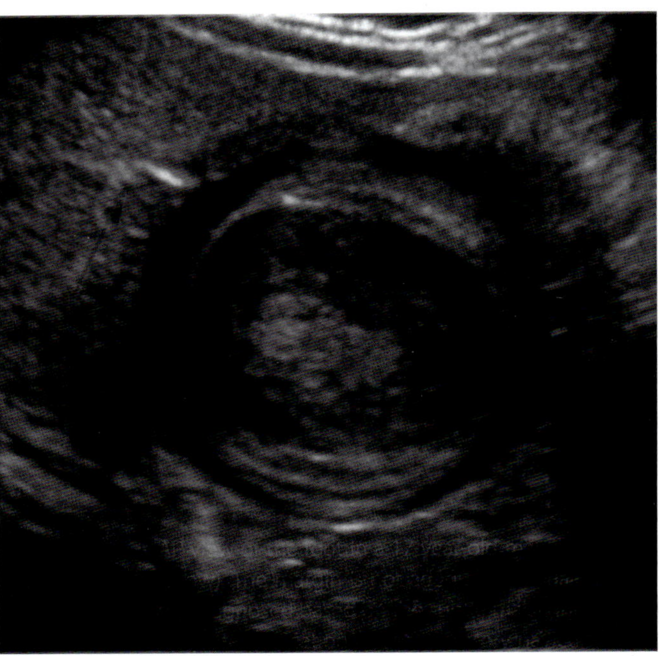

Figure 91.1 Transverse ultrasound image from a 6-month-old boy with intussusception demonstrates the characteristic targetoid pattern of concentric hyperechoic and hypoechoic rings. Subsequent enema revealed an ileocolic intussusception extending to the hepatic flexure.

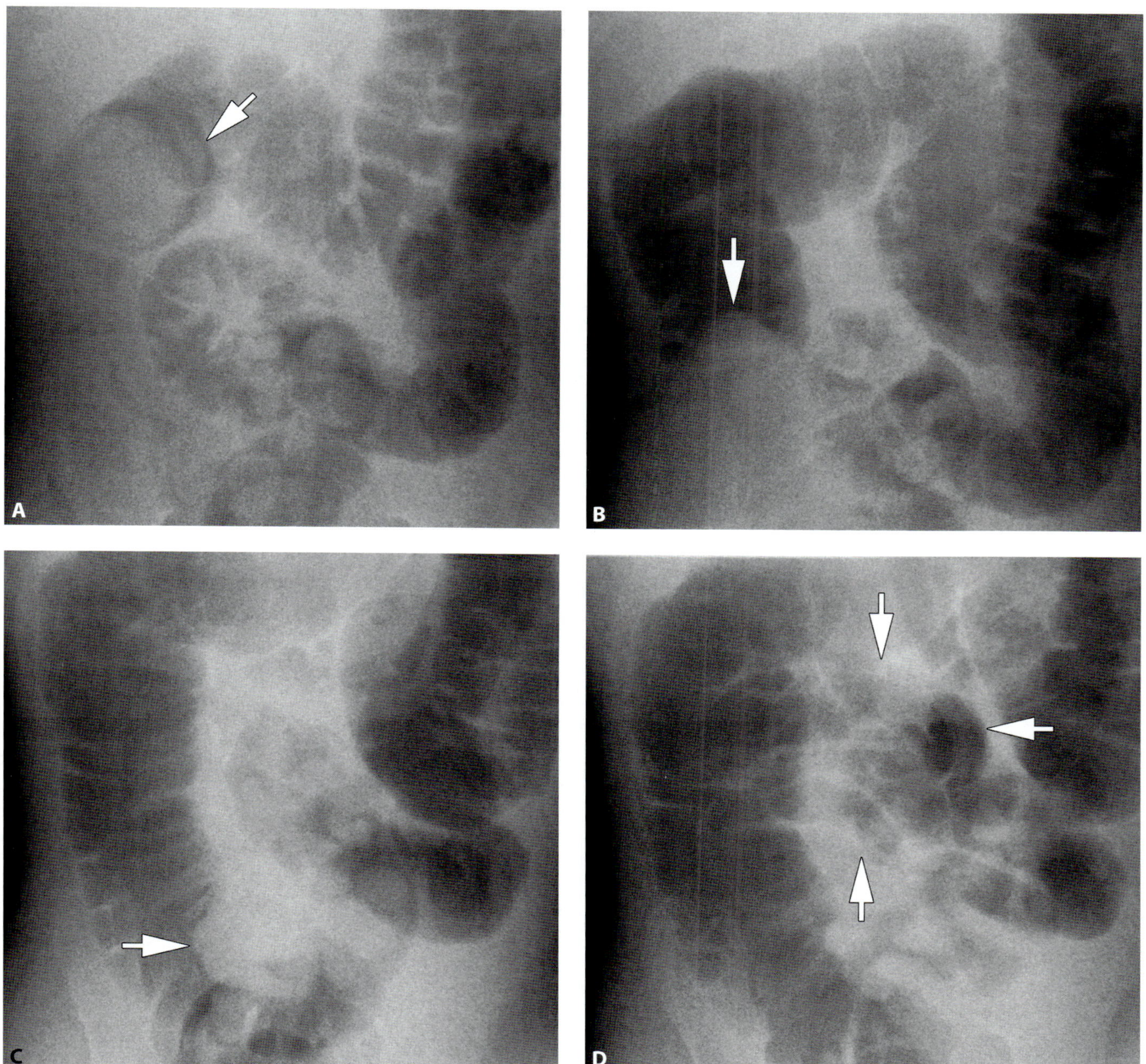

Figure 91.2 A–C. Fluoroscopic spot images of the right abdomen in a 3-year-old boy with successful ileocolic intussusception reduction by air contrast enema. Note retrograde movement of the intussusceptum (arrows in **A–C.**) from the hepatic flexure to the cecum. Note the resemblance to a meniscus. **D.** Final spot image shows new reflux of gas within the distal small bowel (arrows), confirming successful reduction.

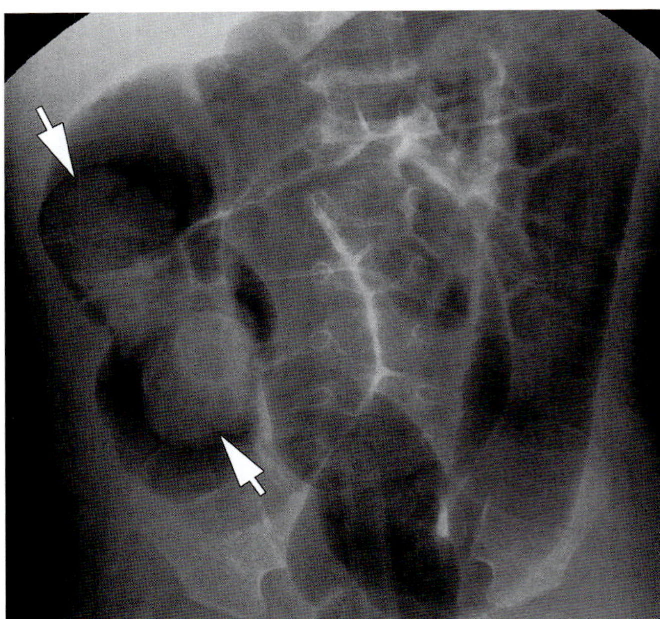

Figure 91.3 Spot image of the abdomen in a 9-month-old boy with an unsuccessful intussusception reduction by air contrast enema. There is a lobulated soft tissue mass (arrows) in the ascending colon indicating the residual intussusception.

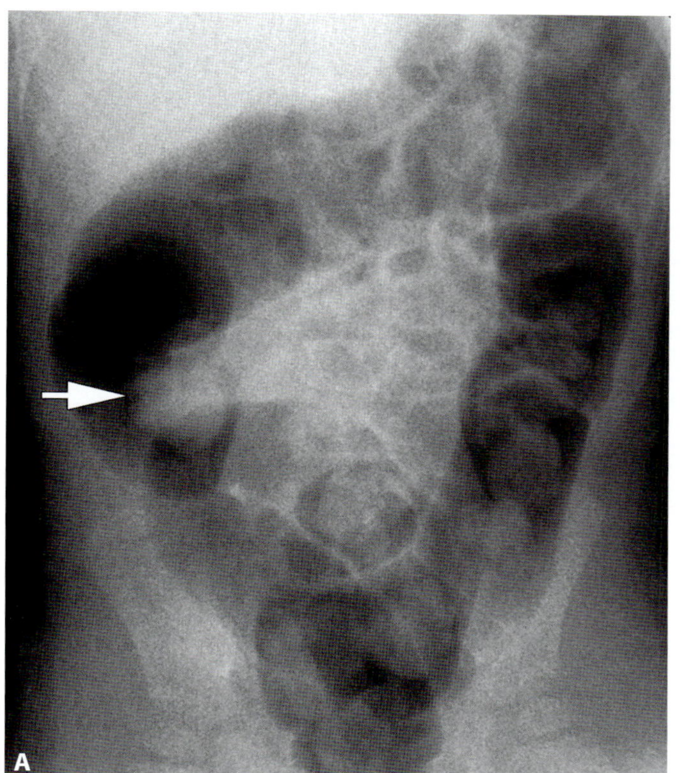

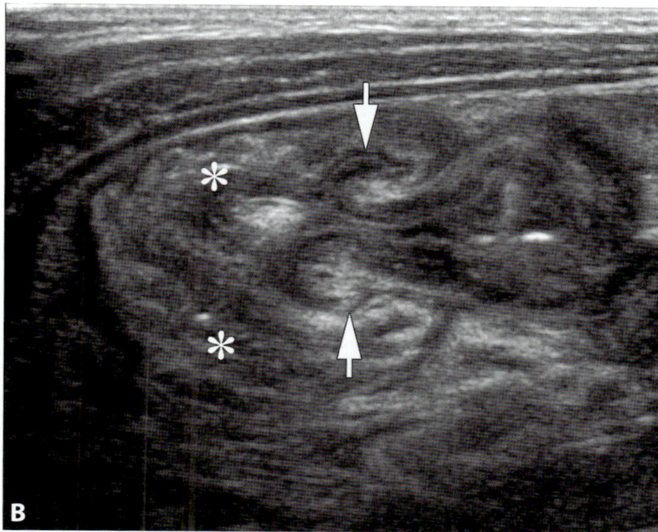

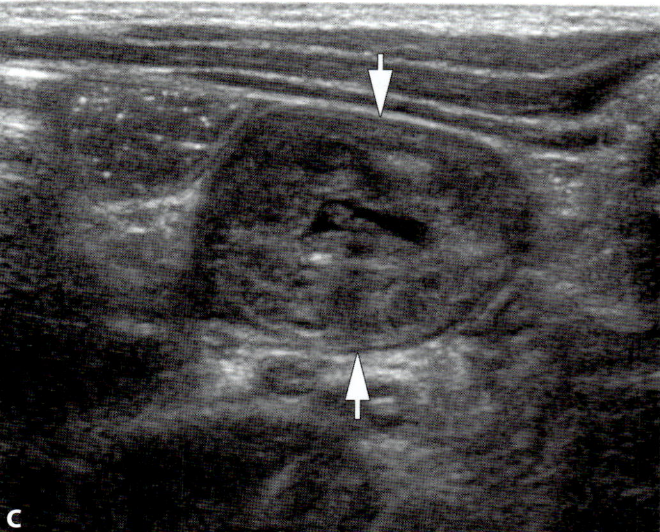

Figure 91.4 **A.** Fluoroscopic spot image of the right abdomen in a 2-year-old boy with ileocecal valve edema. There is a round soft tissue mass in the cecum that may represent either an edematous valve or a small residual intussusception. **B.** Transverse ultrasound image of the right lower quadrant shows the echogenic valve leaflets of the ileocecal valve (arrows) protruding into the cecum (asterisks). **C.** Transverse ultrasound image reveals terminal ileal mural edema (arrows). Note anechoic intraluminal fluid surrounded circumferentially by hypoechoic thickened bowel wall.

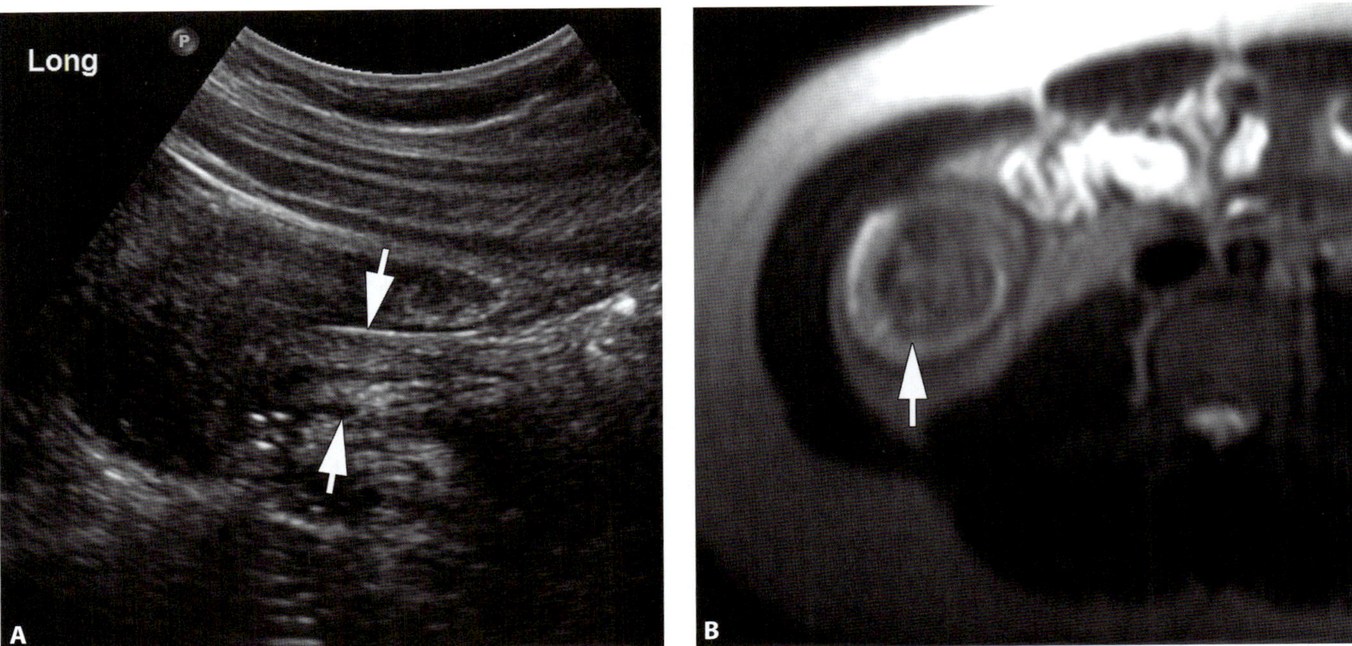

Figure 91.5 **A.** Transverse ultrasound image of the right lower quadrant in a 4-year-old girl with ileocecal valve edema. The echogenic valve leaflets (arrows) are shown longitudinally. **B.** Axial T2-weighted MRI from the same patient shows an isointense soft tissue mass within the cecum, corresponding to a markedly edematous valve.

CASE 92 Ligamentous laxity and intestinal malrotation in the infant

Ramesh S. Iyer

Imaging description

In an otherwise healthy infant with bilious emesis, intestinal malrotation with midgut volvulus is the primary concern. An upper gastrointestinal (GI) series is the reference standard examination to evaluate for intestinal malrotation [1–3]. Barium may be used unless the patient is unstable and bowel ischemia or perforation is suspected, in which case water-soluble contrast is preferred [1]. If the infant cannot tolerate oral contrast, a nasogastric or nasoduodenal tube may be used to rapidly and safely deliver the contrast.

Ascertaining the position of the duodenal-jejunal junction (DJJ) is the primary goal of this evaluation. On a true supine AP projection, the normal DJJ is situated to the left side of the left vertebral pedicle, at or above the level of the pylorus (Figure 92.1) [1–3]. On a lateral view, the normal duodenum passes through the retroperitoneum, affording another opportunity to evaluate for proper rotation [4]. Generally if these criteria are not met then malrotation is suspected. Cecal position may then be assessed through either a delayed radiograph or contrast enema. An abnormally positioned cecum further supports malrotation. The shorter the distance between the DJJ and cecal apex, the shorter the mesenteric vascular pedicle, which then increases the risk for midgut volvulus [3].

Important pitfalls in using these criteria occur in young infants, typically less than six months of age and often with a history of premature births. These babies may exhibit ligamentous laxity with respect to bowel fixation to the retroperitoneum. The DJJ may be inferiorly displaced by a distended stomach, bowel loop, or enteric tube (Figures 92.2 and 92.3). The cecum may also be "high-riding" (Figure 92.4), or situated higher than the iliac fossa. Ligamentous laxity must be considered in young infants with a DJJ and/or cecum that is slightly malpositioned.

The radiologist should also be aware of normal anatomic variations mimicking malrotation, including duodenum inversum, wandering duodenum, and mobile duodenum [5, 6].

It should be noted that the cecum is more variably positioned than the DJJ, rendering the upper GI more useful than enema for malrotation assessment. In cases of surgically proven malrotation, 94–97% of patients had an abnormal DJJ on preoperative imaging compared with 80–87% of cases featuring an abnormal cecal position (Figures 92.5 and 92.6) [1, 2, 7].

Importance

In an infant with bilious emesis, a truly malpositioned DJJ on upper GI equates with intestinal malrotation with possible midgut volvulus. Emergent surgical consultation is warranted. Delayed or missed diagnosis of volvulus may result in bowel infarction and possibly death.

Typical clinical scenario

The classic presentation of malrotation with volvulus is bilious emesis. Up to 90% of cases present in the first year of life, with the vast majority of those patients presenting before one month of age [8, 9]. The majority of infants with bilious emesis, however, show normal intestinal rotation on upper GI. In one prospective study of 63 neonates with bilious emesis, 62% of patients displayed normal intestinal rotation on upper GI [10]. Less commonly malrotation may present in older children with chronic intermittent pain or vomiting, or malnutrition [11].

Differential diagnosis

The false-positive rate for upper GI identifying malrotation ranges up to 15%, while the false-negative rate ranges from 3% to 6% in the literature [2, 8, 12–14]. False positives occur in young infants because these patients feature the highest rates of both ligamentous laxity and true malrotation with midgut volvulus. Normal variant anatomy gives rise to most false-negative interpretations [3, 15]. The discrepancy between false-positive and false-negative upper GI exams reflects the screening nature of this study to identify patients at risk for midgut volvulus, a potentially fatal catastrophe [2].

Teaching point

Ligamentous laxity may account for slightly malpositioned DJJ or cecum in young infants. The DJJ may be displaced by a distended stomach, bowel loop, or enteric tube.

REFERENCES

1. Strouse PJ. Disorders of intestinal rotation and fixation ("malrotation"). *Pediatr Radiol.* 2004;**34**:837–51.
2. Applegate KE. Evidence-based diagnosis of malrotation and volvulus. *Pediatr Radiol.* 2009;**39**(Suppl 2):S161–3.
3. Applegate KE, Anderson JM, Klatte EC. Malrotation of the gastrointestinal tract: a problem solving approach to performing upper GI series. *Radiographics.* 2006;**26**(5):1485–500.
4. Koplewitz BZ, Daneman A. The lateral view: a useful adjunct in the diagnosis of malrotation. *Pediatr Radiol.* 1999;**29**:144–5.
5. Long FR, Mutabagani KH, Caniano DA, *et al.* Duodenal inversion mimicking mesenteric artery syndrome. *Pediatr Radiol.* 1999;**29**:602–4.
6. Beasley SW, de Campo JF. Pitfalls in the radiological diagnosis of malrotation. *Australas Radiol.* 1987;**31**:376–83.

7. Sizemore A, Rabbani KZ, Ladd A, *et al.* Diagnostic performance of the upper gastrointestinal series in evaluation of children with clinically suspected malrotation. *Pediatr Radiol.* 2008;**38**(5):518–28.
8. Torres AM, Ziegler MM. Malrotation of the intestine. *World J Surg.* 1993;**17**:326–31.
9. Filston HC, Kirks DR. Malrotation – the ubiquitous anomaly. *J Pediatr Surg.* 1981;**16**:614–20.
10. Godbole P, Stringer MD. Bilious vomiting in the newborn: how often is it pathologic? *J Pediatr Surg.* 2002;**37**(6):909–11.
11. Spigland N, Brandt ML, Yazbeck S. Malrotation presenting beyond the neonatal period. *J Pediatr Surg.* 1990;**25**:1139–42.
12. Dilley AV, Pereia J, She EC, *et al.* The radiologist says malrotation: does the surgeon operate? *Pediatr Surg Int.* 2000;**16**:45–9.
13. Long FR, Kramer SS, Markowitz RI, *et al.* Intestinal malrotation in children: tutorial on radiographic diagnosis in difficult cases. *Radiology.* 1996;**198**:775–80.
14. Prasil P, Flageole H, Shaw KS, *et al.* Should malrotation in children be treated differently according to age? *J Pediatr Surg.* 2000;**35**:756–8.
15. Long FR, Kramer SS, Markowitz RI, *et al.* Radiographic patterns of intestinal malrotation in children. *Radiographics.* 1996;**16**:547–56.

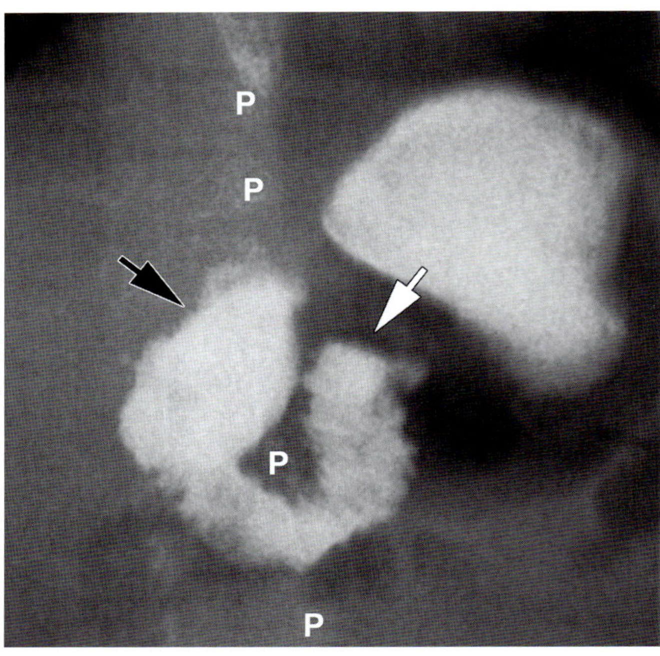

Figure 92.1 Frontal abdominal image from an upper GI series in an 11-month-old girl with abdominal pain demonstrates the normal position of the DJJ. The DJJ (white arrow) is situated at the level of the pylorus (black arrow) and to the left of the left vertebral pedicles (P).

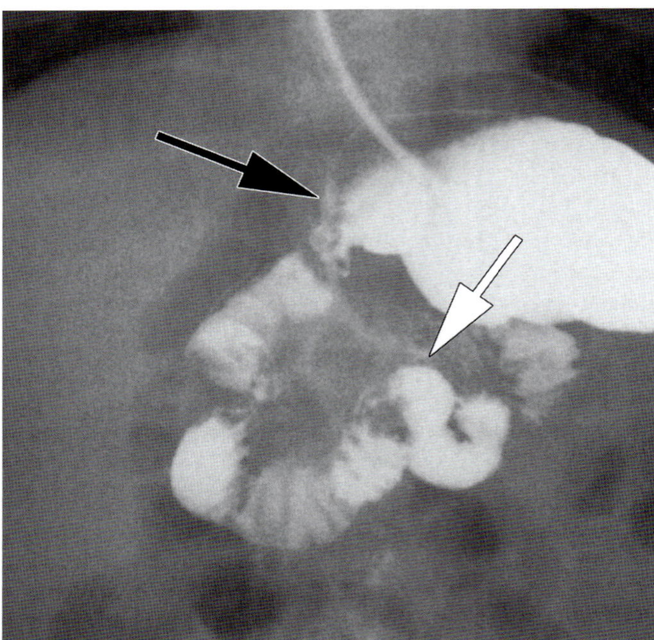

Figure 92.2 Frontal abdominal image from an upper GI series in a 9-month-old boy with ligamentous laxity shows the DJJ (white arrow) situated to the left of the spine but lower than the pylorus (black arrow). This equivocal DJJ position prompted a small bowel follow-through (not shown) which demonstrated a normally situated cecum in the right lower quadrant. The resultant broad mesenteric vascular pedicle does not predispose this child to midgut volvulus. Note the enteric tube that has been retracted into the stomach.

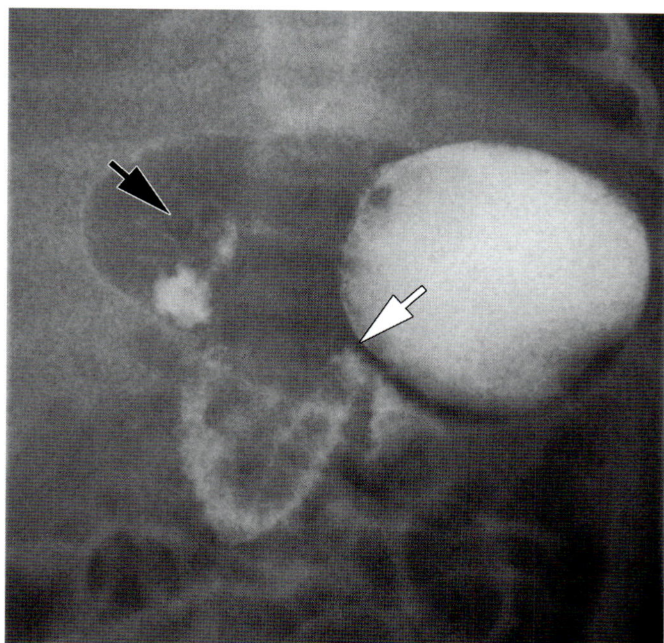

Figure 92.3 Frontal abdominal image from an upper GI series in a 1-month-old girl with ligamentous laxity shows the DJJ (white arrow) displaced inferior to the level of the pylorus (black arrow) by a distended stomach lying transversely.

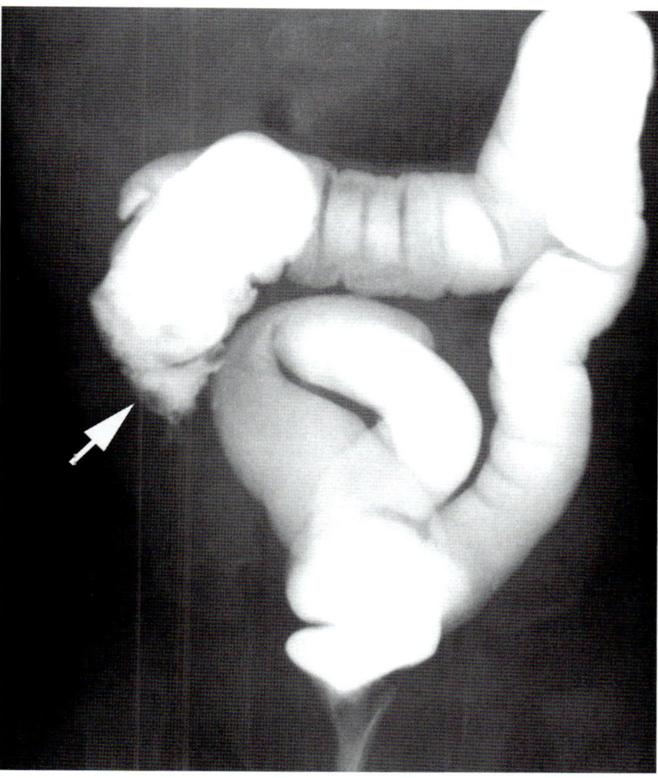

Figure 92.4 Frontal abdominal image from a contrast enema in a 5-month-old boy with ligamentous laxity shows a "high-riding" cecum (arrow) located above the level of the right iliac wing. Note contrast filling the proximal appendix.

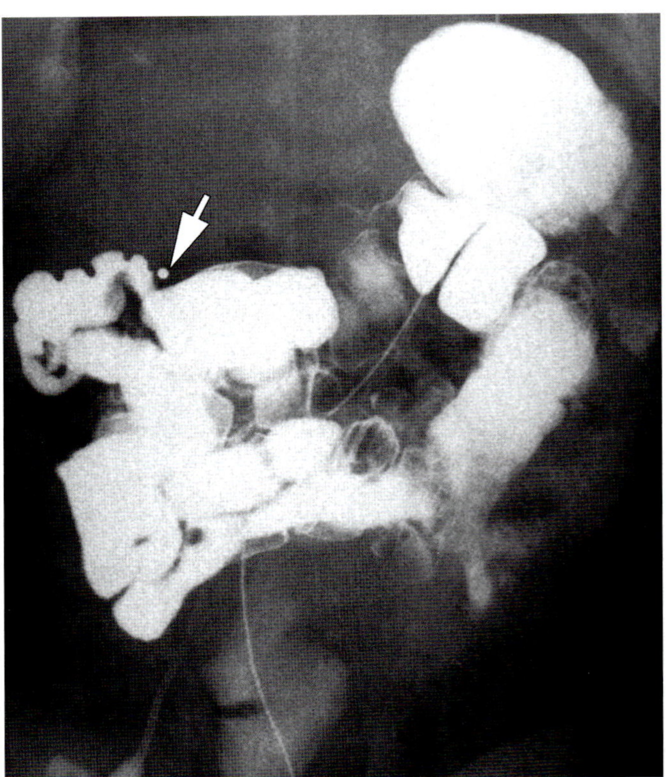

Figure 92.5 Frontal abdominal image from a small bowel follow-through in a 6-day-old girl with malrotation. The ileocecal junction (arrow and round metallic marker) is situated in the right upper quadrant, consistent with malrotation. Contrast this location with the slightly malpositioned cecum in Figure 92.4.

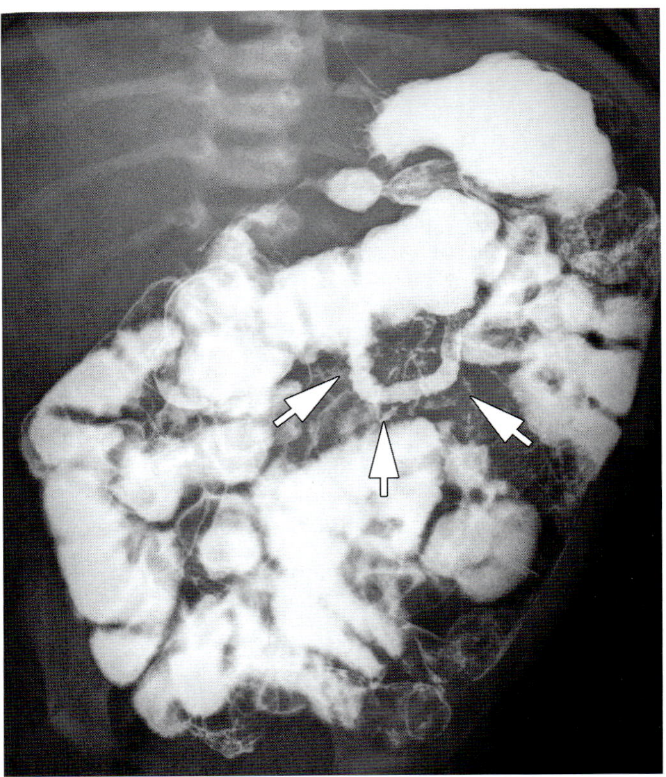

Figure 92.6 Frontal abdominal image from a small bowel follow-through in a 20-day-old girl with malrotation. The cecum is difficult to identify in this image. The appendix is a thin tubular, contrast-opacified structure (arrows) in the left upper quadrant. Based on the location of the appendix the radiologist concluded that the cecum was abnormally located, in keeping with malrotation. As in Figure 92.5, this cecal position results in a short mesenteric vascular pedicle and significantly increases this patient's risk for midgut volvulus.

Hypertrophic pyloric stenosis and pylorospasm

Ramesh S. Iyer

Imaging description

Ultrasound is the preferred imaging modality for diagnosing hypertrophic pyloric stenosis (HPS). The primary sonographic features of HPS include pyloric muscular hypertrophy and channel elongation (Figures 93.1 and 93.2). There is variability in the literature for single wall pyloric muscular thickness diagnostic of HPS, ranging from 3.0 to 4.5 mm [1–6]. Blumhagen and Noble evaluated 326 sonograms in vomiting infants to assess sonographic criteria for HPS diagnosis. They found muscle thickness in HPS patients measured 4.8 +/− 0.6 mm, compared with 1.8 +/− 0.4 mm in normal children [2]. At our institution, single wall thickness exceeding 3.0 mm is consistent with HPS. Similarly, there is a reported range of pyloric channel length consistent with HPS. Most authors consider cutoffs between 14 and 17 mm as diagnostic for HPS [1–6]. Rohrschneider *et al.* reported a 94% accuracy rate using a 15 mm channel length to differentiate normal patients (< 15 mm) from those with HPS (> 15 mm) [5]. In patients with HPS there is often crowding and thickening of the pyloric mucosa, which may protrude into the gastric antrum [1].

Pylorospasm (PS) can mimic HPS clinically as well as on both upper GI (GI) and ultrasound. This condition results in an intermittent narrowing of the pyloric channel, with resultant transient gastric outlet obstruction causing forceful non-bilious emesis. There may be significant overlap in pyloric measurements between HPS and PS. In a 1998 series of 34 patients with HPS by Cohen *et al.*, 18 children had pyloric thicknesses of greater than 4 mm, and 19 children had pyloric lengths greater than 14 mm. Pylorospasm can therefore serve as a major pitfall for diagnosing true pyloric stenosis on ultrasound, particularly when relying on static measurements alone [7, 8].

Differentiating these two conditions necessitates dynamic imaging of the pylorus. Over the course of the exam, the aforementioned measurement criteria for HPS will not persist in the setting of PS. The pyloric appearance may wax and wane, often intermittently normal (Figures 93.3 and 93.4). In HPS, there is fixed mural thickening and elongation of the pylorus throughout the examination. In cases of PS, there is also intermittent transpyloric passage of gas or fluid [7–9]. The pylorus may even be briefly distended with intraluminal contents. In infants with HPS there is no or minimal transpyloric flow, often accompanied by retrograde gastric peristalsis [1, 5].

Importance

The importance in making this distinction lies in management. Surgical pyloromyotomy is required for treatment of HPS, often performed on the same hospital admission after fluid and electrolyte replacement. Pylorospasm on the other hand may be managed medically with antispasmodics [9].

Typical clinical scenario

Hypertrophic pyloric stenosis and PS typically present in young infants aged 2 to 12 weeks, with a strong male predominance of approximately 4:1 [1, 10, 11]. The patient presents with forceful non-bilious emesis after feeding, often described as "projectile." With protracted symptoms there may be blood-tinged vomitus from gastritis [10]. The infant may also be lethargic from dehydration. The palpable "olive-sized" mass in the right upper abdomen that is characteristic for HPS is variably present and highly examiner-dependent [2, 12].

Differential diagnosis

The main differential diagnosis for forceful non-bilious emesis in infants includes HPS, PS, and gastroesophageal reflux. Distinguishing the first two entities is described in the above section. Gastroesophageal reflux is the cause of the majority of non-bilious emesis in infants, and exhibits a normal pylorus on ultrasound. Less common causes of gastric outlet obstruction include malrotation with midgut volvulus and bezoar.

Teaching point

Pylorospasm can mimic HPS on ultrasound, particularly when relying solely on static pyloric measurements. Dynamic pyloric imaging is critical for this distinction. In HPS there is fixed muscular thickening and channel elongation over the entire exam. There will be no or minimal transpyloric passage of gas or fluid. With spasm the pylorus will wax and wane in size, and will feature intermittent transpyloric transit, during the exam.

REFERENCES

1. Hernanz-Shulman M. Infantile hypertrophic pyloric stenosis. *Radiology*. 2003;**227**(2):319–31.
2. Blumhagen JD, Noble HG. Muscle thickness in hypertrophic pyloric stenosis: sonographic determination. *AJR Am J Roentgenol*. 1983; **140**(2):221–3.
3. Blumhagen JD, Maclin L, Krauter D, *et al.* Sonographic diagnosis of hypertrophic pyloric stenosis. *AJR Am J Roentgenol*. 1988; **150**(6):1367–70.
4. Forster N, Haddad RL, Choroomi S, *et al.* Use of ultrasound in 187 infants with suspected infantile hypertrophic pyloric stenosis. *Australas Radiol*. 2007;**51**(6):560–3.
5. Rohrschneider WK, Mittnacht H, Darge K, *et al.* Pyloric muscle in asymptomatic infants: sonographic evaluation and discrimination from idiopathic hypertrophic pyloric stenosis. *Pediatr Radiol*. 1998;**28**(6):429–34.

6. Blumhagen JD. The role of ultrasonography in the evaluation of vomiting in infants. *Pediatr Radiol.* 1986;**16**(4):267–70.

7. Gilet AG, Dunkin J, Cohen HL. Pylorospasm (simulating hypertrophic pyloric stenosis) with secondary gastroesophageal reflux. *Ultrasound Q.* 2008;**24**(2):93–6.

8. Cohen HL, Zinn H, Haller J, *et al.* Ultrasonography of pylorospasm: findings may simulate hypertrophic pyloric stenosis. *J Ultrasound Med.* 1998;**17**:705–12.

9. Swischuk LE, Hayden CK, Jr., Tyson KR. Short segment pyloric narrowing: pylorospasm or pyloric stenosis? *Pediatr Radiol.* 1981;**10**(4):201–5.

10. Aspelund G, Langer JC. Current management of hypertrophic pyloric stenosis. *Semin Pediatr Surg.* 2007;**16**(1):27–33.

11. Mitchell LE, Risch N. The genetics of infantile hypertrophic pyloric stenosis: a reanalysis. *Am J Dis Child.* 1993;**147**:1203–11.

12. Macdessi J, Oates R. Clinical diagnosis of pyloric stenosis: a declining art. *BMJ.* 1993;**306**:553–5.

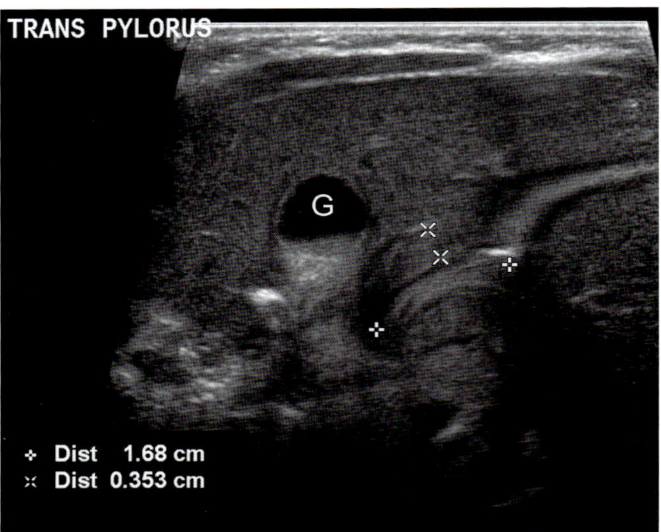

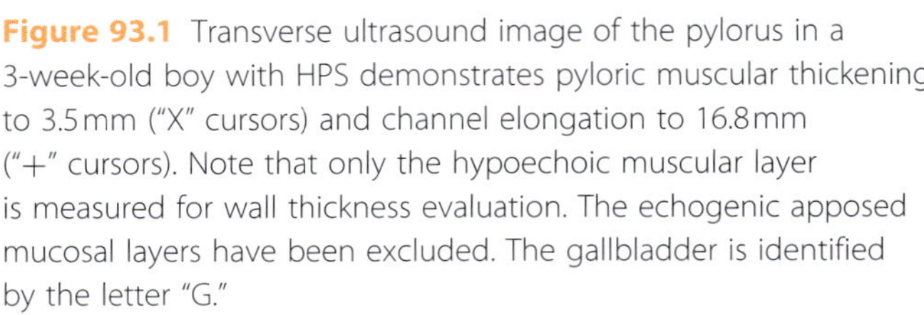

Figure 93.1 Transverse ultrasound image of the pylorus in a 3-week-old boy with HPS demonstrates pyloric muscular thickening to 3.5 mm ("X" cursors) and channel elongation to 16.8 mm ("+" cursors). Note that only the hypoechoic muscular layer is measured for wall thickness evaluation. The echogenic apposed mucosal layers have been excluded. The gallbladder is identified by the letter "G."

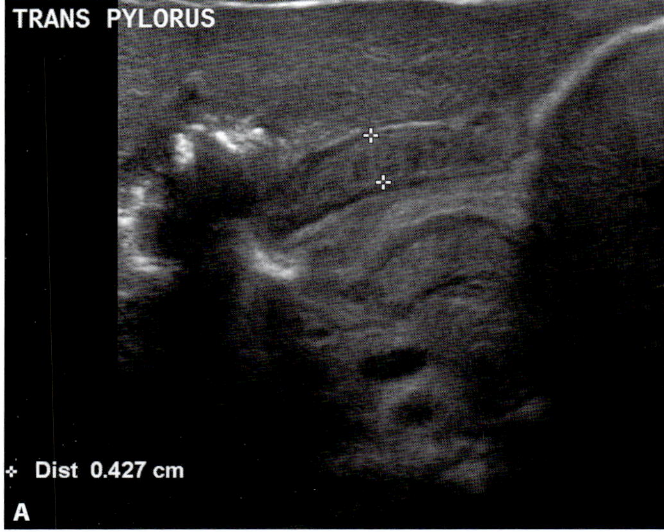

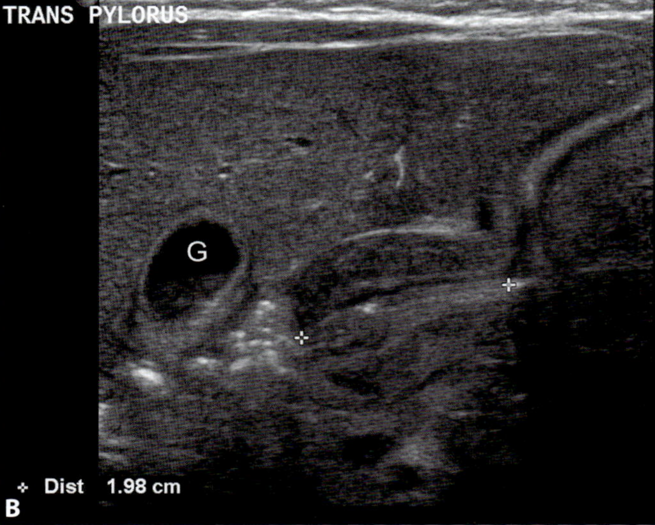

Figure 93.2 **A.** Transverse ultrasound image of the pylorus in a 4-week-old boy with HPS shows mural thickening measuring 4.3 mm ("+" cursors). **B.** Pyloric channel length in this patient measures 19.8 mm ("+" cursors). The gallbladder is identified by the letter "G."

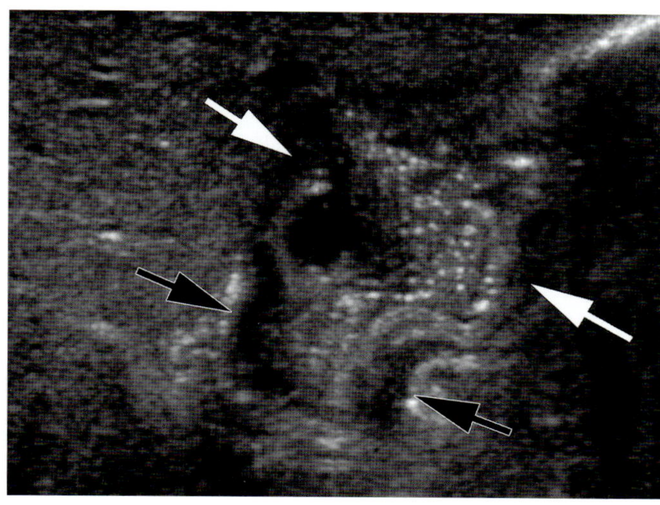

Figure 93.3 Transverse ultrasound image of a normal pylorus in a 7-week-old girl. The pylorus (black arrows) does not exhibit the mural thickening and channel elongation seen in Figures 93.1 and 93.2. The gastric antrum (white arrows) is distended with gas and fluid.

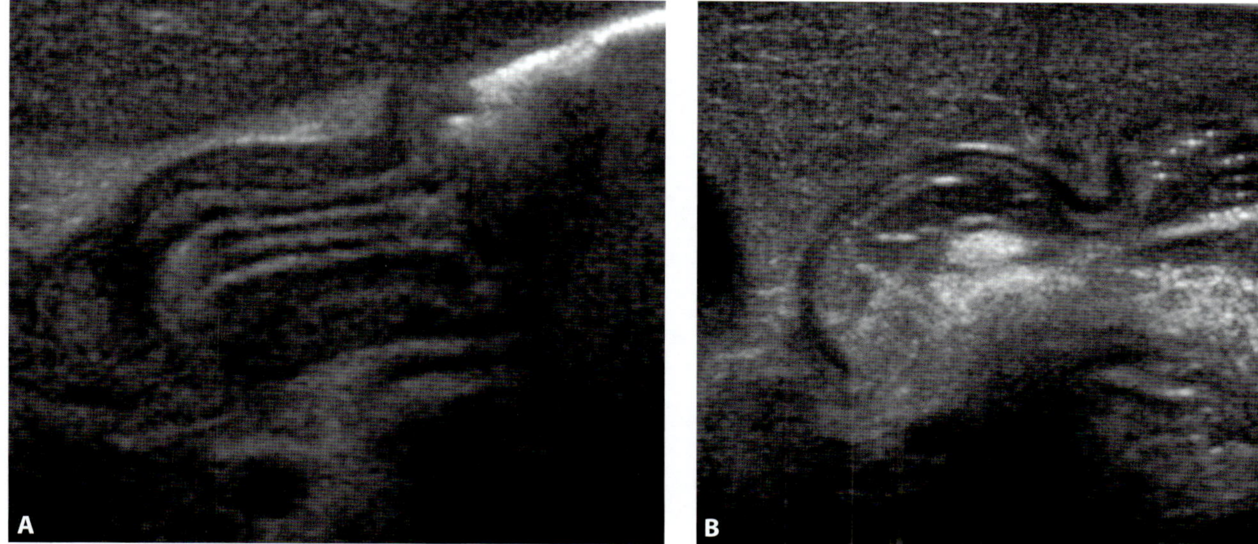

Figure 93.4 **A.** Transverse ultrasound image of the pylorus in a 2-month-old boy with PS shows mural thickening and channel elongation suggestive of HPS. **B.** Image acquired 90 seconds later shows a large bolus of fluid and gas passing through the pylorus.

Retropharyngeal pseudothickening

Ramesh S. Iyer

Imaging description

Initial imaging evaluation for retropharyngeal abscess consists of a lateral radiograph of the cervical soft tissues. In the setting of either retropharyngeal abscess or cellulitis, there is usually marked thickening of the prevertebral soft tissues. The degree of thickening typically exceeds 50% of the anterior-posterior (AP) diameter of the adjacent vertebral body (Figure 94.1) [1]. In severe cases there may be anterior displacement of the airway and/or loss of normal cervical lordosis due to soft tissue expansion. Identification of soft tissue gas is the only method to distinguish abscess from cellulitis alone on radiography. Sensitivity and specificity of lateral neck radiography for retropharyngeal abscess has been reported to be 80% and 100%, respectively [2].

Pseudothickening of the cervical soft tissues results from neck flexion, expiration, or swallowing during radiography. This entity is particularly common in infants and young children, where the neck is short [3]. There is apparent widening of the pre-vertebral soft tissues that mimics abscess or cellulitis (Figure 94.2).

Distinguishing retropharyngeal abscess from pseudothickening requires the lateral neck radiograph to be performed in full extension (or extended to the degree the child can tolerate) during inspiration [1–3]. Along with thickening, a soft tissue configuration with apex anterior convexity is suggestive of true inflammation (Figure 94.1). If there is doubt of technical adequacy, the radiograph should be repeated to ensure full neck extension, inspiration, and true lateral projection.

Importance

Retropharyngeal abscess requires either prompt surgical drainage or antibiotics, depending on abscess size. Potential complications of untreated abscess include airway compromise, spread of infection into the mediastinum, and cervical vascular compromise (e.g., internal jugular venous thrombosis, carotid arterial pseudoaneurysm) [4]. In contradistinction, pseudothickening is a normal finding and a result of suboptimal exam technique.

Typical clinical scenario

Retropharyngeal abscess typically occurs in children younger than 5 years of age. The most common presenting signs and symptoms include neck pain, fever, sore throat, palpable neck mass, stridor, and respiratory distress [1, 5].

Differential diagnosis

Epiglottitis also presents as acute upper respiratory obstruction in a febrile young child. On a lateral neck radiograph there is thickening of the epiglottis and aryepiglottic folds [3, 6]. Croup, or viral laryngotracheobronchitis, usually presents with fever and stridor in children between the ages of 6 months and 3 years. These patients exhibit symmetric subglottic tracheal narrowing and distension of the hypopharynx on the lateral neck radiograph [6]. Lymphatic malformations are infiltrative lesions that may involve the retropharyngeal soft tissues along with anterior and posterior cervical triangles. While swelling is typically chronic, a sudden increase in size may occur from infection or intralesional hemorrhage [7].

Teaching point

Distinguishing between retropharyngeal abscess and pseudothickening requires a lateral neck radiograph in full cervical extension. With proper exam technique, prevertebral soft tissue thickening (greater than half the AP diameter of the adjacent vertebral body) with apex anterior convexity is suggestive of true inflammation. Presence of soft tissue gas distinguishes abscess from cellulitis alone.

REFERENCES

1. Craig FW, Schunk JE. Retropharyngeal abscess in children: clinical presentation, utility of imaging, and current management. *Pediatrics.* 2003;**6**(111):1394–8.
2. Boucher, C, Dorion D, Fisch C. Retropharyngeal abscesses: a clinical and radiologic correlation. *J Otolaryngol.* 1999;**28**(1):134–7.
3. Ludwig BJ, Foster BR, Saito N, *et al.* Diagnostic imaging in nontraumatic pediatric head and neck emergencies. *Radiographics.* 2010;**30**:781–99.
4. Hoang JK, Branstetter BF 4th, Eastwood JD, *et al.* Multiplanar CT and MRI of collections in the retropharyngeal space: is it an abscess? *AJR Am J Roentgenol.* 2011;**196**(4):W426–32.
5. Cmejrek RC, Coticchia JM, Arnold JE. Presentation, diagnosis, and management of deep-neck abscesses in infants. *Arch Otolaryngol Head Neck Surg.* 2002;**128**:1361–4.
6. Robson C, Hudgins P. Pediatric airway disease. In: Som PM, Curtin HD, eds. *Head and neck imaging*, 4th edn. St. Louis, MO: Mosby, 2003:1521–94.
7. de Serres LM, Sie KC, Richardson MA. Lymphatic malformations of the head and neck: a proposal for staging. *Arch Otolaryngol Head Neck Surg.* 1995;**121**(5):577–82.

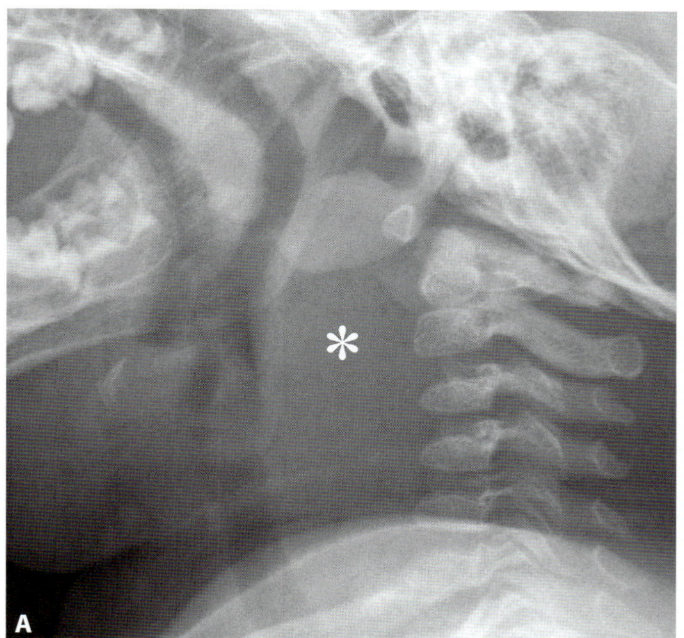

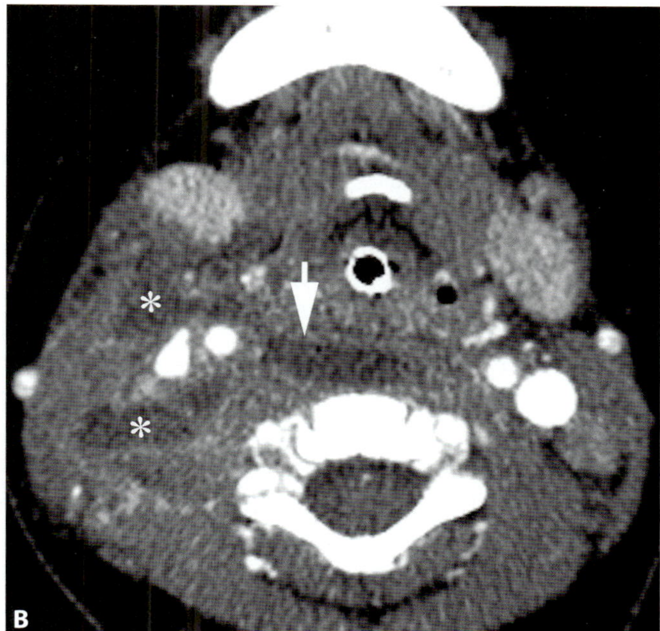

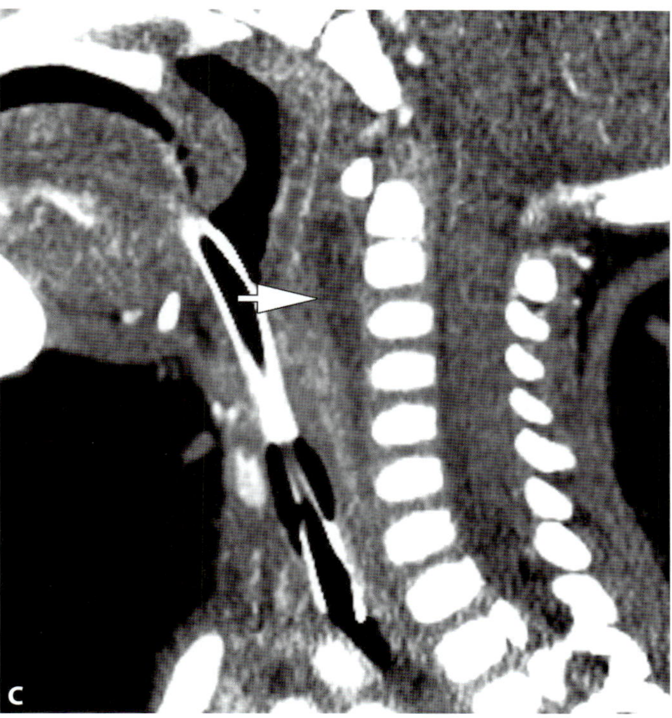

Figure 94.1 **A.** Lateral neck radiograph from a 1-year-old boy with retropharyngeal abscess exhibits marked diffuse thickening of the prevertebral soft tissues (asterisk). Note the apex anterior convex soft tissue thickening and the mild associated cervical kyphosis. **B.** Axial contrast-enhanced neck CT demonstrates multiple rim-enhancing abscesses in the retropharynx (arrow) and surrounding the right common carotid artery and internal jugular vein (asterisks). **C.** Sagittal CT reformation shows the retropharyngeal abscess extending superiorly-inferiorly from C2 through C6 (arrow).

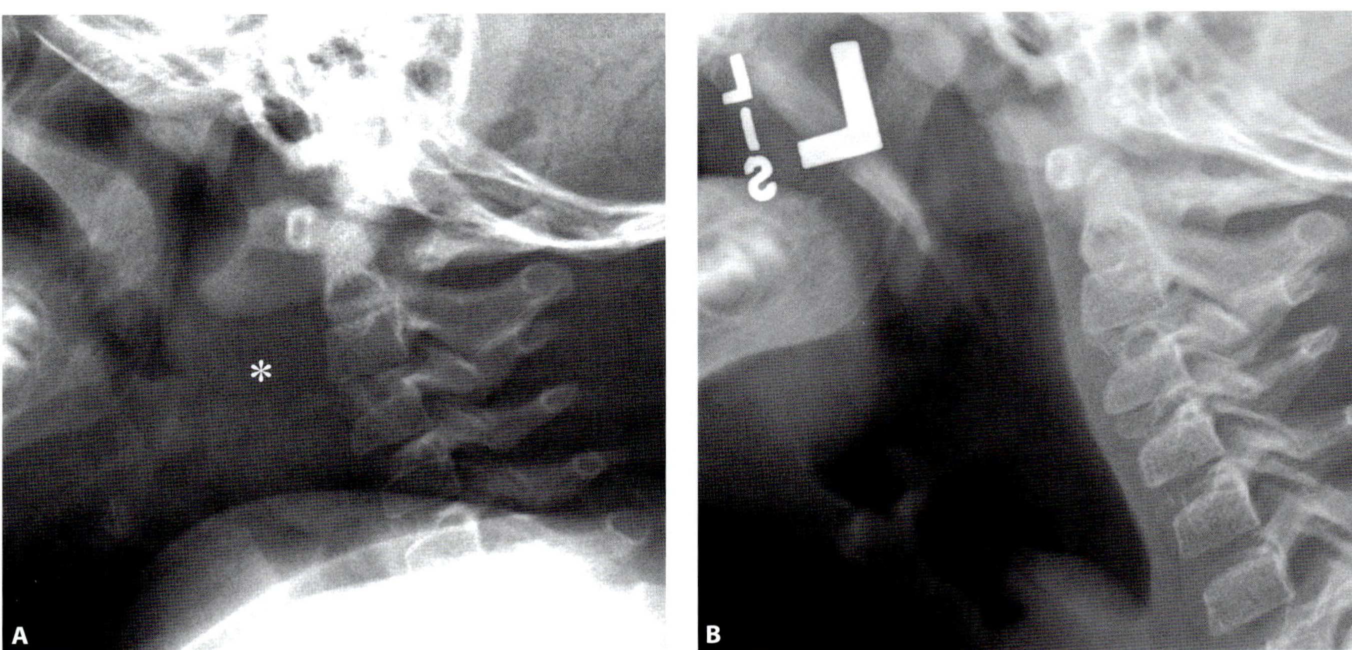

Figure 94.2 **A.** Lateral neck radiograph in a 2-year-old girl with pseudothickening of cervical soft tissues shows apparent prevertebral widening secondary to neck flexion and expiration (asterisk). **B.** Lateral neck radiograph performed one hour later with improved extension and inspiration demonstrates normal prevertebral soft tissues.

Cranial sutures simulating fractures

Ramesh S. Iyer

Imaging description

Skull radiographs may still be performed to evaluate for pediatric calvarial fractures. However, at most facilities, radiography has been replaced by CT due to its superior detection and characterization of fractures and sutures, and assessment of intracranial pathology. Even on CT, calvarial fractures may be difficult to identify because of the thin cortex in children. Three-dimensional shaded surface reconstructions of the skull (3D-CT) are invaluable to evaluate for pediatric head trauma [1–3]. This technique offers exquisite detail in characterizing surface anatomy and the osseous defect(s) in question. MRI provides no significant advantage over CT to distinguish fractures from normal sutures.

Common sutures include the midline sagittal and metopic, and bilateral coronal and lambdoid. Accessory sutures are most common in the parietal and occipital bones. The parietal bone arises from two ossification centers, while the occipital bone ossifies from six centers [1, 2, 4, 5].

Symmetry is the first criterion to differentiate accessory sutures from fractures. Accessory sutures are generally bilateral and symmetric. Fractures are most often unilateral, and if bilateral are usually asymmetric [1]. Bilateral skull fractures are usually the result of high-energy trauma, are often diastatic, and are frequently accompanied by extra-axial hemorrhage. These fractures, though not pathognomonic, should raise suspicion for non-accidental trauma [4].

Skull fractures are sharp lucent straight lines with non-sclerotic margins. Cranial sutures feature multiple interdigitations and a "zig-zag" or corrugated pattern, with sclerotic margins (Figures 95.1 and 95.2) [1–5]. Fractures may cross sutures (Figure 95.3), and if they terminate at a suture may be associated with sutural diastasis (Figure 95.4) [1]. Accessory sutures merge with the adjacent common suture and do not widen the latter. Finally, the presence of overlying soft tissue swelling strongly suggests fracture. Kleinman and Spevak reported 4 mm of scalp swelling overlying calvarial fractures in all 35 children that they evaluated [6]. However, the lack of associated soft tissue swelling does not exclude a fracture [1, 7].

Importance

Distinguishing fractures from cranial sutures is a standard component of assessing pediatric head trauma, particularly in cases of abuse.

Typical clinical scenario

A child presents for imaging following head trauma.

Differential diagnosis

As mentioned above, primary diagnostic considerations include fracture and normal cranial suture. Occasionally prominent calvarial vascular channels may mimic fractures. Vascular channels exhibit sclerotic margins, while the margins of fractures are sharp and non-sclerotic.

Teaching point

Three-dimensional volume rendered views of the cranium are particularly helpful for detection and characterization of skull defects. Normal cranial sutures are typically bilateral and symmetric, while fractures are usually unilateral and when bilateral are asymmetric. Fractures may cross or widen adjacent sutures. While associated soft tissue swelling strongly suggests a fracture, lack of swelling does not exclude it.

REFERENCES

1. Sanchez T, Stewart D, Walvick M, *et al.* Skull fracture vs. accessory sutures: how can we tell the difference? *Emerg Radiol.* 2010;**17**:413–18.

2. Choudhary AK, Jha B, Boal DK, *et al.* Occipital sutures and its variations: the value of 3D-CT and how to differentiate it from fractures using 3D-CT? *Surg Radiol Anat.* 2010;**32**:807–16.

3. Miller AJ, Kim U, Carrasco E. Differentiating a mendosal suture from a skull fracture. *J Pediatr.* 2010;**157**(4):691.

4. Weir P, Suttner NJ, Flynn P, *et al.* Normal skull suture variant mimicking intentional injury. *BMJ.* 2006;**332**(7548): 1020–1.

5. Tharp AM, Jason DR. Anomalous parietal suture mimicking skull fracture. *Am J Forensic Med Pathol.* 2009;**30**(1):49–51.

6. Kleinman PK, Spevak MR. Soft tissue swelling and acute skull fractures. *J Pediatr.* 1992;**121**(5 Pt 1):737–9.

7. Fernando S, Obaldo RE, Walsh IR, *et al.* Neuroimaging of nonaccidental head trauma: pitfalls and controversies. *Pediatr Radiol.* 2008;**38**(8):827–38.

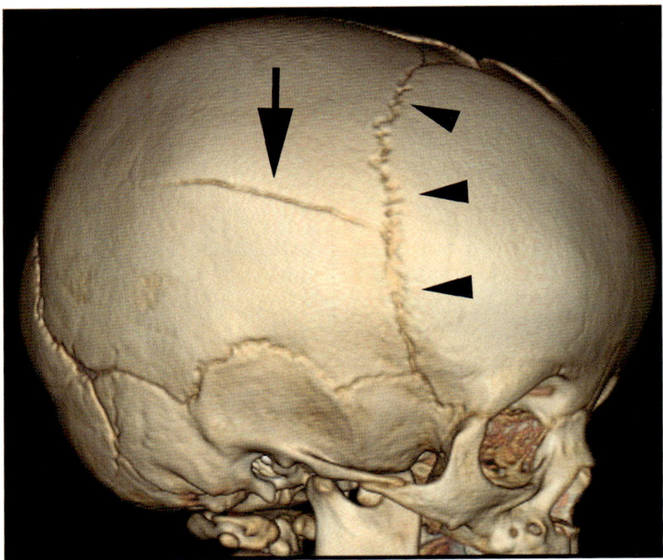

Figure 95.1 Right lateral 3D volume rendered CT image of the calvarium of a 7-month-old boy with a linear non-depressed right parietal fracture. Compare the sharp linear course of the fracture (arrow) with the corrugated appearance of the adjacent right coronal suture (arrowheads).

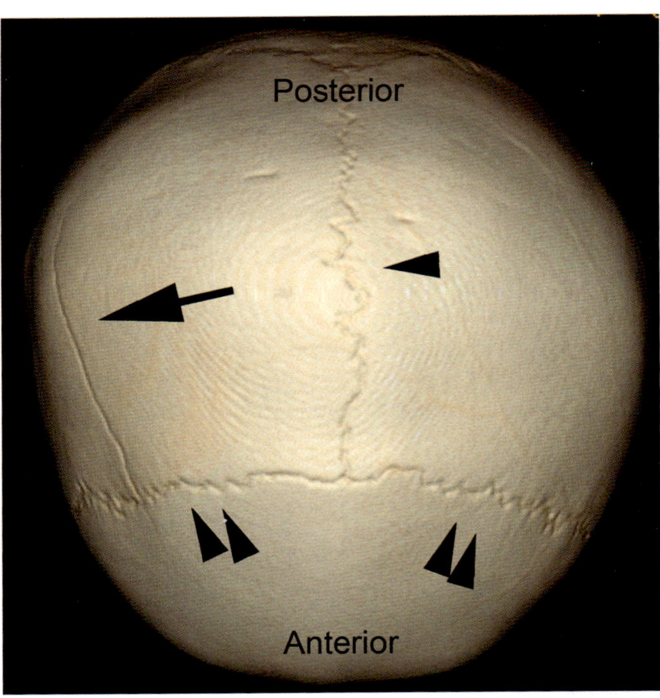

Figure 95.2 Vertex 3D volume rendered CT image of the calvarium of a 4-year-old boy with a linear non-depressed right parietal fracture. The fracture (arrow) is unilateral and linear, and does not feature interdigitations like the sagittal (single arrowhead) and bilateral coronal (double arrowheads) sutures.

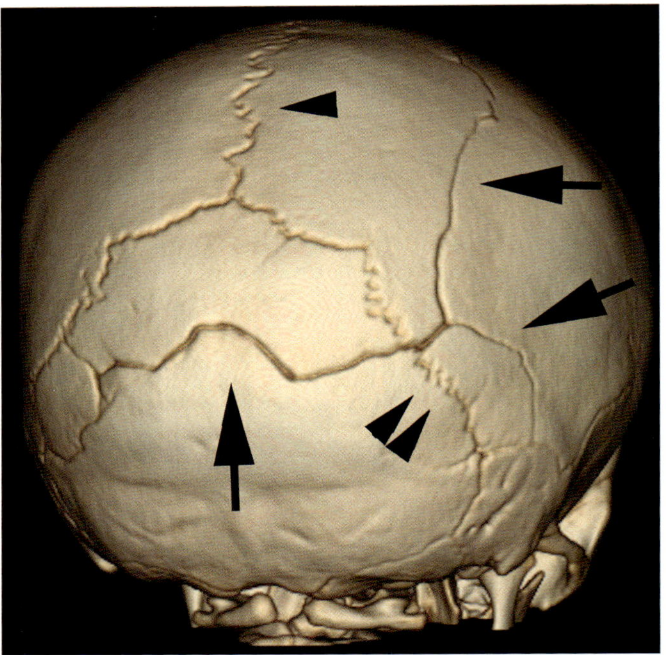

Figure 95.3 Posterior 3D volume rendered CT image of the calvarium of an 8-month-old boy with a stellate occipital fracture extending into the right parietal bone (arrows). Note how the fracture crosses the right lambdoid suture (double arrowheads). Compare the linearity of the fracture planes with the "zig-zag" pattern of the right lambdoid and sagittal sutures (arrowhead). The complex nature of this fracture indicates a high-energy impact, and should raise suspicion of non-accidental trauma.

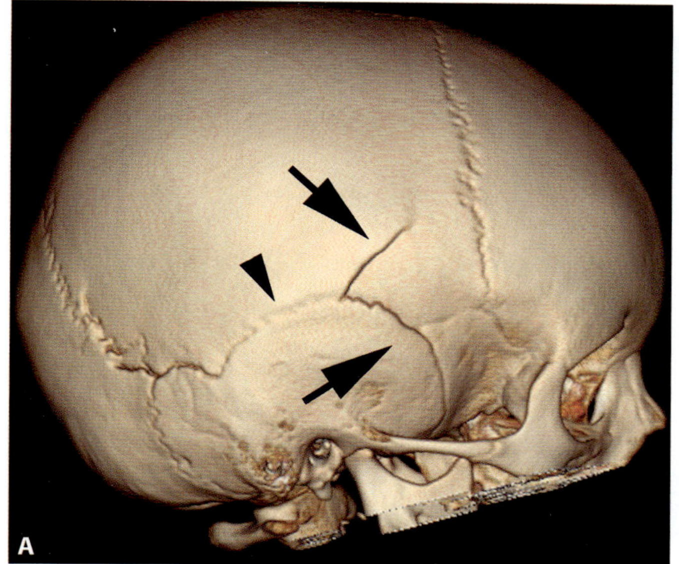

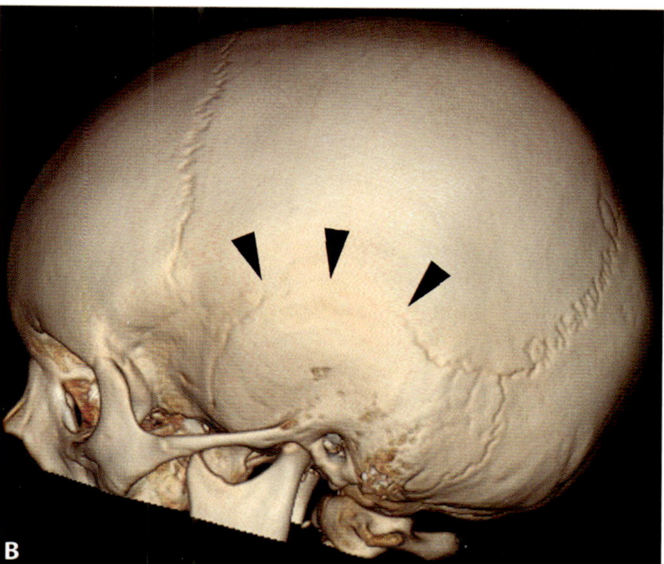

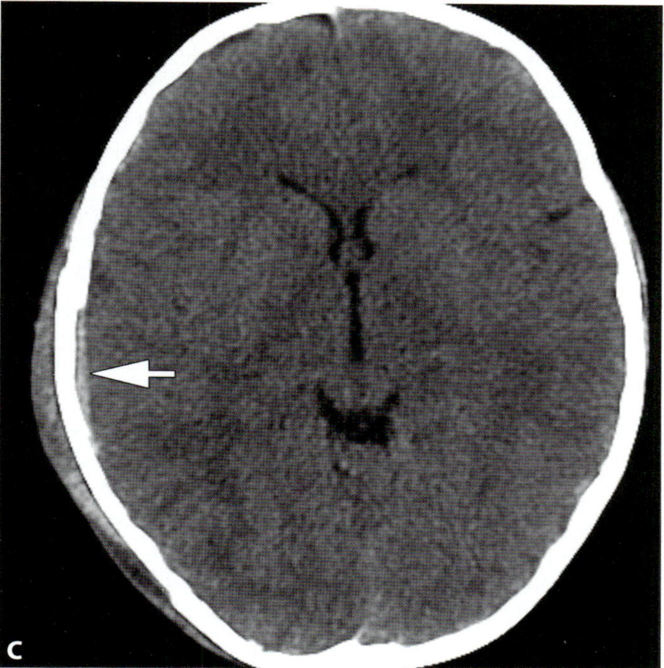

Figure 95.4 **A.** Right lateral 3D volume rendered CT image of the calvarium of a 4-year-old girl with a right parietal fracture widening the anterior portion of the right squamosal suture (arrows), a normal minor cranial suture. The posterior aspect of this suture (arrowhead) remains partially fused. **B.** Left lateral 3D volume rendered image from the same study. Compare the right-sided pathologic sutural widening in **A** with the normal left squamosal suture (arrowheads), also partially fused. **C.** Axial non-contrast CT image demonstrates subdural hemorrhage (arrow) subjacent to the fracture, and a subgaleal hematoma overlying the site.

Systematic review of elbow injuries

Claire K. Sandstrom

Imaging description

Consider the imaging description in the following four scenarios of elbow radiographs performed in children, and try to identify the pitfall illustrated in each case.

Scenario 1: In a child less than 30 months of age, the anterior humeral line does not bisect the ossification center of the capitellum. No joint effusion is present, and no fracture line is detected.

Scenario 2: In a child less than eight years of age, elbow radiographs reveal three ossification centers, one each at the expected location of the capitellum, radial head, and trochlea. There is an elbow effusion.

Scenario 3: In a child younger than 10 years of age, a fracture of the lateral humeral condyle is detected. The distal humeral epiphysis remains partially cartilaginous, but the fracture is not seen in the ossified portions of the epiphysis.

Scenario 4: An adolescent complains of medial elbow pain following trauma. Anterior-posterior (AP) view shows lucency through the base of the medial epicondyle, while the remainder of the elbow is normal.

Importance

At birth, the distal humeral epiphysis, including the epicondyles, is completely cartilaginous. Progressive ossification predisposes children to characteristic injuries of the elbow, and age-directed evaluation can help detect otherwise subtle injuries [1]. Unfortunately, ossification is not always symmetric, so relying solely on comparison to the contralateral side may be problematic and misleading [2].

The capitellum, which typically begins to ossify around 6–12 months of age, can be used to assess for supracondylar humeral fractures. In older children and adults, a line drawn along the anterior cortex of the humeral – the anterior humeral line (AHL) – should pass through the middle third of the capitellum [3]. Failure to do so suggests a supracondylar fracture of the distal humerus with dorsal angulation (Figure 96.1). However, in children less than two-and-a-half years of age, the AHL may normally pass through the anterior third of the small capitellar growth center, as shown in scenario 1 (Figure 96.2) [3]. In these young children, comparison to the contralateral side may be helpful to avoid incorrectly diagnosing a supracondylar fracture.

Epiphyseal ossification at the elbow follows a predictable sequence, though exact age and rate of individual development vary. The mnemonic CRITOE is often quoted and conveys the sequence of appearance of epiphyseal ossification centers: Capitellum, Radius, Internal (medial) epicondyle, Trochlea, Olecranon, and External (lateral) epicondyle (Figure 96.3). The precise age, confirmed by McCarthy and Ogden [2, 4], at which these ossification centers appear is quite variable, but on average they should be visible by the ages given in Table 96.1. CRITOE is most useful when evaluating the elbow in children ranging in age from 3 to 7 years, when the medial epicondyle has begun to ossify but the trochlea remains cartilaginous (Figure 96.4). If an ossification center is seen at the expected location of the trochlea with none at the medial epicondyle, as in scenario 2 (Figure 96.5), medial epicondyle avulsion and intraarticular entrapment is the diagnosis.

Table 96.1. Average ages at which secondary ossification centers of the elbow appear in boys and girls

Ossification center	Boys	Girls
Capitellum	5 months old	4 months old
Radius	3 years old	3 years old
Medial epicondyle	5–7 years old	3–6 years old
Trochlea	8–10 years old	7–9 years old
Olecranon	9 years old	9 years old
Lateral epicondyle	12 years old	11 years old

In children younger than 10 years of age, fractures of the lateral humeral condyle are the most common Salter–Harris grade III and IV injuries (Figure 96.6). A common pitfall is to ascribe Salter–Harris grade II classification to lateral condyle fractures, as only the metaphyseal fracture is visible. However, these fractures nearly always propagate across the cartilaginous trochlear articular surface [2], and therefore should always be considered Salter–Harris grade IV injuries until proven otherwise.

Finally, fusion of the medial epicondylar apophysis typically occurs at 15 years in girls and 18 years in boys. This is much later than fusion of the capitellum, trochlea, and lateral epicondyle to the distal humeral metaphysis, and the normal patent physis should not be confused with a fracture, as in scenario 4 (Figure 96.7). However, the delayed closure of the physis also predisposes the medial epicondyle to injuries in adolescents and young teenagers, and true fractures in this location tend to be grade IV by Salter–Harris classification (Figure 96.8) [1].

Typical clinical scenario

The most common cause of elbow injuries in children is falling on an outstretched arm, though motor vehicle collisions and other high-energy trauma are other important causes of fractures.

Differential diagnosis

As illustrated in the four scenarios above, most common diagnostic dilemmas in the pediatric elbow involve differentiating acute fracture from normal development.

Teaching point

Understanding development of the elbow is vital to recognizing and correctly diagnosing elbow injuries in children. The AHL may be problematic in children under two-and-a-half years of age but is essential for diagnosing subtle supracondylar fractures in older children. CRITOE should be routinely used on elbow radiographs in children 3–7 years of age, particularly to avoid missing medial epiphyseal avulsion. Fractures of the lateral condyle and medial epicondyle should be considered Salter–Harris grade IV until proven otherwise.

REFERENCES

1. Sonin A. Fractures of the elbow and forearm. *Semin Musculoskelet Radiol.* 2000;**4**(2):171–91.
2. McCarthy SM, Ogden JA. Radiology of postnatal skeletal development. V. Distal humerus. *Skeletal Radiol.* 1982;**7**(2):239–49.
3. Rogers LF, Malave S, White H, Tachdjian MO. Plastic bowing, torus and greenstick supracondylar fractures of the humerus: radiographic clues to obscure fractures of the elbow in children. *Radiology.* 1978; **128**(1):145–50.
4. McCarthy SM, Ogden JA. Radiology of postnatal skeletal development. VI. Elbow joint, proximal radius, and ulna. *Skeletal Radiol.* 1982;**9**(1):17–26.

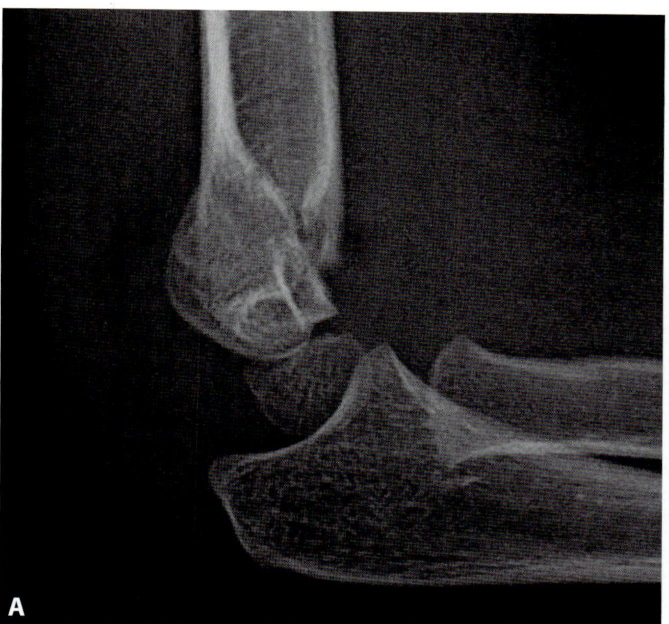

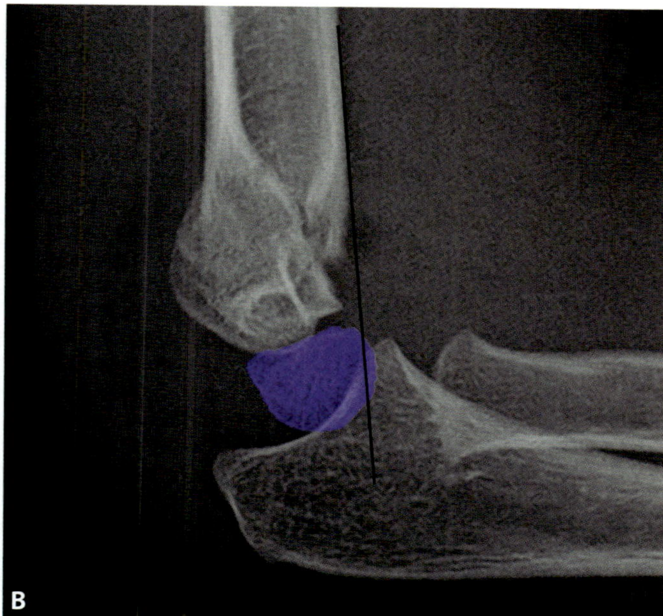

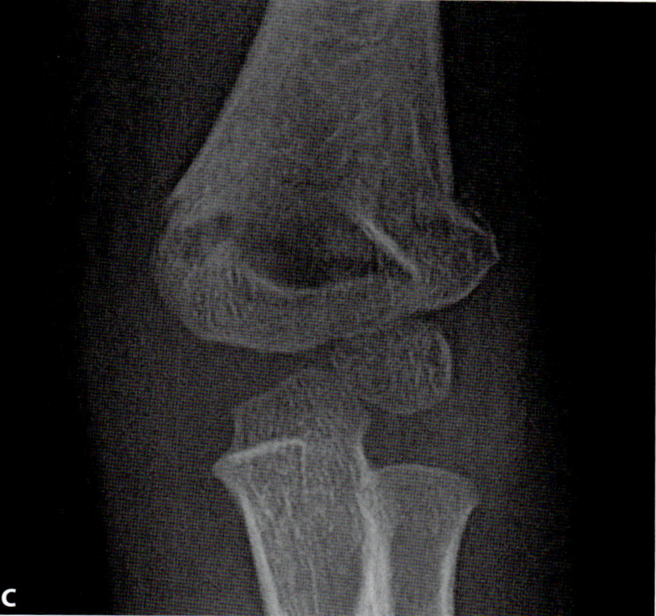

Figure 96.1 **A.** Lateral radiograph of the right elbow in a 4-year-old boy following a fall shows elevation of the anterior and posterior fat pads. **B.** Same lateral elbow radiograph shows that the anterior humeral line (black line) does not intersect the middle third of the capitellum (blue shading), in this case consistent with a supracondylar fracture of the elbow. **C.** Frontal radiograph of the elbow confirms the linear lucency of the supracondylar fracture.

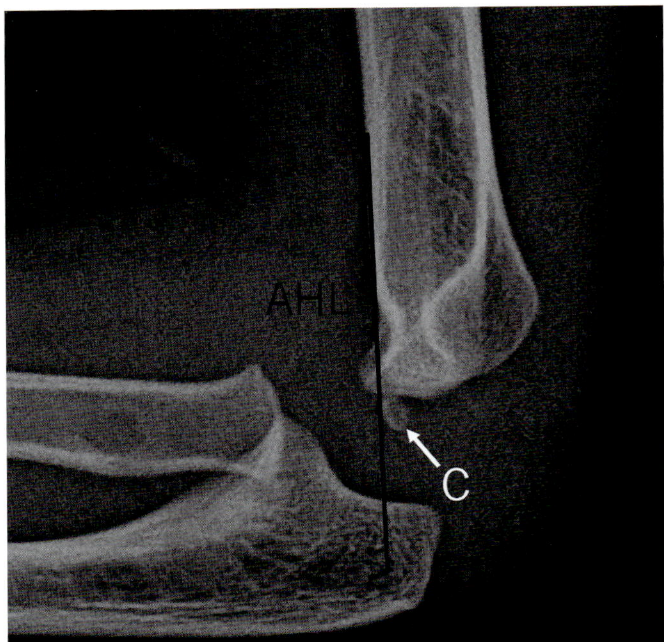

Figure 96.2 (Scenario 1) Lateral radiograph of the right elbow in a 26-month-old boy who fell down stairs shows the anterior humeral line (AHL, black line) passing through the anterior third of the capitellum (C, white arrow), due to the small size the capitellar ossification center. No joint effusion or fracture is present.

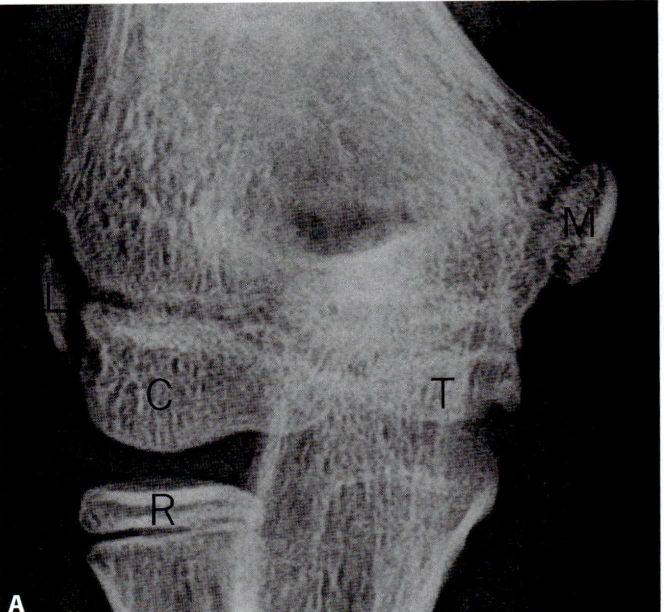

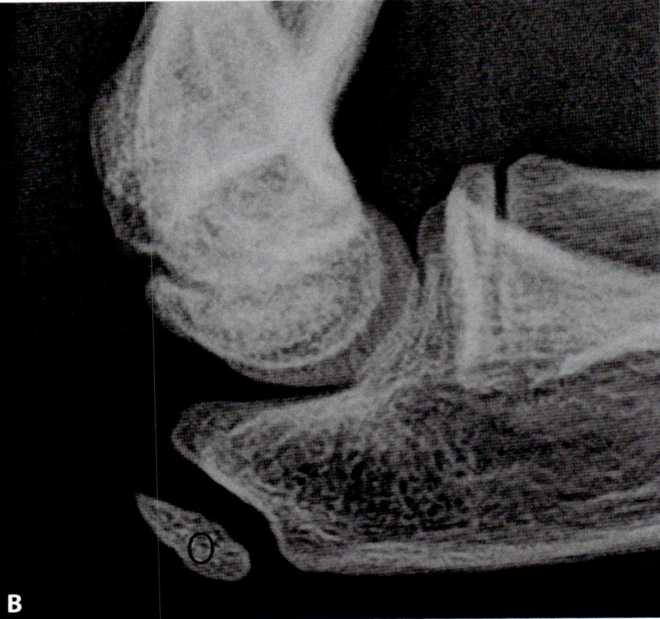

Figure 96.3 A. AP radiograph of the left elbow in a 12-year-old boy with no elbow fracture shows the normal position of the unfused growth centers of the capitellum (C), radius (R), trochlea (T), and medial (M) and lateral (L) epicondyles. **B.** Lateral radiograph of the same elbow shows the normal olecranon (O) ossification center.

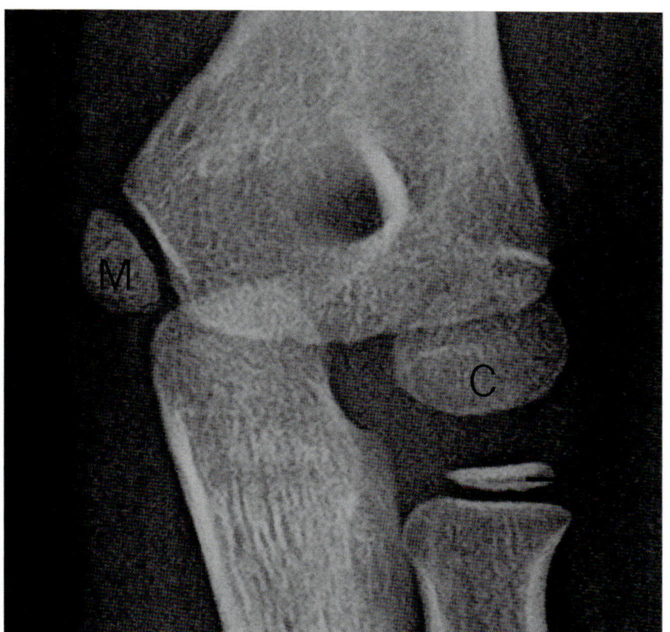

Figure 96.4 AP radiograph of the right elbow in a 5-year-old girl complaining of arm pain shows the normal position of the growth centers of the capitellum (C), radius (R), and medial epicondyle (M).

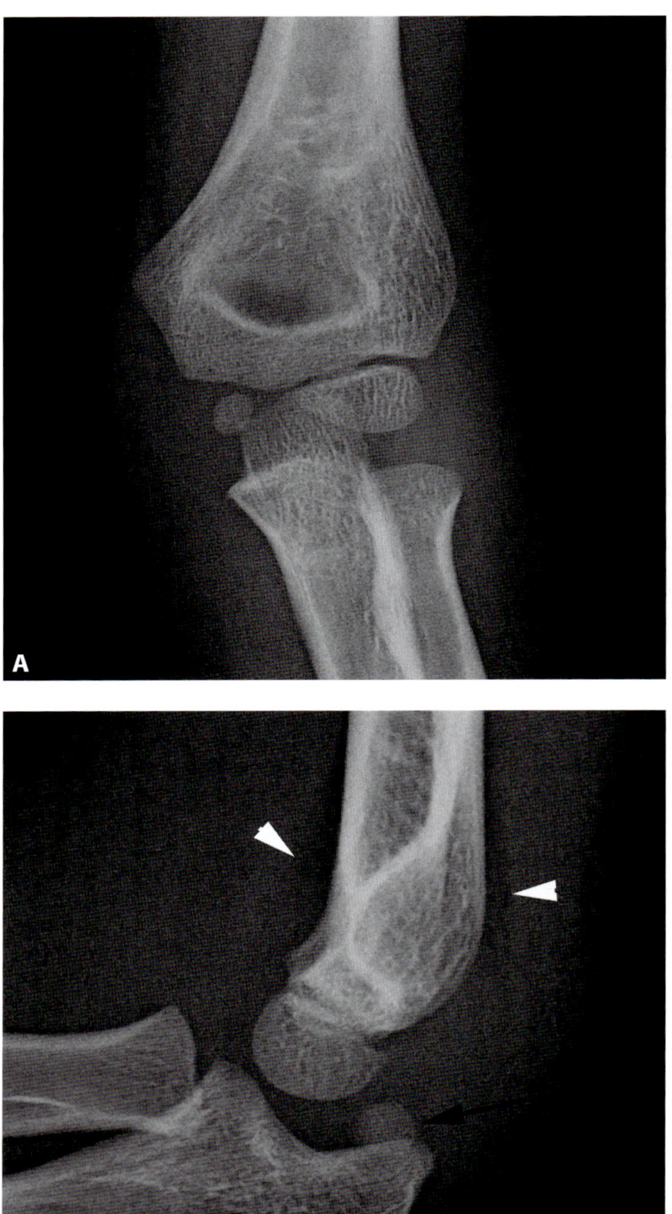

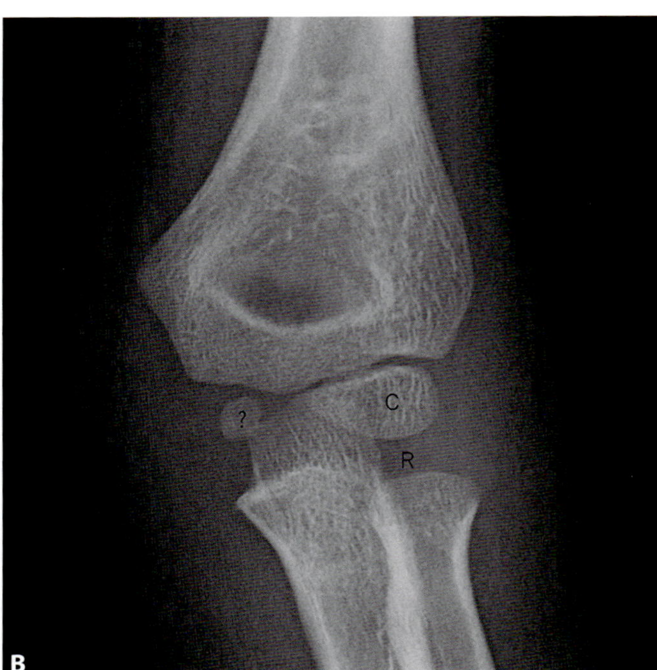

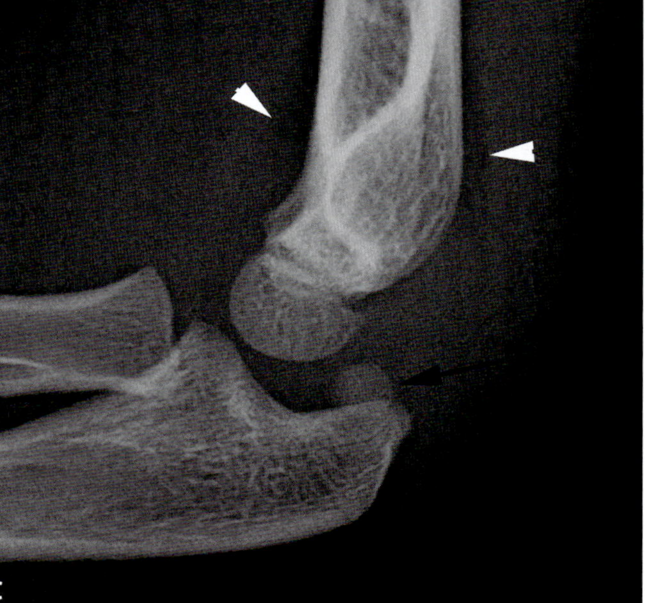

Figure 96.5 (Scenario 2) A. AP radiograph of the right elbow in a 7-year-old boy after falling. Three growth centers are seen, which as labeled in **B.** include the capitellum (C) and radius (R), and a third growth center (labeled "?") which is in the expected location of the trochlea. The diagnosis is medial epicondylar avulsion with intraarticular dislocation. **C.** Lateral radiograph shows the displaced medial epicondylar growth center (arrow). The anterior and posterior fat pads are elevated (arrowheads), compatible with intraarticular fracture.

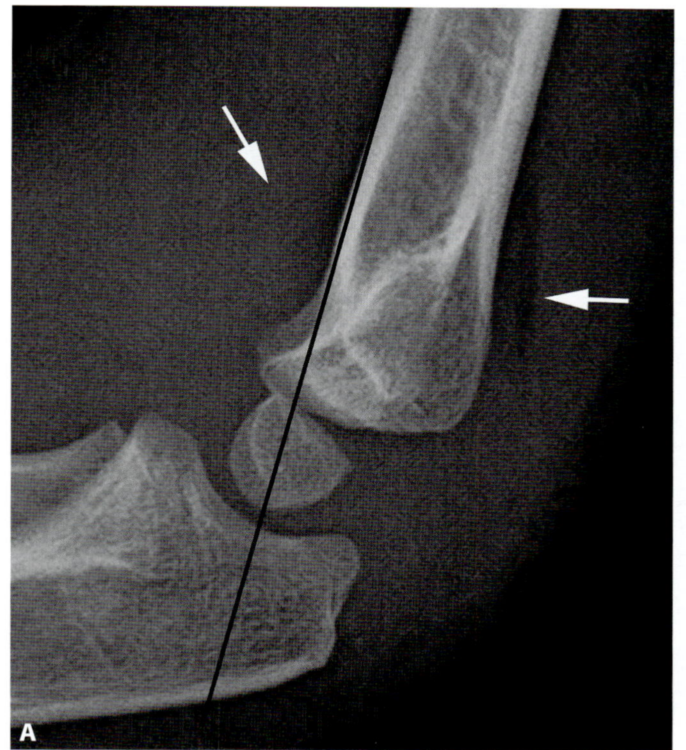

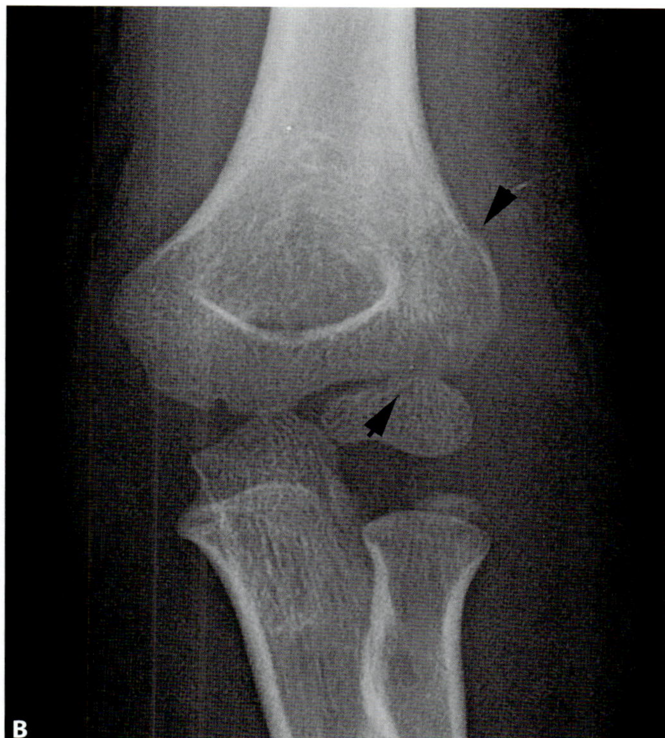

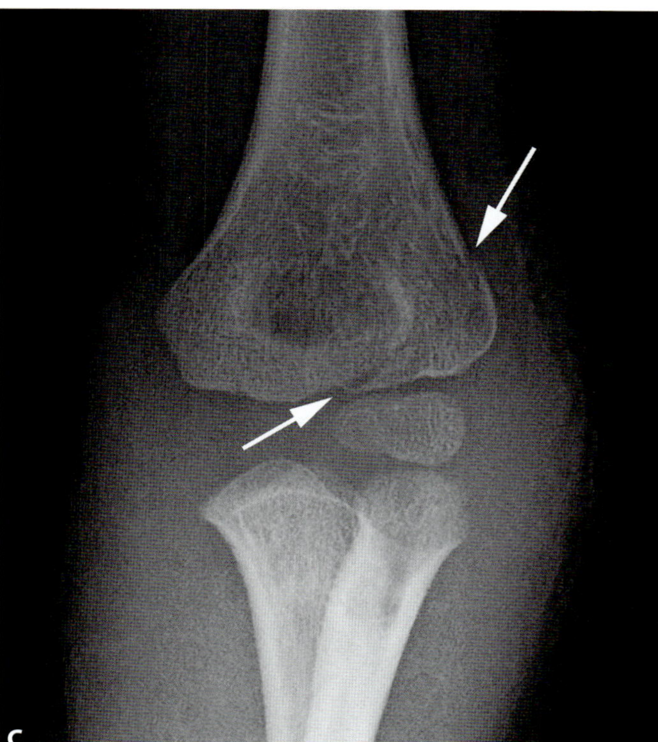

Figure 96.6 (Scenario 3) A. Lateral radiograph of the right elbow in a 4-year-old girl after falling shows elevation of the anterior and posterior fat pads (arrows). The capitellar growth center is in the expected location relative to the anterior humeral line (black line). **B.** The AP view shows a cortical defect and linear lucency (arrowheads) through the lateral humeral condyle. **C.** The fracture is seen to better extent on an oblique view (arrows). This should be described as a Salter–Harris grade IV injury.

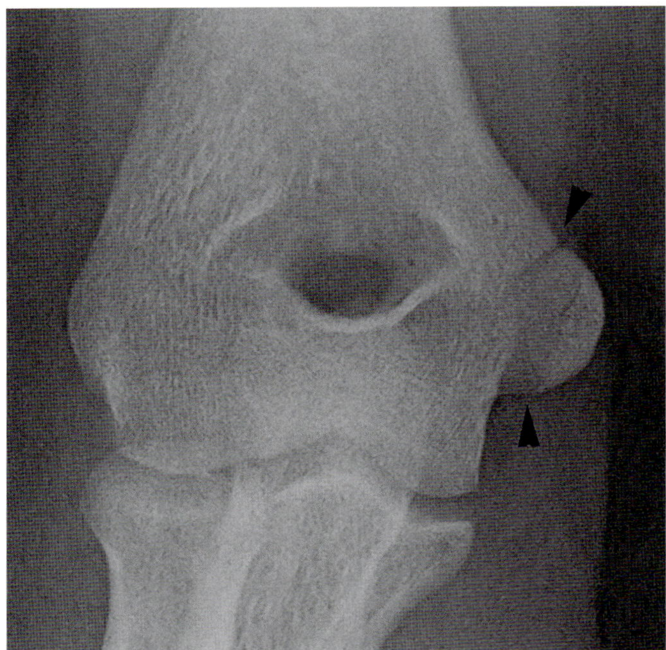

Figure 96.7 (Scenario 4) AP radiograph of the left elbow in a 15-year-old boy after a motor vehicle collision shows lucency at the base of the medial epicondyle (black arrowheads). This was initially thought to represent a fracture before being recognized as a normal physis.

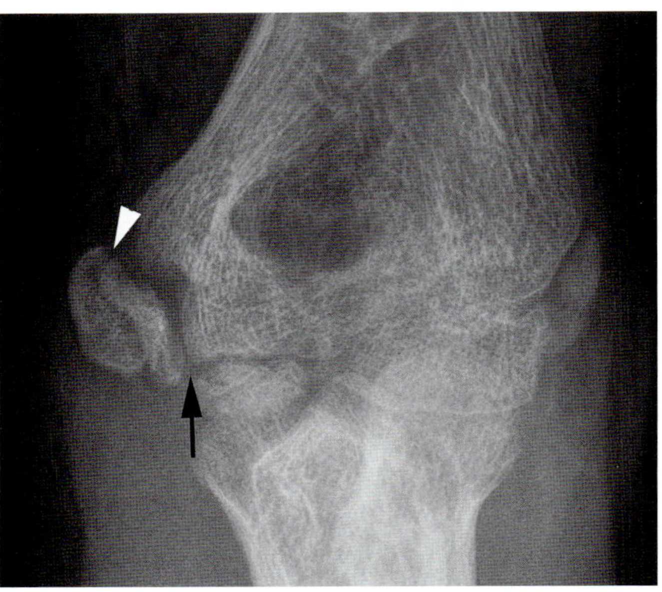

Figure 96.8 AP radiograph of the right elbow in an 11-year-old boy referred for medial epicondyle fracture 1 month earlier. Resorption along the fracture line (black arrow) is seen, as well as disuse osteoporosis. The fracture line passes into the superior aspect of the physis (white arrowhead) and likely across the apophysis, consistent with Salter–Harris type IV injury.

CASE **97**

Pelvic pseudofractures: normal physeal lines

Claire K. Sandstrom

Imaging description

At birth, the primary ossification centers of the ilium, pubis, and ischium converge on the triradiate cartilage at the hip (Figure 97.1). On a poorly positioned radiograph (Figure 97.2), the ischium may appear to be displaced medially relative to the ilium, which should not be mistaken for a fracture through the triradiate cartilage. The triradiate cartilage gradually thins (Figure 97.3), and the roof of the acetabulum may appear irregular in children 7–12 years of age. Particularly when viewed on axial CT images, this should not be mistaken for comminuted acetabular fracture (Figure 97.4). The bony acetabulum fuses around 11 to 14 years of age, achieving its adult appearance slightly earlier in girls than boys [1].

Around puberty, three secondary ossification centers develop around the acetabulum, including the os acetabuli (epiphysis of os pubis along the anterior wall of the acetabulum), epiphysis of the ilium (forms the superior wall of the acetabulum), and a small epiphysis of the ischium (Figures 97.5 and 97.6) [2]. These contribute to the depth of the acetabulum but may be confused with avulsion injuries (Figure 97.7). The os acetabuli may persist into adulthood as a separate, well-corticated ossicle (Figure 97.8).

The body and alae of the sacrum develop from several separate primary ossification centers (Figure 97.1), which typically fuse between one and seven years of age [3]. Cartilage bordering the articular surfaces of the sacroiliac joints in young children makes them appear wider on radiographs than would be normal for an adult (Figure 97.1). Small triangular secondary ossification centers appear around puberty along the anterior sacroiliac joint spaces at the levels of S1 and S3 (Figure 97.6) [4]. These begin to fuse to the lateral os sacrum around 18 years of age.

The ischiopubic synchondrosis is seen until fusion, usually by four to eight years of age. This physis can be quite variable in appearance in terms of width and symmetry in young patients (Figures 97.2, 97.3, 97.9, and 97.10).

Importance

As in the rest of the skeleton, evaluation of the pediatric pelvis can be difficult when one is unfamiliar with the normal and variant appearances of the physes. It is important to recognize that pelvic fractures in children are rare. Rather, the presence of a fracture indicates more severe trauma and an increased likelihood of concomitant, potentially life-threatening injuries [1, 5]. In young children, pelvic ligament disruption and pelvic ring instability is very unusual, and, therefore, pediatric pelvic injuries almost never necessitate internal fixation. Exsanguination from pelvic injuries is similarly unlikely. However, detection of a pelvic fracture should prompt more aggressive workup to evaluate for, and guide treatment of, other systemic injuries that may potentially be fatal, including abdominal organ, lung, and head injuries.

Typical clinical scenario

Blunt trauma to the pelvis occurs in children most frequently as a result of falls, pedestrian-versus-motor vehicle collisions, and motor vehicle collisions.

Differential diagnosis

Following blunt trauma, the differentiation of normal physeal anatomy from acute fracture is of paramount importance, and recognition of typical maturation-related fracture patterns may be helpful. In the immature pelvis before fusion of the triradiate cartilage, isolated fractures of the pubic rami and iliac wings are most commonly seen. After closure of the triradiate cartilage, typically in the early teenage years, the pelvis takes on a mature form and is subject to fracture patterns similar to those seen in adults, namely acetabular fractures and pubic symphysis and sacroiliac joint disruption [1]. Open reduction and internal fixation is typically only necessary in the older child with a skeletally mature pelvis.

Sports-related injuries are typically avulsion fractures of unfused apophyses. These occur most commonly at the ischial tuberosity, anterior inferior iliac spine, anterior superior iliac spine, superior parasymphyseal pubis, lesser trochanter, and iliac crest [6, 7].

Teaching point

In the young child, the triradiate cartilage, wide-appearing sacroiliac joints, sacral synchondroses, and ischiopubic synchondroses may potentially be confused with pelvic injuries. In older children, the appearance of secondary ossification centers, particularly around the acetabulum and anterior sacroiliac joints may be similarly confusing.

REFERENCES

1. Silber JS, Flynn JM. Changing patterns of pediatric pelvic fractures with skeletal maturation: implications for classification and management. *J Pediatr Orthop*. 2002;**22**(1):22–6.
2. Ponseti IV. Growth and development of the acetabulum in the normal child. Anatomic, histological, and roentgenographic studies. *J Bone Joint Surg*. 1978;**60A**(5):575–85.
3. Keats TE, Anderson MW. *Atlas of Normal Roentgen Variants That May Simulate Disease*, 8th edn. Philadelphia: Mosby Elsevier; 2007.

4. Gotz W, Funke M, Fischer G, Grabbe E, Herken R. Epiphysial ossification centres in iliosacral joints: anatomy and computed tomography. *Surg Radiol Anat.* 1993;**15**(2):131–7.

5. Banerjee S, Barry MJ, Paterson JM. Paediatric pelvic fractures: 10 years experience in a trauma centre. *Injury.* 2009;**40**(4):410–13.

6. Rossi F, Dragoni S. Acute avulsion fractures of the pelvis in adolescent competitive athletes: prevalence, location and sports distribution of 203 cases collected. *Skeletal Radiol.* 2001;**30**(3):127–31.

7. Sanders TG, Zlatkin MB. Avulsion injuries of the pelvis. *Semin Musculoskelet Radiol.* 2008;**12**(1):42–53.

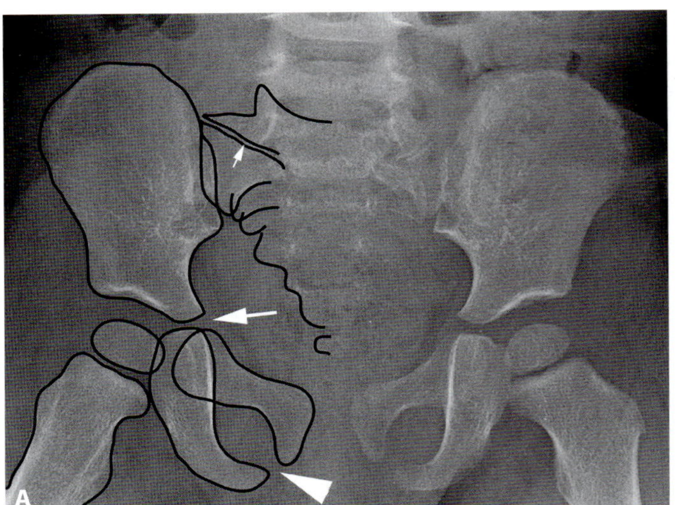

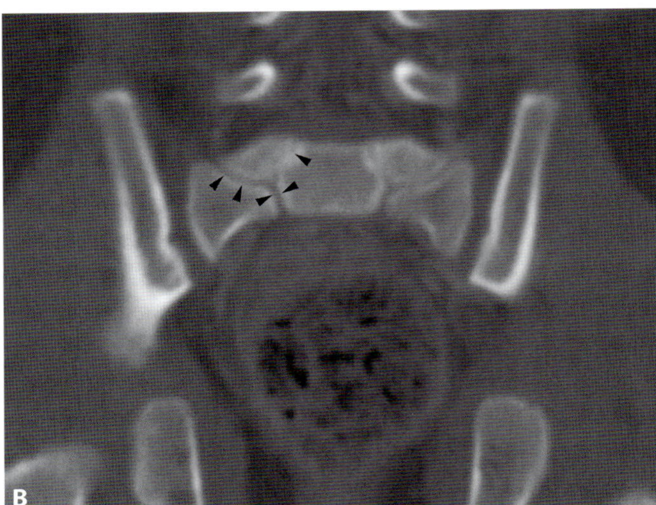

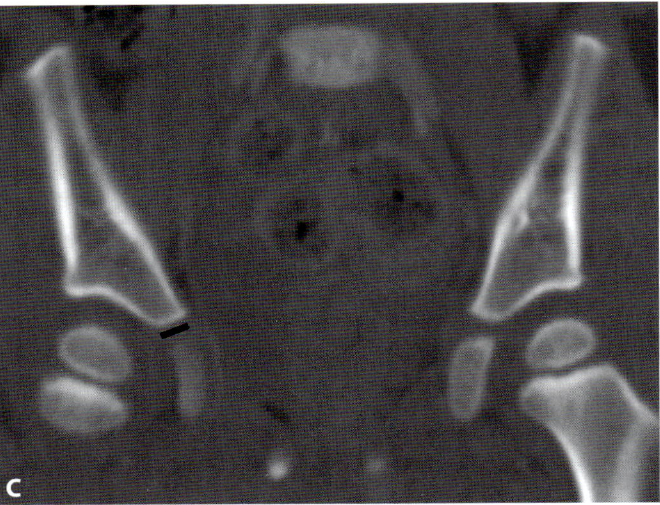

Figure 97.1 A. Anterior-posterior (AP) radiograph of the pelvis in a 21-month-old girl after a 15ft fall shows the primary ossification centers of the right hemipelvis and proximal femur, outlined in black. The normal triradiate cartilage (large arrow), ischiopubic synchondrosis (arrowhead), and sacral synchondroses (small arrow) are shown. Width of the sacroiliac joints is normal and symmetric, though they appear wider than in adults. **B.** Coronal reformation of a non-contrast CT through the sacrum shows multiple ossification centers of the alae and body with normal, symmetric patent physes (arrowheads). **C.** Coronal reformation through the acetabulum shows normal, symmetric triradiate cartilages (black line).

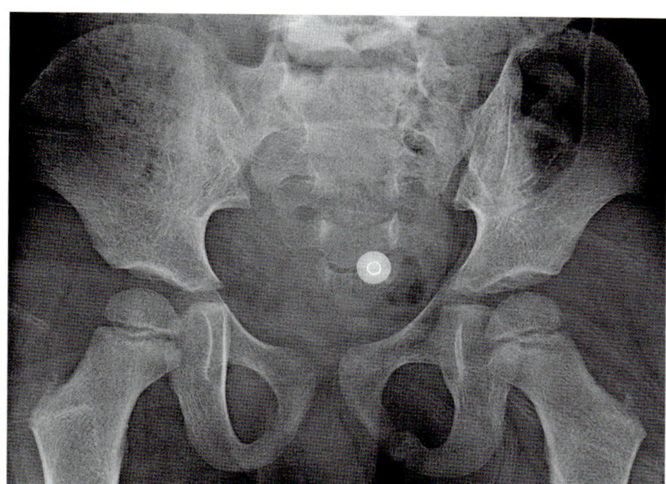

Figure 97.2 AP radiograph of the pelvis in a 4-year-old boy with known subdural hematoma from a fall shows mild rightward rotation of the pelvis, reflected by rounder appearance of the left obturator ring and flared appearance of the right iliac crest. The sacroiliac joints and ischiopubic synchondroses appear asymmetric due to rotation. There is no acute fracture or dislocation.

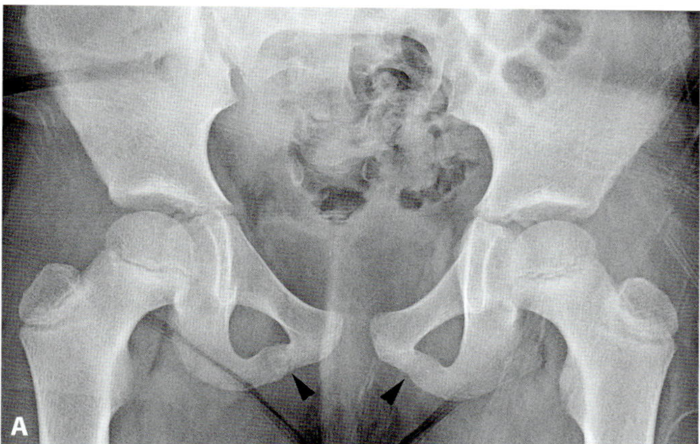

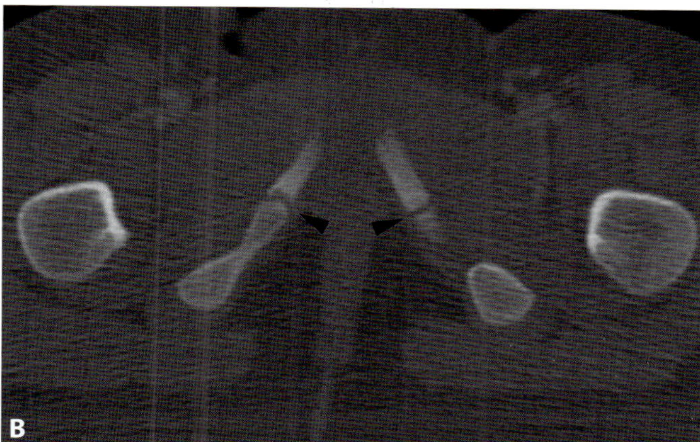

Figure 97.3 **A.** AP radiograph of the pelvis in a 6-year-old girl after a motor vehicle collision shows irregularity of the bilateral ischiopubic rami (black arrowheads) that could be confused with fractures. The triradiate cartilage is open but thinning. **B.** Axial CT of the inferior pelvis in bone windows confirms that these are normal ischiopubic synchondroses (black arrowheads) without acute traumatic abnormality.

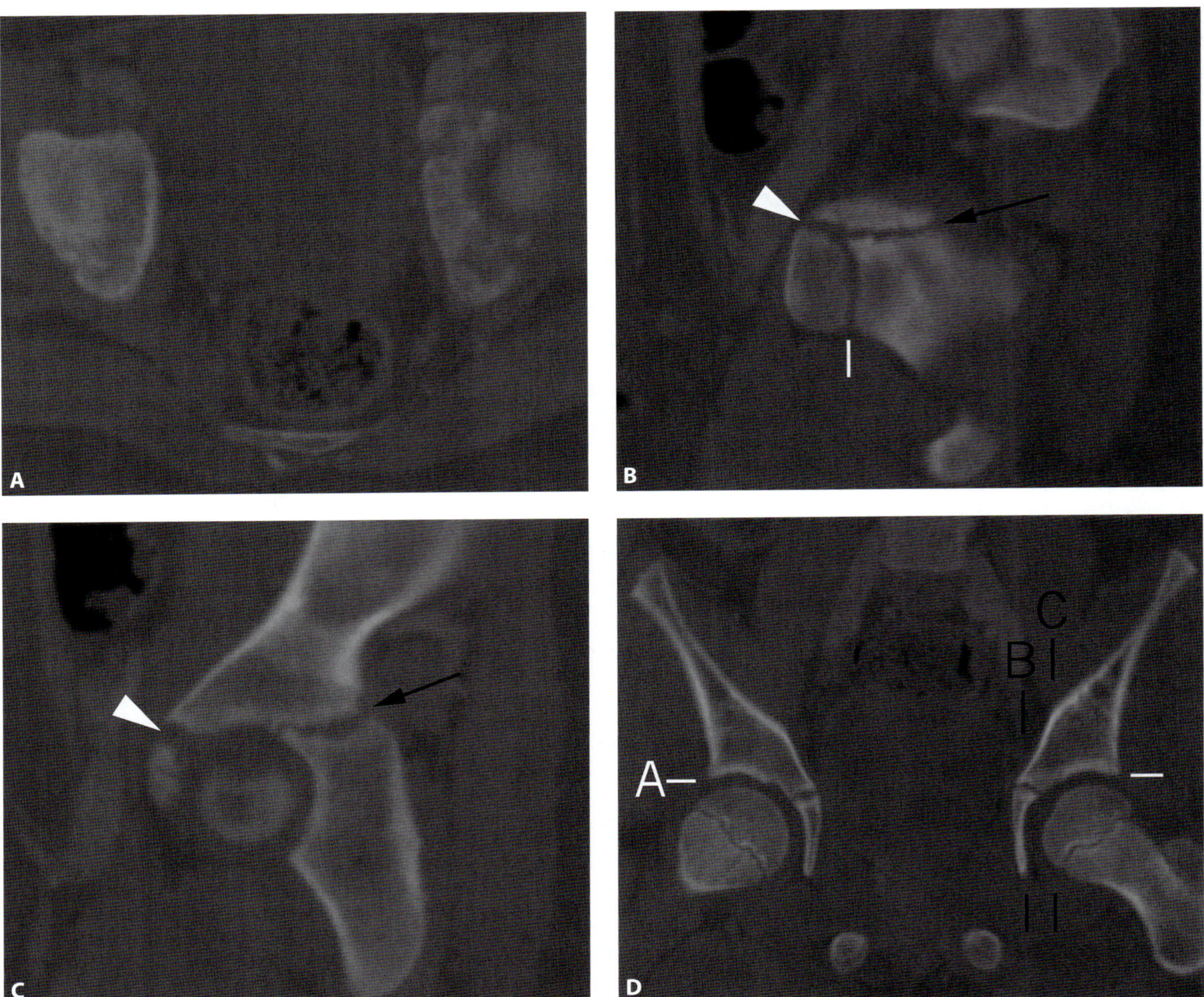

Figure 97.4 A. Axial non-contrast CT of the pelvis in bone windows in an 8-year-old boy following an all-terrain vehicle accident shows the normal fragmented appearance of the left acetabular roof. A similar appearance was also noted on the right on a different slice (not shown). The possibility of acetabular fracture was raised. Sagittal reformations through the medial (**B**) and lateral (**C**) aspects of the left acetabulum show normal appearance of the ilioischial (arrow), iliopubic (arrowhead), and ischiopubic (white line) physes at the hip. **D.** Coronal reformation of the pelvis shows the levels at which figures A, B, and C were obtained.

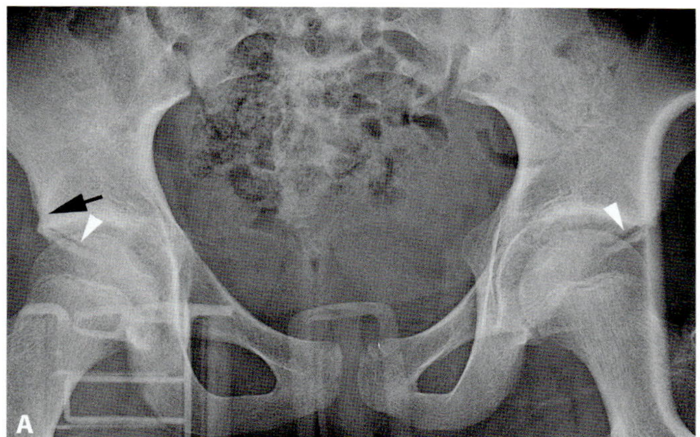

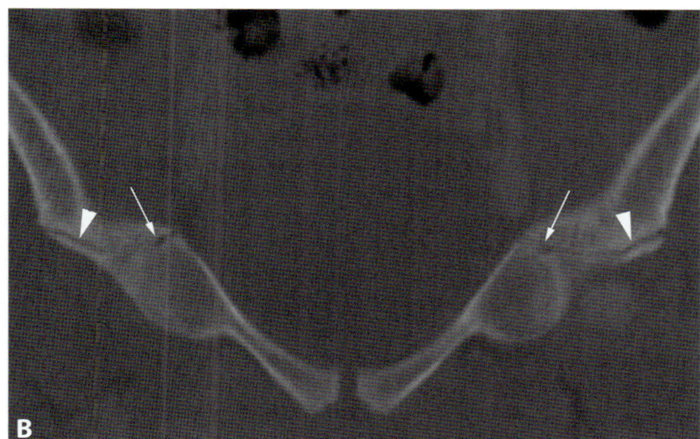

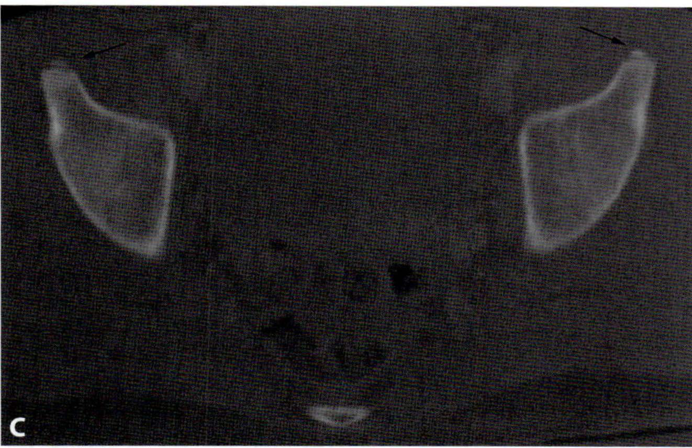

Figure 97.5 **A.** AP pelvis radiograph from an 11-year-old girl involved in a motor vehicle collision shows symmetric linear lucencies through the superior acetabulum (white arrowheads), and a linear lucency through the right ilium (black arrow). The left lateral ilium is obscured by the trauma backboard. **B.** Coronal reformation CT shows the symmetric residual ilioischial physes (arrows) and the secondary ossification centers of the acetabulum (arrowheads). **C.** Axial CT image shows the normal and symmetric secondary ossification centers of the anterior inferior ischial spines (arrows), which gave rise to the lucency in the right ilium in (A).

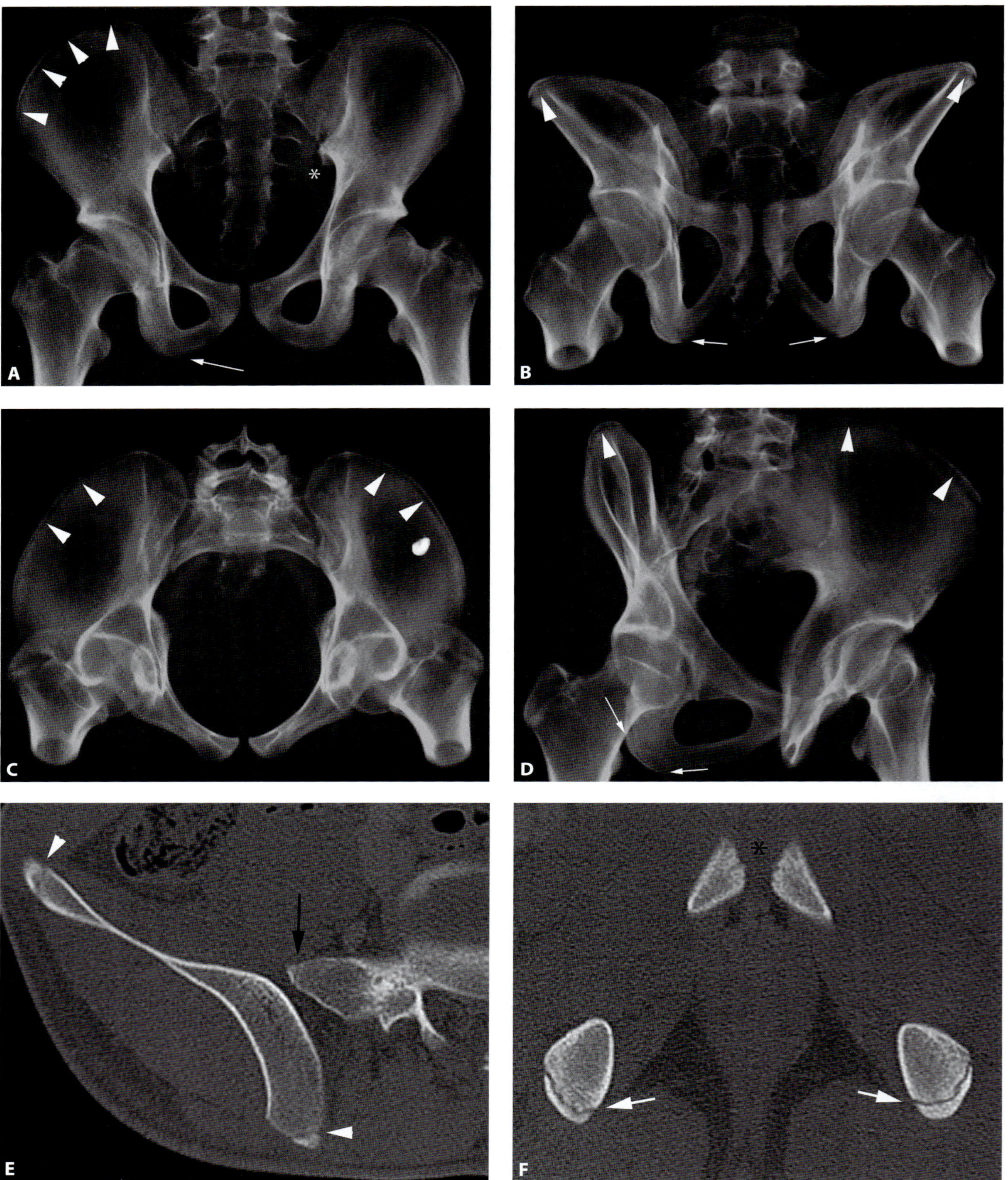

Figure 97.6 **A.** AP 3D volume rendered reconstruction CT of the bony pelvis in a 15-year-old boy following a football injury shows normal developmental anatomy. The right iliac apophysis (arrowheads) is fully ossified but not yet fused, and is symmetric to the left. The ischial apophysis is barely visible on the AP view (arrow). The left sacroiliac joint appears slightly irregular (asterisk), a normal variant in adolescents. Three-dimensional volume rendered views in simulated outlet (**B**), inlet (**C**), and Judet (**D**) projections shows the iliac and ischial apophyses (arrowheads and arrows, respectively) to varying degrees. **E.** Axial image in bone window through the iliac crest shows the anterior and posterior aspects of the iliac apophysis (arrowheads), as well as the secondary ossification center of the sacrum (arrow). **F.** Axial image through the ischial tuberosities shows the unfused ischial tuberosities (arrows). Irregularity of the pubic symphysis (asterisk) is also within normal limits.

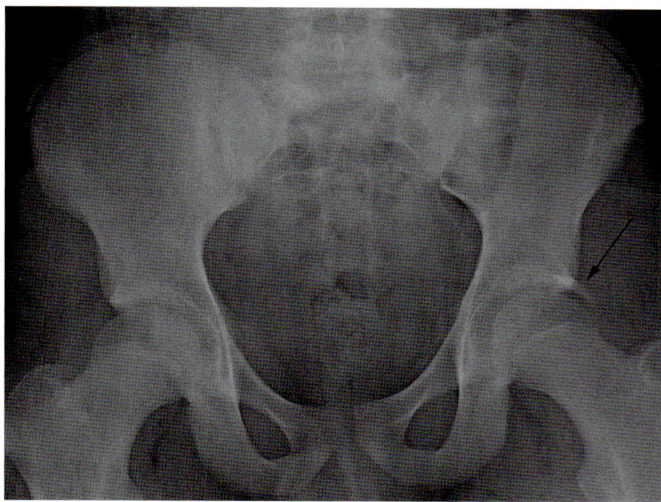

Figure 97.7 AP pelvis radiograph from a 14-year-old boy following a motor vehicle collision shows a bony fragment along the lateral aspect of the left acetabulum (arrow). This should be differentiated from an os acetabuli in this patient with prior posterior hip dislocation and posterior acetabular wall fracture.

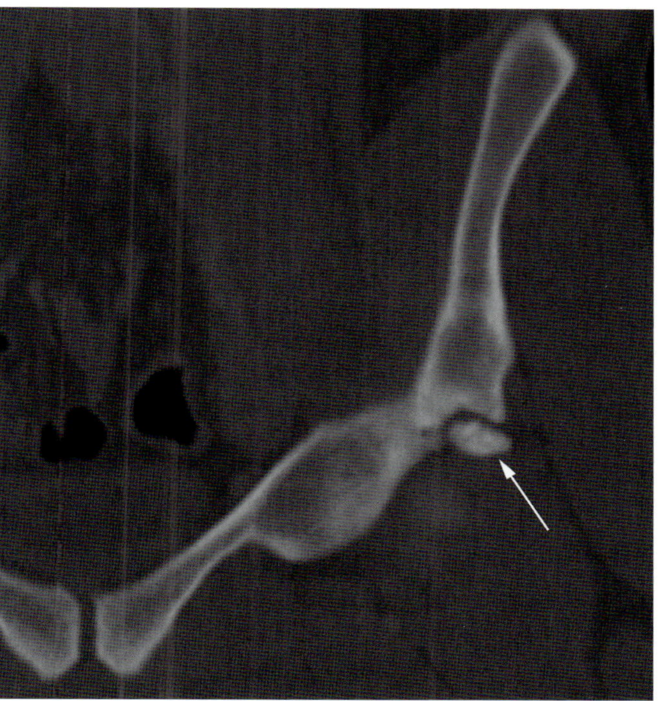

Figure 97.8 Coronal CT reformation in bone window in a 27-year-old man with abdominal pain shows an incidental os acetabuli (white arrow) of the left hip.

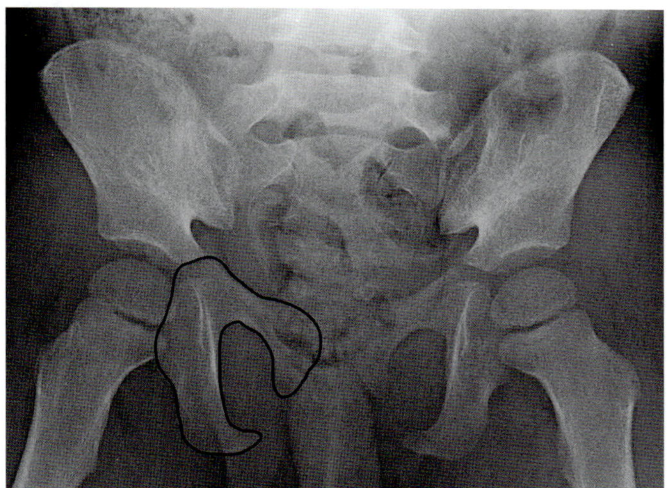

Figure 97.9 AP radiograph of the pelvis in a 2-year-old boy with pain after a fall shows wide but symmetric and normal ischiopubic synchondroses. The right ischium and pubis are traced in black, with only the medial synchondrosis well seen on AP radiograph.

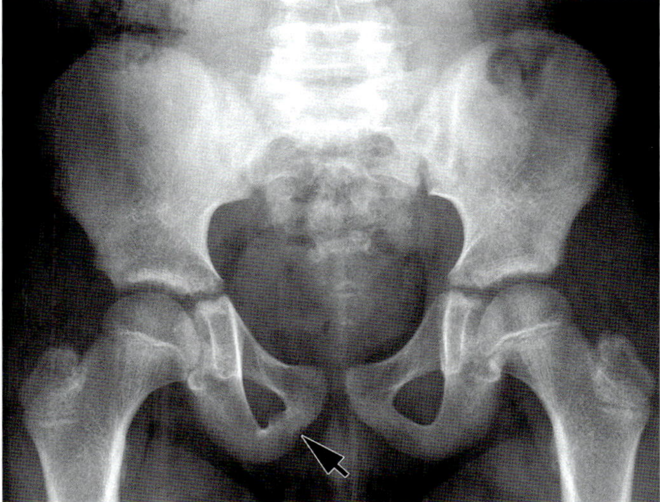

Figure 97.10 AP pelvis radiograph from a 9-year-old boy kicked by a horse shows asymmetry of the right ischiopubic physis (arrow) without acute fracture.

Hip pain in children

Matthew H. Nett

Imaging description

Hip pain in a young child can be due to toxic or transient synovitis, which is a benign entity that will resolve spontaneously with minimal to no treatment, or due to septic arthritis, which is a serious condition that may require emergency surgery to prevent rapid total joint destruction.

The initial evaluation of a limping child with hip pain typically involves radiography of the pelvis and hips. The best indicator of hip effusion is widening of the joint space, but obliteration of para-articular fat pads may also occur [1]. Joint space widening can appear as increased distance between the ossified proximal femoral metaphysis and the radiographic teardrop when compared to the non-affected contralateral hip joint (Figure 98.1A). With joint inflammation there is para-articular soft tissue edema which obliterates the adjacent fat pads including the obturator and gluteus minimus fat pads, with the obturator fat pad being the more important of the two [1].

Ultrasound is more sensitive than radiography for joint effusions. For evaluation of the hip, the patient lies supine with the hip in extension and slight abduction. The transducer is oriented anteriorly along the axis of the femoral neck. Hip effusions are seen as fluid displacing the joint capsule away from the echogenic cortex of the femoral neck (Figure 98.1B). In subtle cases, a difference in joint distention of greater than or equal to 2 mm between the symptomatic and asymptomatic hip has been reported as significant [2]. A normal ultrasound excludes septic arthritis.

Importance

Differentiation of septic arthritis and toxic synovitis can be challenging clinically, but is extremely important as septic arthritis can rapidly lead to joint destruction. Algorithms based on clinical variables have been developed to try to differentiate these two entities; however, these have not been shown to be reproducible at other institutions [3]. Therefore imaging plays a vital role in the evaluation of a limping child with hip pain.

Typical clinical scenario

Evaluation of a limping child is a common clinical scenario in the emergency department or pediatricians' office. While initially the differential diagnosis is extremely broad, it can be quickly narrowed by clinical variables such as age, location of pain, mechanism of injury, presence of fever, etc.

In a limping child under the age of 10 years with hip pain, two of the more common etiologies are septic arthritis and toxic synovitis. Both may present with fever, elevated white blood cell count, and elevated erythrocyte sedimentation rate. The typical age of presentation for toxic synovitis is 4 to 10 years. Septic arthritis is also common within this age group, although it may occur at any age. Transient synovitis may be preceded by an upper respiratory infection [4].

Patients will typically hold the hip in flexion, abduction, and external rotation, which maximizes capsule volume and minimizes strain on the sensitive joint capsule.

Differential diagnosis

The differential diagnosis for a painful hip in a child includes transient or toxic synovitis, septic arthritis, slipped capital femoral epiphysis, and osteonecrosis, either idiopathic (Legg–Calvé–Perthes) or due to corticosteroid use or sickle cell anemia. Toxic synovitis and septic arthritis can both present with a joint effusion, therefore joint aspiration is required. Ultrasound can help to guide aspiration. Legg–Calvé–Perthes typically presents in the 6- to 8-year-old child, whereas slipped capital femoral epiphysis presents at age 10–15 years. Both of these entities have characteristic radiographic findings.

Teaching point

Ultrasound is useful to identify an effusion in a limping child with a painful hip. Because effusions from septic arthritis and toxic synovitis have similar imaging appearances, aspiration must be performed.

REFERENCES

1. Hayden CK, Jr., Swischuk LE. Paraarticular soft-tissue changes in infections and trauma of the lower extremity in children. *AJR Am J Roentgenol.* 1980;**134**(2):307–11.
2. Fessell DP, Jacobson JA, Craig J, *et al.* Using sonography to reveal and aspirate joint effusions. *AJR Am J Roentgenol.* 2000; **174**(5):1353–62.
3. Luhmann SJ, Jones A, Schootman M, *et al.* Differentiation between septic arthritis and transient synovitis of the hip in children with clinical prediction algorithms. *J Bone Joint Surg Am.* 2004;**86**-A(5):956–62.
4. McCarthy JJ, Noonan KJ. Toxic synovitis. *Skeletal Radiol.* 2008; **37**(11):963–5.

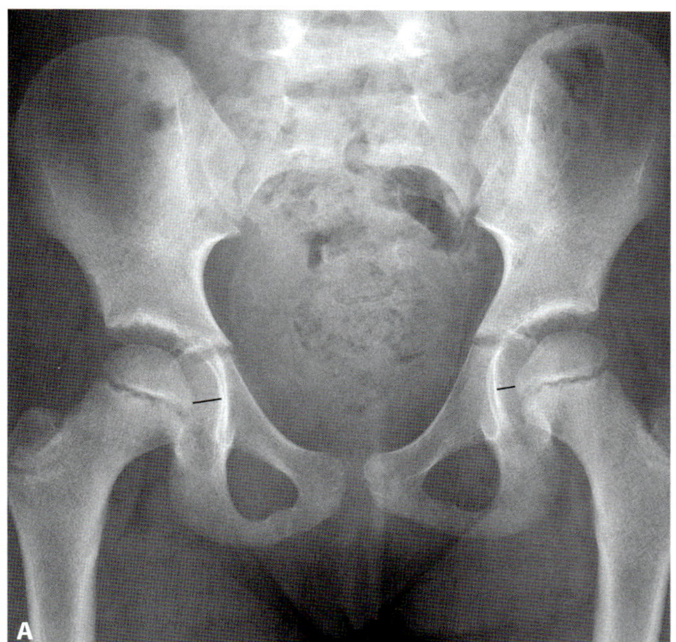

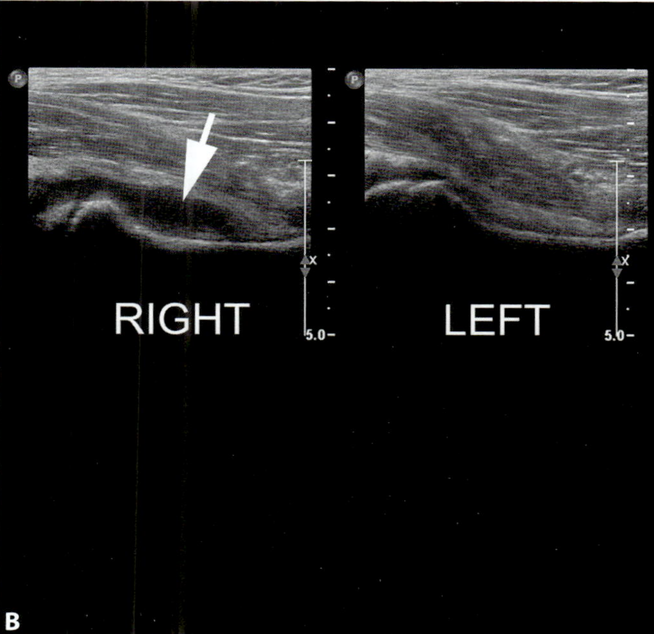

Figure 98.1 **A.** AP radiograph of the pelvis in a 7-year-old girl with hip pain demonstrates a subtle asymmetric increase in the distance between the right femoral head and upper part of the teardrop (shown as black lines). **B.** Ultrasound more distinctly shows the effusion with distension of the joint recess on the right (arrow) compared with the normal left side. This plane is useful for image-guided aspiration.

Common pitfalls in pediatric fractures: ones not to miss

Ramesh S. Iyer

Imaging description

There are numerous pitfalls in pediatric musculoskeletal trauma, in large part due to the progressive ossification of the maturing skeleton. Three fractures unique to pediatric imaging will be discussed here: supracondylar humeral, toddler's type 1, and the classic metaphyseal lesion.

Supracondylar fractures are the most common pediatric elbow fractures, comprising 50–70% of such injuries [1]. These fractures are shown to greatest advantage on the lateral view, usually showing posterior angulation of the distal fragment. These are covered in detail in Case 96.

Non-displaced or hairline spiral fracture of the tibial diaphysis is referred to as toddler's type 1 fracture, the most common subtype (Figure 99.1) [2]. Impaction or buckle fracture of the proximal tibial diaphysis, or toddler's type 2, is a recently described but less common variant [3]. Hairline fractures may be extremely difficult or impossible to identify on standard orthogonal views. Overlying soft tissue swelling may be variably present. If suspicious for this injury, one should perform an additional oblique projection of the lower leg to optimize detection. Toddler's fractures may manifest either as sharp oblique lucent or sclerotic lines, depending on both acuity and projection [2]. If clinical suspicion remains high and three-view tibia/fibula radiographs are negative, a scintigraphic bone scan may be performed (Figure 99.2). These exams feature a wide field of view, do not require anesthesia, and are less expensive than MRI. Bone scans may also identify tarsal fractures, particularly the cuboid and calcaneus, which may mimic tibial injuries in toddlers [4].

Metaphyseal corner fracture, or the so-called classic metaphyseal lesion, is highly specific for non-accidental trauma. Distal femur, proximal and distal tibia, and proximal humerus are the most common sites although any long bone metaphysis may be involved. Imaging appearance of these injuries is highly dependent on the projection. If the metaphyseal fragment is viewed tangential to the physeal plane, then the injury appears as a discrete triangular or "corner" fracture (Figures 99.3 and 99.4). If viewed oblique to this plane, however, the metaphyseal corner fracture may have a crescentic or "bucket-handle" appearance. Bucket-handle fractures may exhibit greater metaphyseal circumferential involvement than pure corner fractures [5–7]. Classic metaphyseal lesions are frequently bilateral and strikingly symmetric [6]. An overview of non-accidental skeletal trauma is provided in Case 101.

As with other fractures, identifying these injuries prompts orthopedic management. Non-displaced supracondylar fractures are treated with splint or cast immobilization, while displaced fractures typically require closed reduction and percutaneous pinning. Delay in diagnosis may uncommonly result in neurovascular compromise acutely, while remote supracondylar injuries may feature malalignment or growth arrest [1]. Treatment for toddler's type 1 fractures may include long-leg casting [8]. Identifying metaphyseal corner fractures may be critical in cases of child abuse. Radiographic evidence of injuries sustained may be grounds for removing such children from dangerous environments.

Typical clinical scenario

The vast majority (>95%) of supracondylar humeral fractures are hyperextension injuries, often the result of a fall on an outstretched hand. Patients present with elbow pain, swelling and limited range of motion. These injuries typically afflict children aged 3–10 years [1].

Toddler's type 1 fractures of the tibia typically present with refusal to bear weight after a twisting injury. As its name suggests, this injury occurs in young children learning to ambulate, often between 6 months and 2 years of age. Occasionally an ankle injury is suspected, prompting ankle radiographs, which can exclude the tibial fracture [2].

Classic metaphyseal lesions occur in infants under 1 year old. They are often clinically occult and are identified on skeletal surveys performed in cases of non-accidental trauma [5–10].

Differential diagnosis

The differential diagnosis for classic metaphyseal lesion is unfortunately broad, including various syndromes, dysplasias, metabolic conditions, and even normal anatomy [10]. Patients with rickets have metaphyseal irregularity, cupping, and fraying along with physeal widening. Metabolic bone disease is a well-documented cause of fractures in premature infants [9]. Both obstetric and iatrogenic injury (e.g., from vigorous physical therapy) may cause metaphyseal fractures identical to those from abuse [6]. Underlying bone disorders including osteogenesis imperfecta and various dysplasias may render the metaphysis more susceptible to fracture, or cause metaphyseal dysgenesias that can simulate a fracture. The normal metaphysis in the young child is fraught with radiologic fracture pitfalls, including irregularity, slight cortical stepoffs, fragmentation, peripheral beaking (Figure 99.5) and/or spurs [9, 10].

Teaching point

Displaced supracondylar fractures are usually posteriorly angulated, causing the anterior humeral line to pass anterior to the mid-capitellum. Non-displaced spiral tibial fractures may require oblique radiographs or even a bone scan for identification. Metaphyseal corner fractures in infants are highly specific for abuse.

REFERENCES

1. Alburger PD, Weidner PL, Randal RB. Supracondylar fractures of the humerus in children. *J Pediatr Orthop.* 1992;**12**:16–19.
2. Swischuk LE, John SD, Tschoepe EJ. Upper tibial hyperextension fractures in infants: another occult toddler's fracture. *Pediatr Radiol.* 1999;**29**:6–9.
3. Connolly LP, Connolly SA. Skeletal scintigraphy in the multimodality assessment of young children with acute skeletal symptoms. *Clin Nucl Med.* 2003;**28**(9):746–54.
4. Boal DK. Metaphyseal fractures. *Pediatr Radiol.* 2002;**32**(7):538–9.
5. Kleinman PK. Problems in the diagnosis of metaphyseal fractures. *Pediatr Radiol.* 2008;**38**(Suppl 3):S388–94.
6. Kleinman PK, Marks SC, Blackbourne B. The metaphyseal lesion in abused infants: a radiologic-histopathologic study. *AJR Am J Roentgenol.* 1986;**146**(5):895–905.
7. Halsey MF, Finzel KC, Carrion WV, *et al.* Toddler's fracture: presumptive diagnosis and treatment. *J Pediatr Orthop.* 2001; **21**(2):152–6.
8. Carroll DM, Doria AS, Paul BS. Clinical-radiological features of fractures of premature infants: a review. *J Perinat Med.* 2007; **35**:366–75.
9. Kleinman PK, Belanger PL, Karellas A, *et al.* Normal metaphyseal radiologic variants not to be confused with findings of infant abuse. *AJR Am J Roentgenol.* 1991;**156**(4):781–3.
10. Kleinman PK, Sarwar ZU, Newton AW, *et al.* Metaphyseal fragmentation with physiologic bowing: a finding not to be confused with the classic metaphyseal lesion. *AJR Am J Roentgenol.* 2009; **192**(5):1266–8.

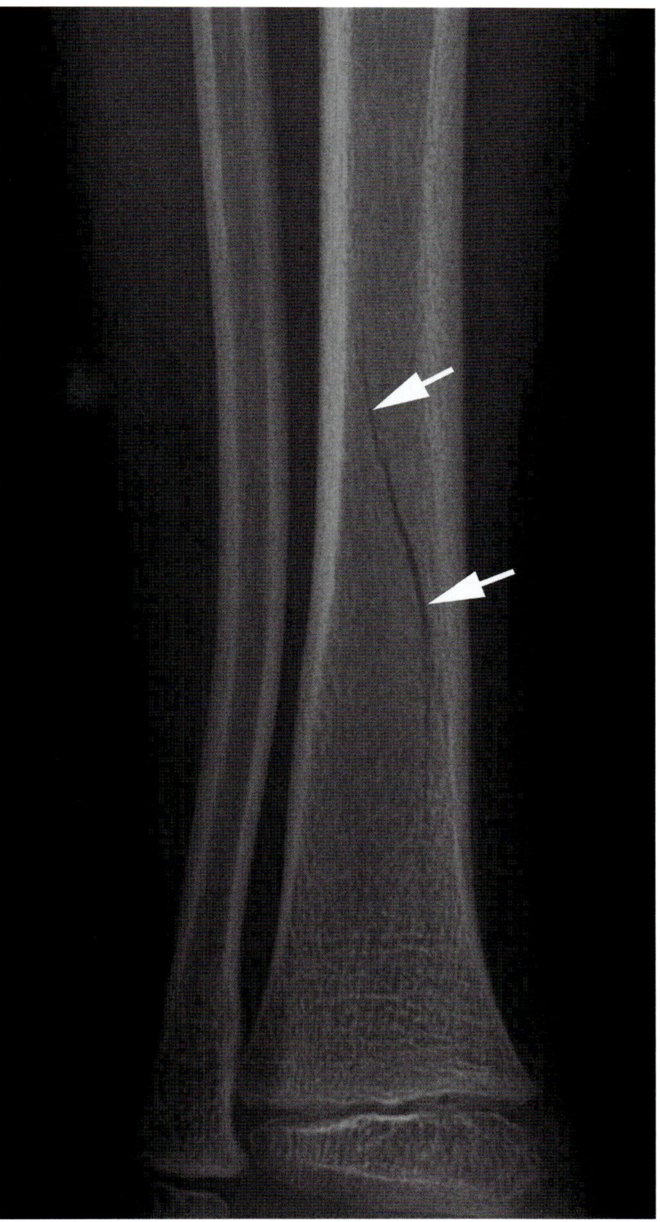

Figure 99.1 AP radiograph of the tibia in a 2-year-old boy with a toddler's type 1 fracture reveals a non-displaced spiral fracture of the mid- and distal tibial diaphysis (arrows).

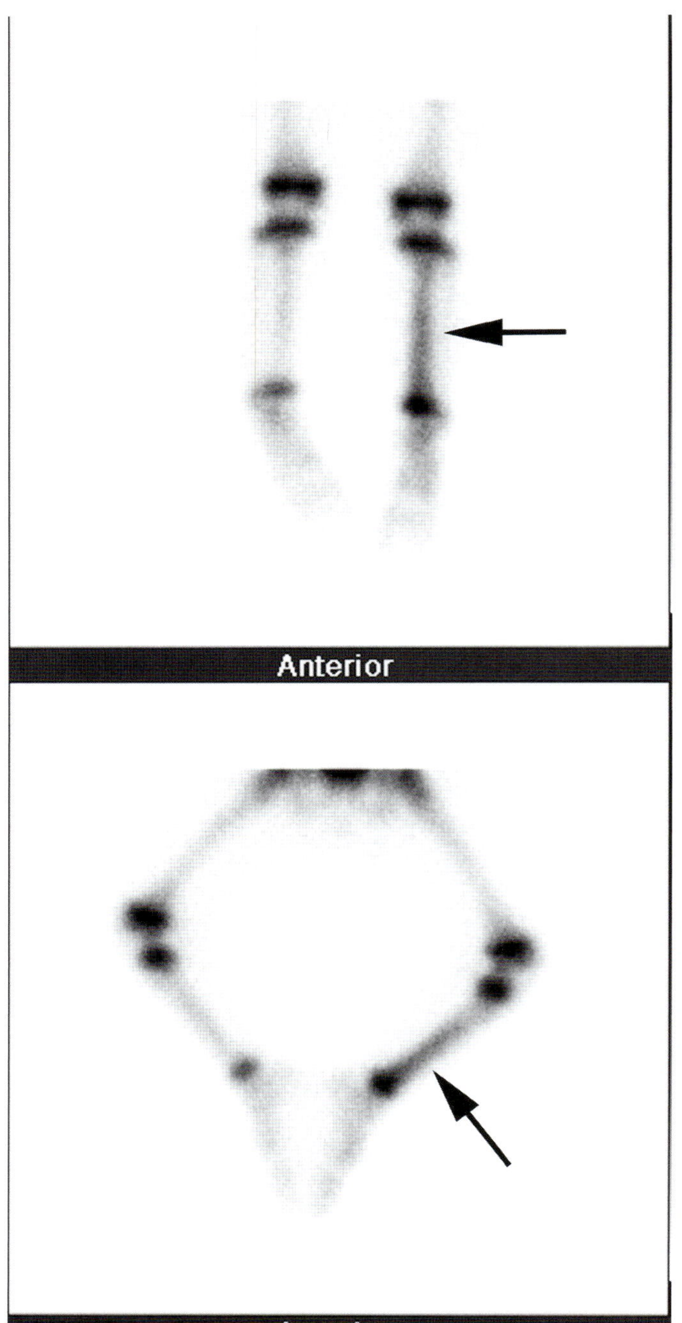

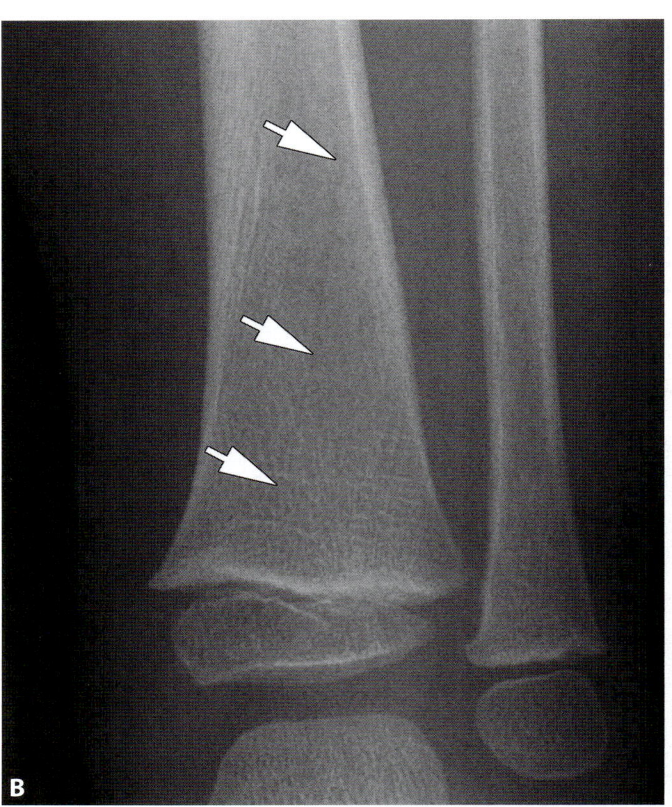

Figure 99.2 **A.** Frontal and lateral images of the lower legs from a bone scintigraphy study in a 2-year-old girl with a toddler's type 1 fracture. Note the mild diffuse radiotracer uptake within the mid- and distal left tibial diaphysis (arrows). **B.** AP radiograph of the left ankle in the same patient shows the subtle non-displaced tibial fracture line (arrows). Radiograph was initially read as normal, which prompted the bone scan.

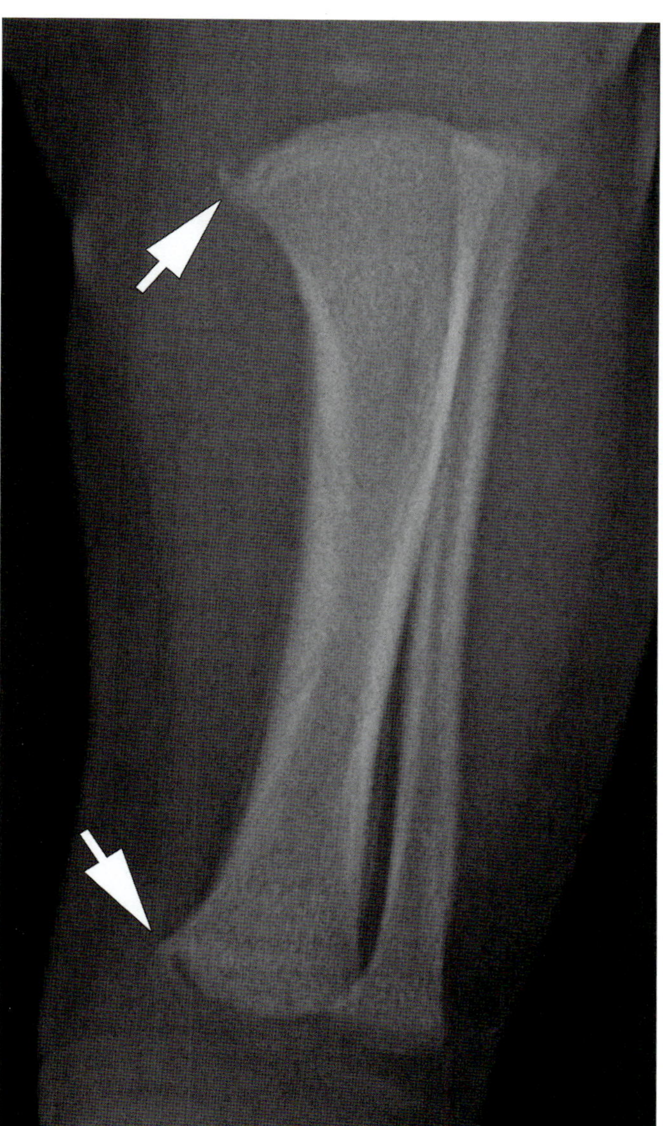

Figure 99.3 Frontal radiograph of the left tibia in a 4-week-old boy with classic metaphyseal lesions of the proximal and distal tibia (arrows). Note the crescentic or "bucket-handle" configuration of the fragments.

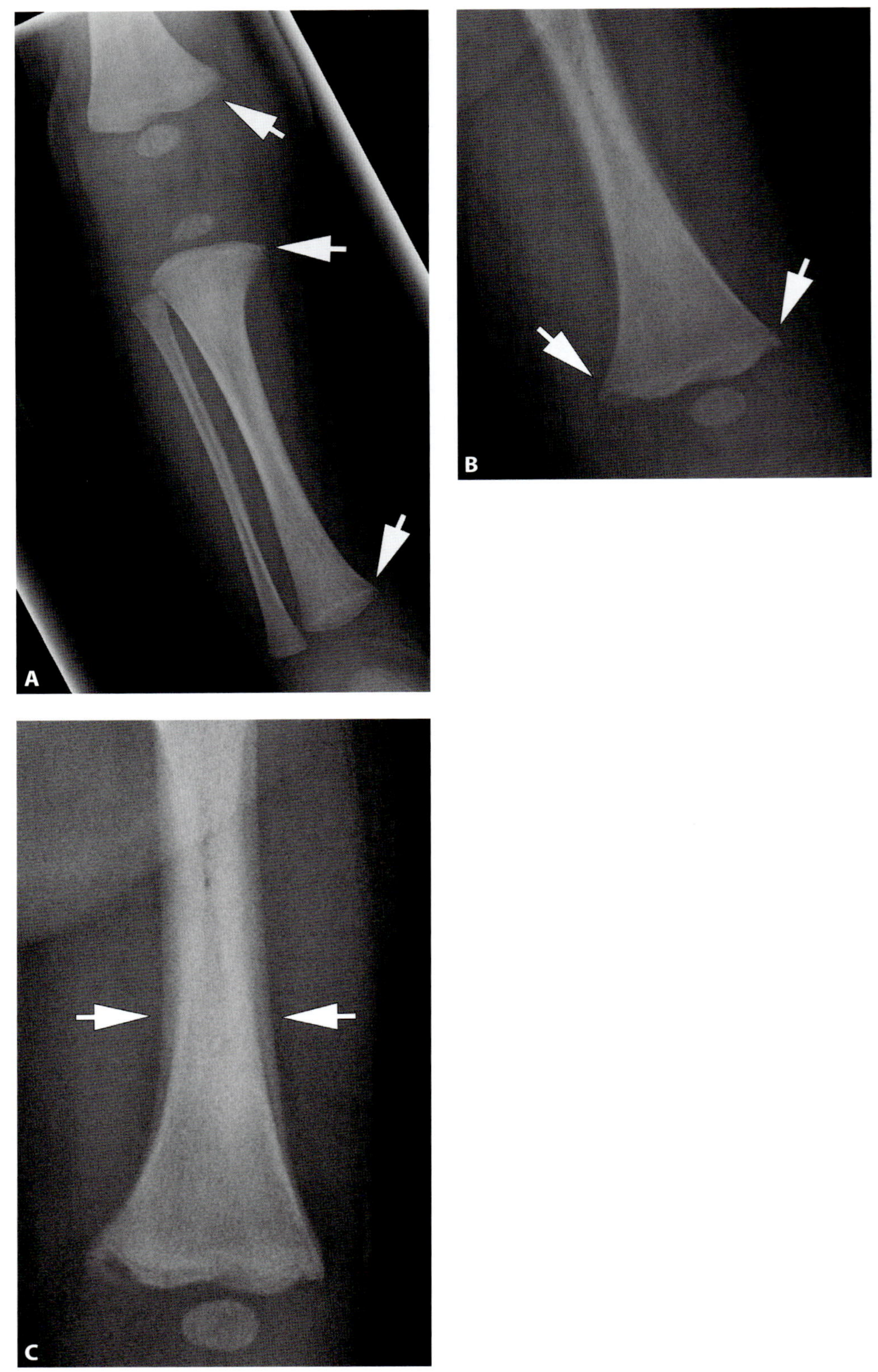

Figure 99.4 A. Frontal radiograph of the right leg in a 5-week-old boy with classic metaphyseal lesions of the distal femur, and proximal and distal tibia (arrows). The proximal tibial lesion has a triangular fragment, representative of a "corner" fracture, while the other fracture fragments have a "bucket-handle" configuration. The same mechanism causes both of these fractures, with the bucket-handle lesion involving a greater metaphyseal circumference. **B.** Frontal radiograph of the left distal femur in the same patient reveals a classic metaphyseal lesion involving both medial and lateral aspects. Frequently these injuries are bilateral and symmetric. **C.** Frontal radiograph of the left femur in the same patient performed one week later. Note the healing distal metaphyseal fracture and the interval periosteal elevation along the femoral diaphysis (arrows).

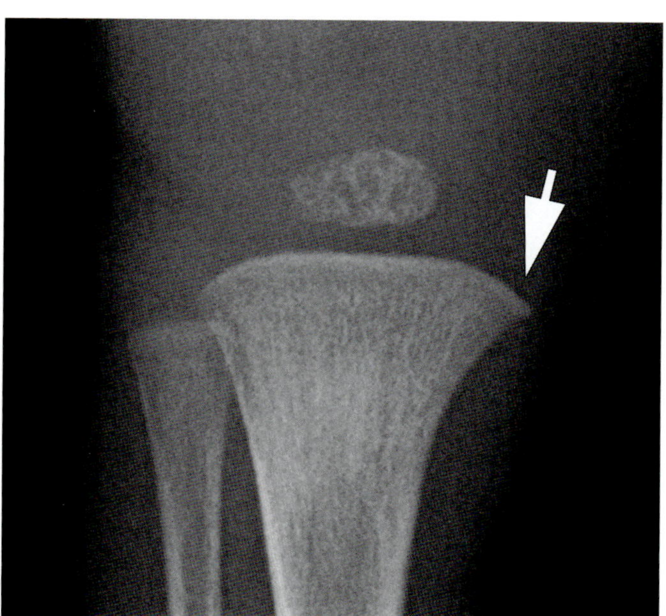

Figure 99.5 Normal frontal radiograph of the proximal tibia in a 4-week-old baby shows beaking of the proximal tibial metaphysis (arrow), which can be mistaken for injury.

CASE 100

Non-accidental trauma: neuroimaging

Ramesh S. Iyer

Imaging description

Head trauma is the leading cause of child abuse fatality and a major factor in long-term morbidity [1, 2]. It is important to image the brain in all suspected cases of inflicted trauma. This is typically performed with a CT head without contrast, for its ability to detect intracranial hemorrhage, cerebral edema, and skull fractures. MRI is occasionally performed to better characterize extra-axial fluid collections, identify parenchymal infarcts or ischemia in the setting of suffocation, and recognize subtle cases of shear injury [3, 4]. Cranial ultrasound does not play a significant role in the initial neuroimaging evaluation of nonaccidental trauma.

Subdural hemorrhage is the most common intracranial manifestation of inflicted trauma [1]. As in adults, acute subdural blood is hyperattenuating on a non-contrast CT, the result of shear forces tearing bridging cortical veins that drain into the dural venous sinuses. Subdural hemorrhage is nonspecific for traumatic injury. Close correlation with available clinical history is warranted when differentiating accidental versus non-accidental trauma. Nonaccidental injury produces hemorrhage that is more frequently bilateral and distant from the site of calvarial fracture, due to shear forces rather than direct coup injury [5]. Acute subdural hemorrhage must also be distinguished from the normal falx cerebri and tentorium cerebelli, which are prominent in infants. The latter are symmetrically dense and thickened on non-contrast CT, while hemorrhage thickens these dural reflections in an asymmetric or nodular fashion [6, 7].

Subdural hemorrhage of varying ages is also suggestive but not pathognomonic of inflicted trauma (Figure 100.1). Subacute and chronic subdural blood may cause diagnostic pitfalls on non-contrast CT. Subacute blood may be isodense to brain parenchyma, rendering subdural hematomas difficult to recognize when they are bilateral and symmetric. MRI may be useful in this setting; subacute hemorrhage, in contradistinction to normal cerebrospinal fluid (CSF), demonstrates intrinsic high T1 and FLAIR signal. Bilateral chronic subdural hemorrhage is isodense to CSF and can be mistaken for large subarachnoid spaces, a benign entity that is common in infants. Chronic subdural hemorrhage features absence or paucity of vessels and local mass effect upon the subjacent cortex; expanded subarachnoid spaces exhibit traversing vessels and lack mass effect [1].

Diffuse cerebral edema may result from severe shear injury, such as in shaken infants, or from asphyxiation. There is bilateral symmetric effacement of ventricles, sulci, and basal cisterns. On CT, the cerebrum is homogeneously hypodense with loss of gray–white matter differentiation. The cerebellum is typically spared, making it relatively hyperdense compared to the swollen cerebrum – the "cerebellar reversal" sign (Figure 100.2) [8].

Skull fractures occur in up to one third of cases of nonaccidental trauma in young children [9, 10]. Fracture characteristics that are suggestive but not diagnostic of inflicted trauma include bilateral involvement and diastasis (Figure 100.3) [1, 10].

Importance

Timely diagnosis of nonaccidental head trauma is essential. Many sustained injuries, including subdural hematomas, may require urgent or emergent neurosurgical intervention. Gliosis and/or brain atrophy may result if left untreated. Imaging evidence of inflicted trauma facilitates social service intervention and removal of the child from a dangerous environment.

Typical clinical scenario

Battered children with head trauma may have a variety of clinical presentations. These include altered mental status, lethargy, irritability, seizures, bruising, facial or scalp swelling, found while down, apnea, or death [11].

Differential diagnosis

Birth-related trauma may appear identical to nonaccidental trauma in the neonatal period. Extra-axial hemorrhage and skull fractures may be seen. Hemorrhage related to parturition typically evolves by 6 weeks, while fractures typically heal by 6 months of age [1, 12]. Injuries of varying ages are not seen in birth-related trauma, and if these are present, child abuse should be strongly considered.

Glutaric aciduria type 1 (GA1) is a rare autosomal recessive metabolic disorder that can mimic nonaccidental head trauma (Figure 100.4). Infants with this condition may present with macrocephaly, seizures, motor and developmental delay. Subdural fluid collections, often of varying age and hemorrhagic, are commonly identified along with brain atrophy and even retinal hemorrhage. Diagnosis of GA1 is achieved through metabolic and genetic screening [13].

Teaching point

Non-accidental head trauma is frequently indistinguishable from accidental or birth-related trauma by imaging. Subdural hemorrhage is the most common manifestation of inflicted head trauma and is suggestive of non-accidental trauma if it is of varying ages, bilateral, and distant from calvarial fractures. Bilateral and diastatic fractures are also suggestive but not diagnostic of non-accidental trauma.

REFERENCES

1. Fernando S, Obaldo RE, Walsh IR, *et al.* Neuroimaging of nonaccidental head trauma: pitfalls and controversies. *Pediatr Radiol.* 2008;**38**:827–38.
2. Keenan HT, Runyan DK, Marshall SW, *et al.* A population-based study of inflicted traumatic brain injury in young children. *JAMA.* 2003;**290**:621–6.
3. Soul JS, Robertson RL, Tzika AA, *et al.* Time course of changes in diffusion-weighted magnetic resonance imaging in a case of neonatal encephalopathy with defined onset and duration of hypoxic-ischemic insult. *Pediatrics.* 2001;**108**:1211–14.
4. Suh DY, Davis PC, Hopkins KL, *et al.* Nonaccidental pediatric head injury: diffusion-weighted imaging findings. *Neurosurgery.* 2001;**49**:309–18.
5. Hymel KP, Rumack CM, Hay TC, *et al.* Comparison of intracranial computed tomographic (CT) findings in pediatric abusive and accidental head trauma. *Pediatr Radiol.* 1997;**27**:743–7.
6. Zimmerman RD, Yurberg E, Russell EJ, *et al.* Falx and interhemispheric fissure on axial CT: I. normal anatomy. *AJR Am J Roentgenol.* 1982;**138**:899–904.
7. Zimmerman RD, Russell EJ, Yurberg E, *et al.* Falx and interhemispheric fissure on axial CT: II. recognition and differentiation of interhemispheric subarachnoid and subdural hemorrhage. *AJNR Am J Neuroradiol.* 1982;**3**:635–42.
8. Han BK, Towbin RB, De Courten-Myers G, *et al.* Reversal sign on CT: effect of anoxic/ischemic cerebral injury in children. *AJR Am J Roentgenol.* 1990;**154**:361–8.
9. King J, Diefendorf D, Apthorp J, *et al.* Analysis of 429 fractures in 189 battered children. *J Pediatr Orthop.* 1988;**8**:585–9.
10. Arnholz D, Hymel KP, Hay TC, *et al.* Bilateral pediatric skull fractures: accident or abuse? *J Trauma.* 1998;**45**:172–4.
11. Duhaime AC, Christian CW, Rorke LB, *et al.* Nonaccidental head injury in infants – the “shaken-baby syndrome”. *N Engl J Med.* 1998; **338**(25):1822–9.
12. Whitby EH, Griffiths PD, Rutter S, *et al.* Frequency and natural history of subdural haemorrhages in babies and relation to obstetric factors. *Lancet.* 2004;**363**:846–51.
13. Hartley LM, Khwaja OS, Verity CM. Glutaric aciduria type 1 and nonaccidental head injury. *Pediatrics.* 2001;**107**(1):174–5.

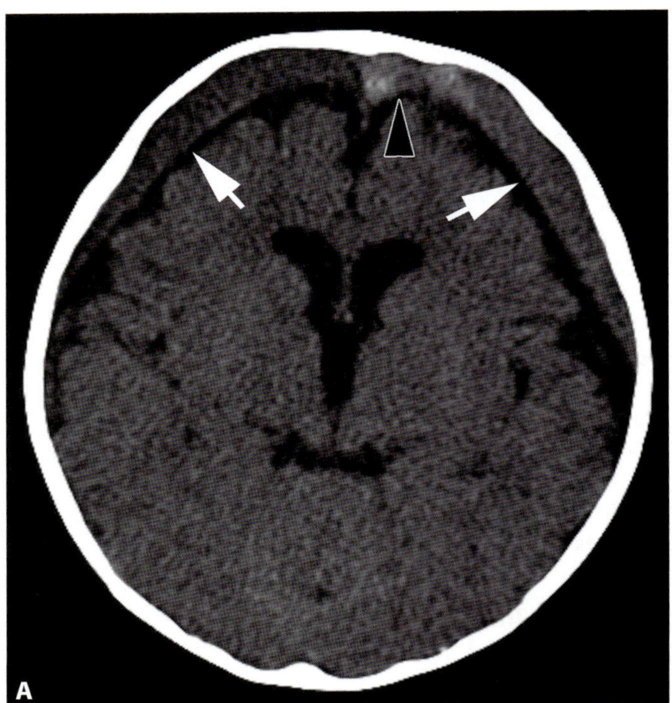

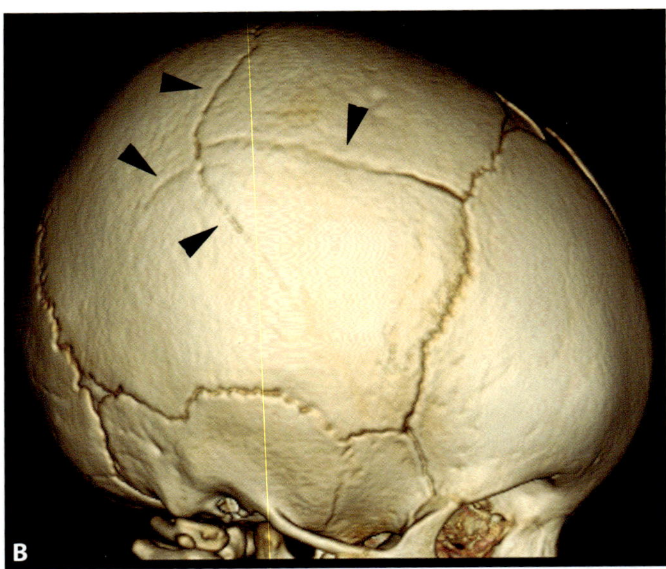

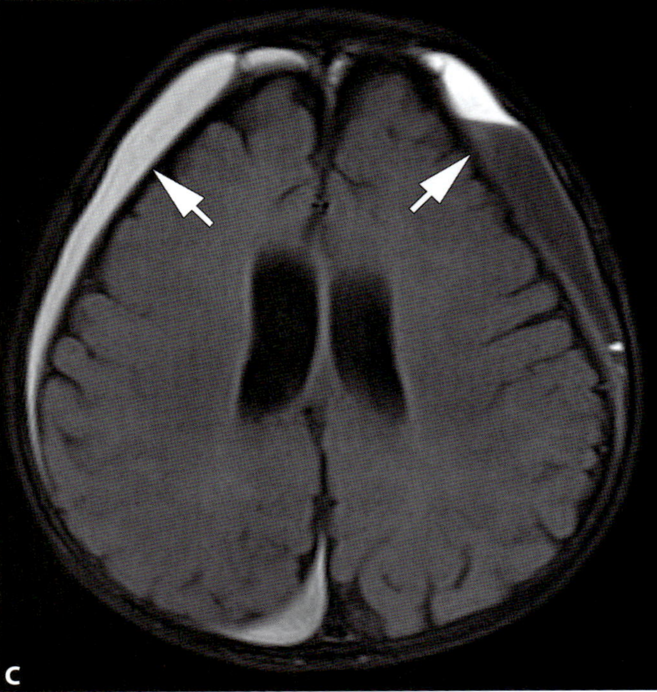

Figure 100.1 **A.** Axial non-contrast CT from a 7-month-old girl with non-accidental head trauma demonstrates bilateral subdural hematomas of varying age (arrows). The left collection is slightly more dense than the right with an acute hemorrhagic component (arrowhead). **B.** Three-dimensional volume rendered view of the calvarium from the same CT shows a stellate non-depressed right parietal fracture (arrowheads). **C.** Axial FLAIR magnetic resonance image exhibits bilateral subdural hemorrhage (arrows). Note the fluid-fluid level in the left collection showing blood products of different ages.

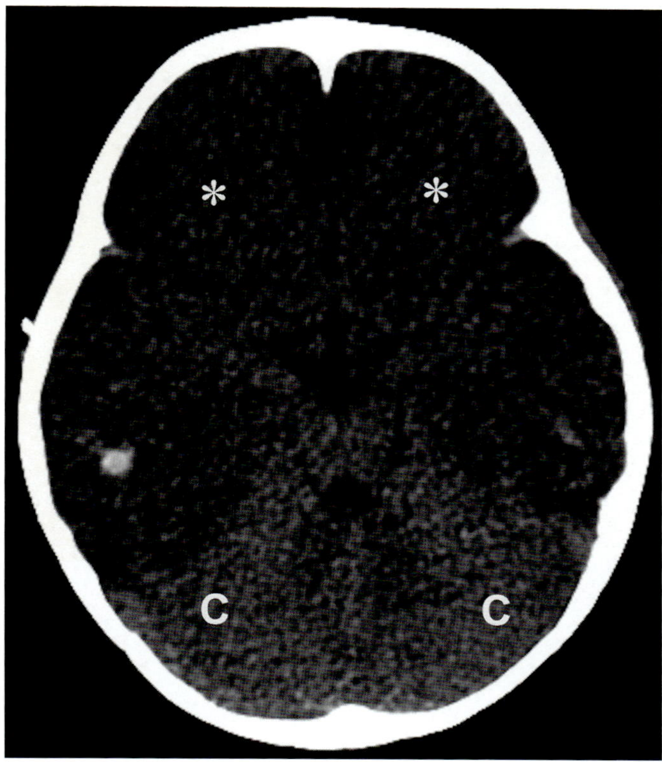

Figure 100.2 Axial non-contrast CT from an 11-month-old girl with inflicted asphyxiation shows diffuse cerebral edema. Bilateral supratentorial parenchyma (asterisks) is diffusely hypodense with loss of gray–white differentiation. The cerebellum ("C") is spared and relatively hyperdense compared to the cerebrum, reflecting the "cerebellar reversal" sign.

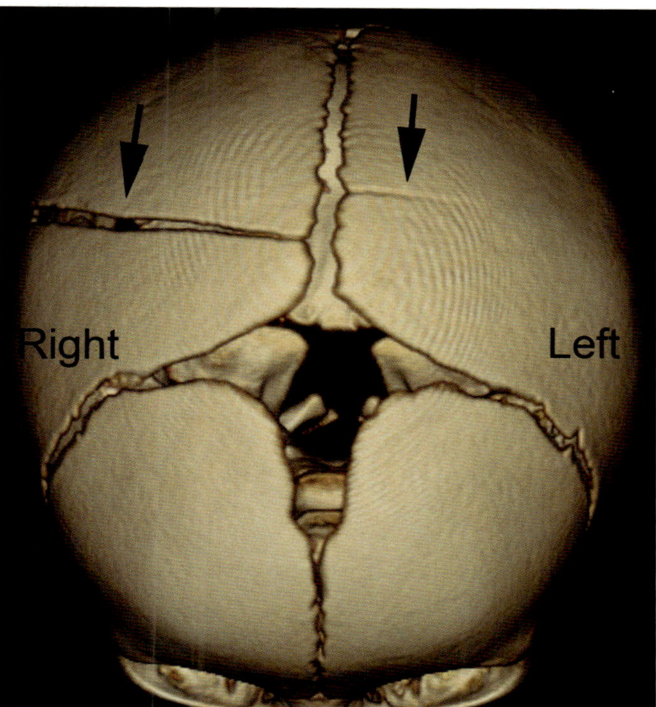

Figure 100.3 Vertex view from a 3D volume rendered CT reconstruction in a 3-month-old boy with nonaccidental head trauma shows bi ateral parietal fractures (arrows). Note fracture diastasis on the right. Bilateral and diastatic fractures are suggestive but not diagnostic of inflicted trauma.

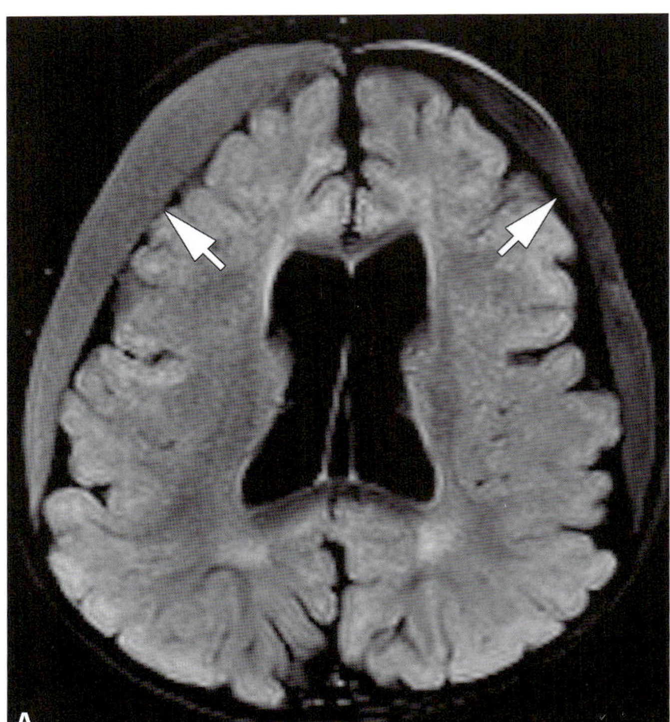

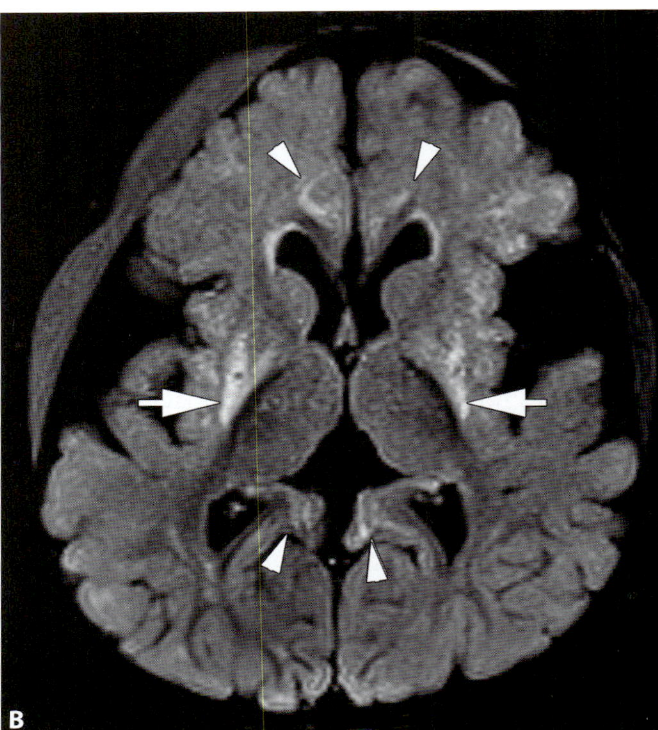

Figure 100.4 **A.** Axial FLAIR MRI of a 1-year-old boy with glutaric aciduria type 1 (GA1). There are bilateral subdural fluid collections of varying ages (arrows), mimicking non-accidental trauma. **B.** There is also high FLAIR signal within bilateral basal ganglia (arrows) and periventricular white matter (arrowheads).

CASE 101 Non-accidental trauma: skeletal injuries

Ken F. Linnau

Imaging description

Skeletal injuries that have a high predictive value for non-accidental trauma (NAT) include metaphyseal corner fractures, posterior rib fractures, scapula fractures, and spinous process fractures. These bones are usually difficult to break. Humeral and femoral shaft fractures, particularly distal shaft fractures, are the most common long bone fractures in NAT and should be treated with suspicion in children less than three years [1–3]. Moreover, the presence of multiple fractures of different ages is highly suspicious for NAT.

Metaphyseal corner fractures of NAT, also referred to as "metaphyseal lesions" are avulsion fractures of an arcuate metaphyseal fragment passing through the primary spongiosa overlying the lucent epiphyseal cartilage. This results in irregularity and fragmentation of the metaphysis (Figure 101.1). When a classic metaphyseal lesion is suspected, two radiographic projections of the affected joint are required to avoid confusion with mild physiologic irregularity of the metaphysis or chronic stress such as in malignancy [4]. Metaphyseal fractures of child abuse are most commonly encountered around the knee or elbow. They are also discussed in Case 99.

Multiple fractures that are new and old in the same patient represent an injury pattern highly specific for child abuse, which can often be observed in the rib cage (Figure 101.2). Rib fractures due to abuse can occur anywhere in the ribs, but are most commonly detected posteromedially adjacent to the costovertebral junction and rib head [5]. Acute rib fractures in young children can easily be missed owing to the lack of periosteal new bone formation, which usually peaks 10–14 days after trauma. The routine addition of oblique views of the chest increases the diagnostic accuracy for acute rib fractures [6].

Other pediatric fractures that are highly specific for abuse include scapular, spinous process, and sternal fractures. Fractures in children that have moderate specificity for abuse include vertebral fractures, digital fractures, and bilateral synchronous fractures.

Other clues which should raise the suspicion of child abuse include occult or unexpected fractures, multiplicity of fractures, injuries that are out of proportion to the given history, and multisystem injuries.

Importance

Skeletal injuries are often the strongest indicators for child abuse and are detected in up to 55% of young children who have been abused [7]. Most of these children are younger than 3 years old and subsequently, clinical history as well as interpretation of symptoms is usually unreliable. Many abusive fractures are occult. For this reason the American College of Radiology recommends skeletal radiographic surveys for all cases of suspected abuse in children younger than 2 years and a more tailored approach for older children based on clinical circumstances [8]. Oblique views of the chest should be included in the skeletal survey to increase accuracy. At times, complementary bone scintigraphy and repeat skeletal survey are indicated to detect occult fractures [7]. At our institution, bone scintigraphy is not used routinely owing to the higher radiation dose and normal physiologic tracer uptake at the metaphysis in children, which may obscure classic metaphyseal lesions (Figure 101.3). "Babygrams," where large parts of the body are radiographed in one exposure have a low sensitivity for metaphyseal corner fractures and high radiation dose, and should not be used.

Although skeletal injuries are rarely life-threatening, detection and dating of skeletal injuries can provide investigators with valuable information which may help in identifying potential perpetrators in the setting of child abuse [9].

Typical clinical scenario

Unfortunately, no fracture can distinguish NAT from accidental injury [3]. Fractures are more common in children than adults, with a peak incidence of sport-related fractures in the 4–9 year age group [10]. By 16 years of age, approximately one-third of children will have had at least one fracture. Fractures unrelated to abuse occur most commonly in the upper limb.

Consequently, fractures caused by NAT represent a very small proportion of the childhood fractures. There is an inverse relation between the age of the child and rate of non-accidental fractures. Therefore, the risk of non-accidental injury is the highest in a child under 3 years with a rib or long bone fracture, including the humerus and femur [3].

When infants and toddlers present with a fracture in the absence of a confirmed cause, NAT should be considered [3].

Differential diagnosis

Accidental long bone fractures such as toddler's fracture or torus fracture usually affect the diaphysis. However, long bone fractures in non-ambulating young children/infants are suspicious.

Rib fractures in infants and young children can be due to osteogenesis imperfecta, birth trauma, or long-term ventilator therapy in prematurity.

Leukemia, round cell tumors, or metabolic bone disease (rickets, scurvy) may mimic child abuse.

Osteogenesis imperfecta results in brittle bones and can lead to multiple fractures of different ages. Usually osteoporosis is present in osteogenesis imperfecta.

The differential diagnosis for metaphyseal corner lesion is discussed in Case 99.

Teaching point

Careful inspection of skeletal survey radiographs should be performed while the patient is still in the emergency department imaging suite. If appendicular findings are equivocal for child abuse, additional projections of the bone or joint in question should be obtained.

Oblique views of the chest increase diagnostic accuracy for the detection of rib fractures.

If skeletal injuries are out of proportion to the history, the suspicion of child abuse should be raised. Communicate directly with the ordering physician.

REFERENCES

1. Shrader MW, Bernat NM, Segal LS. Suspected nonaccidental trauma and femoral shaft fractures in children. *Orthopedics*. 2011;**34**(5):360.
2. Dalton HJ, Slovis T, Heifer RE, *et al.* Undiagnosed abuse in children younger than 3 years with femoral fracture. *Am J Dis Child*. 1990; **144**(8):875–8.
3. Kemp AM, Dunstan F, Harrison S, *et al.* Patterns of skeletal fractures in child abuse: systematic review. *BMJ*. 2008;**337**:a1518.
4. Kleinman PK, Marks SC, Blackbourne B. The metaphyseal lesion in abused infants: a radiologic-histopathologic study. *AJR Am J Roentgenol*. 1986;**146**(5):895–905.
5. Kleinman PK, Marks SC, Jr., Nimkin K, Rayder SM, Kessler SC. Rib fractures in 31 abused infants: postmortem radiologic-histopathologic study. *Radiology*. 1996;**200**(3):807–10.
6. Ingram JD, Connell J, Hay TC, Strain JD, Mackenzie T. Oblique radiographs of the chest in nonaccidental trauma. *Emerg Radiol*. 2000; **7**(1):42–6.
7. Kemp AM, Butler A, Morris S, *et al.* Which radiological investigations should be performed to identify fractures in suspected child abuse? *Clin Radiol*. 2006;**61**(9):723–36.
8. ACR Appropriateness Criteria: Suspected Physical Abuse – Child American College of Radiology; 2009. Available from: http://www.acr.org/SecondaryMainMenuCategories/quality_safety/app_criteria.aspx (accessed April 26, 2012).
9. Section on Radiology; American Academy of Pediatrics. Diagnostic imaging of child abuse. *Pediatrics*. 2009;**123**(5):1430–5.
10. Falvey EC, Eustace J, Whelan B, *et al.* Sport and recreation-related injuries and fracture occurrence among emergency department attendees: implications for exercise prescription and injury prevention. *Emerg Med J*. 2009;**26**(8):590–5.

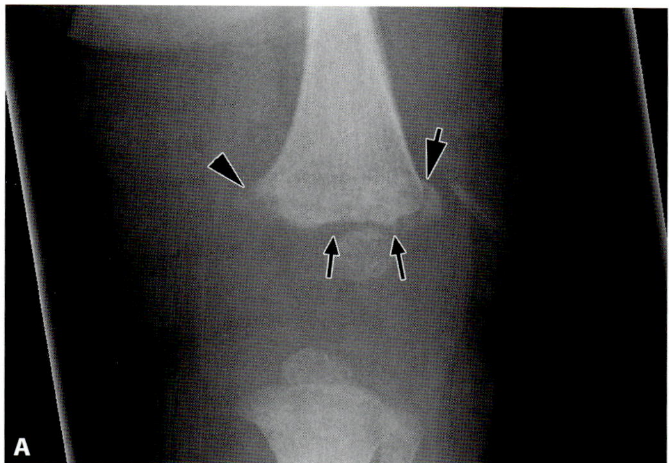

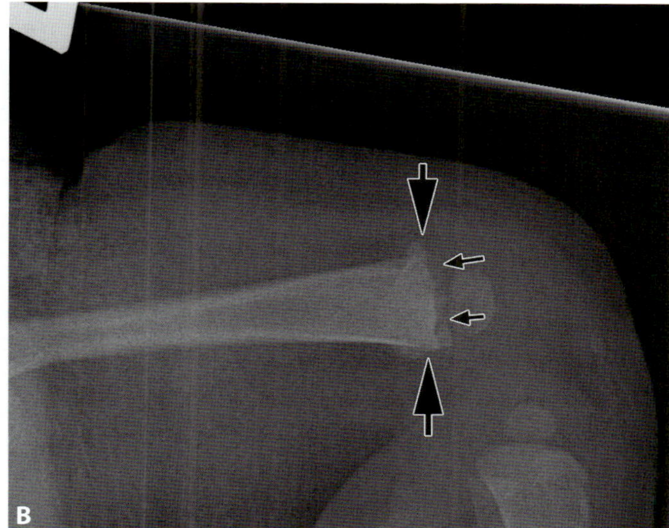

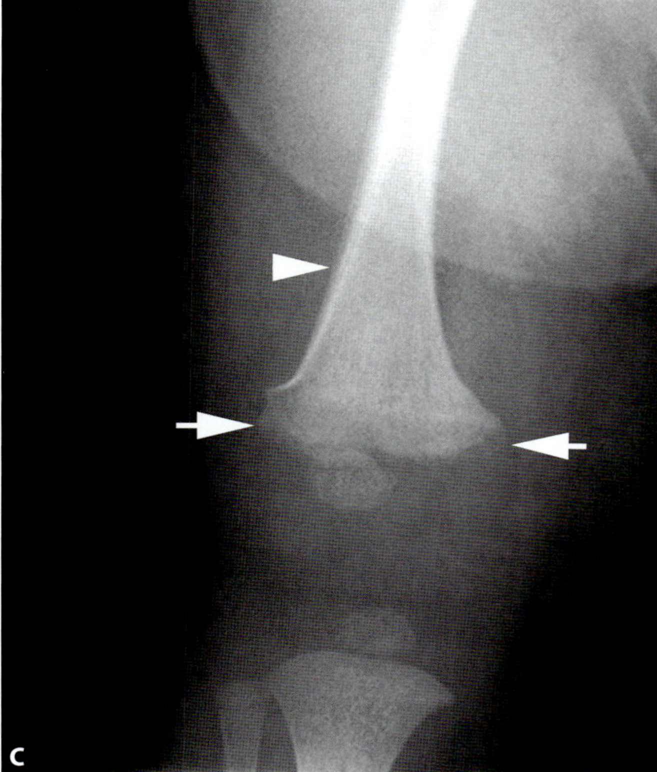

Figure 101.1 Frontal projection of the left femur (**A**) obtained as part of a skeletal survey for suspected child abuse in a 9-week-old boy shows a lucency through the lateral metaphysis of the distal femur (arrow) and irregularity of the medial metaphysis (arrowhead). There is a thin rim of calcification connecting the metaphyseal corners (small arrows), resulting in the "bucket-handle" appearance of this classic metaphyseal lesion, which is near pathognomonic for child abuse. At our institution, the imaging protocol for skeletal survey radiographs only includes frontal projections of the extremities, which are reviewed by a radiologist while the patient is still in the department. If abnormalities are detected, orthogonal projections of the joint are acquired. The subsequently obtained lateral view (**B**) of the left femur confirms the findings (black arrows). On the contralateral right anterior-posterior femur radiograph (**C**) a similar metaphyseal corner lesion (whitearrows) and mild periosteal reaction (white arrowhead) are detected.

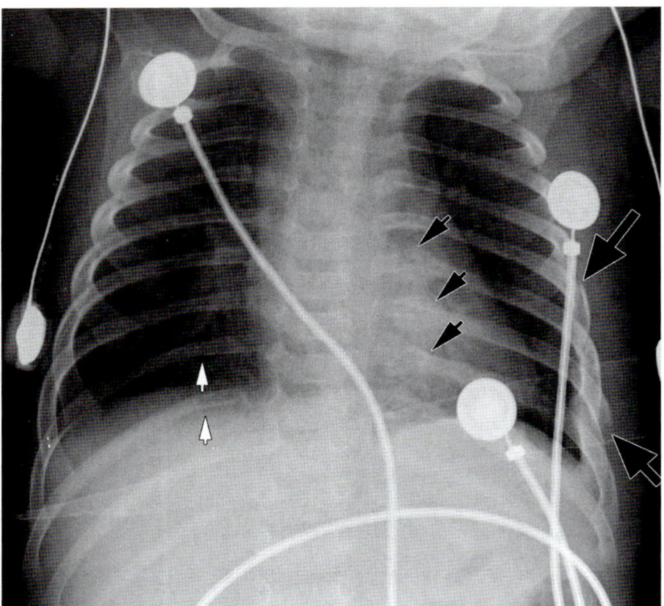

Figure 101.2 Frontal chest radiograph from a 4-month-old boy shows multiple bilateral rib fractures. There are acute fractures of the left sixth through eighth ribs (large black arrows) without periosteal reaction. Cortical disruption and mild angulation without periosteal reaction indicate subtle acute right posterior ninth and tenth rib fractures (small white arrows). Additional posteromedial left rib fractures with exuberant callus formation (small black arrows), which appear more bulbous, are indicative of older injury. The detection of multiple new and old rib fractures is highly specific for child abuse. Oblique chest radiographs increase diagnostic accuracy.

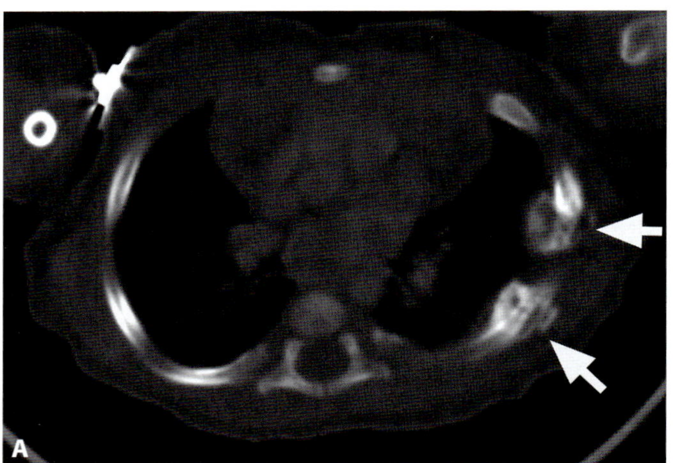

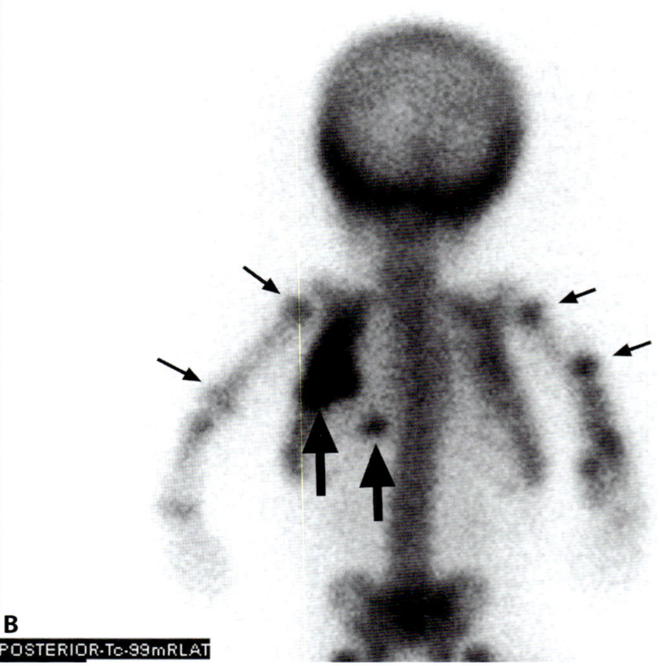

Figure 101.3 Axial CT image (A) of an 8-week-old boy with acute liver laceration secondary to child abuse shows subacute posterolateral left rib fractures with substantial callus formation (arrows). Multiorgan system injury in the absence of a plausible clinical history should always raise concern for child abuse. The posterior projection of the bone scintigraphy (B) shows avid tracer uptake in the left posterolateral ribs confirming multiple rib fractures (large arrows). Mildly increased tracer accumulation in the metaphyses of the bilateral humerus (small arrows) is normal physiologic activity and not due to nonaccidental trauma.

Index